W0106326

Handbook of Microscopic Anatomy

Continuation of Handbuch der mikroskopischen Anatomie des Menschen
Founded by Wilhelm von Möllendorff
Continued by Wolfgang Bargmann

Edited by A. Oksche and L. Vollrath

B.K.B. Berkovitz A. Boyde R.M. Frank H.J. Höhling
B.J. Moxham J. Nalbandian C.H. Tonge

TEETH

With 413 Figures

Springer-Verlag
Berlin Heidelberg NewYork
London Paris Tokyo

Handbook of Microscopic Anatomy
Volume V/6: Teeth

Professor Dr. Drs. h.c. A. Oksche

Institut für Anatomie und Zytobiologie der Justus-Liebig-Universität, Aulweg 123, D-6300 Giessen

Professor Dr. L. Vollrath

Anatomisches Institut der Johannes Gutenberg-Universität, Saarstraße 19–21, D-6500 Mainz

ISBN-13: 978-3-642-83498-1   e-ISBN-13: 978-3-642-83496-7
DOI: 10.1007/978-3-642-83496-7

Library of Congress Cataloging-in-Publication Data. Teeth / edited by A. Oksche and L. Vollrath; contributors, B.K.B. Berkovitz... [et al.]. p. cm. − (Handbook of microscopic anatomy; v. v/6) Includes bibliographies and indexes. 1. Teeth − Histology. 2. Teeth − Growth. 3. Teeth − Ultrastructure. I. Oksche, A. II. Vollrath, Lutz, 1936 −. III. Berkovitz, B.K. B. IV. Series.QM311 .T44 1988 611′.314 − dc1988-29489 CIP.

This work is subject to copyright. All rights are reserved, whether the whole or part of the material is concerned, specifically the rights of translation, reprinting, re-use of illustrations, recitation, broadcasting, reproduction on microfilms or in other ways, and storage in data banks. Duplication of this publication or parts thereof is only permitted under the provisions of the German Copyright Law of September 9, 1965, in its version of June 24, 1985, and a copyright fee must always be paid. Violations fall under the prosecution act of the German Copyright Law.

© Springer-Verlag Berlin Heidelberg 1989

The use of registered names, trademarks, etc. in this publication does not imply, even in the absence of a specific statement, that such names are exempt from the relevant protective laws and regulations and therefore free for general use.

Product liability: The publisher can give no guarantee for information about drug dosage and application thereof contained in this book. In every individual case the respective user must check its accuracy by consulting other pharmaceutical literature.

Typesetting, printing and bookbinding: Universitätsdruckerei H. Stürtz AG, Würzburg

List of Authors

B.K.B. BERKOVITZ, Dr., King's College London, University of London, Department of Anatomy and Human Biology, Strand, GB-London WC2R 2LS

A. BOYDE, Prof. Dr., University College London, Department of Anatomy and Embryology, Gower Street, GB-London WC1E 6BT

R.M. FRANK, Prof. Dr., Université Louis Pasteur, Faculté de Chirurgie Dentaire, 4, Rue Kirschleger, F-67085 Strasbourg-Cedex

H.J. HÖHLING, Prof. Dr., Institut für Medizinische Physik der Universität, Hüfferstr. 68, D-4400 Münster

B.J. MOXHAM, Dr., University of Bristol, The Medical School, Department of Anatomy, GB-Bristol BS8 1TD

J. NALBANDIAN, Dr., University of Connecticut, School of Dental Medicine, 263 Farmington Avenue, Farmington, Connecticut 06032, USA

C.H. TONGE, Emeritus Professor of Oral Anatomy, The University of Newcastle upon Tyne, GB-Newcastle upon Tyne NE1 7RU

Preface

The editors and the publisher are pleased to present another volume in this series of monographs. The topic of teeth was last reviewed within the framework of this Handbook more than fifty years ago, in 1936, by Josef Lehner and Hanns Plenk of Vienna, who wrote a comprehensive treatise on the subject in volume V/3. The introduction of new methods (e.g., transmission and scanning electron microscopy, histochemistry, radioautography, element analysis) and progress in dental research have made an update necessary.

In present times, characterized by scientific specialization and very rapid progress, it is virtually impossible to find a single individual prepared to review a field of research as large as that of teeth. Consequently, several authors were asked to contribute to the present volume. Originally, the intention was to cover the field in one volume. When it became clear that the material had become too extensive for a single volume, and when some authors were forced by external factors to withdraw from the project, it was decided to publish two volumes. In 1986, the volume *Periodontium,* written by Hubert E. Schroeder of Zürich, appeared in this series (volume V/5, 418 pages).

It is not without irony that, in a time seemingly conducive to basic and applied research, potential authors have had to struggle to prevent the closure of their institutions. As a result some researchers could not contribute to the present volume, and in other instances the contributions were delayed. The editors are grateful to the contributing authors and the publisher for their patience. We are confident that both volumes, *Periodontium* and *Teeth,* will become standard sources of scientific information, promoting both basic and applied dental research.

A. OKSCHE
L. VOLLRATH

Contents

Development of Dentine and Pulp
R.M. Frank and J. Nalbandian (With 88 Figures)

Structure and Ultrastructure of Dentine
R.M. FRANK and J. NALBANDIAN (With 95 Figures)

Structure and Ultrastructure of the Dental Pulp
R.M. FRANK and J. NALBANDIAN (With 59 Figures)

Special Aspects of Biomineralization of Dental Tissues
H.J. HÖHLING (With 20 Figures)

XVI Contents

Tooth Development – General Aspects

C.H. Tonge

A. General Arrangement of the Dental Tissues and Their Embryological Derivation

Human deciduous and permanent teeth have pulpal tissue within the dentine of their crowns and roots. Crown dentine is covered by enamel, root dentine by cementum. The periodontal ligament attaches the cementum to the alveolar bone in which the tooth lies. Gingivae surround the crown where it projects into the mouth. The pulp periodontal and gingival tissues have a blood and nerve supply. In craniofacial and dentofacial development the neural crest has a major inductive role, giving origin to ectomesenchyme which migrates to form neural skeletal and connective tissue derivatives (LE DOUARIN 1982, 1984).

Vertebrate embryos have been studied by extirpation, grafting, nuclear labelling, and histochemical techniques, to trace the migratory path of ectomesenchymal cells, identifying areas of neural crest related to specific developments, such as teeth, and the ultimate differentiation of neural crest cells upon reaching their final location. Mesencephalic neural crest cells migrating to the facial area are seen, from many studies, to form the trigeminal ganglion, Meckel's cartilage, the maxilla, mandible and first branchial arch muscles, whilst interference prior to cell migration results in varying degrees of facial deformity (MORRISS and THOROGOOD 1978; JOHNSTON and SULIK 1979; NODEN 1982, 1983; TOSNEY 1982). Extirpation, limited to premigratory mesencephalic neural crest cells from chick embryos, was compensated by contributions from the metencephalic neural crest and the dorsomedian part of the neural tube. Morphogenesis and differentiation proceeded normally and no craniofacial malformations occurred. MCKEE and FERGUSON (1984) concluded, therefore, that premigratory neural crest cells are not necessarily regionally patterned and committed to specific structural developments. Varying facial deformities, reported by earlier workers, were explained by their more radical extirpations, involving both the premigratory neural crest cells and the areas from which the compensatory cells arise. The mechanism initiating compensation and determining its completion remains unknown. If confirmed, these experiments suggest that the precise position of tooth formative neural crest, determined by SELLMAN (1946), was really only the premigratory phase, and that there remained greater flexibility in the potential for differentiation.

LUMSDEN (1984), using an intraocular grafting technique, showed that explanted mouse embryo neural crest, excised from the margins of the rhombencephalic neural plate in combination with homologous mandibular arch epithelium, had an odontogenic potential and that neural crest cell migration was not necessary for differentiation. It appears, therefore, that adequate evidence exists in amphibian, avian, and mammalian studies to show that ectomesenchymal cells of neural crest origin have a primary role in initiating tooth development by means of a succession of epithelio-ectomesenchymal interactions. These interactions give rise to the enamel organ and dental papilla which constitute the tooth germ and from which ameloblasts, enamel, odontoblasts, dentine, and pulp are formed. Around the margins of the dental papilla a condensed layer of ectomesenchymal cells forms the dental sac or follicle, the source of cementoblasts, fibroblasts, and osteoblasts which form cementum, periodontal ligament, and alveolar bone. The gingivae, a specialised part of the epithelial lining of the oral cavity, are probably ectodermal in origin.

B. Formation of Odontogenic Sites in the Developing Mouth

Neural crest cells migrate, as ectomesenchyme, into the maxillary and mandibular first arch processes and interact with the lining epithelium of the primitive oral cavity. Increased mitotic activity of that part of the epithelium lining the opposing borders of the maxillary and mandibular processes (Fig. 1), and the accumulation of squamous cells upon the basal layer, together with evidence of condensation of the underlying mesenchyme, constitute an odontogenic area (TONGE 1966, 1976; SLAVKIN 1974). Further growth results in the formation of a continuous band around the future dental arches, usually called the dental lamina, which later invaginates beneath the surface. According to MOSS-SALENTIJN (1982), the invagination arises passively due to the growth patterns of the surrounding tissues. Individual tooth primordia form before invagination in man and arise, after invagination, directly from the dental lamina in most other species.

I. Interactions Between Odontogenic Epithelium and Ectomesenchyme

Among others, the studies of GYSEL (1970), SLAVKIN (1974), THESLEFF and HURMERINTA (1981) and KOLLAR (1981, 1983) relate to the interactions between odontogenic epithelium and ectomesenchyme in leading to the differentiation of odontoblasts and ameloblasts and the laying down of predentine and enamel matrix. Recent fish, amphibians, reptiles, mammals, primates, and man possess teeth; only birds are toothless. There is significance in ascertaining whether birds have lost all trace of genetic coding associated with odontogenesis.

Experimentally KOLLAR and FISHER (1980) showed that intraocular grafts of chick epithelium, in combination with mouse molar mesenchyme, produced

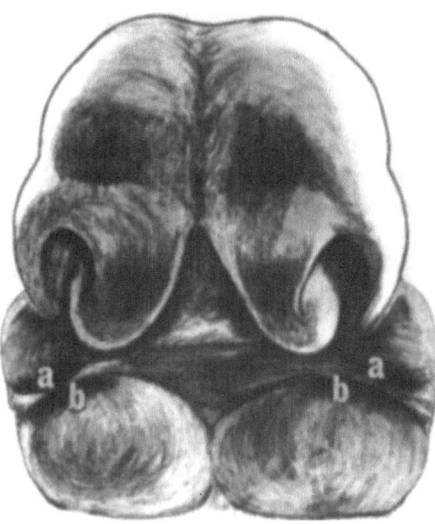

Fig. 1. The origin of odontogenic epithelium; *a* maxillary, *b* mandibular

perfectly formed crowns with differentiated ameloblasts which synthesised enamel matrix. This suggests that in birds the genetic information necessary for enamel synthesis remains quiescent rather than changed or suppressed. Perhaps an altered developmental sequence or an inhibition of the role of neural crest cells in forming odontogenic ectomesenchyme occurs in birds, and reproviding it, in the form of mouse molar mesenchyme, reactivates the quiescent avian oral epithelium.

Epitheliomesenchymal interactions, in odontogenesis, follow definite sequences; the epithelium and the mesenchyme being reciprocal and inter-dependent in the differentiation of secretory odontoblasts and ameloblasts. Ectomesenchymal cells in contact with epithelium induce the epithelium to form the enamel organ, the innermost cells of which reciprocally induce the ectomesenchyme to form the dental papilla and odontoblasts which then induce ameloblast differentiation from the internal enamel epithelium. Enamel synthesis commences once dentine has begun to form.

Examples of the dominant inductive role of dental mesenchyme:

i) Enzymatic separation of odontogenic epithelium and dental mesenchyme prevents tooth formation. Formation proceeds after the recombination of the separated elements (KOLLAR and BAIRD 1969).

ii) Epithelium from a non-dental site, the foot, combined with dental mesenchyme differentiates into ameloblasts which form enamel (KOLLAR and BAIRD 1970).

iii) Dental mesenchyme from a 10-day-old mouse embryo recombined with odontogenic epithelium from an 11–12-day-old embryo fails to differentiate to form a tooth, whereas dental mesenchyme from an 11–12-day-old embryo recombined with odontogenic epithelium from a 10-day-old embryo, forms a tooth. Epitheliomesenchymal interaction occurs at an optimum time of limited duration during odontogenesis (HERITIER 1970).

iv) The shape of a tooth is determined by dental mesenchyme, since incisor epithelium combined with molar mesenchyme forms a molar tooth, and incisor mesenchyme in combination with molar epithelium forms an incisor tooth (KOLLAR and BAIRD 1969). Similarly, mouse molar mesenchyme combined with edentulous diastemal epithelium forms a molar tooth (KOLLAR 1972). Further support for the role of the mesenchyme in promoting epithelial differentiation is provided in the odontogenic studies of RUCH et al. (1973) and THESLEFF (1977). Gradually, the epitheliomesenchymal junction folds to outline the cuspal pattern of the tooth and forms the potential amelodentinal junction.

II. Primordia for the Different Classes of Teeth

It is uncertain when, if at all, the dental mesenchyme becomes committed to form an individual class of tooth: incisor, canine, or molar. In the mouse embryo, from 12 days or earlier, molar teeth form from a single cell mass whose later developmental pattern is already determined (OSMAN and RUCH 1976; OSMAN et al. 1979; LUMSDEN 1979). The timing of human odontogenesis, based on assessments of embryonic age, is more speculative. NERY et al. (1970), in a study of 24–37-day human embryos, reported dental development commencing at 28 days and identified four maxillary and two mandibular arch zones which formed a continuous dental lamina in both jaws at 37 days. TONGE (1969a), when studying human embryos of 27–48 days ovulation age, found that at 28–30 days an odontogenic epithelium of tall columnar cells lined the margins of the maxillary and mandibular processes. At 31–32 days, superficial flattened cells covered the basal epithelium, and a basement membrane separated the epithelium from the underlying condensed mesenchyme. A single layer of cuboidal epithelium covered the non-odontogenic oral mucosa and skin. Later growth centers, consisting of centrally mitosing cells, were formed for the central incisors in both jaws at 36–38 days; lateral incisors, canines, and first molars by 38–40 days; and second molars by 40–44 days (Fig. 2). These centers give rise to deciduous teeth and their successors. Differentiation usually occurs earlier in the male, whilst in both sexes the mandibular anterior region can be identified before a similar differentiation within the maxillary process takes place. The odontogenic epithelial band unites at the mid-line and, as the individual growth centers invaginate, connection with the surface is maintained by the dental lamina which precedes the formation of the vestibular sulcus (Fig. 3). Between the individual primordia, the epithelium and the mesenchyme remain inactive.

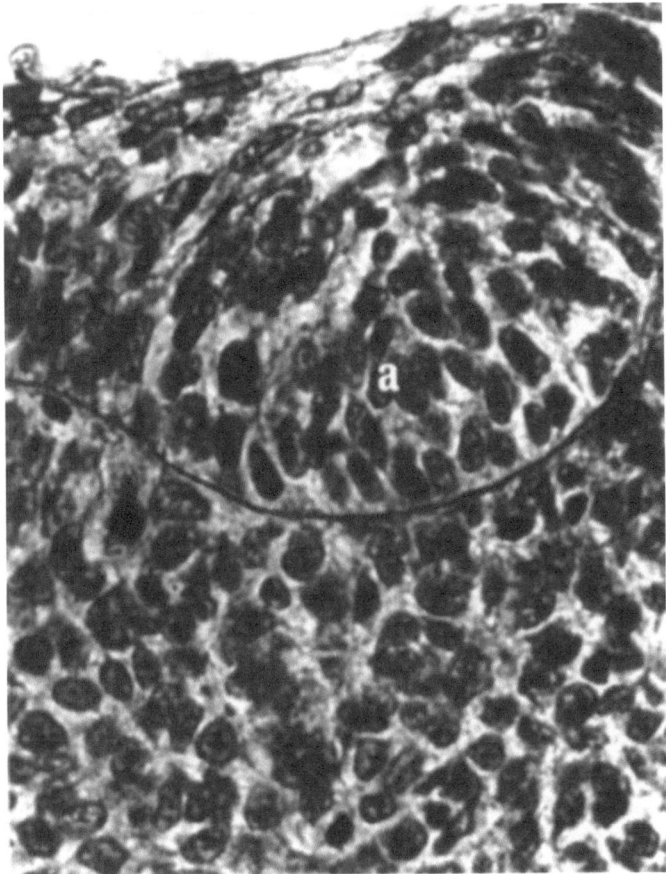

Fig. 2. The lower central incisor odontogenic epithelium (*a*) and the underlying mesenchyme in a human embryo of 34–36 days

III. Early Innervation and Blood Supply

Concurrently with the formation of odontogenic epithelium, the maxillary and inferior alveolar nerves enter the mesenchyme to distribute branches to the facial and oral regions. PEARSON (1977) demonstrated incisive branches of the inferior alveolar nerve reaching the primordia of the lower incisor teeth and the canines, unless separately supplied, while later posterior branches enter the mesenchyme of the molar area. All three superior alveolar nerves contribute to a plexus closely associated with the mesenchyme adjacent to the maxillary odontogenic epithelial band, the anterior supplying the incisor and canine, and the posterior the molar areas. As the enamel organ, dental papilla, and dental follicle form, nerve fibres spread over the outer surface of the dental follicle and external enamel epithelium, but do not initially enter the dental papilla.

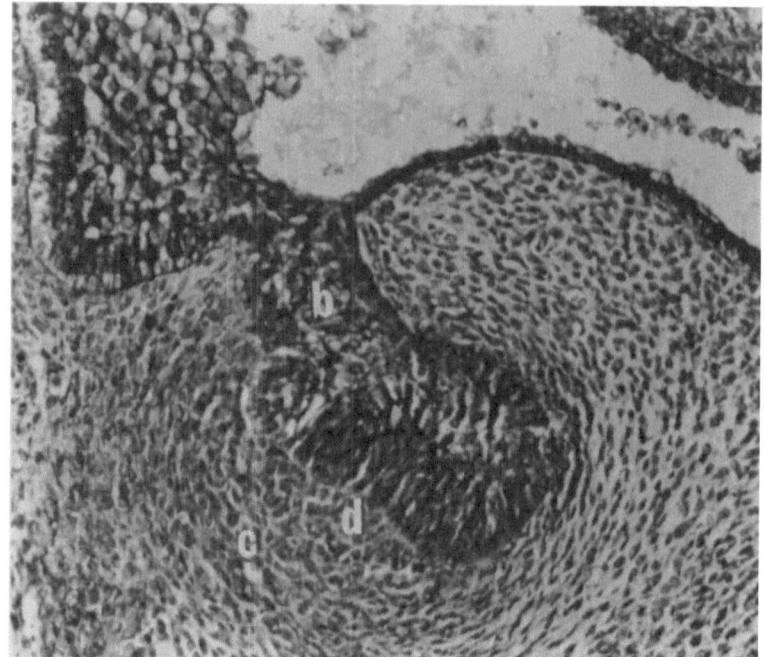

Fig. 3. Coronal section through lower canine area showing lip formation (*a*) and dental lamina (*b*) with the mesenchyme condensing (*c*) around the dental papilla (*d*)

However, their presence cannot be excluded because of the difficulties experienced in studying nerve fibres.

Autonomic nerve fibres are also said to reach the tooth-bearing areas early in development. The postulate that nerves entering the dental mesenchyme may induce the overlying mesenchyme to become odontogenic receives some support from KOLLAR and LUMSDEN (1979) and LUMSDEN (1982), but it seems unlikely that innervation has a sole inductor role. TOBIN (1972) showed that the vascular plexus surrounding developing tooth germs was derived mainly from the inferior alveolar and palatine arteries with a lesser contribution from blood vessels in adjoining tissues. As hard tissues develop, odontoblasts have an increased vascularity on their deeper surface, whilst the plexus on the outer surface of the external enamel epithelium lies nearest to the ameloblasts. A venous plexus, or sinus, is located at the base of the dental papilla.

C. Structural Formations in Mouth Development

The human lower jaw, including the oral floor structures, is identifiable in 5-week-old embryos. The primary palate forms at about the same time and secondary palatal fusion occurs during the seventh and eighth weeks in utero.

I. Tooth Germ, General Morphological and Histological Features of the Enamel Organ and Dental Papilla

As the tooth bud forms a cuboidal cell layer, the external enamel epithelium peripherally, and a low columnar cell layer, the internal enamel epithelium next to the dental papilla, differentiate. Polyhedral cells and intercellular fluid lie between these two layers of the enamel organ. Increased fluid formation and cell separation create the stellate reticulum, although the cells remain joined together by processes ending in desmosomes. The intercellular spaces contain glycosaminoglycans; and the cells, alkaline phosphatase, ribonucleic acid, and glycogen. Expansion of the fluid-filled spaces occurs from the center outwards to the margins of the enamel organ. This leaves transient accumulations of cells, forming the enamel knot, adjacent to the middle of the internal enamel epithelium and the enamel cord. The cord extends from the internal to the external enamel epithelium, which shows, at this point, a slight indentation – the enamel navel – which contains a blood capillary. The stellate reticulum prevents pressure distorting the interface between the internal enamel epithelium and the dental papilla during the shaping of the tooth crown.

The stratum intermedium, consisting of a few layers of cells containing more alkaline phosphatase than the stellate reticulum, persists, overlying the internal enamel epithelium, itself a single layer of short columnar cells with large centrally placed nuclei (Fig. 4). Before they become ameloblasts, the internal enamel epithelial cells are taller and columnar, and hexagonal in cross section; their Golgi apparatus, centrioles, and scattered mitochondria migrate to the distal part of the cell, which is separated by a basement membrane from the dental papilla. The nucleus of the ameloblast is now at the basal end of the cell nearest to the stratum intermedium. Initially, the dental papilla consists of irregularly arranged spindle-shaped cells of varying size. Then those lying next to the basal lamina, between them and the internal enamel epithelium, differentiate into odontoblasts which duly initiate predentine formation as a prerequisite to the ameloblast laying down enamel matrix (Fig. 5).

The interface between the enamel organ and the dental papilla has been studied by transmission and scanning electron microscopy (HURMERINTA and THESLEFF 1981). The fibrillar material next to the basal lamina increases during odontoblast differentiation, providing attachment for their cellular processes; probably aided by more fibronectin being present. Interactions involving the extracellular matrix during odontogenesis (RUCH 1975; GALBRAITH and KOLLAR 1976; THESLEFF and HURMERINTA 1981; RUCH et al. 1984) indicate that collagen deposition and glycoconjugate synthesis are both necessary for normal tooth morphogenesis and that interference with the stability of the extracellular matrix inhibits it. Cyclophosphamide cytotoxic injury induced in the rat incisor (ADATIA 1980) caused a cessation of odontoblast formation and a lower cellularity of ordinary pulpal tissue. Internal enamel epithelial cells atrophied and failed to become ameloblasts. During recovery, mesenchymal cells proliferated, odontoblasts formed, irregular dentine was laid down, followed by a differentiation of stunted internal enamel epithelial cells into ameloblasts. The interdependence

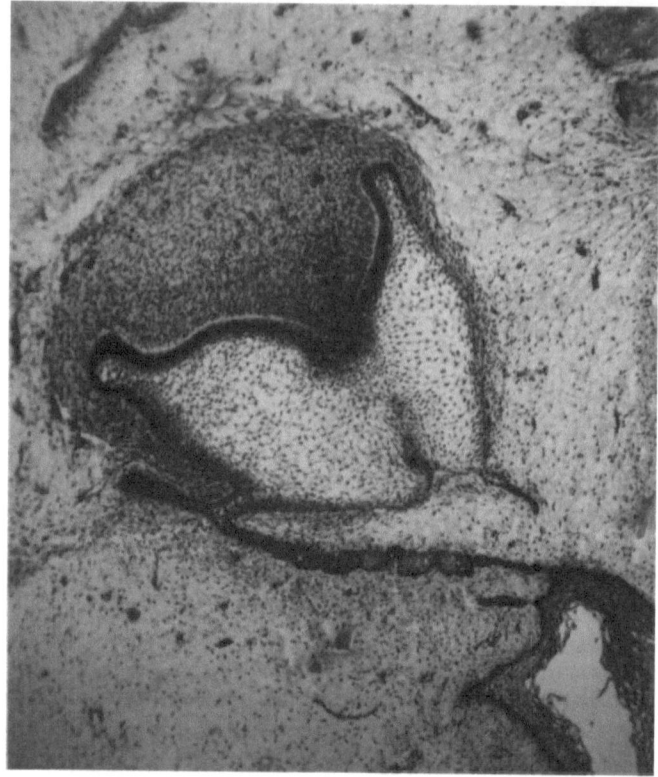

Fig. 4. Sagittal section through the maxillary central deciduous incisor of a 100-day-old human foetus

of the epithelial and mesenchymal cells, in situ, suggests that the nature of the contacts between them, including structural and chemical factors, may be significant in the inductive mechanism.

BURGESS and KATCHBURIAN (1982) and KATCHBURIAN and BURGESS (1983) have shown that contacts or approximations of diverse morphological types appear between the ameloblasts and odontoblasts about the time when the basal lamina is disintegrating. Their membranes never fuse and the contacts never form a well-known type of junction. THESLEFF (1986) has successfully cultured dental papilla cells, thus creating another method whereby their structural and functional features can be studied and related to their morphogenetic properties. Undifferentiated mandibular mesenchyme from 11-day-old mouse embryos, was shown to resemble, in morphology and growth rate, dental papilla cells cultured from 17-day-old mouse embryos; later studies will no doubt add to the understanding of the complexities of odontogenesis.

Dental papilla, apart from odontoblasts, produces a network of fibroblasts with fibers and intercellular ground substance, macrophages, and a reservoir of mesenchymal cells in the course of pulpal development.

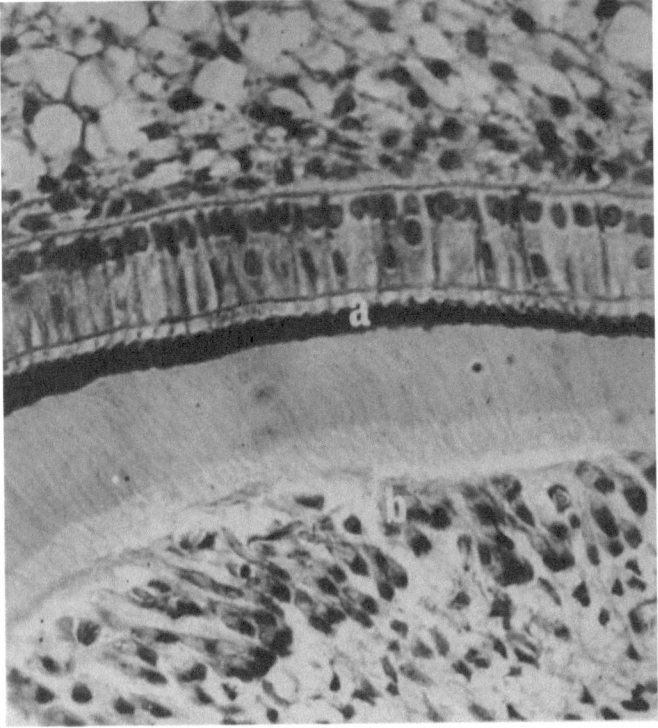

Fig. 5. Early amelogenesis (*a*) and odontogenesis (*b*)

II. Dental Lamina, Enamel Niche and Accessory Laminae

During the formation of the enamel organ, the part of the dental lamina connected to the surface, now non-odontogenic, breaks up by means of a mesenchymal invasion which first divides it into a main lamina and a lateral lamina occupying the intervening enamel niche. This soon disappears, the remnants of the dental lamina persisting as cell clusters, or islands, within the jaws or associated with the gingivae. Permanent successor teeth arise as buds from the deepest part of the dental lamina or adjacent external enamel epithelium lying lingual or palatal to the deciduous tooth germ. A posterior extension of the lamina, without independent connections to the oral epithelium, gives origin to the permanent molar teeth.

III. Labial and Vestibular Laminae

At 38–40 days ovulation age, in man, an epithelial thickening labiobucally to the dental lamina develops and, by disintegrating, a vestibular sulcus forms to separate the lips and cheeks from the alveolar process.

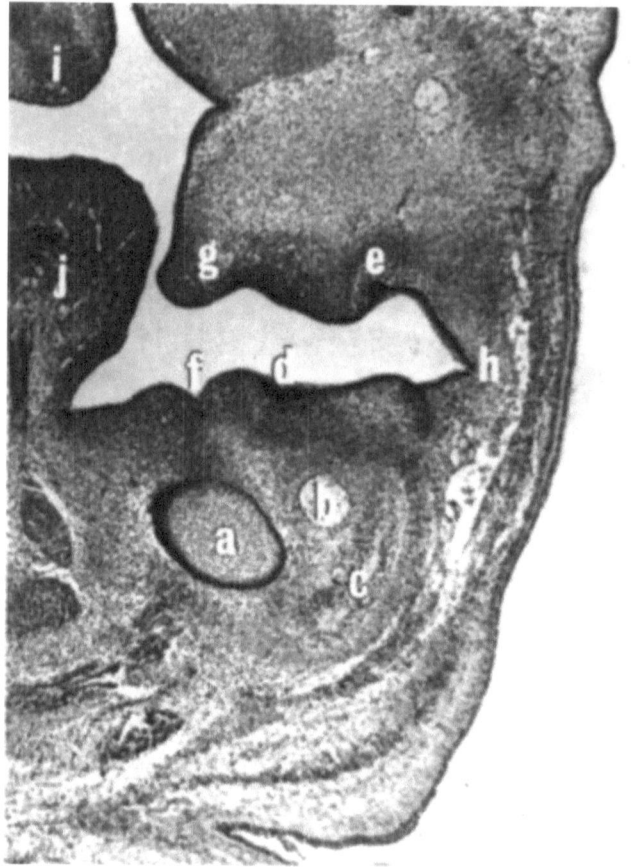

Fig. 6. Coronal section of right side of human embryo of 40–42 days. *a* Meckel's cartilage, *b* inferior alveolar nerve, *c* developing mandible, *d* odontogenic epithelium forming deciduous canine, *e* first deciduous molar and successor, *f* lingual gingival sulcus, *g* palatal shelf, *h* parotid, *i* nasal septum, *j* tongue

IV. Linguoalveolar Sulcus

At 40–42 days ovulation age, the limits of the future human gingivae, both palatally and lingually, can be identified and, in the lower jaw, the linguoalveolar sulcus deepens to separate the tongue from the future alveolus (Fig. 6).

V. Dental Follicle and Early Alveolus

As the tooth germ increases in size, the inner layer of the dental follicle surrounding it forms a vascular capsule of closely interwoven collagen fibers and fibroblasts around the margins of the dental papilla, while the outer part lines the alveolar bone; and a looser arrangement of reticular fibers, undifferentiated mesenchymal cells, macrophages, and blood vessels lies in the middle (TONGE 1971). The neural crest is thought to be the source of both the dental follicle and the dental papilla giving the periodontal tissues a dental origin (TEN CATE 1975). Using ocular grafts from mouse embryos of 16–19 days various recombinations of the dental papilla and the dental follicle with the enamel organ have shown (YOSHIKAWA and KOLLAR 1981) that both had the same capacity for tooth induction initially. However, the dental follicle had a much reduced capacity by 17 days although it retained the capability to form a root, periodontal ligament, and alveolar bone.

D. The Development of the Crown and its Morphology

Many theories attempt to explain the role of gene action in establishing functional morphological structure and the manner in which adaptive modifications evolve. The regional field theory (BUTLER 1939, 1956, 1982) postulates locations along the jaws acted upon by morphogens corresponding to incisor canine and molar fields. Individual tooth crown size and shape, within a given series, being dependent upon its position within the field and its capability of responding differently to the morphogen. This theory which envisages morphogenetic gradients, and also VAN VALEN'S (1970) theory which involves a series of programmed prepatterns to explain tooth morphogenesis, is very difficult to investigate by experimental embryological techniques. OSBORN (1973, 1978) critically analyses the "Field theory" and replaces it with his own "Clone theory". This postulates that a single mass of mesenchymal cells forms a clone for a particular class of teeth and, by repeated cell divisions, a critical mass of tissue is formed from which to initiate the first tooth germ of the class (e.g. molar). An inhibitory zone created around it prevents the formation of a second tooth germ until the clone mass has grown beyond the margins of the first primordium. With repeated cell divisions, the mesenchyme loses its tooth-forming potentiality ultimately limiting the number of teeth formed. Each tooth germ formed is different, its shape and size being intrinsic to the growing tissue. There is no prepatterning other than at the time at which it is formed. Presumably, therefore, the last formed elements are not necessarily simpler than the first formed elements.

LUMSDEN (1979) analyses the various developmental theories formulating a number of axioms, testing the predictions based upon them using the lower molar dentition of the mouse as an experimental model. His results are consistent with the view that gradations of shape and size in individual tooth germs in a sequence, occur as a result of intrinsic time-dependent alterations in the growing cell population which forms them, rather than from an extrinsic control such as a morphogenetic gradient field. Unequal growth of the internal enamel epithelium causes it to fold, in a cervical direction, from its most occlusal part, outlining the cusps, occlusal ridges, marginal ridges and fissures. The border between the internal enamel epithelium and the dental papilla delineates the amelodentinal junction. Crown shape, incisally or occlusally, is outlined prior to amelogenesis, the cervical part of the crown being mapped out whilst occlusal amelogenesis and dentinogenesis proceed.

I. Developmental Integuments of the Enamel Surface

After the completion of amelogenesis, a reduced ameloblast layer is formed as the cells shorten and become cuboidal. This layer is covered externally by the residual two or three layers of flattened cells derived particularly from the stratum intermedium and the external enamel epithelium. The reduced ameloblasts, prior to tooth eruption, lose their separate identity, and the collective name "reduced enamel epithelium" is given to the cellular remnants covering the enamel. The reduced enamel epithelium cells are connected to the enamel surface by hemidesmosomes and a basal lamina, but the so-called structureless primary enamel cuticle lying upon the enamel surface seems to be an artefact. Externally, the reduced enamel epithelium is separated, by a vascular network and the connective tissues of the dental follicle, from the oral mucosa to which it fuses during tooth eruption to form the epithelial attachment or junctional epithelium which persists in the gingival region. Developmental integuments covering the erupted crown enamel quickly disintegrate and any subsequent deposition on the enamel surface constitutes an acquired pellicle or plaque.

E. Development of the Periodontium

Collectively, root cementum, periodontal ligament, alveolar bone, and gingiva make up the periodontium which supports the tooth adapting to its functional requirements. Development of the root, the cementum, the periodontal ligament, and alveolar bone commences before eruption, the gingiva forming progressively while the tooth erupts.

I. Outline of Root Formation

When crown development nears completion, the cervical loop formed from cells of the internal and external enamel epithelia proliferates, growing into the connective tissue for a short distance to form the epithelial root sheath of HERTWIG. The number of roots and the outline of their shape and size for each tooth are determined by the epithelial root sheath. It also induces the underlying cells of the dental papilla to differentiate into odontoblasts which lay down root dentine. The margins of the epithelial root sheath turn inwards at the site of the amelocemental junction to form an epithelial diaphragm from which a root canal, or canals, develops of smaller dimension than the pulp chamber within the crown of the tooth.

Defects in the epithelial root sheath or growth around blood vessels, can produce accessory root canals communicating between the root pulp and the periodontal tissue. These accessory foramina, commonly found in the apical third of a root, also occur in the pulpal floor of interradicular zones of multirooted teeth if the epithelial diaphragm fails to fuse around the tubular extensions outlining these roots. During pre-eruptive movement, a tooth acquires about $1/2$–$3/4$ of its root length. Permanent human single rooted teeth require $1^1/_2$–2 years to develop and multirooted molars up to 3 years. During root formation, the epithelial root sheath and dental papilla cells proliferate; odontoblasts differentiate; dentine is laid down; the connective tissue cells of the dental follicle invade the epithelial root sheath; and cementoblasts, differentiated from dental follicle cells, deposit a layer of cementum upon the surface of the exposed dentine. Rarely, the inner layer of the epithelial root sheath retains an amelogenic capability depositing small enamel pearls on the dentine surface prior to the break-up of the sheath.

II. The Periodontal Ligament

The growing tooth crown lies within a large rounded alveolus, the margins of which are surrounded by compact bone. Extensive remodelling and reconstruction of the alveolar bone occurs throughout root formation and tooth eruption so that the alveolus becomes U or V shaped. The fibers and attachments of the periodontal ligament are also adjusted and re-orientated between the root and the alveolar process. Cellular activity for bone deposition, resorption, cementum formation, fibrous tissue formation, and degradation is provided by osteoblasts, osteoclasts, cementoblasts, and fibroblasts recruited from a reservoir of progenitor mesenchymal cells. TEN CATE and DEPORTER (1975) showed that the fibroblast may synthesise collagen and also phagocytose and degrade it, possibly simultaneously on different surfaces of the cell during the development of the periodontal ligament.

Dentinogingival fibres from just below the amelocemental junction are well organised before tooth eruption, closely followed by transseptal fibers between the cementum of adjacent teeth and the alveolar crest fibers. Other described groups, horizontal, oblique, apical, and interradicular form progressively as the tooth erupts and becomes functional. As active eruption proceeds, an intermediate plexus of intermingling fibres, especially well marked in rodent incisors, can be observed. Variously considered as a zone of adjustment during movement, or as an interlacing of fibers not continuously passing between the alveolar bone and cementum, its role remains obscure since the limited evidence that it is an active site of collagen synthesis is not confirmed by radiobiological studies (SCOTT and SYMONS 1982). In addition to collagen and reticular fibers, oxytalan fibers surround the base of the dental papilla, the cervical loop, and, to a lesser extent, the outer part of the dental follicle. They may be embedded in cementum and some are associated with the walls of blood vessels and lie in a plane parallel to the tooth surface.

III. Epithelial Rests of Malassez

The fenestrated epithelial root sheath persists as the epithelial rests of Malassez within the periodontal ligament observed as a network in tangential sections and as clusters in longitudinal sections. Ultrastructural and histochemical evidence (Valderhaug and Nylen 1966) indicates that human epithelial rests are separated from the connective tissue cells by a basement lamina. Although resting, they are vital epithelial cells which are not glandular and synthesize very little, if any, protein, but may become activated, in a changed environment of low O_2 high CO_2 tension, to proliferate (Grupe et al. 1967). In man, occasionally they are incorporated in the cementum of rapidly growing roots but soon degenerate. In the periodontal tissues they persist throughout life mainly in the cervical region becoming much reduced in numbers.

F. The Developmental Positions of the Teeth in the Jaws

Increasing data, concerning crown-size inter-relationships and interdental spacing, is accumulating. Nutrition, affecting craniofacial development prenatally, in the neonate, and postnatally, is described affecting tooth jaw relationships.

I. The Deciduous Teeth

In studies based on fifteen human embryos aged 6.5–12 weeks, the morphogenesis of maxillary (Burdi and Lillie 1966) and mandibular (Burdi 1968) dental arches has been analysed. Between 6.5 and 8 weeks, the maxillary dental arch, with all the deciduous tooth buds except the second molar which develops separately, was wide and flattened antero-posteriorly. At 7.5–9 weeks, the maxillary dental arch was increased in depth and elongated; conforming, after further growth, to a catenary curve in embryos of 9.5–12 weeks. Between 6.5 weeks the earliest time when central and lateral incisors are present and 8 weeks when all the deciduous teeth are developing, the mandibular dental arch was generally flattened antero-posteriorly. At about $8^1/_2$ weeks an increasing curvature extending from right to left canine tooth germs, marks the earliest time at which the mandibular dental arch conforms to a catenary curve.

Butler (1971) discusses the developmental growth of human tooth germs, including differences in the growth rates of individual teeth which result in differences in interdental spacing. Trenouth (1985) found that between 10–23 weeks, absolute human foetal growth rates were unrelated to changes in shape and relative proportion although correlating highly with the size of the craniofacial region measured. However, by standardising for size, absolute growth rates were converted into relative growth rates which correlated with changes in relative proportion and, thus, shape. Composite reconstructions, made from optically projected measurements of 10-µm frontal maxillofacial sections of 10.5–28-week-old human foetuses, were used by GARN et al. (1979) to analyze tooth crown size and spacing in a manner comparable to that obtained on plaster casts of children. Their results indicated that individual differences in relative tooth size and interdental spacing existed by 10.5–11 weeks. Despite an eight-fold increase in crown dimensions during the 10.5–28-week developmental span, interdental spacing, which was approximately equal to crown size at 11 weeks, decreases slightly but remains relatively constant from about 20 weeks. Tooth movement, alveolar remodelling, mandibular, and maxillary growth are among the factors preventing interdental space decreasing despite the continuing expansion of the teeth, so that overcrowding mainly affects the canine incisor region. Foetuses show variations in tooth size, some having large teeth overall; the deciduous dentition, however, retains its arrangement within the dental arch both prenatally and postnatally, much better than the permanent dentition.

II. The Successional Teeth

Successor incisor, canine, and premolar teeth, repeating the developmental sequence of their deciduous predecessors have formed tooth germs by 30 weeks, but dentinogenesis and amelogenesis do not start before birth. Initially, they occupy the same alveolus as the deciduous teeth, acquiring an independent alveolus and dental follicle when the deciduous teeth erupt postnatally.

A gubernacular cord, comprising fibrous tissue and the epithelial remnants of the successional lamina, passes through a gap in the roof of the permanent tooth alveolus excepting for the premolar, where the attachment is within the deciduous molar bony crypt.

III. The Permanent Molars

Each of the permanent molar teeth arise, in turn, from an extension of the dental lamina growing posteriorly beneath the surface, the first permanent molar being the earliest to differentiate (TONGE 1969b) and developing at a quicker pace than the successional teeth. At 100 days tooth buds are present, at 120 days the cellular elements of the enamel organ and an enamel knot have differentiated, whereas the first premolars are just showing early invagination of the dental papilla within the epithelial bud. At 24 weeks, the tooth germs of the first permanent molars are fully developed, while dentine and, sometimes enamel formation, occurs before birth. Some teeth can be identified radiologically at birth. At 120 days, the permanent molar extension of the dental lamina, showing little evidence of disintegration, extends as a finger-like process from which tooth buds arise for the second permanent molars at about 6 months after birth, and for the third permanent molars during the fourth to fifth years. By 150 days, the dental lamina associated with the permanent first molars breaks up into cellular clusters rather than the keratin-like whorls commonly found elsewhere in relation to the deciduous teeth and their successors.

It would be of interest to know when the dental lamina and its extensions lose their capacity to form teeth. Research examining the effects of ageing on epithelium and on its ability to respond to an inductive stimulus provided by mouse dental papillae (RICHMAN and KOLLAR 1986) indicates that up to 17 days gestation age palatal epithelium is capable of forming an enamel organ. At 18 days, tooth formation did not occur but the epithelium developed a stratum corneum and also keratin-filled cysts. It seems therefore that the disintegrating dental lamina is behaving in a similar manner after it has lost its potentiality for tooth development. There is need for further studies particularly aimed at ascertaining the time-scale of the capability of the dental laminal extension for the permanent molar teeth. Also whether, and to what extent, the long quiescent phase associated with the development of the last tooth of the series, can be influenced by factors other than genetic.

IV. Growth Retardation

A general account of nutrition and craniofacial growth (TONGE 1979) shows that, although there are many human studies, much of the evidence upon which the criteria for prenatal and postnatal growth is based is derived from animal experiments.

Severe calorie and protein deficiencies affected the general and skeletal growth of the pig (MCCANCE 1968) and similarly the rat (WARREN and BEDI 1985). In a series of studies (TONGE and MCCANCE 1965, 1973; LUKE et al. 1979, 1980) pigs fed on protein and on calorie-deficient diets all showed marked retardation of the growth of jaws and teeth, always more severe in the calorie-deficient. Tooth development and eruption was delayed to a relatively lesser extent than jaw development with resultant overcrowding particularly of the permanent molar dentition as well as displacement and malocclusion. There was a prolonged retention of the deciduous dentition with delayed eruption of their successors. The third permanent molar teeth, which are the largest of the pig molars and which were affected throughout the entire period of calorie deficiency, were frequently small, misshapen, or even absent. Nutritional rehabilitation or "catch up" was quicker in the protein-deficient than in the calorie-deficient and persistent malocclusion was similarly less severe. Rehabilitated animals never achieved the same level of growth and dental development as the normal age controls. Their offspring were, however, normal in all respects.

Human physical prenatal and postnatal growth is broadly considered by TANNER (1978). A scientific and medically orientated account of human protein-energy malnutrition is given in the child studies of ALLEYNE et al. (1977). Human studies on dental and craniofacial development, comparing normal with growth-retarded human foetuses, whose gestational ages ranged from 22–42 weeks, indicated that there are significant differences (LUKE 1976; LUKE et al. 1978). The upper and lower jaws and the base of the skull, which normally have a relatively greater growth rate during the second half of gestation, were reduced in size in the growth retarded. The neurocranium and the crowns of the deciduous teeth, whose growth rate is normally either low throughout the second half of gestation or falls during the ten weeks prior to term, were not similarly affected by intrauterine growth retardation.

G. Developmental Anomalies

The reviews of KINGSTON (1985) and POSWILLO (1975) provide guidance to the different syndromes affecting the dentofacial complex. The majority of genetic effects, whether multifactorial, chromosomal, or the action of a single gene, result in extremes of dentofacial normality and give rise to growth patterns to which are added environmental influences which complicate the interpretation of data acquired clinically.

Inherited defects include hereditary enamel hypoplasia and hereditary opalescent dentine. Absence of one or more of both the deciduous and permanent teeth can occur with either both, or only one, dentition being involved in the same individual. Complete absence of teeth (anodontia) is very rare. Missing permanent teeth are commonly the third molars, maxillary lateral incisors, maxillary or mandibular second premolars, mandibular central incisors, and maxillary first premolars. Misshapen and misplaced teeth are found. Submerged and ankylosed human deciduous molars are described (DARLING and LEVERS 1973), most of them having originally erupted and reached occlusion and at a later date becoming submerged.

Other common anomalies include cuspal variations in number and position, and the presence of supernumerary teeth which are found in the incisor or premolar region and may also occur in the deciduous dentition (OOË 1971). Occasional anomalies include dens in dente, which is an invagination into the tooth crown probably due to an 'abnormal proliferation of part of the internal enamel epithelium, or as a result of retarded growth at the site of the invagination.

References

Adatia AK (1980) Direct evidence of epithelialmesenchymal interdependence in situ. J Anat 130:469–478

Alleyne GAO, Hay RW, Picou DI, Stanfield JP, Whitehead RG (1977) Protein-energy malnutrition. Arnold, London

Burdi AR (1968) Morphogènesis of mandibular dental arch shape in human embryos. J Dent Res 47:50–58

Burdi AR, Lillie JH (1966) A catenary analysis of the maxillary dental arch during human embryogenesis. Anat Rec 154:13–20

Burgess AMC, Katchburian E (1982) Morphological types of epithelial-mesenchymal cell contacts in odontogenesis. J Anat 135:577–584

Butler PM (1939) Studies of the mammalian dentition. Differentiation of the postcanine dentition. Proc Zool Soc Lond 109:1–36

Butler PM (1956) The ontogeny of molar pattern. Biol Rev 31:30–70

Butler PM (1971) Growth of human tooth germs. In: Dahlberg AA (ed) Dental morphology and evolution. Univ Chicago Press, Chicago, pp 3–13

Butler PM (1982) Some problems of the ontogeny of tooth patterns. In: Kurten B (ed) Teeth: Form function and evolution. Columbia Univ Press, New York, pp 44–51

Darling AI, Levers BGH (1973) Submerged human deciduous molars and ankylosis. Arch Oral Biol 18:1021–1040

Galbraith DB, Kollar EJ (1976) In vitro utilisation of exogenous pro-collagen by embryonic tooth germs. J Exp Zool 197:135–140

Garn SM, Burdi AR, Babler WJ, Asp R (1979) Crown size arch space relationships during human prenatal dental development. J Dent Res 58:554–559

Grupe HE, Ten Cate AR, Zander HA (1967) A histochemical and radiobiological study of in vitro and in vivo human epithelial cell rest proliferation. Arch Oral Biol 12:1321–1329

Gysel C (1970) Ontogènese de la dent et gradients morphogenetiques. Rev Belge Med Dent 25:313–317

Héritier M (1970) Etude in vitro d'associations hétérochroniques d'épithélium et de mésenchyme odontogènes chez la souris. C R Acad Sci [III] 271:1704–1706

Hurmerinta K, Thesleff I (1981) Ultrastructure of the epithelial-mesenchymal interface in the mouse tooth germ. J Craniofac Genet Dev Biol 1:191–202

Johnston MC, Sulik KK (1979) Some abnormal patterns of development in the craniofacial region. In: Melnick M, Jurgensen R (eds) Developmental aspects of craniofacial Dysmorphology, birth defects. Original articles series, Vol. 11. Liss, New York, pp 23–42

Katchburian E, Burgess AMC (1983) Lysosomes and removal of the basal lamina of ameloblasts in early stages of odontogenesis. Cell Biol Int Rep 7:407–415

Kingston HN (1985) Dental syndromes in medical genetics. Dental Update 12:314–328

Kollar EJ (1972) The development of the integument: spatial temporal and phylogenetic factors. Am Zool 12:125–135

Kollar EJ (1981) Tooth development and dental patterning. In: Connelly (ed) Morphogenesis and pattern formation. Raven, New York, pp 87–102

Kollar EJ (1983) Epithelial mesenchymal interactions in the mammalian integument. Tooth development as a model for instructive induction. In: Fallon JF, Sawyer RH (eds) Praeger, New York, pp 27–49

Kollar EJ, Baird GR (1969) The influence of the dental papilla on the development of tooth shape in embryonic mouse tooth germs. J Embryol Exp Morphol 21:131–148

Kollar EJ, Baird GR (1970) Tissue interactions in embryonic mouse tooth germs. I. Reorganisation of the dental epithelium during tooth-germ reconstruction. J Embryol Exp Morphol 24:159–171

Kollar EJ, Fisher C (1980) Tooth induction in chick epithelium: Expression of quiescent genes for enamel synthesis. Science 207:993–995

Kollar EJ, Lumsden AGS (1979) Tooth morphogenesis: The role of the innervation during induction and pattern formation. J Biol Buccale 7:49–60

Le Douarin NM (1982) The neural crest. Cambridge Univ Press

Le Douarin NM (1984) Cell migrations in embryos. Cell 38:353–360

Luke DA (1976) Dental and craniofacial development in the normal and growth retarded human fetus. Biol Neonate 29:171–177

Luke DA, Stack MV, Hey EN (1978) A comparison of morphological and gravimetric methods of estimating human foetal age from the dentition. In: Butler PM, Joysey KA (eds) Development function and evolution of teeth. Academic, New York London, pp 511–518

Luke DA, Tonge CH, Reid DJ (1979) Metrical analysis of growth changes in the jaws and teeth of normal protein deficient and calorie deficient pigs. J Anat 129:449–457

Luke DA, Tonge CH, Reid DJ (1980) Histology of mandibular bone from normal protein deficient and calorie deficient pigs. J Anat 130:859–865

Lumsden AGS (1979) Pattern formation in the molar dentition of the mouse. J Biol Buccale 7:77–103

Lumsden AGS (1982) The developing innervation of the lower jaw and its relation to the formation of tooth germs in mouse embryos. In: Kurtén B (ed) Teeth: Form function and evolution. Columbia Univ Press, New York, pp 32–43

Lumsden AGS (1984) Tooth forming potential of mammalian neural crest. J Embryol Exp Morphol [Suppl Sept] 68

McCance RA (1968) The effect of calorie deficiencies and protein deficiencies on final weight and stature. In: McCance RA, Widdowson EM (eds) Calorie deficiencies and protein deficiencies. Churchill, London, pp 319–328

McKee GJ, Ferguson MWJ (1984) The effects of mesencephalic neural crest cell extirpation on the development of chicken embryos. J Anat 139:491–512

Morriss GM, Thorogood PV (1978) An approach to cranial neural crest cell migration and differentiation in mammalian embryos. In: Johnston MC (ed) Development in mammals, vol 3. Elsevier, North Holland, pp 363–412

Moss-Salentijn L (1982) Morphological aspects of the growth behaviour of the early dental lamina in the cat and rat. In: Kurtén B (ed) Teeth: Form function and evolution. Columbia Univ Press, New York, pp 7–20

Nery EB, Kraus BS, Croup M (1970) Timing and topography of early human tooth development. Arch Oral Biol 15:1315–1326

Noden DM (1982) Patterns and organisation of craniofacial skeletogenic and myogenic mesenchyme: a perspective. In: Dixon A, Sarnat B (eds) Factors and mechanisms influencing bone growth, Liss, New York, pp 167–203

Noden DM (1983) The role of the neural crest in patterning of avian cranial skeletal connective and muscle tissues. Dev Biol 96:144–165

Ooë T (1971) Three instances of supernumerary tooth germs observed with serial sections of human foetal jaws. Z Anat Entwickl-Gesch 135:202–209

Osborn JW (1973) The evolution of dentitions. Am Scient 61:548–559

Osborn JW (1978) Morphogenetic gradients: fields versus clones. In: Butler PM, Joysey KA (eds) Development function and evolution of teeth. Academic, New York London, pp 171–201

Osman A, Ruch JV (1976) Répartition topographique des mitoses dans l'incisive et la première molaire inférieures de l'embryon de souris. J Biol Buccale 4:331–348

Osman M, Karcher-Djuricic V, Ruch JV (1979) Différenciation in vitro du matériel molaire présomptif de l'embryon de souris. C R Acad Sci [III] 289:149–151

Pearson AA (1977) The early innervation of the developing deciduous teeth. J Anat 123:563–577

Poswillo D (1975) Causal mechanisms of craniofacial deformity. Br Med Bull 31:101–106

Richman JM, Kollar EJ (1986) Tooth induction and temporal patterning in palatal epithelium of fetal mice. Am J Anat 175:493–505

Ruch JV (1975) On odontogenic tissue interactions. In: Slavkin HC, Greulich RC (eds) Extra cellular matrix influences on gene expression. Academic, New York, pp 549–554

Ruch JV, Karcher-Djuricic, Gerber R (1973) Les déterminismes de la morphogenèse et des cytodifférenciations des ébauches dentaires de souris. J Biol Buccale 1:45–56

Ruch JV, Lesot H, Karcher-Djuricic V, Meyer JM (1984) Extracellular matrix-mediated interactions during odontogenesis. Matrices and cell differentiation. Liss, New York, pp 103–114

Scott JH, Symons NBB (1982) Introduction to dental anatomy, 9th edn. Churchill Livingstone, Edinburgh

Sellmann S (1946) Some experiments on the determination of the larval teeth in Amblystoma mexicanum. Odont Tidskr 54:1–128

Slavkin HC (1974) Embryonic tooth formation. A tool for developmental biology. In: Melcher AH, Zarb GA (eds) Oral Sci Rev, vol 4. Munksgaard, Copenhagen, pp 1–136

Tanner JM (1978) Foetus into man. Physical growth from conception to maturity. Open Books, London

Ten Cate AR (1975) Development of the periodontal membrane and collagen turnover. In: Poole DFG, Stack MV (eds) Eruption and occlusion. Butterworth, London, pp 281–289

Ten Cate AR, Deporter DA (1975) The degradative role of the fibroblast in the remodelling and turnover of collagen in soft connective tissue. Anat Rec 182:1–14

Thesleff I (1977) Tissue interactions in tooth development in vitro. In: Karkinen-Jaaskelainen M, Saxen L, Weiss L (eds) Cell interactions in differentiation. Academic, London, pp 191–207

Thesleff I (1986) Dental papilla cells in culture. Comparison of morphology growth and collagen synthesis with two other dental related embryonic mesenchymal cell populations. Cell Differ 18:189–198

Thesleff I, Hurmerinta K (1981) Tissue interactions in tooth development. Differentiation 18:75–88

Tobin CE (1972) Blood supply of human fetal teeth. Am J Anat 131:217–226

Tonge CH (1966) Advances in dental embryology. Int Dent J 16:328–349

Tonge CH (1969a) Time-structure relationship of tooth development in human embryogenesis. J Dent Res 48:745–752

Tonge CH (1969b) La première molaire permanente. Pt 1 Embryologie et anatomie. Edition de la Société Française d'orthopédie dento-faciale. Vitte, Lyon, pp 3–18

Tonge CH (1971) The role of mesenchyme in tooth development. In: Dahlberg (ed) Dental morphology and evolution. University Chicago Press, Chicago, pp 45–58

Tonge CH (1976) Morphogenesis and development of teeth. In: Cohen B, Kramer IRH (eds) Scientific foundations of Dentistry. Heinemann, London, pp 325–334

Tonge CH (1979) Nutrition and craniofacial growth. J R Coll Surg Edinb 24:1–8

Tonge CH, McCance RA (1965) Severe undernutrition in growing and adult animals. 15. The mouth jaws and teeth of pigs. Br J Nutr 19:361–372

Tonge CH, McCance RA (1973) Normal development of the jaws and teeth in pigs and the delay and malocclusion produced by calorie deficiencies. J Anat 115:1–22

Tosney KW (1982) The segregation and early migration of cranial neural crest cells in the avian embryo. Dev Biol 89:13–24

Trenouth MJ (1985) The relationship between differences in regional growth rates and changes in shape during human fetal craniofacial growth. Arch Oral Biol 30:31–35

Valderhaug JP, Nylen MU (1966) Function of epithelial rests as suggested by their ultrastructure. J Periodont Res 1:69–78

Van Valen L (1970) An analysis of developmental fields. Dev Biol 23:456–477

Warren MA, Bedi KS (1985) The effects of a lengthy period of undernutrition on the skeletal growth of rats. J Anat 141:53–64

Yoshikawa DK, Kollar EJ (1981) Recombination experiments on the odontogenic roles of mouse dental papilla and dental sac tissues in ocular grafts. Arch Oral Biol 26:303–307

Tissue Changes During Tooth Eruption

B.K.B. Berkovitz and B.J. Moxham

Tooth eruption is the process whereby a tooth moves axially from its developmental position within the jaws to emerge into the oral cavity. However, eruption is part of a more complex system which involves movements in other planes and which continues beyond the developmental stage to maintain the tooth in its functional position.

The first part of this chapter reviews the tissue changes which take place during eruption until the tooth first attains its functional position. Attention will be focussed upon the periodontal tissues because 1) they are the tissues through which the tooth erupts, 2) they show considerable remodelling to accommodate eruptive movements, and 3) there is evidence that the periodontal ligament is responsible for generating the eruptive force (NESS 1964; BERKOVITZ 1975; MELCHER and BEERTSEN 1977; MOXHAM and BERKOVITZ 1982). Although we will be concerned primarily with information obtained from histological studies, occasional reference will be made to changes in biochemistry and changes in metabolic activity. The second part of the review will attempt to relate the structure of the periodontal ligament to the eruptive mechanism.

A. Tissue Changes During Eruption

There are three main phases in the development of a tooth, the pre-eruptive, eruptive, and intra-oral phases (Fig. 1). The tissue changes which occur during eruption will be considered in relation to these and with reference only to teeth of limited growth.

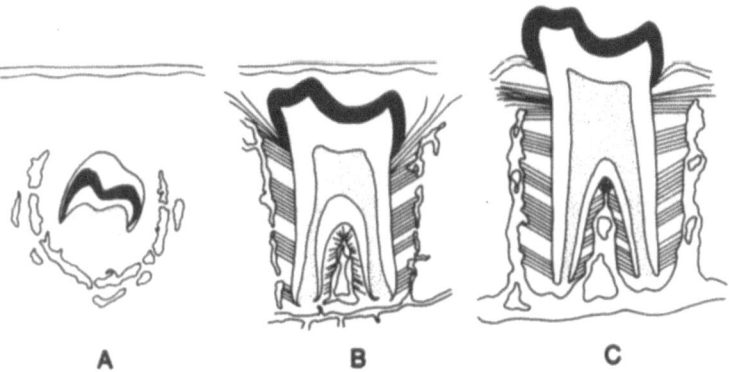

Fig. 1 A–C. Diagram showing the three main phases in the development of a tooth. **A** pre-eruptive phase, **B** eruptive phase, **C** intra-oral phase

I. The Pre-Eruptive Phase

This covers the period from the initial appearance of the tooth germ to the beginning of root formation. During this phase, the tooth remains in its developmental intraosseous location and grows concentrically within its follicle (DARLING and LEVERS 1975). There is little active bodily movement of the tooth axially (THOMAS 1965; DARLING and LEVERS 1975, 1976) but there may be considerable drifting and tilting (BRASH 1928; BAUME 1953; MANSON 1968; RICHARDSON 1978).

Much information is available concerning the histological appearance of the tooth and of its investing connective tissues (the dental follicle) during the pre-eruptive phase. Less is known about the changes occurring in the dental crypt, despite the considerable remodelling it shows to accommodate the growing tooth germ. Early on, resorption is seen over much of the internal surface of the crypt whilst compensatory bone deposition occurs on the external surface (Fig. 2). Later, more localised areas of bone resorption and of bone deposition on the internal surface indicates tilting and drifting of the tooth germs. For example, distolingual deposition of bone within the crypts of monkey permanent mandibular incisors is associated with their forward and outward relocation (BAUME 1953). In addition, it is claimed that backward relocation of cat molars is associated with resorption on the buccal walls of the crypts and deposition of bone on the lingual walls (MANSON 1968). The mechanism responsible for producing tooth movements during the pre-eruptive phase is unknown and it is not clear whether the bone activity is the cause of the movement or is a response to movement produced by a force derived elsewhere.

II. The Eruptive Phase

This begins with root formation and terminates when the crown first emerges through the oral mucosa. It is characterised by axial migration of the tooth which involves its active bodily movement (CARLSON 1944; THOMAS 1965; BJÖRK and SKIELLER 1972; DARLING and LEVERS 1975). Surprisingly little is known about the pattern and rates of eruption during this phase. DARLING and LEVERS (1976) reported that eruption of human mandibular teeth commenced once the roots began to form and they calculated that the eruption rates varied from 1.2 mm/year (for the permanent mandibular third molar) to 3.5 mm/year (for the permanent mandibular second premolar).

The main tissue changes which occur during the eruptive phase involve root development, formation of the periodontal ligament from the dental follicle, adaptation of the alveolar bone, and alterations in the tissues overlying the erupting tooth.

1. The Development of the Periodontal Ligament During the Eruptive Phase

This follows the events associated with root formation. Briefly, root formation involves the downgrowth of an epithelial root sheath from the enamel

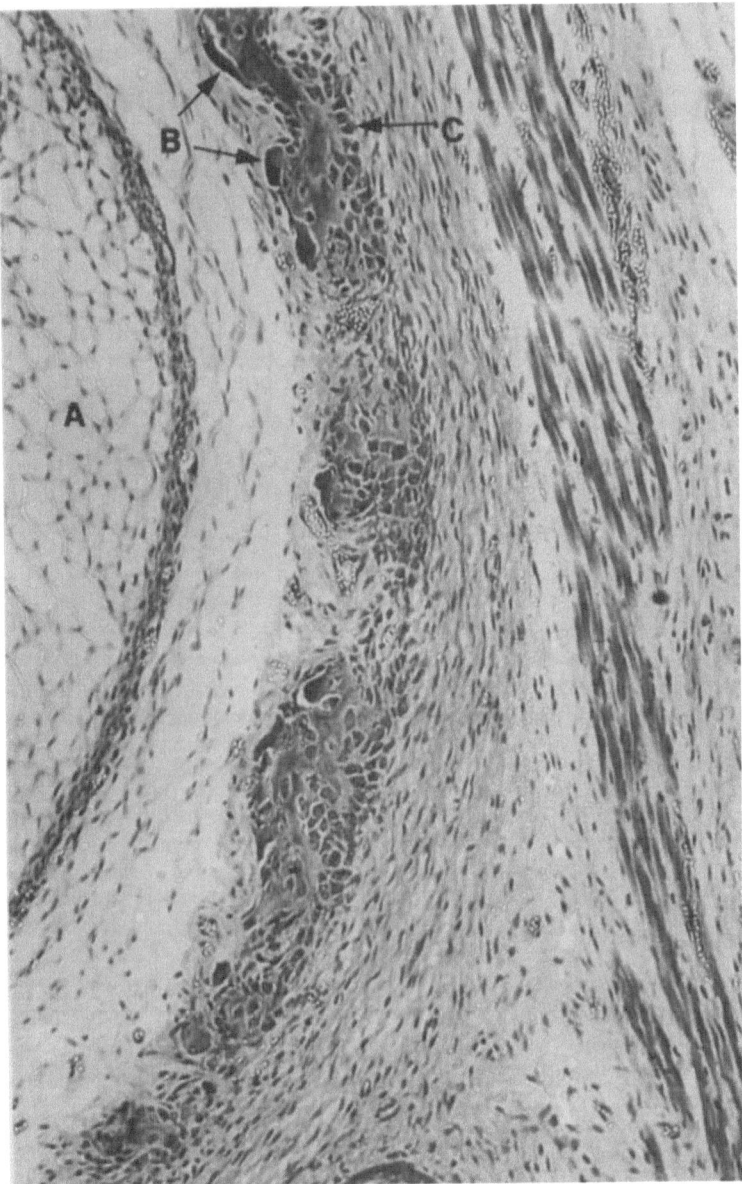

Fig. 2. Photomicrograph of a developing tooth (*A*) showing resorption of the internal wall of the alveolus as evidenced by the presence of osteoclasts (*B*) and deposition of bone externally due to osteoblasts (*C*). × 150

organ of the tooth germ. This sheath induces the formation of root dentine. With loss of continuity of the root sheath, the mesenchymal cells of the dental follicle, immediately adjacent to the root dentine, differentiate into cemento-blasts which commence to form cementum. The subsequent development of

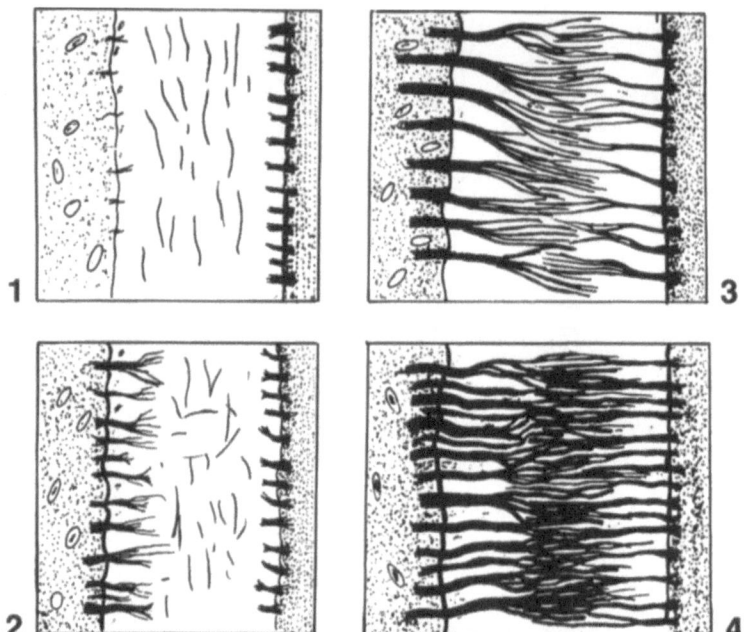

Fig. 3. Diagram summarising the sequence of events in the formation of principal collagen fiber bundles. **1** Fine, brush-like fibers are first seen emanating from cementum. Only a few fibers project from the osteoblast-lined alveolar bone and extend into the non-organized, collagenous elements that occupy the broad central zone of the developing periodontal ligament. **2** Sharpey's fibers, thicker and more widely spaced than those from cementum, emerge from bone to extend toward the tooth and appear to unravel as they arborize at their ends. The closely spaced, cemental fibers are still short, giving the root surface a brush-like appearance. **3** The alveolar fibers extend further into the central zone to join the lengthening cemental fibers. **4** With occlusal function, the principal fibers become classically organized, thicker, and apparently continuous between bone and cementum.
Courtesy of Dr's D.A. Grant and S. Bernick and of the Journal of Periodontology

the periodontal ligament has been described by many authors (SICHER 1923; ORBAN 1927; NOYES et al. 1943; ORBAN 1957; BERNICK 1960; TROTT 1962; TONGE 1963; ECCLES 1964; LEVY and BERNICK 1968; MAGNUSSON 1968; ATKINSON 1972; GRANT and BERNICK 1972; GRANT et al. 1972; PEARSON et al. 1975; BERNICK and GRANT 1982). GRANT and BERNICK (1972) and GRANT et al. (1972) described the sequence leading to the formation of a principal collagen fiber bundle using the cheek teeth in the permanent dentitions of squirrel monkeys and marmosets (Fig. 3). Initially, fine brush-like fibers are seen emanating from cementum. Only a few fibers extend from the alveolar bone into the apparently unorganized, collagenous elements occupying the broad central zone of the developing periodontal ligament. Later, Sharpey fibers (thicker and more widely spaced than those from cementum) emerge from bone and appear to unravel as they arborize at their ends. The closely spaced cemental fibers are still short, giving the root a brush-like appearance. The alveolar fibers extend further into the central zone to join the lengthening cemental fibers. There may be species differences however. In the rat molar, the fibers from alveolar bone appear shortly before those from cementum (TROTT 1962).

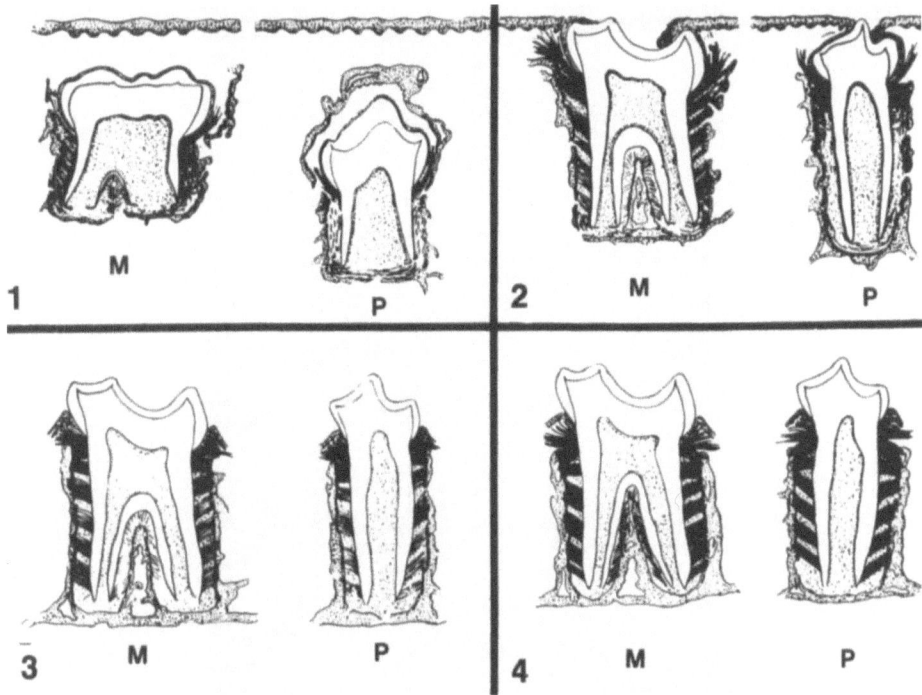

Fig. 4. Diagram illustrating the differences in periodontal fiber formation in primary and secondary succedaneous teeth. **1** With root formation well advanced, the permanent molar (*M*) (tooth without a predecessor) shows principal fibers extending from bone to cementum. The permanent premolar (*P*) (tooth with a predecessor) shows only predentogingival fibers as an organized group. **2** Upon emergence into the oral cavity, the permanent molar shows advanced fiber formation with apparently continuous principal fibers. The premolar shows organized fibers only at the alveolar crest; more apically, the periodontal ligament becomes progressively less organized. **3** With occlusal function, the molar shows complete periodontal fiber apparatus. The premolar shows apparently continuous principal fibers, except near the apex where an intermediate zone is still demonstrable. **4** With continued function, both molar and premolar show classically aligned, and apparently continuous, principal fiber groups. Courtesy of Dr. D.A. Grant and of the Journal of Periodontology

GRANT et al. (1972) studied the sequential histogenesis of the principal collagen fiber bundles of the periodontal ligament. Using permanent molar teeth of the marmoset, they reported that the fibers were formed mainly prior to the emergence of the tooth into the oral cavity (Figs. 4 and 5). At the beginning of root formation, the developing molar appeared to be surrounded only by the loosely organised fibers of the dental follicle. Only the future dentogingival fibers were evident as a distinct bundle. Below, fine fibers were seen to extend from the cementum for but a short distance into the developing ligament. As the root develops, the obliquely-orientated dentoalveolar fibers then become prominent. Others have also reported that the periodontal fibers in molars are present with an oblique alignment before the tooth emerges through the oral mucosa: BERNICK (1960), THOMAS (1965) and MAGNUSSON (1968) for the rat molar, ATKINSON (1972) for the mouse molar, and THOMAS (1965) for the human molar. LEVY and BERNICK (1968) observed a similar sequence in the deciduous teeth of the marmoset.

Fig. 5. Photomicrograph showing the second permanent molar of a marmoset just emerging into the oral cavity. The periodontal ligament is composed of well-formed, obliquely orientated, principal collagen fiber bundles. × 20. Courtesy of Dr. D.A. Grant and of the Journal of Periodontology

A different pattern of development seems to exist for permanent teeth which have deciduous predecessors. GRANT and BERNICK (1972) studied the formation of the periodontal ligament in the premolars of squirrel monkeys (Figs. 4 and 6). Only the presumptive dentogingival fibres were prominent during the eruptive phase, the remaining ligament containing only loosely structured collagenous elements with a few short fibers emerging from the cementum midroot. A similar pattern has been reported for the premolars of the marmoset (GRANT et al. 1972) and for monkey permanent incisors and canines (BERNICK and GRANT 1982). CAHILL and MARKS (1982) observed collagen fibres within the periodontal ligament of erupting premolar teeth in dogs. However, they stated that these fibres passed from the cementum with an almost vertical orientation and did not insert into the alveolar bone until the tooth reached the occlusal plane.

The findings of BERNICK (1960), THOMAS (1965), and MAGNUSSON (1968) appear to conflict with those of other workers. O'BRIEN et al. (1958) and FORMI-COLA and FERRIGNO (1966) have reported that organisation of the periodontal ligament fibers in the rat molar occurs only when the tooth appears in the

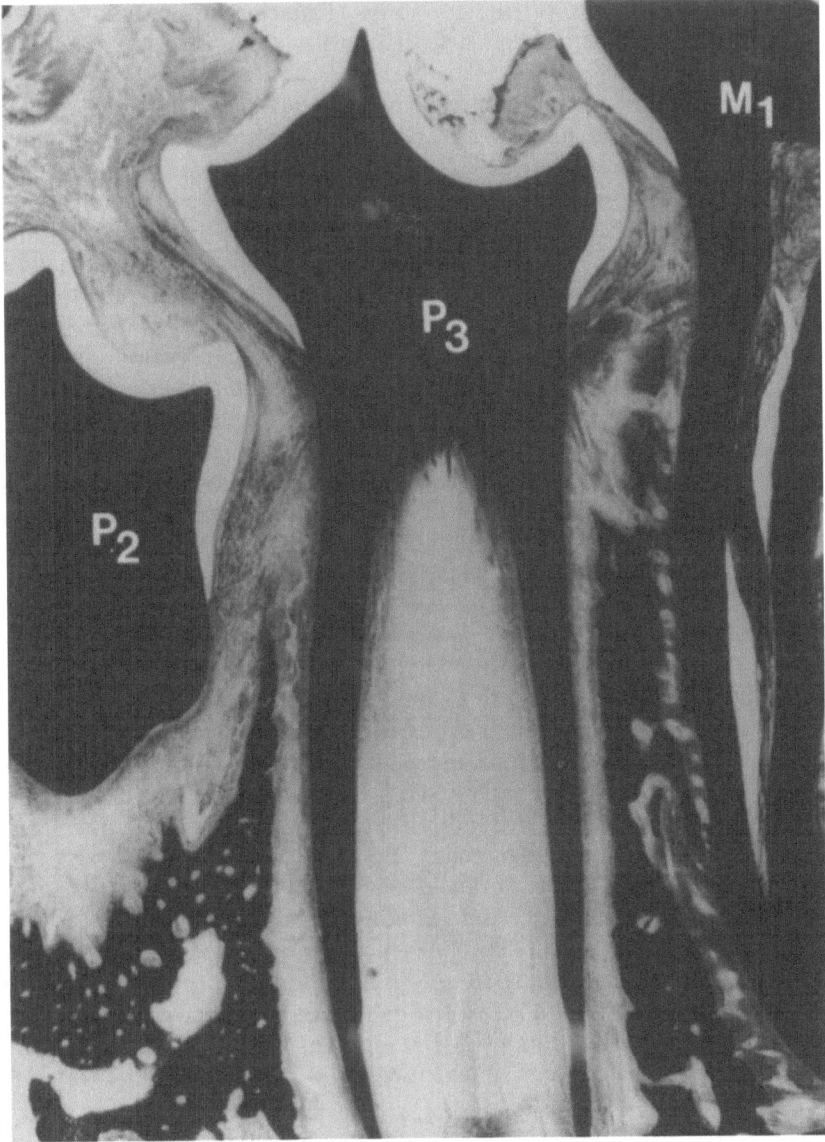

Fig. 6. Photomicrograph showing the emergence into the oral cavity of the third premolar (P_3) of the squirrel monkey. Organized fibers are evident in the coronal and cervical areas. No principal fibers are demonstrable more apically in the periodontal ligament. Distally, the erupted first molar (M_1) shows a classically organized attachment apparatus. P_2 second premolar. × 50. Courtesy of Dr's D.A. Grant and S. Bernick and of the Journal of Periodontology

oral cavity. MAGNUSSON (1968) has observed that a considerable portion of the root in the molar of a macaque monkey is formed before there are obliquely orientated fibers in the periodontal ligament. PEARSON et al. (1975) have reported that unerupted and partially erupted bovine molars show no evidence of organization of the collagen into orientated groups. Thus, a tooth with no predecessor may also lack prominent collagen fiber bundles prior to emergence into the oral cavity.

The above observations are of relevance when considering the proposal that contraction of collagen may be involved in generating the force responsible for tooth eruption (see pages 52–54). This proposal obviously cannot account for the eruption of teeth in the absence of a well-developed collagen fiber system. Even where prominent, orientated collagen fibers have been described in the developing periodontal ligament, care must be taken in interpreting the observations. In many of the studies, silver impregnation techniques have been used to demonstrate collagen. However, structures which appear as distinct argyrophilic collagen fibers at the light microscope level may correspond with areas containing only a few fine collagen fibrils at the electron microscope level (Ten Cate 1972a). Indeed, Ten Cate is of the opinion that the silver impregnation identifies the ground substance-rich, collagen-fibril-poor, extracellular matrix.

There has been controversy concerning the connective tissue immediately beneath the developing root apex, the so-called cushion hammock ligament. This was first described by Sicher (1942a) beneath continuously growing teeth. As a fibrous network with fluid-filled interstices, Sicher (1942a) supposed that it had attachments to alveolar bone and believed that it provided a base where the forces produced by the growing root were resolved into an eruptive force. Although Sicher (1942b) reported that the cushion hammock ligament was absent beneath multi-rooted teeth of limited growth, its existence here was described by Scott (1953). Though subsequent authors have confirmed the presence of a collagenous membrane beneath the developing root (e.g. Hunt 1959; Ness and Smale 1959; Eccles 1961; Ten Cate 1969; Atkinson 1972; Perera 1983), it is not attached to the alveolar wall but merges with the fibers of the developing periodontal ligament (Fig. 7). Perhaps, therefore, it is better termed the 'pulp-limiting membrane'. In a preliminary report on the cellular origin and development of the pulp-limiting membrane, Perera (1983) found no cell types indicative of any special structure and stated that the membrane was only the most apical part of the periodontal ligament developing from the dental follicle. As for its possible function in eruption, root resection and transection studies in continuously growing incisors show that root growth is not responsible for generating eruptive forces (e.g. Berkovitz and Thomas 1969; Moxham and Berkovitz 1974).

Perera and Tonge (1981a) have studied the turnover of collagen in the mouse molar periodontal ligament during the eruptive phase. Their investigations utilised ^{3}H-proline, a material Sodek et al. (1977) have shown is almost entirely incorporated into collagen. Prior to the emergence of the mouse molar into the oral cavity, the turnover rate was high and varied according to site. The rate in the mid-root region and near the apex appeared to be quicker than that near the alveolar crest, having half-lives of ~4.5 days, ~2.5 days, and ~6 days respectively. Regardless of site, collagen turnover occurred throughout the whole thickness of the ligament and there was no evidence of a metabolically active intermediate plexus. Perera and Tonge (1981a) also reported that the rate varied according to the stage of eruption. Although they

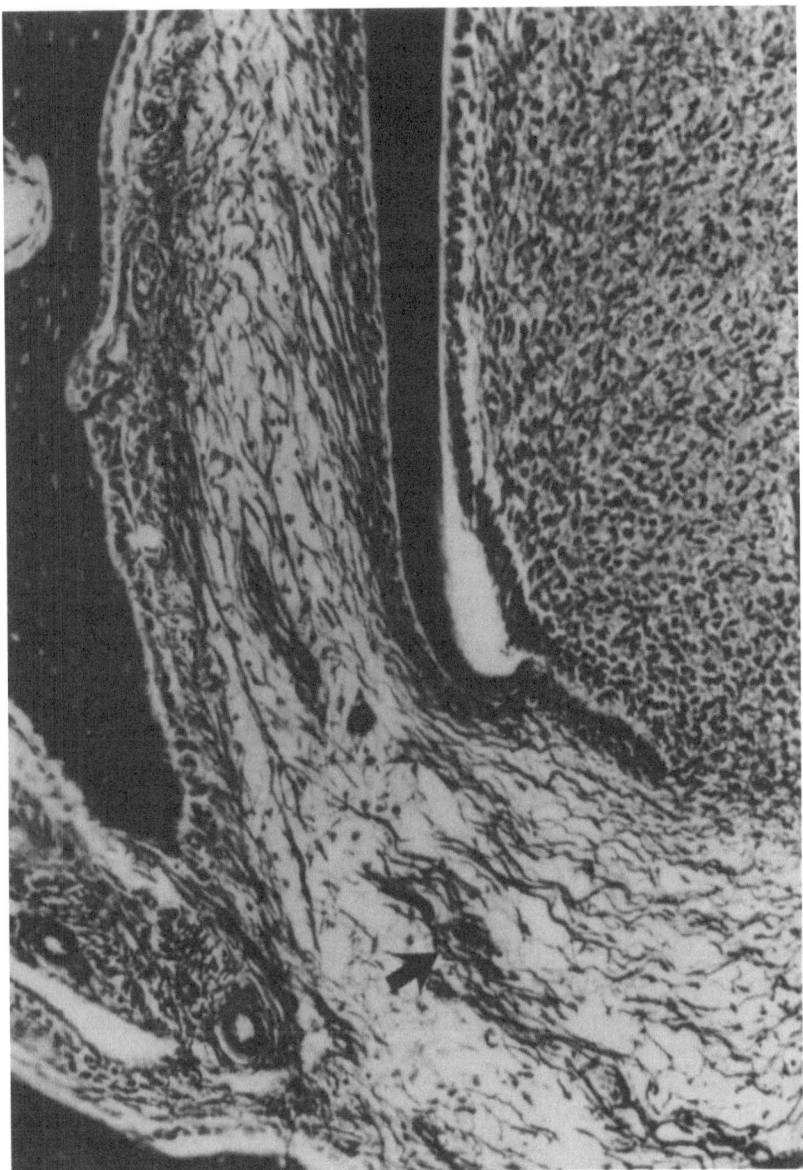

Fig. 7. Photomicrograph of developing root apex showing fibers of pulp-limiting membrane (*arrow*) merging with those of the periodontal ligament. × 200. Courtesy of Dr's D.A. Grant and S. Bernick and of the Journal of Periodontology

claimed that the rate was quickest at the time of emergence, their data for half-lives between 10 and 16 days does not appear to fully confirm this. An earlier qualitative investigation by KAMEYAMA (1973a) provides some information about the initial uptake of ^{3}H-proline but not about its subsequent turnover. As PERERA and TONGE (1981a), he observed that the pattern in the developing rat molar periodontal ligament varied with the location and with the stage of eruption. A change in the labelling pattern appeared to occur first in the cervical region of the developing ligament when the tooth emerged into the oral cavity. KAMEYAMA (1973a) related this change to the observation that periodontal ligament collagen fibers first become organised cervically (e.g. GRANT and BERNICK 1972), though there is no reason to assume that there is a direct relationship.

Few studies have been undertaken on the biochemistry of collagen in the periodontal tissues of erupting teeth. PEARSON et al. (1975) reported that the periodontal ligaments of unerupted and partially erupted bovine molars contain more hexose than those of fully functional molars. The significance of this is unclear although, speculatively, it has been suggested that there may be a relationship in other connective tissues between the levels of hexose and the diameters and the organisation of collagen fibrils (GRANT et al. 1969; MORGAN et al. 1970; PEARSON et al. 1972). From hydroxyproline analysis, GIBSON (1979) – quoted by PEARSON (1982) – determined that the collagen content increased markedly in the periodontal ligaments of bovine incisors during the earliest stages of eruption and more gradually thereafter. In such teeth, prominent collagen fibers were not evident histologically until after the teeth had emerged into the oral cavity. Thus, while the absence of fiber bundles in some erupting teeth seems to argue against a role for collagen in the eruptive mechanism, care must be taken in view of the presence of considerable amounts of soluble collagen. The apparent absence of collagen in a fibrous form may be of relevance to the proposals of KARDOS and SIMPSON (1979, 1980). They suggested that teeth move in response to pressure and that eruptive movements may be explained in terms of the periodontal ligament being a thixotropic gel (a thixotropic material is one which can undergo gel/sol/gel transformations; PRYCE-JONES 1936). If the ligament is indeed thixotropic then the collagen fiber bundles seen in histological preparations are fixation artefacts. It is difficult to reconcile this view with the observation that highly organised fibers are seen in unfixed and undemineralised sections of the periodontal ligament (MASHOUF and ENGEL 1975). Analysis of the degree of cross-linking of collagen in the developing periodontal ligament appears to argue against the hypothesis that collagen contraction generates the eruptive force, at least according to the mechanism proposed by THOMAS (1965, 1967, 1976). THOMAS claimed that the development of covalent cross-links within the collagen generates a tensional force leading to eruption. PEARSON et al. (1975) demonstrated that the collagen of the periodontal ligament was subsequentially cross-linked before eruption commenced and that there was no apparent increase in cross-links relative to the hydroxyproline content of the tissue during the eruptive process.

Less research has been conducted on the cells of the developing periodontal ligament than on the collagen fibers. The transplantation studies of TEN CATE et al. (1971), TEN CATE and MILLS (1972), and FREEMAN et al. (1975) indicate that all the cells of the periodontal ligament can be derived from the investing layer of the dental follicle immediately adjacent to the tooth germ. From an ultrastructural study of developing mouse molars, FREEMAN and TEN CATE (1971) observed that prior to root formation, the cells of the investing layer

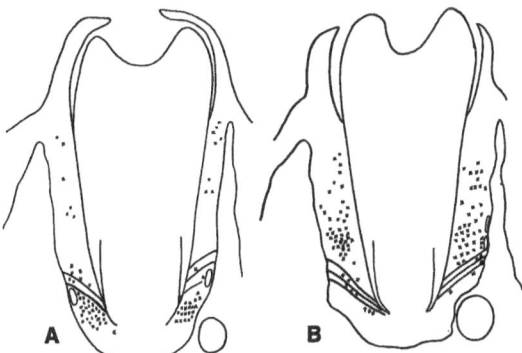

Fig. 8 A, B. Diagrams showing the distribution of ^{3}H-thymidine-labelled cells in developing mouse molar periodontal ligament. **A** 3 hours after labelling, the majority of cells (74%) occur in the unorganised tissue of the dental follicle immediately in advance of the forming periodontal ligament (*the oblique lines*). **B** 72 hours after labelling, the majority of cells (78%) are located within the formed periodontal ligament. Courtesy of Dr. A.R. Ten Cate and of Archives of Oral Biology

of the dental follicle resembled undifferentiated fibroblasts, containing few cytoplasmic organelles but with accumulations of glycogen. With the onset of root formation, there was an abrupt transition from the loose connective tissue containing undifferentiated fibroblasts to a more organized tissue consisting of actively functioning fibroblasts. The functioning fibroblasts showed an increase in organelles (particularly those associated with protein synthesis and secretion) and a loss of glycogen. Indeed, the authors claimed that they could trace the origin of periodontal fibroblasts and cementoblasts to the cells of the investing layer of the dental follicle. FREEMAN and TEN CATE (1971) also suggested that periodontal fibroblasts may develop from perivascular mesenchymal cells. Using ^{3}H-thymidine, TEN CATE (1972b) reported that most labelled cells initially were found near the blood vessels close to the surface of the alveolar bone in the unorganized tissue of the dental follicle immediately below the developing periodontal ligament (Fig. 8A). Electron microscopic examination of this area revealed that cell division was confined to perivascular cells. With continued root formation, the labelled cells were seen to be incorporated into the periodontal ligament (Fig. 8B). The source of the perivascular cells has not yet been determined. Investigations concerning the origin of developing fibroblasts in other sites of the body suggest that they may arise locally or have a haematogenous origin (for reviews see VAN WINKLE 1967; MELCHER and EASTOE 1969). TEN CATE (1972b) believes that the perivascular cells in the developing periodontal ligament do not have a haematogenous origin. This belief stems from his transplantation studies (TEN CATE et al. 1971; TEN CATE and MILLS 1972) which indicate that, as all the supporting tissues of the tooth are derived from the investing layer of the dental follicle, so must the perivascular cells. However, this claim assumes that normal development follows the same pattern as that seen during transplantation. Indeed, it is conceivable that, although the investing layer has the potential to form all the supporting tissues, this is only realized under abnormal conditions.

It has been proposed that an apico-occlusal migration of fibroblasts in the periodontal ligament is responsible for generating the eruptive force (see pages 54–60). Consequently, the population kinetics and migratory activity of fibroblasts in the developing periodontal ligament is of importance. To date, few studies have investigated these features in the periodontal tissues of an erupting tooth prior to emergence into the oral cavity. TEN CATE (1972b) observed that over a period of 3 days there was a shift in cells labelled with ^{3}H-thymidine from an initial site in the apical region of the developing mouse molar to a more occlusal position in the forming periodontal ligament (Fig. 8). More recently, PERERA and TONGE (1981b) found that the labelling index varied in different regions of the ligament in the erupting mandibular first molar of the mouse. During the first 12 h after injection of ^{3}H-thymidine, the labelling index was greatest in the apical third of the developing periodontal ligament, supporting the concept that this is a progenitor zone. There was a subsequent increase in the percentage of labelled fibroblasts in the middle and cervical zones (after 2 and 3 days respectively). These findings were interpreted by the authors as indicating that fibroblasts migrate from the apical zone into the middle and cervical zones. However, the middle and cervical thirds of the developing ligament also had measurable, though smaller, labelling indices during the first 12 h. PERERA and TONGE concluded, therefore, that the periodontal fibroblasts maintained some proliferative activity as they migrated, albeit at a decreased level. Alternatively, it is possible that the progenitor zone is not restricted to the apical region. PERERA and TONGE (1981b) further reported that the proliferative and migratory activities were highest at 12 days when the teeth were emerging into the oral cavity. At 20 days (when the teeth were fully erupted), such activities were considerably decreased. Thus, it would appear that fibroblast proliferation and migration in the developing periodontal ligament are related to the stage of eruption. PERERA and TONGE (1981c) reported that their overall data best fitted a steady state system as the kinetic model applicable to the apical progenitor zone of the periodontal ligament. Accordingly, the population size at this position remains unaltered, though there would be a dynamic equilibrium between cell input (by cell birth or by inward migration) and cell output (by cell death or by outward migration). However, not all the data were consistent with this view. When the results were analyzed using the Gilbert Computer Program which assumes a stationary population (GILBERT 1972), there was a failure of fit. PERERA and TONGE (1981c) claimed that this could be accounted for by cell migration from the apical zone to other zones. The authors also estimated the time course for the various phases of the cell cycle for periodontal ligament fibroblasts in the apical zone, but again assuming a steady state system. To explain the increased rate of fibroblast production during the most active phase of eruption, a reduction in cell cycle parameters (particularly DNA synthesis time and cell generation time) was invoked.

A number of difficulties arise in interpreting data of this kind. Firstly, the underlying assumption is that periodontal fibroblasts are all derived locally from the apical progenitor zone. As previously mentioned, however, some fibroblasts may have a haematogenous origin. Secondly, it is not possible to distinguish cell populations of periodontal fibroblasts from those of other connective

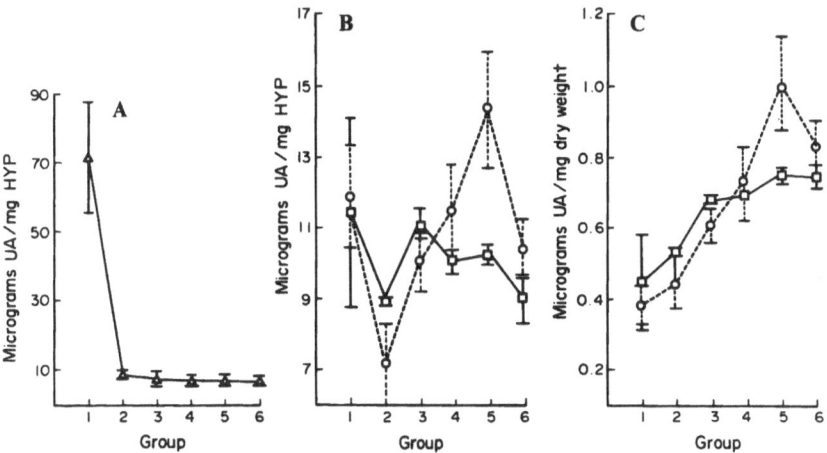

Fig. 9A–C. Graphs showing glycosaminoglycans in the developing periodontal ligament. Unerupted and erupted bovine incisors were divided into six groups. Group *1* being the youngest with only a dental follicle surrounding the unerupted crown and Group *6* being the oldest with the teeth fully erupted. Eruption had just occurred in Group *5* specimens. *UA* hexuronic acid. *Hyp* hydroxyproline. **A** Results for hyaluronate. **B** and **C** Results for dermatan sulphate (□——□) and PG1 proteoglycan (o-----o). Courtesy of Dr. C.H. Pearson

tissue cell types (e.g. cementoblasts and osteoblasts). Thirdly, because the root may be growing downwards and the alveolar process upwards, it is not possible to identify, with certainty, homologous areas of periodontal ligament from one developmental stage to the next. Consequently, care must be taken in interpreting an apparent shift of labelled tissue as evidence of migration. Finally, even if fibroblasts move, one cannot determine whether movement is active or whether the fibroblasts are carried passively with other connective tissue elements in response to tooth movement.

There is little information about changes in the ground substance of the periodontal ligament during eruption. PEARSON et al. (1975) analyzed separated glycopeptides in the tissues of unerupted, partially erupted, and fully erupted bovine incisors. They found that there is less insoluble, non-collagenous glyco-proteins in the mature ligament and that hyaluronic acid progressively decreases relative to chondroitin sulphate during eruption. PEARSON (1982) has described experiments conducted by GIBSON (1979) which showed variations in the quantity of proteoglycans within bovine periodontal tissues during eruption. There was a marked decrease in hyaluronate during the interval between the pre-eruptive stage when the tooth was surrounded by the dental follicle and the stage where a periodontal space was discernible though without identifiable collagen fibres. No further change in hyaluronate levels occurred during subsequent stages of development (Fig. 9a). On a dry weight basis, the amounts of the sulphated galactosaminoglycan components of both the major proteoglycans in bovine periodontal ligament slightly increased as eruption proceeded. Some differences between them were discernible however (Figs. 9b and 9c). With respect to dermatan sulphate (and thus presumably proteodermatan sul-

phate proteoglycan), the increase was not apparent when expressed relative
to the collagen content. Indeed, there seemed to be a decrease leading up to
the emergence of the tooth into the oral cavity. For the PG1 proteoglycan
containing CS(DS) hybrid (a proteoglycan characteristic of the periodontal ligament ground substance (Pearson and Gibson 1982)), on the other hand, the
increase remained when expressed relative to the collagen content. The possible
significance of these findings for the eruptive mechanism is discussed later (pages
60–61).

There is a paucity of information concerning the development of other periodontal ligament structures during the eruptive phase. This is regrettable in view
of the fact that some of these structures (e.g. oxytalan fibers, periodontal vasculature) have been implicated in the eruptive process.

Oxytalan fibers resemble pre-elastin fibres ultrastructurally and histochemically. They stain with aldehyde fuchsin only after oxidation of the tissue
(Fullmer 1960; Rannie 1963). Oxytalan fibres have been found in human
teeth when about 2 mm of root dentine has been formed (Fullmer 1959, 1967).
As the epithelial root sheath looses its continuity, oxytalan fibers become incorporated into cementum, initially at the cervical margin and gradually progressing
towards the root apex.

With regard to the innervation of the periodontal ligament, Fearnhead
(1967) has reported that nerve fibers can be demonstrated in the pulp but not
in the periodontal ligament prior to eruption. With root formation and eruption,
nerves grow from the adjacent alveolar bone into the developing ligament, most
of them accompanying periodontal vessels. Levy and Bernick (1968) and Bernick (1960) have shown that the sensory innervation in the teeth of marmosets
and rats is not established until the ligament has become fully organised at
the time of eruption.

The development of the vasculature of the periodontal ligament of monkey
permanent teeth has been described by Cutright (1970). The first sign of vascular development was the appearance of an encircling plexus of vessels within
the alveolar bone. This was connected to the periodontal vessels of the overlying
deciduous tooth. The encircling plexus was seen to give rise to pulp vessels.
In addition, a periodontal plexus arose by "a thinning and compression of
the vessels of the encircling plexus, forming a dense network of flattened vessels
within the periodontal membrane". Direct vascular connections were noted
between the periodontal plexuses of neighbouring teeth. Kindlová (1970) has
described the development of blood vessels in the gingivae of rat molars. She
noted that they were derived from vascular networks from the enamel organ
and the alveolar mucosa. Vessels then spread apically to supply the developing
periodontal ligament. Prior to eruption, the vascular bed of the periodontal
ligament was found to be more uniform than in fully erupted teeth, indicating
that its pattern may be influenced by function.

It has been suggested that changes in vascular permeability in the periapical
connective tissues of developing teeth can be related to their eruptive behaviour.
Magnusson (1968) described dense accumulations of tissue fluid (termed effusions) beneath the growing roots of erupting molars in monkeys and rats
(Fig. 10). These effusions were observed using PAS and PTAH stains. Magnus-

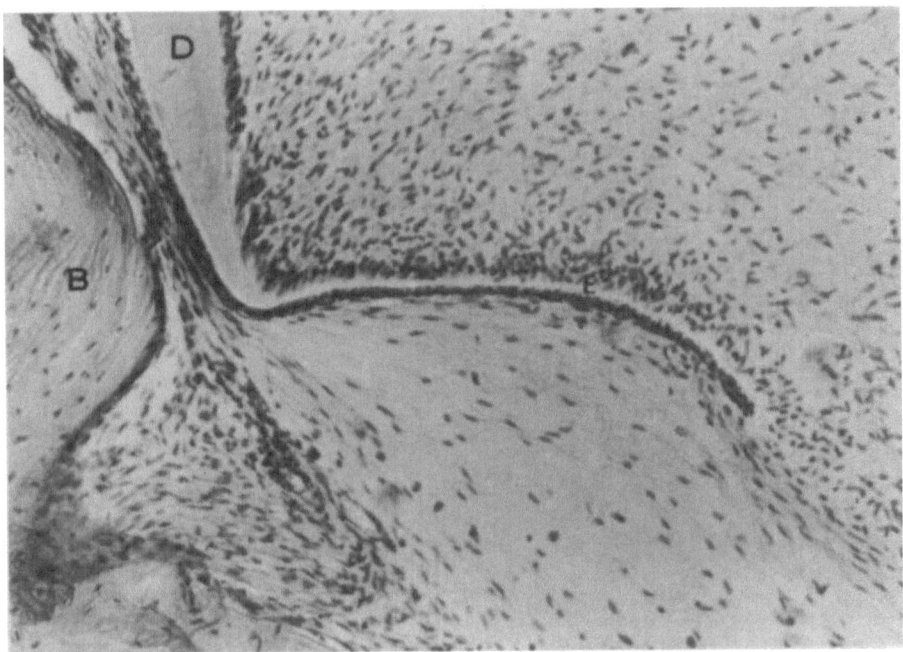

Fig. 10. Photomicrograph of developing root apex of a monkey molar tooth showing "effusions" in the tissue beneath the epithelial root sheath (*E*). *D* Root dentine; *B* alveolar bone. × 150. Courtesy of the University of Umeå

SON (1973) reported that radioactive fibrinogen injected into rats intracardially became incorporated rapidly into the effusions beneath the erupting molars and incisors, supporting the view that they are vascular in origin. As the effusions were induced when the growing root was situated close to the fundus of the alveolar bone, MAGNUSSON (1968) thought that they then forced the root and bone apart, thereby contributing to eruptive migration and enabling further growth.

2. Adaptation of the Alveolar Bone to Eruption

MANSON (1967, 1968) used tetracycline to study the changes in alveolar bone which occur with eruption of the teeth in a variety of animals. In addition to activity associated with relocation of the tooth in the growing jaw, he observed resorption on both the lingual and facial walls of the crypt above the points of maximum convexity of the crown of the erupting tooth. The amount of resorption was usually greater on the facial wall. Bone deposition occurred below the maximum convexity and was greater on the lingual surface where rapid formation resulted in woven bone being frequently found. Resorption was the predominant activity in the fundus of the crypt. On occasions where deposition was found (most frequently in relation to the erupting canine), it could be related to relocation of the crypt within the growing mandible and

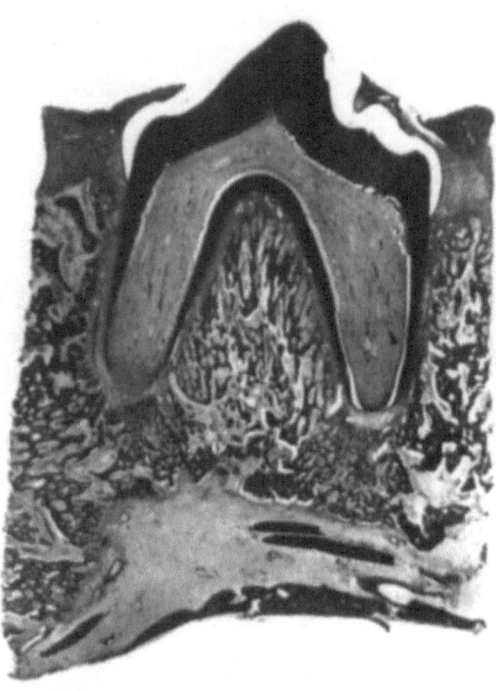

Fig. 11. Photomicrograph of erupting premolar in the dog illustrating evidence of active bone deposition beneath the root apex. Note also the deposition of bone in the interradicular region. × 5. Courtesy of Dr. D.R. Cahill and of Anatomical Record a publication of Wistar Press

not to eruption. Manson concluded that deposition of bone at the fundus could not be the direct cause of tooth eruption.

The findings of Cahill (1969, 1970) and Cahill and Marks (1980, 1982) differ from those of Manson. They observed considerable amounts of bone being deposited beneath the roots of erupting premolar teeth in dogs (Fig. 11). Cahill and Marks (1980) concluded that tooth eruption is the result of bone resorption above the occlusal surface of a developing tooth with bone deposition beneath its root(s). As experiments involving surgical resection and transection of continuously growing rodent incisors show that alveolar bone deposition cannot be responsible for eruption in such teeth (Bryer 1957; Kostlán et al. 1960; Berkovitz and Thomas 1969; Berkovitz 1971; Moxham and Berkovitz 1974; Pitaru et al. 1976), Cahill and Marks (1982) claim that this type of tooth is atypical and not suitable for studying the eruptive mechanism. They believe that the non-continuously growing dentition in dogs is most suitable, corresponding to that found in humans. However, the research of Manson (1967, 1968) indicates that differences even exist between teeth of non-continuous growth. Indeed, the absence of bone deposition beneath erupting rat molars has also been observed by O'Brien et al. (1958) and Kameyama (1973b). It is conceivable that these differences may be related to whether the teeth have predecessors. CAHILL and MARKS used premolars whereas MANSON, O'BRIEN et al. and KAMEYAMA studied teeth without predecessors. Differences between teeth with and without predecessors have been noted already with regard to the development of the collagen fibers within the periodontal ligament. Concerning the claim that the dog dentition corresponds to that of the human, LEVERS (personal communication) has observed that bone activity at the base of alveolar crypts in human mandibular teeth shows considerable variation. For example, the base of the crypt moves occlusally during the most active phase of eruption for the second premolar and second permanent molar, whereas it resorbs throughout eruption for the first and third permanent molars.

Two further points should be mentioned in considering whether alveolar bone deposition is responsible for eruption. Firstly, there is the problem of distinguishing cause and effect. It is not yet possible to determine whether bone deposition beneath an erupting tooth is causing eruption or is a response to eruption produced by another agent. Secondly, fundic bone deposition may be related to the distance the tooth has to erupt. If this distance is greater than the length of the root then, clearly, fundic bone deposition is necessary to maintain the normal dimensions of the periodontal tissue in this region (THOMAS 1965; KENNEY and RAMFJORD 1969). This may explain the differences between teeth with and without predecessors, for teeth with predecessors develop deep within the jaws and will thus have to erupt greater distances.

Bone deposition around the alveolar crest is conspicuous during the eruptive phase and coincides with an important growth period for the jaws. This deposition also has been implicated in providing an eruptive force by producing tension on fibers of the periodontal ligament (WESTIN 1942) or, in the case of multi-rooted teeth (Fig. 11), by generating pressure in the region of the root bifurcation (SICHER 1942b). It is seen that deposition of bone around the alveolar crest can occur independently of tooth eruption in cases of submerging deciduous human molars and following immobilization of erupting hamster molars (GREGG 1965). As the tooth prior to eruption lies within the alveolar crypt, the distance that the tooth erupts far exceeds the amount of alveolar bone deposited at the alveolar crest. Furthermore, human teeth exhibiting retarded eruption can erupt following surgical exposure into an established arch without a localized overgrowth of the alveolus. MANSON (1968) concluded that alveolar bone growth was unrelated to tooth development as, in the cat dentition, the amount of deposition above the forming permanent molar was approximately the same as that around the fully developed deciduous molars and as that around the oral surface of the diastema. He regarded such deposition as part of the vertical growth of the ramus.

3. Alterations in the Tissues Overlying the Erupting Tooth

As the tooth approaches the oral mucosa, the external enamel epithelium covering the crown actively proliferates into the overlying connective tissue (McHUGH 1961; TOTO and SICHER 1966). TOTO and SICHER (1966) and MELCHER (1967) have suggested that the epithelial cells secrete enzymes capable of collagen degradation. MELCHER (1967) reported that as the collagen fibers degenerate they lose their birefringence and their affinity for stains. Although it appeared that the degeneration preceded epithelial proliferation, it was not possible to determine whether the cells proliferate into the area after the collagen fibers have been removed by some other agency, or whether they mediate the removal. In silver impregnated sections, MELCHER (1967) also observed a network of reticulin in areas where collagen had been removed. He postulated that at least part of the reticulin network was newly laid down. Depolymerisation of muco-polysaccharides has also been detected in the connective tissue overlying erupt-ing teeth (ENGEL 1951). TEN CATE (1971) observed that fibrocytes in the overly-ing connective tissue cease fibrillogenesis, actively take up extracellular material and synthesize acid hydrolases; these changes leading eventually to disintegra-tion of the cells (Fig. 12). He further suggested that the reduced enamel epithelial cells aid the removal of breakdown products. There may be a relationship be-tween degeneration of the connective tissue and pressure exerted by the underly-ing erupting tooth, ischaemia having been proposed as a contributory factor (SCHOUR 1960). That pressure alone is not responsible is indicated by the finding that there is always evidence of new collagen formation in this region (FULLMER 1961).

If the erupting tooth has a deciduous predecessor, resorption and shedding of the deciduous tooth must take place to provide an eruptive pathway. NANDA (1969) has conducted a radiographical survey to determine rates of resorption of deciduous teeth in a large sample of Indian children aged 6–12 years. No sexual differences were detected but the rates varied with age and site (Fig. 13). The rates reported by NANDA (1969) are considerably slower than those reported by FANNING (1961) and MOORREES et al. (1963) from smaller surveys conducted on American children.

For a deciduous incisor or canine, root resorption initially occurs on the lingual surface adjacent to the developing permanent tooth. With subsequent movement and relocation of the teeth in the growing jaws, the developing perma-nent tooth comes to lie directly beneath the deciduous tooth and further resorp-tion occurs from the apex. For a deciduous molar, root resorption often com-mences on the inner surfaces where the permanent premolars initially develop. The premolars later come to lie beneath the roots of the deciduous molar and further resorption occurs from the root apices. The shift in position of the deciduous tooth relative to the permanent successor may account for the inter-mittent nature of root resorption.

Resorption of deciduous teeth results from the activity of osteoclast-like cells. These cells have been termed odontoclasts. However, the evidence suggests that all "-clast" cells associated with vertebrate mineralised tissues are similar morphologically (YAEGER and KRAUCUNAS 1969; FREILICH 1971; ADDISON 1979). In addition, osteoclasts in vitro will resorb cementum, dentine, enamel, biological calcite or non-biological calcite (BOYDE et al. 1983; JONES et al. 1984). The "-clast" cells are usually large and multinucleated (there may be several hundred nuclei, although most have between ten and twenty). The cell surface in contact with the mineralised tissue has a "brush border" and the cytoplasm here has a characteristic foamy appearance. Ultrastructurally, the brush border

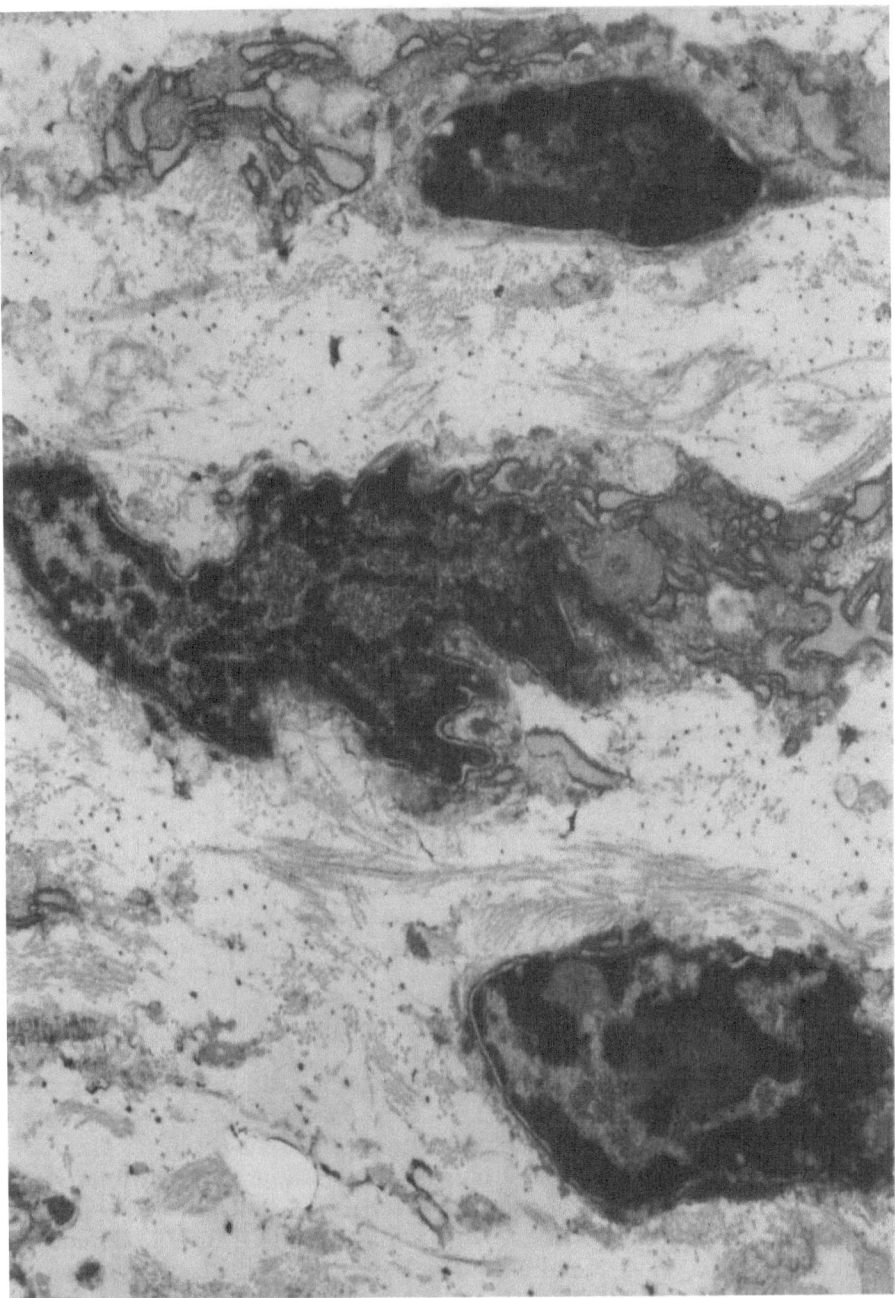

Fig. 12. Electron micrograph of the connective tissue overlying an erupting rat molar showing degenerating fibrocytes with pyknotic nuclei. × 17000

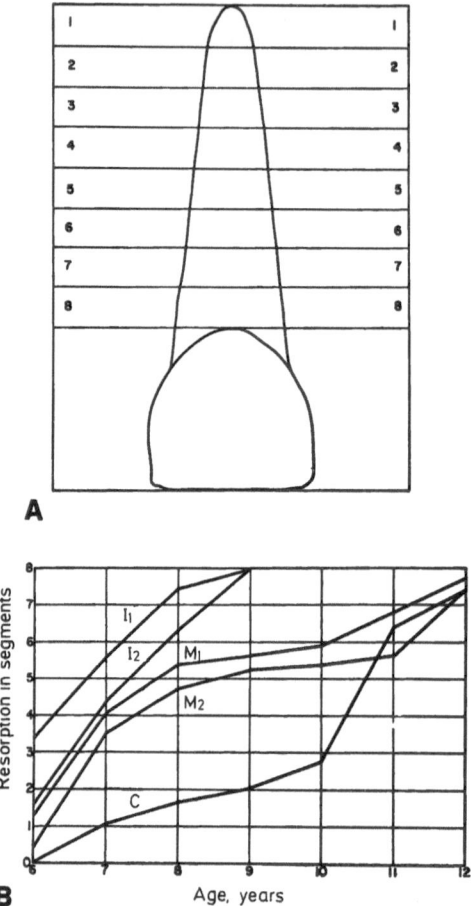

Fig. 13. A Diagram showing arbitrary division of deciduous root into 8 equal segments used for measuring root resorption. **B** Graph showing rates of root resorption of the mandibular deciduous teeth. Courtesy of Dr. R.S. Nanda and of Archives of Oral Biology

is comprised of many tightly packed microvilli (Fig. 14). In odontoclasts, the spaces between the microvilli are 2–3 μm deep. Mineral crystals may be found within the spaces but may possibly be a reprecipitation artefact. The tooth surface shows a zone of decreased mineral content approximately 1 μm wide. The cell contains many mitochondria and vesicles of different sizes and types but endoplasmic reticulum is scarce (FURSETH 1968). The origin of odontoclasts is unknown. However, it is probable that, like osteoclasts, they arise from the fusion of precursor cells of the monocyte/macrophage type (e.g. JEE and NOLAN 1963; GOTHLIN and ERICSSON 1973a and b; JOTEREAU and LE DOUARIN 1978; TINKLER et al. 1981). Indeed, MARKS and CAHILL (1983) found mononuclear cells (with abundant cytoplasm, euchromatic nuclei and prominent nucleoli) in a juxtavascular location in the dental follicle overlying unerupted permanent premolars in the dog. These cells increased in number immediately preceding

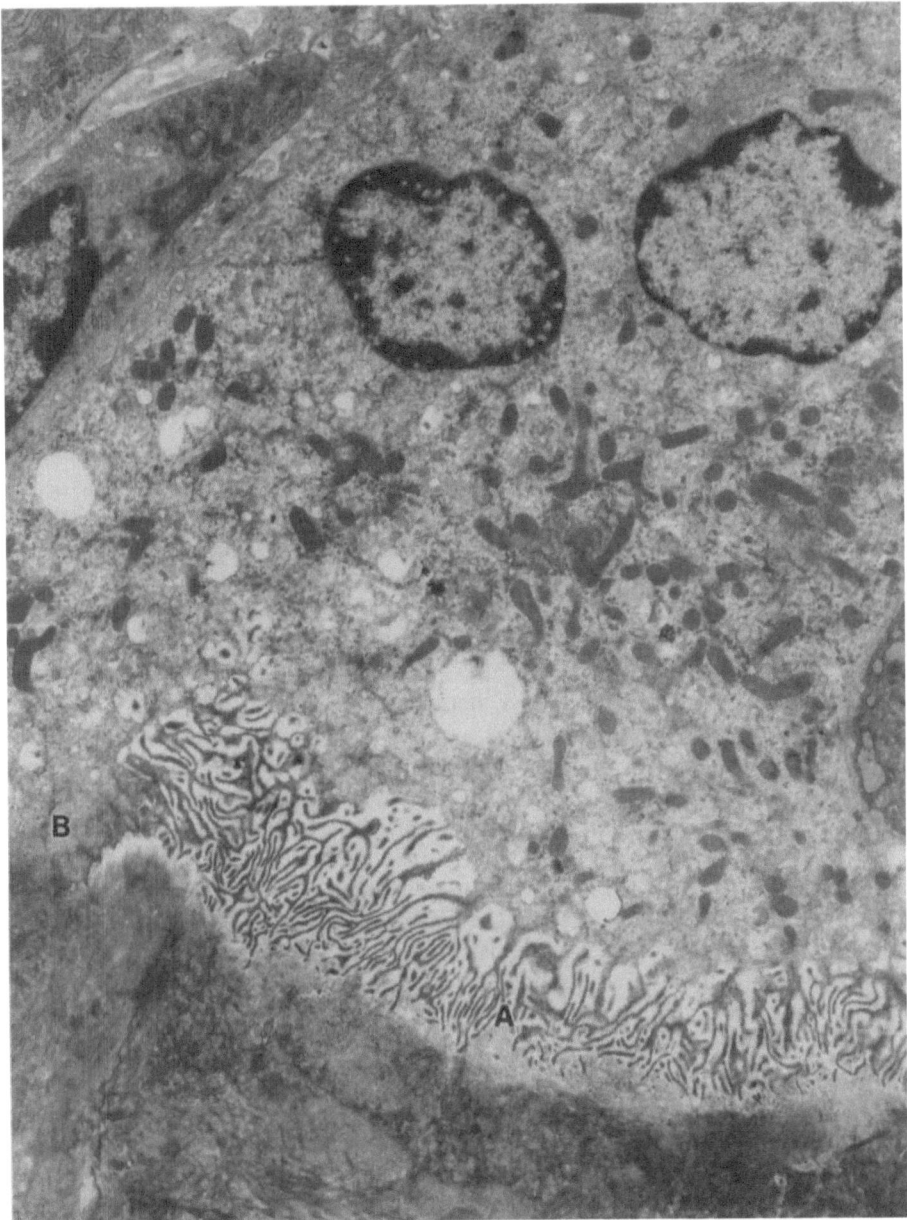

Fig. 14. Electron micrograph of a demineralised section of alveolar bone showing part of a multinuc-leated osteoclast. The brush border (*A*) has numerous microvilli whilst there is a lack of microvilli in the annular zone (*B*). There are numerous vesicles within the cytoplasm but little endoplasmic reticulum. × 7000

and during tooth eruption in parallel with an increase in osteoclasts on the adjacent crypt wall. Recently, the view has been put forward that interactions with other cell types may be important in the differentiation of osteoclasts (e.g. BARON et al. 1983). The experiments of ALI et al. (1984) support this view. Where monocyte-enriched human blood cells were seeded on to sperm whale dentine for up to 20 days, although multinucleate giant cells were formed, there was no morphological evidence of resorptive activity. In contrast, preparations containing known osteoclasts (derived from bone) resorbed the same substrate within hours.

There has been much debate about the precise role of -clast cells in the resorption of calcified tissues. What information we do have mainly concerns osteoclasts. It is thought that they a) secrete substances which degrade both inorganic and organic components and b) phagocytose and digest materials intracellularly (e.g. GOTHLIN and ERICSSON 1976; MELCHER 1978). Unlike periodontal fibroblasts, intracellular collagen profiles have not been observed within osteoclasts, suggesting that the bulk of the digestion of bone collagen occurs extracellularly. This may be related either to the presence of a localised extracellular microenvironment (delineated by the annular zone of the osteoclast (Fig. 14)) or to differences between bone and periodontal ligament collagen. It has even been suggested that the -clast cell is only involved in the initial stages of resorption and that the final degradation of collagen is undertaken by adjacent fibroblasts (BEERTSEN et al. 1978; HEERSCHE and DEPORTER 1979). However, JONES et al. (1984) have demonstrated that osteoclasts can resorb demineralised dentine in vitro.

Figure 15 depicts the histological appearance of a resorbing deciduous tooth. The connective tissue between the crown of the erupting tooth and the resorbing root appears to be highly vascular. This connective tissue in vitro has collagenolytic activity, particularly where it is rich in odontoclasts (MORITA et al. 1970; WOESSNER and CAHILL 1974). The odontoclasts lie in concavities in the resorbing root surface (Howship's lacunae). The lacunae are ovoid and give the resorbing tissue a honeycombed appearance (FURSETH 1968). BOYDE and LESTER (1967) have shown that the lacunae meet beneath the surface of the resorbing root, producing considerable undermining of the tissues. They also found that peritubular dentine remains proud of the resorbing surfaces, its mineral component somehow being selectively protected. BOYDE and JONES (1979) have estimated the size of resorption lacunae in mammalian calcified tissues using SEM stereophotogrammetry. Lacunae in bone were observed to be smaller on average than those in resorbing teeth but a wide range of size was evident in all tissues. This may be related to the number of nuclei and ruffled borders, the nature of the substratum being resorbed, and the freedom of lateral movement of the cell (S.J. JONES, personal communication). A greater proportion of nearly spherical lacunae were found within resorbing deciduous teeth whilst anisotropy was more common in bone where many lacunae resemble paths ("snail track lacunae") rather than pits (Fig. 16). BOYDE and JONES (1979) noted that the dimensions of the resorption lacunae were often considerably smaller than the known dimensions for -clast cells and they therefore deduced that only small portion(s) of a cell may be actively engaged in resorption at any one time.

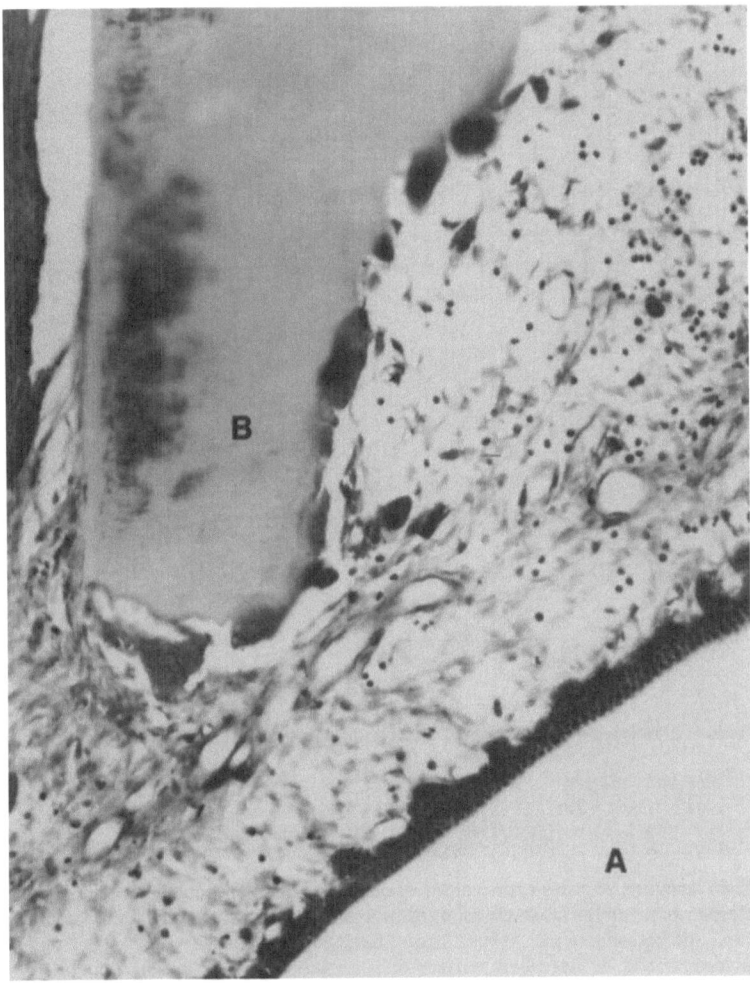

Fig. 15. Photomicrograph of a demineralised section of the jaw of a ferret (*Putorius putorius*) showing part of the crown of an erupting premolar tooth (*A*) beneath the resorbing root of its deciduous predecessor (*B*). The base of the root and much of the pulpal surface of the deciduous tooth are occupied by odontoclasts lying in Howship's lacunae. × 200

The pulp may play an active role in the physiological resorption of deciduous teeth. KRONFELD (1932) showed that odontoclasts resorbing the deciduous teeth of cats and dogs were found on the walls of the pulp chambers and that the teeth were "hollowed out" from within. This has been confirmed by CAHILL (1974) in dogs. It had been thought that internal resorption in human teeth is seen only where the pulp shows pathological changes (KRONFELD 1932; HARNDT 1948; ZEROSI 1965). However, WEATHERELL and HARGREAVES (1966) found signs of resorption at the surface of the pulp in sound deciduous teeth. FURSETH (1968) observed that resorption rarely occurred in the pulp chamber but was seen frequently in the root canals. This may be related to the different

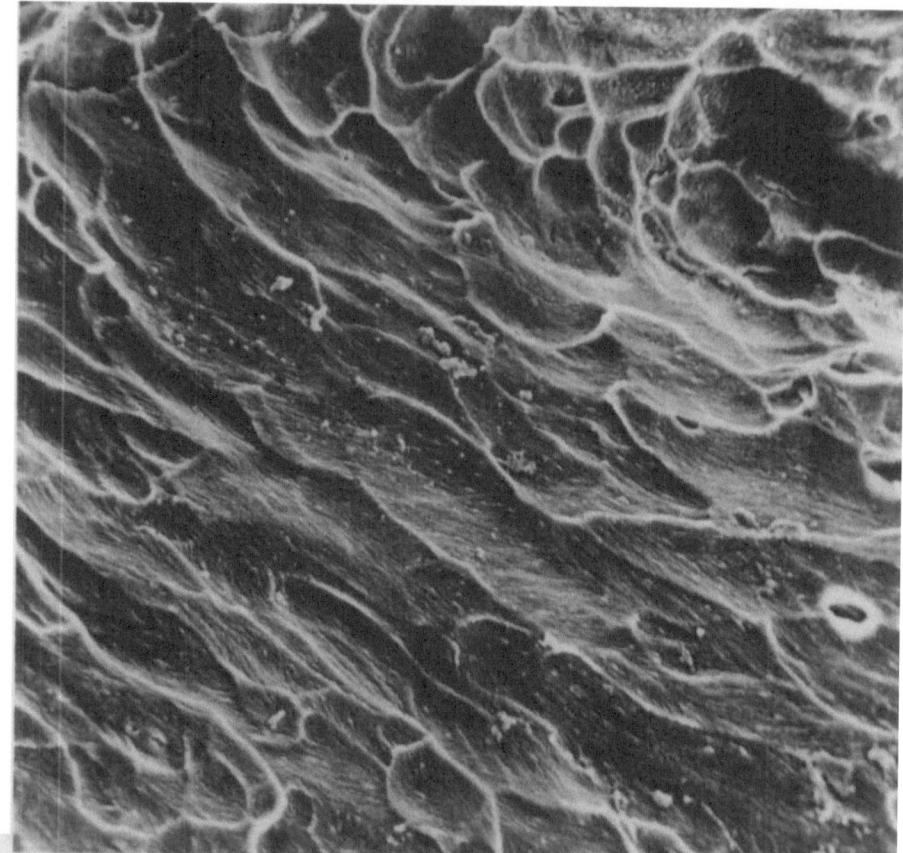

Fig. 16 A, B. Scanning electron micrographs of resorbing bone and dentine. **A** Surface of resorbing human alveolar bone on the labial aspect of the maxilla. Note both the long "snail track" resorption lacunae and the resorption pits. The collagen fibers of the lamellar bone are seen in the walls of the lacunae. × 650. **B** Resorbing surface of a human deciduous tooth. Compared with bone, the lacunae are larger and more rounded. Dentinal tubules are seen as small circular openings in the embayed dentine. × 525. Courtesy of Dr. S.J. Jones

patterns of resorption seen in different human teeth (Ten Cate 1980). Single-rooted teeth usually are shed before the root has been resorbed completely and there is an intact odontoblast layer without any odontoclasts. On the other hand, the roots of deciduous molars are more often resorbed completely before exfoliation and the odontoblast layer is replaced by odontoclasts. It would appear, therefore, that internal resorption is not a general feature of the shedding of human teeth. The internal resorption seen in cats and dogs may be related to the very rapid rate of resorption, the teeth being shed in just a few weeks (Fig. 15).

The factors which initiate, and subsequently influence, root resorption are not fully understood. Pressure from the erupting permanent tooth is generally

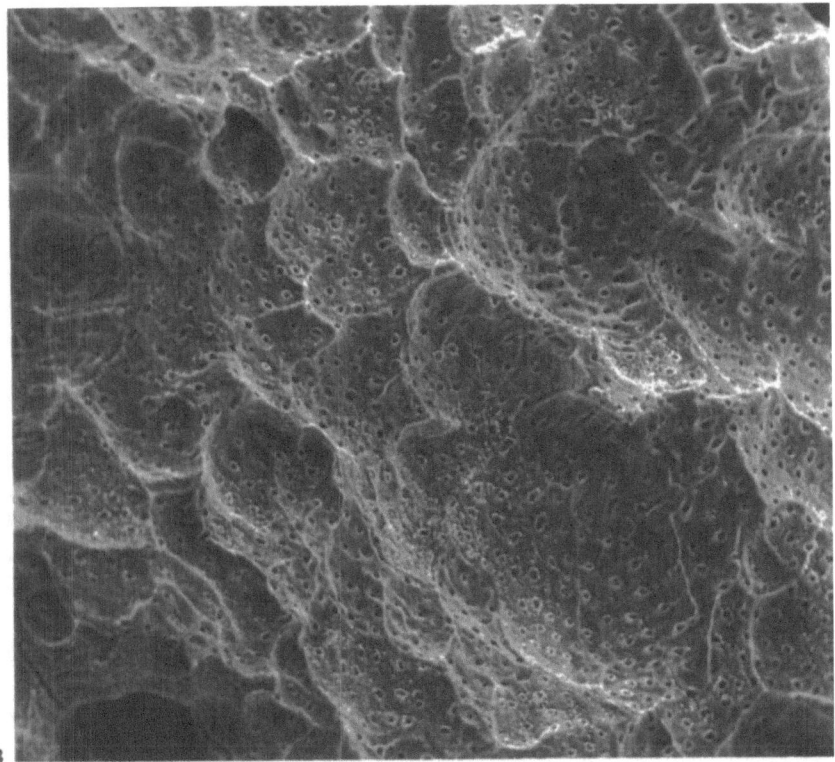

B

thought to be important. To assess this, OBERSZTYN (1963) studied the effects of removing the permanent incisor and canine tooth germs in dogs. Resorption of their deciduous predecessors still occurred, though this was delayed. He did not consider the possibility, however, that the surgical procedures interfered with the process. Nevertheless, his findings are consistent with the clinical observation that shedding of a deciduous tooth is retarded where the successor is congenitally absent or occupies an abnormal position within the jaw. OBERSZTYN (1963) also presented evidence that the influence of the permanent tooth was exerted not solely by its presence but by its position directly adjacent to the deciduous predecessor. Where the direction of eruption of a permanent tooth was changed by osteotomy, a delay in the resorption of the deciduous tooth resulted. OBERSZTYN's conclusions have been extended by CAHILL and MARKS (1982) who implicated the follicle of the underlying permanent tooth rather than the crown itself as being responsible for the resorptive process. It has been suggested that increased masticatory loads affect the pattern and rate of deciduous tooth resorption (WESTIN 1942; BERNICK et al. 1949, 1951). The experiments of OBERSZTYN (1963) provide further evidence for this. He showed that, if the deciduous teeth of dogs were splinted following the removal of the developing permanent teeth, there was less root resorption compared with removal of the permanent teeth alone.

Root resorption may be influenced by inflammation. Nowick (1958), Kli-mecka-Zakiewicz (1960) and Obersztyn (1963) have reported that inflammation speeds resorption. Kantorowicz (1929), on the other hand, claimed that resorption was delayed. Although internal resorption seems to occur in human teeth only when the pulp tissues are inflamed (Kronfeld 1932; Harndt 1948; Zerosi 1965), some may be found in root canals containing sound tissues (Furseth 1968).

Resorption is not continuous but alternates with periods of rest or repair (e.g. Kronfeld 1932; Westin 1942; Sicher 1962; Furseth 1968). A reversal line separates the repair tissue from the underlying dental tissues (Furseth 1968). The repair tissue resembles cellular cementum. Furseth (1968) observed that the formative cells had the fine structure of cementoblasts, that there were lines resembling the incremental lines in cementum and that there was a zone of uncalcified repair tissue homologous to precementum. However, differences were noted between the repair tissue and cementum. Firstly, the width of its uncalcified zone (15 µm) was greater than that reported by Selvig (1965) for precementum (4 µm). Secondly, its degree of mineralisation was less (as judged by electron density). Thirdly, its crystals were smaller. Fourthly, calcific globules were present, suggesting that mineralisation was not proceeding evenly. These differences may be related to the speed of formation of the repair tissue. Where this is very slow, the repair tissue cannot be distinguished histologically or in its mineralisation pattern from primary cementum. However, where the repair tissue is formed rapidly (as in most resorbing deciduous teeth), in most respects it resembles woven bone (Jones and Boyde 1972).

A further feature associated with the eruption of permanent teeth is the gubernaculum. Initially, each deciduous tooth and its developing permanent successor share a common alveolar crypt, the permanent tooth germ being situated lingually. With continued growth and eruption of the deciduous tooth, the permanent tooth comes to lie within its own crypt near to the root apex of the deciduous tooth. However, the crypt of the permanent tooth is not complete, there being a canal (the gubernacular canal) in its roof through which the dental follicle is attached to the overlying oral mucosa. Permanent molars lack deciduous predecessors but also have a gubernaculum (Scott 1948; Car-ollo et al. 1971; Hodson 1971). The gubernacular canal contains a structure called the gubernacular cord. The cord is comprised of a central strand of epithelium (derived from the dental lamina) surrounded by connective tissue (Fig. 17). Carollo et al. (1971) reported that the connective tissue is organised into inner and outer layers. Collagen fibers of the inner layer showed greater organisation and ran mainly parallel to the long axis of the epithelium. In the outer layer, the collagen fibers were fewer and less organised. Differences between the layers were also discerned with respect to the vasculature, the vessels in the outer layer being larger.

It has been suggested that the gubernacular cords play an active role in the movements of the erupting permanent teeth through the jaws (Scott 1948, 1953, 1967). During eruption, the gubernacular cords decrease in length but increase in thickness (Scott and Symons 1971; Cahill 1974; Cahill and Marks 1982). In addition, they become less dense (Scott and Symons 1971).

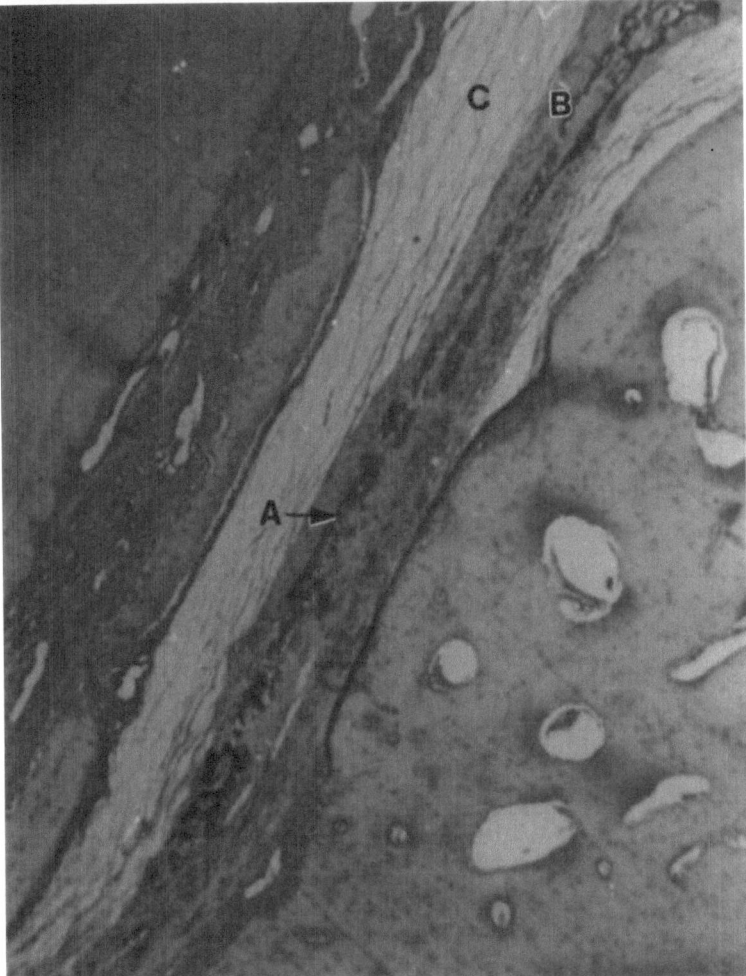

Fig. 17. Photomicrograph showing the gubernacular cord lying in the gubernacular canal. The cord contains a central strand of epithelium (*A*) surrounded by inner (*B*) and outer (*C*) connective tissue layers. × 50

Whether the gubernaculum provides an eruptive pathway or is actively engaged in pulling the tooth out has not been established, though it cannot be implicated in the process of eruption once the tooth has breached the mucosa. The observation that there was disintegration of the epithelium and loss of connective tissue organisation within the gubernaculum just before eruption of the tooth led CAROLLO et al. (1971) to conclude that the eruptive mechanism did not reside in the gubernacular cord. Furthermore, surgical removal of the cord does not prevent eruption of the permanent tooth (CAROLLO et al. 1971; CAHILL and MARKS 1980).

CAHILL (1969) has studied the eruptive pathway of the permanent premolars in dogs under different experimental conditions. When the premolars were

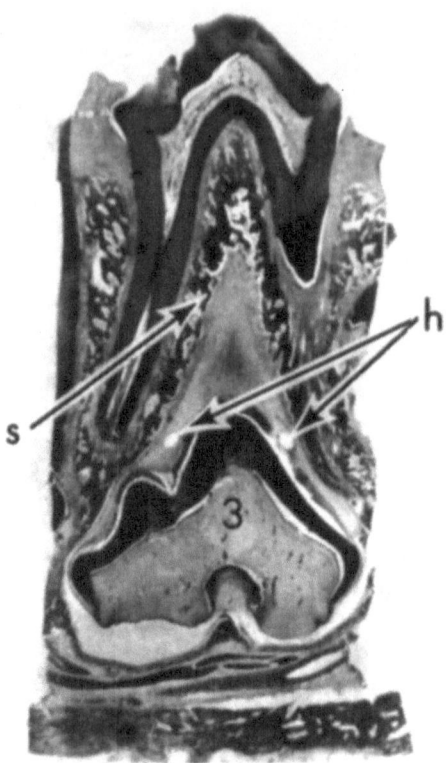

Fig. 18. Photomicrograph showing the crown of the third premolar (*3*) of a dog which had been impacted for two weeks by metal wires. An eruptive pathway has still formed above the tooth due to osteoclastic resorption along the crypt wall (*s*). *h* Holes for impaction wires. × 5. Courtesy of Dr. D.R. Cahill and of Anatomical Record a publication of Wistar Press

pinned to prevent their eruption, bone resorption still occurred above them to form a normal pathway for eruption (Fig. 18). Thus, continuous and direct pressure from an erupting tooth is not essential for the formation of an eruptive pathway. These findings led CAHILL (1969) to suggest that osteoclastic activity by the dental follicle (including the gubernaculum) aids directly in enlarging the pathway.

CAHILL and MARKS (1980, 1982) observed that removal of developing dog premolars without major disturbance of the dental follicle did not prevent the creation of an eruptive pathway, there being resorption of alveolar bone above and deposition of bone in the base of the empty bony crypt. However, when the tooth and surrounding follicle were removed and only the crown replaced, there was no eruption of the tooth and no enlargement of the eruptive pathway. When the crown was replaced by a metal or silicone replica and the dental follicle retained, the replica "erupted" with the formation of an eruptive pathway, with root resorption and exfoliation of the deciduous predecessor and with substantial alveolar bone deposition at the base of the crypt. CAHILL and

MARKS concluded that tooth eruption is a series of metabolic events in alveolar bone characterised by bone resorption and formation which is co-ordinated by the dental follicle and to which the tooth does not contribute. It is to be remembered that in the normal situation the connective tissues of the dental follicle develop into those of the periodontal ligament.

III. The Intra-Oral Phase

During this phase, the tooth erupts through the oral mucosa into the oral cavity and continues to move axially to attain its functional position. Maximum eruption rates are approximately 1–2 mm/month (BURKE and NEWELL 1958; BURKE 1963; BERKOVITZ and BASS 1976; SMITH 1978).

Tooth emergence may be facilitated by recession of the mucosa around the erupting tooth. This is termed passive eruption to distinguish it from active eruption where crown exposure is due to bodily movement of the tooth. BERKOVITZ and BASS (1976) have demonstrated that gingival retraction may contribute significantly to exposure of the clinical crown of human maxillary permanent third molars. SMITH (1978) has recorded the changing depths of the gingival crevice during eruption. The crevices of teeth which had just penetrated the mucosa were found to exceed 7 mm in depth, but became more shallow as eruption continued. This reduction was associated partly with active eruptive movements and partly with gingival recoil. The patterns of eruption and of gingival recoil were characterized initially by rapid changes followed by slower ones (though considerable variations were found).

Much information is available concerning the changes which take place in the epithelial tissues of the oral mucosa as the tooth emerges. This leads to the establishment of the junctional epithelium (for review see SCHROEDER and LISTGARTEN 1977).

Several studies have been concerned with the changes which occur in the periodontal ligament during the intra-oral phase. The appearance of collagen fiber groups in the developing periodontal ligament during the eruptive phase has been discussed previously (pages 22–28). The evidence suggests that the pattern of development relates to whether the tooth has a predecessor. LEVY and BERNICK (1968) and GRANT et al. (1972) have reported that the periodontal ligaments of monkey teeth lacking predecessors show advanced fiber formation at the time of emergence into the oral cavity. This is supported by the findings of O'BRIEN et al. (1958), BERNICK (1960), THOMAS (1965), FORMICOLA and FERRIGNO (1966), MAGNUSSON (1968) and ATKINSON (1972). However, the work of PEARSON et al. (1975) indicates that partially erupted bovine molars do not have well organised fiber groups. GRANT and BERNICK (1972), GRANT et al. (1972) and BERNICK and GRANT (1982) have shown that monkey permanent teeth with deciduous predecessors only show organised fibers around the alveolar crest at the time of emergence into the oral cavity. Even when the teeth first reach the occlusal plane, principal fibers still do not cross the ligament in the apical region (Fig. 19).

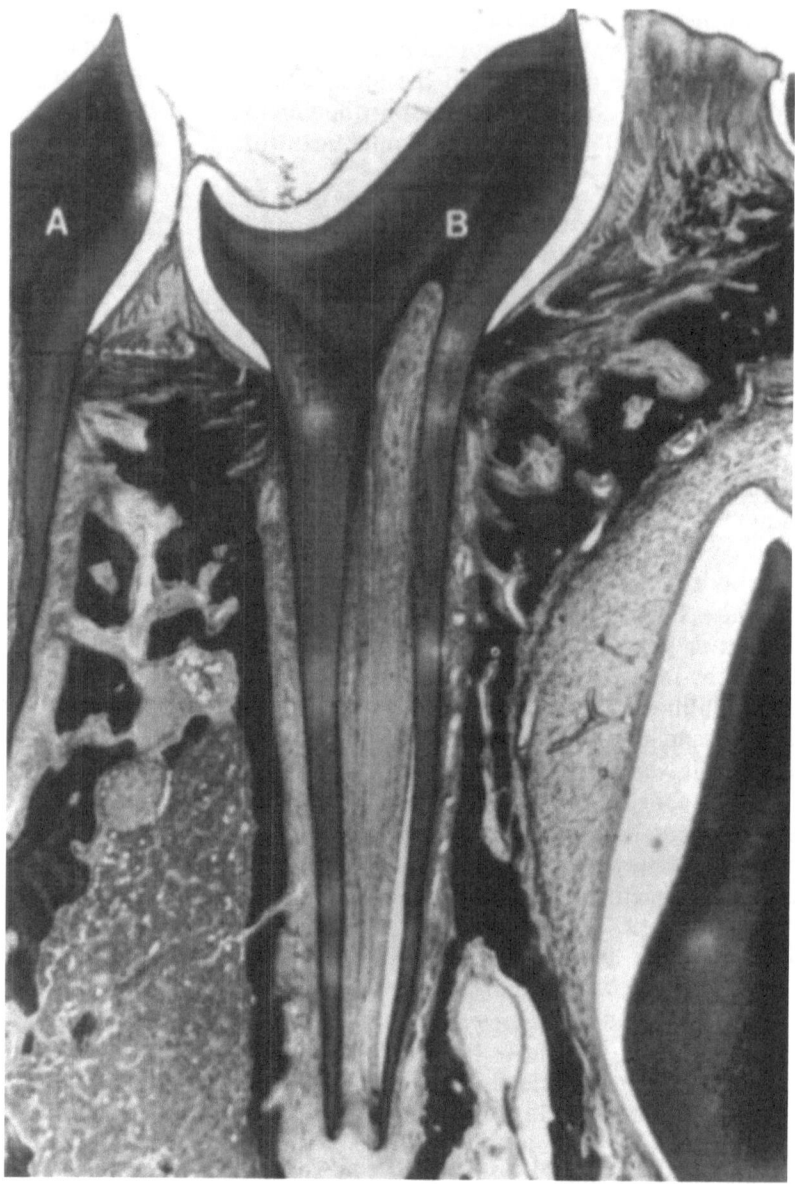

Fig. 19. Photomicrograph showing two premolars in the squirrel monkey in closely succeeding stages of eruption from first occlusal contact (tooth *B*) to full articulation (tooth *A*). The dentogingival and transseptal fiber groups are well developed on tooth *A*. On tooth *B*, obliquely orientated fibers course from cementum to bone at the cervical third of the root while more apically the ligament becomes progressively less mature. × 60. Courtesy of Dr's D.A. Grant and S. Bernick and of the Journal of Periodontology

There is evidence of a change in the obliquity and in the dimensions of the principal collagen fibers in the periodontal ligament during the intra-oral phase. MAGNUSSON (1968), BERNICK and GRANT (1982) and CAHILL and MARKS (1982) have reported that the inclination of the oblique principal fibers decreases during eruption. BERNICK and GRANT (1982) observed that fiber bundles in monkey premolars thicken with function. This also has been reported for rat molars (BERNICK 1960; TROTT 1962) but not for mouse molars (ATKINSON 1972). It is to be hoped that future work involving quantification will provide more information. BERKOVITZ et al. (1985) found few structural differences between the periodontal ligaments of teeth (rat molars) which had just emerged into the oral cavity and of teeth which had been functional for a considerable period of time. No differences in oxytalan fibers were seen but collagen fibril diameters were slightly smaller in the erupting tooth. Preliminary observations indicate, however, that just prior to emergence the ligament is more cellular and the fibroblasts possess fewer intracellular collagen profiles (CHANDRASEKERA and BEYNON 1983). MOXHAM et al. (unpublished data) have observed large numbers of fenestrated capillaries in the periodontal ligament of the erupting rat molar. Indeed, preliminary data show that there may be more fenestrations in the erupting tooth than in the fully functional tooth. This may be significant in view of the proposal that the periodontal vasculature plays a role in eruption (for review see MOXHAM and BERKOVITZ 1982).

Concerning the ground substance, GIBSON (1979) – quoted by PEARSON (1982) – reported that no differences in hyaluronate and dermatan sulphate levels were evident during the intra-oral phase. However, the PG1 proteoglycan levels in the periodontal ligaments of bovine incisors which had just erupted into the oral cavity were higher than those in the ligaments of fully erupted teeth (Fig. 9). The possible significance of this is discussed on pages 60–61.

PERERA and TONGE (1981 a, b, c) have compared collagen turnover and cell kinetics in the periodontal ligaments of the emerging and functional mouse molar. They claim that collagen turnover is most rapid and fibroblast proliferative and migratory activities greatest at the time of crown emergence. KAMEYAMA (1973a) reported that there was a change in the uptake of ^{3}H-proline in the cervical part of the developing periodontal ligament as the tooth erupted into the oral cavity. These studies have been discussed on pages 28–30, 32–33.

B. Mechanisms of Tooth Eruption

There is considerable controversy concerning the mechanism by which the force(s) responsible for tooth eruption is (are) generated. Most investigators have adopted a physiological rather than an histological approach to examine the eruptive mechanism. The reader is referred to previous reviews for details of the physiological evidence (e.g. MELCHER and BEERTSEN 1977; MOXHAM and BERKOVITZ 1982). Our intention is to discuss the findings of recent histological studies in an attempt to relate the structural features of the periodontal ligament with the proposed agents causing eruption.

In contrast to investigations concerning the tissue changes which take place during eruption, research on the eruptive mechanism has involved almost exclusively the use of continuously growing incisors of rodents and lagomorphs. This is primarily because they erupt continuously at high rates. However, care must be taken before extrapolating from these studies to the mechanism involved in teeth of limited growth. Although it is believed that the mechanisms are similar, few experiments have been undertaken to assess this. Whereas propylthiouracil retards eruption in both tooth types (PAYNTER 1954; GARREN and GREEP 1955), the teeth react differently to thyroxine (HOSKINS 1927) and to hexamethonium (MOXHAM and BERKOVITZ 1983; BERKOVITZ and MOXHAM 1984).

Recently, three possible causative agents have attracted most attention:

i. Contraction of collagen in the periodontal ligament.
ii. Contraction/migration of fibroblasts in the periodontal ligament.
iii. Vascular/fluid pressures in the tissues around or beneath the tooth.

I. The Collagen Contraction Hypothesis

This proposes that tractional forces are set up within the oblique fiber system of the periodontal ligament during collagen maturation (SHRIMPTON 1960; THOMAS 1967). To date, the various explanations offered for the underlying cause of contraction are considered to be biochemically untenable (BAILEY 1976). The results of experiments using substances which specifically inhibit the formation of intramolecular cross-links in collagen (lathyrogens) also argue against the collagen contraction hypothesis. For, even though there is disruption of the periodontal ligament collagen, lathyritically affected teeth maintained free of the bite erupt at high rates (SARNAT and SCIAKY 1965; BERKOVITZ et al. 1972; TSURUTA et al. 1974; MICHAELI et al. 1975; THOMAS 1976).

Little information relating to the collagen contraction hypothesis can be gained from an analysis of periodontal ligament structure. The oblique arrangement of many periodontal ligament collagen fibers would allow extrusion if contraction did occur. GATHERCOLE and KELLER (1982) have suggested that the formation of crimps within the collagen fibrils of the periodontal ligament (Fig. 20) may lead to their shortening. However, there is no evidence that a force can be generated at the crimp and continuously erupting teeth do not possess crimping in the necessary orientation for the contraction of the crimp to result in extrusion. Further histological evidence against the collagen contraction hypothesis comes from research which shows that some teeth can erupt in the absence of a well organised fiber system (see pages 26–28).

Study of the ultrastructure of collagen indicates that there is a relationship between fibril diameters and the function of a connective tissue (e.g. PARRY et al. (1978) claim that there is a correlation between the ultimate tensile strength of a connective tissue and the size of its collagen fibrils). It might be expected that collagen fibril diameters would differ in periodontal tissues of teeth having different eruptive behaviours. This is not the case, at least for the rat dentition.

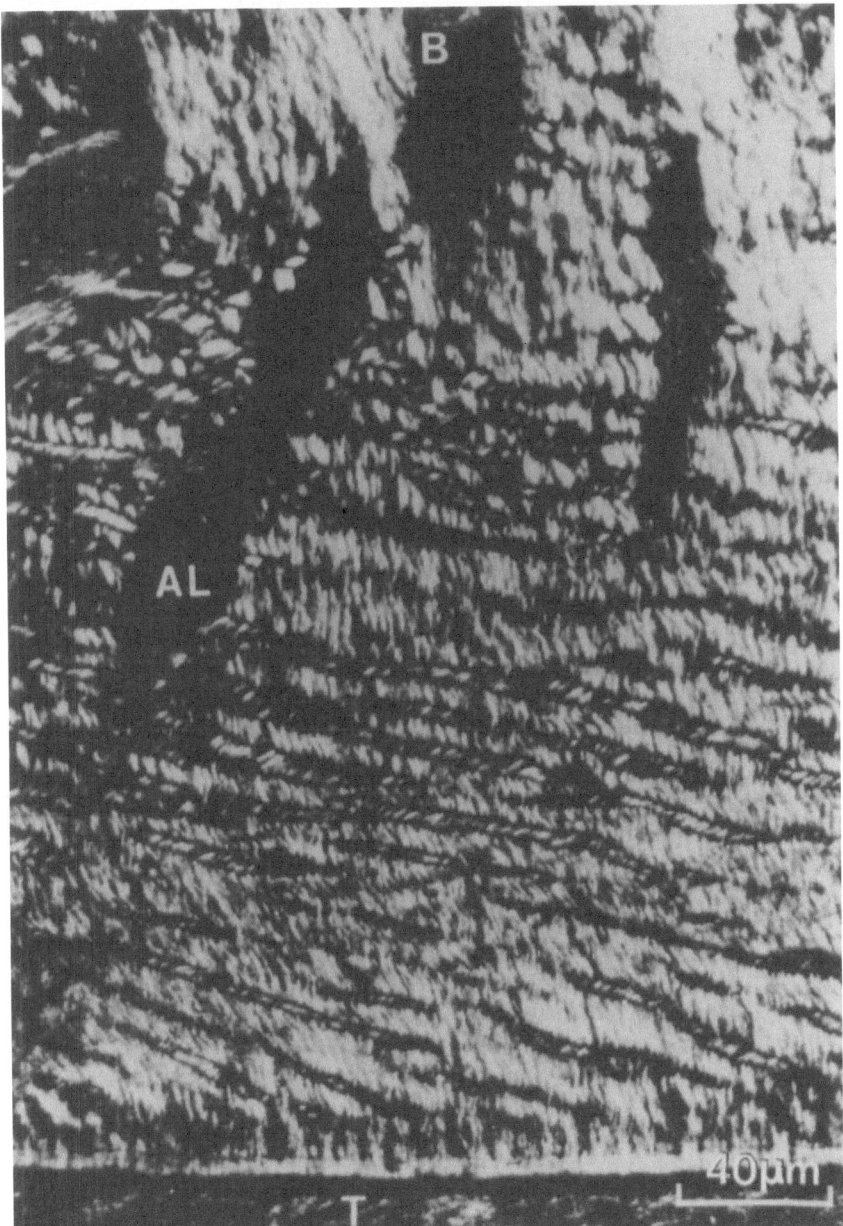

Fig. 20. Photomicrograph of transverse section of rabbit incisor periodontal ligament viewed in polarised light. The horizontal lines in the ligament, especially in the alveolar-related portion (*AL*), are evidence of crimping of the collagen fibers. *T* Tooth; *B* alveolar bone. × 500. Courtesy of Dr. P. Sloan

Firstly, there is no difference between collagen fibrils in the continuously erupting incisor and in the erupted molar (Berkovitz et al. 1981). Secondly, there is no difference between fibrils in the continuously erupting incisor maintained in occlusion (impeded) or maintained free of occlusion (unimpeded), despite the rate of eruption being three times greater in the unimpeded state (Shore et al 1982). Thirdly, there is no difference between fibrils in the periodontal ligament of the continuously erupting incisor and in the adjacent, but unattached, enamel-related connective tissue (Berkovitz et al. 1981).

II. The Periodontal Fibroblast Contraction/Motility Hypothesis

Evidence from tissue culture suggests that fibroblast traction may be important in connective tissue morphogenesis (Bellows et al. 1982a; Stopak and Harris 1982; Oster et al. 1983; Lewis 1984). The possibility that fibroblast traction in the periodontal ligament is also responsible for eruption has recently attracted considerable attention. Two cellular activities have been implicated – fibroblast contraction and fibroblast motility. Because of the difficulty of devising experiments to alter selectively any potential contractile or motile properties of periodontal ligament fibroblasts, evaluation of the periodontal contraction/motility hypothesis has relied upon interpretation of the functional significance of fibroblast morphology and upon the behaviour of fibroblasts in vitro.

Ness (1967) suggested that periodontal ligament fibroblasts generate the eruptive force by a process similar to that found in a contracting wound. Wound contraction appears to result from the activities of transformed fibroblasts, termed myofibroblasts. The myofibroblast is structurally distinct (Fig. 21), possessing a crenulated nucleus, a prominent basal lamina, pinocytotic vesicles, and intercellular attachment sites (Gabbiani 1979; Lipper et al. 1980). In addition, there are prominent and extensive arrays of microfilament bundles throughout the cytoplasm which react immunologically with anti-actin (Hirschel et al. 1971; Gabbiani 1979). The important characteristic of myofibroblasts is that they react pharmacologically as smooth muscle (e.g. Majno et al. 1971; Gabbiani 1979). There is, however, neither morphological nor pharmacological evidence for the existence of myofibroblasts within the periodontal ligament. Although Azuma et al. (1975) claimed that there is a subpopulation of myofibroblasts within the periodontal ligament, their illustrations do not show the structurally distinct features characteristic of this cell type. Furthermore, myofibroblasts have not been identified in the periodontal ligament by any other worker.

It is possible that periodontal ligament fibroblasts can contract even though they do not have the specific morphology of myofibroblasts. Some evidence for this derives from the work of Bellows et al. (1981, 1982a). They showed that periodontal ligament fibroblasts cultured on collagen gels produced gel compaction and that sufficient tension was developed to bring together fragments of tooth and bone initially placed some distance apart within the gel. This contraction was inhibited by colcemid and cytochalasin D, suggesting that microtubules and microfilaments were involved, organelles present in periodon-

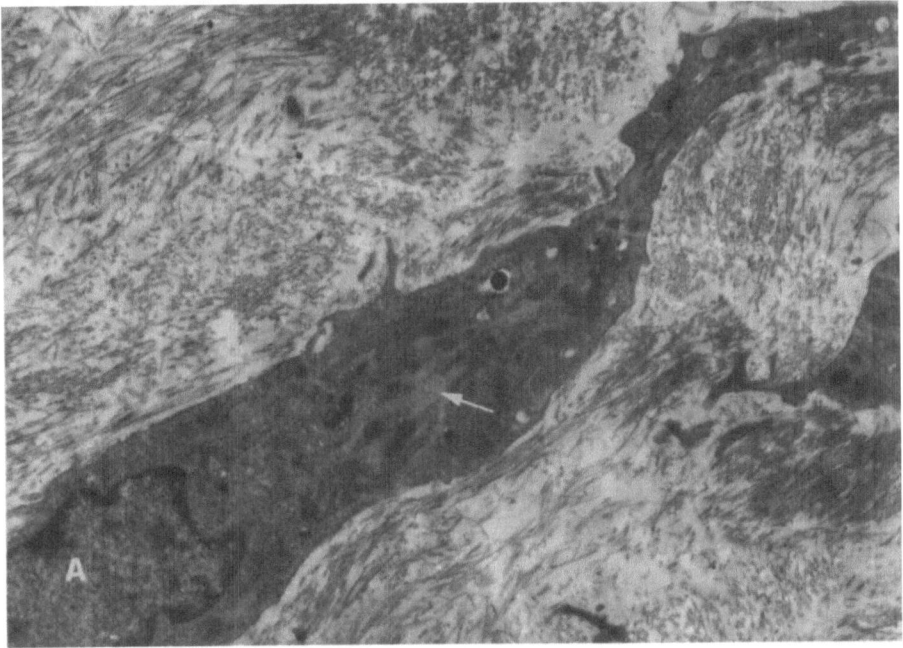

Fig. 21. Electron micrograph of a myofibroblast from the pubic symphysis of a pregnant mouse. Note the crenulated nucleus (*A*) and prominent bundles of microfilaments (*arrow*) throughout the cytoplasm. × 7500

tal ligament fibroblasts in vivo (BEERTSEN et al. 1974; SHORE and BERKOVITZ 1979). BELLOWS et al. (1981, 1982a) extrapolated from the results of their in vitro experiments to suggest that a similar contractile activity can occur in fibroblasts in vivo to effect tooth eruption. Some criticisms can be levelled at this conclusion:

Firstly, contracting periodontal ligament fibroblasts appear spindle-shaped and highly polarised. Ultrastructurally, they are similar to myofibroblasts, possessing thick cell coats, considerable amounts of microfilamentous material, little rough endoplasmic reticulum, numerous structures resembling gap junctions, and occasionally crenulated nuclei (Fig. 22) (BELLOWS et al. 1982a, 1982b). In all these features, they differ from periodontal fibroblasts in vivo which have an irregular disc-shape and a cytoplasm containing considerable amounts of rough endoplasmic reticulum and microfilamentous material primarily in the form of stress fibers beneath the cell membrane (BEERTSEN et al. 1974; BEERTSEN and EVERTS 1977; SHORE and BERKOVITZ 1979; BERKOVITZ 1981; BERKOVITZ and SHORE 1982). Gap junctions are also infrequent in vivo where the more common type of intercellular contact is the simplified desmosome (SHORE et al. 1981). Significantly, as contraction of the collagen gel in vitro ceases, the morphology of the cultured fibroblasts changes to resemble periodontal ligament fibroblasts in vivo. The cells assume a more rounded mor-

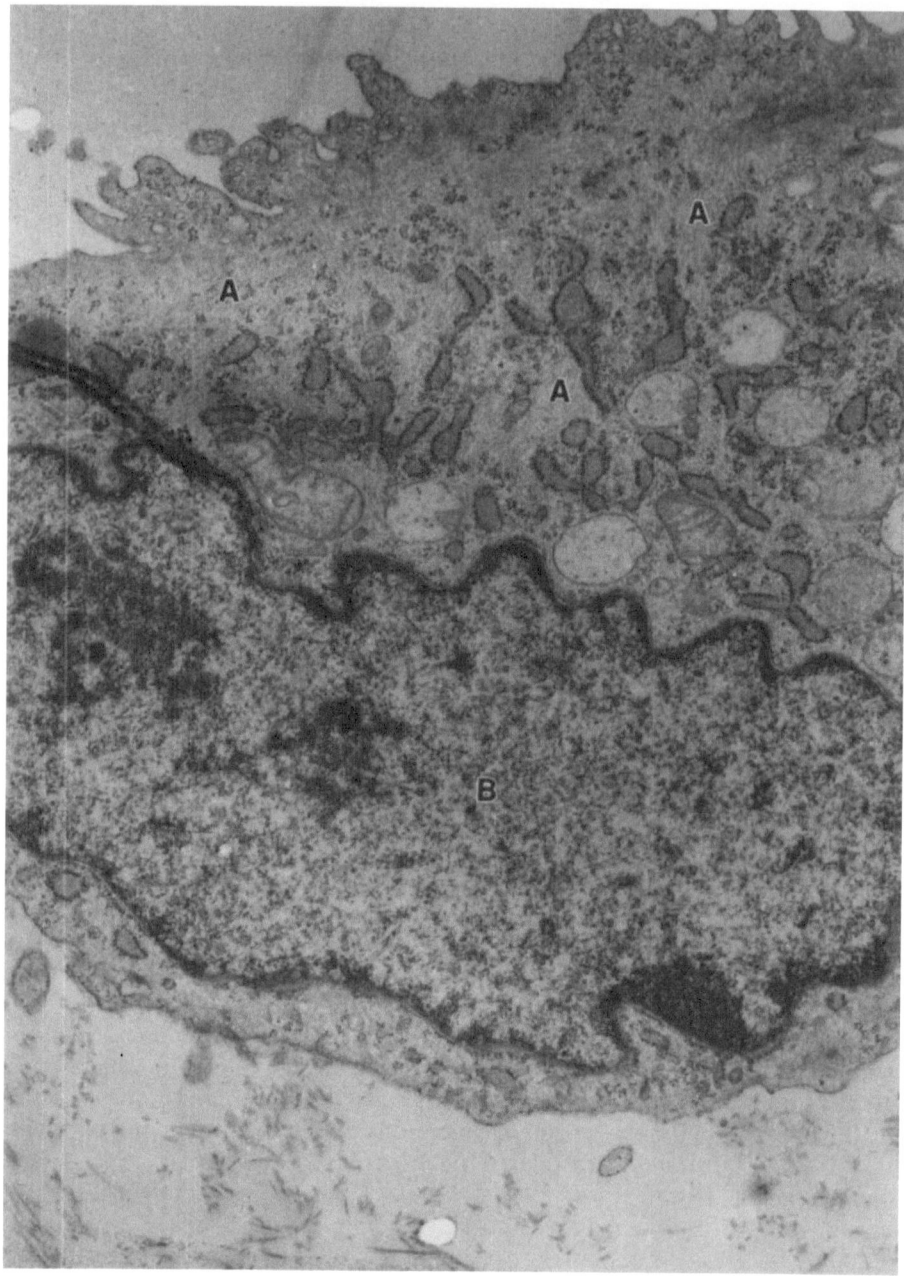

Fig. 22. Electron micrograph of periodontal ligament fibroblast from a contracting collagen gel. The cell contains considerable amounts of microfilamentous material (*A*), a crenulated nucleus (*B*) but little rough endoplasmic reticulum. × 19000

phology, exhibit extensive endoplasmic reticulum and show few gap junctions but more desmosome-like contacts (BELLOWS et al. 1982b).

Secondly, it might be expected that periodontal ligament fibroblasts of teeth showing differences in their eruptive behaviour would show differences in their intracellular organelles (particularly in microfilaments). However, no differences were apparent between periodontal fibroblasts in erupting and erupted rat molars (BERKOVITZ et al. 1985), or in normal and immobilised rat incisors (SHORE et al. 1985).

Thirdly, it has been shown that the appearance of stress fibers within fibroblasts in vitro corresponds to the period after maximum contractility (CURATOLO et al. 1982). Thus, that periodontal ligament fibroblasts in vivo contain stress fibers (BEERTSEN et al. 1974; SHORE and BERKOVITZ 1979) suggests that they have passed any main contractile phase, a view not entirely consistent with the notion that they continue to maintain eruption. Indeed, there is no evidence for subpopulations of periodontal ligament fibroblasts distinguished by the presence or absence of stress fibers.

It has been proposed that fibroblast migration rather than contraction is responsible for tooth eruption (NESS 1970; BEERTSEN et al. 1974). BEERTSEN et al. (1974) claim that periodontal ligament fibroblasts near the tooth are polarised (elongated in the direction of eruption) and that the eruptive force is somehow produced as a result of unidirectional cell migration in an occlusal direction. They implicated the arrays of microfilaments and microtubules present within periodontal fibroblasts as being involved in cell motility. These organelles have been implicated in cell locomotion in vitro (VASILIEV et al. 1970; SPOONER et al. 1971). Indeed, BEERTSEN et al. (1974) claimed that periodontal ligament fibroblasts in vivo resembled morphologically fibroblasts migrating in vitro.

A number of authors have shown that periodontal ligament fibroblasts migrate occlusally (e.g. CHIBA 1968; BEERTSEN 1973, 1975; ZAJICEK 1974; BEERTSEN and EVERTS 1977; PERERA and TONGE 1981b). Not all the cells migrate however. It has been proposed that in the continuously-growing rodent incisor a zone of shear occurs in the middle of the periodontal ligament such that the inner, tooth-related zone moves with the tooth whereas the outer, alveolar-related zone remains behind (MELCHER 1967; BEERTSEN 1973; BEERTSEN and EVERTS 1977). BEERTSEN and EVERTS (1977) showed that the rate of migration in the rat incisor can correspond with the rate of eruption. When eruption was doubled by removing the tooth from the bite (unimpeded eruption), fibroblast migration in the tooth-related zone of the periodontal ligament was also doubled. The possibility exists, however, that cell migration is an effect of eruption and not its cause. Consequently, much of the argument has centred on whether periodontal ligament fibroblasts migrate actively. In contrast to the findings of BEERTSEN and EVERTS (1977), ZAJICEK (1974) reported that periodontal fibroblasts move approximately three times faster than the tooth in a region ~20 µm from the tooth surface. He claimed that this could not be explained by passive movement of the fibroblast with the erupting tooth. However, it is still possible to account for this by assuming deformation of collagen fibers during eruption (SHORE and BERKOVITZ 1978).

Actively migrating fibroblasts in vitro are markedly polarised in shape and in the arrangement of microfilaments and microtubules (Bard and Hay 1975; Hay 1982). However, Shore and Berkovitz (1979) and Berkovitz (1981) have shown that fibroblasts within the periodontal ligament of the rat incisor are flattened irregular discs and are not elongated and unidirectionally polarised (Fig. 23). The lack of polarity of shape does not preclude the possibility that the cells possess an internal polarity. Garant and Cho (1979) reported that the nuclei within the fibroblasts of the transseptal fibers above the periodontal ligament show polarity. 63 per cent of cells were considered to be polarized, the nuclei in half these cells being sited at the mesial end of the cell and in half at the distal end. These data were considered as evidence that the cells were migrating actively, a process which the authors related to the elaboration and orientation of new collagen fibrils. Beertsen et al. (1979) found that the centrioles of fibroblasts within the tooth-related zone of the periodontal ligament of the rodent incisor were preferentially located in the occlusally directed part of the cytoplasm. Polarity was less pronounced in the alveolar-related zone. That there might not be a simple relationship between polarity of organelles and unidirectional movement is apparent from the same study of Beertsen et al. (1979). Using the *fully erupted* molar of the rat, they found that the Golgi region was preferentially situated in that part of the cell directed towards the alveolar wall and the occlusal plane. Furthermore, Trelstad et al. (1967) showed that the Golgi apparatus in migrating mesenchymal cells is located in the trailing part of the cytoplasm. Beertsen et al. (1979) concluded by suggesting that structural polarity of periodontal ligament fibroblasts may be associated with secretion and/or phagocytosis of collagen. Indeed, microfilaments and microtubules are known to be involved in endocytosis, exocytosis and protein transport (e.g. Ehrlich and Bornstein 1972; Allison 1973).

It has been shown that migration rates of fibroblasts in vitro are reduced by collagen, dense connective tissues being associated with slower rates (Armstrong and Armstrong 1980). The maximum rate of movement observed within a collagenous framework is approximately 60 µm/h (Bard and Hay 1975). For an unimpeded rat incisor erupting at 1 mm/day, periodontal ligament fibroblasts would need to migrate at 40 µm/h. The only direct evidence of periodontal ligament fibroblast migration rates in vitro suggests, however, that their maximum rate of movement is only 12.5 µm/h (Brunnette et al. 1977).

If periodontal ligament fibroblasts are actively motile, they might be expected to show ultrastructural differences in teeth exhibiting different eruptive behaviour. For example, there may be different energy requirements (and thus differences in mitochondrial volume) and variations in the amounts of microfilaments and microtubules. On comparing periodontal ligament fibroblasts in erupting and erupted rat molars (Berkovitz et al. 1985). in normal and immobilised rat incisors (Shore et al. 1985) and in normal and unimpeded rat incisors (Shore et al. 1982), the only finding which appeared to lend support to the fibroblast motility hypothesis was an increase of microtubules in the unimpeded tooth. In another study, Berkovitz and Shore (1978) investigated the structure of the fibroblasts adjacent to the enamel organ of the rat incisor, cells which

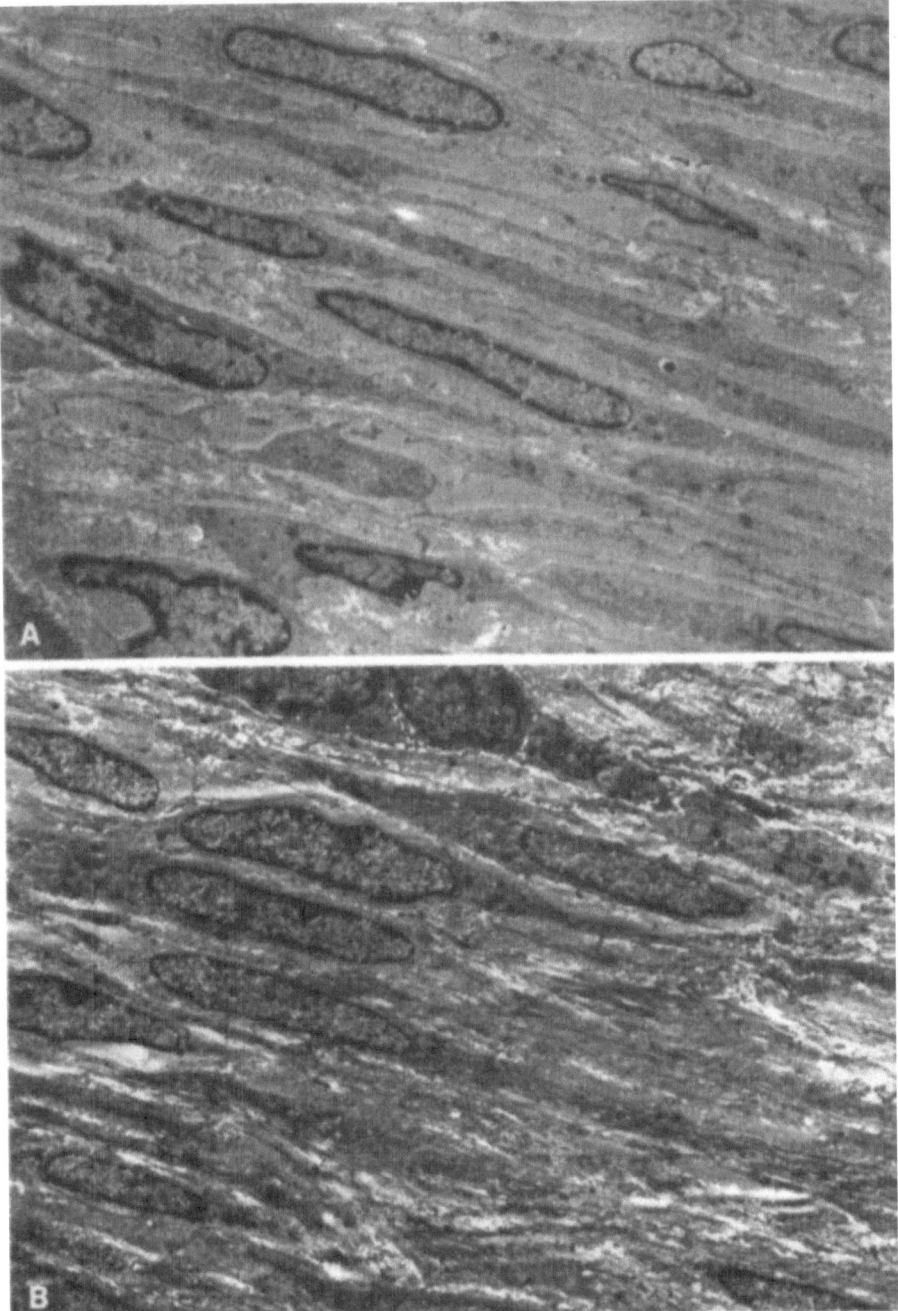

Fig. 23A, B. Electron micrographs of periodontal ligament of rat incisor. × 3800. **A** Section along the longitudinal axis of the tooth. **B** Section along the transverse axis of the tooth. The elongated appearance of the periodontal fibroblasts in both planes of section is taken to indicate that the cells are disc-shaped

have never been implicated in the generation of an eruptive force. It was reported that they did not differ from fibroblasts in the adjacent periodontal ligament.

Even if periodontal fibroblasts were shown to contract or actively migrate in vivo, a satisfactory explanation has to be provided to account for the way the force is transmitted to the tooth to effect eruption. Of importance in this respect is the need to know the position of the zone of shear, yet this is in dispute (e.g. BEERTSEN and EVERTS 1977; BERKOVITZ et al. 1980). It may be that a tensional force is generated which relies upon the close relationship between fibroblasts and collagen or between the fibroblasts themselves (e.g. the close adherence produced by junctional complexes) for its transmission. However, that teeth with periodontal ligaments weakened by the administration of lathyrogens erupt at high rates (e.g. BERKOVITZ et al. 1972) argues against the idea that tension is generated through the collagen. It is indeed an unusual feature that fibroblasts in the periodontal ligament are joined by intercellular contacts (SHORE et al. 1981). Nevertheless, teeth showing differences in eruptive behaviour do not show consistent changes in the quantity of junctional complexes and, perhaps more importantly, erupting and immobilised teeth not erupting possess similar numbers of contacts (SHORE et al. 1982, 1985; BERKOVITZ et al. 1985). Indeed, fibroblasts in the enamel-related connective tissue of rat incisors have more intercellular contacts than those in the periodontal ligament (SHORE et al. 1981).

III. The Vascular/Tissue Fluid Pressure Hypothesis

CONSTANT (1900) first suggested that the eruptive force was generated by blood pressure. This idea subsequently has been developed by invoking the Starling hypothesis, the eruptive force being said to be derived from the hydrostatic pressures around or beneath the tooth. Hydrostatic pressure may be exerted either by fluid which is free in the tissues and/or by fluid which is bound to fibers or ground substance. The ground substance, through its hyaluronate and sulphated proteoglycans, appears to be the major water-binding constituent of a connective tissue (SCHUBERT and HAMERMAN 1968; BENTLEY 1970; MELCHER and WALKER 1976; BETTELHEIM and BRADY 1979; PEARSON 1982). GUYTON (1972) has shown that interstitial gel removed from a tissue and placed in an electrolyte medium can swell considerably (up to 50%) and will exert much pressure against any physical barrier that attempts to prevent its swelling. That the periodontal ligament, and tooth position, may be affected by electrolyte balance is indicated by the work of TYLER and BURN-MURDOCH (1976). Using incisor teeth in dissected rat mandibles, they reported that solutions of varying electrolyte content perfused into the periodontal ligament produced predictable intrusive and extrusive tooth movements depending upon whether the solutions were hyper- or hypo-tonic.

PEARSON (1982), reviewing the work of GIBSON (1979), reported that the composition of the ground substance in the periodontal ligaments of bovine incisors changed at various stages of tooth eruption (Fig. 9). If proteoglycans are involved in the mechanism of tooth eruption, it seems more likely to be

PG1 than proteodermatan sulphate because of the quantitative changes and the probable locations of the proteoglycans in the tissue. An interfibrillar location of PG1 would allow a fuller expression of its ability to influence the osmotic pressure and swelling of the tissue compared with proteodermatan sulphate which is more intimately associated with collagen fibrils (e.g. PEARSON 1982). The latter association might partly "tie up" the dermatan sulphate chains and reduce their effect on tissue swelling similar to that observed when glycosaminoglycans in umbilical cord or cornea are allowed to interact with polycations (GELMAN and SILBERBERG 1976; COMPER and LAURENT 1978). Hyaluronate, like PG1, will be less affected by interactions with collagen. However, the content of this glycosaminoglycan is almost constant during eruption (Fig. 9). Thus, PG1 may control the internal osmotic pressure of the periodontal ligament during tooth eruption. Furthermore, small changes in concentration of glycosaminoglycans and proteoglycans may have a considerable effect on osmotic pressures because their behaviour is "non-ideal" (COMPER and LAURENT 1978; URBAN et al. 1979). Any physical swelling of the tissue is likely to be opposed by the elastic contribution of the collagen fibers (COMPER and LAURENT 1978). However, the development of the collagen fibers shows considerable variation in erupting teeth (see pages 24–28).

Although it is known that nervous vasomotor activity or vasoactive drugs can influence the behaviour of an erupting tooth (MOXHAM 1979, 1981; MYHRE et al. 1979; AARS 1982a, b), histological studies have provided little information on the role of the vasculature during tooth eruption. However, some information has been published by BRYER (1957), MAGNUSSON (1968, 1973) and MOXHAM et al. (1985).

BRYER (1957) studied the effects of nutritional disturbances and surgical interference on the eruption of the rodent incisor, relating the resulting changes in vascularity of the periodontal ligament to eruption rate. He reported that a reduction in eruption (following Vitamin A deficiency, rickets, Vitamin D toxicity, cobalt administration, and various surgical interferences with the tissues and blood supply to the teeth) was associated with a reduced vascularity. An increase in eruption (during semistarvation and after localized sympathectomy) appeared to be associated with an increased vascularity. However, BRYER's studies relied upon a subjective assessment of vascularity without quantification. Furthermore, although Vitamin C deficiency reduces eruption rates, it is associated with hypermia and oedema of the pulp (HÖJER and WESTIN 1925; KEY and ELPHICK 1931; MACLEAN et al. 1939; BERKOVITZ 1974).

MAGNUSSON (1968, 1973) observed periapical effusions beneath the molars of rats and monkeys (Fig. 10). He claimed these effusions were induced by changes in the permeability of the periapical vessels. He further suggested that the effusions separate the tooth from the bone and thereby create the eruptive migration (see pages 34–35). However, the precise nature of such effusions awaits clarification as does the question of whether they can generate a force sufficient to effect tooth eruption.

CORPRON et al. (1976) and FRANK et al. (1976) have reported that some capillaries within the periodontal ligament are fenestrated (Fig. 24). As there may be considerable differences in permeability between fenestrated and contin-

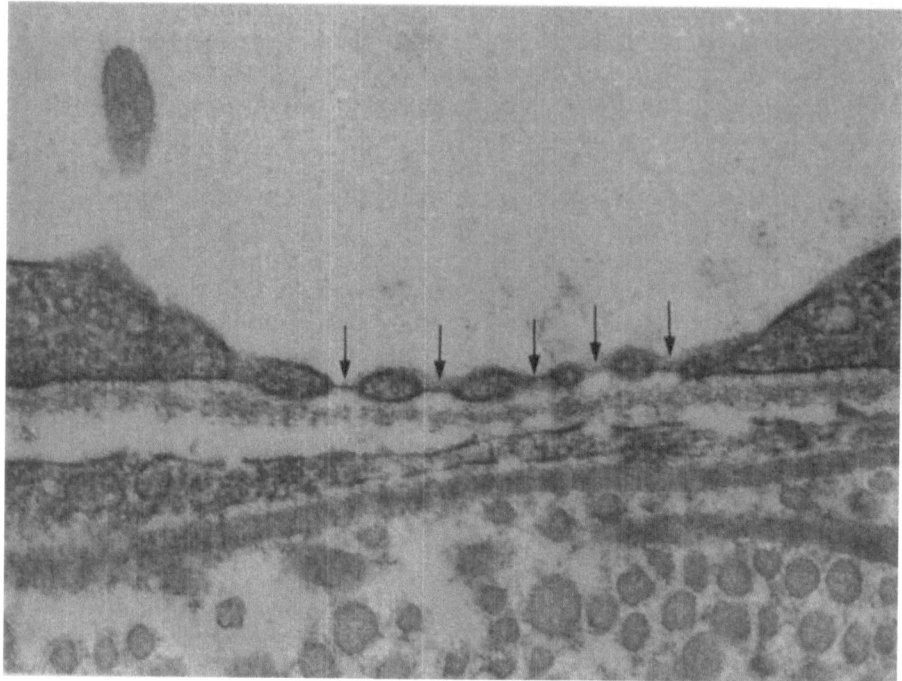

Fig. 24. Electron micrograph of fenestrations (*arrows*) in a capillary in the rat incisor periodontal ligament. × 70000

uous endothelia (e.g. Haraldsson et al. 1982), Moxham et al. (1985) quantified the numbers of fenestrated capillaries in the periodontal ligament of the continuously erupting rat incisor. They found that there are $\sim 3.5 \times 10^6$ fenestrations per mm^3 at the base of the tooth whereas more occlusally there are $\sim 1 \times 10^6$ fenestrations per mm^3. Such differences may provide a basis for pressure differentials along the length of the tooth. A comparison of the number of fenestrated capillaries has also been made between an erupting and an erupted tooth of limited growth. Moxham et al. (unpublished data) found that, whilst the periodontal ligament of the erupted rat molar contains a greater volume of capillaries, the ligament of the erupting tooth appears to have a greater percentage of its capillary surface area occupied by fenestrations and contains a greater number of fenstrations per unit volume of tissue ($\sim 10 \times 10^6$ fenestrations per mm^3 in the erupted molar, $\sim 30 \times 10^6$ fenestrations per mm^3 in the erupting molar). Care must be taken, however, about attributing a rôle for the fenestrations in generating the tissue fluid pressures in the periodontal ligament. Although generally it is assumed that fenestrations facilitate permeability, it is possible that the fenestral diaphragms provide restrictive barriers in some capillary beds. Recently, Bankston and Milici (1983) have demonstrated that fenestral diaphragms can exhibit various binding and permeability patterns to the same size cationic tracer (polycationic ferritin). This they attributed to differences in luminal glycosaminoglycans. Consequently, the charge of the glycocalyx covering the fenestral diaphragm may play a rôle in the permeability characteristics of fenestrated capillaries. Of course, to obtain a proper understanding of the functioning and control of the periodontal ligament vasculature we need some measure of filtration and diffusion capacities. Because of the difficulty of isolating the periodontal tissues, techniques for doing this are not yet available.

References

Aars H (1982a) The influence of sympathetic nerve activity on axial position of the rabbit incisor tooth. Acta Physiol Scand 116:417–421

Aars H (1982b) The influence of vascular β-adrenoceptors on the position and mobility of the rabbit incisor tooth. Acta Physiol Scand 116:423–428

Addison WC (1979) The distribution of nuclei in human odontoclasts in whole cell preparations. Arch Oral Biol 23:1167–1171

Ali NN, Jones SJ, Boyde A (1984) Monocyte-enriched cells on calcified tissues. Anat Embryol (Berl) 170:169–175

Allison AC (1973) The rôle of microfilaments and microtubules in cell movement, endocytosis and exocytosis. In: Ciba Symposium 14. Locomotion of tissue cells. Elsevier, Amsterdam, pp 110–143

Armstrong MT, Armstrong PB (1980) The role of extracellular matrix in cell motility in fibroblast aggregates. Cell Motil Cytoskeleton 1:99–112

Atkinson ME (1972) The development of the mouse molar periodontium. J Periodont Res 7:255–260

Azuma M, Enlow DH, Fredrickson RG, Gaston LF (1975) A myofibroblastic basis for the physical forces that produce tooth drift and eruption, skeletal displacement of sutures, and periosteal migration. In: McNamara JA (ed) Determinants of mandibular form and growth. University of Michigan, Ann Arbor, pp 179–207

Bailey AJ (1976) In: Poole DFG, Stack MV (eds) The eruption and occlusion of teeth. Butterworths, London, pp 306–307

Bankston PW, Milici A (1983) A survey of the binding of cationic ferritin in several fenestrated capillary beds: Indication of heterogeneity in the luminal glycocalyx of fenestrated diaphragms. Microvasc Res 26:36–48

Bard JBL, Hay ED (1975) The behaviour of fibroblasts from the developing avian cornea. J Cell Biol 67:400–418

Baron R, Vignery A, Horowitz M (1983) Lymphocytes, macrophages and the regulation of bone remodelling. In: Peck WA (ed) Bone and mineral research, Ann 2. Elsevier, Amsterdam, pp 175–243

Baume LJ (1953) The development of the lower permanent incisors and their supporting bone. Am J Orthod 39:526–544

Beertsen W (1973) Tissue dynamics in the periodontal ligament of the mandibular incisor of the mouse. A preliminary report. Arch Oral Biol 18:61–66

Beertsen W (1975) Migration of fibroblasts in the periodontal ligament of the mouse incisor as revealed by autoradiography. Arch Oral Biol 20:659–666

Beertsen W, Everts V (1977) Site of remodelling of collagen in periodontal ligament of mouse incisor. Anat Rec 189:479–498

Beertsen W, Everts V, Van den Hooff A (1974) Fine structure of fibroblasts in the periodontal ligament of the rat incisor and their possible role in tooth eruption. Arch Oral Biol 19:1087–1098

Beertsen W, Brekelmans M, Everts V (1978) The site of collagen resorption in the periodontal ligament of the rodent molar. Anat Rec 192:305–318

Beertsen W, Everts V, Brekelmans M (1979) Unipolarity of fibroblasts in rodent periodontal ligament. Anat Rec 195:535–544

Bellows CG, Melcher AH, Aubin HE (1981) Contraction and organization of collagen gels by cells cultured from periodontal ligament, gingiva and bone suggest functional differences between cell types. J Cell Sci 50:299–314

Bellows CG, Melcher AH, Aubin JE (1982a) Association between tension and orientation of periodontal ligament fibroblasts and exogenous collagen fibres in collagen gels in vitro. J Cell Sci 58:125–138

Bellows CG, Melcher AH, Bhargava V, Aubin JE (1982b) Fibroblasts contracting three-dimensional collagen gels exhibit ultrastructure consistent with either contraction or protein secretion. J Ultrastruct Mol Struct Res 78:178–192

Bentley JP (1970) The biological role of the ground substance mucopolysaccharides. Adv Biol Skin 10:103–121

Berkovitz BKB (1971) The effect of root transection and partial root resection on the unimpeded eruption rate of the rat incisor. Arch Oral Biol 16:1033–1043

Berkovitz BKB (1974) The effect of vitamin C deficient diet on eruption rates for the guinea pig lower incisor. Arch Oral Biol 19:807–811

Berkovitz BKB (1975) Mechanisms of tooth eruption. In: Lavelle CLB (ed) Applied physiology of the mouth. Wright, Bristol, pp 99–123

Berkovitz BKB (1981) A critique of the fibroblast migration hypothesis of tooth eruption with a note on the tissue fluid pressure hypothesis. In: Barrer HG (ed) Orthodontics: The state of the art. University of Philadelphia, pp 239–255

Berkovitz BKB, Bass TB (1976) Eruption rates of human upper third molars. J Dent Res 55:460–464

Berkovitz BKB, Moxham BJ (1984) Differences in the eruptive behaviour of teeth of continuous and limited growth. J Dent Res 63:513

Berkovitz BKB, Shore RC (1978) The ultrastructure of the enamel aspect of the rat incisor periodontium in normal and root resected teeth. Arch Oral Biol 23:681–689

Berkovitz BKB, Shore RC (1982) Cells of the periodontal ligament. In: Berkovitz BKB, Moxham BJ, Newman HN (eds) The periodontal ligament in health and disease. Pergamon, Oxford, pp 25–50

Berkovitz BKB, Thomas NR (1969) Unimpeded eruption in the root-resected lower incisor of the rat with a preliminary note on root transection. Arch Oral Biol 14:771–780

Berkovitz BKB, Migdalski A, Solomon M (1972) The effect of the lathyritic agent aminoacetonitrile on the unimpeded eruption rate in normal and root-resected rat lower incisors. Arch Oral Biol 17:1755–1763

Berkovitz BKB, Shore RC, Sloan P (1980) Histology of the periodontal ligament of rat mandibular incisors following root resection, with special reference to the zone of shear. Arch Oral Biol 25:235–244

Berkovitz BKB, Weaver ME, Shore RC, Moxham BJ (1981) Fibril diameters in the extracellular matrix of the periodontal connective tissues of the rat. Connet Tissue Res 8:127–132

Berkovitz BKB, Shore RC, Moxham BJ (1985) Ultrastructural studies on the developing periodontal ligament, proceedings of Inserm colloquium: In: Belcourt A, Ruch JV (eds) Tooth morphogenesis and differentiation. INSERM 125:545–556. Strasbourg

Bernick S (1960) The organization of the periodontal membrane fibres of the developing molars of rates. Arch Oral Biol 2:57–63

Bernick S, Grant DA (1982) Development of the periodontal ligament. In: Berkovitz BKB, Moxham BJ, Newman HN (eds) The periodontal ligament in health and disease. Pergamon, Oxford, pp 197–213

Bernick S, Rutherford R, Rabinowitch B (1949) Microscopic studies of the teeth of a 6-year old boy: II Tooth absorption. Anat Rec 105:249–265

Bernick S, Rutherford R, Rabinowitch B (1951) The role of the epithelial attachment in tooth resorption of primary teeth. Oral Surg 4:1444–1450

Bettelheim FA, Brady E (1979) Hydration and proteoglycan content of rat skin. In: Schaur R et al. (eds) Glycoconjugates. Thieme, Stuttgart, pp 662–664

Björk A, Skieller V (1972) Facial development and tooth eruption. An implant study at the age of puberty. Am J Orthod 62:339–383

Boyde A, Jones SJ (1979) Estimates of the size of resorption lacunae in mammalian calcified tissues using SEM stereophotogrammetry, SEM/1979/II, 393–402, SEM Inc, AMF O'Hare, Illinois

Boyde A, Lester KS (1967) Electron microscopy of resorbing surfaces of dental hard tissues. Z Zellforsch 83:538–547

Boyde A, Ali NN, Jones SJ (1983) Computer aided measurement of resorbtive activity of isolated osteoclasts. Proc RMS 18:357

Brash JC (1928) The growth of the alveolar bone and its relation to the movements of the teeth including eruption. Int J Orthod 14:196–223, 283–293, 398–405, 487–494, 494–504

Brunette DM, Kanoza RJ, Marmary Y, Chan J, Melcher AH (1977) Interactions between epithelial and fibroblast-like cells in cultures derived from monkey periodontal ligament. J Cell Sci 27:127–140

Bryer LW (1957) An experimental evaluation of the physiology of tooth eruption. Int Dent J 7:432–478

Burke PH (1963) Eruptive movements of permanent maxillary central incisor teeth in the human. Proc R Soc Med 56:513–515

Burke PH, Newell DJ (1958) A photographic method of measuring eruption of certain human teeth. Am J Orthod 44:590–602

Cahill DR (1969) Eruptive pathway formation in the presence of experimental tooth impaction in puppies. Anat Rec 164:67–78

Cahill DR (1970) The histology and rate of tooth eruption with and without tremporary impaction in the dog. Anat Rec 166:225–238

Cahill DR (1974) Histological changes in the bony crypt and gubernacular canal of erupting permanent premolars during deciduous premolar exfoliation. J Dent Res 53:786–791

Cahill DR, Marks SC Jr (1980) Tooth eruption: evidence for the central role of the dental follicle. J Oral Pathol 9:189–200

Cahill DR, Marks SC Jr (1982) Chronology and histology of exfoliation and eruption of mandibular premolars in dogs. J Morphol 171:213–218

Carlson H (1944) Studies on the rate and amount of eruption of certain human teeth. Am J Orthod 30:575–588

Carollo DA, Hoffman RL, Broadie AG (1971) Histology and function of the dental gubernacular cord. Angle Orthod 41:300–307

Chandrasekera MS, Beynon AD (1983) Ultrastructural stereological analysis of the developing rat molar periodontal ligament. J Dent Res 62:429

Chiba M (1968) Movement during unimpeded eruption of the position of cells and of material incorporating tritiated proline in the lingual periodontal membrane of the mandibular incisors of adult male mice. J Dent Res 47:986

Comper WD, Laurent TC (1978) Physiological function of connective tissue polysaccharides. Physiol Rev 58:255–315

Constant TE (1900) The eruption of teeth. Int Dent Congr 2:180–192

Corpron KE, Avery JK, Morawa AP, Lee SD (1976) Ultrastructure of capillaries in mouse periodontium. J Dent Res 55:551

Curatolo L, Chaponnier C, Donati MB, Morasca L, Gabbiani G (1982) Actin organisation and fibrin-clot retractile activity of cultured mouse fibroblasts. Cell Tissue Res 223:665–673

Cutright DE (1970) The morphogenesis of the vascular supply to the permanent teeth of *Macaca rhesus*. Oral Surg 30:284–291

Darling AI, Levers BGH (1975) The pattern of eruption of some human teeth. Arch Oral Biol 20:89–96

Darling AI, Levers BGH (1976) The pattern of eruption. In: Poole DFG, Stack MV (eds) The eruption and occlusion of teeth. Butterworths, London, pp 80–96

Eccles JD (1961) Studies in the development of the periodontal membrane: The apical region of the erupting tooth. Dent Practr Dent Rec 11:153–157

Eccles JD (1964) The development of the periodontal membrane in the rat incisor. Arch Oral Biol 9:127–133

Ehrlich HP, Bornstein P (1972) Microtubules in transcellular movement of procollagen. Nature 238:257–260

Engel MB (1951) Some changes in the connective tissue ground substance associated with the eruption of the teeth. J Dent Res 30:322–330

Fanning EA (1961) A longitudinal study of tooth formation and root resorption. N Z Dent J 57:202–217

Fearnhead RW (1967) Innervation of dental tissues. In: Miles AEW (ed) Structural and chemical organization of teeth, vol 1. Academic Press, London, pp 247–281

Formicola AJ, Ferrigno PD (1966) An autoradiographic study of the developing periodontium of the rat. Periodontics 4:297–301

Frank RM, Fellinger E, Steuer P (1976) Ultrastructure du ligament alvéolo-dentaire du rat. J Biol
    Buccale 4:295–313
Freeman E, Ten Cate AR (1971) Development of the periodontium: An electron microscope study.
    J Periodont 42:387–395
Freeman E, Ten Cate AR, Dickinson J (1975) Development of a gomphosis by tooth germ implants
    in the parietal bone of the mouse. Arch Oral Biol 20:139–140
Freilich LS (1971) Ultrastructure and acid phosphatase cytochemistry of odontoclasts: Effects of
    parathyroid extract. J Dent Res 50:1047–1055
Fullmer HM (1959) Observations on the development of oxytalan fibres in the periodontium of
    man. J Dent Res 38:510–518
Fullmer HM (1960) A comparative histochemical study of elastic, preelastic and oxytalan connective
    tissue fibres. J Histochem Cytochem 8:290–295
Fullmer HM (1961) A histochemical study of periodontal disease in the maxillary alveolar processes
    of 135 autopsies. J Periodont 32:206–218
Fullmer HM (1967) The development of oxytalan fibres. In: Anderson DJ, Eastoe JE, Melcher
    AH, Picton DCA (eds) The mechanism of tooth support. Wright, Bristol, pp 72–75
Furseth R (1968) The resorption processes of human deciduous teeth studied by light microscopy,
    microradiography and electron microscopy. Arch Oral Biol 13:417–431
Gabbiani G (1979) The role of contractile proteins in wound healing and fibrocontractive diseases.
    Methods Achiev Exp Pathol 9:187–206
Garant PR, Cho M-I (1979) Cytoplasmic polarisation of periodontal ligament fibroblasts. J Periodont
    Res 14:95–106
Garren L, Greep RO (1955) Effect of thyroid hormone and propylthiouracil on eruption rate of
    upper incisor teeth in rats. Proc Soc Exp Biol Med 90:652–655
Gathercole LJ, Keller A (1982) Biophysical aspects of the fibres of the periodontal ligament. In:
    Berkovitz BKB, Moxham BJ, Newman HN (eds) The periodontal ligament in health and disease.
    Pergamon, Oxford, pp 103–117
Gelman RA, Silberberg A (1976) The effect of a strongly inter-acting macromolecular probe on
    the swelling and exclusion properties of loose connective tissue. Conn Tissue Res 4:79–90
Gibson GJ (1979) Proteoglycans of the periodontal ligament. PhD Thesis, University of Alberta
Gilbert CW (1972) The labelled mitosis curve and the estimation of the parameters of the cell
    cycle. Cell Tissue Kinet 5:53–63
Gothlin G, Ericsson JLE (1973a) Electron microscopic studies on the uptake and storage of thorium
    dioxide molecules in different cell types of fracture callus. Acta Pathol Microbiol Immunol Scand
    [A] 81:523–543
Gothlin G, Ericsson JLE (1973b) On the histogenesis of the cells in fracture callus. Electron micro-
    scopic autoradiographic observations in parabiotic rats and studies on labelled monocytes. Vir-
    chows Archs [B] 12:318–329
Gothlin G, Ericsson JLE (1976) The osteoclast. Clin Orthop Rel Res 120:201–231
Grant D, Bernick S (1972) The formation of the periodontal ligament. J Periodontol 43:17–25
Grant D, Bernick S, Levy BM, Dreizin S (1972) A comparative study of periodontal ligament
    development in teeth with and without predecessors in marmosets. J Periodontol 43:162–169
Grant ME, Freeman IL, Schofield JD, Jackson DS (1969) Variations in the carbohydrate content
    of human and bovine polymeric collagens from various tissues. Biochim Biophys Acta 177:682–
    685
Gregg JM (1965) Immobilization of the erupting molar in the Syrian hamster. J Dent Res 44:1219–
    1226
Guyton AC (1972) Compliance of the interstial space and the measurement of tissue pressure. Pflügers
    Arch [Suppl] 336:S1–S20
Haraldsson B, Rippe B, Moxham BJ, Folkow B (1982) Permeability of fenestrated capillaries in
    the isolated pig pancreas, with effects of bradykinin and histamine, as studied by simultaneous
    registration of filtration and diffusion capacities. Acta Physiol Scand 114:67–74
Harnot E (1948) Milchzahnstudien. I. Die Resorption der Milchzahnwurzel unter physiologischen
    und pathologischen Bedingungen. Dt Zahn-Mund-u-Kieferheilk 11:12–36
Hay ED (1982) Interaction of embryonic cell surface and cytoskeleton with extracellular matrix.
    Am J Anat 165:1–12

Heersche JNM, Deporter DA (1979) The mechanism of osteoclastic bone resorption: A new hypothesis. J Periodont Res 14:266–267

Hirschel BJ, Gabbiani G, Ryan GB, Majno G (1971) Fibroblasts of granulation tissue. Immunofluorescent staining with antismooth muscle serum. Proc Soc Exp Biol Med 138:466–469

Hodson JJ (1971) The gubernaculum dentis. Dent Practit 21:423–428

Höjer A, Westin G (1925) Jaws and teeth in scorbutic guinea-pigs. Dent Cosmos 67:1–24

Hoskins MM (1927) The effect of acetylthryoxin on the teeth of newborn rats. Proc Soc Exp Biol Med 25:55–57

Hunt AM (1959) A description of the molar teeth and investing tissues of normal guinea-pigs. J Dent Res 38:216–231

Jee WSS, Nolan PD (1963) Origin of osteoclasts from the fusion of phagocytes. Nature 200:225–226

Jones SJ, Boyde A (1972) A study of human root cementum surfaces as prepared for and examined in the scanning electron microscope. Z Zellforsch 130:318–337

Jones SJ, Boyde A, Ali NN (1984) The resorption of biological and nonbiological substrates by cultured avian and mammalian osteoclasts. Anat Embryol (Berl) 170:247–256

Jotereau FV, Le Douarin NM (1978) The developmental relationship between osteocytes and osteoclasts: A study using the quail-chick nuclear marker in endochondral ossification. Dev Biol 63:253–265

Kameyama Y (1973a) An autoradiographic investigation of the developing rat periodontal membrane. Arch Oral Biol 18:473–480

Kameyama Y (1973b) The pattern of alveolar bone activity during development and eruption of the molar in the rat. J Periodont Res 8:179–191

Kantorowicz FA (1929) Klinische Zahnheilkunde. Meusser, Berlin, p100

Kardos TB, Simpson LD (1979) A theoretical consideration of the periodontal membrane as a collagenous thixotropic system and its relationship to tooth eruption. J Periodont Res 14:444–451

Kardos TB, Simpson LD (1980) A new periodontal membrane biology based upon thixotropic concepts. Am J Orthod 77:508–515

Kenney EB, Ramfjord SP (1969) Patterns of root and alveolar-bone growth associated with development and eruption of teeth in rhesus monkeys. J Dent Res 48:251–256

Key KM, Elphick GK (1931) A quantitative method for the determination of vitamin C. Biochem J 25:888–897

Kindlova AM (1970) The development of the vascular bed of the marginal periodontium. J Periodont Res 5:135–140

Klimecka-Zakiewicz A (1960) Wplyw chorób miarzgi i ozebnej zebow mlecznych na resorpcje ich korzeni. Czas Stomat 13:729–735

Kostļán J, Thořová J, Škach M (1960) Erupce llodavého zubu po resekci jeho růstové zóny. Čslká Stomat 6:401–410

Kronfeld R (1932) The resorption of the roots of deciduous teeth. Dent Cosmos 74:103–120

Levy BM, Bernick S (1968) Development of organization of the periodontal ligament of deciduous teeth in marmosets (Calithrix jacchus). J Dent Res 47:27–33

Lewis J (1984) Morphogenesis by fibroblast traction. Nature 307:413–414

Lipper S, Kahn LB, Reddick RL (1980) The myofibroblast. Pathol Annu 15:409–441

Maclean DL, Sheppard M, McHenry EW (1939) Tissue changes in ascorbic acid deficient guinea-pigs. Br J Exp Pathol 20:451–457

Magnusson B (1968) Tissue changes during molar tooth eruption. Trans R Schs Dent Stockh Umeå 13:1–122

Magnusson B (1973) Autoradiographic study of erupting teeth in rats after intracardial injection of ^{131}I-fibrinogen. Scand J Dent Res 81:130–134

Majno G, Gabbiani G, Hirschel BJ, Ryan GB, Statkov PR (1971) Contraction of granulation tissue in vitro: Similarity to smooth muscle. Science 173:548–550

Manson JD (1967) Bone changes associated with tooth eruption. In: Anderson DJ, Eastoe JE, Melcher AH, Picton DCA (eds) The mechanism of tooth support. Wright, Bristol, pp 98–101

Manson JD (1968) A comparative study of the postnatal growth of the mandible. Kimpton, London

Marks SC Jr, Cahill DR (1983) The cytology of the dental follicle and adjacent alveolar bone during tooth eruption. Am J Anat 168:277–289

Marks SC Jr, Cahill DR (1984) Experimental study in the dog of the nonactive role of the tooth in the eruptive process. Arch Oral Biol 29:311–322

Mashouf K, Engel MB (1975) Maturation of periodontal connective tissue in the newborn rat incisor. Arch Oral Biol 20:161–166

McHugh WD (1961) The development of the gingival epithelium in the monkey. Dent Practr Dent Rec 11:314–324

Melcher AH (1967) Changes in connective tissue covering erupting teeth. In: Anderson DJ, Eastoe JE, Melcher AH, Picton DCA (eds) The mechanism of tooth support. Wright, Bristol, pp 94–97

Melcher AH (1978) Biological process in resorption, deposition and regeneration of bone. In: Stahl SS (ed) Periodontal surgery, biological basis and technique. Thomas, Springfield, pp 99–120

Melcher AH, Beertsen W (1977) The physiology of tooth eruption. In: McNamara JA (ed) The biology of occlusal development. Monographs in craniofacial growth, No 7. University of Michigan, Ann Arbor, pp 1–23

Melcher AH, Eastoe JE (1969) The connective tissues of the periodontium. In: Melcher AH, Bowen WH (eds) Biology of the periodontium. Academic, London, pp 161–343

Melcher AH, Walker TW (1976) The periodontal ligament in attachment and as a shock absorber. In: Poole DFG, Stack MV (eds) The eruption and occlusion of teeth. Butterworths, London, pp 183–192

Michaeli Y, Pitaru S, Zajicek G, Weinreb MM (1975) Role of attrition and occlusal contact in the physiology of the rat incisor: IX Impeded and unimpeded eruption in lathyritic rats. J Dent Res 54:891–898

Moorrees CFA, Fanning EA, Hunt EE Jr (1963) Formation and resorption of three deciduous teeth in children. Am J Phys Anthropol 21:205–213

Morgan PH, Jacobs HG, Segrest JP, Cunningham LW (1970) A comparative study of glycopeptides derived from selected vertebrate collagens. J Biol Chem 245:5042–5048

Morita H, Yamashiya H, Shimizu M, Sasaki S (1970) The collagenolytic activity during root resorption of bovine deciduous teeth. Arch Oral Biol 15:503–508

Moxham BJ (1979) The effects of some vasoactive drugs on the eruption of the rabbit mandibular incisor. Arch Oral Biol 24:681–688

Moxham BJ (1981) The effects of section and stimulation of the cervical sympathetic trunk on eruption of the rabbit mandibular incisor. Arch Oral Biol 26:887–891

Moxham BJ, Berkovitz BKB (1974) The effects of root transection on the unimpeded eruption rate of the rabbit mandibular incisor. Arch Oral Biol 19:903–909

Moxham BJ, Berkovitz BKB (1982) The periodontal ligament and physiological tooth movements. In: Berkovitz BKB, Moxham BJ, Newman HN (eds). The periodontal ligament in health and disease. Pergamon, Oxford, pp 215–247

Moxham BJ, Berkovitz BKB (1983) Continuous monitoring of the position of the ferret mandibular canine tooth to enable comparisons with the continuously-growing rabbit incisor. Arch Oral Biol 28:477–481

Moxham BJ, Shore RC, Berkovitz BKB (1985) Fenestrated capillaries in the connective tissues of the periodontal ligament. Microvasc Res 30:116–124

Myhre L, Preus HR, Aars H (1979) Influences of axial load and blood pressure on the position of the rabbit's incisor tooth. Acta Odontol Scand 37:153–159

Nanda RS (1969) Root resorption of deciduous teeth in Indian children. Arch Oral Biol 14:1021–1030

Ness AR (1964) Movement and forces in tooth eruption. In: Staple PH (ed) Advances in oral biology, vol 1. Academic, London, pp 33–75

Ness AR (1967) Eruption – a review. In: Anderson DJ, Eastoe JE, Melcher AH, Picton DCA (eds) The mechanisms of tooth support. Wright, Bristol, pp 84–88

Ness AR (1970) Eruption 70. Apex J University College Hosp dent Soc 4. 4: 23–27

Ness AR, Smale DE (1959) The distribution of mitoses and cells in the tissues bounded by the socket wall of the rabbit mandibular incisor. Proc R Soc B 151:106–128

Nowik IO (1958) Srokakh rassasyvaniya korniei molochnykh zubov. Stomatologiia (Mosk) 37:3–6

Noyes FB, Schour I, Noyes HJ (1943) In: Oral histology and embryology. Lea & Febiger, Philadelphia, p 170

Obersztyn A (1963) Experimental investigation of factors causing resorption of deciduous teeth. J Dent Res 42:660–674

O'Brien C, Bhaskar SN, Brodie AG (1958) Eruptive mechanism and movement in the first molar of the rat. J Dent Res 37:467–484

Orban BJ (1927) Embryology and histogenesis, Fortschritte der Zahnheilkunde. Mische J (ed) Oral histology and embryology, vol 3, 1st edn. Mosby, St Louis, p 749 (cited in Orban BJ 1944)

Orban BJ (1957) Oral histology and embryology, 4th edn. Mosby, St Louis, p 185

Oster GF, Murray JD, Harris AK (1983) Mechanical aspects of mesenchymal morphogenesis. J Embryol Exp Morphol 78:83–125

Parry DAD, Barnes GRG, Craig AS (1978) A comparison of the size distribution of collagen fibrils in connective tissues as a function of age and a possible relation between fibril size distribution and mechanical properties. Proc R Soc Lond [Biol] 203:305–321

Paynter KJ (1954) The effect of propylthiouracil on the development of molar teeth of rats. J Dent Res 33:364–376

Pearson CH (1982) The ground substance of the periodontal ligament. In: Berkovitz BKB, Moxham BJ, Newman HN (eds) The periodontal ligament in health and disease. Pergamon, Oxford, pp 119–149

Pearson CH, Gibson GJ (1982) Proteoglycans of bovine periodontal ligament and skin: occurrence of different hybrid-sulphated galactosaminoglycans in distinct proteoglycans. Biochem J 201: 27–37

Pearson CH, Happey F, Naylor A, Turner RL, Palframan J, Shentall RD (1972) Collagens and associated glycoproteins in the human intervertebral disc. Ann Rheum Dis 31:45–53

Pearson CH, Wohllebe M, Carmichael DJ, Chovelon A (1975) Bovine periodontal ligament. An investigation of the collagen, glycosaminoglycan and insoluble glycoprotein components at different stages of tissue development. Conn Tissue Res 3:195–206

Perera KAS (1983) Cellular origin and developmental sequence of the 'pulplimiting' (pulp delineating) membrane. J Dent Res 62:438

Perera KAS, Tonge CH (1981a) Metabolic turnover of collagen in the mouse molar periodontal ligament during tooth eruption. J Anat 133:359–370

Perera KAS, Tonge CH (1981b) Fibroblast cell proliferation in the mouse molar periodontal ligament. J Anat 133:77–90

Perera KAS, Tonge CH (1981c) Fibroblast cell population kinetics in the mouse molar periodontal ligament and tooth eruption. J Anat 133:281–300

Pitaru S, Michaeli Y, Zajicek G, Weinreb MM (1976) Role of attrition and occlusal contact in the physiology of the rat incisor. The part played by the periodontal ligament in the eruptive process. J Dent Res 55:819–824

Pryce-Jones J (1936) Some fundamental aspects of thixotropy. J Oil and Colour Chem Assoc 19:295– 337

Rannie I (1963) Observations on the oxytalan fibres of the periodontal membrane. Trans Euro Orthodont Soc 39:127–136

Richardson M (1978) Pre-eruptive movements of the mandibular third molar. Angle Orthod 48:187– 193

Sarnat H, Sciaky I (1965) Experimental lathyrism in rats. Effect of removing incisal stress. Periodontics 3:128–134

Schour I (1960) Noyes' oral histology and embryology, 8th edn. Kimpton, London

Schroeder HE, Listgarten MA (1977) Fine structure of the developing epithelial attachment of human teeth. In: Wolsky E (ed). Monographs in developmental biology, vol 2, 2nd edn. Karger, Basel

Schubert M, Hamerman D (1968) A primer on connective tissue biochemistry. Lea and Febiger, Philadelphia

Scott JH (1948) The development and function of the dental follicle. Br Dent J 85:193–199

Scott JH (1953) How teeth erupt. Dent Practr Dent Rec 3:345–350

Scott JH (1967) Dento-facial development and growth. Pergamon, Oxford

Scott JH, Symons NBB (1971) Introduction to dental anatomy, 6th edn. Livingstone, London

Selvig KA (1965) The fine structure of human cementum. Acta Odontol Scand 23:423–441

Shore RC, Berkovitz BKB (1978) Model to explain differential movement of periodontal fibroblasts. Arch Oral Biol 23:507–509

Shore RC, Berkovitz BKB (1979) An ultrastructural study of periodontal ligament fibroblasts in relation to their possible role in tooth eruption and intracellular collagen degradation in the rat. Arch Oral Biol 24:155–164

Shore RC, Berkovitz BKB, Moxham BJ (1981) Intercellular contacts between fibroblasts in the periodontal connective tissues of the rat. J Anat 133:67–76

Shore RC, Moxham BJ, Berkovitz BKB (1982) A quantitative comparison of the ultrastructure of the periodontal ligaments of impeded and unimpeded rat incisors. Arch Oral Biol 27: 423–430

Shore RC, Berkovitz BKB, Moxham BJ (1985) The effects of preventing movement of the rat incisor on the structure of its periodontal ligament. Arch Oral Biol 30:221–228

Shrimpton BA (1960) Dynamics of eruption. NZ Dent J 56:122–124

Sicher H (1923) Bau und Funktion des Fixationsapparates der Meerschweinchenmolaren. Oral histology and embryology, 1st edn. Mosby, St Louis, (Z Stomat 21:580, cited in Orban BJ 1944)

Sicher H (1942b) Tooth eruption: The axial movement of continuously growing teeth. J Dent Res 21:201–210

Sicher H (1942b) Tooth eruption: Axial movement of teeth of limited growth. J Dent Res 21:395–402

Sicher H (1962) Orban's oral histology and embryology. In: Sicher H (ed) 5th edn. Mosby, St Louis, pp 321–330

Smith RG (1978) A clinical study into the depth of the so-called gingival crevice of some erupting teeth of humans. MDS Thesis, Bristol University

Sodek J, Brunette DM, Feng J, Herrsche JNM, Limeback HF, Melcher AH, Ng B (1977) Collagen synthesis is a major component of protein synthesis in the periodontal ligament in various species. Arch Oral Biol 22:647–653

Spooner BS, Yamada KM, Wessels NK (1971) Microfilaments and cell locomotion. J Cell Biol 49:595–613

Stopak D, Harris AK (1982) Connective tissue morphogenesis by fibroblast traction, I Tissue culture observations. Dev Biol 90:383–398

Ten Cate AR (1969) The development of the periodontium. In: Melcher AH, Bowen WH (eds) Biology of the periodontium. Academic, London, pp 53–89

Ten Cate AR (1971) Physiological resorption of connective tissue associated with tooth eruption. J Periodont Res 6: 168–181

Ten Cate AR (1972a) Developmental aspects of the periodontium. In: Slavkin HC, Bavetta LA (eds) Developmental aspects of oral biology. Academic, London, pp 309–324

Ten Cate AR (1972b) Cell division and periodontal ligament formation in the mouse. Arch Oral Biol 17:1781–1784

Ten Cate AR (1980) Shedding of deciduous teeth. In: Bhaskar SN (ed) Orban's oral histology and embryology, 9th edn. Mosby, St Louis, pp 386–403

Ten Cate AR, Mills C (1972) The development of the periodontium. The origin of the alveolar bone. Anat Rec 173:69–78

Ten Cate AR, Mills C, Solomon G (1971) The development of the periodontium. An autoradiographic and transplantation study. Anat Rec 170:365–380

Thomas NR (1965) The process and mechanism of tooth eruption. PhD Thesis, University of Bristol

Thomas NR (1967) The properties of collagen in the periodontium of an erupting tooth. In: Anderson DJ, Eastoe JE, Melcher AH, Picton DCA (eds) The mechanisms of tooth support. Wright, Bristol, pp 102–106

Thomas NR (1976) Collagen as the generator of tooth eruption. In: Poole DFG, Stack MV (eds) The eruption and occlusion of teeth. Butterworths, London, pp 290–301

Tinkler SMB, Linder JE, Williams DM, Johnson NW (1981) Formation of osteoclasts from blood monocytes during 1 α-OH Vit D-stimulated bone resorption in mice. J Anat 133:389–396

Tonge CG (1963) The development and arrangement of the dental follicle. Trans Eur Orthod Soc 118–126

Toto PD, Sicher H (1966) Eruption of teeth through the oral mucosa. Periodontics 4:29–32

Trelstad RL, Hay ED, Revel JP (1967) Cell contact during early morphogenesis in the chick embryo. Dev Biol 16:78–106

Trott JR (1962) The development of the periodontal attachment in the rat. Acta Anat (Basel) 51:313–328

Tsuruta M, Eto K, Chiba M (1974) Effect of daily or 4-hourly administrations of lathyrogens on the eruption rates of impeded and unimpeded mandibular incisors of rats. Arch Oral Biol 19:1221–1226

Tyler DW, Burn-Murdoch R (1976) Tooth movements in an in vitro model system. In: Poole DFG, Stack MV (eds) The eruption and occlusion of teeth. Butterworths, London, pp 302–304

Urban JPG, Maroudas A, Bayliss MT, Dillon J (1979) Swelling pressures of proteoglycans at the concentrations found in cartilagenous tissues. Biorheol 16, pp 447–464

Van Winkle W Jr (1967) The fibroblast in wound healing. Surg Gynecol Obstet 124:369–386

Vasiliev JM, Gelfand IM, Domnina LV, Ivanovna CY, Komm SG, Olshevskaja LV (1970) Effect of colcemid on the locomotory behaviour of fibroblasts. J Embryol Exp Morphol 24:625–640

Weatherell JA, Hargreaves JA (1966) Effect of resorption on the fluoride content of human deciduous dentine. Arch Oral Biol 11:749–756

Westin C (1942) Über Zahndurchbruch und Zahnwechsel. Z Mikrosk Anat Forsch 51:393–470

Woessner JF Jr, Cahill DR (1974) Collagen breakdown in relation to tooth eruption and resorption in the dog. Arch Oral Biol 19:1195–1201

Yaeger JA, Kraucunas E (1969) Fine structure of the resorptive cells in the teeth of frogs. Anat Rec 164:1–14

Zajicek G (1974) Fibroblast cell kinetics in the periodontal ligament of the mouse. Cell Tissue Kinet 7: 479–492

Zerosi C (1965) Su alcuni aspetti morfologici della matrice fibrillare della polpa di denti decidui in riassorbimento. Schweiz Monatsschr Zahnmed 75:123–126

Development of Dentine and Pulp

R. M. FRANK and J. NALBANDIAN

A. Introduction

Dentine is the first of the calcified dental tissues to be deposited during tooth embryogenesis. Its development involves cellular and extracellular activities which take place primarily in the dental papilla, the precursor of the dental pulp. In the early formative stages, important epithelio-mesenchymal interactions occur between the enamel organ and the dental papilla, notably between the cells of the internal dental epithelium and the peripheral cells of the dental papilla.

From developmental, structural, and functional points of view, dentine and pulp constitute an undissociable tissue complex. Embryologically, dentine is produced by the pulp. Structurally, certain of the pulpal elements such as odontoblasts and nerve fibers extend into the dentine. The two tissues are closely interrelated functionally as well. Metabolically, the pulp nourishes the avascular dentine and irritation factors affecting the dentine produce a reaction in the dental pulp. The dentine and pulp are also directly involved, as an integrated complex, in all phenomena related to tooth sensitivity. And finally, reparative processes are the result of cellular activities in the pulp, expressed as new dentine deposition in the adult tooth.

Subsequent to the fundamental contribution of LEHNER and PLENK (1936) regarding the structure and development of the teeth, significant progress has been achieved in our understanding of dentine and pulp, with the advent of a number of important methodological advances. Improved fixation, demineralization, and embedding methods have allowed preparation of better decalcified and nondecalcified sections for light microscopy. Advances have also accrued through new technology in the fields of histochemistry, autoradiography, electron microscopy, biophysics, immunology and biochemistry.

From a historical perspective, it is interesting to note that, following his invention of the microscope, Anthony VAN LEEUWENHOEK (1675) was, himself, the first to describe the tubular structure of dentine. The dentinal tubules were then rediscovered independently by VON PURKINJE (1835, cited by FRÄNKEL 1835, and by RETZIUS 1837). John HUNTER (1778) was among the pioneers who recognized that dentine was a specific hard tissue different from bone, and CUVIER (1805) gave the name of "ivory" to this calcified substance, formerly considered to be an "osseous tooth substance". The fibrillar nature of the content of the dentinal tubule was initially described by Sir John TOMES (1856), whereas VON EBNER (1875) noted the presence of collagen fibers in the intertubular dentine.

In a well documented review, Baume (1980) analyzed the developments that led to the recognition of the fundamental organization and structure of dentine. He described four historical periods. From 1835 to 1851, the tubular structure of dentine was demonstrated. Between 1852 to 1863, the relationship between pulp and dentine was recognized. During the period from 1863 to 1868, the structures around the dentinal tubules as well as the nerves were discovered and discussed, and from 1867 to 1906, the ontogeny and phylogeny of dentine was scrutinized. For those interested in dentine and pulp from a historical perspective, the reviews of Lehner and Plenk (1936) and Baume (1980) are highly recommended.

B. The Dental Papilla

The development of the dental papilla, part of the tooth germ and formative organ of the dentine and pulp, will be considered briefly. Descriptions of the formation of this dental primordium were made early in the 19th century (Serres 1817). The first sign of human tooth development is a proliferation of the oral epithelium, observed in the 6–7-week-old embryo (11–15 mm). This gives rise to the dental lamina which takes the form of a continuous band of thickened oral epithelium in the region of the future dental arches.

With further development, the dental lamina extends into the underlying mesenchyme (Fig. 1). Subsequently lateral, lingual or palatal epithelial thickenings develop and form the tooth buds. The basal portion of each tooth bud becomes indented to form an epithelial cap-shaped structure (Fig. 2). In the concavity of the epithelial cap, a condensation of mesenchymal cells will constitute the dental papilla. The cap continues its growth and ultimately differentiates into a four-layered enamel organ which assumes a bell shape. At this point (Figs. 3 and 4), the dental papilla is separated from the enamel organ by a basement membrane and a narrow cell-free zone is found between this basement membrane and the mesenchymal cells (Fig. 5).

It is widely held that the mesenchymal cells of the dental papilla originate from the cranial neural crest. In fact, such an origin has only been clearly established in Amphibia (Stone 1926; Raven 1932; Sellman 1946). Using methods involving surgical removal of the cephalic neural crest and grafting, associated with tritiated thymidine labeling, Chibon (1967) demonstrated that in amphibians, the cells of the dental papilla originated in the neural crest. At present, however, direct experimental evidence for this origin is still lacking with respect

Fig. 1. Dental lamina of a human fetus. O Oral cavity; M Mesenchyme; G Oral epithelium. × 105

Fig. 2. Developing tooth germ of a human embryo illustrating the enamel organ at the early cap stage. Along the concave side of the enamel organ (EO), a condensation of mesenchymal cells gives rise to the dental papilla (DP). M Mesenchyme; O Oral cavity; G Oral epithelium. × 55

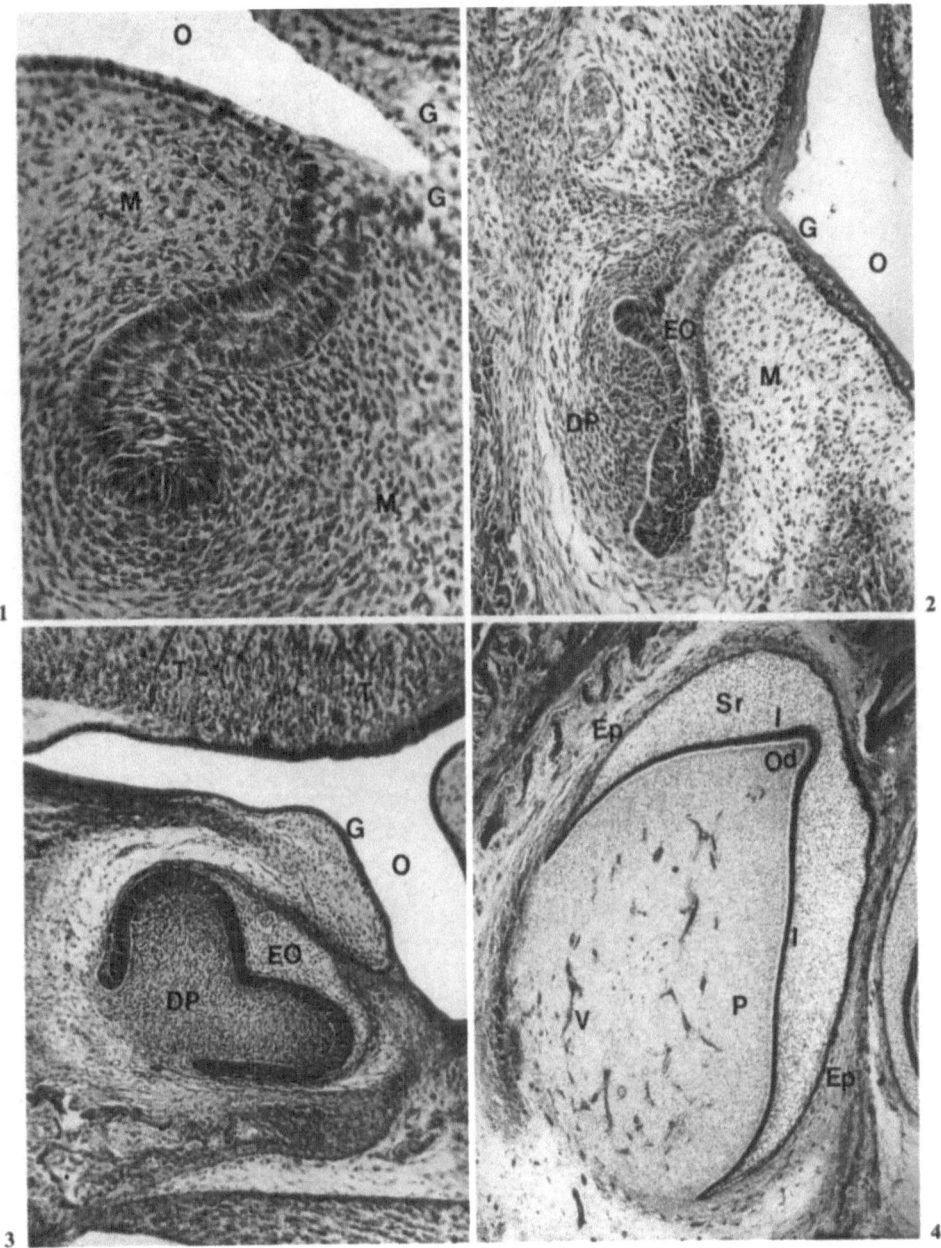

Fig. 3. Developing molar tooth germ of a rabbit embryo at the bell stage. *EO* Enamel organ, with a well-developed internal dental epithelium; *DP* Dental papilla; *O* Oral cavity; *G* Oral epithelium; *T* Tongue. ×60

Fig. 4. Developing human tooth germ (7-month-old embryo) at the bell stage. *P* Pulp; *Od* Odontoblasts; *V* Blood vessels; *Sr* Stellate reticulum; *I* Internal dental epithelium; *Ep* External dental epithelium. ×40

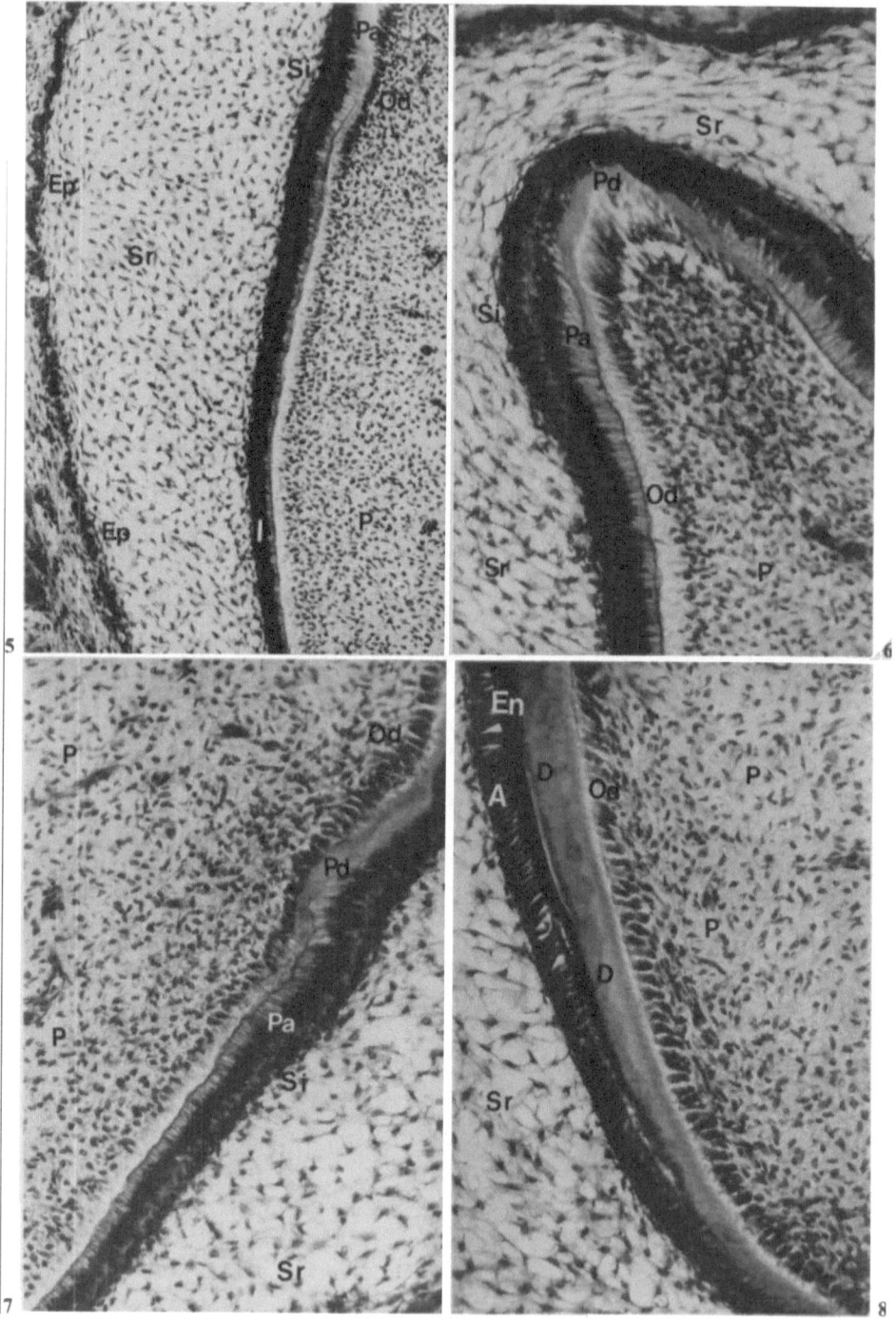

to mammalian species. DUNLAP et al. (1984), by means of a light microscopic immunoperoxidase technique, attempted to localize in the developing dental papilla of 18-day-old rat fetuses a protein unique to neural tissue (the so-called S-100 protein) to serve as a marker of neural crest cells. No marker protein was identified in the cells of the dental papilla. However, further studies along these lines, such as may be possible when monoclonal antibodies become available for more accurate localization of marker proteins, will be necessary before definite conclusions can be reached.

The cells of the dental papilla (Figs. 6–14) have irregular oval shapes and are separated by variable but significant amounts of extracellular space (Figs. 9, 12–14). The nuclei are very large compared to the very scant amounts of cytoplasm and they have an homogeneous distribution of chromatin. Denser regions correspond to nucleoli (Figs. 9, 13, 14). The nuclei are circumscribed by a double nuclear membrane with nuclear pores (Fig. 10). The outer nuclear membrane, with its adherent ribonucleoprotein particles, is in continuity with the rough-surfaced endoplasmic reticulum (Figs. 12–14), the latter organelle being moderately well developed. The surrounding cytoplasm presents a poorly developed Golgi apparatus in addition to the endoplasmic reticulum, some mitochondria, ribosomes, and a cytoskeleton composed of sparsely distributed microfilaments and microtubules. Occasionally, a typical stereocilium with a basal body and centrioles can be observed (Fig. 11).

A continuous trilaminar plasma membrane circumscribes the cells of the dental papilla. In localized areas, desmosome-like intercellular junctions can be observed (Fig. 9, lower right). In other cases, the plasma membranes of adjacent cells are parallel to each other and separated by very narrow spaces (Fig. 9). The content of the extracellular compartment varies in its complexity (Figs. 9 and 12). It often consists of an amorphous matrix rich in proteoglycans and glycoproteins (Figs. 9, 13 and 14), but typical collagen fibrils with 640 Å periodicity can be found occasionally (Fig. 12). In addition to these mature collagen fibrils, fibrils with smaller diameters and fewer crossbandings have been observed between the cells of the dental papilla, suggesting a progressive maturation of the collagen fibrils (FRANK 1965).

Fig. 5. Developing human tooth germ (7-month-old embryo). Bell stage. *Od* Odontoblasts; *P* Pulp; *I* Internal dental epithelium; *Pa* Preameloblasts; *Si* Stratum intermedium; *Sr* Stellate reticulum; *Ep* External dental epithelium. × 150

Fig. 6. Commencement of predentine (*Pd*) formation at a cuspal region of a deciduous incisor. Developing human tooth germ (7-month-old embryo). *Od* Odontoblasts; *P* Pulp; *Pa* Preameloblasts; *Si* Stratum intermedium; *Sr* Stellate reticulum. × 180

Fig. 7. Commencement of predentine (*Pd*) formation in the lateral coronal region of a deciduous incisor. Developing tooth germ (7-month-old embryo). *Od* Odontoblasts; *P* Pulp; *Pa* Preameloblasts; *Si* Stratum intermedium; *Sr* Stellate reticulum. × 180

Fig. 8. Commencement of enamel (*En*) and dentine (*D*) formation in a deciduous incisor. Developing human tooth germ (7-month-old embryo). *Od* Odontoblasts; *P* Pulp; *A* Ameloblasts; *Sr* Stellate reticulum. × 205

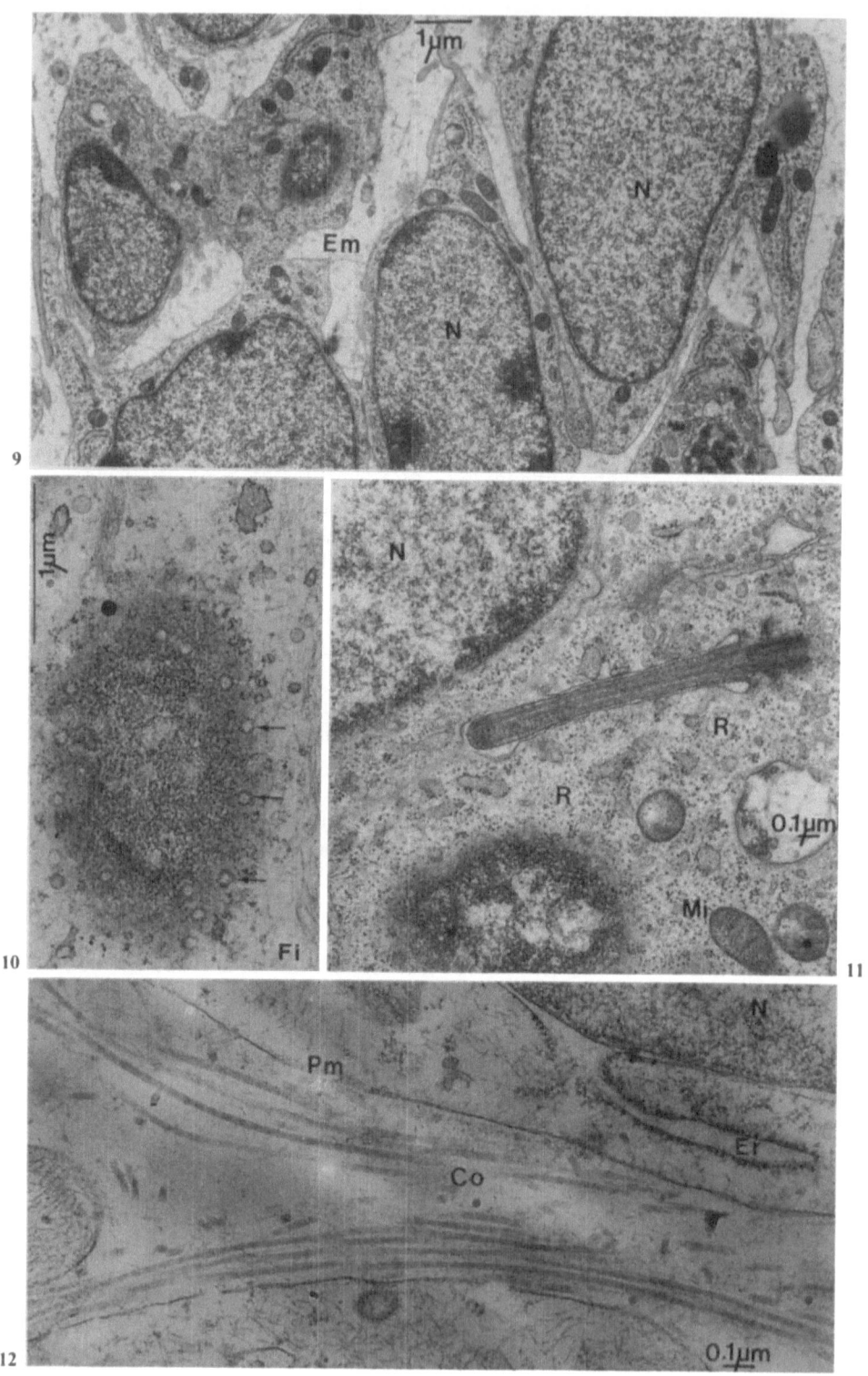

Blood capillaries with characteristic ultrastructural features (continuous array of endothelial cells, peripheral pericytes, and well-differentiated basement membrane) can be found in the central part of the dental papilla in the early stages of development. At the bell stage, nerve fibers enter the dental papilla before odontoblasts have differentiated, but innervation remains rudimentary until birth. At the time of birth and in the ensuing period, a sudden increase in innervation is observed in the deciduous teeth (SCOTT and SYMONS 1982).

C. The Internal Dental Epithelium and the Preameloblasts

When describing the early histological features associated with dentinogenesis, it becomes clear that the differentiation of the peripheral cells of the dental papilla into odontoblasts, occurring initially in the cuspal areas (Figs. 5–8), is preceded by events that take place at the bell stage in the enamel organ. These events include the transformation of the cells of the internal dental epithelium into preameloblasts, as part of a complex sequence of interactions, the so-called epithelio-mesenchymal interactions, which will be described in detail later. At present a brief description of the enamel organ is in order (Figs. 3–19).

At the bell stage (Figs. 3–5, 7, 8) the enamel organ has differentiated into four distinct layers. From its surface, inward, these are the external dental epithelium, the stellate reticulum, the stratum intermedium, and the internal dental epithelium. The external dental epithelium is composed of a single layer of cuboid cells separated from the surrounding oral mesenchyme by an outer basement membrane. Up to 8 hemi-desmosomes per cell have been observed along the basal plasma membranes of the external epithelial cells (FRANK 1968a). A large nucleus with nucleoli, occupying the center of the cell, is surrounded by cytoplasm containing small amounts of rough-surfaced endoplasmic reticulum, a poorly developed Golgi apparatus, lipid droplets, a few mitochondria, glycogen, numerous ribonucleoprotein particles, and an abundant cytoskeleton consisting primarily of intermediate microfilaments, but with some microtu-

Fig. 9. Mesenchymal cells of the dental papilla in a mouse molar tooth germ. Note the large size of the nuclei (*N*). *Em* Extracellular matrix

Fig. 10. Tangential section of the nucleus of a mesenchymal cell from the dental papilla of a molar tooth germ in a newborn cat, showing pores in the nuclear membrane (*arrows*). *Fi* Microfilaments

Fig. 11. Presence of a stereocilium in a mesenchymal cell of the dental papilla of a mouse molar tooth germ. *N* Nucleus; *Mi* Mitochondria; *R* Free ribosomes

Fig. 12. Dental papilla of a molar tooth germ in a newborn cat. Typical cross-striated collagen fibrils (*Co*) are visible in the extracellular space between two mesenchymal cells. *Pm* Plasma membrane; *N* Nucleus; *Er* Endoplasmic reticulum. (From FRANK 1965, courtesy of J Micros)

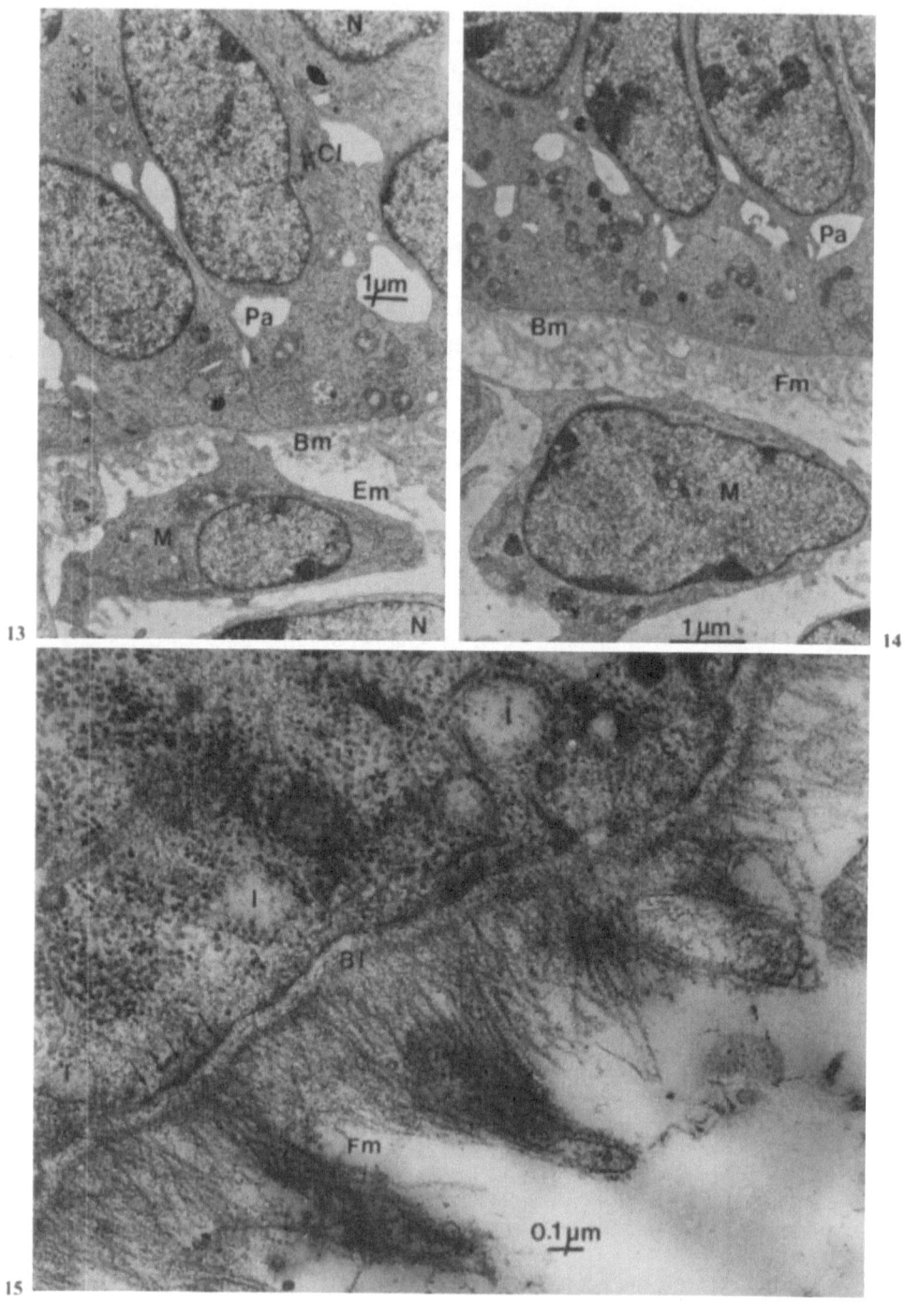

bules. Each cell is bound by a plasma membrane. Adjacent cell membranes make contact through typical desmosomes and gap junctions. These types of cell-to-cell attachments are also found between the cells of the external dental epithelium and the stellate reticulum.

Between the internal and the external dental epithelium, two other zones, the stellate reticulum and the stratum intermedium, are found. The stratum intermedium consists of two or three rows of somewhat flattened cells, separated from each other by small intercellular spaces and arranged so that their long axes are perpendicular to the cells of the internal dental epithelium. Numerous microvilli interdigitate with those of adjacent cells. The cells of the stellate reticulum nearest the stratum intermedium have this same general arrangement, but as the central part of the stellate reticulum is approached their orientation becomes more random and the intercellular spaces increase in size. Each cell of the stellate reticulum assumes a star-shaped form with a central body containing a large nucleus and long, branching cytoplasmic extensions (Fig. 20).

According to PANNESE (1960) and FRANK and NALBANDIAN (1967), the general cytomorphological characteristics of the stellate reticulum and the stratum intermedium are similar, the most notable difference being the large intercellular spaces in the former *vs.* the very much smaller ones in the latter. Both cell types have large nuclei and centrioles. The Golgi complex is relatively well-developed but endoplasmic reticulum and mitochondria are scarce. Numerous glycogen and ribonucleoprotein granules are scattered throughout the cytoplasm, along with some lipid droplets. The cytoskeleton consists of microtubules and intermediate microfilaments, often grouped in bundles. Desmosomes (FRANK and NALBANDIAN 1967) and gap junctions (GARANT 1972) can be found between the cells of the stratum intermedium, the stellate reticulum, and the internal dental epithelium.

Using tritiated thymidine, HUNT and PAYNTER (1963) showed that the stratum intermedium serves as a continuous source of stellate reticulum cells. Taking in account the presence of a well-developed Golgi apparatus with a poorly developed endoplasmic reticulum, KALLENBACH (1978) proposed that a relatively high rate of polysaccharide synthesis and a low rate of protein synthesis occurred in the stellate reticulum. Aside from secretory activities, the latter layer seems to be able to store and concentrate some metabolic products such as glycos-

Fig. 13. Epithelial-mesenchymal interface in a mouse tooth germ. The internal dental epithelium, at an early preameloblast (*Pa*) stage, is separated from mesenchymal cells (*M*) by a basement membrane (*Bm*). *N* Nucleus; *Cl* Centriole; *Em* Extracellular matrix

Fig. 14. Preameloblasts (*Pa*) at an early stage of differentiation, separated from the mesenchymal cells (*M*) by a lamina densa (*Bm*) and associated filamentous material (*Fm*). Mouse tooth germ

Fig. 15. Basement membrane separating a cell of the internal dental epithelium (*I*) from the extracellular matrix of the dental papilla in a tooth germ of a human embryo. The *arrows* indicate a thickening of the epithelial plasma membrane reminiscent of a hemi-desmosome. *Fm* lamina rara interna, with its filamentous material extending to the lamina densa (*Bl*)

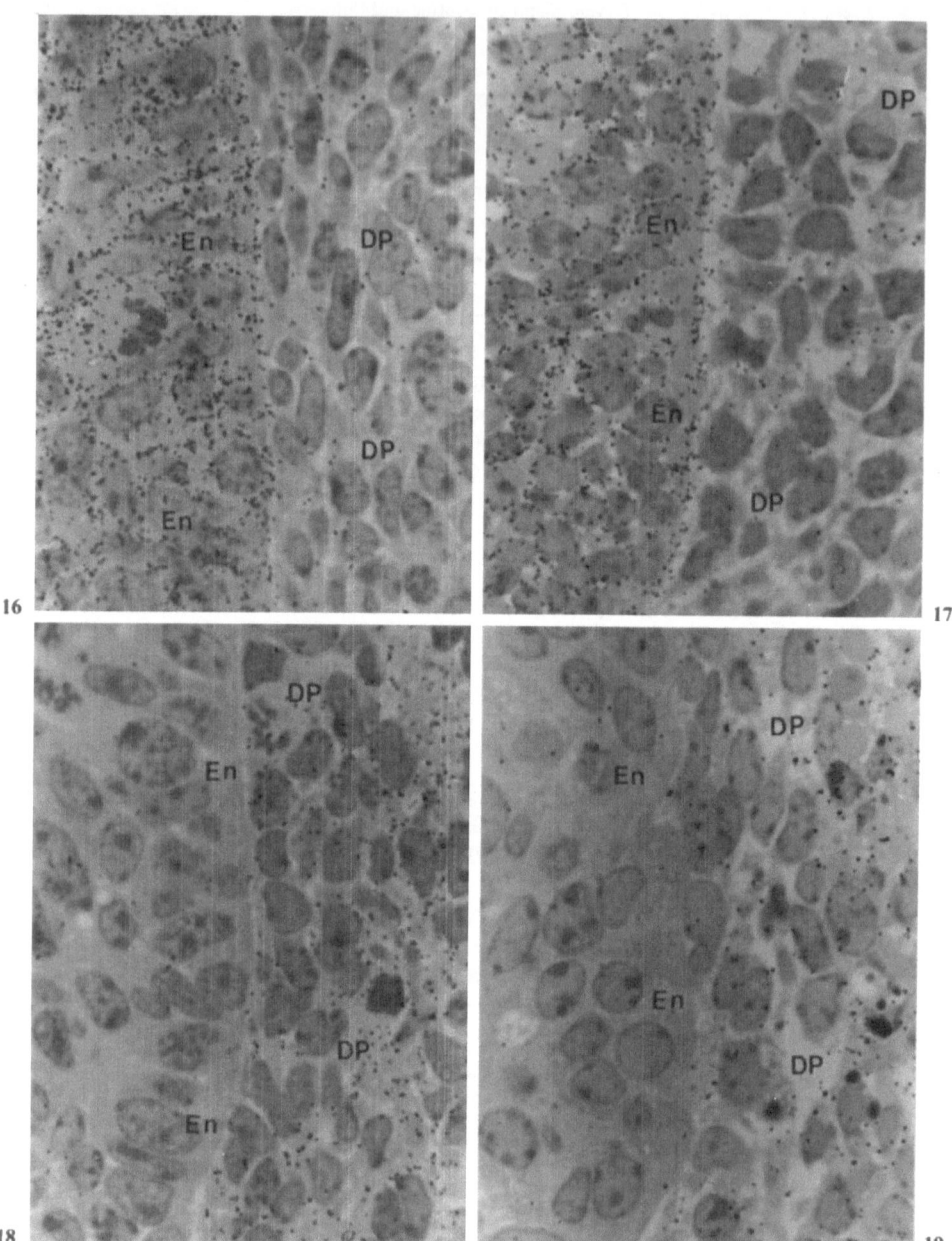

Fig. 16. Light microscopic autoradiogram of a ³H-glucosamine-labeled enamel organ (*En*) associated with a non-labeled dental papilla (*DP*) and cultivated for 9.5 hours. More silver grains are observed over the enamel organ than over the dental pulp. (From FRANK et al. 1979, courtesy of J Biol Buccale). × 490

Fig. 17. Light microscopic autoradiogram of a ³H-glucosamine-labeled enamel organ (*En*) cultivated in association with a non-labeled dental pulp (*DP*) for 19 hours. The labeling in the dental papilla has not increased and the enamel organ is still labeled, with some increase in the number of silver grains located in the intercellular spaces of the stellate reticulum and along the epithelial-mesenchymal interface. (From FRANK et al. 1979, courtesy of J Biol Buccale). × 490

aminoglycans (Fig. 20) in the intercellular spaces (FRANK et al. 1979). Such is not the case for the dental papilla.

The internal dental epithelium consists of a layer of tall cells separated from the dental papilla by a basement membrane (Figs. 13–15). In the cat, at an early stage when the cells of the internal dental epithelium are adjacent to undifferentiated mesenchymal cells of the dental papilla, they are about 2 to 3 µm wide and 10–12 µm high (PANNESE 1962). Dimensions are similar in the mouse tooth germ (Figs. 13 and 14). In the human embryo, these cells seem to be larger, i.e. about 25–30 µm long and 4–5 µm in diameter (FRANK and NALBANDIAN 1967). Initially, their oval nuclei are located in the half of the cell adjacent to the basement membrane. Mitotic figures have been described (NYLEN and SCOTT 1960). One of the striking cytoplasmic characteristics is the large number of ribonucleoprotein granules disseminated in the cytoplasm, contrasting with the scarcity of endoplasmic reticulum. In the human and cat embryo, mitochondria are evenly but sparsely distributed throughout the cell. The Golgi zone is poorly developed in the early stages of differentiation. Fine intermediate microfilaments and a few microtubules, similar to those observed in the other layers of the enamel organ, constitute the cytoskeleton of the cell.

In longitudinal section, the outlines of the cells of the internal dental epithelium are rather straight in the young enamel organ and the usual distance between two adjacent membranes is about 200 Å. Some larger intercellular spaces are found here, as well as between the internal dental epithelium and the stratum intermedium. Along the basement membrane, the plasma membranes are parallel to the *lamina densa*, at a distance of about 300 Å (Fig. 15). In the human embryo (Fig. 15), typical hemi-desmosomes have been observed in this location (FRANK and NALBANDIAN 1967). Desmosomes and gap junctions (GARANT 1972) are found between the plasma membranes of adjacent cells of the internal enamel epithelium.

With advancing development, the cells of the internal enamel epithelium are transformed into preameloblasts (SCHOUR 1960) which show, in addition to an increase in size, a proximal migration of their nuclei in the direction of the stratum intermedium. There is still a scarcity of rough-surfaced endoplasmic reticulum and an abundance of ribonucleoprotein granules. At this time, the Golgi apparatus is observed in the proximal juxta-nuclear cytoplasm (i.e. on the side of the nucleus towards the stratum intermedium). Before predentine and dentine formation, the Golgi apparatus of the preameloblast is always

Fig. 18. Light microscopic autoradiogram of a ³H-glucosamine-labelled dental pulp (*DP*) associated with a non-labeled enamel organ (*En*) and cultivated for 9.5 hours. A diffuse labeling is observed over the dental papilla. The enamel organ is essentially free of silver grains. (From FRANK et al. 1979, courtesy of J Biol Buccale). × 490

Fig. 19. Light microscopic autoradiogram of a ³H-glucosamine-labeled dental papilla (*DP*) associated with a non-labeled enamel organ (*En*) and cultivated for 19 hours. The dental papilla is covered with fewer silver grains. (From FRANK et al. 1979, courtesy of J Biol Buccale). × 490

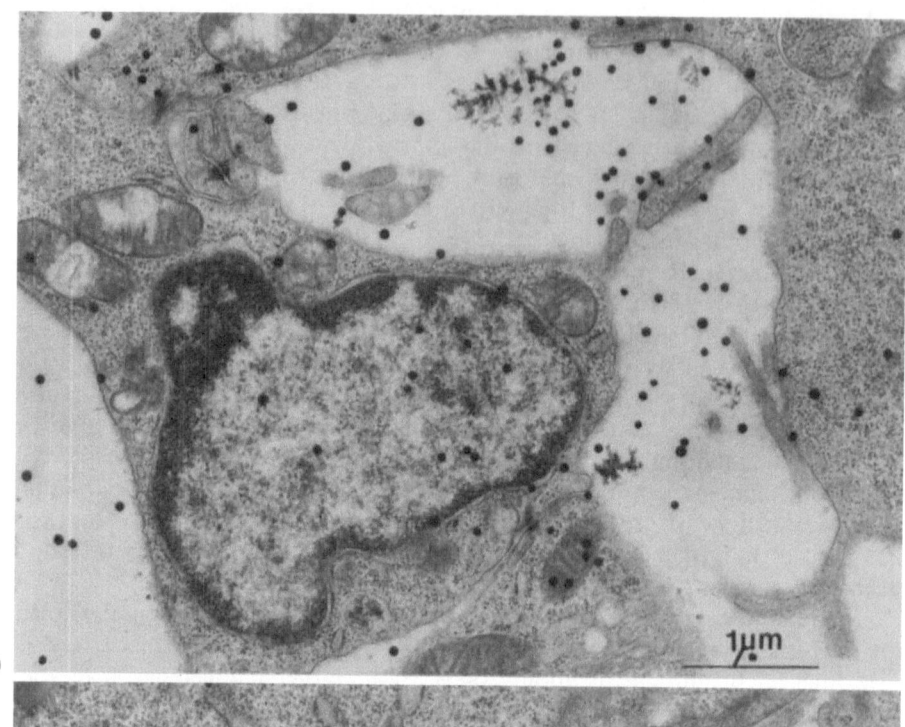

20

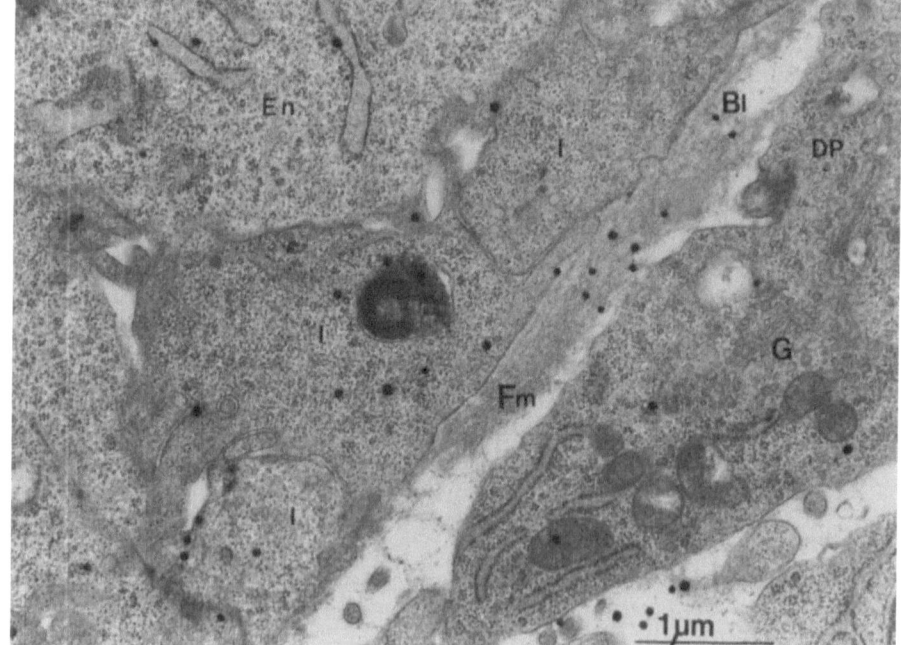

21

seen in this position. Even during initial dentine formation, but before the appearance of enamel, this cytoplasmic organelle remains in a proximal position (KEREBEL and GRIMBERT 1958; FRANK and NALBANDIAN 1967). Just prior to enamel formation, however, the Golgi apparatus migrates around the nucleus to a distal juxta-nuclear position (i.e. on the side of the nucleus facing the dental papilla). The occurrence of this phenomenon coincides with the differentiation of preameloblasts into a secretory ameloblast.

In addition to migration and enlargement of the Golgi complex, differentiation of the ameloblast is accompanied by a significant growth in size and by an extensive proliferation of the granular endoplasmic reticulum, closely associated with an increase in the number and size of mitochondria. The secretory pole of the cell shows development of a specialized process, the ameloblastic process or Tomes' process, with formation of a terminal bar apparatus at both the proximal and the distal ends of cells. At this stage, specific membrane-limited secretion granules are elaborated in the Golgi apparatus.

D. The Dental Basement Membrane

The cells of the internal dental epithelium are separated from the dental papilla by a basement membrane which, in light microscopy, appears as a continuous [PAS-positive sheet (RUCH et al. 1983)]. Various terms have been applied to the basement membrane and its components. Basal lamina and basement membrane are sometimes used interchangeably. The basement membrane was described in electron microscopy as consisting of an electron-dense structure, the *lamina densa*, separated from the plasma membrane of the epithelial cells by a *lamina lucida* (NYLEN and SCOTT 1960; PANNESE 1962; FRANK and NALBAN-DIAN 1967). Adjacent to the *lamina densa*, on the dental papilla side, a less distinct layer, the *lamina diffusa*, was sometimes described.

More recently, a slightly different terminology has been used (HASCALL and HASCALL 1981; HAY 1981). Accordingly, the dental basement membrane consists of a central *lamina densa*, composed of a compact sheet of collagen, mainly

Fig. 20. Electron microscopic autoradiogram of a ^{3}H-glucosamine-labeled enamel organ cultivated for 9.5 hours in association with a non-labeled dental pulp. A great number of silver grains are found in the large intercellular spaces of the stellate reticulum. Also note intracellular labeling over the nucleus, the cytoplasm and some mitochondria. (From FRANK et al. 1979, courtesy of J Biol Buccale)

Fig. 21. Electron microscopic autoradiogram of a ^{3}H-glucosamine-labeled enamel organ (*En*) cultivated for 9.5 hours in association with a non-labeled dental pulp. Note the labeling of the filamentous material (*Fm*) on the pulpal side adjacent to the basal lamina (*Bl*). The cytoplasm of the internal dental epithelium (*I*) is diffusely labeled. Some silver grains are seen in the intercellular spaces of the dental papilla (*DP*). *G* Golgi apparatus. (From FRANK et al. 1979, courtesy of J Biol Buccale)

type IV, associated with other glycoproteins (Fig. 15). This central *lamina densa*, measuring 30–60 nm in width, is separated from the cells of the internal dental epithelium by a less electron-dense zone, the *lamina rara externa*, and from the underlying connective tissue by the *lamina rara interna*. Under the electron microscope, small filaments (less than 10 nm in diameter) associated with very fine granules, are adherent to the *lamina densa* and extend through the *lamina rara interna* (Figs. 13–15). These small filaments do not present any periodic cross-striation and are oriented perpendicularly to the *lamina densa*. Such is the ultrastructure of the basement membrane in the early stages of development. With differentiation of odontoblasts and elaboration of predentine matrix, cross-striated collagen fibrils as well as matrix vesicles appear in close contact with the *lamina densa* and the *lamina rara interna* (Nylen and Scott 1958; Slavkin et al. 1977; Garant 1978; Orams 1978).

The composition of the dental basement membrane has been extensively studied by light and electron microscopic cytochemistry and immunocytochemistry, autoradiography, and biochemistry. Dental basement membrane contains collagenous and non-collagenous glycoproteins and closely associated glycosaminoglycans.

Basement membrane collagen has been termed type IV collagen (Kefalides 1973) with the following chain structure $[\alpha_1(IV)]3$. However, Glanville et al. (1979) have suggested the presence of two genetically different chains. Type IV collagen has been identified in the dental basement membrane by indirect immunofluorescence, with light microscopy (Lesot et al. 1978, 1981; Thesleff et al. 1979, 1981). With similar techniques, Cournil et al. (1979) described the presence of type IV procollagen in the same location. In addition, the dental basement membrane contains type I and III collagen (Lesot et al. 1978, 1981; Lesot 1980; Thesleff et al. 1979, 1981). Type III procollagen was found in the basement membrane during the bud and cap stages in embryonic mouse tooth germs, but apparently disappeared with the differentiation of odontoblasts (Thesleff et al. 1979).

Two types of non-collagenous glycoproteins have been demonstrated in the dental basement membrane: fibronectin and laminin. Fibronectin, a major component of connective tissue matrix (Linder et al. 1975; Stennman and Vaheri 1978; Ruoslahti 1981), has been detected in the dental basement membrane by indirect immunofluorescence in the early stages of development, but it seems to disappear from this location with the differentiation of the odontoblasts (Thesleff et al. 1979; Linde et al. 1982a). Laminin, a non-collagenous glycoprotein extracted from a transplantable mouse tumor producing basement membrane (Timpl et al. 1979) was also found in the dental basement membrane (Lesot 1980; Thesleff et al. 1980; Lesot et al. 1981).

A third class of extracellular matrix molecules has been found in basement membranes. These are proteoglycans (PG) which contain glycosaminoglycans (GAG) such as chondroitin sulfate, keratan sulfate and heparan sulfate. They are linked to protein and occur as monomers associated with basal laminae. When fixed with ruthenium red, PG monomers appear as small granules in the *lamina rara externa* and *interna* (Hay 1983). Using antibodies to heparan sulfate, Thesleff and Pratt et al. (1980) demonstrated the presence of proteoglycans in the dental basement membrane, using indirect immunofluorescence. Hyaluronate, chondroitin-4 and -6 sulfates as well as heparan sulfate were demonstrated by LAU (1983) and LAU and RUCH (1983).

It would appear from the above that the collagenous and non-collagenous glycoproteins, along with the proteoglycans, form an integrated complex that constitutes the basement membrane structure.

The tissue origin of the various components of the dental basement membrane has been investigated by culture techniques using dissociated and/or reassociated tooth germs. A new *lamina densa* and associated matrix were redeposited in trypsin-isolated enamel organs when cultured in vitro following removal of the original basement membrane. (SLAVKIN et al. 1975; OSMAN and RUCH 1980). Type IV collagen, laminin, and some glycosaminoglycans seem to be epithelially derived (FRANK et al. 1979; OSMAN and RUCH 1981a; HURMERINTA 1982; RUCH et al. 1983), whereas type I and type III collagen are synthesized by the dental mesenchyme (LESOT and RUCH 1979; LESOT et al. 1981). Mesenchymal origin is also the case for chondroitin-4 and -6 sulfates (LAU 1983) and for fibronectin.

Light and electron microscope radioautography was performed (FRANK et al. 1979) after ^{3}H-glucosamine labeling of trypsin-isolated enamel organ and dental papilla from embryonic mandibular first molars of Swiss mice (Figs. 16–21). The labeled fragment was reassociated with its cold counterpart and cultured for 9.5 h or 19 h. In the case of enamel organ labeling, numerous silver grains were observed in the intercellular spaces of the stellate reticulum at both time intervals (Figs. 16, 17, 20), indicating a concentrating and storing role of the latter. After dental papilla labeling, a significant decrease in labeling was noted between the two experimental times, suggesting that ^{3}H-glucosamine is not stored in the dental papilla (Figs. 18 and 19). A clear labeling of the basement membrane area and of the adjacent aperiodic filamentous material found on the pulpal side (Fig. 20) was only observed after enamel organ labeling with ^{3}H-glucosamine, suggesting an epithelial origin of the glycosaminoglycans found in the dental basement membrane (FRANK et al. 1979).

E. Epithelio-Mesenchymal Interactions

The importance of epithelio-mesenchymal interactions during tooth development became particularly apparent with improvements in tissue and organ culture techniques. Tooth germs were found to be easily dissociated into enamel organ and dental papilla with the use of either proteases or cation chelating agents. With trypsin, the basement membrane was destroyed. This was not the case with ethylene diamine tetraacetic acid (EDTA) solutions; rather, the basement membrane remained adherent to the dental papilla. When the two separate dental tissues were recombined and cultured in vitro, they differentiated and produced tooth organs comparable to the intact explant culture.

The fundamental role of these interactions was readily recognized because both dental epithelium and dental papilla isolated with trypsin failed to differentiate (HUGGINS et al. 1934; KOCH 1967; KOLLAR and BAIRD 1969; RUCH and KARCHER-DJURICIC 1971; SLAVKIN 1974). Transfilter induction between enamel organ and dental papilla (KOCH 1967) was demonstrated when the two tissues were cultured in juxtaposition to a Millipore filter (0.45 μm pore size and 25 μm thick). Dentine matrix was present on one side of the filter whereas on the other side an enamel-like substance was deposited. When the culture components were separated by cellophane or by a Millipore filter 70 to 80 μm in thickness, no tissue differentiation was observed.

In relation to the very first initiating events leading to the differentiation of the dental lamina, PEARSON (1977) and KOLLAR and LUMSDEN (1979) suggested that the distribution of nerves might influence the epithelial downgrowth in the stomodeal mesenchyme. Several investigations (KOLLAR 1972; RUCH et al. 1973; SLAVKIN 1974) have underlined the specificity of the dental papilla which, in tissue culture, is able to induce enamel organ differentiation from various non-dental epithelia (eg. snout, foot pad, skin epithelium, etc.). Along these lines RUCH et al. (1973) performed associations between 11-day-old mandibular dental papilla of Swiss mouse embryos and either oral or foot pad epithelium, and between 11-day-old oral epithelium and foot or gut mesenchyme. The dental papilla was a prerequisite for tooth development.

Concerning the determination of coronal tooth shape, DRYBURGH (1967) and MILLER (1969), using heterotopic reassociations of epithelial and mesenchymal components of presumptive molar and incisor-bearing parts of developing mouse jaws, grown on the chick chorio-allantois, concluded that the morphology of the tooth germ seemed to be under the control of the enamel organ.

However, using similar technics, KOLLAR and BAIRD (1969, 1970a and b) as well as HERITIER and DEMINATTI (1970) came to a different conclusion, i.e., that crown morphogenesis was directed by the dental papilla. In other words, when the dental papilla of a mouse incisor tooth germ is recombined with the enamel organ of a molar tooth germ, the enamel organs assumes the shape of the incisor and vice versa. Therefore, it appears that the dental papilla not only induces the development of a tooth but also determines its shape.

Functional odontoblasts are post-mitotic cells. After the last mitosis, the daughter cells which are in contact with the basement membrane overtly differentiate within a few hours. As early as 1887, VON BRUNN (1887) suggested that odontoblasts differentiate only in the presence of internal dental epithelium. The role of the internal dental epithelium and the preameloblasts on the differentiation of odontoblasts was documented by transfilter associations between enamel organ and dental papilla, and by in vitro cultured iso- and heterochronal associations between dental papilla and internal dental epithelium (RUCH et al. 1984).

It was shown that 1) heterotypic cell interactions were necessary for cytological and functional differentiation of odontoblasts; 2) that recombinants of polarizing odontoblasts and isochromal internal dental epithelium allowed maintenance of odontoblast polarization and predentine secretion; and 3) that odontoblast differentiation can be a labile stage and that in associations of polarizing odontoblasts and younger internal epithelium, the odontoblasts can depolarize in a few hours, incorporate ^{3}H-thymidine and divide again. Later, daughter cells would become post-mitotic again, repolarize and secrete predentine.

It has been suggested that the action of the internal dental epithelium on odontoblast differentiation is mediated by the basement membrane. It is important to note that during odontoblast differentiation in vivo a continuous basement membrane is constantly present between the preameloblasts and the dental papilla. This basement membrane therefore prevents direct cellular contact between the internal dental epithelium and the differentiating mesenchymal cells. However cellular contacts have been demonstrated in electron microscopy between the mesenchymal cell processes and the basement membrane and associated structures (Figs. 13 and 15). When such contacts are inhibited by interposing a nucleopore filter in cultures of reassociated enamel organ and dental papilla, odontoblast differentiation does not occur (THESLEFF 1978).

The role of the basement membrane was further confirmed by studies utilizing EDTA-dissociated mouse tooth germs (OSMAN and RUCH 1981 b; SLAVKIN et al. 1982). In this case the basement membrane is retained in its normal association with the dental papilla. When dental papillae derived with this technique were cultured alone for 24 h on plasma clots, post-mitotic odontoblasts which were present polarized and secreted predentine. With trypsin-isolated dental papillae, absence of a basement membrane, and similar cultural conditions, a dedifferentiation of odontoblasts was observed. The basal lamina seems to be necessary for the maintenance of polarized odontoblasts, but the precise nature of the biochemical interactions involved is presently unclear.

THESLEFF and PRATT (1980), treating tooth germ explants with tunicamycin, observed a marked decrease in the distribution of fibronectin with drastic reductions in the dental mesenchyme and the basement membrane. This treatment inhibited odontoblast differentiation and consequently these authors postulated that fibronectin reduction in the basement membrane, accompanied by surface alterations in the mesenchymal cells, may prevent the cell-matrix interactions necessary for odontoblast differentiation. Along these lines, interesting observations were made by MAGLOIRE et al. (1981) who observed a differentiation of peripheral odontoblast-like cells in explant cultures of human dental pulp. Using specific antibodies, light microscopic immunofluorescence and immunocytochemical electron microscopy, they showed that these "odontoblasts" synthesized, simultaneously, type I and type III collagen. Therefore differentiation of cultured pulpal cells to odontoblast-like cells seems possible, even in the absence of basement membrane material. A possible inductive role of fibronectin was thus postulated.

The terminal differentiation of odontoblasts, characterized by polarization of organelles, will be described in detail in a subsequent section. This differentiation is accompanied by cell membrane modifications, including an increase in adenylate-cyclase activity (OSMAN et al. 1981) and redistribution of concanavalin A-binding sites (MEYER et al. 1981). In addition, cytoskeletal modifications have been noted. For example, vimentin initially present throughout the cytoplasm becomes gradually restricted to the apical part of the odontoblast during polarization (LESOT et al. 1982).

The terminal differentiation of ameloblasts, with changes from preameloblasts to secretory cells, requires integrity of the cytoskeleton. The behavior
of ameloblasts cultured in iso- and heterochronal association with dental papillae (with and without predentine) and preameloblasts suggests that the terminal
differentiation of ameloblasts can only be achieved after a minimum of cell
cycles (Ruch et al. 1972).

Another important prerequisite for ameloblast differentiation is the appearance of a layer of predentine in the extracellular areas delimited by the basement
membrane and the recently differentiated odontoblastic processes. This predentine layer will undergo mineralization and an initial thin shell of dentine will
be formed concomitantly. With these initial stages of predentine and dentine
formation the basement membrane becomes discontinuous and disappears.

Direct contact can thus be established between preameloblasts and predentine, and/or odontoblastic cell processes. These associations will be described
in detail later, but presently it should be noted that these close-range relationships and direct cell contacts seemed to be coincident with initial dentine mineralization, cessation of cell division within the internal dental epithelium and
cell polarization that occurs as the preameloblast elongate into ameloblast.

The importance of predentine and dentine in ameloblast differentiation has
been experimentally demonstrated by Ruch et al. (1984). These workers isolated
extracellular dental matrixes such as those of predentine, dentine and enamel
after EDTA treatment and then associated them with enamel organs in tissue
culture. Preameloblasts associated with the epithelial side of predentine or dentine, polarized and secreted enamel, whereas preameloblasts associated with
the mesenchymal side of these tissues or with enamel, failed to differentiate.
These observations underline the importance of predentine and dentine in ameloblast differentiation. The significance of predentine collagen was further demonstrated through the use of a bacterial collagenase in tissue culture (Ruch
et al. 1972). The enzyme did not disturb the differentiation of the mouse molar
odontoblast but inhibited the differentiation of the ameloblast in a reversible
manner. Wolters (1978) on the basis of radioautographic experiments suggested
that collagen could play a stabilizing influence on ameloblast differentiation.

From this review of the epithelio-mesenchymal interactions that occur between the dental papilla and components of the enamel organ, it appears that
some inroads have been made towards understanding the complex interconnec-
tions that exist. Progress has been achieved primarily on morphological grounds
and is attributable largely to the development of tissue culture techniques. Considerable study will be necessary to decipher the more subtle molecular basis
for these complex and reciprocating cellular events that take place along the
basement membrane.

Saxen et al. (1976) tried to summarize the possible hypotheses for transmission of the inductive signals responsible for epithelio-mesenchymal interactions.
The following mechanisms were considered: a) extracellular diffusion of molecules; b) extracellular matrix-mediated instructions; c) cell-surface-mediated
interactions of complementary structures or molecules of iso- or heterotypic
cells; and d) cell-junction-mediated interactions.

In the early stages of mineralization of calcified tissues, matrix vesicles, 50–100 nm in diameter, were described in extracellular matrices and were thought to play a major role in the initiation of apatite crystal growth. These vesicles related to mineralization have been described in the earliest increments of predentine laid down (BERNARD 1972; KATCHBURIAN 1972). It was thought later that, besides acting as primary sites of mineral nucleation in predentine, the earlier formed predentinal matrix vesicles were directly involved in epithelio-mesenchymal interactions (SLAVKIN 1972; KARDOS and HUBBARD 1982). The appearance of these initial matrix vesicles at the time of cellular transition (pre-odontoblast-odontoblast; preameloblast-ameloblast) suggested a possible role in heterotypic cell communication. Demonstration of RNA in some vesicles offered the possibility of informational change (SLAVKIN 1972, 1974), while the presence of major histocompatible antigens on their surfaces afforded some specificity to the process through self recognition (SLAVKIN et al. 1977). Later, after the finding of collagenase activity associated with a matrix vesicle fraction, it was proposed that matrix vesicles may indirectly induce epithelio-mesenchy-mal interaction by mediating basal lamina breakdown, thereby allowing hetero-typic cell contact to occur. Such hypotheses need to be substantiated experimen-tally. WOLTERS and VAN MULLEM (1978), by culturing enamel organs and dental papillae on opposite sides of Millipore filters, concluded that matrix vesicles were not responsible for initiation of ameloblast differentiation.

The molecular understanding of epithelio-mesenchymal interactions will cer-tainly be the next step for investigations leading to the elucidation of the precise role of various intracellular and extracellular biochemical constituents. The importance of fetal growth factors such as adenylatecyclase (OSMAN et al. 1981 c) or transferrin (THESLEFF and PARTANEN 1984) have been recognized.

The epithelio-mesenchymal interactions in the tooth germs can be summa-rized as follows: 1) dental papilla controls histogenesis; 2) enamel organ main-tains dental papilla; 3) basal lamina preserves dental papilla; and 4) only the enamel organ can maintain the dental papilla.

F. The Odontoblast and the Odontoblastic Process

As has been stated previously, the differentiation of odontoblasts requires the presence of preameloblasts. The most peripheral cells of the dental papilla, initially those located in the portion corresponding to the cuspal or incisal tip, will gradually elongate (Fig. 22). They become postmitotic and polarize. At this stage, epithelio-mesenchymal interactions seem to be mediated via the basement membrane and it is important to note that no direct cell contacts are observed between preameloblasts and differentiating odontoblasts. Such cell contacts arise later.

In the light microscope, progressive differentiation of an odontoblast layer is observed at the preameloblast stage, adjacent to the apex formed by indenta-

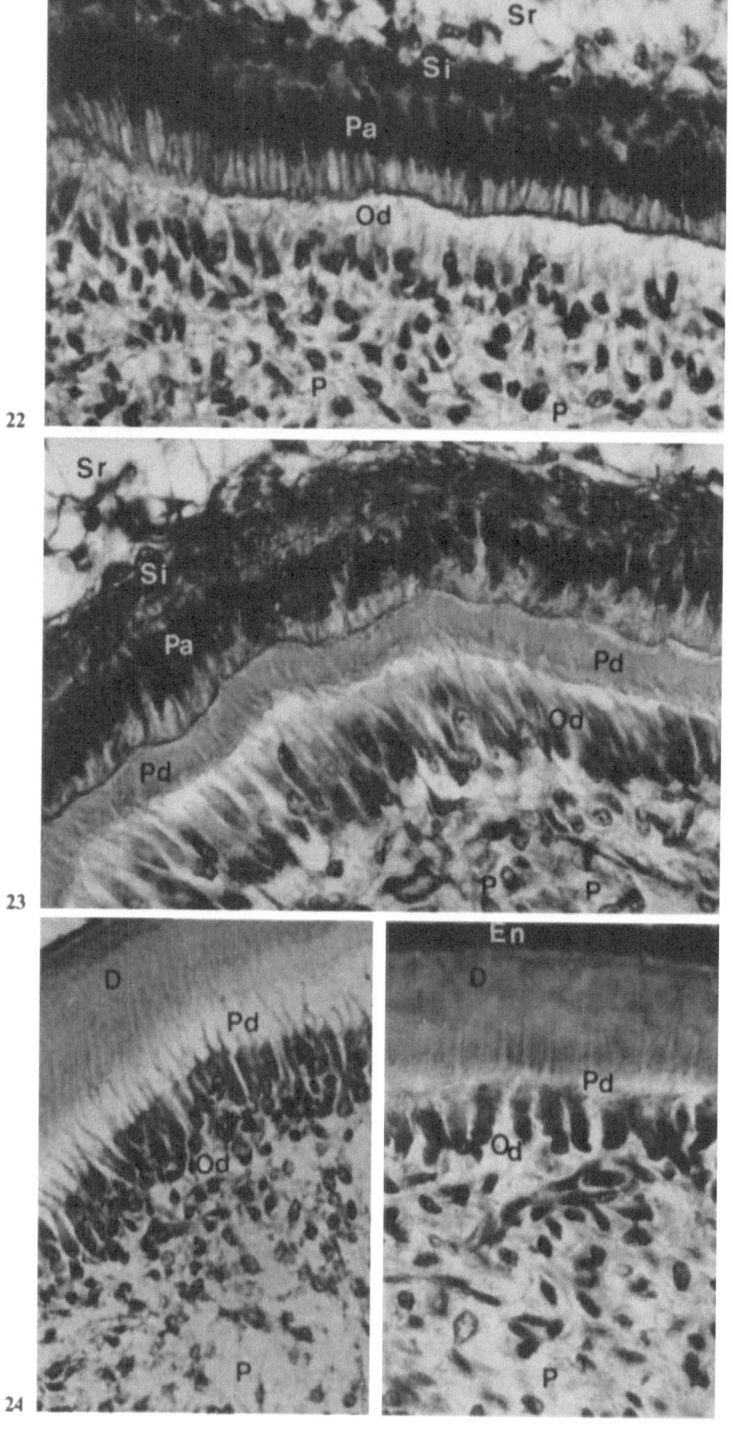

tion of the internal dental epithelium (Fig. 22). From this location odontoblast differentiation and dentine formation proceed down the cuspal slopes. In multicusped teeth, these events begin independently along each of the future cusp slopes, and the dentine of adjacent formative centers becomes fused. When fully-developed odontoblasts (with odontoblastic processes) have differentiated, a thin layer of predentine is deposited along the basement membrane (Fig. 23). Upon mineralization, an initial layer of calcified matrix dentine results (Fig. 24) and will become part of the so-called peripheral mantle dentine. After elongation of the odontoblastic processes and migration of the odontoblast cell bodies towards a more central (pulpal) position, an additional increment of predentine is formed (Fig. 24). It is at this stage that initial deposition of enamel is observed over the peripheral part of the calcified dentine (Fig. 25).

The various steps in the differentiation of the cells of the dental papilla into fully formed odontoblasts will now be described histologically and ultrastructurally. First, however, it seems advisable to clarify the terminology related to the polarity of the odontoblast. The portion of the cell adjacent to the internal dental epithelium has been called the distal or apical end, whereas the part facing the pulp has been referred to as the basal or proximal end. Having in mind the unquestionable orientation of the odontoblast, it seems reasonable to refer to the end of the cell which faces the internal dental epithelium (the extremity where predentine and dentine are laid down) as the distal end, and this terminology will be followed here. However, the portion of the cell itself will be referred to as the "formative part". Likewise the terms "proximal" end and "non-formative" part will be used.

I. Differentiation of the Odontoblasts

This stage of development consists of the progressive growth and elongation of the oval mesenchymal cells, located adjacent to the internal dental epithelium, into tall cylindrical cells with long cell processes. Events will transform the undifferentiated mesenchymal cells into highly polarized post-mitotic cells, assuming important secretory activities. The cells undergoing these complex phenomena have often been called "preodontoblasts". They withdraw from the cell cycle, assume an epithelial arrangement, polarize and finally secrete predentine.

Fig. 22. Progressive differentiation of odontoblasts (*Od*) along a layer of preameloblasts (*Pa*) in a tooth germ of a 7-month-old human embryo. *Si* Stratum intermedium; *Sr* Stellate reticulum; *P* Pulp. × 550

Fig. 23. Predentine (*Pd*) formation in a human tooth germ (6-month-old embryo). *Od* Odontoblasts; *Pa* Preameloblasts; *Si* Stratum intermedium; *Sr* Stellate reticulum; *P* Pulp. × 550

Fig. 24. Dentine (*D*) and predentine (*Pd*) formation in a human tooth germ (7-month-old embryo). *Od* Odontoblasts and their processes; *P* Pulp. × 330

Fig. 25. Enamel (*En*), dentine (*D*) and predentine (*Pd*) formation in a human tooth germ (5-month-old embryo). *Od* Odontoblasts; *P* Pulp. × 395

In the early stages of development, the preodontoblasts have a relatively high nuclear: cytoplasmic ratio (Fig. 26). The cytoplasm of these cells is relatively poor in endoplasmic reticulum and mitochondria. Differentiation of the Golgi apparatus is observed on the non-formative side of the cell in close proximity to the nucleus (Fig. 26). This organelle consists of a few dictyosomes associated with various types of vesicular elements. Polyribosomes as well as microfilaments are distributed across the cytoplasm. With elongation of the cell, the nucleus remains in the non-formative part of the cell (Fig. 27). Significant growth in size is accompanied by a dramatic increase in the endoplasmic reticulum and in the number of mitochondria which become closely associated with each other.

The young differentiating odontoblast of the new-born cat exhibits multivesicular bodies in the nucleus (Frank 1969), up to three bodies per nucleus (Fig. 28). These consist of a trilaminar outer membrane containing a great number of small vesicles, each with their own trilaminar limiting membrane and amorphous contents. The limiting membranes of the multivesicular bodies were seen to be fused and in continuity with the internal nuclear membrane (Fig. 29). Their contents were therefore in a space continuous with the perinuclear space. In fact, numerous vesicles were found in the perinuclear space and even in the lumen of dilated cisternae of rough-surfaced endoplasmic reticulum (Fig. 30). Finally, similar vesicles were observed in the intercellular spaces between adjacent young odontoblasts (Figs. 31 and 32).

It is of course difficult to speculate about the significance of these multivesicular formations on morphological grounds alone. Membrane-limited spherical bodies containing varying numbers of small vesicles, 10 nm to 100 nm in diameter, have been described in the cytoplasm of young odontoblasts by Meyer et al. (1977) who also identified multivesicular bodies in the extracellular spaces between preodontoblasts and preameloblasts. As mentioned earlier, matrix vesicles located between the basement membrane and the odontoblasts were considered to be acting either as inducing agents of epithelio-mesenchymal interactions at the time of transition of the odontogenic cells or as primary sites of mineral nucleation in predentine (Slavkin 1972; Kardos and Hubbard 1982). Further investigations will be necessary to determine if the described extracellular vesicles (Figs. 31 and 32) are related to these types of activities.

The elongation of the preodontoblasts leads to a palisade arrangement of cells (Fig. 33). The nuclei take up eccentric positions in the non-formative part of the cells. Concomitant with the growth in size, a notable proliferation is observed in organelles such as the rough-surfaced endoplasmic reticulum and the mitochondria. The cisternae of the granular endoplasmic reticulum flatten and become parallel to the long axis of the cells. The Golgi apparatus, differentiating initially at the non-formative part of the cells (Fig. 26) migrates around the sides of the nuclei (Fig. 34) and finally reaches the formative part of the cells where it occupies a juxta-nuclear position (Fig. 27). The polarization of the cells is characterized by the eccentric position of the nuclei and the presence of the Golgi apparatus on the formative side. Both these organelles are surrounded by an abundant rough-surfaced endoplasmic reticulum and associated mitochondria and polyribosomes (Fig. 27). This cell polarity requires the presence and integrity of a well-developed cytoskeleton made up of intermediary

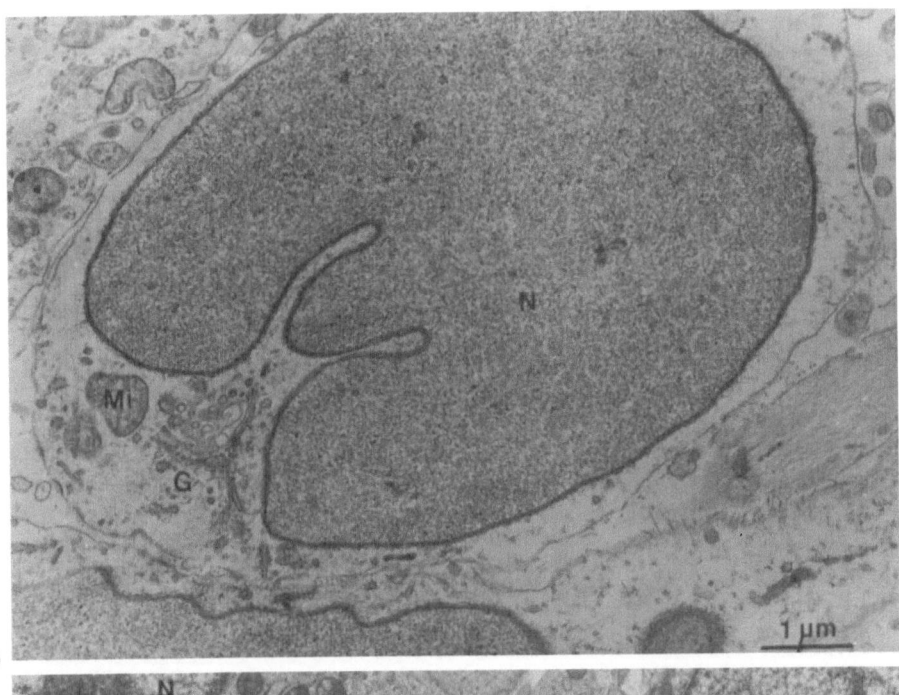

26

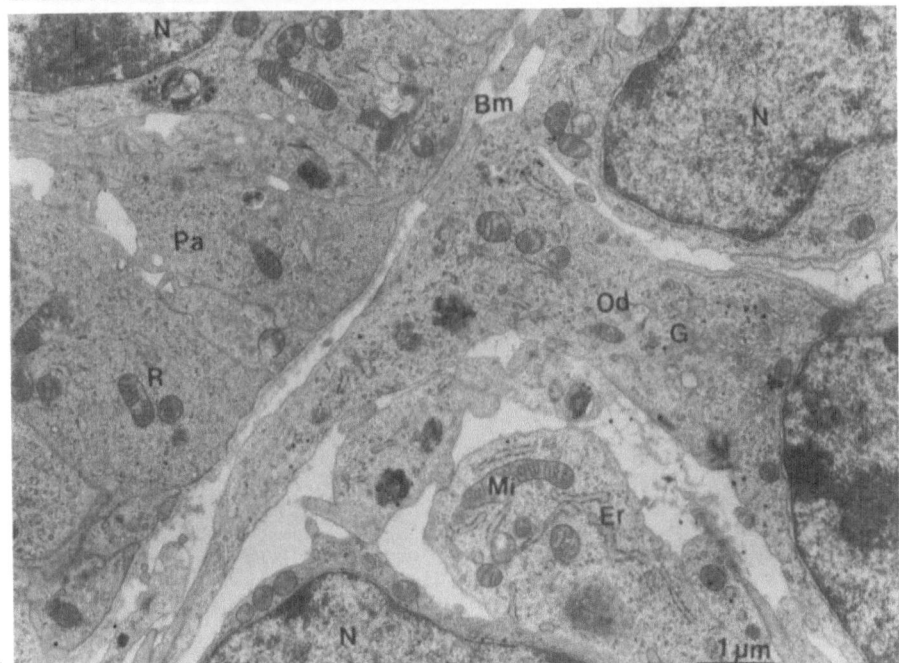

27

Fig. 26. Differentiating mesenchymal cell located adjacent to the basement membrane (not shown) in a dental papilla of a molar tooth germ (newborn cat). Note the great size of the nucleus (*N*) and the differentiating Golgi apparatus (*G*). *Mi* Mitochondria

Fig. 27. Differentiating mouse odontoblast (*Od*) along a layer of preameloblasts (*Pa*). *Bm* Basement membrane; *G* Golgi apparatus; *N* Nucleus; *Mi* Mitochondria; *Er* Endoplasmic reticulum; *R* Ribosomes

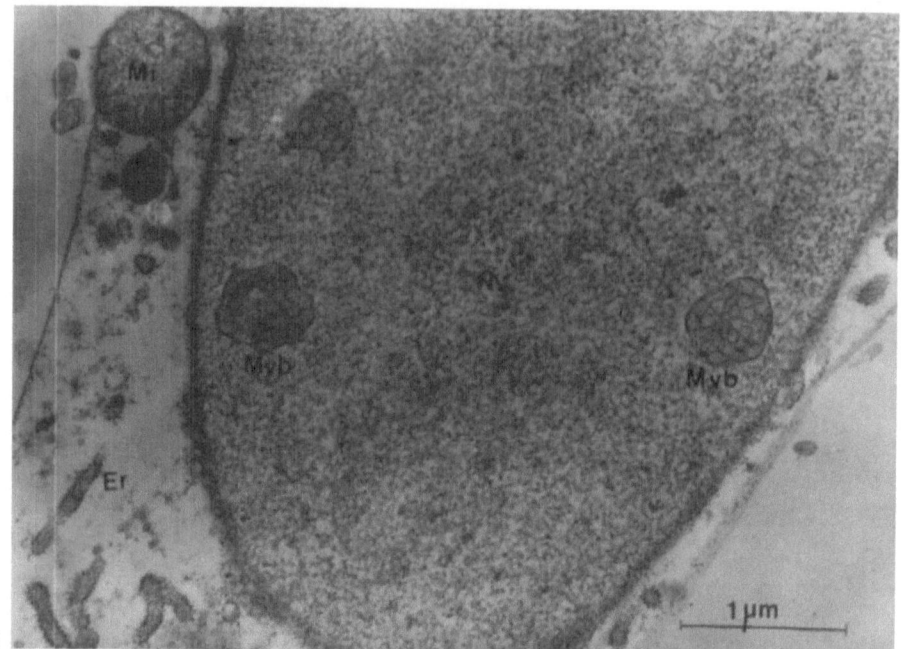

28

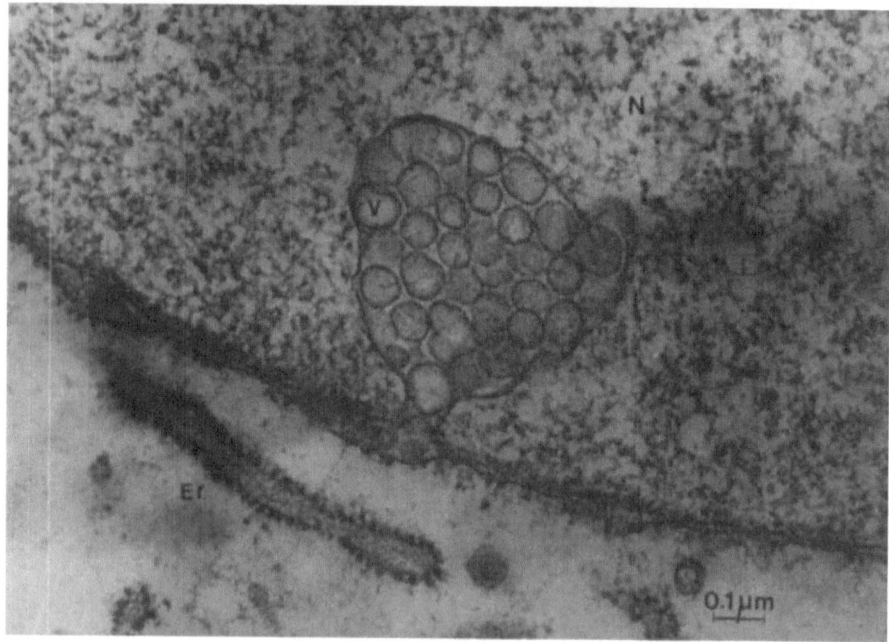

29

Fig. 28. Differentiating odontoblast in a dental papilla of a molar tooth germ (newborn cat). Presence of 3 multivesicular bodies (*Mvb*) in the nucleus (*N*). *Mi* Mitochondria; *Er* Endoplasmic reticulum. (From FRANK 1969, courtesy of Z Zellforsch)

Fig. 29. Multivesicular body containing numerous vesicles (*V*) in a young differentiating cat odontoblast. The limiting membrane of the multivesicular body is in continuity with the inner nuclear membrane and some vesicles are located in the perinuclear space between inner and outer nuclear membranes. *Er* endoplasmic reticulum; *N* Nucleus. (From FRANK 1969, courtesy of Z Zellforsch)

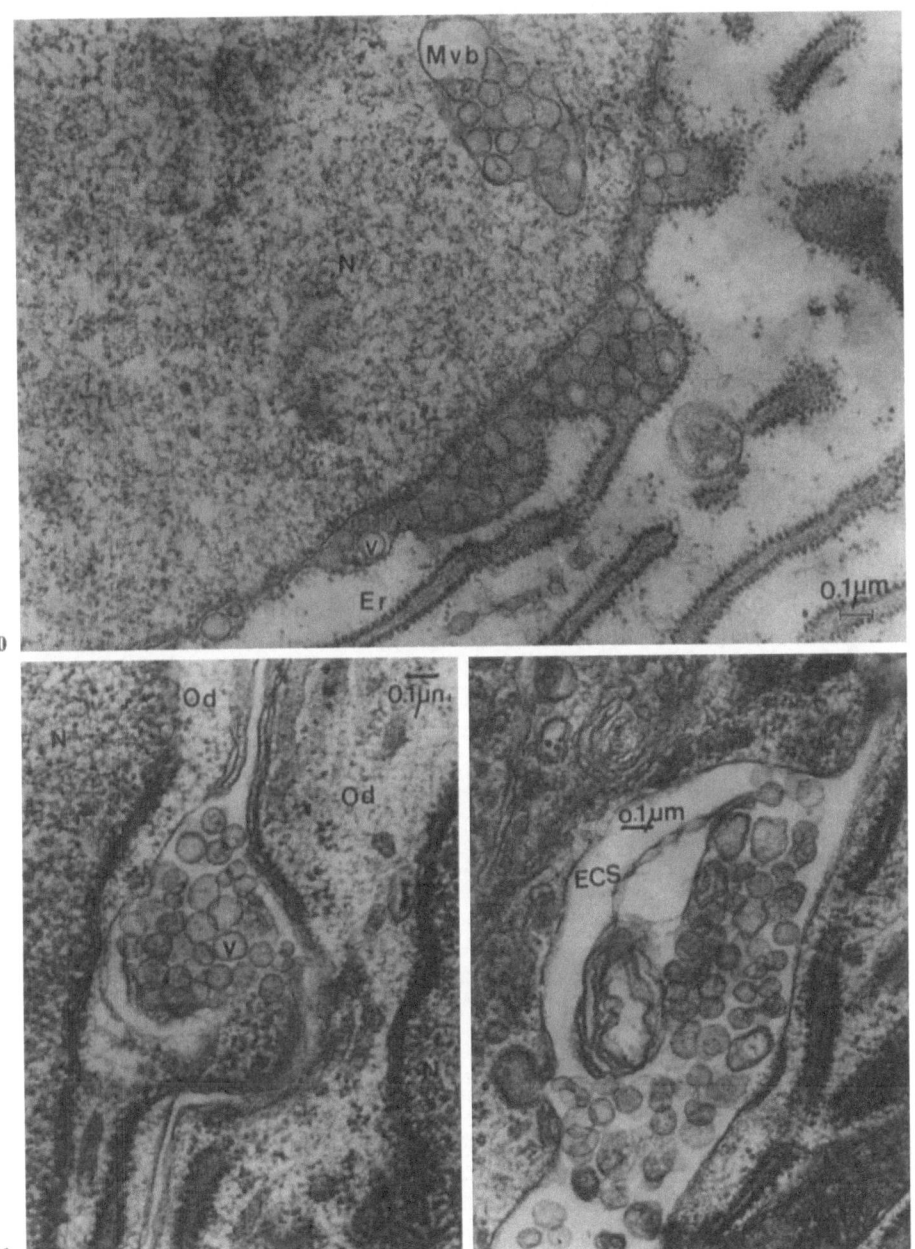

Fig. 30. Presence of numerous vesicles (*V*), identical to those contained in a multivesicular body (*Mvb*) of the nucleus (*N*), in the perinuclear space between inner and outer nuclear membranes. This space is continuous with the lumen of the endoplasmic reticulum (*Er*). Differentiating cat odontoblast. (From FRANK 1969, courtesy of Z Zellforsch)

Fig. 31. Presence of vesicles (*V*) in the intercellular space located between two differentiating cat odontoblasts (*Od*) at the level of their nuclei (*N*). (FRANK 1969, courtesy of Z Zellforsch)

Fig. 32. Numerous vesicles in the intercellular space located laterally between two differentiating cat odontoblasts. *ECS* Extracellular space. (FRANK 1969, courtesy of Z Zellforsch)

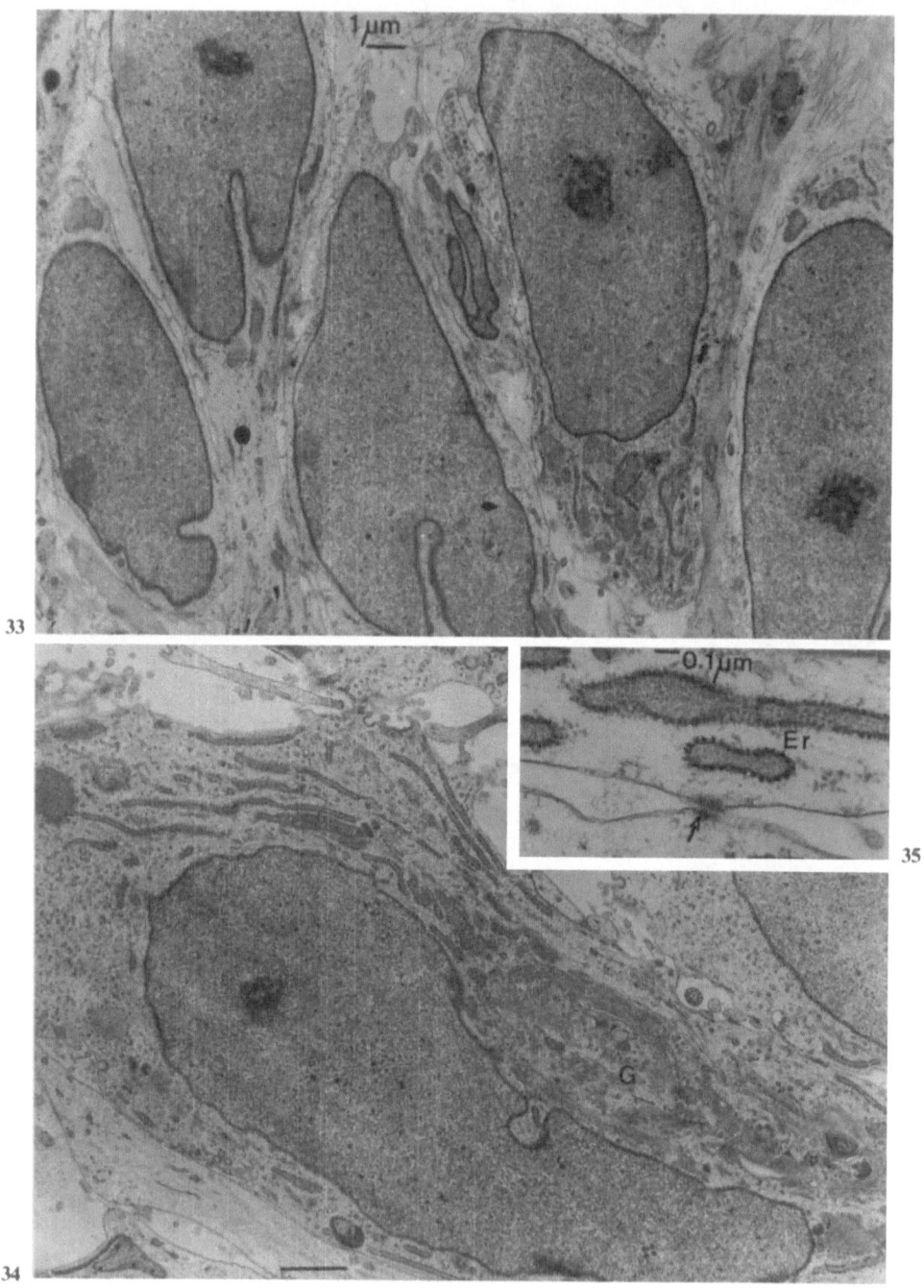

Fig. 33. Peripheral cells of the dental papilla of a cat molar tooth germ, located along the internal dental epithelium. These cells are in the process of differentiating into odontoblasts and are assuming an elongated shape

microfilaments (vimentin), microfilaments and microtubules which are oriented parallel to the long axis of the cell.

The localization of the various cytoskeletal components has been examined in tooth germs by indirect immunofluorescence, using specific antibodies (Lesot et al. 1982). In these studies, vimentin present throughout the cell became progressively restricted to the distal part of the odontoblasts during polarization. It has also been shown with injections of colchicine or vinblastine, interacting with microtubules, or of cytochalasin B, interfering with microfilaments, that the integrity of the cytoskeleton is necessary for the polarization of the odontoblast and secretion of predentine (Kudo 1975; Ruch et al. 1975; Miake et al. 1982).

II. The Cell Body of the Differentiated Odontoblast

The differentiated odontoblast (Figs. 35–45) is a highly polarized secretory cell directly associated with dentinogenesis. The name "odontoblast" was first proposed by Waldeyer (1865). This cell consists of an elongated cell body possessing a long, slender process with lateral branches. The odontoblastic cell body lies adjacent to a collagenous organic matrix, the predentine, from which dentine arises following calcification. The odontoblastic processes, which will be described separately below, extend through the predentine, reaching dentinal tubules that permeate the calcified dentine (Figs. 46–50).

The odontoblast cell bodies are oriented parallel to each other in a palissade arrangement, generally with very little intercellular space (Figs. 35–40). Larger localized, intercellular spaces can occasionally be observed (Fig. 37). The length of the cell bodies is quite variable among the different species. In human and cat embryos, dimensions of 40 to 80 µm long and 4 to 8 µm wide have been observed, with oval to round contours (Figs. 38, 39).

The nucleus of the odontoblasts is located in an eccentric position in the non-formative part of the cell body (Figs. 36, 37). Nuclear chromatin is present mainly as dispersed euchromatin filaments with only moderate masses of dense chromatin, heterochromatin, adjacent to the inner nuclear membrane. The presence of a nuclear matrix has been demonstrated after extraction of chromatin. This matrix, composed of actin, keratan, and other proteins, has been shown to be the site of chromatin DNA replication and RNA attachment (Penman et al. 1983)

One or several prominent nucleoli are present in the odontoblast nuclei. Autoradiographic evidence shows nucleoli to be the sites of extensive RNA

Fig. 34. Elongated shape of a differentiating odontoblast. The basement membrane and the internal dental epithelium are located on the lower right side of the micrograph. Molar tooth germ of a newborn cat. *G* Golgi apparatus

Fig. 35. Desmosome-like junction (*arrow*) along the lateral sides of two young odontoblasts. Newborn cat. Molar tooth germ. *Er* Endoplasmic reticulum

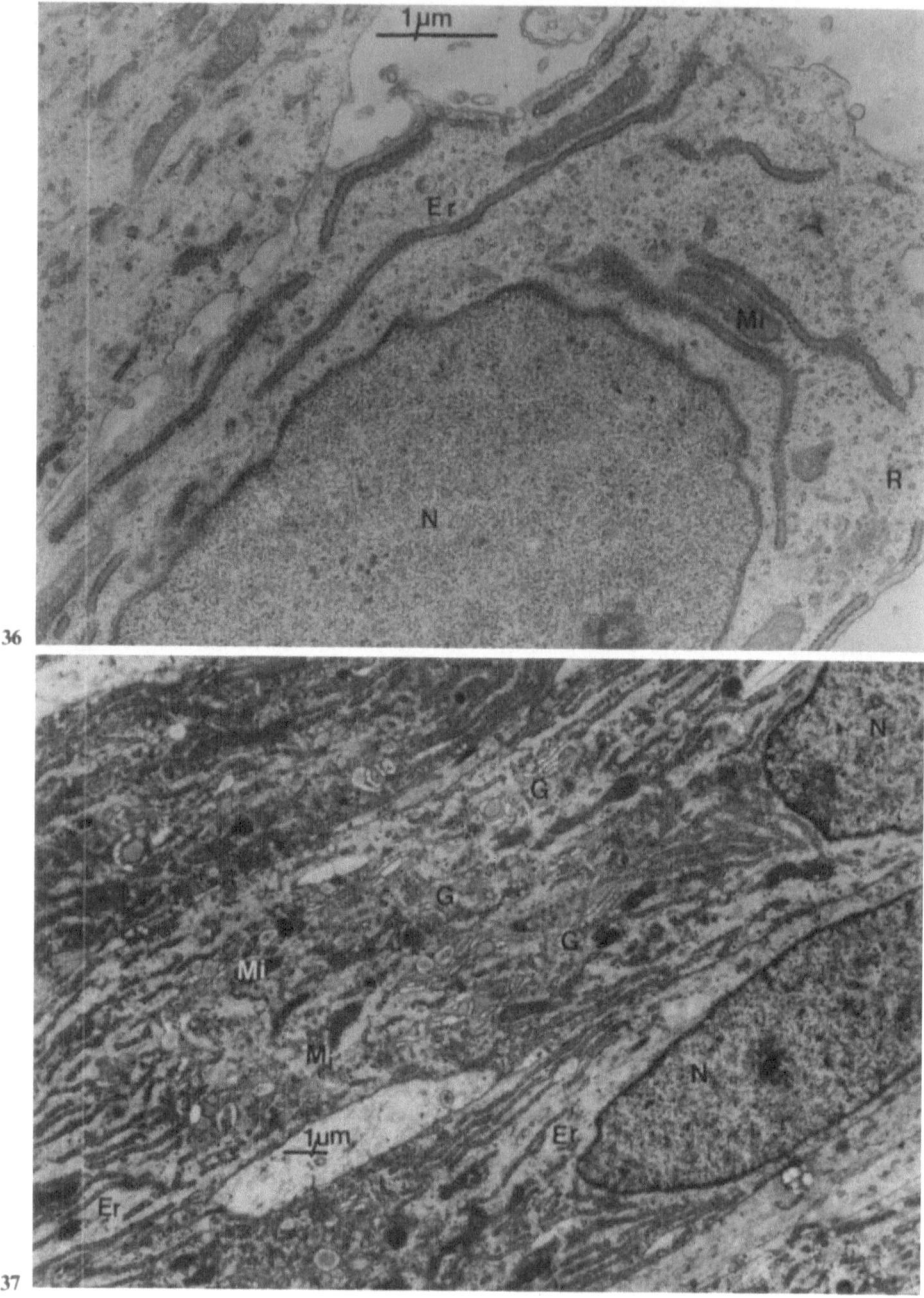

Fig. 36. Non-formative end of a newly differentiated odontoblast. *N* Nucleus; *Mi* Mitochondria; *R* Ribosomes; *Er* Endoplasmic reticulum. Newborn cat. Molar tooth germ

Fig. 37. Formative end of an odontoblast in the newborn cat. *N* Nucleus; *Er* Endoplasmic reticulum; *G* Golgi apparatus; *Mi* Mitochondria

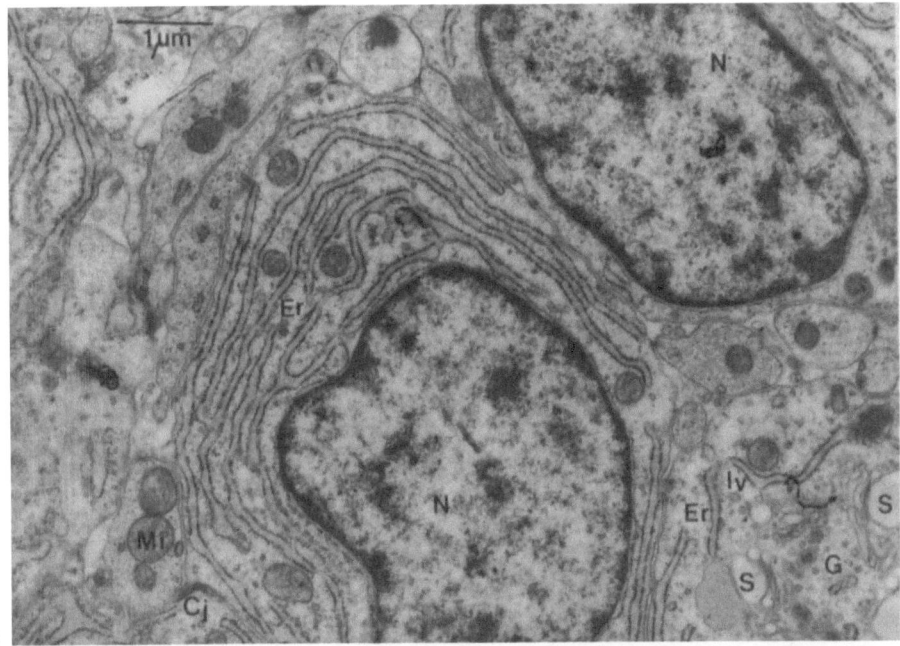

38

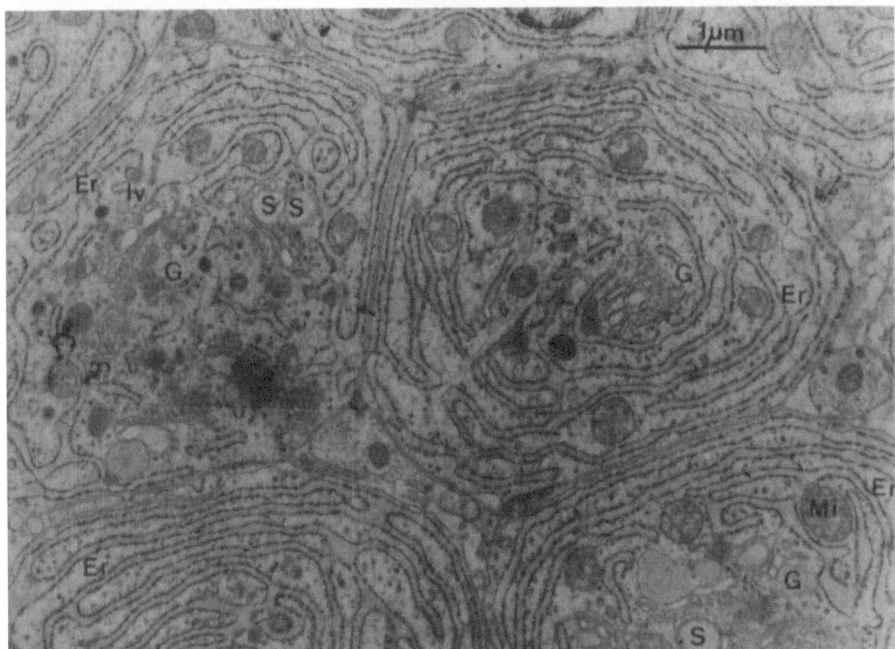

39

Fig. 38. Transverse section through odontoblasts in a tooth germ of a newborn cat, at the level of the nucleus (*N*). *Er* Endoplasmic reticulum; *Iv* Intermediary vesicle; *G* Golgi apparatus; *S* Golgi saccules; *Mi* Mitochondria; *Cj* Cell junction

Fig. 39. Transverse section through odontoblasts in a tooth germ of a newborn cat, at the level of the Golgi apparatus (*G*) which is completely surrounded by endoplasmic reticulum (*Er*) and associated mitochondria (*Mi*). Note the various forms of vesicles and vacuoles related to the Golgi apparatus. *Iv* Intermediary vesicles; *S* Golgi saccule

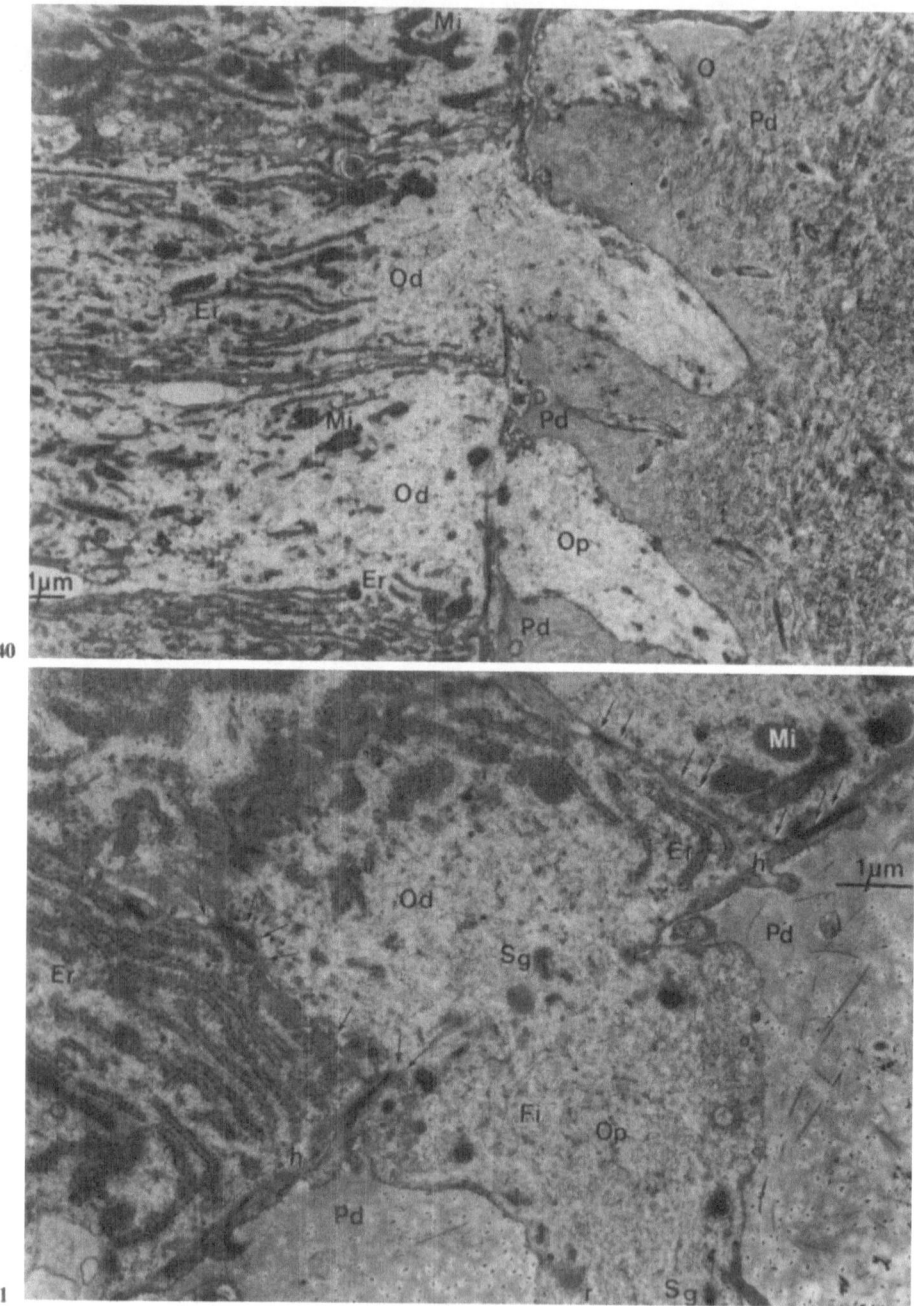

Fig. 40. Formative part of several odontoblasts (*Od*) showing continuity with the odontoblastic process (*Op*) in longitudinal section. *Pd* Predentine; *Er* Endoplasmic reticulum; *Mi* Mitochondria

Fig. 41. Higher magnification of the formative part of an odontoblast (*Od*) and its process (*Op*). Presence of numerous microfilaments (*Fi*), some of which are horizontally oriented (*h*) adjacent to the predentine (*Pd*). A junctional complex between the distal parts of two odontoblasts is indicated by *arrows*. *Er* Endoplasmic reticulum; *Sg* Elongated secretory vesicles; *Mi* Mitochondria

synthesis. Nucleoli are composed of zones of different densities. In addition to chromatin strands, there are areas rich in granules (about 150 Å in diameter) and regions that are primarily fibrillar or amorphous. No membrane delimits the nucleolus. However, contacts between nucleoli and the inner nuclear membrane are frequently seen.

The nucleus is surrounded by a flattened envelope. The outer surface of the envelope facing the cytoplasm, has ribonucleoprotein granules attached to it and is continuous with the rough-surfaced endoplasmic reticulum (Fig. 30). The inner nuclear membrane facing the nuclear contents lacks ribosomes. Heterochromatin is often aggregated along this surface. Nuclear pores occur at intervals along the envelope and appear as roughly circular or polygonal areas where the inner and outer membranes of the nuclear envelope are fused. The pores, measuring 500–800 Å across, serve as routes of nuclear-cytoplasmic interchange.

In the odontoblasts that are actively engaged in synthesizing and secreting proteins, a particularly highly developed rough-surfaced endoplasmic reticulum (RER) is found, closely associated with a great number of mitochondria (Figs. 36–40, 42). Both of these organelles are found in the non-formative part of the cell, around the nucleus, around the Golgi apparatus and throughout the entire formative part of the cell. The RER appears as flattened cisternea aligned parallel to the long axis of the cell. Some smooth-surfaced, i.e., ribosome-free, endoplasmic reticulum can be found. Ribosomes are the intracellular sites of protein synthesis. They appear in electron microscopy as spherical to ellipsoidal bodies, 150–250 Å in diameter, consisting of a large and a small sub-unit. Bound ribosomes are attached to the external lipoprotein component of the membranes of the endoplasmic reticulum, whereas the so-called free ribosomes are found in the cytoplasm. Both types of ribosomes form polyribosomes with messenger RNA and both play an important role in protein synthesis. From recent work to be reviewed later, it appears that the free polyribosomes are, in fact, bound to a three-dimensional cytoplasmic network, apparent in cells treated with a non-ionic detergent (PENMAN et al. 1983).

As has been already noted by BEAMS and KING (1933), the mitochondria are distributed throughout the odontoblast cell body in close association with the endoplasmic reticulum (Figs. 36–39). They can assume various shapes. Generally they are bean-shaped, but can be globular or elongated (Fig. 36). Mitochondria are the major sites of ATP production. They are linked to oxygen consumption and play important roles in many metabolic pathways, including the breakdown and synthesis of carbohydrates, fats, and amino acids. This organelle is delineated by an outer membrane separated from an inner membrane by a space. The inner membrane is folded into a number of cristae. In the odontoblast, as well as in other cells associated with mineralized tissues such as osteoblasts (FRANK 1979), dense intramitochondrial granules are seen in the matrix. These dense granules were found to consist of calcium phosphates, and diffraction studies confirmed the presence of hydroxyapatite. As will be demonstrated later, mitochondria play an important role in the intracellular storage and regulation of calcium.

In a juxtanuclear position on the formative side of the odontoblast, a prominent Golgi apparatus is seen, surrounded by endoplasmic reticulum (Figs. 37–39,

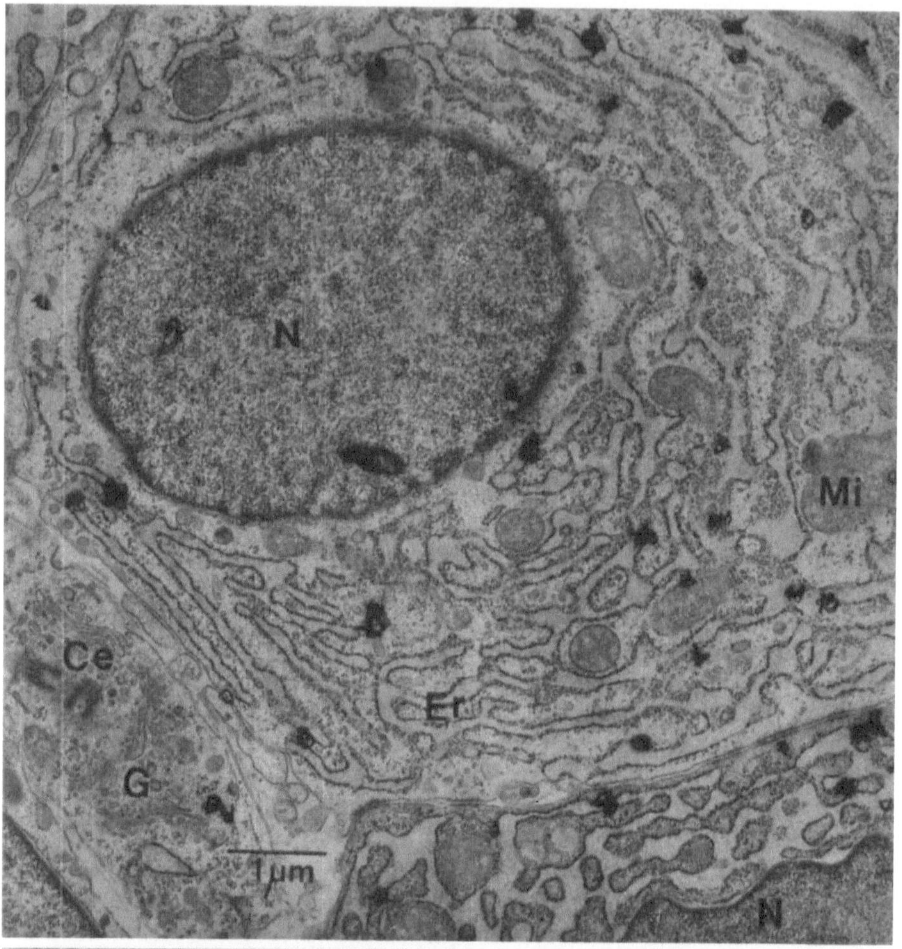

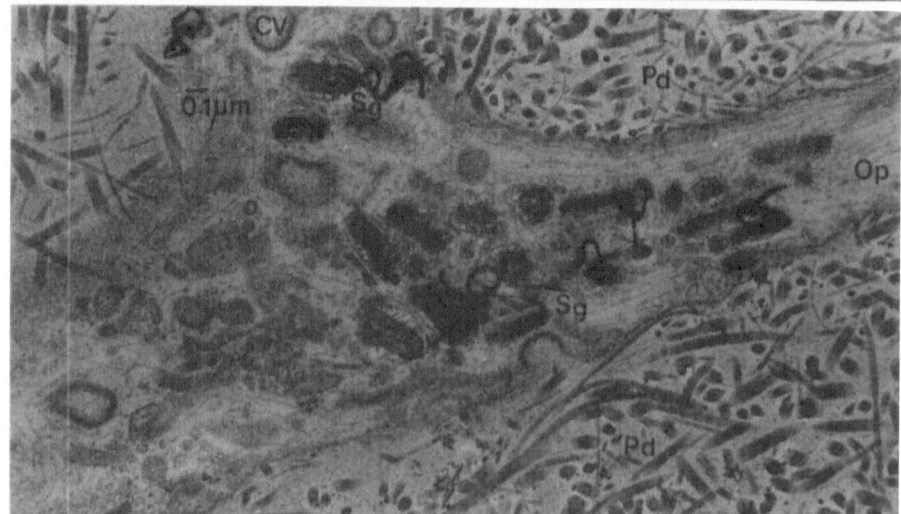

42

43

44, 55). The Golgi apparatus occupies a cylindrical region located in the center of the cell near the nucleus (Figs. 39, 55). The surface of the "cylinder" is in contact with a specialized portion of endoplasmic reticulum called transitional endoplasmic reticulum, lacking bound ribosomes. The walls of the Golgi cylinder are made up of two or three circular stacks of about four to a half-dozen dictyosomes, consisting of flattened membrane-bound cisternae. These flattened sacs tend to be bowed and are piled up parallel to the long axis of the cell. On transverse sections, the lateral sides of the dictysosomes are dilated peripherally. At the edges of the lamellated sacs near the expanded peripheries, saccules as well as various granules, vesicles, and vacuoles are observed. Since the stacks of dictyosomes are bowed, they present a convex outer face, the so-called cisface, which is adjacent to the transitional endoplasmic reticulum (ROTHMAN 1981). Intermediate vesicles are observed at these levels. The internal face of the stack of dictyosomes is concave and has been termed trans-face. Saccules, vesicle, vacuoles and newly formed secretion granules are associated with the trans-face but not the cis-face and occupy the center of the Golgi apparatus. ROTHMAN (1981) has proposed that the Golgi apparatus consists of two attached but different organelles, the cis and trans Golgi. These two compartments may act sequentially to refine the protein export of the endoplasmic reticulum by removing escaped proteins. Purification would occur in sequential stages in the cis portions. The trans Golgi consisting of the inner one or two cisternae, may be the receiver that collects from the cis Golgi only its most refined fraction for later distribution to specific locations throughout the cell (ROTHMAN 1981).

In the center of the tubulo-cylindrical stacks of dictyosomes, an accumulation of various vesicles, vacuoles, bodies and elongated secretory vesicles is observed adjacent to the trans Golgi. Coated vesicles of various sizes, similar to the coated vesicles initially described by ROTH and PORTER (1964), have been observed by a number of authors (FRANK 1968a; REITH 1968a; MATTHIESSEN and VON BÜLOW 1970; NAGAI 1970; GARANT 1978; FRANK 1979; TOMINAGA et al. 1984). Smooth-surfaced vesicles, Golgi saccules, vacuoles with an amorphous content, as well as multivesicular bodies have also been seen. Elongated cylindrical or rod-shaped secretory vesicles are sometimes found in the central part of the Golgi apparatus (FRANK 1968a; REITH 1968a; FRANK 1970; MATTHIESSEN and VON BÜLOW 1970; WEINSTOCK 1972; WEINSTOCK and LEBLOND 1974; GARANT 1978; FRANK 1979; SASAKI et al. 1984; TOMINAGA et al. 1984). The secretory vesicles contain bundles of parallel filaments oriented lengthwise and in, addition they may appear as dense granular particles arranged with

Fig. 42. Cross section of secreting odontoblasts at the nuclear level (N) 5 minutes after intravenous injection of ^{3}H-proline. Numerous silver grains are located over the endoplasmic reticulum (Er). Note the closely associated mitochondria (Mi). Ce A pair of centrioles; G Golgi apparatus. (From FRANK 1979, courtesy of Int Rev Cytol)

Fig. 43. Dense, elongated secretory vesicles (Sg) in the odontoblast process (Op) of the predentine (Pd) 1 hour after intravenous injection of ^{3}H-proline. Note the absence of silver grains over the coated vesicles (CV). (From FRANK 1979, courtesy of Int Rev Cytol)

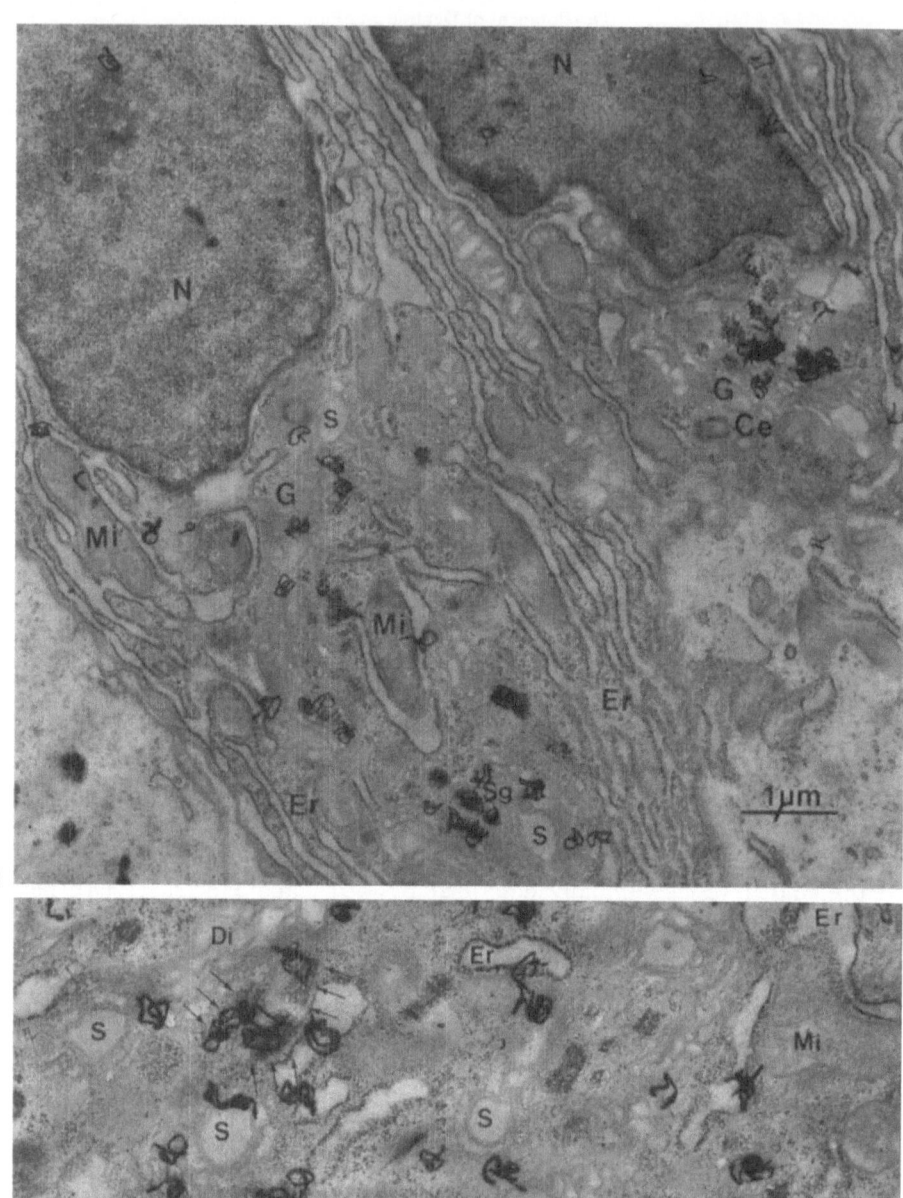

Fig. 44. Labeling of the juxtanuclear Golgi apparatus (*G*) of odontoblasts in the newborn cat 1 hour after intravenous injections of ^{3}H-proline. *N* Nucleus; *Er* Endoplasmic reticulum; *Mi* Mitochondria; *S* Golgi saccules; *Sg* Secretory granules; *Ce* Centriole. (From FRANK 1979, courtesy of Int Rev Cytol)

Fig. 45. Golgi apparatus of an odontoblast 1 hour after intravenous injection of ^{3}H-proline. Numerous dense, elongated, rod-shaped secretory vesicles (*arrow*) are covered with silver grains, as are the Golgi saccules. *Mi* Mitochondria; *Di* Dictyosomes; *Er* Endoplasmic reticulum; *S* Golgi saccules. (From FRANK 1979, courtesy of Int Rev Cytol)

transverse periodic bands or in more or less compact groupings without distinct periodicity (Figs. 51–53). A variety of intermediate stages of formation between distended Golgi saccules and the elongated secretory vesicles have been described (Reith 1968a; Weinstock and Leblond 1974; Leblond and Weinstock 1976). Weinstock (1972) tried to determine the nature of the dense particles found in the secretory vesicles. They were not dissolved with EDTA or nitric acid treatment, indicating an organic composition. It is interesting to note that in cultures of explanted human pulp tissue, peripheral odontoblast-like cells elaborated similar secretory granules with typical periodic striations (Magloire 1983).

As early as 1968, Katchburian and Holt thought that lysosomes may play an important role in the secretory process associated with dentine formation. They described the presence of acid phosphatase in Golgi cisternae, in vacuoles and even in secretion granules. The presence of acid phosphatase was confirmed in Golgi saccules, multivesicular bodies, and in coated and secretory granules by Nagai (1970), Larsson and Bloom (1973), Kajikawa and Kakihara (1974), Garant (1978), and Sasaki et al. (1982).

A pair of centrioles is present in the odontoblast in the area of the Golgi apparatus, near the nucleus (Frank 1968a; Garant et al. 1968). The two centrioles lie with their long axes perpendicular to each other. Each is made up of nine sets of tube-like structures arranged as the walls of a cylinder. Each set is a triplet composed of three closely associated tubular elements. These triplets are embedded in an amorphous matrix. The complex ultrastructure of the centrioles was first described by De Harven and Bernhard (1956). One centriole in the odontoblast serves as a basal body and gives rise to a cilium which may or may not extend into the interodontoblastic space. Since no central fibrils are present in the cilium, the latter is suspected to be nonmotile (Garant 1978).

An important part of the odontoblast cell body is the structured cytoskeleton. It consists of microtubules, microfilaments and the so-called intermediary microfilaments, located between the previously described organelles. This cytoskeleton contributes to cell shape and cell polarity, as already mentioned, after the use of colchicine, cytochalasin B, or vinblastine. It provides a framework for certain motile processes both inside the cell and for the cell as a whole. Before describing the various parts of the cytoskeleton, it may be valuable to discuss the nature of the cytoplasmic matrix of the cell body, long considered as an unstructured medium defined as a "cytosol". In fact, it appears recently that the so-called cytosol is not a sol but rather a semi-solid scaffold defined as a "microtrabecular lattice" different from the cytoskeleton (Porter et al. 1983). The microtrabecular lattice, which probably is present in the odontoblast, consists of slender strands which connect all the better-known components of the cytoplasm to form a space-filling lattice. Similar three-dimensional cytoplasmic networks have been demonstrated in cultured fibroblasts exposed to nonionic detergents such as Triton X100 (Penman et al. 1983). It was thus shown that the so-called free polyribosomes were in fact all bound to this cytoplasmic network which was retained after Triton extraction.

The cell shape, polarity and motility are associated with the presence of special structures, i.e. microtubules and microfilaments. The name "microtubule" was coined by Slautterback (1963). Microtubules are assembled into elongated tubular structures with an average exterior diameter of 24 nm and an indefinite length. In the odontoblast cell body, as well as in the cell process, they are oriented parallel to the long axis of the cell. They are capable of rapid changes in length by assembly and disassembly of their subunit protein molecules or tubulins (Dustin 1978), or of building other complex protein assemblies such as centrosomes, centrioles and cilia. Microtubules can be disrupted by high pressure, low temperature, or by drugs such as colchicine and vinblastine which bind to tubulin molecules. The centrosome which contain the centrioles is considered as the organizing center of the cytoplasmic microtubules (McIntosh 1983).

Microfilaments of actin are also present in the odontoblast and its process, with diameters in the range of 5 to 8 nm. They play a role in cell shape asymmetry and cell movements. Cytochalasin B can produce a disruption of microfilaments and interfere with cytoplasmic movements and secretion phenomena.

Another family of cytoplasmic fibrous proteins are involved in cytoskeletal functions including determination and maintenance of cell shape. Five major subclasses of intermediate filaments with diameters of 7 to 12 nm have been described: 1) keratin (tono) filaments occurring in epithelial cells and cells of epithelial origin; 2) desmin filaments seen mainly in muscle cells; 3) vimentin filaments in many differentiating cells; 4) glial filaments in astrocytes; and 5) neurofilaments in neurons (Lazarides and Granger 1983). In the differentiating odontoblasts, intermediate microfilaments have been identified by Lesot et al. (1982) with the use of specific antibodies and indirect immunofluorescence. Vimentin is initially present throughout the entire odontoblast, but during cell polarization it becomes progressively restricted to the distal part of the odontoblasts.

At this point we will proceed to a description of the various types of cell junctions that interconnect neighboring odontoblasts. The cell body of the odontoblast, which may appear straight or show microvillosities, is limited by a continuous trilaminar cell membrane. It is important to note that the intercellular spaces between adjacent odontoblasts are generally narrow (Figs. 35–39) and only a few collagen fibrils are found there. Sometimes blood capillaries have been seen penetrating between the odontoblastic cell bodies (Adams 1962; Frank 1968a; Boyde et al. 1978), but nerve fibrils have not been found between secretory odontoblasts (Frank 1968a).

Desmosome-like junctions along the lateral faces of the odontoblasts (Figs. 35, 61–63) were recognized by Frank (1968a), Garant (1978), Goldberg (1983), and Tominaga et al. (1984). These junctions were described as desmosome-like because they consisted of dense plaques found along the cytoplasmic sides of the inner leaflet of adjacent odontoblast cell plasmalemmas, but the thin and dense intercellular discs, found in typical epithelial desmosomes, were missing.

Conflicting data are found in the electron microscopic literature concerning the exact location and nature of other intercellular junctions connecting adjacent odontoblasts or connecting odontoblasts and fibroblasts of the sub-odontoblastic layer. In fact, for the study of intercellular junctions to be definitive, two additional techniques are important. By the first method, the permeability of the intercellular junctions is tested through the use of lanthanum hydroxide as initially described by REVEL and KARNOVSKY (1967). A second reliable method is the freeze-fracture technique by which a thin platinum-carbon replica made from the frozen and fractured cell junction area is observed in a transmission electron microscope.

Gap junctions, also called nexus or *macula communicans*, are specialized regions allowing direct intercellular communications between adjacent cells through functional units, the so-called connexons. Gap junctions have been described between the lateral faces of adjacent odontoblasts as well as between odontoblasts and fibroblasts (HOLLAND 1975; COX et al. 1976; GARANT 1978; GOLDBERG 1983; KÖLING 1983; TOMINAGA et al. 1984; CALLE 1985; ZAKI and WEBER 1985).

Along the most distal parts of the odontoblast cell bodies adjacent to predentine, the presence of junctional complexes assures cohesiveness of the odontoblast layer. Each junctional complex is made of an alignment of 3 different types of junctions. From proximal to distal these consist of a desmosome-like junction, as described previously; an intermediary junction or *zonula adherens*, according to the classification of FARQUHAR and PALADE (1963); and a distal junction about which there are contradictory observations, adjacent to predentine. TANAKA (1980) injected lanthanum intravascularly and observed penetration of this tracer into the periodontoblastic spaces of dentinal tubules in developing rat molar teeth. He therefore concluded that there were no true tight junctions between odontoblasts. GOLDBERG (1983) also noted the absence of tight junctions. However, SASAKI et al. (1982) described tight junctions of the macular type continuous with gap junctions. After intravenous injections of horseradish peroxidase in newborn rats, these workers found that the distal junctional complex is not an effective barrier against horseradish peroxidase penetration. These particles pervaded the intercellular spaces as well as the predentine.

Using intrarterial injection of horseradish peroxidase in Sprague-Dawley rats, SATTELBERG and TURNER (1984) claimed to have found tight junctions of the *zonula occludens* type and observed, in contrast to SASAKI et al. (1982), that the reaction product never extended beyond the most distal intercellular junction of the odontoblasts. In their case, dentine and predentine were totally devoid of reaction product. The presence of tight junctions between the odontoblasts of rat incisors has also been described by BISHOP (1985) after perfusion with physiological solutions containing lanthanum nitrate. Lanthanum was completely halted at the apical ends of the odontoblast bodies and was also found in the predentine. Using freeze-fracture techniques, KÖLING (1983) denied the existence of tight junctions at this level but found gap junctions of various shapes and sizes with intercellular distance reduced to about 20 nm. Using the same technique of freeze-fracture on human tooth germs, CALLE (1984, 1985) localized zonular tight junctions and gap junctions at the distal end of the odontoblastic cell bodies. Further work is needed to clarify the nature of the most distal junction of the complex.

It is interesting, to note, however, that at the cytoplasmic level of these most distal junctions, a distinct terminal web is observed, consisting of a dense bundle of microfilaments running across the proximal ends of the odontoblast cell bodies at a right angle to their axes (Figs. 40, 41, 59, 60). This terminal web has been well described by GARANT (1978). The transverse filaments marking the limit between the odontoblast body and its process insert into a dense material adjacent to the cell membrane at the site of the most distal portion of the junctional complex. In fact, it appears from various longitudinal sections of the odontoblasts, that these transverse bundles of filaments are discontinuous in the central part of the cell (Fig. 59). This is not the case in more lateral sections (Fig. 60). The terminal web therefore has a configuration of a diaphragm with a central opening. The distal junctional complex and the terminal web appear to be parts of an attachment apparatus that serves to fasten adjacent odontoblasts to each other near the junction between the predentine and the interodontoblastic spaces.

III. The Odontoblastic Process

The odontoblastic process is a direct continuation of the cell body of the odontoblast and, as would be expected, it is limited by a trilaminar cell membrane. The junction between the cell body proper and the odontoblastic process is recognizable by the terminal web (Figs. 40, 41, 59) as well as a striking reduction in cell diameter.

Using the polyene antibiotic, filipin, as a probe to visualize cholesterol in freeze-fracture replicas, GOLDBERG and ESCAIG (1984a) demonstrated numerous cholesterol-rich domains in the cell membrane of the young rat odontoblastic process, whereas only a few deformations were observed on the plasma membrane of the cell body, an exception being the regions occupied by junctional complexes.

The odontoblastic processes have numerous side branches leaving the main stem at right angles or at various acute angles with the distal extension (Figs. 49, 55). These offshoots themselves may branch and connect with the lateral

Fig. 46. Longitudinal section of an odontoblastic process (*Op*) in the predentine (*Pd*) of a newborn cat 1 hour after intravenous injection of ³H-proline. Silver grains are located over dense, rod-shaped secretory vesicles (*Sg*). The coated vesicles (*V*) are not labeled. *D* Dentine. (From FRANK 1976, courtesy of J Microsc Biol Cell)

Fig. 47. Longitudinal section of an odontoblastic process (*Op*) in a dentinal tubule near the predentine-dentine junction, 1 hour after intravenous injection of ³H-proline. Some uncalcified collagen fibrils (*F*) are seen in the intertubular dentine. Within the odontoblastic process, dense, elongated, secretory vesicles (*Sg*) are covered with silver grains. Note the presence of microfilaments (*Fi*) and microtubules in the odontoblastic process. (From FRANK 1979, courtesy of Int Rev Cytol)

Fig. 48. Odontoblasts (*Od*), odontoblastic processes (*Op*) and predentine (*Pd*) in a newborn cat 24 hours after intravenous injection of ³H-proline. Most of the silver granules are located over the collagenous matrix of predentine

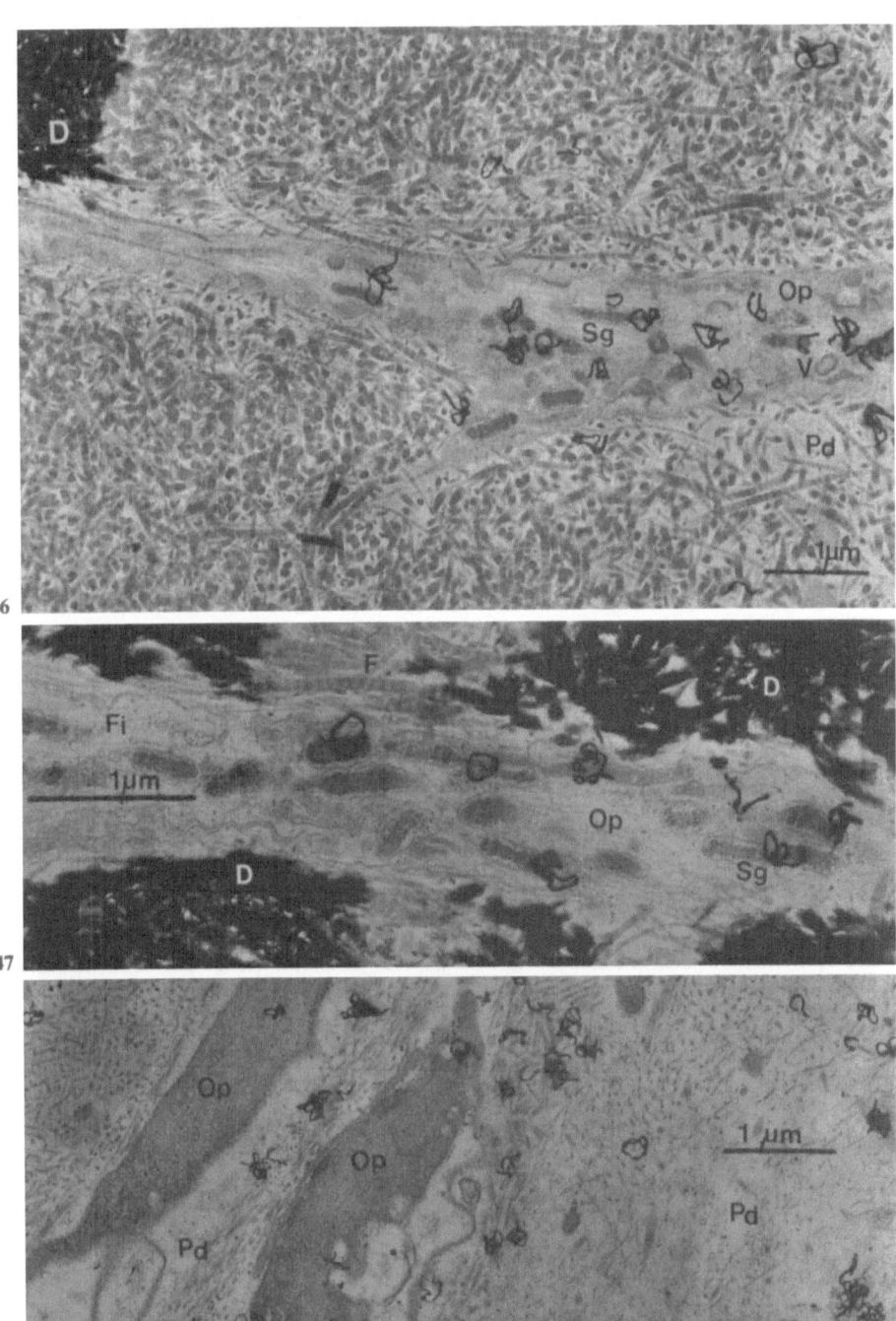

46

47

48

branches of other odontoblasts. Microfilaments oriented parallel to the cell membranes are constant features of the side branches, whereas in the main odontoblastic process, microtubules are observed together with microfilaments (Fig. 61). LESOT et al. (1982) demonstrated the presence of intermediate filaments (vimentin) in the process with the use of indirect immunofluorescence.

In addition to the cytoskeleton, the odontoblastic process contains numerous elongated secretory vesicles which, on longitudinal section, appear as cylindrical, oval or rod-shaped organelles (Figs. 43, 46, 47, 49, 51, 53, 54, 60, 75). In transverse sections, these secretory vesicles appear to be round (Figs. 41, 43, 52). Occasionally, fusion between the limiting membrane of a secretory granule and the cell membrane of the process can be observed (Fig. 49) and parts of the vesicular content are then observed in the extracellular matrix, without cellular disruption.

Numerous coated vesicles of various sizes are also present (Figs. 43, 49, 50, 54). They are often aligned along the cell membrane of the process in regions where invagination of the plasma membrane related to the formation of coated vesicles can be observed (Figs. 49, 50). Autoradiographic findings have indicated that the coated vesicles are related to endocytic phenomena (FRANK 1979). SASAKI et al. (1984) studied the endocytic activity of the odontoblasts by freeze-fracture replication of the odontoblastic processes in young kittens. Freeze-fracture replicas observed in transmission electron microscopy revealed depressions with particles on the plasma membranes of the proximal parts of the odontoblastic processes. Comparison with thin sections suggested that these repressions represented sites of endocytosis. GOLDBERG (1983) and SASAKI et al. (1984) concluded that the odontoblasts, especially in the proximal parts of their

Fig. 49. Odontoblastic process (*Op*) in the predentine (*Pd*) of a newborn cat. Both types of vesicles seen show fusion of their membranes with the plasma membrane. Elongated, rodshaped vesicles (*Sg*) contain a dense filamentous material with dark granules. In the area marked by three *arrows*, dark granules are observed in the extracellular space, with fusion of the secretory vesicle membrane and the plasma membrane. Coated vesicles (*CV*) are located in the odontoblastic process as well as along the plasma membrane. *Op 2* Lateral branch of the odontoblastic process. (From FRANK 1979, courtesy of Int Rev Cytol)

Fig. 50. Coated vesicles (*CV*) in the odontoblastic process (*Op*) of a newborn cat. Note the continuity of the membranes of the coated vesicles with the plasma membrane. *Pd* predentine

Fig. 51. Longitudinal section of an elongated secretory vesicle in an odontoblast of a newborn cat. Note the density of both the filamentous material and the overlying dense granules. Periodicity is observed in the transverse arrangement of the latter

Fig. 52. Transverse section of an elongated secretory vesicle in an odontoblast of a newborn cat

Fig. 53. Longitudinal section of an elongated secretory vesicle in an odontoblast of a newborn cat. There is no apparent periodicity in the vesicle contents. *Fi* microfilament

Fig. 54. Elongated secretory vesicles (*Sg*) in the odontoblast process 1 hour after intravenous ^{45}Ca injection. Silver grains appear in the cytoplasm of the process, but the vesicles are not labelled. *Pd* Predentine. (From FRANK 1979, courtesy of Int Rev Cytol)

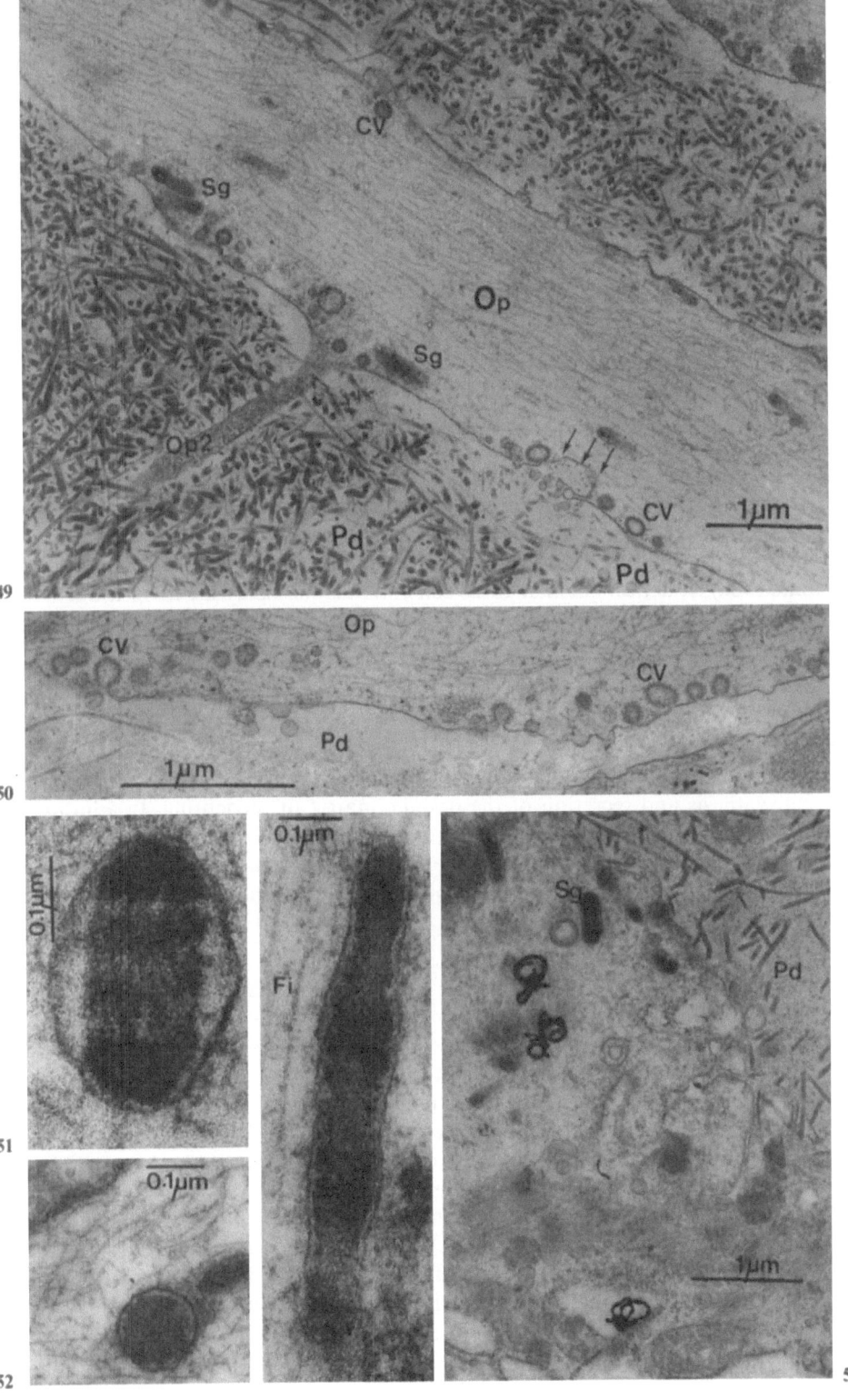

49

50

51

52

53, 54

processes, play a significant functional role in the modification of predentine matrix during early dentinogenesis.

Besides secretory vesicles and coated vesicles, polyribosomes and multivesicular bodies were observed in the odontoblastic process during dentinogenesis. Mitochondria have also been noted, but infrequently at this level (Fig. 64), by Herold and Kaye (1966) and Frank (1968a).

As stated earlier, during the differentiation of preodontoblasts into odontoblasts, a continuous basement membrane is present between the preameloblasts and the dental papilla. However, when an initial layer of predentine has differentiated between the odontoblastic processes, long finger-like projections of the preameloblast extend into the predentine through localized interruptions of the basement membrane (Frank and Nalbandian 1967). These projections of the preameloblasts or young ameloblasts appear to penetrate deeply into the predentine, and come close to the odontoblastic cells (Slavkin and Bringas 1976; Katchburian and Burgess 1977; Meyer et al. 1977; Slavkin et al. 1977). The contacts between the processes of ameloblasts and odontoblasts assume a variety of appearances but their membranes have not been observed to fuse or form any recognized specific type of cellular junction (Burgess and Katchburian 1982). It is thought that these cell contacts may play a role in inductive mechanisms related to cytodifferentiation of the odontogenic cells.

IV. The Metabolism of the Odontoblast

During dentinogenesis, the odontoblast and its process has under its control the biosynthesis and secretion of the organic matrix of predentine. In addition it seems that this cell is involved in the metabolism of inorganic elements such as those involving calcium transfer. Finally, the odontoblast assumes an important lysosomal function and plays an active role in modifying the newly secreted predentine matrix.

1. Collagen Biosynthesis

A predominant part of the organic matrix of predentine and dentine is made up of collagen fibrils. The shape of the native collagen molecule and most of its properties are determined by the triple-helical regions which comprize more than 95% of the molecule. These regions consist of three separate chains (α-chains), each of which contains approximately 1000 amino acids twisted in the form of a left-handed helix (Ramachandran and Ramakrishnan 1976). These three helical chains are then wrapped around one another in a high-order ropelike fashion to produce the tight triple-helical structure of the molecule. This conformation is stabilized by interchain hydrogen bonds.

Folding of the component chains into the proper helical conformation requires that glycine be present at every third amino acid residue (approximately 333 residues per chain). The chains are therefore composed of a series of triplet Glycine-X-Y sequences in which X or Y can be any amino acid. Frequently X is proline and Y is hydroxyproline, each occurring at about 100 sites per chain. The presence of other amino acids at the X and Y positions does not seem to follow any recognizable or repeating pattern. The only other amino acid with a known functional significance is lysine. This amino acid and its enzymatically modified forms participate in covalent cross-linking between chains and molecules and as sites of sugar attachment (LINSENMEYER 1981).

In addition to the triple helical region, collagen molecules also have short (about 20 amino acids), non-helical peptide region at the NH_2- and COOH-terminal ends of each component chain (PIEZ 1976). These terminal extension peptides play important functional roles, for they represent the sites where hydroxylysine-derived intramolecular and intermolecular cross-links are formed when molecules are arranged in native fibrils. The cross-links stabilize the molecular arrangement within fibrils and, depending on the type of cross-links formed, they greatly decrease the solubility of the molecules.

The precursor form of the predentine and dentine collagen molecule (found in the extracellular matrix) is called procollagen, collagen molecules with a much higher molecular weight. In the procollagen molecule, each chain has large additional extension peptides, called propeptides at both NH_2- and COOH-terminals. At least two different enzymes seem to be involved in the normal processing of procollagen to collagen, one to remove the NH_2-terminal propeptide piece and a second to remove the COOH-terminal one. These enzymes are called procollagen proteases (LAPIERRE et al. 1971).

Improved extraction and analytical procedures have shown that there are at least seven or even nine different genetic types of collagen in a whole organism and others are being discovered (LINSENMAYER 1981). In predentine and dentine, the exclusive presence of type I collagen seemed to have been demonstrated (BUTLER 1984a). From a biochemical point of view, the most striking characteristics of this type of collagen is that each molecule contains two genetically different types of α-chains: one is designated as the $\alpha1$ type I, or $\alpha1(I)$ chain and the second as the $\alpha2$ type I, or $\alpha2(I)$ chain. Each type I molecule is composed of two $\alpha1(I)$ chains and one $\alpha2(I)$ chain, the complete molecule being abbreviated: $[\alpha1(I)]_2\alpha2(I)$.

In the mid-1970s, a form of collagen known as the type I trimer was observed in cell cultures (MAYNE et al. 1975). This collagen, classified in the type I family, has no $\alpha2(I)$ chains, but was shown to consist of three $\alpha(I)$ chains. Although its presence is accepted, the level of type I trimer in dentine has not been unequivocally established (BUTLER 1984a). It seems that the synthesis of type I-trimer is an integral and important process in the formation of dentine. The functional significance of type I-trimer is unknown. However, the observations that it might be degraded prior to cross-linking (SODEK and MANDELL 1982), and that cells seem to produce relatively large quantities that are incorporated into the final product (MUNKSGAARD et al. 1978; WOHLLEBE and CARMICHAEL 1978; DIMUZIO et al. 1981), may suggest that this molecular species is in some way involved in the maturation process of collagen fibrils in predentine (BUTLER 1984a).

Two types of morphological approaches have clearly demonstrated that the odontoblast plays a major role in the biosynthesis of dentine collagen. These include light and electron microscope radioautography conducted with labeled amino acids and immunocytochemical investigations using antibodies to type I-procollagen.

Several studies using light microscope radioautography indicated that the precursors of the organic dentine matrix were elaborated in the odontoblasts (for reviews see A. WEINSTOCK 1972a, b; LEBLOND and WEINSTOCK 1976; FRANK 1979). The intracellular mechanisms associated with organic biosynthesis were far from clear, however. Concerning collagen biosynthesis, the majority of standard electron microscope studies on morphological grounds alone, indicated that the role of the odontoblasts was a major one. This notion was initially confirmed by a series of electron microscope radioautographic investigations using ^{3}H-proline (FRANK 1970b; WEINSTOCK and LEBLOND 1974; LEBLOND and WEINSTOCK 1976).

By following electron microscope radioautography intravenous injection of tritiated proline, electron microscope radioautography, FRANK (1970b) studied collagen biosynthesis in the odontoblast of newborn cats and confirmed the optical level findings of YOUNG and GREULICH (1963) and GREULICH and SLAVKIN (1965). No cells of the dental papilla showed such intense labeling as the odontoblast at 5 minutes, 30 minutes or 1 hour after intravenous injection. The sparse labeling of pulpal cells and of the lateral intercellular spaces of the odontoblast layer observed in this work does not give credibility to a pulpal origin for predentine collagen, although a few collagen fibrils were found in these locations (FRANK 1968a).

Initial incorporation of ^{3}H-proline into the odontoblast occurs in the rough-surfaced endoplasmic reticulum (RER) (REITH 1968a; FRANK 1970b; WEINSTOCK and LEBLOND 1974; FRANK 1979). Five minutes after intravenous injection into newborn cats, maximal labeling is observed over the rough-surfaced cisternae (Figs. 42 and 56). These data suggest that the RER and the ribosomes are the sites of synthesis of the polypeptide chains of collagen precursors. The RER shows a progressive decrease in silver grains at 30 minutes, 1 hour and 24 hours after intravenous injection (Fig. 56). In a measure proportional to the decrease in radioactivity in the ergastoplasm, a progressive increase is observed in the Golgi apparatus, where the most dense labeling is reached 1 hour after injection. Thus all the conditions required for a transfer of the synthesized polypeptide chains from RER to Golgi zones seem to be met (Fig. 56).

Fig. 55. Schematic overview of the odontoblast (*Od*) and the odontoblastic process (*Op*) in predentine (*Pd*) and dentine (*D*). A periodontoblastic space (*S 1*) occurs between the odontoblastic process (*Op*) and the calcified tubular wall of the dentine. Secretory granules (*Sg*) elaborated in the Golgi apparatus (*G*) are found in both the body (*Od*) and the process (*Op*) of the odontoblast from which they discharge their content into the extracellular matrix through a membrane fusion phenomenon (*double arrows*). N Nucleus; *Er* Endoplasmic reticulum; *Mi* Mitochondria; *Co* collagen fibrils; *CV* coated vesicles 2 Lateral branching of the odontoblastic process; *Mt* Microtubules

Fig. 56. Distribution of radioactivity, expressed as the number of silver grains per 100 μm², in the various organelles of the odontoblasts, the cytoplasm of the odontoblastic processes, their lateral interodontoblastic spaces, and predentine in the newborn cat after intravenous injection of ^{3}H-proline. (FRANK 1979, courtesy of Int Rev Cytol)

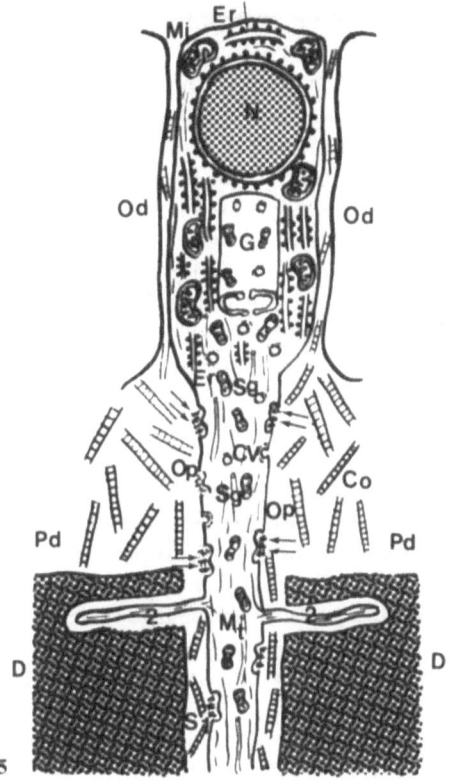

55

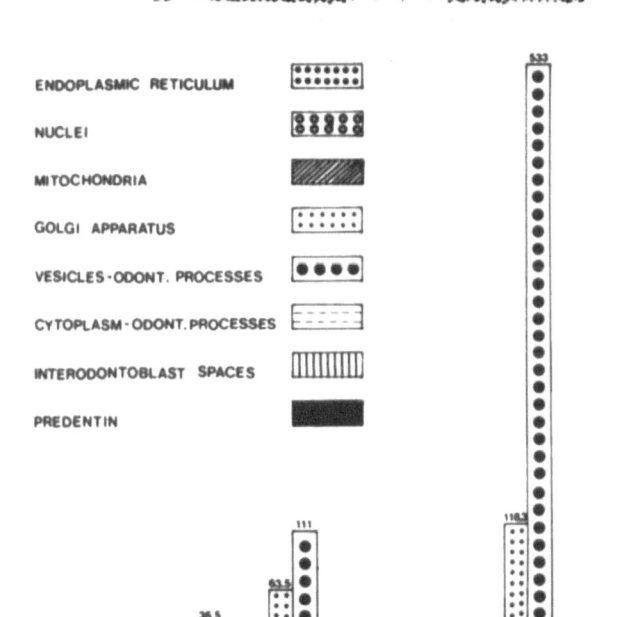

ENDOPLASMIC RETICULUM

NUCLEI

MITOCHONDRIA

GOLGI APPARATUS

VESICLES - ODONT. PROCESSES

CYTOPLASM - ODONT. PROCESSES

INTERODONTOBLAST SPACES

PREDENTIN

56

5'

30'

1h

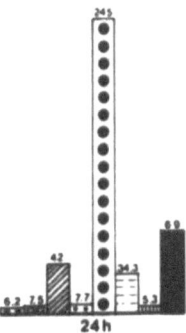

24h

In the Golgi apparatus 1 hour after intravenous injection, numerous silver grains are seen over flattened dictyosomes and distended saccules, as well as over the elongated secretory vesicles previously described (Figs. 44 and 45) and which presumably arose by budding from the dictyosomes (FRANK 1970b). Between the Golgi apparatus and the odontoblastic process, numerous labeled secretory vesicles are found. The collagen precursors packaged in the secretory vesicles at the level of the Golgi apparatus are transferred to the odontoblastic process. A great number of vesicles located in the odontoblastic process are covered by silver grains, at the level of the predentine (Figs. 43 and 46) and the dentine (Fig. 47). A prominent labeling of the predentine collagenous matrix is observed 24 hours after intravenous injection (Figs. 48 and 56); at this time dentine is only slightly labeled. These observations demonstrate that the collagen precursors synthesized in the RER are transferred to the Golgi apparatus where they are packaged in the form of secretory vesicles. The latter migrate to the odontoblastic process. Taking into account the radioautographic findings (Fig. 56), together with the morphological observations (Fig. 49) of secretory vesicles at the periphery of the odontoblastic process with their long axes oriented approximately parallel to the cell membrane, membrane fusion, and the presence of granular content of the vesicles in the extracellular matrix, it can be concluded that the secretory vesicles discharge their contents into the predentine by exocytosis.

No radioactivity has been observed over the coated vesicles present in the odontoblast and the odontoblastic process (Figs. 49 and 50). It is highly probable that these vesicles are involved in endocytosis as suggested by REITH (1968a), and KATCHBURIAN and HOLT (1968).

The above sequence of events in collagen biosynthesis was confirmed with intravenous injection of ^{3}H-proline in young rats (WEINSTOCK and LEBLOND 1974; LEBLOND and WEINSTOCK 1976). By 2–5 minutes after injection the label was restricted to the endoplasmic reticulum. Presumably the label corresponded to pro-α-chains located within the cisternae. By 10 minutes, the silver grains appeared in the Golgi apparatus where spherical saccules with entangled threads were marked. At 20 minutes, label was detected within cylindrical portions of the Golgi saccules containing parallel threads. At the 30-minute interval, a concentration of ^{3}H-proline was observed over the secretory granules between the Golgi apparatus and the odontoblastic process where WEINSTOCK and LEBLOND (1974) observed membrane fusions between the secretory granules and the plasma membrane followed by exocytosis. By 90 minutes and 4 hours, a significant labeling of the predentine was observed.

Recently, similar intracellular biosynthetic pathways for collagen biogenesis were described in the fibroblast of the rat foot pad, using the same technic (MARCHI and LEBLOND 1983, 1984). After intravenous injection of ^{3}H-proline, a radioactive peak appeared over the RER at 4 minutes, over the intermediate vesicles and associated tubules at 10 minutes, the spherical distensions of cis-side Golgi saccules at 20 minutes, the cylindrical distensions of trans-side saccules between 40 and 60 minutes, and the secretory granules at 60 minutes. It is proposed that the succession of peaks corresponds to the migration pathway of collagen precursor proteins within fibroblasts, i.e., the proteins synthesized in the RER are delivered by intermediate vesicles and/or tubules to the spherical distensions of trans-side saccules and finally are carried by secretory granules to the extracellular space. Intermediate vesicles and tubules are also present in odontoblasts, and hence the radioautographic observations described above appear to be applicable to the secretion phenomena in dentine.

However, a number of arguments have been raised against these radioautographic studies using ^{3}H-proline as a precursor to label collagen polypeptides since proline can obviously be associated with other cellular activities than collagen synthesis. An immunocytochemical approach using antibodies to type I-procollagen has provided confirmation of the initial radioautographic findings.

COURNIL et al. (1979) and KARIM et al. (1979), with the aid of procollagen I antibodies produced in rabbits, were able to localize procollagen I in rat odontoblasts using immunoperoxidase electron microscopic techniques. Within odontoblasts, the immunostaining indicative of procollagen I antigenicity was moderate in RER, strong in spherical and cylindrical distensions of the Golgi apparatus, intense in secretory vesicles, and variable in lysosomal structures. In predentine, the immunostaining was intense close to the odontoblast layer but decreased gradually in a distal direction. It was thus concluded that procollagen I (and/or substances endowed with similar antigenicity such as pro-α(I) chains and procollagen fragments) are present 1) along the intracellular pathways of collagen precursors where its concentration gradually increases to reach a maximum in secretory vesicles; 2) in predentine into which it is released from the vesicles for transformation into non-immunoreactive collagen I; and 3) in lysosomal structures where some of it is hydrolyzed.

It can thus be concluded that collagen secretion by the odontoblast involves synthesis of procollagen I in the RER, transport from the RER to the Golgi complex, packaging in the Golgi region and translocation of secretory vesicles. Procollagen I is progressively transformed into collagen type I in the predentine.

Some of the intracellular transport processes of procollagen can be perturbed experimentally by local anesthetics, colchicine, vinblastine, cytochalasin B, uncouplers of oxidative phosphorylation and the Na+ ionophore monensin (OLSEN 1981). As has already been mentioned, colchicine, vinblastine and cytochalasin B, interfering with the cytoskeleton on the odontoblast and its process, have been extensively used in the study of the epithelio-mesenchymal interactions and dentinogenesis. KUDO (1975) studied the effects of subcutaneous injections of colchicine, on the incorporation of ^{3}H-proline into odontoblasts, in the rat. It appeared that the drug did not appreciably affect the incorporation of ^{3}H-proline but affected the secretion of the dentinal matrices by interfering with the structure and function of microtubules in these cells.

CHO and GARANT (1984) studied the effect, in young mouse incisors, of a proline analog, L-a-azetidine-2-carboxylic acid (LACA), known to interrupt the transport of procollagen from rough endoplasmic reticulum to the Golgi apparatus. The effect of LACA on protein matrix secretion of odontoblasts and ameloblasts was followed by light and electron microscopic radioautography after injection of ^{3}H-glycine. Whereas no apparent effect was observed on ameloblasts, LACA inhibited the secretion of dentine matrix with accumulation of ^{3}H-glycine labeled procollagen in the cisternae of the RER. Golgi cisternae appeared not to be affected; however all Golgi saccules and the secretory vesicles disappeared within 2 hours after LACA administration. The inhibition of the translocation of procollagen from RER to the Golgi apparatus provided a means to determine what components of the odontoblast Golgi complex were related to the packaging of exportable protein. The disappearance of Golgi saccules in LACA treated cells, indicated that their presence is a requirement for active collagen secretion (Cho and Garant 1984).

2. Glycoprotein and Fibronectin Biosynthesis

Among non-collagenous proteins, several decidedly acidic and some less acidic glycoproteins have been described in dentine (Butler 1984b). These glycoproteins are synthesized by odontoblasts and rather rapidly transported to the mineralization front, as visualized by radioautography.

The biosynthesis of glycoprotein in dentine of young rats, was studied, after an intravenous injection of ^{3}H-fucose, with electron microscope radioautography (A. Weinstock et al. 1972). By 5–10 minutes after injection, the radioactivity was restricted to the Golgi apparatus of the odontoblast, where the carbohydrate component was incorporated in the forming glycoprotein. By 35 minutes, silver grains were observed over the elongated secretory vesicles located in the Golgi zone as well as in the odontoblastic process and there were a few silver grains in the adjacent predentine. By 4 hours the radioactive reaction was present on the dentine side of the predentine-dentine junction. These observations indicate that ^{3}H-fucose is added to forming glycoprotein in the Golgi apparatus and packaged into secretory vesicles by which the fucose-containing glycoprotein can be transported to the odontoblastic process and subsequently discharged into the predentine.

In a study of rat dentinogenesis in which thiocarbohydrazide-silver proteinate staining was combined with electronmicroscopy, Takagi and co-workers (1981) confirmed the packaging of glycoproteins into secretory vesicles and subsequent release into the extracellular matrix.

The behavior of substances labelled with either ^{3}H-proline of ^{3}H-fucose injected intravenously, were compared in the rat incisor with the use of light microscope radioautography (Warshawsky and Josephsen 1981). With ^{3}H-proline, the odontoblast bodies were labelled at early intervals. They secreted a layer of intensely labeled predentine which by days 1 and 2 was converted into mineralized dentine. Odontoblastic processes were not labeled in the dentine formed prior to injection. With ^{3}H-fucose, the cell bodies were also labeled at the earlier intervals and the newly formed glycoproteins were incorporated into the predentine. In short, the glycoproteins were incorporated into the dentine at the calcification front, but, unlike the behavior of labeled proteins, by 1 or 2 days the labeled glycoproteins appeared along the entire length of the odontoblastic process.

The value of ^{3}H-fucose as a specific radioautographic marker for glycoprotein is now widely recognized (Cho and Garant 1985). Studies using light and electron microscopic radioautography have confirmed that ^{3}H-fucose is rapidly incorporated into the Golgi apparatus of mouse odontoblasts. ^{3}H-fucose is subsequently incorporated into secretion granules similar to those that contain collagen precursors.

Once extruded from the cell, ^{3}H-proline or ^{3}H-fucose labeled material behave differently. The labeled glycoprotein accumulates at the predentine-dentine junction, within 4 hours, especially in the mineralized matrix just distal to the junction (A. WEINSTOCK et al. 1972). Labeled collagen does not reach this junction until 20 to 30 hours later (WEINSTOCK and LEBLOND 1974; WARSHAWSKY and JOSEPHSEN 1981). It was therefore suggested that the glycoprotein may have a role in the mineralization process.

In the studies of CHO and GARANT (1985) an association of ^{3}H-fucose with the terminal web microfilaments and the adjacent region of the plasma membrane of the odontoblasts was noted. It was stated that addition of new membrane glycoprotein to the odontoblast process can occur via fusion of secretory granule membranes in the predentine region of the process. Such a mechanism would facilitate the growth of the odontoblast process until it reached its maximum length.

Another type of high molecular weight glycoprotein, fibronectin, is probably synthesized by preodontoblasts and odontoblasts (LESOT et al. 1981; THESLEFF et al. 1980). Fibronectin, which is an important constituent of basement membranes, is also considered to be a mediator of adhesion between cells and extracellular matrix. The distribution of fibronectin was studied during dentinogenesis in rat incisors with the use of indirect immunofluorescence (LINDE et al. 1982b). Fibronectin was localized in the odontoblast layer at the level where the cell processes leave the cell bodies and where the odontoblasts adhere to each other at junctional complexes. In addition, fibronectin was found in the dental pulp and the predentine at certain stages of dentinogenesis. It was present in predentine during mantle dentine formation but was absent from predentine during further (circumpulpal) dentine formation.

3. Proteoglycan Biosynthesis

Proteoglycans are large macromolecules consisting of a central protein core to which are linked several carbohydrate side chains. The glycosaminoglycans (GAG) side chains are composed of repeating disaccharide units consisting of one uronic acid and one hexosamine residue. The GAG chains may be sulfated and the sulfate groups together with the uronic acid carboxyl groups confer a high negative charge density to the proteoglycans. The specific composition, the high proportion of carbohydrate relative to protein, the anionic character and the large molecular size motivate the segregation of proteoglycans into a category separate from glycoproteins in general.

Recent biochemical studies on the nature of dentine proteoglycans indicate that their structure differs sharply from those of cartilage (BUTLER 1984b). The GAG chains of dentine proteoglycans contain mostly chondroitin-4-sulfate, with lesser amounts of chondroitin-6-sulfate and iduronic acid-containing polymers (HJERPE et al. 1983). The major proteoglycans from predentine and dentine also consist of small core proteins with few chondroitin sulfate chains (RAHEMTULLA et al. 1984).

The presence of GAG in odontoblasts, predentine, and dentine has been recognized on the basis of light microscopic studies employing histochemical stains such as toluidine blue and alcian blue or the dialyzed iron method. Using ultrastructural cytochemistry with the aid of ruthenium red, Nygren et al. (1976) observed positive reactions on the elongated secretory vesicles and in the cell coat of the odontoblasts. Treatment with hyaluronidase prior to staining with ruthenium red abolished the staining of the vesicles, but not that of the cell coat.

Takagi et al. (1981) confirmed the results of Nygren et al. (1976) and observed the presence of GAG in Golgi saccules and secretory vesicles within odontoblasts. Sulfated glucoconjugates were also detected in predentine.

Sulfated proteoglycan formation was traced with ^{35}S-sulfate (Leblond and Weinstock 1976). The label was observed over the elongated secretory vesicles 30–35 minutes after injection. The radioactive material was released into the predentine and subsequently reached the predentine-dentine junction by 4 hours, as in the case of the fucose label.

Thus from the foregoing it becomes evident that the secretory vesicles of the odontoblasts are able to carry not only procollagen, but also glycoprotein, proteoglycan, and other molecules. However, it is unclear whether all these components are transferred simultaneously in one and the same secretory vesicle or if each is carried by its own unique species of vesicle.

4. Biosynthesis of γ-Carboxyglutamic Acid (GLA)-Containing Proteins

The vitamin K-dependent amino acid, γ-carboxyglutamic acid (Gla), was found initially in prothrombin and other blood coagulation proteins. It has subsequently been identified in bone (Hauschka et al. 1975; Price et al. 1976). The bone Gla-protein or osteocalcin has received considerable attention since it has the ability to bind cations such as Ca^{2+}, due to its two closely fixed carboxyl groups. The Gla-containing protein was first demonstrated in rat incisor dentine by Linde et al. (1980, 1982a), where it was found to be one of the most abundant soluble non-collagenous proteins. DiMuzio et al. (1983) demonstrated, in organ culture, that the Gla-containing protein is synthesized by the odontoblasts. By means of electron microscopic immunocytochemistry it has been possible to demonstrate components which cross-react with antibodies to dentine Gla-protein within the odontoblasts and its processes (Linde 1984). Observations indicated that Gla-proteins are secreted into the dentine at the mineralization front (Linde 1984).

5. Phosphoprotein Biosynthesis

Phosphoproteins constitute the major non-collagenous protein fraction from the dentine of all mammalian species. The first evidence of the existence of phosphoprotein in dentine was obtained by Veis and Schleuter (1964) and Schleuter and Veis (1964). The dentine phosphoproteins, referred to as phosphoryns (DiMuzio and Veis (1978) are relatively well characterized but they present biochemical differences among species (BUTLER 1984b). Bovine phosphoprotein from unerupted teeth appears to be a single component of 155000 daltons with about 46% phosphoserine and 45% aspartic acid (STETLER-STEVENSON and VEIS 1983). In the rat incisor (BUTLER et al. 1983) the phosphoryns are more complex, consisting of three families of proteins which are highly phosphorylated (40% phosphoserine), moderately phosphorylated (25% phosphoserine) and minimally phosphorylated (5–7% phosphoserine). The reasons for the species differences are not known, but the similar amino acid composition suggests that the phosphoryns have the same functional significance (BUTLER 1984b). Dentine phosphoproteins seemed to have an important role in the mineralization process.

Radioautographic studies strongly suggest that the phosphoryns are synthesized by the odontoblasts and secreted at the mineralization front. At the optical level, using intravenous ^{33}P-phosphate and ^{3}H-serine, as well as ^{3}H-proline for purposes of comparison, WEINSTOCK and LEBLOND (1973) noted, in the rat incisor, that within 30 minutes the labeled phosphorus, serine and proline were taken up by odontoblasts and deposited into predentine. The proline label remained in the predentine at least 4 hours after injection, whereas the labeled phosphorus and serine were displaced to the dentine side of dentine-predentine junction as early as 90 minutes after injection. These findings were confirmed by biochemical studies of the phosphoprotein content of bovine predentine and mineralized dentine (JONTELL and LINDE 1983). Very low amounts of phosphoprotein were found in predentine compared to dentine.

Localization of phosphoproteins by indirect immunofluorescence, in mouse molars, indicated that perhaps this protein is transported to the mineralization front by the odontoblastic processes (McDOUGALL et al. 1983).

6. Calcium and Phosphate Metabolism

Dentine is the result of calcification of an organic matrix, the predentine. The calcification process is characterized by accumulation of calcium and phosphate ions in this cell-elaborated organic matrix. Initiation of calcification is first observed, with subsequent growth of apatite crystals. This section will consider the transfer mechanisms and routes taken by calcium and phosphate ions as they traverse from the blood capillary lumens located in the dental papilla, across the odontoblast layer and to the dentine. It is important to realize that significant amounts of calcium and phosphate must pass between and/or through the odontoblasts in order to reach the mineralization front.

The transfer of ^{45}Ca to dentine seems to be extremely rapid. MUNHOZ and LEBLOND (1974) using intravenous injections of ^{45}Ca, in young rats, coupled with light microscope radioautography, noted a maximum reaction at the predentine-dentine junction between 30 seconds and 5 minutes after injection, with a gradual decrease towards the dentino-enamel junction. No radioactivity was found over the odontoblasts. These observations, demonstrating a direct extracellular pathway from blood vessels to dentine, were confirmed, in part, by NAGAI and FRANK (1974) following intravenous injection of ^{45}Ca, in newborn cats. A comparison was made between the cellular and extracellular compartments of the pulpo-dentinal organ by electron microscope radioautography. For purposes of analysis, the radioactivity over both the odontoblast and its process was grouped under the collective term of "odontoblasts" (Fig. 57). It was shown that 5 minutes after intravenous injection, the extracellular spaces of the dental papilla were the most intensely labeled, followed by the fibroblasts. At 30 minutes, the highest radioactivity was noted in the pulpal fibroblasts (Fig. 57). At 1 hour, the odontoblast and the dentine were the most densely labeled compartments, whereas at 6 hours the dentine had the maximum number of silver grains. Apart from intracellular radioactivity noted over the odontoblast and its process, a significant part of the radioactive calcium diffused directly from the intercellular spaces of the dental papilla via the spaces located laterally between the odontoblasts, finally reaching the dentine. It is interesting to note that after 5 minutes the radioactivity was slightly higher in the predentine, than in the dentine; at 30 minutes it was almost equal at these sites and became strikingly higher in dentine at 1 hour and even more so at 6 hours (Fig. 57).

Two routes of almost equal importance have thereby been demonstrated for radioactive calcium transfer through the layer of odontoblasts (Fig. 57). In the direct route, the calcium starting from the capillary lumen follows the

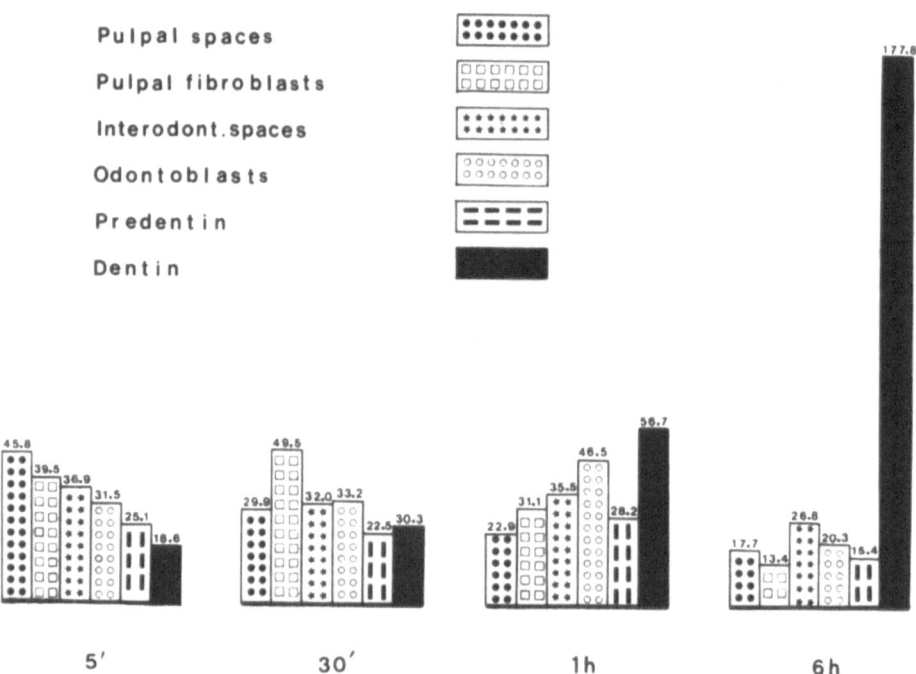

Fig. 57. Concentration of ^{45}Ca radioactivity expressed as number of silver grains per 100 μm² in the intracellular region of the dental papilla (pulpal spaces), in the pulpal fibroblasts, in the odontoblasts and their processes, in the lateral interodontoblastic spaces, and in the predentine and dentine, at various time intervals following injection. (Frank 1979, courtesy of Int Rev Cytol)

extracellular spaces of the dental papilla, the lateral interodontoblastic spaces, and the predentine to finally reach the dentine. In the indirect route, after passage through the extracellular spaces of the dental papilla, the calcium follows a transcellular route through the odontoblast.

The intracellular transfer of ^{45}Ca through the odontoblast body was visualized as a progressive increase of radioactivity over various cellular compartments over the first hour, followed by a decrease after 6 hours (Fig. 58). Among the cellular organelles, the mitochondria were richest in ^{45}Ca, confirming their importance in calcium storage and regulation. The endoplasmic reticulum, Golgi apparatus and nuclei accumulated calcium, subsequently transferred to predentine and dentine through the odontoblast process (Fig. 58). At 6 hours, a diffuse labeling was observed in the intertubular dentine, all along the dentinoenamel junction. The radioactivity was clearly more concentrated in the dentine as a whole (177.8 grains/1000 μm^2) than in enamel (38.5 grains/1000 μm^2). Under the experimental conditions used, calcium transfer via the elongated secretory vesicles was not observed.

The importance of mitochondria in the regulation and concentration of calcium and phosphate ions in cells of mineralizing tissues, such as odontoblasts, osteoblasts and chondroblasts has been reported in several instances (FRANK 1979). As early as 1965, BAUD and DUPONT described the presence of dense granules, 300–500 Å in diameter, in osteocyte mitochondria. These granules containing calcium and phosphates have been observed in odontoblast mito-

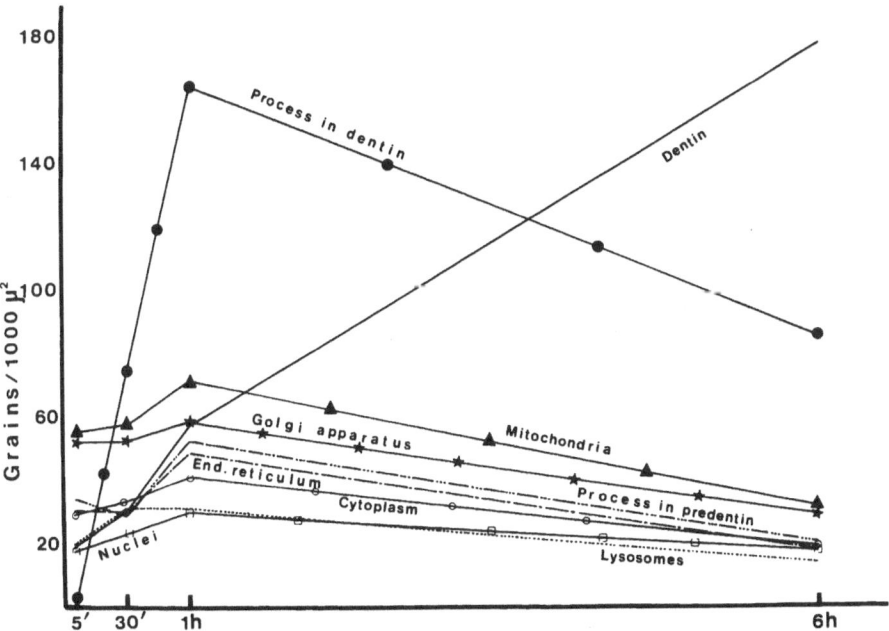

Fig. 58. Concentration of radioactivity expressed as number of silver grains per 100 μm^2 in the odontoblast, its process and the predentine at various time intervals following injection. (FRANK 1979, courtesy of Int Rev Cytol)

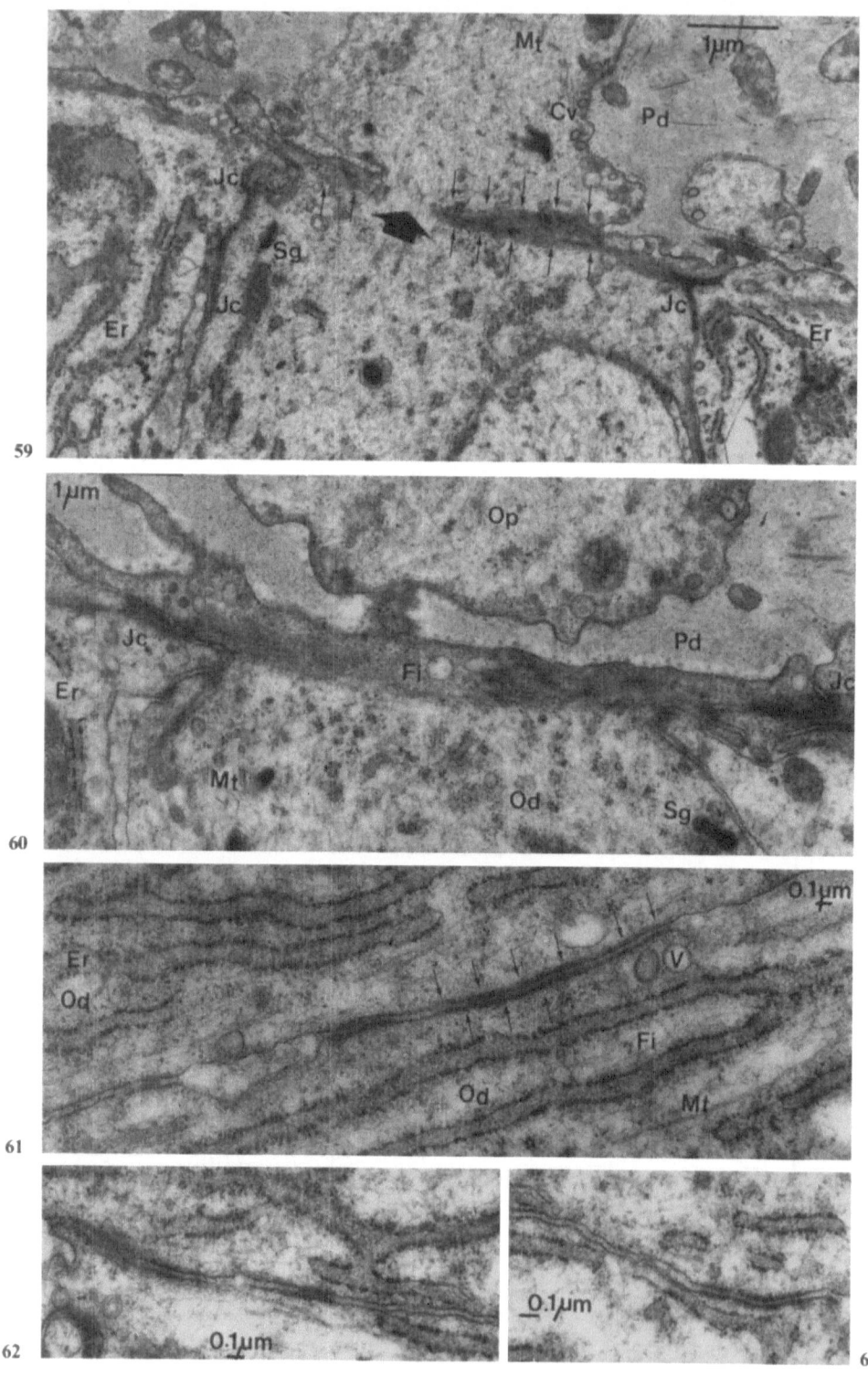

chondria, and recently GOLDBERG and ESCAIG (1984b) reported improved preservation of intramitochondrial granules in the odontoblast by rapid freezing and freeze-substitution fixation.

After intraperitoneal injection of $^{45}CaCl_2$ in rats, HÖHLING and FROMME (1984) observed an accumulation of ^{45}Ca in the endoplasmic reticulum after 10 minutes. At 30 minutes, most of the ^{45}Ca was still localized in the endoplasmic reticulum and some in the central Golgi zone. Considerable ^{45}Ca activity could be demonstrated in the odontoblastic process at 60 minutes. At 120 minutes, the radioactivity was observed within the mineralized dentine.

A number of cytochemical studies have been directed towards localizing calcium in odontoblasts. At the light microscope level, KASHIWA and SIGMAN (1966) demonstrated the presence of calcium granules in the odontoblast process, histochemically. Using the potassium pyroantimonate technique, OZAWA et al. (1972) noted with the electron microscope, reaction products in the mitochondria, Golgi apparatus, dense granules, and interodontoblastic spaces. With the same technique, FRAZIER and NYLEN (1972) reported electron-dense precipitates in flattened sacs and vesicles of the Golgi region as well as in secretory vesicles. The reaction product was also found between cells, whereas the cytoplasm, mitochondria, and endoplasmic reticulum of the odontoblast were virtually free of reaction product. Using potassium pyroantimonate, REITH (1976) found, in odontoblasts of developing rat molar teeth, positive reactions in the elongated secretory vesicles situated in the Golgi apparatus and at the distal pole of the cells. Tissues pretreated to allow the escape of diffusible ions showed a positive calcium reaction only in the secretory vesicles.

However, the specificity of the potassium pyroantimonate technique has been seriously questioned since the potassium pyroantimonate-osmium reaction does not lead to reliable calcium localization. Serious dislocations occur during Ca^{2+} trapping such that heterogeneous nucleation may occur during fixation and embedding (VAN IREN et al. 1979). APPLETON and MORRIS (1979) used pyro-

Fig. 59. Odontoblasts during dentinogenesis in a newborn cat, with junctional complex (*Jc*) on the lateral sides of the formative end of the odontoblast body. A bundle of transverse microfilaments (*thin arrows*) delineates the odontoblast body, below, from its process above. It is discontinuous in the central part of the cell (*large arrow*). *Sg* Secretory granule; *Er* Endoplasmic reticulum; *Cv* Coated vesicles; *Mt* Microtubules and microfilaments; *Pd* Predentine

Fig. 60. Longitudinal section of an odontoblast (newborn cat) passing through the cell body (*Od*) and the odontoblastic process (*Op*) in the peripheral part of the cell. A substantial bundle of parallel microfilaments (*Fi*) is seen adjacent to the odontoblast plasma membrane facing the predentine (*Pd*) and joining the two lateral junctional complexes (*Jc*). *Sg* Secretory granule; *Mt* Microtubule; *Er* Endoplasmic reticulum

Fig. 61. The junctional complex at the formative end of the odontoblast cell body (*Od*). Dense material (*arrows*) is seen along the inner leaflets of adjacent plasma membranes. *Fi* Microfilaments; *Mt* Microtubules; *V* Vesicle; *Er* Endoplasmic reticulum

Figs. 62 and 63. Details of the junctional complex at the formative end of the odontoblast cell body. Note localized narrowing of the intercellular space and accumulation of dense material along the inner leaflet of the plasma membranes

antimonate together with energy dispersive analysis, to localize calcium in odontoblast of 4-day-old rats. Calcium was found within the Golgi apparatus as well as in elongated secretory vesicles. Boyde and Reith (1977) studied the calcium content of freeze-sectioned developing dentine using energy dispersive X-ray emission analysis and found high concentration of calcium in the distal secretory pole of the odontoblasts. Calcium and phosphorus levels were high in the odontoblastic process. By electron microprobe analysis of odontoblasts, Ozawa et al. (1981) found a peak of CaKα radiation over the endoplasmic reticulum.

The intracellular localization and transport of Ca^{2+} within odontoblasts is further suggested by the presence, in mantle predentine, of matrix vesicles, the primary loci for mineralization, to be described in detail later (for review see Bonucci 1984). Histochemical and biochemical analysis of the content of matrix vesicles indicates that they accumulate calcium and phosphate ions and, by implication, that these inorganic elements transit through the cell.

The presence of a Ca^{2+} pump in intracellular vesicles of rat incisor odontoblasts was suggested by Linde and Granström (1980) and Granström and Linde (1981). These workers demonstrated the presence of Ca^{2+}-activated ATPase in the membrane of intracellular vesicles. They incubated a microsomal fraction of the odontoblast with $^{45}Ca^{2+}$ and found that the characteristics of the energy dependent intravesicular accumulation of calcium were such that it should be presumed that Ca^{2+}-ATPase was involved. A fraction containing high alkaline phosphatase activity, with a high Ca^{2+} accumulating ability, was isolated by Granström (1984) from similar intracellular vesicles of rat incisor odontoblasts. He suggested a correlation between alkaline phosphatases in intracellular vesicles and their relation to extracellular vesicles and to the mineralization process.

It must be noted, however, that our present knowledge about the basic intracellular mechanisms governing calcium and phosphate ion transport through the odontoblasts and its processes is still limited. Several questions need to be addressed. The possible role and function of ATPases, vitamin D dependent calcium and/or phosphate binding proteins, the cytoskeleton and many other factors need clarification. It would be interesting to elucidate the role of calmodulin in dentinogenesis. Calmodulin is a multifunctional modulator protein that mediates the Ca^{2+} concentration in a variety of cellular processes and is believed to be a calcium pump (Cheung 1980).

7. Degradative Activities of the Odontoblasts

A variety of degradative activities occur in the odontoblasts during dentinogenesis. Reviewing current work on intracellular degradation of newly synthesized collagen in cultured fibroblasts, Bienkowski (1984) estimated that approximatively 15% of the collagen synthesized by this cell is broken down and by two distinct pathways of degradation. *Basal degradation* functions continuously and independently of collagen synthesis. The basal degradation mechanism would be located in the distal region of endoplasmic reticulum (smooth ER) or the cis-region of the Golgi complex, with enzymes capable of attacking collagenous peptides located in one of these organelles. The second pathway, the so-called *enhanced degradation,* occurs in the case of abnormal collagen synthesis. It is postulated that the latter process results from the interaction between newly synthesized collagen molecules and an apparatus, probably located in the Golgi, which recognizes abnormal collagen structures and directs molecules either towards sites for packaging into secretory vesicles or towards the site of degradation. Enhanced degradation is mediated by lysosomal proteases and occurs in lysosomes (BIENKOWSKI 1984). At present, it is not known if these mechanisms are operating in odontoblasts, but lysosomal functions as defined by DE DUVE (1978) are certainly present in odontoblasts. By electron microscopic immunostaining, KARIM et al. (1979) observed the presence of procollagen I in multivesicular and dense lysosomal bodies of the rat incisor odontoblast. Two possible explanations for this lysosomal involvement were proposed by KARIM et al. (1979): 1) the diversion of a fraction of collagen precursors from the normal biosynthetic pathway to the lysosomal pathway or 2) the reincorporation by phagocytosis of collagenous material recently released into predentine. The second hypothesis can be eliminated, based on the study of MARCHI and LEBLOND (1983) who, after intravenous injection of ^{3}H-proline in rats, observed the label over the lysosomal system of fibroblasts as early as 20 minutes, when no label had yet been released from the cells. MARCHI and LEBLOND (1983) concluded that a fraction of the type I collagen precursors were deviated from the path leading to secretory granules and entered the lysosomal system, presumably to be destroyed.

Cytochemical localization of lysosomal enzymes in odontoblasts has been studied at the ultrastructural level. KATCHBURIAN and HOLT (1968) noted the presence of acid phosphatase in the Golgi cisternae and vacuoles of odontoblasts in 3- to 4-day-old rats. Enzyme activity occurred in dense round vesicles in the distal part of the odontoblast as well as in the odontoblastic process. These workers concluded that lysosomes may play a role in the secretory processes associated with dentine formation and moreover that transport of material seemed to be occurring in both directions across the cell membrane, suggesting that lysosomes are also involved in digestion of material resorbed from the surrounding medium.

The presence of acid phosphatase in the Golgi complex of odontoblasts was confirmed by NAGAI (1970), TAKUMA and NAGAI (1971), LARSSON and BLOOM (1973), and SASAKI et al. (1982, 1984). Positive acid phosphatase reactions were observed by NAGAI (1970) in granular endoplasmic reticulum, probably related to enzyme production, as well as in agranular endoplasmic reticulum, Golgi elements, and coated vesicles. Typical cytosegresomes, digesting part of the cytoplasm, were also described. In addition, acid phosphatase was observed in elongated secretory vesicles (NAGAI 1970) suggesting a possible means of regulating the secretory process or an exocytic discharge of hydrolytic enzymes implicated in degradation of extracellular structures. A lysosomal function of the elongated secretory vesicles of the odontoblast has been discussed by KAJIKAWA and KAKIHARA (1974).

Along these lines, LINDE and PERSLIDEN (1977) demonstrated the presence of a lysosomal acid proteinase with high activity in odontoblasts and predentine.

This enzyme was identified as cathepsin D. NYGREN et al. (1979) localized cathepsin D in lysosomal vesicles in the Golgi region of the odontoblasts, as
well as in elongated secretory vesicles of the odontoblastic process, and as electron dense precipitates in the predentine matrix. According to these authors,
transport of cathepsin occurs from the Golgi zone to the odontoblastic process
in vesicular structures. The localization of high cathepsin D activity in predentine is strongly suggestive for degradative enzymatic function of the proteoglycans.

Using a gelatin film substrate technique, BETTI and KATCHBURIAN (1982)
demonstrated a proteolytic activity in developing predentine and stated that
odontoblasts are concerned with the breakdown of components of the organic
matrix as well as with resorption and digestion of breakdown products during
dentinogenesis. Studying odontoblasts in rat incisors after intravenous injection
of horseradish peroxidase (HRP), freeze-fracture replication and acid phosphatase staining, SASAKI et al. (1982) observed in the electron microscope large
coated vesicles, about 0.1 μm in diameter, as well as small coated vesicles, about
0.05 μm in diameter, distributed from the distal end of the odontoblast to the
odontoblast process. Some of the large coated vesicles were continuous with
the cell membrane and manifested coated pits. In freeze-fracture replication,
particle-rich depressions corresponding to those pits occurred in the cell membrane P face. Taken into cells by means of the large coated vesicles, HRP
was subsequently incorporated into dense bodies. These dense bodies received
hydrolytic enzymes (acid phosphatase) from small coated vesicles that are primary lysosomes. SASAKI et al. (1984) confirmed the high endocytic activity of
odontoblastic processes in young kittens. These depressions, commonly revealed
by freeze-fracture replication in the plasma membranes of the proximal parts
of the odontoblast processes, underline the significant functional role played
by the odontoblasts in the modification of predentine matrix during early dentinogenesis.

It can be concluded that the cellular activities of the odontoblasts are highly
complex. Beyond elaboration of the organic matrix of predentine and control
of calcium and phosphate ion transfer, the odontoblast plays a major role
in degradative processes involving autophagic, endocytic, and exocytic phenomena.

G. Predentine

Formation of predentine and dentine begins in the dental papilla of the
late bell stage, in the region where indentation of the inner dental epithelium
has produced a fold and where preameloblasts and ameloblasts have differentiated. These are locations which correspond to cusp tips and incisal edges.

After the initial differentiation of a layer of odontoblasts in these sites
(described in detail in the previous sections), the plasma membranes of the
odontoblasts adjacent to the enamel organ display several short processes

(Figs. 13, 14). One of these stubby processes persists and becomes accentuated to form the odontoblastic process. In the extracellular spaces limited by the basement membrane and the plasma membranes of the odontoblasts and their extensions, a first layer of predentine is laid down. As more predentine matrix is formed, the odontoblasts begin to retreat towards the center of the dental papilla. A progressive elongation of the odontoblastic process is observed, concurrent with further growth in thickness of predentine and dentine. In the developing canine of the cat, the initial calcification which will transform predentine into dentine is observed when the predentine is about 75 µm wide (SILVA and KAILIS 1972). Elaboration of predentine is a continuous process not only during the formation of the primary coronal and radicular dentine, but also persisting in the adult, erupted tooth.

The organic matrix of predentine is composed principally of collagen fibrils (Fig. 66). The collagen appears to be exclusively type I collagen, although minor amounts of type I-trimer is recognized but not unequivocally established (BUTLER 1984a). With respect to the metabolic function of the odontoblast, it has been shown that collagen precursors elaborated intracellularly are secreted within the predentine as type I procollagen (COURNIL et al. 1979; KARIM et al. 1979). Procollagen was found in the predentine adjacent to the odontoblastic cell bodies, whereas in the predentine near the predentine-dentine junction, the type I procollagen had been transformed into type I collagen.

Predentine contains, in addition, some unique non-collagenous protein macromolecules, most of which have already been reviewed. These include proteoglycans (PG) and glycosaminoglycans (GAG), but also glycoproteins, γ-carboxyglutamic (Gla)-containing proteins, phosphoproteins, and serum proteins (albumin, transferrin, immunoglobulins and 2-HS-glycoprotein) as well (BUTLER 1984b).

PG and GAG have been localized repeatedly in predentine by ultrastructural cytochemistry. With ruthenium red staining, these proteins appear as fine granules, either associated with or not associated with collagen fibrils (NAGAI et al. 1974; NYGREN et al. 1976). With cuprolinic blue, the proteoglycans are visualized as ribbon-like structures with radiating filaments (GOLDBERG 1983; GOLDBERG and SEPTIER 1983). Biochemical analysis has demonstrated a high concentration of PG and GAG in the predentine, with a significant decrease after mineralization (LINDE 1973, 1984).

The glycoproteins of predentine have a tendency to be more concentrated along the mineralization front (MARTENS 1968; GOLDBERG et al. 1978; TAKAGI et al. 1981). Using indirect immunofluorescence with light microscopy, LINDE et al. (1982b) localized fibronectin in the pulp as well as in the mantle predentine. Fibronectin was absent from circumpulpal predentine and the dentine in its entirety. The presence of fibronectin in predentine and its absence in dentine was also reported by THESLEFF (1978) and THESLEFF et al. (1981). Using monoclonal antibody immunofluorescence, CONNOR et al. (1984) noted fibronectin labeling in predentine as well as in the lumen of dentinal tubules. The pulp was not labeled, nor were the odontoblasts and their processes, except for the region in contact with the predentine. These contradictory observations need clarification.

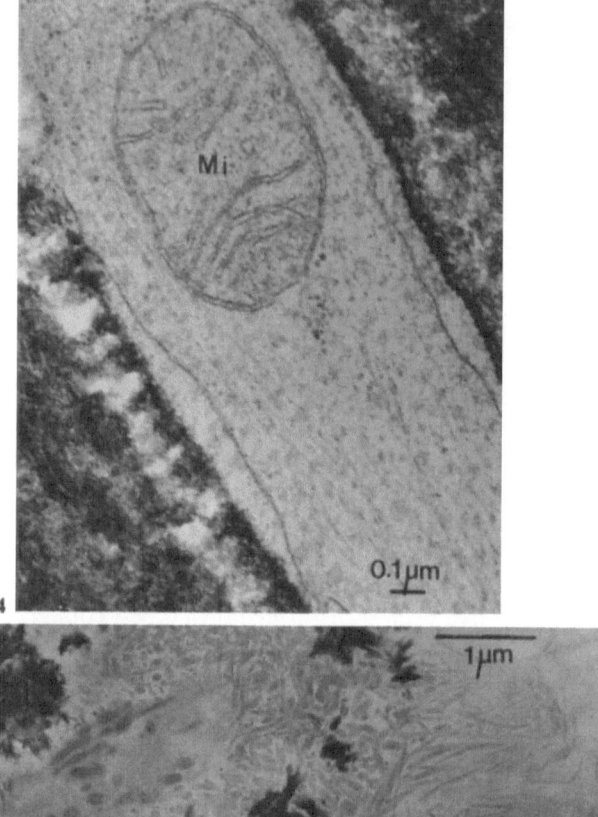

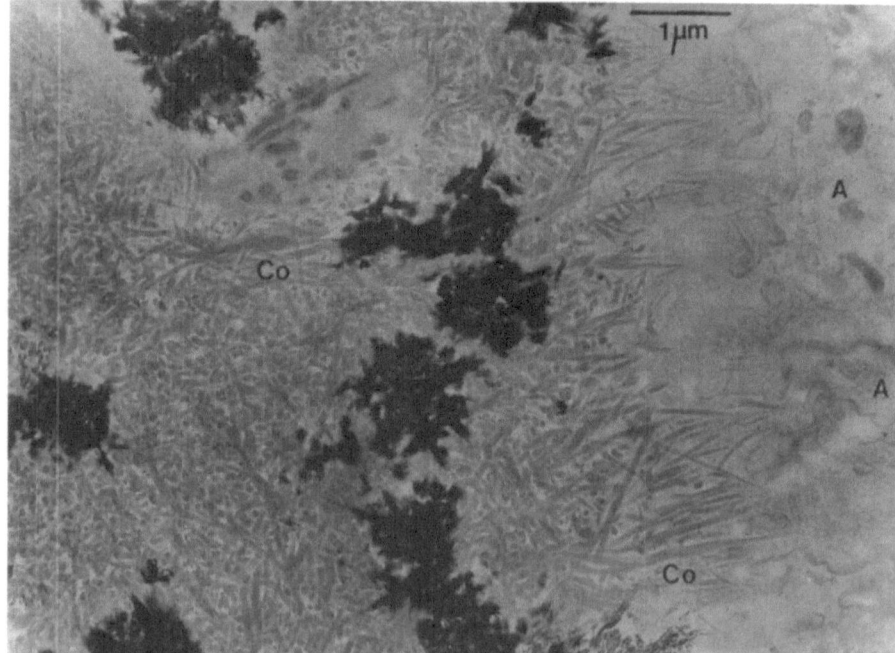

Fig. 64. Presence of a mitochondrion in the odontoblastic process within a dentinal tubule of a newborn cat. *Mi* microfilaments

Fig. 65. Initial mineralization of the predentine layer adjacent to the ameloblast layer (*A*). Apatite crystal deposits appear as irregularly disseminated dense foci in the collagenous matrix. Apparent absence of matrix vesicles. *Co* Collagen fibrils

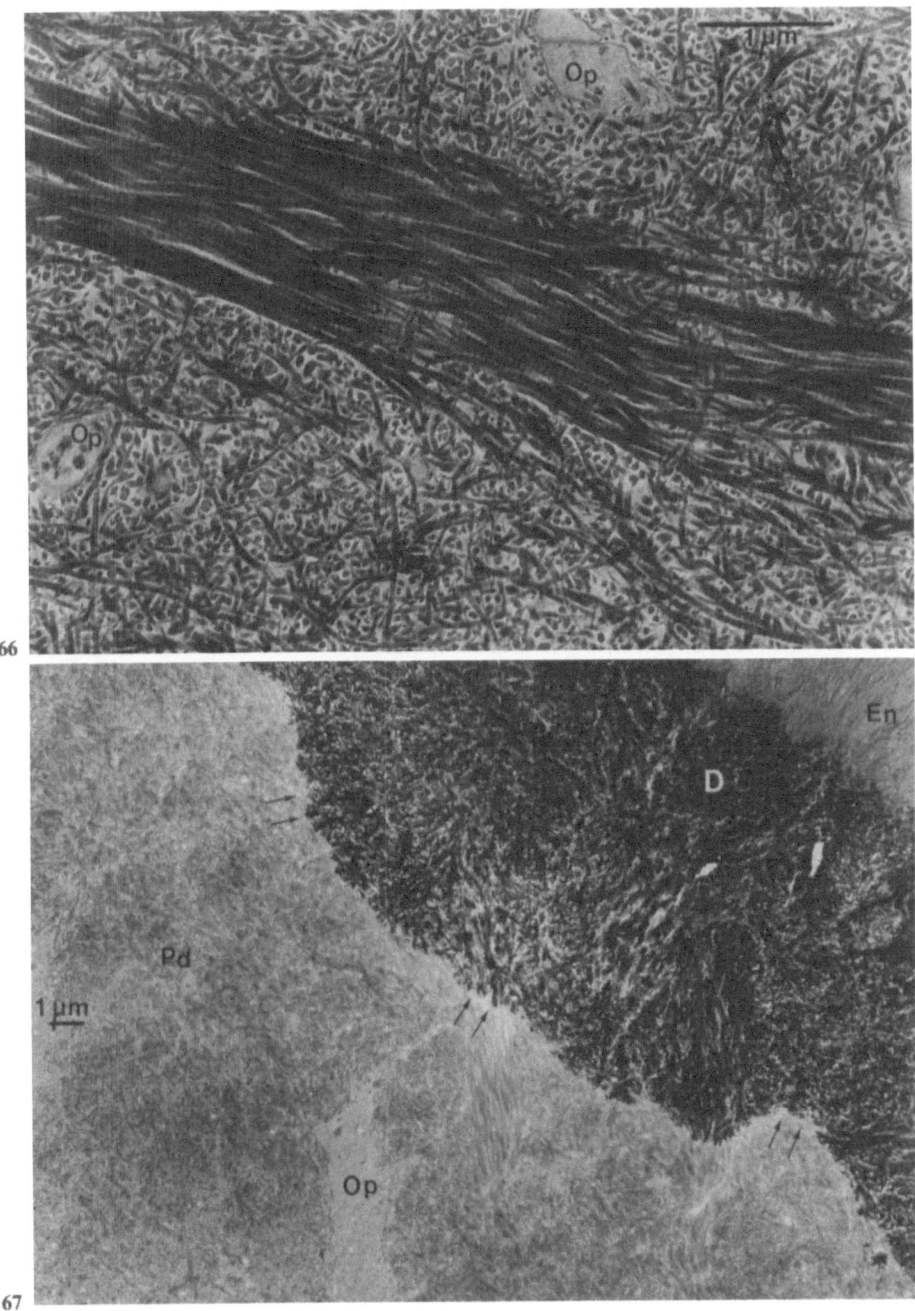

Fig. 66. A large bundle of collagen fibrils in the predentine layer in a molar of a newborn cat. *Op* Odontoblastic process

Fig. 67. First calcified layer of mantle dentine (*D*) adjacent to calcifying enamel (*En*). A sharply defined calcification front (*arrows*) delineates this initial dentine layer from predentine (*Pd*). *Op* Odontoblastic process

It has already been mentioned that γ-carboxyglutamic acid (GLA)-containing proteins as well as phosphoproteins are secreted into the predentine where they are then found primarily near the mineralization front. The levels of lipids are much higher in predentine than in dentine. In the latter, lipids are minor components, comprizing about 0.2% to 0.3% of the dry weight (Wuthier 1984). These lipids, primarily the phospholipids, seem to be involved in the mineralization process.

Since the time of Weidenreich (1925) the first thin layer of dentine formed has been called "mantle dentine", whereas the principal, central part has been known as "circumpulpal dentine". Structural differences are evident when comparing the peripheral and central regions of the dentine. In mantle predentine, along with a collageneous and non-collageneous organic matrix, matrix vesicles are found. The latter are absent in circumpulpal dentine. In the initial stages of dentinogenesis, when the first predentine layer deposited along the basement membrane in the *lamina rara interna* is only about 35 μm wide, the continuity of the basement membrane is interrupted by occasional long, thin preameloblast processes that extend across the future dentinoenamel junction (Silva and Kalis 1972). As the protrusions become longer and more numerous, the basement membrane is disrupted and eventually disappears. Heterotopic cell contacts between cell processes of odontoblasts and ameloblasts thus arise (Frank and Nalbandian 1967; Lester 1969; Silva and Kalis 1972; Sisca and Provenza 1972; Slavkin and Bringas 1976; Katchburian and Burgess 1977; Meyer et al. 1977). The processes of these cell types come into close approximation with each other, but their membranes never fuse or form any known type of cell junction (Burgess and Katchburian 1982). As already pointed out, these contacts may play a role in the inductive mechanisms governing the cytodifferentiation of odontogenic cells.

In the mantle predentine, the first collagen bundles deposited have a fan-like arrangement, perpendicular to the basement membrane (Fig. 65). Large fibrils with this disposition were first observed under the electron microscope by Nylen and Scott (1958, 1960). The origin of these collagen fibrils has been a matter of debate ever since the work of Von Korff (1906) who thought that they arose from the pulpal cells in the subodontoblastic zone, rather than from the odontoblasts. This issue will be covered in greater detail in the section on mantle dentine.

In the circumpulpal predentine, a network of finer collagen fibrils is observed. These are interwoven and generally smaller and more closely packed than those in the mantle predentine (Figs. 41, 43, 46, 49, 73). They are generally aligned at right angles to the odontoblast processes. No matrix vesicles are present here (Figs. 40, 66).

H. The Process of Mineralization

The process of dentine mineralization is characterized by a progressive growth of apatite crystals in a collagenous matrix. In the initial stages of mantle dentine formation, mineralization is associated with matrix vesicles. These vesicles, however, are absent during the formation of circumpulpal dentine. All the calcified tissues of the skeleton and teeth share a common inorganic crystalline phase similar to hydroxyapatite: $Ca_{10}(PO_4)_6(OH)_2$. WATSON and AVERY (1954), JOHANSEN and PARKS (1962), and JOHANSEN (1964) were among the first to describe the mature apatite crystals in dentine as having a platelet form. By high resolution transmission electron microscopy, it has been possible to resolve periodic lattice images in the dentine apatite crystals (SELVIG 1970; SPECTOR 1975; VOEGEL and FRANK 1977; FRANK and VOEGEL 1978). In these demonstrations the represented planes are nearly parallel to the electron beam, hence the images obtained can be considered as projections onto the screen of the various planes of the crystal. The periodic lattice images or fringes has allowed determination of the precise size and shape of the crystals in relation to their crystallographic axes. The visualization of three sets of lattice planes equivalent to the (100) planes, forming 60° angles and with 8.2 Å equidistances, has allowed precise determination of the plane of crystal sectioning. In other words, when hexagonal fringe patterns are visualized within the dentine crystal, it can only mean that a section was made strictly perpendicular to the c-axis of the crystal. It was on such sections that the plate-like shape of the dentine apatite crystal was confirmed (VOEGEL and FRANK 1977; FRANK and VOEGEL 1978).

It is probable that the development of the flattened hexagonal prism shape of the mature apatite crystal of dentine is a gradual process. However, detailed high resolution images of crystals in the initial stages of crystal formation are, as yet, a missing link in this story. The crystalline growth is probably similar to the steps observed in enamel, where initially a conspicuous growth occurs along the length (c-axis) and along the width of the crystal followed by a slow growth in thickness. The growth appears to be related to a progressive piling up of unit cells (DACULSI and KEREBEL 1978; WEISS et al. 1981).

It has been proposed that the central dark line observed in thin sections of enamel and dentine crystallites is the site of crystal initiation formed by the incorporation of carbonate (MARSHALL and LAWLESS 1981; NAKAHARA 1982). According to KAKEI and NAKAHARA (1983) the carbonate ions are supplied at the calcification front by means of carbonic anhydrase activity which is localized in areas of initial mineralization in developing dentine. Another possible explanation of the dark line in biological apatite crystals is the presence of a twin boundary with a twin plane parallel to the ($1\bar{1}00$) crystallographic planes (BRÈS et al. 1986).

Studying the mineralization front in rat incisor predentine with electron probe X-ray microanalysis, HÖHLING et al. (1967, 1984) noted a high content of phosphorus before mineralization, suggesting that the phosphate groups are incorporated into the organic predentine matrix at an early stage. The next phase leading to mineralization entailed a sharp increase in the calcium weight-fraction, while the phosphorus weight-fraction changed only slightly. Initially the Ca/P ratios remained far below those of the mineral being formed in dentine in few microns away. Only in later stages of mineralization did the Ca and P concentrations increase in parallel.

Although hydroxyapatite represents the primary mineral component of dentine, considerable interest exists regarding the presence and distribution of other calcium-phosphate minerals which may act as precursors of the apatite phase. Among such minerals amorphous calcium phosphate, octocalcium phosphate and brushite have been suggested as being precursors in in vitro and in vivo conditions (Posner and Tannenbaum 1984).

The most widely accepted hypothesis to account for maturational changes in the X-ray diffraction characteristics of calcified tissue mineral has been "the amorphous calcium phosphate theory" (Quinaux and Richelle 1967; Termine 1972) which postulated that an initial amorphous calcium phosphate solid phase is deposited in the organic matrix and gradually converted to poorly crystalline hydroxyapatite. However, Glimcher (1984), based on X-ray radial distribution function analysis and ^{31}P nuclear magnetic resonance studies, concluded that a solid phase of amorphous calcium phosphate did not exist in any significant degree in bone mineral.

Based on theoretical, physical, and biochemical considerations, octocalcium phosphate $(Ca_8H_2(PO_4)_6 \cdot 5H_2O = OCP)$ has been proposed as the first crystalline phase of calcium phosphate occurring during mineralization (Brown 1962, 1965). Octocalcium phosphate, having close structural similarity with hydroxyapatite, is presumed to be transformed into the latter phase by hydrolysis. An indication of the presence of octocalcium phosphate during apatite crystal growth in amelogenesis has been provided by high resolution electron microscopy of the periodic lattice fringe patterns (Weiss et al. 1981).

Roufosse et al. (1979) demonstrated brushite to be a major component of the earliest solid mineral in embryonic chick bone. Glick (1982) using differential density centrifugation demonstrated the presence of brushite in mineralizing dentine. Further work is needed regarding the possible role of brushite in the formation of hydroxyapatite.

In the intertubular dentine, the apatite crystals appear to be deposited between, at the surface, and within the collagen fibrils. Most of the crystals seem to be arranged with the same axial periodicity as the collagen molecules and there exists a parallelism between the c-axis of the apatite crystals and the long axes of the collagen fibrils. Accordingly, when dentine is mineralized, the inorganic crystal deposition occurs in an ordered fashion with respect to the periodic cross-striations of the collagen. In bone and dentine, the apatite crystals located in the collagen fibrils are initially deposited in the "holes" existing in the collagen fibrils and later spread to "pores" within the same fibrils (Glimcher 1982). Low angle X-ray diffraction and neutron diffraction studies (White et al. 1977; Berthet-Colominas et al. 1979) have confirmed that the mineral particles are initially deposited within the collagen holes and later within the pores, eventually filling up all the available intermolecular spaces within a collagen fibril with a solid phase of calcium phosphates.

Despite the significant progress made, the exact biochemical mechanisms involved in the deposition of apatite crystals during mineralization are poorly understood and agreement on the physico-chemical conditions governing mineralization of calcified tissues, including dentine has not been reached. Some of the notable observations and concepts regarding mineralization will now be reviewed.

I. Matrix Vesicles

In the early stages of mantle predentine formation, matrix vesicles, 50–200 nm in diameter, have been described in the collagenous organic matrix. It has already been mentioned in the section on epithelio-mesenchymal interactions that these matrix vesicles were thought to play a role in heterotypic cell communication between odontoblasts and ameloblasts (SLAVKIN 1972; KARDOS and HUBBARD 1982). Fundamentally, however, they have been considered as primary sites of mineral nucleation, and a number of authors have described them in mantle predentine during the initial mineralization stages of dentine (CROISSANT 1971; BERNARD 1972; EISENMANN and GLICK 1972; SISCA and PROVENZA 1972; SLAVKIN et al. 1972; KATCHBURIAN 1972; LARSSON and BLOOM 1973; NYGREN et al. 1976; ORAMS 1978; TAKAGI et al. 1981; KARDOS and HUBBARD 1982; KATCHBURIAN and SEVERS 1982; GOLDBERG and SEPTIER 1983; BONUCCI 1984; KOGAYA and FURUHASHI 1985).

The matrix vesicles are thought to be produced by the odontoblasts and their processes. This would allow some degree of cellular control over the mineralization process. KATCHBURIAN and SEVERS (1982), using a freeze-fracture electron microscopic technique, showed that some vesicles arise by budding off from the odontoblasts surface. The matrix vesicles are limited by a trilaminar membrane and have an amorphous content. Biochemical and histochemical studies have shown that the vesicles contain glycoproteins and a variety of lipids such as free fatty acids, cholesterol, sphingomymelin, glycolipids, and phosphatidylserine (BONUCCI 1984). Various enzyme activities are found in these vesicles: alkaline phosphatase, pyrophosphatase, adenosine triphosphatase, adenylcyclase, and phosphodiesterase, as well as ç-AMP. A high calcium content has also been reported. The outer layer of the matrix vesicles is coated with acidic proteoglycans (KOGAYA and FURUHASHI 1985).

Mineralization starts inside the matrix vesicle as a single needle-like crystal, presumably of hydroxyapatite. The growth of this initial crystal, followed by further growth of closely associated needle-shaped crystals, results in rupture of the vesicle membrane. Further mineralization then spreads circumferentially to form small noduli that eventually fuse into the initial front of mantle dentine. Later the crystals become associated with collagen. Whereas matrix vesicles are noted to serve as early loci of calcification in mantle dentine, they are not observed during the calcification of circumpulpal dentine. Similar observations have been made for bone and cartilage.

II. Alkaline Phosphatases

ROBISON (1923) described phosphatase (later called alkaline phosphatase) in various calcifying sites and proposed that they acted upon hexose phosphate esters, thus liberating phosphorus. The solubility product of calcium phosphate would be exceeded and precipitation of calcium salts could occur. In spite of the many shortcomings in this hypothesis, the chief one being the scarcity of hexose phosphate esters, the notion was popular for over 30 years, until it became clear that mineralization was not just a precipitation phenomenon. The recognition that various phosphatases were present in non-calcifying tissues also reduced the credibility of this explanation.

The presence of phosphatases in odontoblasts, matrix vesicles and loci of calcification has stimulated various biochemical and ultrastructural investigations (Linde 1982). Aside from the function considered in the original Robison theory, several other roles have been ascribed to alkaline phosphatase. Since alkaline phosphatase hydrolyzes a number of phosphate esters, some of them known to poison mineral formation in vitro, it has been suggested that the normal function of alkaline phosphatase would be to remove these from the mineralization area (Neuman and Neuman 1958).

In addition to alkaline phosphatase, a separate adenosine triphosphate (ATP) hydrolyzing enzyme was demonstrated in rat incisor odontoblasts (Linde and Magnusson 1975). This Ca^{2+} and Mg^{2+}-activated ATPase was biochemically separated from odontoblasts, together with alkaline phosphatase (Linde and Grännström 1978). The existence of a Ca^{2+}-activated ATPase in odontoblasts implies a role for this enzyme in the turnover of Ca^{2+} ions during the mineralization process. However, as stated by Linde (1982) in a general review, it can be concluded that the function which alkaline phosphatase may play in biological calcification is unclear.

III. The Concept of Epitaxy

The term epitaxy was introduced by Royer (1928), Professor of Mineralogy at the University of Strasbourg, and was intended to describe the oriented growth of two different crystals, one over the other. A number of conditions had to be satisfied in epitaxy: the lattice spacings in the two crystal faces in contact had to be identical or in a simple numerical ratio; ions in the growing crystal had to be of the same sign as the host crystal; and chemical bonding of crystal elements had to be identical in the two crystals.

In 1958, Neuman and Neuman applied the mineralogical concept of epitaxy to biological mineralization and postulated that apatite crystal growth could be initiated on collagen fibrils by a process of epitaxy. The calcium phosphate precipitation theory of Robison (1923) was replaced by a theory of epitactic nucleation. An organic crystalline macromolecule should be present in the mineralizing matrix. This preexisting macromolecule would induce the seeding or nucleation of the very first unit cells of an inorganic calcium phosphate crystallite. Glimcher (1960) considered that collagen was the macromolecule on which apatite was nucleated. Reconstituted collagen, often originating from tissues that do not normally calcify, nucleated hydroxyapatite crystals, in vitro, provided that the collagen had a 640 Å periodicity. Glimcher (1960) and numerous other workers persisted with the view that during in vivo calcification, crystallites were deposited in close association with collagen fibrils (Fig. 75).

However, the initial deposition of hydroxyapatite crystals in matrix vesicles during dentine, bone and cartilage calcification, as well as difficulties fitting biological mineralization to the precise conditions defined by ROYER (1928) for epitaxy, have cast some doubts on this concept. In recent years alternative concepts have been emphasized.

IV. The Role of Various Non-Collagenous Proteins

For a number of years, many investigators have contended that the proteoglycans (PG) and glycosaminoglycans (GAG) secreted by osteoblasts and odontoblasts are lost and removed from the mineralizing tissue prior to the mineralization process (LINDE 1982, 1984).

By means of electron microprobe measurements, it was shown that the concentration of sulfur, presumably representing sulfated PG, is lower in rat incisor dentine than in predentine and that the decrease in sulfur occurs slightly beyond the mineralization front (NICHOLSON et al. 1977). A decrease in PG at the dentine-predentine mineralizing front has also been demonstrated by electron microscopy using cationic dyes to visualize the PG (TAKAGI et al. 1981; GOLDBERG and SEPTIER 1983).

The mineralization of dentine appears to be correlated with complete loss of chondroitin-6-sulfate and partial loss of chondroitin-4-sulfate. Together with the observation that predentine contains significantly more PG than dentine, this suggests that some selective biodegradation and removal of PGs-containing chondroitin-6-sulfate occurs when predentine is transformed to dentine (ENGFELDT and HJERPE 1972).

Degradation of PG has been suggested to be due to enzymatic activity. LINDE and PERSLIDEN (1977) demonstrated a high cathepsin D activity in isolated odontoblasts. By electron microscopic immunochemistry, cathepsin D was localized in intracellular odontoblast lysosomes as well as in the predentine matrix, where PG degradation occurs (NYGREN et al. 1979). In addition it seems that glycosidases capable of degrading GAGs are also present in the odontoblast-predentine region (ENGSTRÖM et al. 1976).

BUTLER (1984b) raised questions about the validity of the studies suggesting a loss of PG at the mineralization front, mainly because one is not fully assured that the reagents used actually fix proteins within the mineralized areas to the same extent as those in unmineralized regions. Using injections of ^{35}S-sulfate into young rats, BUTLER (1984b) quantitated the amount of labeled proteoglycans in unmineralized and mineralized portions of incisors. The biochemical data suggest that 35–45% of the newly synthesized PG is removed from the tissue at a time consistent with that period in which mineralization occurs but do not show conclusively that this loss is related to calcification. Additional data are needed, according to BUTLER (1984b) to show a causal relationship between PG removal and deposition of minerals.

As already mentioned, γ-carboxyglutamic acid-containing proteins (Gla), identified in bone as osteocalcin (HAUSCHKA et al. 1975), have been demonstrated in dentine by LINDE et al. (1980, 1982a). Due to its ability to bind Ca^{2+}, the Gla-protein is considered to play a role in the onset of mineralization and the nucleation of apatite crystals.

Dentine phosphoproteins referred to as phosphoryns (DiMuzio and Veis 1978) are generally considered as directing calcification of dentine and bone, but the exact mechanism of involvement is a matter for speculation. Veis et al. (1982) believe that free and collagen-bound phosphoryns bind large amounts of calcium ions with relatively high affinity and enhance the local concentration of calcium and phosphate. This local effect thus nucleates, facilitates and maintains growth of apatite crystals. Lee et al. (1983) demonstrated that an isolated bovine dentine phosphoprotein binds both calcium and inorganic orthophosphate ions and therefore has the requisite physical-chemical properties necessary to facilitate the heterogeneous nucleation of $CaPO_4$ solid phase from solution during mineralization.

A role of mineralization inhibitor has been demonstrated in vitro for a high molecular weight phosphoryn by Termine and Coon (1976). For Davis and Cavanagh (1982), it is not the dentine phosphoryn, but an associated low molecular weight polypeptide fragment which is a potent inhibitor of collagen mineralization in vitro.

V. The Role of Lipids

The first indication that lipids might have a special role in dentine formation was the observation made by Irving (1958, 1959) that a zone of intense sudanophilia is present at the predentine-dentine junction. This sudanophilia was shown to result from the presence of acidic phospholipids which, because of their affinity for Ca^{2+} were found at the site of mineral formation (Wuthier and Irving 1964; Shapiro et al. 1966). Biochemical investigations have shown that the composition of phospholipids extracted from dentine before and after demineralization was strikingly different (Shapiro et al. 1966; Prout et al. 1973; Wuthier 1984).

The acidic phospholipids, phosphatidylserine, phosphatidylinositol, and phosphatidic acid were tightly complexed with the dentine mineral and could not be extracted unless the tissue were demineralized. The possible role of lipids in mineralization, centers on the ability of certain anionic phospholipids to bind Ca^{2+} with moderate affinity and selectivity (Cotmore et al. 1971; Wuthier 1984). Binding of Ca^{2+} to these acidic phospholipids is enhanced markedly by the presence of inorganic orthophosphate, resulting in the formation of chemically stable phospholipid-calcium-inorganic orthophosphate complexes (Cotmore et al. 1971). In fact, phosphatidylserine, a lipid with strong affinity for Ca^{2+} is consistently found in lipid-calcium-inorganic orthophosphate complexes isolated from calcifying tissues, as well as in matrix vesicles (Peress et al. 1974; Wuthier 1984).

As mentioned previously, matrix vesicles are observed in the early stages of mantle dentine formation and it is highly probable that part of the sudanophilia seen at the predentine-dentine junction at this stage can be related to these matrix vesicles. However, at later stages of circumpulpal dentinogenesis, matrix vesicles are no longer visible and the intensity of the lipid reaction is still observed at the mineralization front. It is therefore unlikely that matrix vesicles alone are responsible for the sudanophilia observed at the predentine-dentine junction.

On the other hand, in vitro studies have repeatedly shown that acidic phospholipids alone or complexed as proteolipids or as lipid-calcium-inorganic orthophosphate complexes are able to cause nucleation of hydroxyapatite from metastable calcium phosphate solutions (ENNEVER et al. 1984; WUTHIER 1984). Treatment of phosphatidylserine, the lipid-calcium-inorganic orthophosphate complexes or even matrix vesicles with phospholipases caused either marked inhibition or total blockage of mineralization induced in vitro by these structures (WUTHIER 1984).

Notwithstanding the important advances described above, it should be apparent from this review that the factors involved in dentine mineralization are only partially understood and further work is necessary to clarify our concepts about this complex aspect of dentine formation.

J. The Development of Coronal Dentine

Since the writings of WEIDENREICH (1925), as mentioned earlier, the initial layer of dentine formed has been referred to as mantle dentine, whereas the central part comprising the bulk of the tooth has been known as circumpulpal dentine. The development of coronal dentine follows an incremental pattern. Formation of a layer of organic matrix, the predentine, is followed by its mineralization, resulting in a layer of dentine. With continuation of these sequential phenomena dentine formation proceeds towards completion. It has been estimated, in man, that the rate of dentine deposition in deciduous teeth is about 4 μm in 24 hours, but the rate can vary somewhat in different parts of the tooth (SCHOUR and PONCHER 1937a). Initially, the so-called intertubular dentine is formed, delineating the dentinal tubules that contain the odontoblastic processes. This is followed by development of peritubular dentine which fills in and narrows the lumens of the dentinal tubules. Dentine starts to form in the tip of the crown region, developing in thickness and paralleling the dentinoenamel junction in a centripetal, pulpal direction. According to SILVA and KAILIS (1972), initial calcification starts when predentine is about 75 μm thick.

I. Intertubular Dentine

1. Mantle Dentine and the Dentinoenamel Junction

Mantle dentine, the initial zone of dentine deposited, is very thin. According to TEN CATE (1980) and JONES and BOYDE (1984), it is not greater than a few tenths of a μm in thickness. The existence of this layer was confirmed by polarized light microscopy (SCHMIDT and KEIL 1971) and by electron microscopy. The mantle dentine contains coarse fibrils oriented perpendicular to the dentinoenamel junction (Fig. 65) while the circumpulpal dentine contains finer collagen fibrils which lie more or less parallel to this junction (Fig. 65).

The origin of the first formed, coarse collagen fibrils of mantle dentine has been a matter of debate ever since VON KORFF (1905) described some conspicuous argyrophilic fibers, now known as VON KORFF's fibers, in the mantle region. These fibers stained after silver impregnation and appeared to arise in the pulp deep to the odontoblasts, often passing in a spiral course between the odontoblasts and fanning out to form the initial mantle dentine matrix. They have sometimes been described as "corkscrew" fibers. Because VON KORFF's fibers appear to be present only during early dentinogenesis, the concept of a dual origin of dentine matrix collagen arose, a concept whereby mantle dentine collagen is related to VON KORFF's fibers originating from subodontoblastic pulpal cells, while circumpulpal dentine collagen is elaborated by the odontoblasts. Using light and electron microscopy, TEN CATE et al. (1970) have indicated that the presence of argyrophilic fibers between odontoblasts is probably an artifact caused by the binding of silver to the ground substance in this location.

With transmission electron microscopy, some investigators did not observe collagen fibrils between the odontoblasts (LENZ 1959; TEN CATE 1978), while others found a few bundles (FRANK 1965, 1968a; NALBANDIAN 1968; REITH 1968; WHITTAKER and ADAMS 1972). It should be noted that the mere finding of collagen fibrils between the odontoblasts does not, in any case, afford proof that the collagen 1) is of pulpal origin and 2) will be incorporated into mantle dentine. With respect to the latter possibility, JONES and BOYDE (1984) considered it unlikely that such collagen could pass through the junctional complexes of the odontoblasts adjacent to the predentine matrix.

Along these lines, FRANK (1970b) observed a discrete labeling of the subodontoblastic pulpal cells with electron microscope autoradiography, after intravenous injection of ^{3}H-proline. However, this was unimpressive when compared to the high ^{3}H-proline uptake of the odontoblasts in early dentinogenesis. Finally, the genetic typing of collagens brought additional arguments for the exclusive odontoblastic origin of dentine collagen. Only type I collagen has been found in dentine. LECHNER and KALNITSKY (1981) isolated type I and type III collagen in the pulp of bovine teeth undergoing active, moderate and inactive dentinogenesis. The proportions of type I and type III collagen were determined by quantitative electrophoresis: type III comprised a constant 45 per cent of the total collagen at each stage. These findings indicate that pulpal collagen cannot be the immediate precursor of dentine collagen.

Proponents of the VON KORFF's fiber concept might argue that type I collagen could be selectively incorporated into dentine, whereas type III is excluded. However, this possibility was ruled out by LECHNER and KALNITZSKY (1981) who found that the type III/type I ratio in the pulp remains constant, even though the total content of pulp collagen increases from 15 or 24 per cent during development.

Other characteristics of the organic matrix of mantle dentine, as already mentioned, are the presence of matrix vesicles and small aperiodic filaments extending perpendicularly from the basement membrane in the *lamina rara interna*. With specific antibodies, LINDE et al. (1982b) demonstrated the presence of fibronectin exclusively in the mantle dentine and its absence in circumpulpal dentine.

The first dentine to mineralize is the mantle dentine below the tips of the cusps. Mineralization starts in relation to the matrix vesicles, with a progressive growth of apatite crystals that lead to the formation of inorganic noduli or globules which eventually become confluent. It is recognized that collagen will be secondarily involved in the mineralization process. On anorganic specimens examined with scanning electron microscopy, the globular or calcospherite feature of intertubular dentine mineralization is clearly displayed and the smallest and more numerous globules are observed in the peripheral region of dentine (JONES and BOYDE 1984).

The mineralization of circumpulpal intertubular dentine is continuous with that of the mantle dentine which, after all, is only a fraction of a micron thick (Figs. 65, 67, 70, 72). In the initial stages of dentine mineralization, the outer surface of the mineralized mantle dentine faces the enamel organ with its young differentiating ameloblasts and typical Tomes' processes (Fig. 70). In human dentine, cross-striated collagen fibrils have been observed in the extracellular spaces delineated by the calcified dentine matrix and the folds of the plasma membranes of the Tomes' processes, prior to the secretion of enamel matrix (WATSON 1960; RÖNNHOLM 1962; FRANK and NALBANDIAN 1967). These uncalcified collagen fibrils have also been observed during early coronal dentinogenesis in newborn cats (Figs. 70 and 71). They often appear to be oriented perpendicular to the surface of mantle dentine and occasionally seemed to be in continuity with already calcified dentinal collagen fibrils (Fig. 71).

The origin of these uncalcified collagen fibrils on the external surface of coronal mantle dentine is a matter of speculation. They could represent portions of the so-called VON KORFF's fibers which were not initially calcified or they could be collagen synthesized on the surface of already calcified mantle dentine from precursors exported by way of odontoblastic processes which are in very close proximity to these areas (Fig. 72).

Whatever the situation may be, the content of ameloblastic secretory granules located in the Tomes' process (Fig. 70) is subsequently discharged into the extracellular spaces by a process of exocytosis involving membrane fusion (FRANK and NALBANDIAN 1967). The secreted enamel matrix appears to intermingle with the uncalcified collagen fibrils, and mineralization of the area leads to the formation of the dentinoenamel junction and to the first increments of enamel (Fig. 72). The close relationship of uncalcified collagen fibrils and enamel matrix contributes to the physical cohesivenes of the dentinoenamel junction. With transmission electron microscopy, differentiating enamel apatite crystals can be recognized there (Fig. 69), since they are much longer than the dentinal apatite crystals. Various stages of dentinoenamel junction development, characterized by the appearance of inner enamel and subsequent formation of prismatic enamel, are illustrated in Figs. 67–72. The scalloped contour of the external surface of the dentinoenamel junction is particularly apparent in light microscopy (Fig. 79).

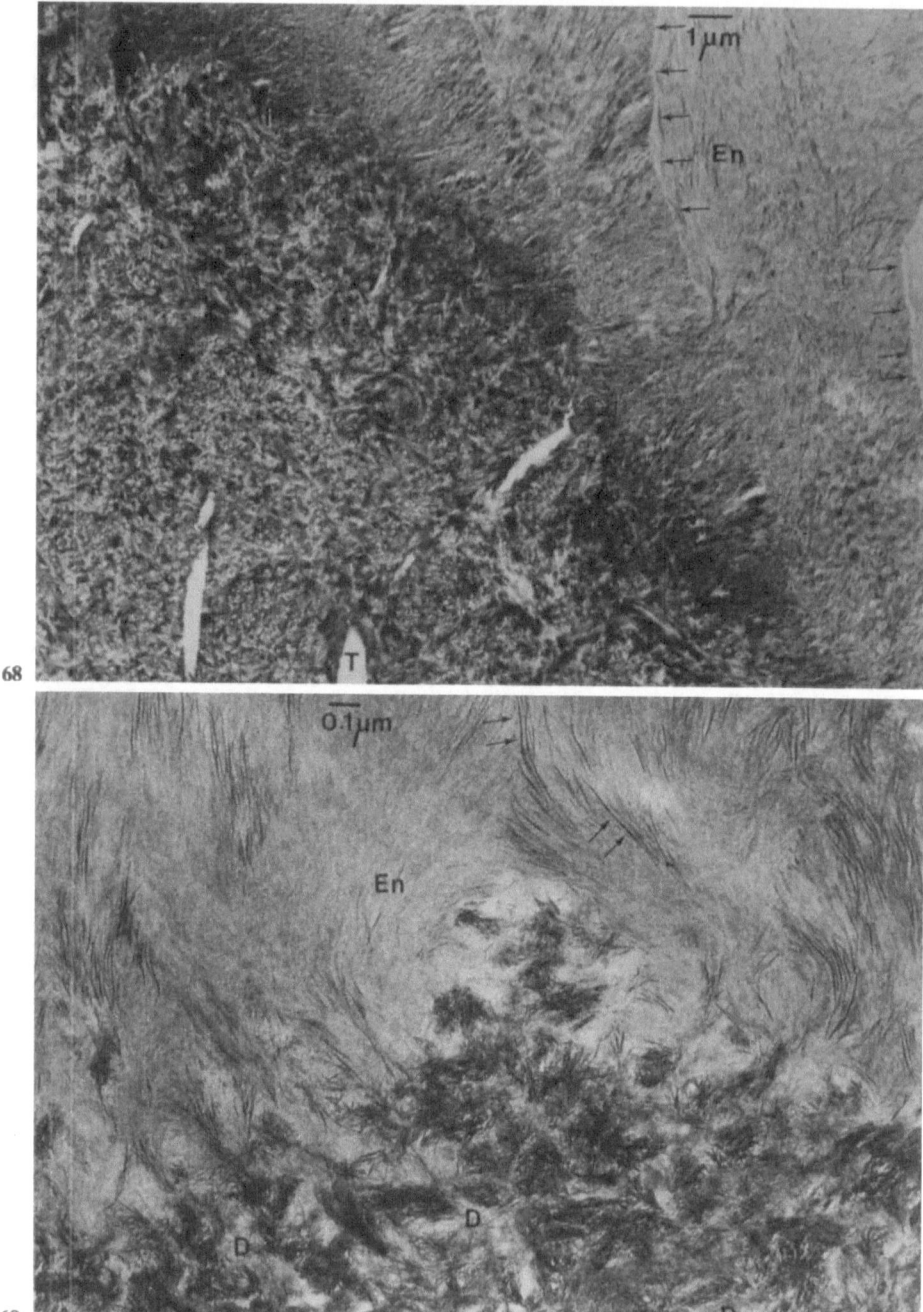

Fig. 68. Developing dentinoenamel junction in a central incisor of a 5-month-old human fetus. The intertubular dentine on the left contains dentinal tubules (*T*). Sheathes (*arrows*) delineate the enamel prisms (*En*)

Fig. 69. Higher enlargement of the dentino-enamel junction in a central incisor of 5-month-old human fetus. Mantle dentine (*D*) at this stage is more strongly calcified than the adjacent enamel (*En*), and the developing apatite crystals of the dentine are generally much shorter than the developing enamel crystal (*arrows*)

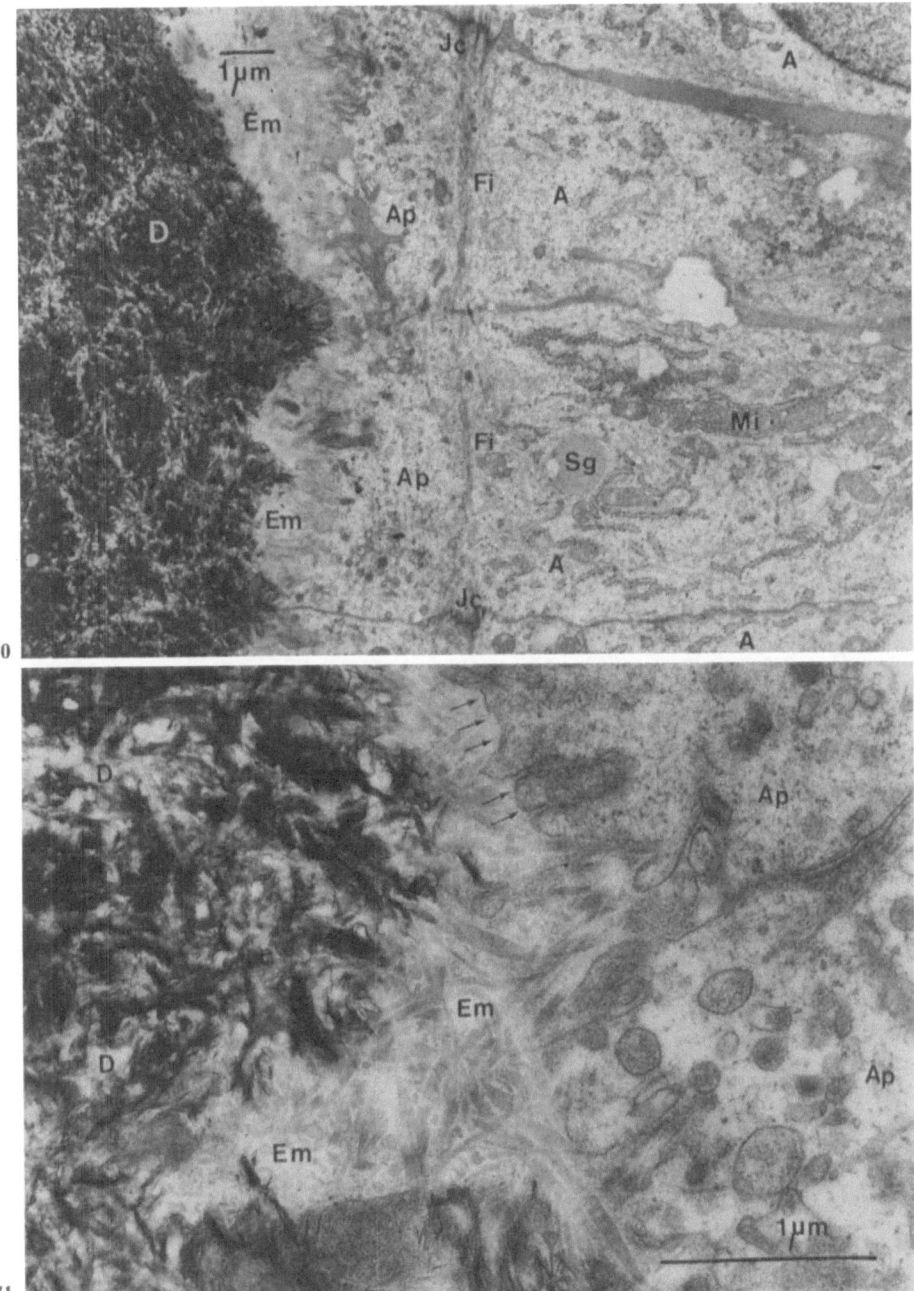

Fig. 70. Overview of the initial calcified layer of dentine (*D*) adjacent to the ameloblast layer (*A*) before the appearance of calcified enamel. Newborn cat molar. *A* Ameloblast cell body; *Sg* Secretory granule; *Mi* Mitochondria; *Jc* Junctional complex; *Ap* Ameloblast (Tomes') process; *Em* Extracellular matrix; *Fi* Bundle of tonofilaments

Fig. 71. Higher enlargement of enamel matrix (*Em*) located between the external surface of calcified mantle (*D*) and the ameloblast (Tomes') process (*Ap*). Uncalcified collagen fibrils are seen in the extracellular matrix. The plasma membrane of Tomes' process is folded (*arrows*). Newborn cat molar

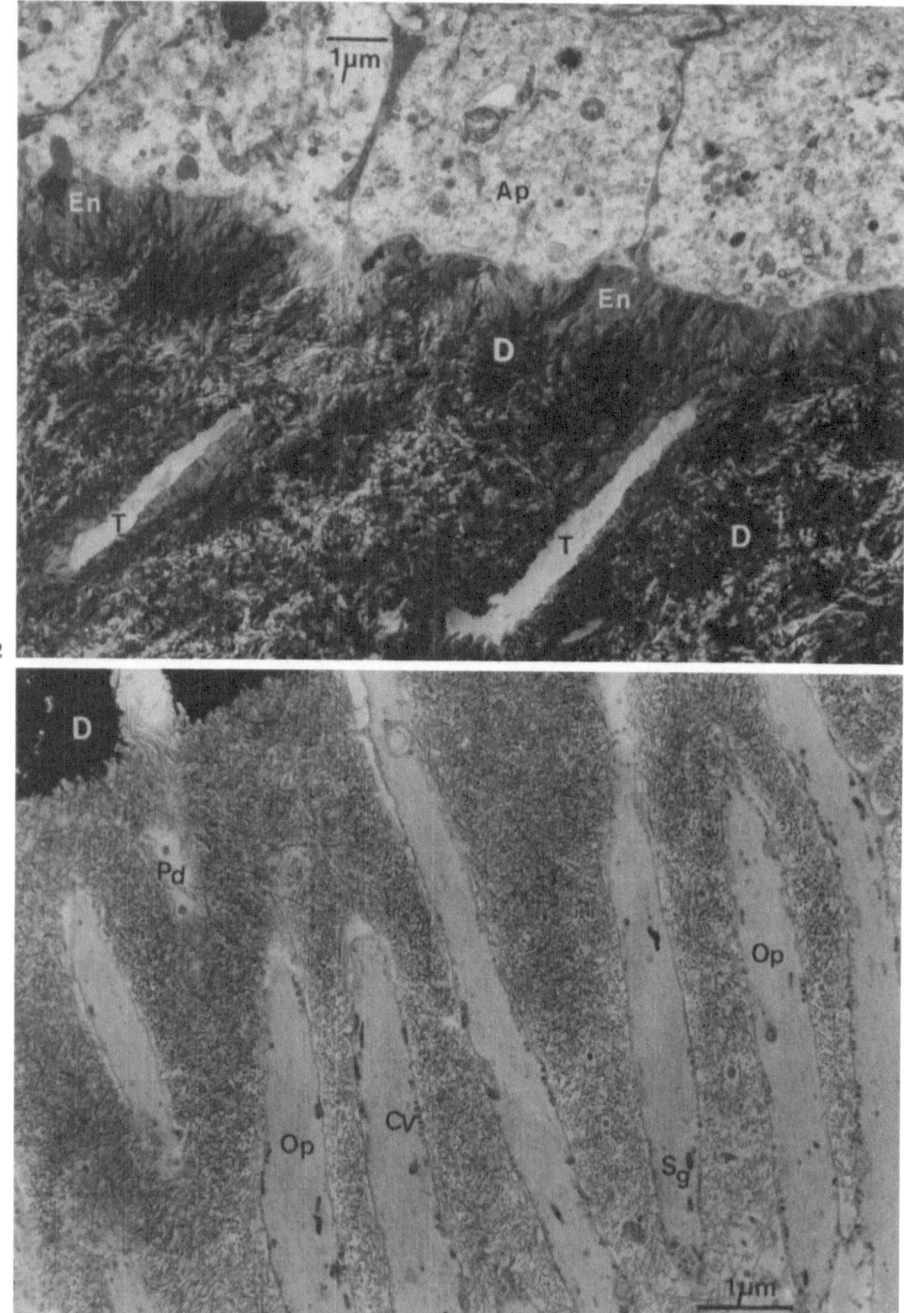

Fig. 72. Presence of a small layer of calcifying enamel (*En*) over the mineralized mantle dentin (*D*) in a molar of a newborn cat. *Ap* ameloblast (Tomes') process; *T* Dentinal tubule

Fig. 73. Dentine (*D*) and predentine (*Pd*) in an incisor of a newborn cat. The odontoblastic process (*Op*) contains numerous elongated secretory granules (*Sg*) as well as coated vesicles (*CV*)

2. Circumpulpal Dentine

The mineralization of the circumpulpal coronal dentine is a process directly continuous with the mineralization of the thin layer of mantle dentine. As already mentioned, circumpulpal dentine forms the bulk of the tooth. Matrix vesicles do not participate in its mineralization (Figs. 65, 67, 73–77). The collagen fibrils are smaller and largely oriented perpendicular to the dentinal tubules, although they are also interwoven in a transverse direction. Bundles of collagen fibrils grouped in parallel can also be observed in circumpulpal dentine (Figs. 66, 67).

A progressive growth of apatite crystals is observed in circumpulpal dentine. These crystals will finally assume the shape of platelets, or more precisely, flattened, hexagonal prisms, as demonstrated by high resolution electron microscopy (VOEGEL and FRANK 1977; FRANK and VOEGEL 1978). They will be deposited within, at the surface of, and between the collagen fibrils (Fig. 75). The apatite crystal growth proceeds in the circumpulpal dentine in spherical masses known as calcospherites, constituting individual calcification foci which will finally fuse (Fig. 65). Evidence from polarized light microscopy suggests that a proportion of the crystals in a calcospherite spread radially from the growth center (SCHMIDT 1969). With scanning electron microscopy of anorganic dentine, the size and shape of the calcospherites were clearly visualized (JONES and BOYDE 1984) and it was shown that in human dentine the calcospherites are smallest, most numerous and most spherical close to the peripheral limits of dentine. They increase in size and decrease in incidence in the pulpal direction. In the outer third of dentine, the shape of the completed calcospherites is a parabolic cone extending pulpward. The calcospherites in the inner half of the dentine are again smaller, yet not so small as those of the initial layer (JONES and BOYDE 1984). The globular character of circumpulpal dentine calcification is also clearly visible when the pulp is stripped away and the calcification front is observed by scanning electron microscopy (LESTER and BOYDE 1967). In human dentine this feature can be seen on either longitudinal (Fig. 80) or transverse (Fig. 81) decalcified sections prepared for light microscopy.

Incomplete fusion of calcospherites occurs in the outer third of the dentine. Consequently a series of interglobular spaces of Czermack results, generally aligned parallel to the dentinoenamel junction. Normally the interglobular spaces are relatively discrete in number and size. A significant increase in interglobular dentine is seen in cases of abnormal dentine mineralization such as vitamin D deficiency, vitamin D-resistant rickets or severe fluorosis (NIKIFORUK and FRASER 1979). Experimentally, abnormally large areas of interglobular dentine can be produced in chronic calcitonin deficiency (KLINE and THOMAS 1977) and in parathyroidectomized rats (SCHOUR et al. 1937b; BERNICK 1969).

A rhythmic pattern of deposition has been demonstrated in coronal as well as in root dentine. By tetracycline labeling, a circadian rhythm of dentine deposition can be seen associated with Von Ebner's incremental lines (KAWASAKI et al. 1977). In addition, ROSENBERG and SIMMONS (1980) identified, in rabbits injected with lead acetate during the day or at night, ultradian and infradian rhythms. The control of the rhythmic functioning of odontoblasts in dentine

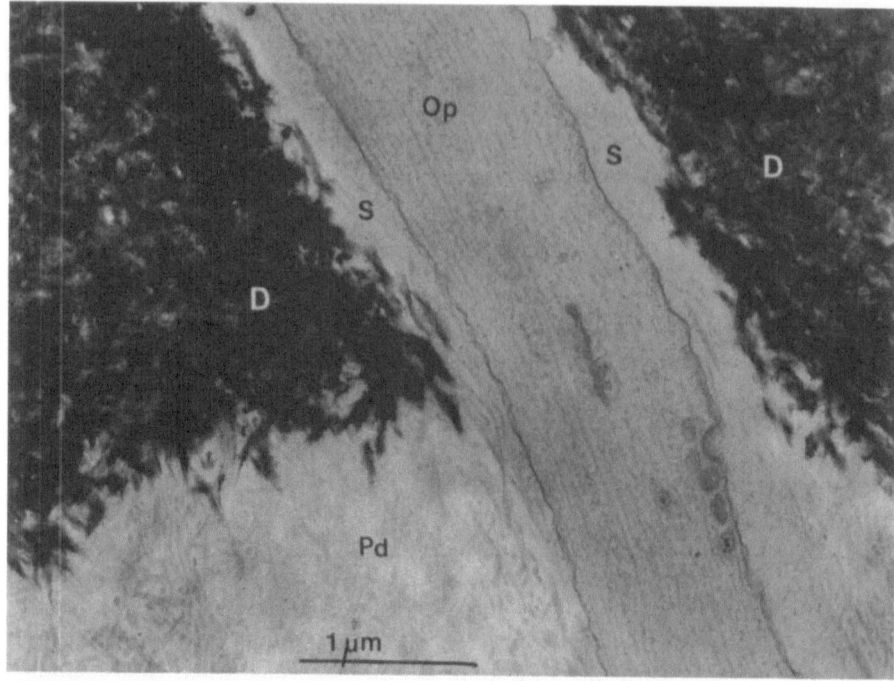

74

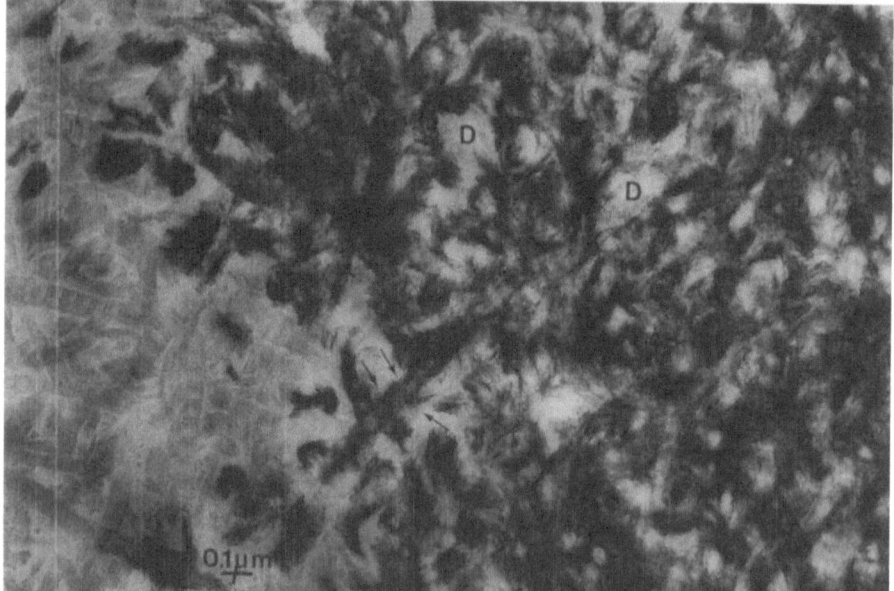

75

Fig. 74. View of the dentine-predentine junction in a molar of a newborn cat. A periodontoblastic space (*S*) is interposed between the odontoblastic process (*Op*) and the calcified intertubular dentine (*D*) forming the wall tubule. *Pd* Predentine

Fig. 75. Higher enlargement of the dentine-predentine junction. Apatite crystal deposition in dentine (*D*) seems to be deposited along collagen fibrils (*arrows*). Numerous scattered areas are still devoid of apatite deposition

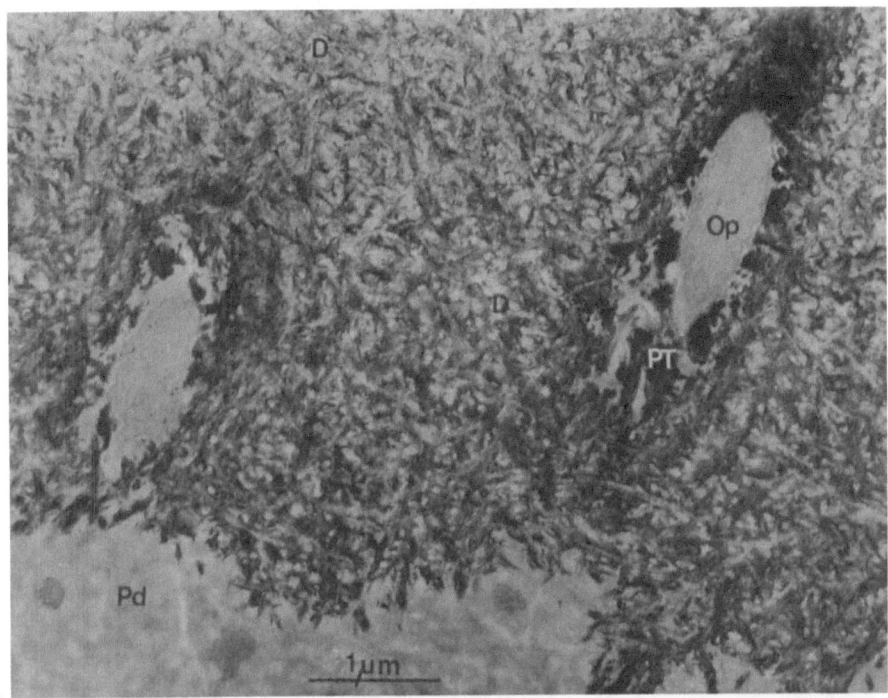

Fig. 76. Predentine-dentine junction in the newborn cat showing early formation of peritubular dentine *(PT)*. *Pd* Predentine; *D* Intertubular dentine; *Op* Odontoblastic process

Fig. 77. Presence of peritubular dentine *(PT)* in a transverse section of developing dentine. Incisor tooth germ of a 5-month-old fetus. *D* Intertubular dentine

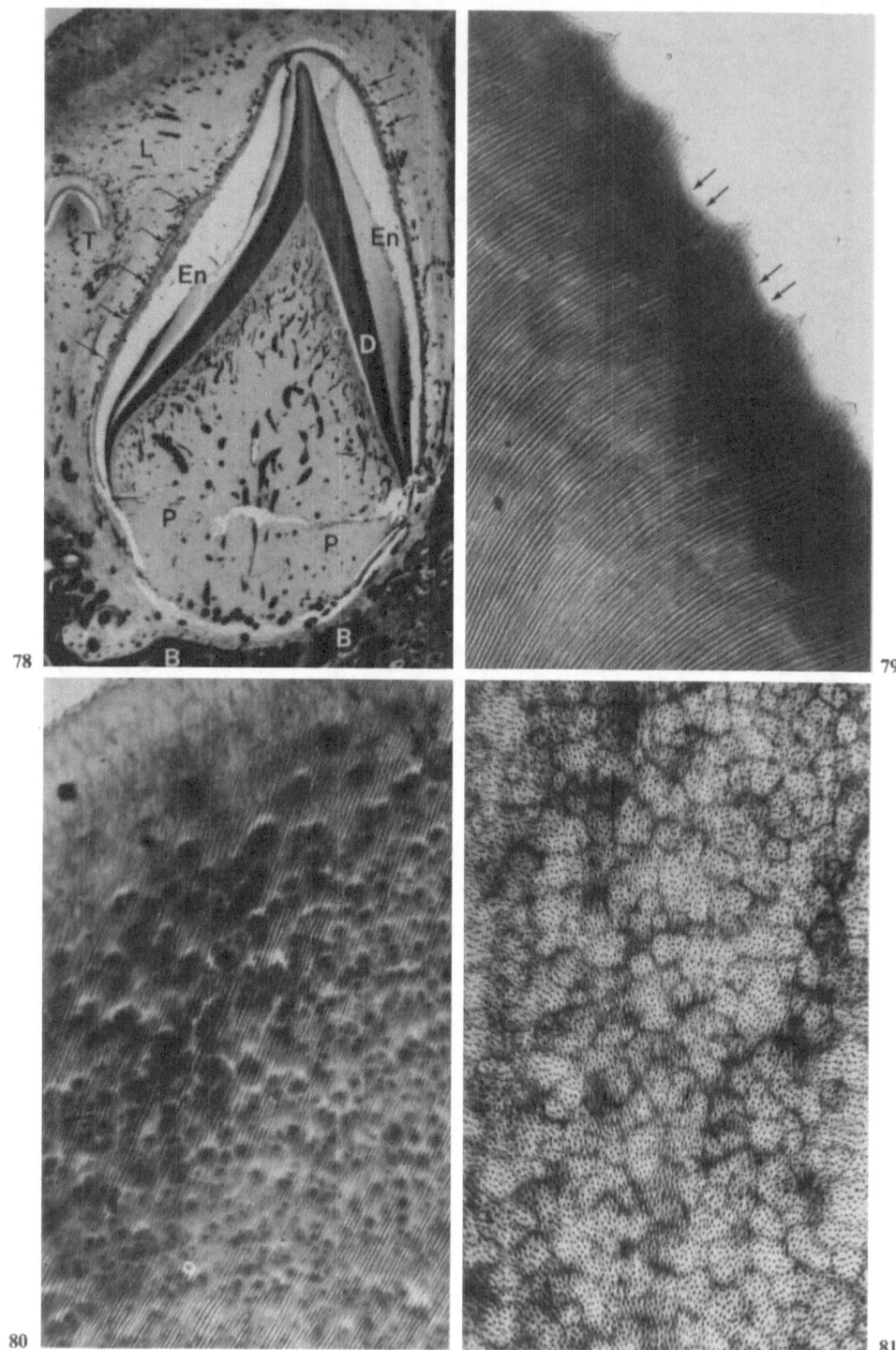

deposition seems to be a complex phenomenon which is only partially understood. Bovine growth hormone was found to increase dentine production slightly in normal rats and significantly in hypophysectomized rats (HANSSON et al. 1978). Along the same lines, ASHIDA (1983) demonstrated that the rhythmicity of daily dentine deposition in rats is influenced by experimental lesions of hypothalamic nuclei.

Proceeding in opposite directions from the dentinoenamel junction, coronal dentine develops in a centripetal, pulpal direction whereas enamel is laid down in a centrifugal direction towards the future crown surface. Formation of both hard tissues will progress until the cervical region is reached. A bucco-lingual section of a human deciduous lower incisor tooth germ (Fig. 78) shows the respective shape, size and relationship of the enamel, dentine and pulp at this stage, prior to development of root dentine.

II. Peritubular Dentine

When initially formed in the course of mineralization of mantle and circumpulpal dentine, the dentinal tubules are delineated by intertubular dentine and no peritubular dentine is present (Figs. 68, 72, 74). According to SYMONS (1961), the formation and calcification of the intertubular and peritubular matrices appear to be separate stages in dentine development.

In the initial formation of the dentinal tubule (Fig. 74), an organic periodontoblastic region or "space" separates the odontoblastic process from the intertubular dentine walls. This space contains a finely ground granular substance and a few collagen fibrils, often arranged parallel to the odontoblastic process. The periodontoblastic space is in direct continuity with the predentine matrix (Fig. 74).

The development of peritubular dentine, in fact, takes place in an *intratubular* location and occurs through centripetal mineral deposition, i.e. from the intertubular dentine walls towards the center of the tubule. This mineralization occurs at the expense of the periodontoblastic space which, as a result, becomes narrowed (Fig. 76). On transverse sections observed in transmission electron mi-

Fig. 78. Development of enamel (*En*) and coronal dentine (*D*) in a human deciduous tooth germ. The enamel is in contact with a reduced enamel organ (*arrows*). *B* Alveolar bone; *P* Pulp with blood vessels; *L* Lamina propria; *T* Tooth germ of the permanent incisor. × 12

Fig. 79. Longitudinal decalcified section of peripheral adult human coronal dentine (light microscopy). Presence of numerous dentinal tubules. Note the scalloped contours of the dentino-enamel junction (*arrows*). × 125

Fig. 80. Longitudinal decalcified section of peripheral adult human coronal dentine, stained with hematoxylin-eosin and showing the structure of interglobular dentine. × 320

Fig. 81. Transverse decalcified section of radicular human adult human dentine stained with hematoxylin-eosin and showing cross-cut dentinal tubules as well as a honey-combed network in the intertubular dentine related to the globular pattern of calcification. × 350

croscopy (Figs. 76 and 77), the peritubular dentine appears as an electron-dense annular region. When completely formed, peritubular dentine is about 40% more highly mineralized than intertubular dentine (Höhling et al. 1972). According to Lester and Boyde (1968), who observed peritubular dentine in fracture planes of dentine treated with ethylene diamine, the mineral fraction is made up of equidiameter mineral particles of roughly spherical or polygonal shape, forming a regular mosaic pattern in all planes of fracture. Lefèvre et al. (1976) studied peritubular dentine using secondary ion mass spectrometry and found higher levels of calcium phosphate and magnesium ions than in intertubular dentine, the magnesium reaching 1% by weight. Through selective electron diffraction, these authors showed that witlockite was present in peritubular dentine. With high resolution electron microscopy of thin sections, Schroeder and Frank (1985) demonstrated the presence of hydroxyapatite crystals with a typical flattened hexagonal shape.

The exact nature of the organic matrix of peritubular dentine is not well known. It is commonly supposed that it has the same composition as the non-collagenous component of intertubular dentine but at least one report indicates that this is not so (Goldberg et al. 1978). It contains virtually no collagen fibrils or only very few (Takuma and Eda 1966).

The amount of peritubular dentine varies greatly in human teeth. It is not found in interglobular dentine. It varies also greatly between species. It is, for example, particularly plentiful in the cow, horse and elephant, but scarce in the rabbit (Jones and Boyde 1984).

K. The Development of Root Dentine

Fundamentally, root dentinogenesis differs only slightly from coronal dentinogenesis. Differentiation of odontoblasts in the radicular region is likewise the end result of epithelio-mesenchymal interactions, but the epithelial cells in this case are derived from Hertwig's epithelial root sheath rather than from the internal dental epithelium.

From the cervical loop of the enamel organ, an initial epithelial proliferation, known as the epithelial diaphragm, grows horizontally from the junction between the ameloblast layer and the external dental epithelium. This diaphragm, part of the Hertwig's sheath, consists of a double layer of epithelial cells surrounded by a basement membrane. It extends around and encloses most of the papilla (Fig 86). During the subsequent development of the root, the diaphragm remains spatially at the same level and root development is accompanied by movement of the tooth germ in an oral direction. Root development continues in an apical direction by incremental dentine deposition until root formation is completed (Figs. 82–85). For the permanent dentition, eruption is accompanied by progressive resorption of the roots of the deciduous teeth (Figs. 82, 83).

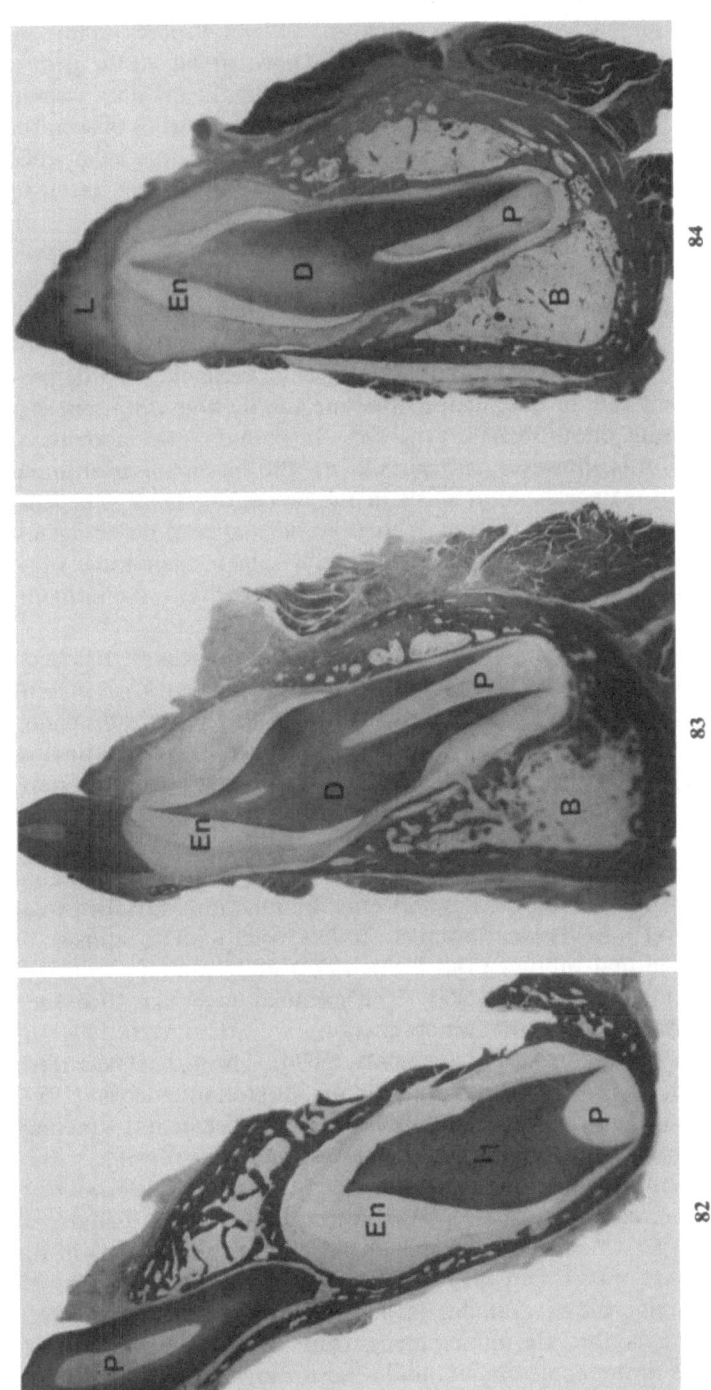

Fig. 82. Decalcified section through the lower jaw of a human fetus. The tooth germ of the permanent lower central incisor, located in a lingual position with regard to the deciduous central incisor, shows evidence of early root formation. *En* Enamel lost during decalcification; *P* Pulp. ×3,5

Fig. 83. Extensive resorption of the root of a deciduous central incisor with development and early eruption of the underlying permanent central incisor. *En* Enamel lost during decalcification. *P* Pulp; *D* Dentine; *B* Lower jaw. ×3,5

Fig. 84. After shedding of its deciduous predecessor the permanent central lower incisor approaches the oral epithelium. Nearly complete development of the root has occurred. Decalcified section of the lower jaw of a human fetus. *L* Gingival lamina propria; *En* Enamel lost during decalcification; *P* Pulp; *D* Dentine; *B* Lower jaw. ×3,5

Hertwig's root sheath consists of two layers of epithelial cells separated from the surrounding mesenchyme by a basement membrane. It plays an important role in differentiation of the odontoblasts. In mice molar tooth germs, the cells of the inner epithelial layer, which face the dental papilla, assume a cuboidal form and do not acquire the ultrastructural characteristics of secretory cells (Andujar et al. 1984; Heritier et al. 1984). The outer epithelial cells of Hertwig's sheath, which face the dental follicular mesenchyme, are flattened cells with large nuclei.

The root odontoblasts differentiate along the basement membrane adjacent to the inner cuboidal cells of Hertwig's sheath through a similar sequence of events as those described for coronal odontoblasts. The ultrastructural features of the odontoblasts during root development are also identical to those of coronal odontoblasts (Furseth 1971). Initially a layer of radicular mantle predentine is laid down between the basement membrane and the root odontoblasts, with their differentiating odontoblastic processes. In root mantle predentine, the coarse collagen fibrils, however, are parallel to the basement membrane (Ten Cate 1978, 1980). Mineralization starts in the predentine layer. The cells of Hertwig's sheath become discontinuous in these initial stages of dentinogenesis and, in fact, Andujar et al. (1984) noted that the basement membrane adjacent to initial root predentine lost its positive immunoreactivity to laminin and type IV collagen.

The root mantle dentine seems to manifest species differences. It has been reported to be absent in rat, rabbit and guinea pig incisors (Moss 1974), present in mouse molars (Ten Cate 1978), present as a thick (8 to 15 µm) radiopaque layer in the dog (Schackelford 1971) and to be 15–30 µm thick in primates (Lavelle et al. 1977). After deposition of root mantle dentine, dentinogenesis proceeds as in the crown by formation of intertubular and then peritubular dentine.

A hypomineralized layer, the so-called granular layer of Tomes, is found in the external root dentine along the dentinocemental junction. Various explanations have been given to its development. It has been said to consist of dilated and curved dentinal tubules (Ten Cate 1972) or to contain epithelial cells, at least in rat molars (Lester 1969). The granular layer has also been suggested to be a zone of deficient mineralization (Schackelford 1971) or of incomplete fusion of calcospherites (Furseth 1974). Using back-scattered electron imaging with scanning electron microscopy, Boyde and Jones (1983) followed the mineralization in the vicinity of the dentinocemental junction. According to their observations, numerous small calcospherites forming within the initial root predentine constitute the leading feather-edge of the mineral part of the developing root. An uneven distribution of these, particularly at the level at which larger calcospherites begin to develop, result in gaps in the mineral part of dentine which can never be bridged and would later be an element of the hypomineralized granular layer of Tomes. As it recedes from the dentinocemental junction, the mineralizing front of the dentine changes with fewer, larger and flatter coalescing mounds. According to Boyde and Jones (1983), the granular layer in the dentine is an inconstant and variable feature but, where present, is hypomineralized and interglobular.

In multirooted teeth, the floor of the pulp chamber is formed by the interradicular portion of dentine which is generally thought to develop as a continuation of the coronal dentine. Dentinogenesis proceeds cervically and then horizontally following the development of the Hertwig's root sheath. This results in the development of two (in premolars and mandibular molars) or three (in maxillary molars) interradicular dentine processes. Subsequently the ends of the processes unite and form the initial floor of the pulp chamber (ORBAN and MÜLLER 1929; FUJITA 1978). An alternative view, however, suggests that the first interradicular dentine develops from a separate mineralization center or centers and that only later does it unite with the coronal dentine (JÖRGENSEN 1950). Using tetracycline labeling, OOE and GOHDO (1984) demonstrated isolated mineralization centers for the interradicular dentine of human mandibular second deciduous molars, indicating that this region in man, unlike the rodent, does not develop as a direct continuation of coronal dentine.

A controversial structure has been described, between the cementum and the dentine, initially called "the hyaline layer of HOPEWELL-SMITH" (1903), then later "intermediate cementum" (BLACKWOOD 1957) and "enameloid" (FEARNHEAD 1979). This seems to be a prerequisite for the formation of the cementum proper and has been considered either as a variety of dentine (BRADFORD 1967; KAWASAKI 1975) or as a form of cementum (HOPEWELL-SMITH 1920; BLACKWOOD 1957; OWENS 1973). LINDSKOG and HAMMARSTRÖM (1984) have expressed the view that this region in monkeys consists of a mineralized enamel matrix formed by the epithelial cells of the root sheath.

L. The Development of the Pulp

The pulp is derived directly from the dental papilla. The coronal pulp is initially delineated, followed by gradual formation of the radicular pulp (Figs. 83 and 84), the last formed pulpal area being at the apex (Fig. 85). The peripheral cells of the pulp are the odontoblasts (Fig. 87), derived from undifferentiated mesenchymal cells through a sequence of changes involving epithelio-mesenchymal interactions.

The dental papilla has already been described with respect to structural characteristics. It consists of undifferentiated mesenchymal cells (Fig. 9), separated by large extracellular spaces, showing a few cross-striated collagen fibrils (Fig. 12). These mesenchymal cells will gradually be transformed into pulpal fibroblasts (Fig. 87).

Although type I collagen is the major fibrous component of differentiating pulp, it has been demonstrated that type III collagen constitutes a substantial portion of the pulpal collagen (LINDE 1985). With maturity, the total collagen content of pulp increases, but type III collagen remains a constant 45 per cent of the total collagen in bovine pulp specimens with active, moderate or minimal dentinogenesis (LECHNER and KALNITZSKY 1981). These data suggests that all the pulp collagen (both type I and III) remains in the pulp and is compressed

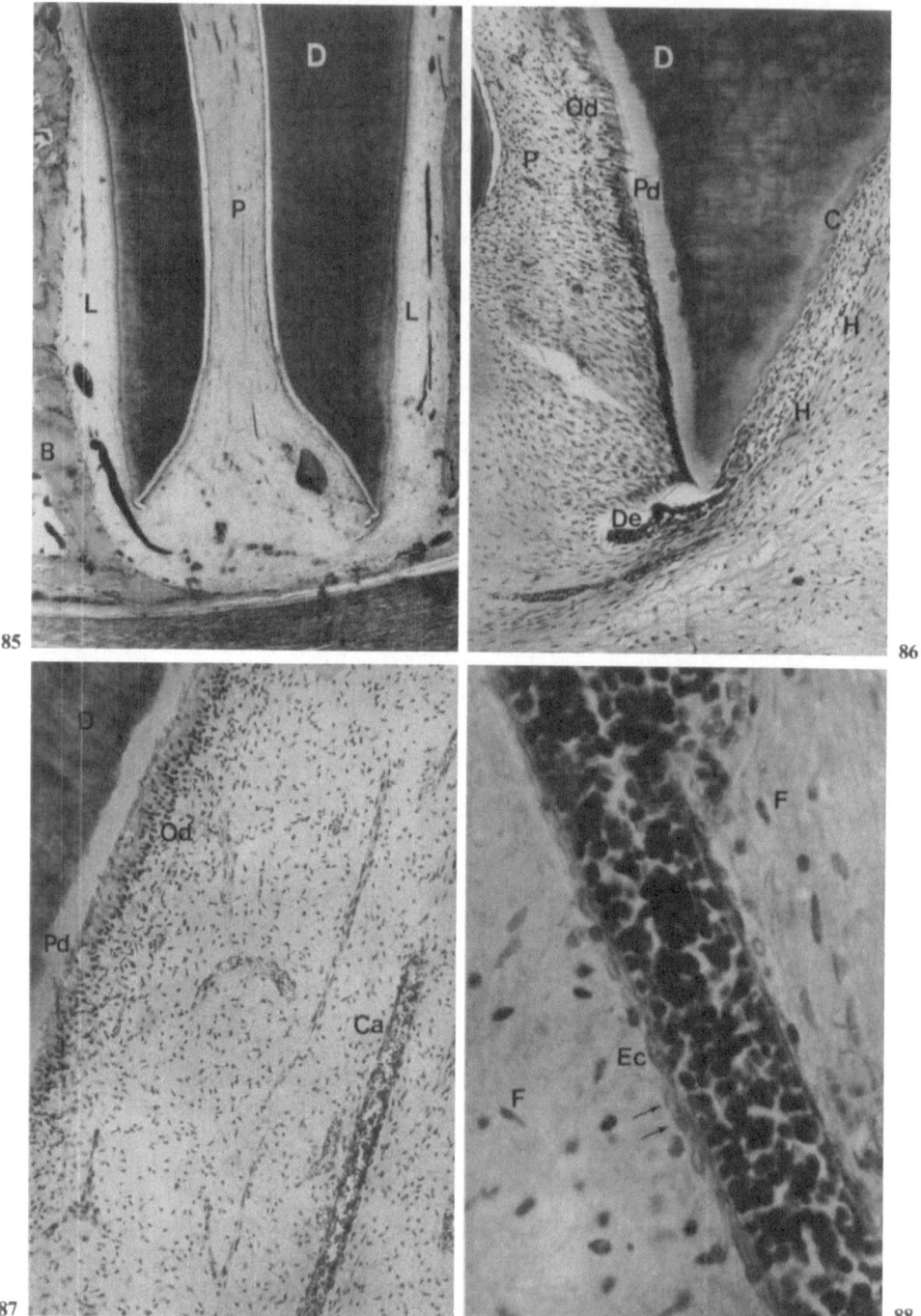

Fig. 85. Terminal stage of root development in the apical region of a human deciduous tooth.
P Pulp; *D* Root dentine; *L* Periodontal ligament; *B* Alveolar bone ×16

into a smaller volume as the pulp cavity becomes narrower during dentinogenesis. With light and electron microscope immunochemistry, MALGOIRE et al. (1982) studied the distribution of type I and III collagen in the pulpal tissue of human developing teeth. Whereas type I collagen corresponded to typical cross-striated fibrils, type III collagen appeared as fine branched filaments or electron dense material distributed throughout the tissue, in close association with the plasma membranes of fibroblasts.

Proteoglycans are present in the pulp, and it has been shown that the total glycosaminoglycan content decreases in the pulp as dentinogenesis diminishes (LINDE 1973). Whereas pulps undergoing active dentinogenesis contained large amounts of chondroitin sulfate, pulps in erupted teeth in which dentinogenesis was completed, showed hyaluronate accounting for more than half of the GAGs. Dermatan sulfate was present in large quantities whereas chondroitin sulphate constituted a minor fraction (LINDE 1973). In an immunofluorescent study using a monoclonal antibody, CONNOR et al. (1984) studied the localization of fibronectin in rat molars and incisors. The connective tissue of the pulp was not labeled and the odontoblasts were labeled for fibronectin only in the region in contact with predentine. In explanted human pulps cultured in vitro, MAGLOIRE (1983) found a high fibronectin activity in the odontoblast-like peripheral cells compared to the central pulpal cells. The vascular supply of developing pulp was studied in human fetuses of various ages by DE SAUNDERS and RÖCKERT (1967), using microradiography after injection with radiopaque solutions. They demonstrated an intradental vascular plexus which consisted of larger blood vessels, of the order of arterioles and venules, occupying the central region and united by a rich capillary network occupying the peripheral zone of the dental papilla, immediately beneath the cusps of the developing crown. Differentiation of the capillary network precedes the appearance of enamel and mineralization of the crown (DE SAUNDERS and RÖCKERT 1967). During crown formation in mongrel dogs, TAKAHASHI (1985) noted that, in the central portion of the pulp, numerous arterioles and venules follow a generally longitudinal course along the tooth axis with arterioles running up towards the pulp horn. The venules are larger than the corresponding arterioles and bend and curve along their path.

In the developing pulp of the cat and man, a subodontoblastic capillary plexus is described, with occasional loops penetrating the odontoblast layer. Many of the capillaries have short, thorn-like irregularities projecting from their outer surfaces (KRAMER 1960). In rodents and cats, the peripheral capillary plexus is a dense network often lying among the odontoblasts, just internal

Fig. 86. Root development in a human deciduous tooth. *De* Epithelial diaphragm; *H* Hertwig's epithelial root sheath; *C* Cementum; *Pd* Predentine; *D* Dentine; *P* Pulp; *Od* Odontoblasts. × 210

Fig. 87. The pulp during root development in a human deciduous tooth. *D* Dentine; *Pd* Predentine; *Od* Odontoblasts; *Ca* Blood capillary. × 90

Fig. 88. Higher enlargement of the pulp of a human deciduous tooth during root formation. Blood capillary lined by endothelial cells (*Ec*) and pericytes (*arrows*). *F* Fibroblast. × 835

to the predentine (Adams 1962). During root development capillaries lie at some distance from the periphery of pulp (Figs. 87 and 88).

Two types of capillaries are found in developing pulp. The *continuous* capillaries consist of continuous layer of endothelial cells, united by intercellular junctions. The endothelial cells are bound on their external surfaces by a basement membrane which also envelops the peripheral pericytes. The *fenestrated* capillaries have very thin endothelial cells and exhibit circular transcellular openings, the fenestrae, with a diameter of approximately 600 to 800 Å. Each fenestration is usually closed by a single-layered diaphragm displaying a central lamina, continuous with the plasmalemma of the endothelial cells. The peripheral basement membrane of this type of capillary is continuous over the fenestrations. The fenestrated capillaries allow faster exchange between the content of the vessel and the connective tissue matrix.

The innervation of the developing human pulp was studied by Fearnhead (1967). This work showed that in the deciduous central incisor tooth germ (8th week of fetal life at the cap stage), nerves are found in the base of the dental papilla but do not enter its substance. One or two pioneer nerve fibers penetrate the dental papilla at an advanced bell stage of development. When dentinogenesis occurs, a few terminal branches may be found to be directed towards the odontoblasts. At this stage, a rich innervation is found in the follicle. In a number of human specimens, Fearnhead (1967) noted that the dental papilla of the mandibular deciduous canine tooth germ appears to receive its pioneer nerve fibers before the other teeth.

In developing mouse molars, Corpron and Avery (1973) found that major nerves appeared in the pulp organ and extended to the basal region of the odontoblasts by the 9th day after birth. At 15 days, a subodontoblastic neural plexus developed and by 25 days isolated nerves penetrated into the predentine of the pulpal horns. Development of sensory innervation in forming rat molars at 3 weeks to 1 year of age was followed after labeling of dental nerves by axonal transport of radioactive protein injected into the Gasserian ganglion (Byers 1980). The labeled nerves were visualized by light microscopic autoradiography. In addition to nerve growth, Byers (1980) found that two processes determine adult dentinal nerve location: 1) enclosure of nerves within circumpulpal dentinal tubules during the last few weeks of dentinogenesis, beginning at the tip of the pulp horn and spreading to include most coronal dentine; and 2) gradual loss of the nerves near the tip of the cusp because of dental attrition and replacement by non-innervated, reparative dentine. Several days before a molar erupts, nerves at the tip of the cusp can already be observed in dentine. Two to three weeks later, when dentinogenesis at the cusp slows down, most of the innervation for that region has been established and nerves extend up to 160 μm into dentine.

Fried and Hildebrand (1981a, b) followed the developmental growth of pulpal axons in primary and permanent mandibular incisors of kittens and cats. In primary incisors, cells resembling Schwann cells preceded the first arrival of pulpal axons 1 week postnatally, whereas myelination was initiated during the second week. Two months after birth, the primary incisors were fully grown and each pulp contained about 100 axons. Between 10 and 20% of these were myelinated and ranged in size from 1 to 5 μm. In permanent mandibular incisors of cats, the first axons entered the incisor pulp about 2 months after birth. Cells resembling Schwann cells, but lacking a relationship to axons were also observed at this time. Axonal ingrowth and maturation continued until 7 months, at which time the pulp contained 92-394 axons, 81–88% of which were unmyelinated.

With the end of the development of the apical dentine, the formation of *primary* dentine is completed. The dentine formed after root formation is complete, and when the tooth is in function, is referred to as *secondary* dentine.

References

Adams D (1962) The blood supply to developing dentine. Arch Oral Biol 7:773–774

Andujar MB, Magloire H, Hartman DJ, Grimaud JA (1984) Modifications of Hertwig's epithelial cells during developing mouse molar tooth: ultrastructural, histochemical and immunohistochemical studies. In: Belcourt AB, Ruch JV (eds) Tooth morphogenesis and differentiation, vol 125. Inserm, Paris, pp 535–544

Appleton J, Morris DC (1979) An ultrastructural investigation of the role of the odontoblast in matrix calcification using the potassium pyroantimonate osmium method for calcium localization. Arch Oral Biol 24:467–475

Ashida O (1983) Hypothalamic nuclei related to circadian rythmicity in dentinogenesis of the rat incisor. Bull Kanagawa Dent Coll 11:15–27

Baud CA, Dupont DH (1965) La structure microscopique des ostéocytes en rapport avec leur fonction. In: Richelle LJ, Dallemagne MJ (eds) Proc Eur 2nd Symp Calcified tissues. Collection Colloques. University Liège, pp 31–37

Baume LJ (1980) The biology of pulp and dentine: a historic, terminologic, taxonomic, histological, biochemical, embryonic and clinical survey. Monographs in Oral Science. Karger, Basel

Beams HW, King RL (1933) The Golgi apparatus in the developing tooth with special reference to polarity. Anat Rec 57:29–39

Bernard GW (1972) Ultrastructural observations of initial calcification in dentin and enamel. J Ultrastruct Mol Struct Res 41:1–17

Bernick S (1969) Histochemical study of dentin in parathyroidectomized rats. J Dent Res 48:1251–1257

Berthet-Colominas C, Miller A, White SW (1979) Structural study of the calcifying collagen in turkey leg tendons. J Mol Biol 134:431–445

Betti F, Katchburian E (1982) Proteolytic activity of developing dentine of rat tooth terms revealed by the gelatin-film substrate technique. Arch Oral Biol 27:891–896

Bienkowski RS (1984) Intracellular degradation of newly synthesized collagen. Coll Relat Res 4:399–412

Bishop MA (1985) Evidence for tight junctions between odontoblasts in the rat incisor. Cell Tissue Res 239:137–140

Blackwood HJJ (1957) Intermediate cementum. Br Dent J 102:345–350

Bonucci E (1984) Matrix vesicles: their role in calcification. In: Linde A (ed) Dentin and dentinogenesis, vol I. CRC Press, Boca Raton, pp 135–154

Boyde A, Jones S (1983) Mineralization in the vicinity of the cement-dentine junction. Proc Anat Soc Great Britain Ireland 136:642–643

Boyde A, Reith EJ (1977) Qualitative electron probe analysis of secretory ameloblasts and odontoblasts in the rat incisor. Histochemistry 50:347–354

Boyde A, Reith EJ, Jones SJ (1978) Intercellular attachments between calcified collagenous tissue forming cells in the rat. Cell Tissue Res 191:507–516

Bradford E (1967) Microanatomy and histochemistry of dentin. In: Miles AEW (ed) Structural and chemical organization of teeth, vol II. Academic, New York London, pp 3–34

Bres EF, Voegel JC, Barry JC, Frank RM (1986) Theoretical image simulation of dark contrast line in twinned apatite bicrystals and its possible correlation with the chemical properties of human dentine and enamel crystals. Biophys J 50:1185–1193

Brown WE (1962) Crystal structure of octocalcium phosphate. Nature 196:1048–1050

Brown WE (1965) A mechanism for growth of apatitic crystals. In: Stack MV, Fearnhead RW (eds) Tooth enamel. Wright, Bristol, pp 11–14

Brunn A von (1887) Über die Ausdehnung des Schmelzorgans und seine Bedeutung für die Zahnbildung. Arch Mikr Anat 29:367–382

Burgess AMC, Katchburian E (1982) Morphological types of epithelial mesenchymal cell contacts in odontogenesis. J Anat 135:577–584

Butler WT (1984a) Dentin collagen: chemical structure and role in mineralization. In: Linde A (ed) Dentin and dentinogenesis, vol II. CRC Press, Boca Raton, pp 37–53

Butler WT (1984b) Matrix molecules of bone and dentin. Coll Relat Res 4:297–307

Butler WT, Bhown M, Di Muzio MT, Cothran WC, Linde A (1983) Multiple forms of rat dentin phosphoproteins. Arch Biochem Biophys 225:178–186

Byers MR (1980) Development of sensory innervation of dentin. J Comp Neurol 191:413–427

Calle A (1984) Intercellular junctions between human odontoblasts revealed by freeze-fracture. In: Belcourt AB, Ruch JV (eds) Tooth morphogenesis and differentiation, vol 125. Inserm, Paris, pp 167–171

Calle A (1985) Intercellular junctions between human odontoblasts. A freeze-fracture study after demineralization. Acta Anat (Basel) 122:138–144

Cheung WY (1980) Calmodulin. An introduction. In: Cheung WY (ed) Calcium and cell function, vol I. Academic, New York, pp 1–12

Chibon P (1967) Etude expérimentale par ablations, greffes et autoradiographie de l'origine des dents chez l'amphibien unrodèle Pleurodeles Waltii Michah. Arch Oral Biol 12:745–755

Cho MI, Garant PR (1984) Comparative radioautographic study of the effect of L-azetidine-2-carboxylic acid on matrix secretion and Golgi of the mouse incisor. Calcif Tissue Int 36:409–420

Cho MI, Garant PR (1985) Radioautographic analysis of (^{3}H)-fucose utilization by mouse odontoblasts with emphasis on intracytoplasmic and plasma membrane glycoproteins. Archs Oral Biol 30:110–120

Connor NS, Aubin JE, Melcher AH (1984) The distribution of fibronectin in rat tooth and periodental tissues: an immunofluorescence study using a monoclonal antibody. J Histochem Cytochem 32:565–572

Corpron RE, Avery JK (1973) The ultrastructure of intradental nerves in developing mouse molars. Anat Rec 175:585–606

Cotmore JM, Nichols G Jr, Wuthier RE (1971) Phospholipid-calcium phosphate complex: enhanced calcium migration in the presence of phosphate. Science 172:1339–1341

Cournil I, Leblond CP, Pomponio J, Hand AR, Sederlof L, Martin GR (1979) Immunohistochemical localization of procollagens. I. Light microscopic distribution of procollagen I, III and IV antigenicity in the rat incisor tooth by the indirect peroxidase-antiperoxidase method. J Histochem Cytochem 27:1059–1069

Cox CF, Avery JK, Lee SD, Tomaro AA, Simmons TA (1976) Ultrastructural characterization of junctional complexes between odontoblasts. J Dent Res 55: B86 (abstract) (special issue)

Croissant RD (1971) Isolation of an intercellular matrix RNA-protein complex during odontogenesis. J Dent Res 50:1065–1071

Cuvier G (1805) Leçons d'anatomie comparée. Dix septième leçon des dents. Crochars et Fantin (Paris) 3:103–200

Daculsi G, Kerebel B (1978) High resolution electron microscopic study of human enamel crystallites: size, shape and growth. J Ultrastruct Res 65:163–172

Davis NR, Cavanagh JC (1982) Hard tissue mineralization inhibitors. In: Veis A (ed) The chemistry and biology of mineralized connective tissues. Elsevier North Holland, New York, pp 489–491

Dimuzio, MT, Veis A (1978) Phosphoryns, major noncollagenous proteins of rat incisor dentin. Calcif Tissue Res 25:169–178

Dimuzio MT, Bhown M, Butler WT (1981) Odontoblast, dentin organ cultures: the biosynthesis of matrix protein. In: Veis A (ed) The chemistry and biology of mineralized connective tissues. Elsevier, North Holland, New York, pp 427–431

Dimuzio MT, Bhown M, Butler WT (1983) The biosynthesis of α-carboxyglutamic acid-containing proteins in rat incisor odontoblasts in organ culture. Biochem J 216:249–257

Dryburgh L (1967) Epigenetics of early tooth development in the mouse. J Dent Res 46:1264 (abstract)

Dunlap C, Williams C, Barker B, Hoff R (1984) An investigation of S-100 protein in embryonic dental papillae of rats. Oral Surg Oral Med Oral Pathol 58:575–578

Dustin P (1978) Microtubules. Springer, Berlin

Duve C de (1978) An integrated view of lysosome function. In: Berlin RD, Herrmann, Lepow IH, Tanzer JM (eds) Molecular basis of biological degradative processes. Academic, New York, pp 25–38

Ebner V von (1875) Über den feineren Bau der Knochensubstanz. Sitzgsber Akad Wiss Wien Math Naturwiss Kl III 72:1–90

Eisenman DR, Glick PL (1972) Ultrastructure of initial crystal formation in dentin. J Ultrastruct Mol Struct Res 4:18–28

Engfeldt B, Hjerpe A (1972) Glycosaminoglycans of dentine and predentine. Calcif Tissue Res 10:152–159

Engström C, Linde A, Persliden B (1976) Acid hydrolases in the odontoblast-predentin region of dentinogenically active teeth. Scand J Dent Res 84:76–81

Ennever J, Riggan JL, Vogel JJ (1984) Proteolipid and calcification in vitro. Cytobios 39:151–157

Farquhar MG, Palade GE (1963) Junctional complexes of various epithelia. J Cell Biol 17:375–412

Fearnhead RW (1967) Innervation of dental tissues. In: Miles AEW (ed) Structural and chemical organization of teeth, vol I. Academic Press, New York, pp 247–281

Fearnhead RW (1979) Matrix-mineral relationships in enamel tissues. J Dent Dres 58 B:909–916

Frank RM (1965) Microscopie électronique de la genèse de collagène dans la papille dentaire. J Microsc 4:43–56

Frank RM (1968a) Etude ultrastructurale de la dentinogenèse et de l'amelogénèse. Thèse doct 3ème cycle science odont. Université Louis Pasteur, Strasbourg

Frank RM (1968b) Ultrastructural relationship between the odontoblast, its process and the nerve fibre. In: Symons NBB (ed) Dentine and pulp: their structure and reactions. Livingstone, Edinburgh, pp 115–145

Frank RM (1969) Mise en évidence de corps multivésiculaires intra-nucléaires au niveau des odontoblastes jeunes. Z Zellforsch 95:310–316

Frank RM (1970b) Etude autoradiographique de la dentinogenèse en microscopie électronique à l'aide de la proline tritiée chez le chat. Arch Oral Biol 15:583–596

Frank RM (1979) Electron microscope autoradiography of calcified tissues. Int Rev Cytol 56:183–253

Frank RM, Nalbandian J (1967) Ultrastructure of amelogenesis. In: Miles AEW (ed) Structural and chemical organization of teeth, vol I. Academic, New York, pp 399–466

Frank RM, Voegel JC (1978) Dissolution mechanisms of the apatite crystals during dental caries and bone resorption. In: Berlin RD, Herrmann H, Lepow IH, Tanzer JM (eds) Molecular basis of biological degradative processes. Academic, New York, pp 277–311

Frank RM, Osman M, Meyer JM, Ruch JV (1979) ^{3}H-glucosamine electron microscope autoradiography after isolated labeling of the enamel organ or the dental papilla followed by reassociated toothgerm culture. J Biol Buccale 7:227–241

Fränkel L (1835) De penitiori dentium humanorum structura observationes. Diss Vratislaviae, Pressburg

Frazier PD, Nylen MU (1972) Biophysical studies of calcium transport in mineralizing tissues. In: Takeuchi T, Ogawa K, Fujita S (eds) 4th Int Congress Histochem Cytochem. Japan Scient Press, Kyoto, pp 91–92

Fried K, Hildebrand C (1981a) Pulpal axons in developing, mature and aging feline permanent incisors. A study by electron microscopy. J Comp Neurol 203:23–36

Fried, K, Hildebrand C (1981b) Developmental growth and degeneration of pulpal axons in feline primary incisors. J Comp Neurol 203:37–51

Fujita Y (1978) Formation of interradicular portion of mouse first maxillary molars. Jap J Oral Biol 20:1–8

Furseth R (1971) The fine structure of odontoblast/predentin area in the root. Scand J Dent Res 79:141–150

Furseth R (1974) The structure of peripheral dentine in young human premolars. Scand J Dent Res 82:557–561

Garant PR (1972) The demonstration of complex gap junctions between the cells of the enamel organ with lanthanum nitrate. J Ultrastruct Mol Struct Res 40:333–348

Garant PR (1978) Microanatomy of the oral mineralized tissues. In: Shaw JH, Sweeney EA, Cappuccino CC, Meller SB (eds) Textbook of oral biology, 1st edn. Saunders, Philadelphia, pp 181–225

Garant PR, Szabo G, Nalbandian J (1968) The fine structure of mouse odontoblasts. Arch Oral Biol 13:857–876

Glanville RW, Rauter A, Fietzek PP (1975) Isolation and characterization of a native placental basement membrane collagen and its component chains. Eur J Biochem 95:383–389

Glick PL (1982) Identification of mineral fractions in developing rat incisors. In: Veis A (ed) The chemistry and biology of mineralized tissues. Elsevier, North Holland New York, pp 309–311

Glimcher MJ (1960) Specificity of the molecular structure of organic matrices in mineralization. In: Sognnaes RF (ed) Calcification in biological systems. Am Assoc Adv Sciences, Washington, pp 421–487

Glimcher MJ (1982) On the form and function of bone: from molecules to organs. Wolff's law revisited. In: Veis A (ed) The chemistry and biology of mineralized connective tissues. Elsevier, North Holland New York, pp 617–673

Glimcher MJ (1984) Recent studies of the mineral phase in bone and its possible linkage to the organic matrix by protein-bound phosphate bonds. Philos Trans R Soc Lond [Biol] 304:479–508

Goldberg M (1983) "Protéoglycanes de la dentine et de l'émail: interrelations avec les composants matriciels. Etudes histochimiques et ultrastructurales"; Thèse Doct d'Etat Sciences Nat Université, Paris 6

Goldberg M, Escaig F (1984a) Distribution of filipin-cholesterol complexes in rat incisor odontoblasts. J Biol Buccale 12:171–180

Goldberg M, Escaig F (1984b) Improved preservation of intramitochondrial granules in rat-incisor odontoblasts by rapid freezing and freeze substitution fixation. Arch Oral Biol 29:295–301

Goldberg M, Septier D (1983) Electron microscopic visualization of proteoglycans in rat incisor predentine and dentine with cuprolinic blue. Arch Oral Biol 38:79–83

Goldberg M, Genotelle-Septier D, Weill R (1978) Glycoprotéines et protéoglycanes dans la matrice prédentinaire et dentinaire chez le rat: étude ultrastructurale. J Biol Buccale 6:75–90

Gränström G (1984) Further evidence of an intravesicular Ca^{2+} pump in odontoblasts from rat incisors. Arch Oral Biol 29:599–606

Gränström G, Linde A (1981) ATP dependent uptake of Ca^{2+} by a microsomal fraction from rat incisor odontoblasts. Calcif Tissue Int 33:125–128

Greulich RC, Slavkin HC (1965) Amino acid utilization in the synthesis of enamel and dentin matrices as visualized by autoradiography. In: Leblond CP, Warren KK (eds) The use of radioautography in investigating protein synthesis. Symposium International Society Cell Biology. Academic Press, New York 4:199–214

Hansson LI, Stenström A, Thorngren KG (1978) Effect of pituitary hormones on dentin production in maxillary incisors in the rat. Scand J Dent Res 86:80–86

Harven de E, Bernhardt W (1956) Etude au microscope électronique de l'ultrastructure du centriole chez les vertébrés. Z Zellforsch 45:378–498

Hascall VC, Hascall GK (1981) Proteoglycans. In: Hay ED (ed) Cell biology of extracellular matrix, vol 2. Plenum, New York, pp 39–63

Hauschka PV, Lian JB, Gallop PM (1975) Direct identification of the calcium-binding amino acid, carboxyglumate, in mineralized tissue. Proc Natl Acad Sci USA 72:3925–3929

Hay ED (1981) Cell biology of extracellular matrix. Plenum, New York

Hay ED (1983) Cell and extracellular matrix: their organization and mutual dependence. In: McIntosh JR, Satir BH (ed) Modern cell biology, vol II. Liss, New York, pp 509–548

Heritier M, Deminatti M (1970) Rôle du mésenchyme odontogène dans l'orientation de la morphogenèse coronaire des dents chez la souris. CR Ac Sciences (Paris) 271:851–853

Heritier M, Bailly Y, Robert JL (1984) Données nouvelles sur l'ultrastructure de la gaine de Hertwig chez la souris. In: Belcourt AB, Ruch JV (eds) Tooth morphogenesis and differentiation, vol 125. Inserm, Paris, pp 527–534

Herold RC, Kaye H (1966) Mitochondria in odontoblastic process. Nature 210:108–109

Hjerpe A, Antonopoulos J, Engfeldt B, Wikström B (1983) Analysis of dentine glycosaminoglycans using high performance liquid chromatography. Calcif Tissue Int 35:496–501

Höhling HJ, Fromme HG (1984) Cellular transport and accumulation of calcium and phosphate during dentinogenesis. In: Linde A (ed) Dentin and dentinogenesis, vol II. CRC Press, Boca Raton, pp 1–15

Höhling HJ, Hall T, Boyde A (1967) Electron probe X-ray microanalysis of mineralization in rat incisor peripheral dentine. Naturwissenschaften 54:617–618

Höhling HJ, Steffens H, Heuck F (1972) Untersuchungen zur Mineralisierungsdichte im Hartgewebe mit Protein-Polysaccharid bzw. mit Kollagen als Hauptbestandteil der Matrix. Z Zellforsch 134:283–296

Holland GR (1975) Membrane junctions on cat odontoblasts. Arch Oral Biol 20:551–552

Hopewell-Smith A (1903) The histology and patho-histology of the teeth and associated parts. The Dental Manufacturing Company, London

Hopewell-Smith A (1920) Concerning human cementum. J Dent Res 2:59–75

Huggins CB, McCarrol MD, Dahlberg AA (1934) Transplantation of tooth germ elements and the experimental heterotopic formation of dentin enamel. J Exp Med 60:199–210

Hunt AM, Paynter KJ (1963) The role of cells of the stratum intermedium in the development of the guinea pig molar. Arch Oral Biol 8:65–78

Hunter J (1778) The natural history of human teeth, 2nd edn. Johnson, London

Hurmerinta K (1982) Autoradiographic visualization of glycoproteins and glycosaminoglycans in the epitheliomesenchymal interface of developing mouse tooth germ. Scand J Dent Res 90:278–285

Iren F van, Essen-Joolen L van, Duyn Schouten P van der, Boers van der, Sluijs P, Bruijn WC de (1979) Sodium and calcium localization in cells and tissues by precipitation with antimonate: a quantitative study. Histochemistry 63:273–294

Irvin JT (1958) A histological stain for newly calcified tissues. Nature 181:704–705

Irving JT (1959) A histological staining method for site of calcification in teeth and bone. Arch Oral Biol 1:89–96

Johansen E (1964) Microstructure of enamel and dentin. J Dent Res 43:1007–1020

Johansen E, Parks HF (1962) Electron microscopic observations of sound human dentin. Arch Oral Biol 7:185–193

Jones SJ, Boyde A (1984) Ultrastructure of dentin and dentinogenesis. In: Linde A (ed) Dentin and dentinogenesis, vol I. CRC Press, Boca Raton, pp 81–134

Jontell M, Linde A (1983) Non-collagenous proteins from dentinogenically active bovine teeth. Biochem J 214:769–776

Jörgensen KD (1950) Macroscopic observations on the formation of the subpulpal wall. Odont Tidskr 2:82–103

Kajikawa K, Kakihara S (1974) Odontoblasts and collagen formation: an ultrastructural and autoradiographic study. J Electron Microsc (Tokyo) 23:9–17

Kakei M, Nakahara H (1983) Ultrastructural localization of carbonic anhydrase activity in developing enamel and dentin of the rat incisor. Jap J Oral Biol 25:1129–1133

Kallenbach E (1978) Fine structure and the stratum intermedium, stellate reticulum and outer enamel epithelium in the enamel organ of the kitten. J Anat 126:247–260

Kardos TB, Hubbard MJ (1982) Are matrix vesicles anoptic bodies. In: Dixon AD, Sarnat BG (eds) Factors and mechanisms influencing bone growth. Liss, New York, pp 45–60

Karim A, Cournil I, Leblond CP (1979) Immunohistochemical localization of procollagens. II. Electron microscopic distribution of procollagen I antigenicity in the odontoblasts and predentin of rat incisor teeth by a direct method using peroxidase linked antibodies. J Histochem Cytochem 27:1070–1083

Kashiwa HK, Sigman MD Jr (1966) Calcium localized in odontogenic cells of rat mandibular teeth by the glyoxal bis (2-hydroxyanil) method. J Dent Res 45:1796–1799

Katchburian E (1972) Membrane-bound bodies as initiators of mineralization of dentine. Am J Anat 116:285–302

Katchburian E, Burgess AMC (1977) Fine structure of contacts between ameloblasts and odontoblasts in the rat tooth germ. Arch Oral Biol 22:551–553

Katchburian E, Holt SJ (1968) Ultrastructural studies on lysosomes and acid phosphatase in odontoblasts. In: Symons NBB (ed) Dentine and pulp: their structure and reactions. Livingstone, Edinburgh, pp 43–57

Katchburian E, Severs NJ (1982) Membranes of matrix vesicles in early developing dentine. A freeze fracture study. Cell Biol Int Rep 6:941–950

Kawasaki K (1975) On the configuration of incremental lines in human dentin as revealed by tetracycline labeling. J Anat 119:61–66

Kawasaki K, Tanaka S, Ishikawa T (1977) On the incremental lines in human dentine as revealed by tetracycline labelling. J Anat 123:427–436

Kefalides NA (1973) Structure and biosynthesis of basement membranes. Int Rev Connect Tissue Res 6:63–104

Kerebel B, Grimbert L (1958) Histogenèse de l'émail. Rev Franc Odonto-Stomat 5:1093–1124

Kline LW, Thomas NR (1977) The role of calcitonin in the calcification of dentin matrix. J Dent Res 56:862–865

Koch W (1967) In vitro differentiation of tooth rudiments of embryonic mice. J Exp Zool 165:155–170

Kogaya Y, Furuhashi K (1985) Ultrastructural distribution of glycosaminoglycans associated with matrix vesicle-mediated calcification in mouse progenitor predentine. Calcif Tissue Int 37:36–41

Köling A (1983) "Membrane structures in the human pulp-dentin region. An electron microscopic investigation of permanent teeth using the freeze fracture technique"; Doctoral thesis. Universitet Centraltryckeriet, Uppsala

Kollar EJ (1972) Histogenic aspects of dermal-epidermal interactions. In: Slavkin HC, Bavetta LA (eds): Developmental aspects of oral biology. Academic, New York

Kollar EJ, Baird G (1969) The influence of the dental papilla on the development of tooth shape in embryonic mouse tooth germs. J Embryol Exp Morphol 24:131–148

Kollar EJ, Baird G (1970a) Tissue interactions in embryonic mouse tooth germs. I. Reorganization of the dental epithelium during tooth-germ reconstruction. J Embryol Exp Morphol 24:159–171

Kollar EJ, Baird G (1970b) Tissue interactions in embryonic mouse tooth germs. II. The inductive role of the dental papilla. J Embryol Exp Morphol 24:173–186

Kollar EJ, Lumsden AGS (1979) Tooth morphogenesis: the role of the innervation during induction and pattern formation. J Biol Buccale 7:49–60

Korff K von (1905) Die Entwicklung des Zahnbeines und Knochengrundsubstanz der Säugetiere. Arch Mikrok Anat Entwicklsmechanik 67:1–17

Kramer IRH (1960) The vascular architecture of the human dental pulp. Arch Oral Biol 2:177–189

Kudo N (1975) Effect of colchicine on the secretion of matrices of dentine and enamel in the rat incisor: an autoradiographic study using (^{3}H)-proline. Calcif Tissue Res 18:37–46

Lapiere CM, Lenaers A, Kohn LD (1971) Procollagen peptidase: an enzyme exicising the coordination peptides of procollagen. Proc Natl Acad Sci USA 68:3054–3058

Larsson A, Bloom GD (1973) Studies on dentinogenesis in the rat. Fine structure of developing odontoblasts and predentine in relation to the mineralization process. Z Anat Entwickl Gesch 139:227–246

Lau EC (1983) Glycosaminoglycanes dans les ébauches dentaires de souris et leurs constituants dissociés. Etude biochimique, histochimique et autoradiographique. Thèse doct, 3ème cycle. Sciences Université de Strasbourg, Strasbourg

Lau EC, Ruch JV (1983) Glycosaminoglycans in embryonic mouse teeth and the dissociated dental constituents. Differentiation 23:234–242

Lavelle CLB, Shellis RP, Poole DFG (1977) Evolutionary changes to the primate skull and dentition. Thomas, Springfield/Illinois

Lazarides E, Granger BL (1983) Transcytoplasmic integration in avian erythrocytes and striated muscle: the role of intermediate filaments. Modern cell biology, vol IIG. Liss, New York, pp 143–162

Leblond CP, Weinstock M (1976) A comparative study of dentin and bone formation. In: Bourne GH (ed) The biochemistry and physiology of bone, 2nd edn, vol IV. Academic, New York, pp 516–562

Lechner JH, Kalnitsky G (1981) The presence of large amounts of type III collagen in bovine dental pulp and its significance with regard to the mechanism of dentinogenesis. Arch Oral Biol 26:265–273

Lee SL, Glonek T, Glimcher MJ (1983) ^{31}P nuclear magnetic resonance spectroscopic evidence for ternary complex formation of fetal dentin phosphoprotein with calcium and inorganic ortho-phosphate ions. Calcif Tissue Int 35:815–818

Lefèvre R, Frank RM, Voegel JC (1976) The study of human dentine with secondary ion microscopy and electron diffraction. Calcif Tissue Res 19:251–261

Lehner J, Plenk H (1936) Die Zähne. In: Möllendorff W von (ed) Hdbuch Mikr Anat Menschen, Bd V. Springer, Berlin, pp 449–708

Lenz H (1959) Elektronenmikroskopische Untersuchungen der Dentinentwicklung. Dtsche Zahn Mund Kieferheilk 30:367–381

Lesot H (1980) Caractérisation des types de collagène de l'ébauche dentaire et localisation par immunofluorescence indirecte des collagènes, de la laminine et de la fibronectine; Thèse doct état es sciences. Université de Strasbourg, Strasbourg

Lesot H, Ruch JV (1979) Analyse des types de collagènes synthétisés par l'ébauche dentaire et ses constituants dissociés chez l'embryon de souris. Biol Cell 34:23–38

Lesot H, Mark K von der, Ruch JV (1978) Localisation par immunofluorescence des types de collagène synthétisés par l'ébauche dentaire chez l'embryon de souris. CR Ac Sciences (Paris) [série D] 286:765–768

Lesot H, Osman M, Ruch JV (1981) Immunofluorescent localization of collagens, fibronectin and laminin during terminal differentiation of odontoblasts. Dev Biol 82:371–381

Lesot H, Meyer JM, Ruch JV, Weber K, Osborn M (1982) Immunofluorescent localization of vimentin, prekeratin and actin during odontoblast and ameloblast differentiation. Differentiation 21:133–137

Lester KS (1969) The incorporation of epithelial cells by cementum. J Ultrastruct Res 27:63–87

Lester KS, Boyde A (1967) Electron microscopy of predentinal surfaces. Calcif Tissue Res 1:44–54

Lester KS, Boyde A (1968) The surface morphology of some crystalline components of dentine. In: Symons NBB (ed) Dentine and pulp: their structure and reactions. Livingstone, London, pp 197–219

Leuwenhoek A van (1675) Microscopical observations on the structure of teeth and other bones. Phil Trans Martyn (London) 10:1002–1003

Linde A (1973) A study of the dental pulp glycosaminoglycans from permanent human teeth and rat and rabbit incisors. Arch Oral Biol 18:49–59

Linde A (1982) On enzymes associated with biological calcification. In: Veis A (ed) The chemistry and biology of mineralized connective tissue. Elsevier, North Holland New York, pp 559–570

Linde A (1984) Non-collagenous proteins and proteoglycans in dentinogenesis. In: Linde A (ed) Dentin and dentinogenesis, vol II. CRC Press, Boca Raton, pp 55–92

Linde A (1985) The extracellular matrix of the dental pulp and dentin. J Dent Res 64:523–529

Linde A, Gränström G (1978) Odontoblast alkaline phosphatases and Ca^{2+} transport. J Biol Buccale 6:293–308

Linde A, Magnusson B (1975) Inhibition studies of alkaline phosphatase in hard tissue-forming cells. J Histochem Cytochem 23:342–347

Linde A, Persliden B (1977) Cathepsin D activity in isolated odontoblasts. Calcif Tissue Res 23:33–38

Linde A, Bhown M, Butler WT (1980) Noncollagenous proteins of dentin. A reexamination of proteins from rat incisor dentin utilizing techniques to avoid artifacts. J Biol Chem 255:5931–5942

Linde A, Bhown M, Cothran WC, Höglund A, Butler WT (1982a) Evidence for several carboxyglutamic acid containing proteins in dentin. Biochim Biophys Acta 704:235–239

Linde A, Johansson S, Jonsson R, Jontell M (1982b) Localization of fibronectin during dentinogenesis in rat incisor. Arch Oral Biol 27:1069–1073

Linder E, Vaheri A, Ruoslahti E, Wartiovaara J (1975) Distribution of fibroblast surface antigen in the developing chick embryo. J Exp Med 142:41–49

Lindskog S, Hammarström L (1984) Enzyme histochemistry of Hertwig's epithelial root sheath in monkeys. In: Belcourt AB, Ruch JV (eds) Tooth morphogenesis and differentiation. Inserm, Paris, 125:527–534

Linsenmayer TF (1981) Collagen. In: Hay ED (ed) Cell biology of extracellular matrix. Plenum, New York, pp 5–37

Magloire H (1983) Elaboration de la trame organique prédentinaire in vitro. Ultrastructure, cytochimie, immunochimie; Thèse doct sciences odont. Université Lyon, Lyon

Magloire H, Joffre A, Grimaud JA, Herbage D, Couble ML, Chavrier C, Dumont J (1981) Synthesis of type I collagen by human odontoblast-like cells in explant culture: light and electron microscope immunotyping. Cell Mol Biol 27:429–435

Magloire H, Joffre A, Grimaud JA, Herbage D, Couble ML, Chavrier C (1982) Distribution of type III collagen in the pulp parenchyma of the human developing tooth. Light and electron microscope immunotyping. Histochemistry 74:319–328

Marchi F, Leblond CP (1983) Collagen biosynthesis and assembly into fibrils as shown by ultrastructural and ^{3}H-proline radioautographic studies on the fibroblasts of the rat foot pad. Am J Anat 168:167–197

Marchi F, Leblond CP (1984) Radioautography characterization of successive compartments along the rough endoplasmic reticulum-Golgi pathway of collagen precursors in foot pad fibroblasts of (^{3}H)proline-injected rats. J Cell Biol 98:1705–1709

Marshall AF, Lawless KR (1981) TEM study of the central dark line in enamel crystallite. J Dent Res 60:1173–1182

Martens P (1968) Human dentinogenesis with special regard to the formation of peritubular crown dentine and zones in fetal deciduous and unabraded permanent teeth. Scand J Dent Res 76:5–169 (suppl)

Matthiessen ME, Bülow von FA (1970) The ultrastructure of human fetal odontoblasts. Z Zellforsch Mikrosk Anat 105:569–578

Mayne R, Vail MS, Miller EJ (1975) Analysis of changes in collagen biosynthesis that occur when chick chondrocytes are grown in 5-bromo-2'-deoxyuridine. Proc Natl Acad Sci USA 72:4511–4515

McDougall M, Zeichner-daved M, Slavkin HC (1983) In situ localization of dentine phosphoprotein during murine tooth organ development. Calcif Tissue Int 35:663 (abstract)

McIntosh JR (1983) The centrosome as an organizer of the cytoskeleton. In: McIntosh JR, Satir BH (eds) Modern cell biology, vol II. Liss, New York, pp 115–142

Meyer JM, Fabre M, Staubli A, Ruch JV (1977) Relations cellulaires au cours de l'odontogenèse. J Biol Buccale 5:107–119

Meyer JM, Stäubli A, Ruch JV (1981) Ultrastructural localization of concanavalin A binding sites on the surface of differentiating odontoblasts. Biol Cell 42:193–196

Miake Y, Yanagisawa T, Takuma S (1982) Electron microscopic study on the effect of vinblastine on young odontoblasts in rat incisor. J Biol Buccale 10:319–330

Miller W (1969) Inductive changes in early tooth development. I. A study of mouse tooth development on the chick chorio-allantois. J Dent Res 48:719–726

Moss ML (1974) Studies of dentine. I. Mantle dentin. Acta Anat 87:481–490

Munhoz COG, Leblond CP (1974) Deposition of calcium phosphate into dentin and enamel as shown by radioautography of sections of incisor teeth following injection of ^{45}Ca into rats. Calcif Tissue Res 15:221–235

Munksgaard EC, Rhodes M, Mayne R, Butler WT (1978) Collagen synthesis and secretion by rat odontoblasts in organ culture. Eur J Biochem 82:609–617

Nagai N (1970) Ultrastructural localization of acid phosphatase in odontoblasts of young rat incisors. Bull Tokyo Dent Coll 11:85–120

Nagai N, Frank RM (1974) Electron microscopic autoradiography of ^{45}Ca during dentinogenesis. Cell Tissue Res 155:513–523

Nagai N, Takuma S, Goto Y, Ogiwara H (1974) Electron microscopy of dentine and predentine of developing rat molars stained with ruthenium red. J Biol Buccale 2:73–83

Nakahara H (1982) Electron microscopic studies of the lattice image and central dark line of crystallites in sound and carious human dentine. Bull Josai Dent Univ 11:209–215

Nalbandian J (1968) The ultrastructure of odontoblasts and dentinogenesis. In: Finn SB (ed) Biology of the dental pulp organ. A symposium. University of Alabama Press, Tuscaloosa, pp 195–210

Neuman WF, Neuman NW (1958) The chemical dynamics of bone mineral. University of Chicago Press, Chicago

Nicholson WAP, Ashton BA, Höhling HJ, Quint P, Schreiber J, Ashton K, Boyde A (1977) Electron microprobe investigations into the processes of hard tissue formation. Cell Tissue Res 177:331–339

Nikiforuk G, Fraser D (1979) Etiology of enamel hypoplasia and interglobular dentin: the roles of hypocalcemia and hypophosphatemia. Metab Bone Dis Rel Res 2:17–23

Nygren H, Hansson HA, Linde A (1976) Ultrastructural localization of proteoglycans in the odontoblast-predentin region of rat incisor. Cell Tissue Res 168:277–287

Nygren H, Persliden B, Hansson HA, Linde A (1979) Cathepsin D: ultra-immunohistochemical localization in dentinogenesis. Calcif Tissue Int 29:251–256

Nylen MU, Scott DB (1958) An electron microscopic study of the early stages of dentinogenesis. Public Health Serv Publ No 613, Washington/DC

Nylen MU, Scott DB (1960) Electron microscopic studies of odontogenesis. J Indiana Dent Assoc 39:406–421

Olsen BR (1981) Collagen biosynthesis. In: Hay ED (ed) Cell biology of extracellular matrix. Plenum Press, New York, pp 139–177

Ooë T, Gohdo S (1984) The development of the human interradicular dentine as revealed by tetracycline labelling. Arch Oral Biol 29:257–262

Orams HJ (1978) The ultrastructure of tissues at the epithelial mesenchymal interface in developing rat incisors. Arch Oral Biol 23:39–44

Orban B, Müller E (1929) The development of the bifurcation of multirooted teeth. J Am Dent Assoc 16:297–319

Osman M, Ruch JV (1980) Secretion of basal lamina by trypsin-isolated embryonic mouse molar epithelia cultured in vitro. Dev Biol 75:467–470

Osman M, Ruch JV (1981a) ^{3}H-glucosamine and ^{3}H-proline radioautography by embryonic mouse dental basement membrane. J Craniofac Genet Dev Biol 1:95–108

Osman M, Ruch JV (1981b) Behavior of odontoblasts and basal lamina of trypsin or EDTA-isolated mouse dental papillae in short-term culture. J Dent Res 60:1015–1027

Osman M, Meyer JM, Stäubli A, Ruch JV (1981c) Cytochemical localization of adenylate cyclase in embryonic mouse molars. Acta Histochem 68:91–102

Owens PDA (1973) Mineralization in the roots of human deciduous teeth demonstrated by tetracycline labeling. Arch Oral Biol 18:889–897

Ozawa H, Yajima, T, Kobayashi S (1972) An investigation of the methods for studying calcium localization by means of electron microscopy. J Niigata Shigakai 2:29–42

Ozawa H, Yamada H, Yamamoto T (1981) Ultrastructural observations on the location of lead and calcium in the mineralizing dentine of rat incisors. In: Ascenzi A, Bonucci E, Bernard B de (eds) Matrix vesicles. Wichtig, Milan, pp 179–183

Pannese E (1960) Observations on the ultrastructure of the enamel organ. I. Stellate reticulum and stratum intermedium. J Ultrastruct Res 3:372–400

Pannese E (1962) Observations on the ultrastructure of the enamel organ. III. Internal and external enamel epithelia. J Ultrastruct Res 6:186–204

Pearson AA (1977) The early innervation of the developing deciduous teeth. J Anat 123:563–577

Penman S, Capco DG, Fey EG, Chatterjee P, Reiter T, Ermish S, Wan K (1983) The three-dimensional structural networks of cytoplasm and nucleus: function in cells and tissue. In: McIntosh JR, Satir BH (eds) Modern cell biology, vol II. Liss, New York, pp 385–405

Peress NS, Anderson HC, Sajdera SW (1974) The lipids of matrix vesicles from bovine fetal epiphyseal cartilage. Calcif Tissue Res 14:275–283

Piez KA (1976) Primary structure. In: Ramanchandran GN, Reddi AH (ed) Biochemistry of collagen. Plenum, New York, pp 1–44

Porter KR, Beckerle M, McNiven M (1983) The cytoplasmic matrix. In: McIntosh JR, Satir BH (eds) Modern cell biology, vol II. Liss, New York, pp 259–302

Posner AS, Tannenbaum PJ (1984) The mineral phase of dentin. In: Linde A (ed) Dentin and dentinogenesis, vol 2. CRC Press, Boca Raton, pp 17–36

Price PA, Otsuka AS, Poser JW, Kirstaponis J, Raman N (1976) Characterization of a α-carboxyglutamic acid-containing protein from bone. Proc Natl Acad Sci USA 73:1447–1451

Prout RES, Odugata AA, Trig FC (1973) Lipid analysis of rat enamel and dentine. Arch Oral Biol 18:373–380

Quinaux N, Richelle LJ (1967) X-ray Diffraction and infrared analysis of bone: specific gravity fractions in the growing rat. Isr J Med Sci 3:667–690

Rahemtulla F, Prince CW, Butler WT (1984) Isolation and partial characterization of proteoglycans from rat incisors. Biochem J 218:877–885

Ramachandran GN, Ramakrishnan CF (1976) Molecular structure. In: Ramachandran GN, Reddi AH (eds) Biochemistry of collagen. Plenum, New York, pp 45–85

Raven CP (1932) Zur Entwicklung der Ganglienleiste. I. Die Kinematik der Ganglienleistenentwicklung bei den Urodelen. Arch Entwickl Mech Org 125:210–291

Reith EJ (1968a) Ultrastructural aspects of dentinogenesis. In: Symons NBB (ed) Dentine and pulp: their structure and reaction. Livingstone, Edinburgh, pp 19–57

Reith EJ (1968b) Collagen formation in developing molar teeth of rats. J Ultrastruct Res 21:383–414

Reith EJ (1976) The binding of calcium with the Golgi saccules of the rat odontoblast. Am J Anat 147:267–272

Retzius A (1837) Bemerkungen über den inneren Bau der Zähne mit besonderer Berücksichtigung auf den im Zahnknochen vorkommenden Röhrenbau. Arch Anat Physiol 4:486–571

Revel JP, Karnovsky MJ (1967) Hexagonal array of subunits in intercellular junctions of the mouse heart and liver. J Cell Biol 33:C7–C12

Robison R (1923) The possible significance of hexose phosphoric esters in ossification. Biochem J 17:286–293

Rönnholm E (1962) An electron microscopic study of the amelogenesis in human teeth. I. The fine structure of the ameloblasts. J Ultrastruct Res 6:229–248

Rosenberg GD, Simmons DJ (1980) Rhythmic dentinogenesis in the rabbit incisor: circadian, ultradian and infradian periods. Calcif Tissue Int 32:29–44

Roth TF, Porter KL (1964) Yolk protein uptake in the oocyte of the mosquito *Aedes aegypti* L. J Cell Biol 20:313–332

Rothman JE (1981) The Golgi apparatus: two organelles in tandem. Science 213:1219

Roufosse AH, Landis WJ, Sabine WK, Glimcher MJ (1979) Identification of brushite in newly deposited bone mineral from embryonic chicks. J Ultrastruct Res 68:235–255

Royer L (1928) Recherches expérimentales sur l'épitaxie ou orientation mutuelle de cristaux d'espèces différentes. Bull Soc Franc Mineral 51:7–159

Ruch JV, Karcher-Djuricic V (1971) Mise en évidence d'un rôle spécifique de l'épithélium adamantin dans la differentiation et le maintien des odontoblastes. Ann Embryol Morph 4:359–366

Ruch JV, Karcher-Djuricic V, Gerber R (1972) Quelques aspects du rôle de la prédentine dans la differentiation des adamantoblastes. Arch Anat Microsc Morphol Exp 61:127–138

Ruch JV, Karcher-Djuricic V, Gerber R (1973) Les déterminismes de la morphogenèse et des cytodifférenciations des ébauches dentaires de souris. J Biol Buccale 1:45–56

Ruch JV, Karcher-Djuricic V, Staübli A, Fabre M (1975) Effects de la cytochalasine et de la colchicine sur les cytodifférentiations dentaires in vitro. Arch Anat Microsc Morphol Exp 64:113–134

Ruch JV, Lesot H, Karcher-Djuricic V, Meyer JM, Mark M (1983) Epithelial-mesenchymal interactions in tooth germs: mechanisms of differentiation. J Biol Buccale 11:173–193

Ruch JV, Lesot H, Karcher-Djuricic V, Meyer JM (1984) Extracellular matrix-mediated interactions during odontogenesis. In: Kemp RB, Hinchliffe JR (eds) Matrices and cell differentiation. Liss, New York, pp 103–114

Ruoslahti E (1981) Fibronectin. J Oral Pathol 10:3–13

Sasaki T, Ishida I, Higashi S (1982) Ultrastructure and cytochemistry on old odontoblasts in rat incisors. J Electron Microsc (Tokyo) 31:378–388

Sasaki T, Tominaga H, Higashi S (1984) Endocytic activity of kitten odontoblasts in early dentinogenesis. 1. Thin section and freeze-fracture study. J Anat 138:485–492

Sattelberg C, Turner DF (1984) Anatomical evidence for the existence of zonula occludens between pulpal odontoblasts. J Dent Res 63:225 (abstract)

Sauderns RL (de) CH, Röckert HOE (1967) Vascular supply of the dental tissues including lymphatics. In: Miles AEW (ed) Structural and chemical organization of teeth, vol I. Academic, New York, pp 199–245

Saxen L, Karkinen-Jaaskelännen M, Lehtonen E, Nordling S, Wartlovaara J (1976) Inductive tissue interactions. In: Poste G, Nicolson GL (eds) Inductive tissue interactions. North Holland, Amsterdam

Schakelford JM (1971) The structure of Tomes' granular layer in dogs premolar teeth. Anat Rec 170:357–368

Schleuter RJ, Veis A (1964) The macromolecular organization of dentin matrix collagen II. Periodate degradation and carbohydrate cross-linking. Biochemistry 3:1657–1665

Schmidt WJ (1969) Die kristalline Struktur der Globuli im Zahnbein des Menschen. Z Zellforsch 97:313–320

Schmidt WJ, Keil A (1971) Polarizing microscopy of dental tissues, 1st edn. Pergamon, New York

Schour I (1960) Noyes' oral histology and embryology, 8th edn. Lea and Febiger, Philadelphia

Schour I, Poncher HG (1937a) The rate of apposition of human enamel and dentin as measured by the effects of acute fluorosis. Am J Dis Child 54:756–776

Schour I, Chandler SB, Tweedy WR (1937b) Changes in the teeth following parathyroidectomy. Am Pathol 13:945–970

Schroeder L, Frank RM (1985) High-resolution transmission electron microscopy of adult human peritubular dentine. Cell Tissue Res 242:449–451

Scott JH, Symons NBB (1982) Introduction to dental anatomy, 9th edn. Churchill and Livingstone, Edinburgh London

Sellman S (1946) Some experiments on the determination of the larval teeth in *Amblystoma mexicanum*. Odont Tidskr 54:1–69

Selvig KA (1970) Periodic lattice images of hydroxyapatite crystals in human bone and dental hard tissues. Calcif Tissue Res 6:227–228

Serres A (1817) Essai sur l'Anatomie et la Physiologie des Dents. Mequignon-Marvis ed Paris

Shapiro IM, Wuthier RE, Irving JT (1966) A study of the phospholipids of bovine dental tissues. I. Enamel and dentin. Arch Oral Biol 11:501–512

Silva DG, Kailis DG (1972) Ultrastructural studies on the cervical loop and the development of the amelo-dentinal junction in the cat. Arch Oral Biol 17:279–289

Sisca RF, Provenza DV (1972) Initial dentin formation in human deciduous teeth. An electron microscope study. Calcif Tissue Res 9:1–16

Slautterback DB (1963) Cytoplasmic microtubules. I. *Hydra*. J Cell Biol 18:367–388

Slavkin HC (1972) Intercellular communication during odontogenesis. In: Slavkin HC, Bavetta LA (eds) Developmental aspects of oral biology. Academic, New York

Slavkin HC (1974) Embryonic tooth formation. A tool for developmental biology. In: Melcher AH, Zarb GA (eds) Oral sciences reviews, vol IV. Munksgaard, Copenhagen

Slavkin HC, Bringas P (1976). Epithelial-mesenchyme interactions during dentinogenesis. IV. Morphological evidence for direct heterotopic cell-cell contact. Dev Biol 50:428–442

Slavkin HC, Bringas P, Croissant R, Bavetta LA (1972) Epithelial-mesenchymal interactions during odontogenesis. Dev 1:139–161

Slavkin HC, Matosian P, Wilson P, Brigas P, Mino W, Croissant RD, Guenther H (1975) Epithelial specific extracellular matrix influences on mesenchyme collagen biosynthesis in vitro. In: Slavkin HC, Greulich RC (eds) Extracellular matrix influences on gene expression. Academic, New York, pp 237–251

Slavkin HC, Trump GN, Brownell A, Sorgente N (1977) Epithelial mesenchymal interactions: mesenchymal specificity. In: Lash JW, Burger MM (eds) Cell and tissue interactions, vol 32. Raven, New York, pp 29–46

Slavkin HC, Cummings F, Bringas P, Honig LS (1982) Epithelial derived basal lamina regulation of mesenchymal cell differentiation. In: Weber R, Burger A (eds) Proc Int Soc Develop Biol, IX Congress. Liss, New York

Sodek J, Mandell SM (1982) The synthesis and maturation of type I, type V and 1(I)-trimer collagen in rat predentine in vivo. In: Veis A (ed) The chemistry and biology of mineralized connective tissue. Elsevier, North Holland New York, pp 19–21

Spector M (1975) High resolution electron microscope study of lattice images in biological apatites. J Microsc 103:55–62

Stenman S, Vaheri A (1978) Distribution of major connective tissue protein, fibronectin in normal tissues. J Exp Med 147:1054–1064

Stetler-Stevenson WG, Veis A (1983) Bovine dentin phosphoryn: composition and molecular weight. Biochemistry 22:4326–4335

Stone LS (1926) Further experiments on the extirpation and transplantation of mesectoderm in *Amblystoma punctatum*. J Exp Zool 44:95–131

Symons NBB (1961) A histochemical study of the intertubular and peritubular matrices in normal human dentine. Arch Oral Biol 5:241–250

Takagi M, Parmley RT, Denys FR (1981) Ultrastructural localization of complex carbohydrates in odontoblasts, predentin and dentin. J Histochem Cytochem 29:747–758

Takahashi K (1985) Vascular architecture of dog pulp using corrosion resin cast examined under a scanning electron microscope. J Dent Res 64:579–584

Takuma S, Eda S (1966) Structure and development of the peritubular matrix of dentin. J Dent Res 45:683–692

Takuma S, Nagai N (1971) Ultrastructure of rat odontoblasts in various stages in their development and maturation. Arch Oral Biol 16:993–1011

Tanaka T (1980) The origin and localization of dentinal fluid in developing rat molar teeth studied with lanthanum as a tracer. Arch Oral Biol 25:153–162

Ten Cate AR (1972) An analysis of Tomes' granular layer. Anat Rec 172:137–149

Ten Cate AR (1978) A fine structural study of coronal and root dentinogenesis in the mouse: observations on the so-called von Korff fibres and their contribution to mantle dentine. J Anat 125:183–197

Ten Cate AR (1980) Oral histology, development, structure and function. Mosby, St Louis

Ten Cate AR, Melcher AH, Pudy G, Wagner D (1970) The nonfibrous nature of the von Korff fibers in developing dentine. A light and electron microscope study. Anat Rec 168:491–523

Termine JD (1972) Mineral chemistry and skeletal biology. Clin Orthop 85:207–241

Termine JD, Coon KM (1976) Inhibition of apatite formation by phosphorylated metabolites and macromolecules. Calcif Tissue Res 22:149–157

Thesleff I (1978) Role of the basement membrane in odontoblast differentiation. J Biol Buccale 6:241–249

Thesleff I, Hurmerinta K (1981) Tissue interactions in tooth development. Differentiation 18:75–88

Thesleff I, Partanen AM (1984) Growth factors and tooth morphogenesis. In: Belcourt A, Ruch JV (eds) Tooth morphogenesis and differentiation, vol 125. Inserm, Paris, pp 73–85

Thesleff I, Pratt RM (1980) Tunicamycin-induced alterations in basement membrane formation during odontoblast differentiation. Dev Biol 80:175–185

Thesleff I, Stenman S, Vaheri A, Timpl R (1979) Changes in the matrix proteins, fibronectin and collagen during differentiation of mouse tooth germ. Dev Biol 70:116–126

Thesleff I, Barrach HJ, Foldart JM, Vaherl A, Pratt RM, Martin GR (1981) Changes in the distribution of type IV collagen, laminin, proteoglycan and fibronectin during mouse tooth development. Dev Biol 81:182–192

Timpl R, Rohde H, Robey PG, Rennard SI, Foidart JM, Martin GR (1979) Laminin. A glycoprotein from basement membranes. J Biol Chem 254:9933–9937

Tomes J (1856) On the presence of fibrils of soft tissue in the dentinal tubes. Philos Trans R Soc Lond [Biol] 146:515–522

Tominaga H, Sasaki T, Higashi S (1984) Ultrastructural changes in odontoblasts during early development. Bull Tokyo Dental Coll 25:9–26

Veis A, Schleuter RJ (1964) The macromolecular organization of dentine matrix collagen. I. Characterization of dentine collagen. Biochemistry 3:1650–1657

Veis A, Stetler-Stevenson W, Takagi Y, Sabsay B, Fullerton R (1982) The nature and localization of the phosphorylated proteins of mineralized dentin. In: Veis A (ed) The chemistry and biology of mineralized connective tissues. Elsevier, North Holland New York, pp 377–387

Voegel JC, Frank RM (1977) Ultrastructural study of apatite crystal dissolution in human dentine and bone. J Biol Buccale 5:181–194

Waldeyer W (1865) Über den Ossifikationsprocess. Arch Mikr Anat 1:354–375

Warshawsky H, Josephsen K (1981) The behavior of substances labeled with ^{3}H-proline and ^{3}H-fucose in the cellular processes of odontoblasts and ameloblasts. Anat Rec 200:1–10

Watson ML (1960) The extracellular nature of enamel in the rat. J Biophys Biochem Cytol 7:489–492

Watson ML, Avery JK (1954) The development of the hamster lower incisor as observed by electron microscopy. Am J Anat 95:109–162

Weidenreich F (1925) Über den Bau und die Entwicklung des Zahnbeins in der Reihe der Wirbeltiere. Knochenstudien IV. Teil. Z Anat Entwickl Gesch 76:218–260

Weinstock A (1972a) Matrix development in mineralizing tissues as shown by radioautography: formation of enamel and dentin. In: Slavkin HC, Bavetta LA (eds) Developmental aspects of oral biology. Academic, New York, pp 201–242

Weinstock A (1972b) Elaboration of enamel and dentin matrix glycoproteins. In: Bourne GH (ed) The biochemistry and physiology of bone, vol II. Academic, New York, pp 121–154

Weinstock A, Weinstock M, Leblond CP (1972) Radioautographic detection of ^{3}H-fucose incorporation into glycoprotein by odontoblasts and its deposition at the site of the calcification front in dentin. Calcif Tissue Res 8:181–189

Weinstock M (1972) Collagen formation – Observations on its intracellular packaging and transport. Z Zellforsch 129:455–470

Weinstock M, Leblond CP (1973) Radioautographic visualization of the deposition of a phosphopro-tein at the mineralization front in the dentin of the rat incisor. J Cell Biol 56:838–845

Weinstock M, Leblond CP (1974) Synthesis, migration and release of precursor collagen by odonto-blasts as visualized by radioautography after (^{3}H)-proline administration. J Cell Biol 60:92–127

Weiss MP, Voegel JC, Frank RM (1981) Enamel crystallite growth: width and thickness study related to the possible presence of octocalcium phosphate during amelogenesis. J Ultrastruct Res 76:286–292

White SW, Hulmes DJS, Miller A, Timmins PA (1977) Collagen-mineral axial relationship in calcified turkey leg tendon by X-ray and neutron diffraction. Nature 266:421–425

Whittaker DK, Adams DA (1972) Electron microscopic studies on von Korff fibers in the human developing tooth. Anat Rec 174:175–189

Wohllebe M, Carmichael DJ (1978) Type I-trimer and type I collagen in neutral-salt soluble lathyritic rat dentin. Eur J Biochem 92:183–188

Wolters JML (1978) The transfilter transmission of ^{3}H-proline labelled material in cultured rat tooth germs. Arch Oral Biol 23:51–55

Wolters JML, Mullem PJ van (1978) Electron microscopy of epithelio-mesenchyme intercellular communication in trans-filter cultures of rat tooth germs. Arch Oral Biol 22:705–707

Wuthier RE (1984) Lipids in dentinogenesis. In: Linde A (ed) Dentin and dentinogenesis, vol II. CRC Press, Boca Raton, pp 93–106

Wuthier RE, Irving (1964) Lipids in developing calf bone. J Dent Res 43:814–815

Young RW, Greulich RC (1963) Distinctive autoradiographic patterns of glycine incorporation in rat enamel and dentin matrices. Arch Oral Biol 8:509–521

Zaki AE, Weber DF (1985) Fine structure of intercellular junctions in human odontoblasts. J Dent Res 64:237 (abstract)

Structure and Ultrastructure of Dentine

R.M. Frank and J. Nalbandian

A. Basic Anatomy

As mentioned by Baume (1980) in his excellent review of pulp and dentine, the histological study of dentine began when Anthonie Van Leuwenhoek (1675), the inventor of the microscope, first noted the tubular structure in this tissue. Leuwenhoek's drawings of tooth sections failed to show a distinction between dentine and enamel, however, and he reported to the Royal Society of London "... that the whole tooth was made up of very small straight and transparent pipes. Six or seven hundred of these pipes put together, I judge, exceed not the thickness of one hair of man's beard".

Dentine is a mineralized connective tissue which forms the bulk and frame of the teeth, both deciduous and permanent, and provides their basic shapes (Figs. 1 and 2). In the crown it is covered by enamel, and in the root, by cementum. Peripherally, dentine circumscribes the dental pulp and constitutes the calcified walls of this highly vascularized and innervated loose connective tissue. In fact, as was demonstrated in the previous chapter, dentine and pulp consist of an undissociable complex, since dentine is produced by peripheral pulpal cells, the odontoblasts, cells which have processes that extend into the dentinal tubules. Embryologically, metabolically, and functionally the tissues are closely related to each other.

A large number of dentinal tubules extend through the dentine from the pulpal surface to the dentinoenamel junction (Figs. 3, 4) or to the cementodentinal junction, usually with an S-shaped curvature (Fig. 4). The dentinal tubules are separated by a calcified intercellular matrix, the intertubular dentine, and they are walled by a calcified matrix with unique properties, the so-called peritubular dentine. The inorganic phase of dentine consists mainly of hydroxyapatite crystals and the organic phase is composed largely of collagen. Dentine is slightly harder than bone and cementum but softer than enamel, and on radiographs, dentine is more radiolucent (darker) than enamel.

Root dentine is sometimes permeated by accessory canals which connect the pulp and periodontal ligament. These may occur laterally, near the apex, or in furcation regions (Fig. 2). In deciduous molars, a relatively high frequency of accessory canals has been noted in the interradicular dentine constituting the floor of the pulp chamber (Winter 1962; Vermot-Gaud 1967).

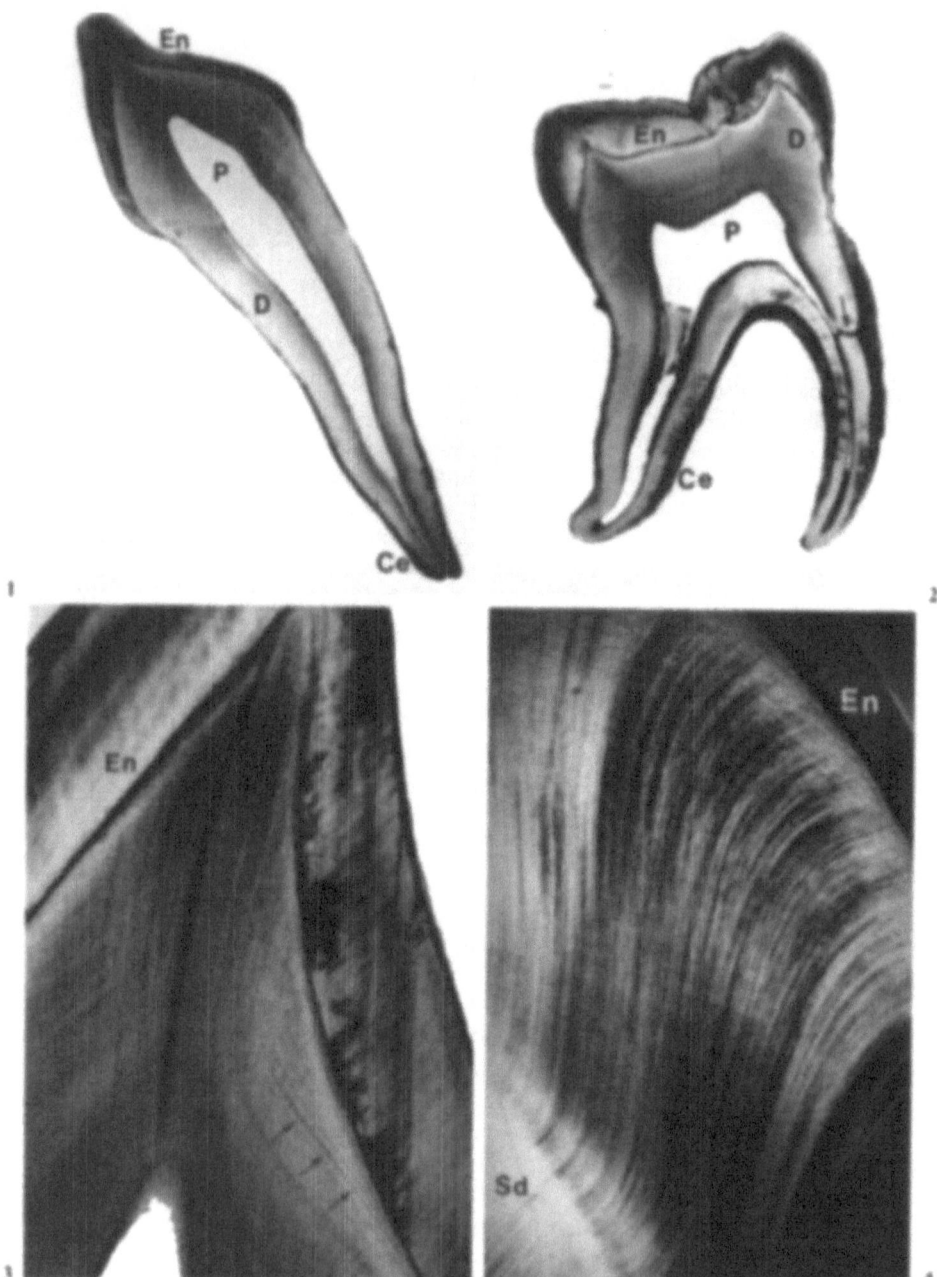

Fig. 1. Ground section of a human deciduous incisor. *En* enamel; *D* Dentine; *P* Pulp area; *Ce* Cementum. × 3

Fig. 2. Ground section of a human permanent molar. *En* Enamel. *D* Dentine; *Ce* Cementum; *P* Pulp area with an accessory canal (*arrow*). × 2,2

B. Physical and Chemical Characteristics

Dentine is light yellow in color. It has a high degree of elasticity, the average value for the elastic modulus being 1.79×10^6 lbs/in^2 (SCOTT and SYMONS 1982). It might be expected a priori that the elastic constants of dentine would depend on tubule orientation. However, detailed study has revealed no significant effect from this variable (BRADEN 1976). Furthermore, extensive measurements indicate that elastic constants are unaffected by age, sex, tooth type, and location within the teeth.

Enamel is much harder than dentine by a factor of about four to five. The mean Knoop hardness number (KHN) expressed in kg/mm^2 and measured by the load necessary to indent 1 mm^2 of surface with a calibrated diamond is about 63KHN (BRADEN 1976). The average value of the breaking stress of dentine is 38,800 lbs/in^2 (SCOTT and SYMONS 1982).

Dentine has electrical properties: like enamel, it is a dielectric. However, both piezo- and pyro-electricity have been demonstrated in dentine whereas such effects do not occur in enamel (ATHENSTAEDT 1971). Piezo-electricity is the generation of electrical charges by mechanical stress on a substance. Hydroxyapatite is not a piezo-electric crystal, and hence the source of the piezo-electricity in dentine is of obvious interest. Since it is not the mineral phase, collagen seems to be the probable source of the effect. Dentine is also pyro-electric, as are other tissues (ATHENSTAEDT 1971). Pyro-electricity is the generation of electric charges due to heating.

Mature dentine is composed of approximately 70% inorganic material, 20% organic material and 10% water by weight. On a volume basis, it is approximately 47% inorganic material, 32% organic material and 21% water (ARMSTRONG and BREKHUS 1937; LEFEVRE and MANLY 1938).

Chemical data show that the mineral phase of dentine is primarily calcium phosphate. A Ca/P molar ratio of 1.969 ± 0.062 was calculated by DRIESSENS (1982) for dentine mineral on the basis of data from the literature on the contents of Ca, PO$_4$, Na and Mg. The highest concentrations of total phosphate were found in peripulpal area of coronal dentine (LAIKKO and LARMAS 1979). Root dentine contained less total phosphate than crown dentine. The amount of total phosphate increased significantly with increasing age of the teeth. Carbonate in dentine, as in bone, is of the order of 3 to 4% expressed as CO_3^- (POSNER and TANNENBAUM 1984). Average contents of Mg, Na, and K on a weight basis are respectively, 1.2, 0.4, and 0.1 per cent (GRÖN 1978; DRIESSENS 1982).

Fig. 3. Ground section of a human permanent incisor illustrating the structure of dentinal tubules as well as contour lines of VON EBNER (*arrows*). *En* Enamel; *Re* Retzius lines. × 11

Fig. 4. Ground section of a human permanent canine showing the curvature of dentinal tubules. *Sd* Secondary dentine. *En* Enamel. × 65

WEATHERELL and ROBINSON (1973) list over 45 elements which are found in trace amounts in dentine mineral.

It is generally recognized that the crystal structure of the calcium phosphate mineral in dentine is that of hydroxyapatite: $Ca_{10}(PO_4)_6(OH)_2$. Whether the mineral in normal dentine is in a single phase is an issue that needs further investigations. The crystal structures of the calcium phosphate biominerals can generally be modeled as derivatives (principally via atomic substitutions and disordering) of one of five model compounds (YOUNG and BROWN 1982):

$Ca_{10}(PO_4)_6(OH)_2$	(hydroxyapatite)
$Ca_8H_2(PO_4)_6 \cdot 5H_2O$	(octocalcium phosphate)
β-$Ca_3(PO_4)_2$ and its modification	(whitlockite)
$CaHPO_4 - 2H_2O$	(brushite)
$CaHPO_4$	(monetite)

A certain proportion of dentine mineral consists of water absorbed onto the apatite surface or in interstices between crystals. About 8 to 16% of dentine on an air-dried basis consists of water which can be removed by heating to 120° C (WEATHERELL and ROBINSON 1973). Most of this water is surface-adsorbed, as only very small quantities of water are found within the apatite structure.

Certain ions are specific apatite seekers and can substitute in the apatite structure. Examples of these are: Cl^-, F^-, Mg^{2+}, Sr^{2+}, Fe^{2+}, Zn^{2+}, and Pb^{2+}.

The size of the apatite crystal in dentine has been determined by line broadening measurements on X-ray diffraction patterns, and a dimension of about 250 Å along the c-axis has been found (TOVBORG JENSEN et al. 1948; TRAUTZ et al. 1953).

By electron microscopy, WATSON and AVERY (1954) and JOHANSEN and PARKS (1962) were among the first to describe the mature apatite crystals in dentine as platelets. JOHANSEN (1964) reported the length of the plate-like crystals in intertubular dentine to be 500–600 Å with a somewhat lesser width and thickness of 20–35 Å. By using high-resolution electron microscopy, upon simultaneous visualization of three lattice plane sets equivalent to (100) planes (forming 60° with each other and with an 8.2 Å equidistance), one is sure to have made a transverse section of an apatite crystal strictly perpendicular to the c-axis of the crystal. With this method, VOEGEL and FRANK (1977) and FRANK and VOEGEL (1978) were able to confirm the flattened hexagonal prism shape of the mature apatite cystals of human dentine. They found a mean width to thickness ratio of 3.6±0.14, with a mean width of 364.5A±14.50 and mean thickness of 103.3A±2.7 for intertubular dentine crystals.

The organic portion of dentine is predominantly made up of collagen which constitutes 90–95% of the weight of the entire organic material. The presence of type I collagen, exclusively, is now recognized although the presence of minor amounts of type I-trimer is accepted but not unequivocally established (BUTLER 1984a). In addition, the organic matrix of dentine contains non-collagenous proteins such as proteoglycans, phosphoproteins, γ-carboxyglutamic acid (Gla)-containing proteins, serum proteins, glycoproteins, and fibronectin (BUTLER 1984b). The nature and function of these non-collagenous proteins has been discussed in the chapter on development of dentine and pulp. Dentine also contains lipids and citrate.

C. Classification of Different Types of Dentine

The specific characteristics of dentine as compared to bone were recognized as early as 1778 by John Hunter and a significant change in the French terminology from "osseous tooth substance" into "ivory" was initiated by CUVIER (1805).

The initial peripheral layer formed in coronal as well as in radicular dentine, as described in the chapter on dentine and pulp development, was referred to as the mantle dentine by WEIDENREICH (1925). The development of this very thin layer is followed by the formation of the so-called circumpulpal dentine, constituting the main mass of coronal and radicular dentine.

Orthodentine is the normal tubular dentine composed of predentine, tubules with their contents, and peritubular and intertubular dentine. Orthodentine has been distinguished from osteodentine, vasodentine and plicidentine (BRADFORD 1967). Osteodentine, often found in fish teeth, has more resemblance to bone than to mammalian dentine. Vasodentine is a special type of circumpulpal dentine containing blood vessels and especially capillaries. In plicidentine, found in elasmobranch fish, the tubules radiate perpendicularly to the boundary of a highly convoluted tubules (TOMES 1898).

Apart from the above forms which relate to species differences, it is the classification of KUTTLER (1959) into primary, secondary, and tertiary dentine that has been recognized classically for mammalian dentine. Primary dentine is the regular normal dentine, most of which is formed prior to eruption. Primary dentine is produced during the formation period up to the completion of the apical tooth region. Secondary dentine comprises that portion of the regular orthodentine region which is in continuity with primary dentine and which is produced circumpulpally during the later periods of tooth life. A line of demarcation and often a change in tubule curvature is usually apparent at the junction of primary and secondary dentine. Secondary dentine is formed in response to the slight irritating effects of normal biological function.

The notion of tertiary dentine was proposed by KUTTLER (1959) as a replacement for a series of confusing conceptual terms for irregular secondary dentine, such as irritation dentine, reparative dentine, irregular dentine, reaction dentine, replacement dentine, defense dentin, etc. Tertiary dentine is a more or less irregular dentine produced locally in reaction to noxious stimuli such as abrasion, mechanical, chemical or thermal stresses, cavity and crown preparations, erosion, caries, etc. Unlike primary and secondary dentine, tertiary dentine is only produced in the area directly affected by the stimulus. This type of dentine occurs with irregular tubules and tubules can even be absent. Calcification can be deficient and cellular inclusions can be found that result in the occurrence of spaces (STANLEY and WHITE 1966).

D. The Odontoblasts and Their Processes

The odontoblasts are post-mitotic cells. They constitute a layer of elongated cells at the periphery of the pulp, cells which extend as long cellular processes, the odontoblastic processes, into the dentinal tubules (Figs. 5, 6, 15). The odon-

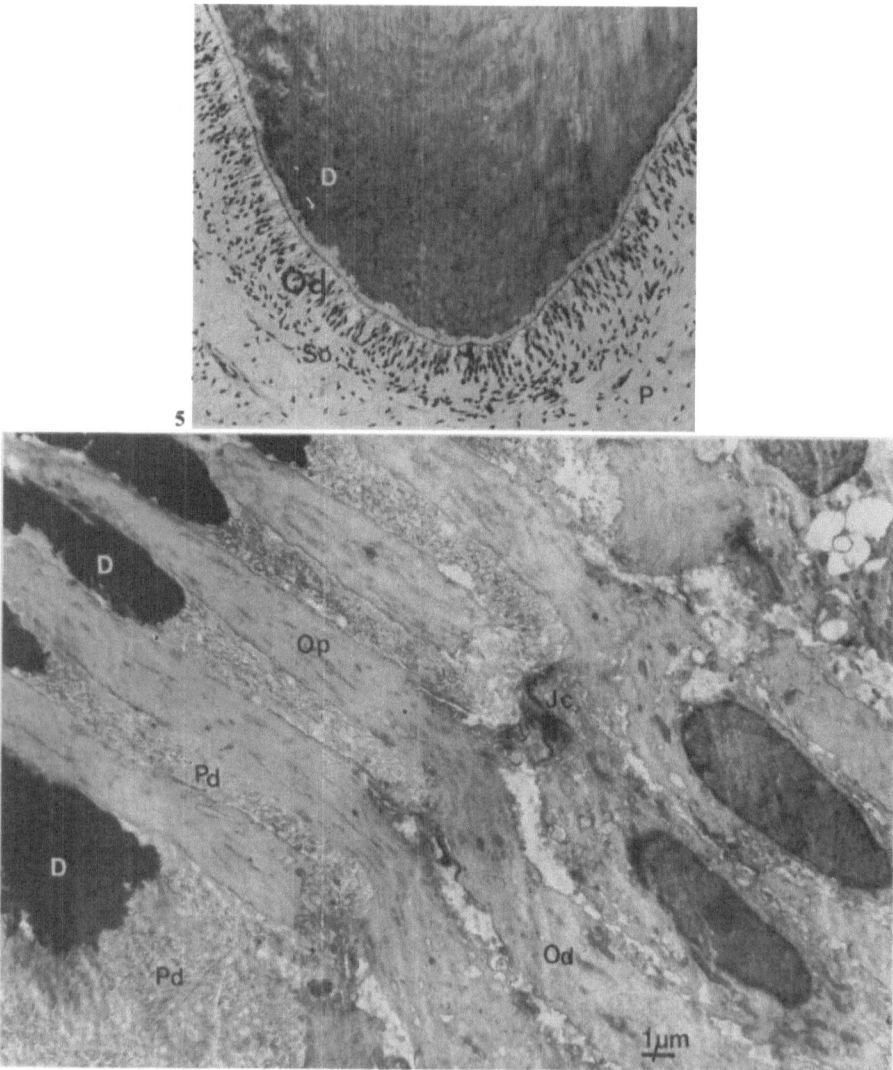

Fig. 5. Decalcified section of dentine and pulp underlying a premolar fissure and showing a well-differentiated odontoblastic layer (*Od*) located along a thin predentine layer. *D* Dentine with straight tubules; *So* Subodontoblastic layer; *P* Pulp. × 160

Fig. 6. Longitudinal non-decalcified section with odontoblasts (*Od*) and their processes (*Op*), predentine (*Pd*) and dentine as (*D*) observed in transmission electron microscopy. *Jc* Junctional complex

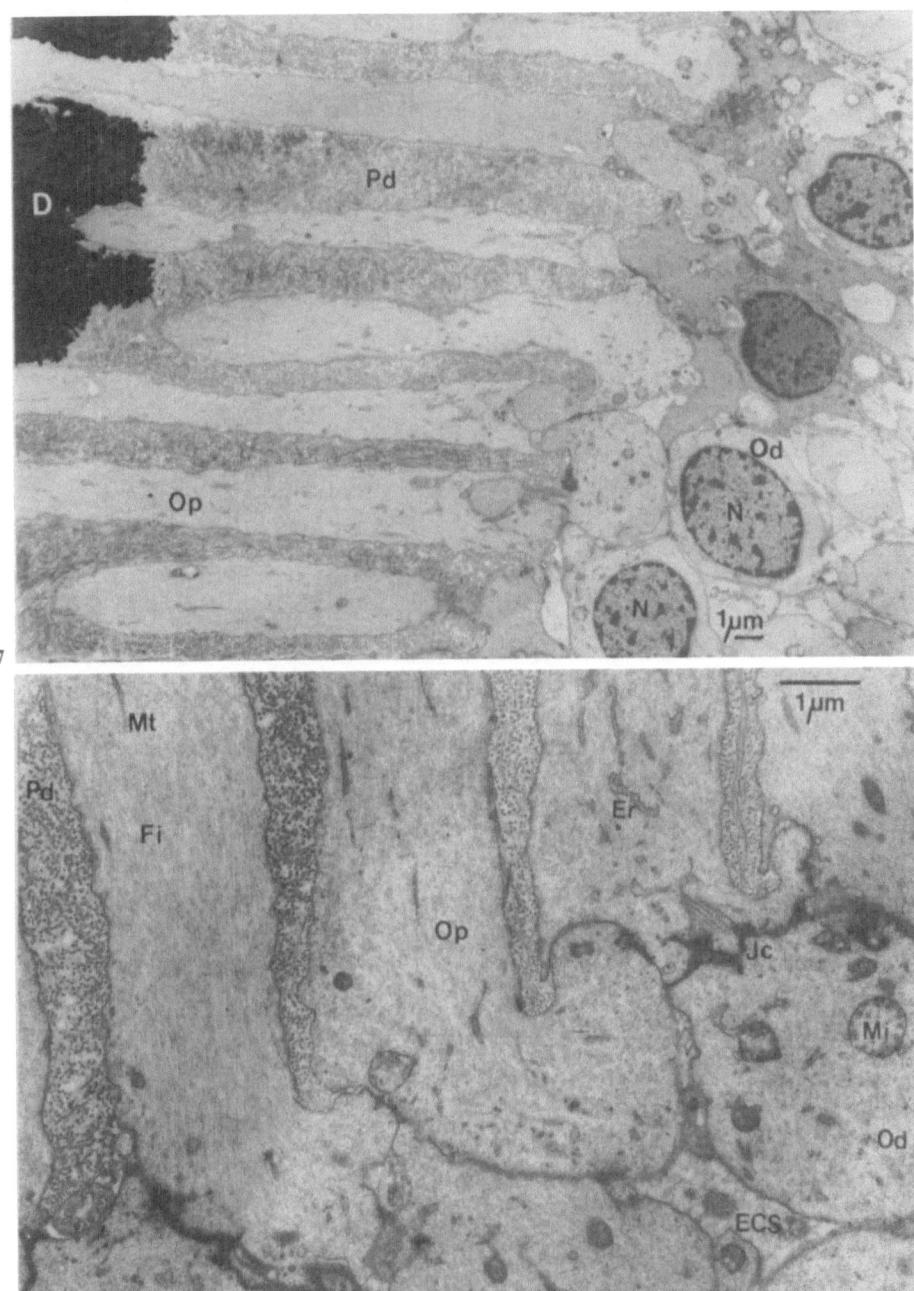

Fig. 7. Non-decalcified section showing human odontoblasts (*Od*), Predentine (*Pd*) and dentine (*D*). *N* Nucleus of an odontoblast; *Op* Odontoblastic process

Fig. 8. Human adult odontoblasts (*Od*) and their processes (*Op*). *Pd* Predentine; *Fi* Microfilaments; *Mt* Microtubule; *Er* Endoplasmic reticulum; *Mi* Mitochondria; *ECS* Extracellular space containing a bundle of cross-cut collagen fibrils; *Jc* Junctional complex

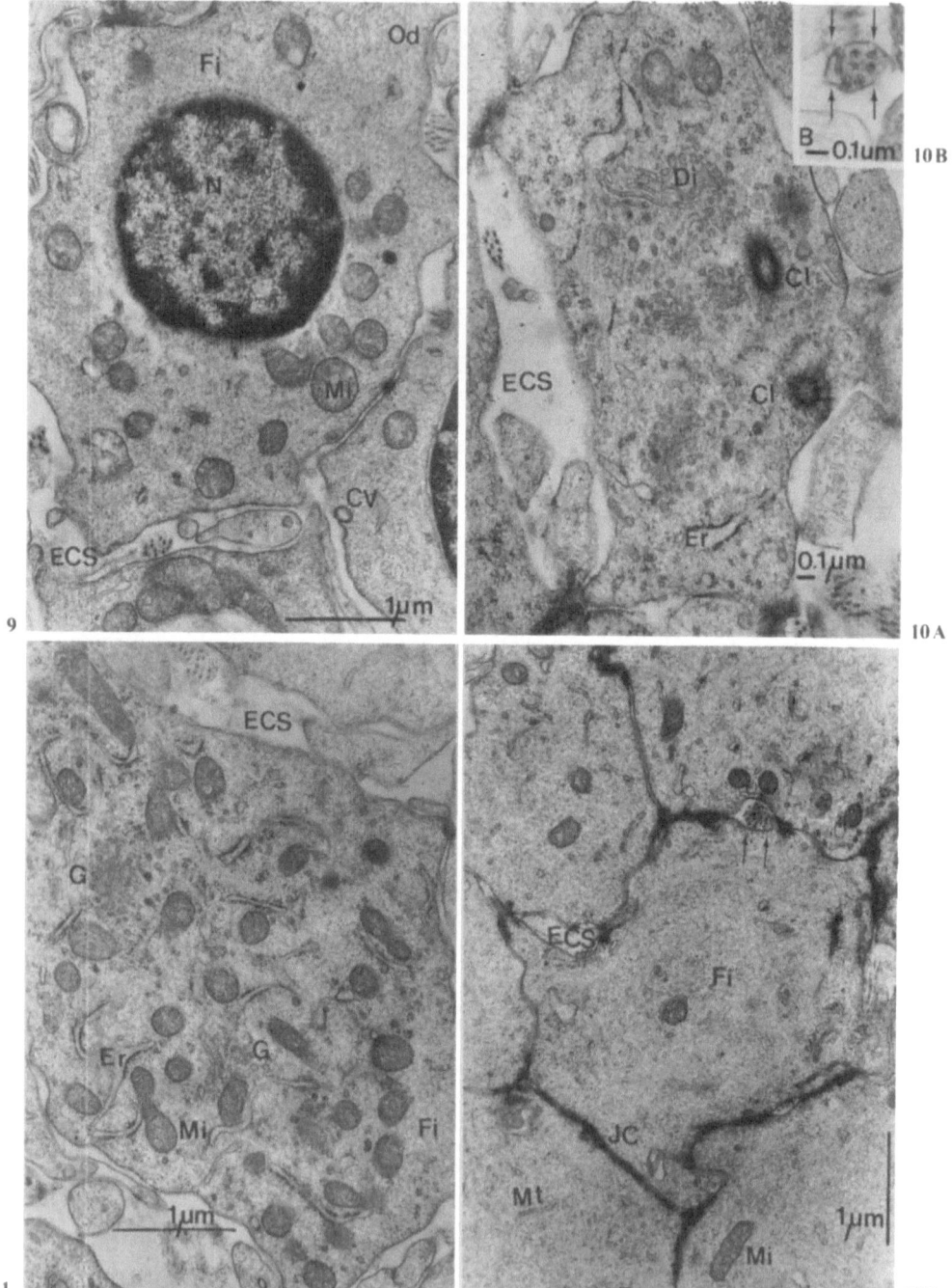

Fig. 9. Transverse section of the adult human odontoblasts (*Od*). *N* Nucleus of an odontoblast; *Fi* Microfilaments; *Mi* Mitochondria; *CV* Coated vesicle; *ECS* Extracellular space. (From FRANK 1966b, courtesy of Springer)

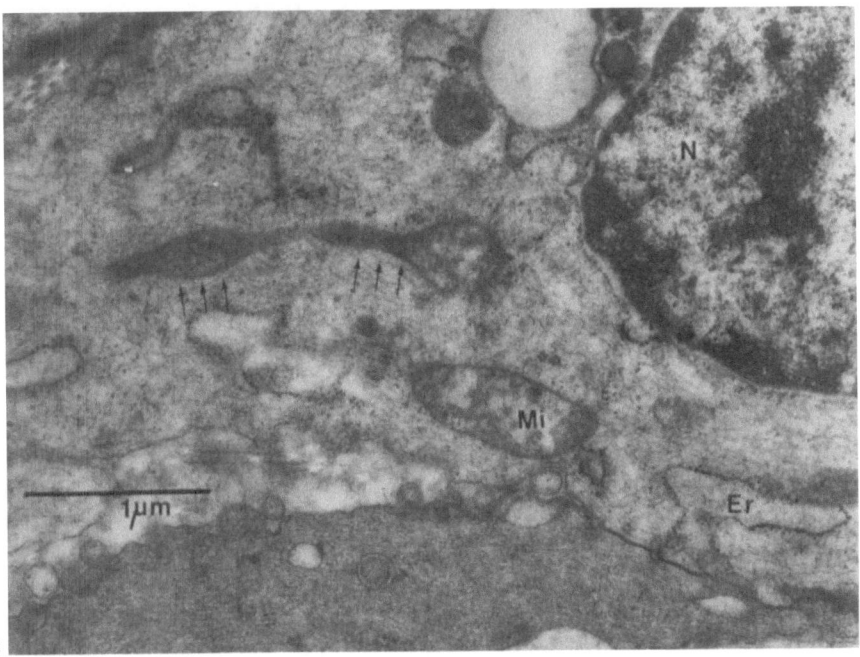

Fig. 12B. Longitudinal section through the distal part of a human odontoblast. A dumbbell-shaped organelle (*arrows*) is indicative of intracellular collagen resorption. *N* nucleus; *Er* Endoplasmic reticulum; *Mi* Mitochondria

toblasts form a single palisade layer adjacent to the predentine, but due to their irregular lengths an impression of a multilayered configuration results (Figs. 5–7). The cell bodies of the coronal odontoblasts are larger in size than those of the root odontoblasts and generally have an elongated cylindrical shape measuring about 25 to 60 µm in length and 4 to 7 um in diameter. Some assume a pear-shaped outline. In the root, the odontoblast cell bodies are shorter and more cuboid (TEN CATE 1980).

Fig. 10A. Transverse section of the adult human odontoblasts at the level of the Golgi apparatus. *Di* Dictyosomes; *Cl* Centriole; *Er* Endoplasmic reticulum; *ECS* Extracellular space. (From FRANK 1966a, courtesy of Archives of Oral Biology)

Fig. 10B. Cross section of a cilium (*arrows*) of an adult human odontoblast located in the space between odontoblasts

Fig. 11. Cross section of a human odontoblast at the level of the distal part of the cell, between the nucleus and the odontoblast process. *G* Part of the Golgi apparatus; *Er* endoplasmic reticulum; *Mi* mitochondria; *Fi* Microfilaments; *ECS* Extracellular space. (From FRANK 1966, courtesy of Archives of Oral Biology)

Fig. 12A. Cross section through the distal part of odontoblast cell bodies, adjacent to predentine. *Mi* Mitochondria; *Mt* Microtubules; *Fi* Microfilaments; *JC* Junctional complex; *ECS* Extracellular space; *Arrows* Cross-cut collagen fibrils. (From FRANK 1966a, courtesy of Archives of Oral Biology)

I. The Odontoblast Cell Body

In the adult human tooth, various structural appearances have been observed among the odontoblasts. For convenience these can be classified into 3 types. The first is that of a cell actively engaged in synthetic activities, exhibiting a well-developed Golgi apparatus surrounded by abundant endoplasmic reticulum and numerous mitochondria, an appearance similar to that exhibited by the odontoblasts during active primary dentinogenesis (Figs. 10A, B, 11). The second and most common is that of a resting cell with a moderately developed Golgi apparatus and a less extensive network of endoplasmic reticulum and mitochondria (Figs. 6–9, 12A). A third category is represented by cells exhibiting lysosome-like bodies and engaged in resorptive activity (Figs. 12B, 13) (Frank 1966b, 1968a).

These various cell forms will be described in detail. All have in common an oval nucleus located in an eccentric position in the part of the cell adjacent to the pulp (non-formative part). The nucleus is made up of dispersed euchromatin with moderate amounts of dense heterochromatin adjacent to the inner nuclear membrane (Figs. 6, 7, 9, 12B). In the nucleus of adult rat odontoblasts, Jessen (1967) found an elongated inclusion consisting of 6 to 7 parallel filaments.

One to four prominent nucleoli are present in the odontoblast nuclei. According to Ivanyi (1972), ring-shaped nucleoli typical for reversible inhibited RNA synthesis are found in human mature odontoblasts, an indication that these cells are able to respond to adequate stimulation by the synthesis of new RNA. This is in agreement with the view that the odontoblasts, after the initial deposition of primary dentine, do not cease their activity and are able to produce secondary or tertiary dentine. The nucleus is surrounded by a flattened envelope. A small space, the socalled perinuclear space, separates the outer nuclear membrane from the inner nuclear membrane. The outer membrane surface has ribonucleoprotein granules attached to it and is continuous with the rough-surfaced endoplasmic reticulum. A smooth inner nuclear membrane circumscribes the nuclear content. Nuclear pores, 500–800 Å in diameter, occur at intervals along the envelope.

Near the distal side of the nucleus, in the area of the Golgi apparatus, a pair of centrioles (Fig. 10A) has been observed in adult odontoblasts (Frank 1966a; Jessen 1967; Garant et al. 1968). One centriole acts as a basal body and gives rise to a cilium which may (Fig. 10B) or may not extend into the interodontoblastic space.

In odontoblasts actively engaged in synthesis, a highly developed rough-surfaced endoplasmic reticulum is observed closely associated with numerous mitochondria (Fig. 11). These organelles are distributed throughout the entire odontoblastic cell body, surrounding the nucleus and the Golgi apparatus. The endoplasmic reticulum is composed of lipoprotein membranes delimiting interconnecting channels and taking the form of flattened sacs known as cisternae. The outer nuclear membrane is in continuity with these cisternae. Rough-surfaced endoplasmic reticulum is studded with ribosomes and smooth-surfaced (ribosome-free) endoplasmic reticulum can also be found. The cisternae of the endoplasmic reticulum are aligned parallel to the long axis of the odontoblast

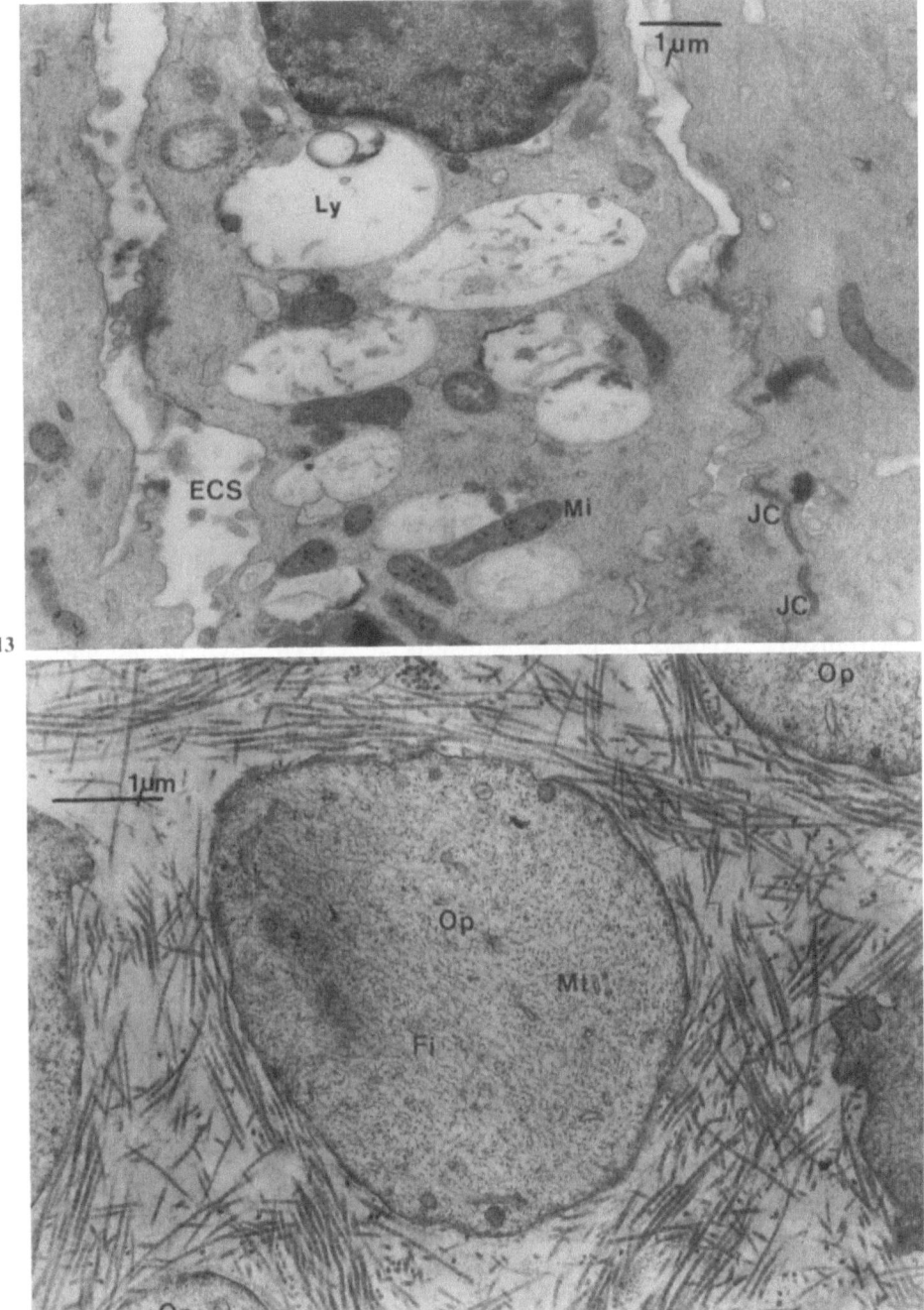

Fig. 13. Longitudinal section of human odontoblasts. Note the large vacuoles containing cellular remnants (*Ly*), probably related to phagolysosomes. Dense granules are seen in the mitochondria (*Mi*). *JC* Distal junctional complex; *ECS* Extracellular space. (From FRANK 1968a, courtesy of Livingstone)

Fig. 14. Transverse section through odontoblastic processes (*Op*) in predentine of a human premolar. The collagen fibrils are interwoven between the processes. *Mt* Microtubule; *Fi* Microfilaments

cell body. In addition to the bound ribosomes, the so-called free ribosomes are found in the cytoplasm, but from recent work it appears that these free ribosomes are bound to a three-dimensional cytoplasmic network which has been demonstrated after treatment with nonionic detergents (Penman et al. 1983).

The mitochondria closely associated with the endoplasmic reticulum are located primarily in the cell body of the odontoblast and exhibit the usual shape and ultrastructural detail. Delineated by an outer membrane, they are separated from the inner membrane by a small space. The inner membrane is folded inward to form several cristae. In the central matrix of the mitochondria, dense granules (Fig. 13) consisting of calcium phosphate accumulation can be found. The presence of hydroxyapatite has been confirmed in these granules by electron-diffraction studies. As has been described in the chapter on dentine development, mitochondria play an important role in the intracellular calcium metabolism of the odontoblasts as well as in many other metabolic pathways (Frank 1979).

Another prominent organelle in actively synthesizing odontoblasts is the Golgi apparatus, occupying a cylindrical region located on the formative side of the nucleus in the center of the cell body. Its general form is made clear on transverse sections of the odontoblasts at a level distal to the nucleus (Fig. 10A, B). The external surface of the cylinder is in contact with the transitional endoplasmic reticulum which lacks bound ribosomes. The walls of the Golgi apparatus cylinder consists of two or three stacks of about four to six dictyosomes consisting of flattened cisternae. Laterally, the ends of these dictyosomes are dilated and in that vicinity, saccules, various granules, vesicles and vacuoles are found. The convex outer face of the stacks of dictyosomes, called the "cis-face", is adjacent to the transitional endoplasmic reticulum (Rothman 1981) and intermediate vesicles are observed there. The concave internal face or trans-face of the dictyosomes has associated with it an accumulation of various saccules, smooth-surfaced vesicles, coated vesicles of various sizes, and multivesicular bodies. In mature human odontoblasts, elongated secretory vesicles containing longitudinally oriented filaments and periodically arranged granular particles such as were seen associated with dentinogenesis (Frank 1970; Weinstock and Leblond 1974; Garant 1978; Frank 1979; Sasaki et al. 1984) were not observed either in the odontoblast cell body or its process.

A substantial cytoskeleton contributes to the cell shape, polarity and metabolism of the cell body of the odontoblast; it consists of microtubules, microfilaments and intermediary filaments, described in detail in the previous chapter on dentine and pulp development. Microtubules composed of tubulin have an average diameter of 24 nm, and microfilaments of actin, ranging in diameter from 5 to 8 nm, are generally oriented parallel to the long axis of the odontoblast cell body. Further work is needed to precisely identify the location of intermediary filaments (vimentin) in the adult odontoblast. In the cytoplasmic matrix lipid granules (Harris and Griffin 1969) and glycogen deposits in rosette-like arrangement can be seen (Frank 1966a, b). It was formerly thought that the cytoplasmic matrix was an unstructured medium referred to as the "cytosol". It has recently been shown, in fact, that this matrix is a semisolid scaffold, defined as a microtrabecular lattice different from the cytoskeleton (PORTER et al. 1983). Such a fine three-dimensional cytoplasmic network is probably also present in the adult odontoblast. In fact, the odontoblast of the mature tooth engaged in synthesizing activities has many structural similarities to the odontoblast engaged in active dentinogenesis. The reader is referred to the previous chapter on dentine and pulp development for further details. The principal difference between these types of odontoblasts is the absence in the mature odontoblast and its process of the typical elongated secretory vesicles present during dentinogenesis.

Most commonly, the odontoblast cell body in adult dentine has the structure of a resting cell. The endoplasmic reticulum and the Golgi apparatus are moderately developed in this case (Figs. 6, 12 A). Around the nucleus, sparsely distributed cisternae of endoplasmic reticulum and a few mitochondria are observed among the prominent cytoskeletal elements consisting predominantly of microfilaments (Figs. 6, 9).

The third common structural appearance of the odontoblast cell body is evident by virtue of the presence of a relatively large number of lysosome-like bodies in the area of the Golgi apparatus (Fig. 13) indicating that the cell is engaged in a resorptive process. Large vacuoles containing various types of granules, vacuoles, membranes and cellular debris are seen in the odontoblastic process as well as in the cytoplasm of the distal part of the cell next to the nucleus. These heterogeneous vacuoles can be quite large, attaining a width of several microns. The presence of hydrolytic enzymes typical of lysosomal activity has been demonstrated by conventional histochemistry (TEN CATE 1967) and by electron microscopy (KATCHBURIAN and HOLT 1968; NAGAI 1970; TAKUMA and NAGAI 1971; LARSSON and BLOOM 1973; GRÄNSTRÖM et al. 1978). Using intravenously injected horseradish peroxidase (HRP), SASAKI et al. (1982) demonstrated in the older odontoblasts of rat incisors that HRP is taken into the cell through the odontoblastic process via large coated vesicles, 0.1 μm in diameter. HRP then migrates into dense bodies which, by fusion with primary lysosomes (small coated vesicles, 0.05 μm in diameter), are exposed to hydrolytic enzymes. In this way, adsorbed materials are digested intracellularly in the odontoblast.

For certain connective tissue cells such as the fibroblast, it has been shown that collagen can undergo phagocytosis (TEN CATE 1972a; BEERSTEN et al. 1974; GARANT 1976; TEN CATE et al. 1976; FRANK et al. 1977). Collagen fibrils appear intracellularly in membrane-bounded phagosomes. The lysosomal system of the fibroblast is then mobilized so that lysosomes fuse with phagosomes. Collagen is thus degraded intracellularly in typical phagolysosomes (TEN CATE 1980). To date, membrane-bounded profiles of intracellular collagen have been reported only in pulpal cells of human teeth undergoing dental caries (TORNECK 1978).

However, MAGLOIRE and DUMONT (1976) observed, in explanted primary cultures of mature human dental pulps, the differentiation of peripheral cells with structural characteristics of odontoblasts, including a cilium, highly developed endoplasmic reticulum and a typical Golgi apparatus with secretory vesicles. They noted, in addition, the presence of a special type of organelle associated with intracellular collagen degradation. This dumbbell-shaped intracyto-plasmic body located in the Golgi apparatus consisted of two round or oval electron-dense masses united by a cross-striated collagen fibril. Similar bodies have also been observed in vivo in fibroblasts of the periodontal ligament en-gaged in intracellular collagen destruction (Ten Cate and Deporter 1974). This peculiar organelle had never been described in the adult odontoblast. Re-cently, we were able to disclose such dumbbell-shaped bodies in the Golgi appa-ratus of a human odontoblast in adult dentine (Fig. 12B) in the absence of any pathological condition. It seems, therefore, that the odontoblast may be involved in collagen turnover via intracellular collagen degradation. Further investigation is necessary to assess the importance of this phenomenon.

The cell body of the odontoblast is limited by a trilaminar cell membrane which is continuous around the odontoblastic process, including its secondary or side branches. The lateral surfaces or sides of the cell bodies are generally separated by narrow intercellular spaces (Figs. 6–9, 11), but larger spaces up to one micron can be seen locally (Figs. 6, 10A, B, 11). A few collagen fibrils as well as some nerve fibers can be observed there. The latter structures will be described in the chapter on pulp when dealing with the total innervation of the dentine-pulp complex. Intercellular spaces between odontoblasts are either straight or irregular depending on the presence or absence of cellular microvil-losities. Sometimes blood capillaries are seen between the odontoblast cell bod-ies.

Various types of intercellular junctions have been described between the fibroblasts of the subodontoblastic pulpal region and the odontoblasts, as well as between adjacent cell surfaces of the odontoblasts. These junctions were initially described in thin sections with transmission electron microscopy. This approach, however, is not precise enough for positive identification of intercellu-lar junctions. Three additional techniques have been used: lanthanum hydroxide labeling and freeze-fracture replicas allow the differentiation of desmosome, gap and tight junctions, while intravascular injections of lanthanum salts or horseradish peroxidase test the permeability of the intercellular spaces between odontoblasts.

Desmosome-like junctions (Fig. 9) have been observed between odontoblasts as well as between odontoblasts and fibroblasts of the dental pulp (Jessen 1967; Frank 1968a; Holland 1975; Garant 1978; Goldberg 1983; Iguchi et al. 1984; Tominaga et al. 1984). In thin section, these junctions consist of the apposition of two unit cell membranes separated by a space of 30 to 70 nm. Electron-opaque material condensed on the intracytoplasmic sides of the plasma membranes resemble attachment plaques. The intercellular region is occupied by a somewhat electron-opaque amorphous material but intercellular structures seen in typical epithelial desmosomes are not obvious between odontoblasts (Fig. 9). In freeze-fracture replicas, no cluster of intramembranous particles characteristic of typical desmosomes was found on either the PF (protoplasmic face) or the EF (extracellular face) of these junctional regions.

Gap junctions, *nexus*, or *macula communicans* have been described between adjacent odontoblasts as well as between odontoblasts and fibroblasts of the dental pulp (Holland 1975, 1976a; Garant 1978; Goldberg 1983; Köling

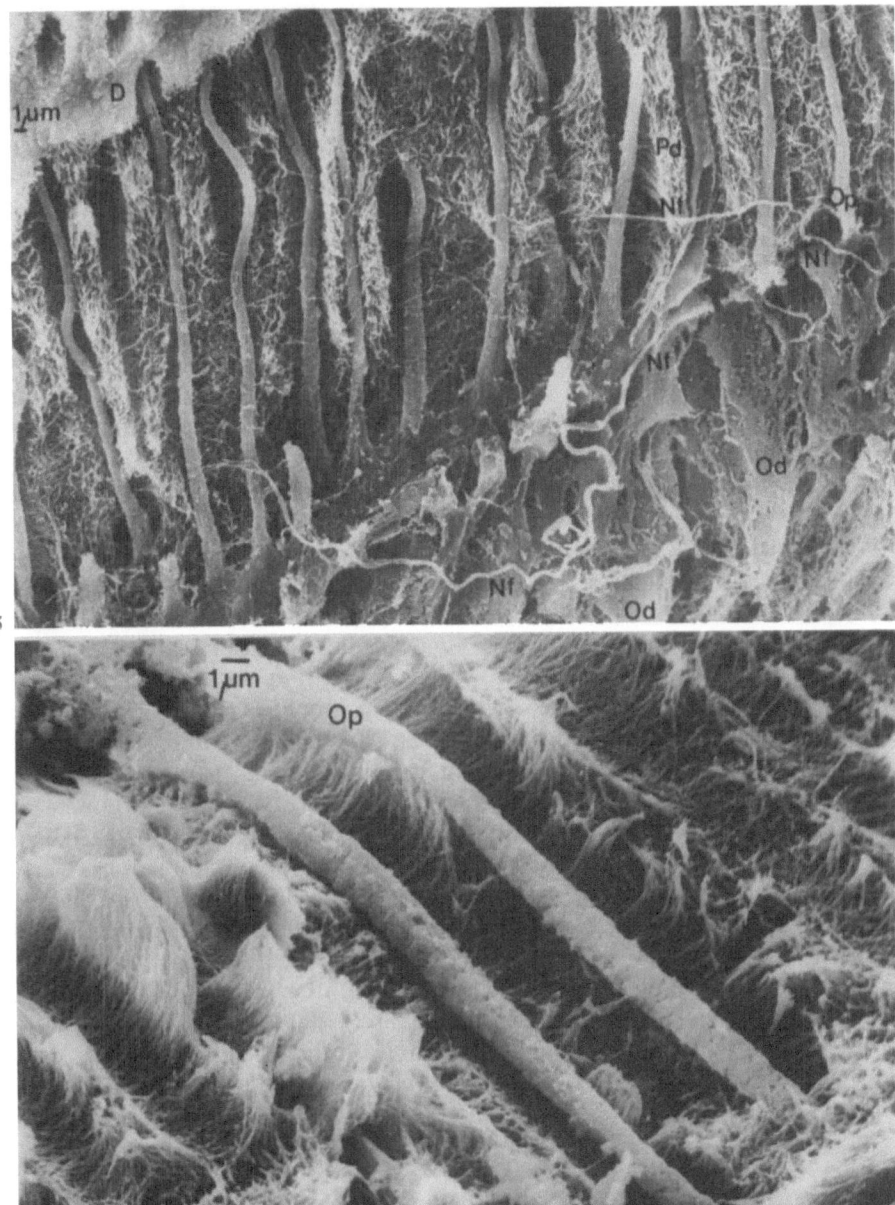

Fig. 15. Odontoblastic layer (*Od*) with odontoblastic processes (*Op*) in predentine (*Pd*) of a human premolar observed in scanning electron microscopy. Ramified nerve fibrils (*Nf*) are running almost transversely across the odontoblast layer and the predentine. *D* Dentine. (RAPP, R. and FRANK, R.M., unpublished data)

Fig. 16. Human adult predentine showing a longitudinal view of the odontoblastic processes (*Op*) in scanning electron microscopy. Note the transverse orientation of the collagen fibrils of predentine. (RAPP, R. and FRANK, R.M., unpublished data)

1983; Igushi et al. 1984; Tominaga et al. 1984; Calle 1985). The gap junctions consisted of a modification of the two outer leaflets of adjacent plasma membranes, resulting in a central dense line. The two opposed unit-membranes are separated by a gap of 2–4 nm which can be penetrated by lanthanum hydroxide. In freeze-fracture replicas, numerous round and ellipsoid gap junctions can be seen on the plasma membranes of the odontoblasts. Their mean diameter is 0.4 ± 0.11 μm (Igushi et al. 1984). On the PF, these junctions were recognized as a cluster of closely packed particles measuring 8–10 nm in diameter. On the EF face, this type of junction appeared as an ordered array of distorted hexagonal pits. Gap junctions allow direct intercellular communications between adjacent odontoblasts through functional units, the so-called connexons.

Finally, around the distal end of the odontoblast cell body, adjacent to the predentine, the presence of a distal intercellular junctional complex is generally recognized. It appears in thin section as a succession of various types of intercellular junctions (Figs. 6–8, 12A, 13, 17–19). The exact nature of the various junctions constituting the junctional complex is a matter of debate and contradictory observations are found in the literature as will be shown in the following lines.

Each junctional complex is made up of an alignment of different types of junctions. From proximal to distal there appears to be a desmosome-like junction, as described previously, and an intermediary junction (or *zonula adherens* according to the classification of Farquhar and Palade 1963). The nature of the most distal junction adjacent to predentine is controversial.

A first group of morphological investigations made on thin sections and freeze-fracture replicas revealed no tight junctions in these areas. Goldberg (1983) reported *zonula adherens*, whereas Köling (1983) found gap junctions in this site. With generally the same techniques, Calle (1985) reported gap junctions as well as tight junctions to be part of the junctional complex. Similarly Iguchi et al. (1984) observed tight junctions at the distal ends of rat incisor odontoblasts, arranged to form small *maculae* rather than belt-like *zonulae* and thought it not likely that the junction contributed to a barrier function.

Similarly, contradictory results were obtained after intravascular injections of lanthanum nitrate or horseradish peroxidase (HRP). Tanaka (1980), after the injection of lanthanum in young rats, observed the label in the interodontoblastic spaces, the predentine, and the tubules, concluding that no true tight junctions were present between the odontoblasts. With intravascular lanthanum

Fig. 17. Longitudinal section of the distal junctional complex of two adjacent human odontoblast cell bodies (*Od*) near the predentine (*Pd*)

Fig. 18. Longitudinal section of the distal intercellular space between two adjacent human odontoblasts (*Od*) near the predentine. Presence of a junctional complex

Fig. 19. Enlargement of part of a distal junctional complex between two human odontoblasts

Fig. 20. End of a human dentinal tubule (*T1*) near the dentino-enamel junction in scanning electron microscopy. Secondary (*T2*) and tertiary (*T3*) branchings can be observed. (Rapp, R. and Frank, R., unpublished data)

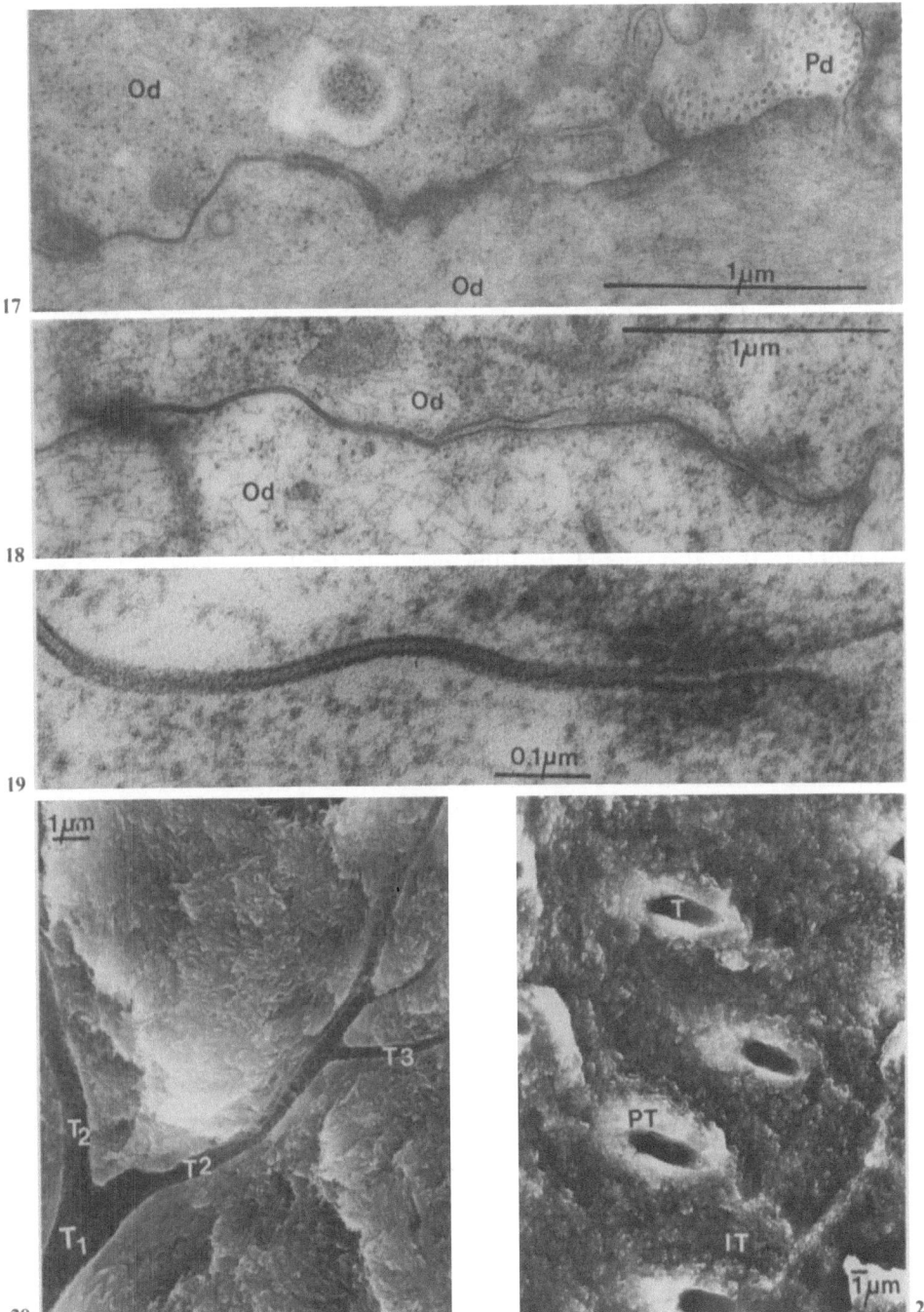

Fig. 21. Peritubular dentine (*PT*) in cross-sectioned human dentinal tubules (*T*) observed in scanning electron microscopy. *IT* Intertubular dentine. (RAPP, R. and FRANK, R., unpublished data)

nitrate, Bishop (1985) obtained exactly opposite results: in the rat incisor the lanthanum was stopped abruptly at the level of the distal junctional complex and was not found in predentine, leading him to conclude in favor of a continuous belt of tight junctions.

With intravenous injection of HRP, the distal junctional complex failed to show an effective barrier against penetration into the predentine and dentinal tubules (Sasaki et al. 1982). These authors concluded that the junctional complex consisted of macular tight junctions and gap junctions. However, in Sprague-Dawley rats, vascular HRP did not pass the junctional complex (Sattelberg and Turner 1984). The latter observation supported the presence of the *zonula occludens* configuration. Further work is needed to clarify these discrepancies.

At the junction between the odontoblasts and the odontoblastic processes, Bradford (1950) described a pulpo-dentinal membrane constituting a line of demarcation between odontoblasts and predentine in light microscopy. He proposed that von Korff's fibers may form the pulpo-dentinal membrane. Reith (1968) stated that the pulpo-dentinal membrane is related to the presence of transverse intracellular filaments which unite the most distal junctions of the odontoblast. It is also possible that this pulpo-dentinal membrane in light microscopy arises from oblique sectioning of odontoblast bodies, as seen in Figs. 7 and 8. In fact, the distinct band of transverse filaments which run across the distal ends of the odontoblasts at right angles to their axes has been described as a terminal web (Garant 1978). They insert into an electron-dense material adjacent to the inner leaflet of the plasma membrane and are thus part of an attachment apparatus that serves to fasten adjacent odontoblasts to each other along the predentine.

In the chapter on dentine and pulp development, we have shown that the terminal web has the configuration of a diaphragm with a central opening. However, it should be mentioned that in human odontoblasts, the transverse terminal web is not always present (Figs. 7, 8).

II. The Odontoblastic Process

The odontoblastic process is a direct cytoplasmic extension of the odontoblast cell body and is limited, as the latter, by a continuous trilaminar plasmalemma (Figs. 6–8, 14–16). The terminal web, when present, marks the junction between the cell body and the odontoblastic process. Quite abruptly at its origin, the odontoblastic process is narrower than the odontoblast cell body, i.e. about 2 to 4 μm in width. The process extends through the predentine and then penetrates the dentinal tubule.

Detailed knowledge of the odontoblastic process initially came through the use of transmission electron microscopy. Dentine was first studied with replica methods, then by direct examination of decalcified sections of tissue fixed with buffered osmic acid solutions (Gerould 1944; Scott and Wyckoff 1947; Helwig and Menke 1949; Syrrist 1949; Menke 1950; Scott and Kennedy 1950; Mennehöh and Fahnenbrock 1951; Pease 1951; Rouiller 1951; Syrrist and

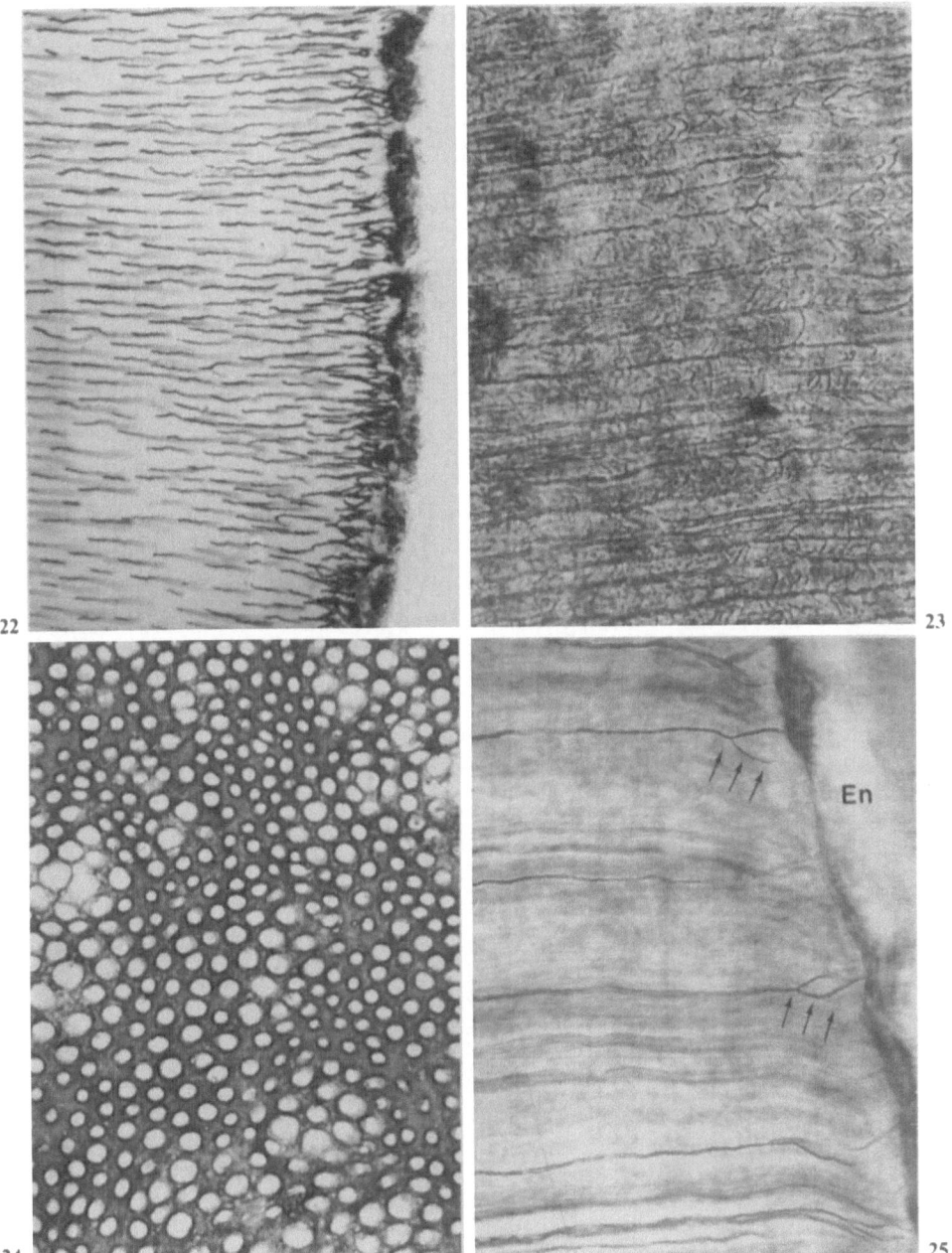

Fig. 22. Nearly longitudinal section of decalcified peripheral human dentine near the dentino-enamel junction (right) as observed in light microscopy. × 215

Fig. 23. Longitudinal section of decalcified human adult dentine stained with silver nitrate and showing the major dentinal tubules as well as numerous lateral branchings. × 460

Fig. 24. Transverse section of decalcified human adult coronal dentine. The peritubular dentine has been lost in the decalcification process. Note the different sizes of the dentinal tubules. × 850

Fig. 25. Longitudinal ground section of human coronal dentine showing the terminal branchings of the tubules (*arrows*). *En* Enamel. × 500

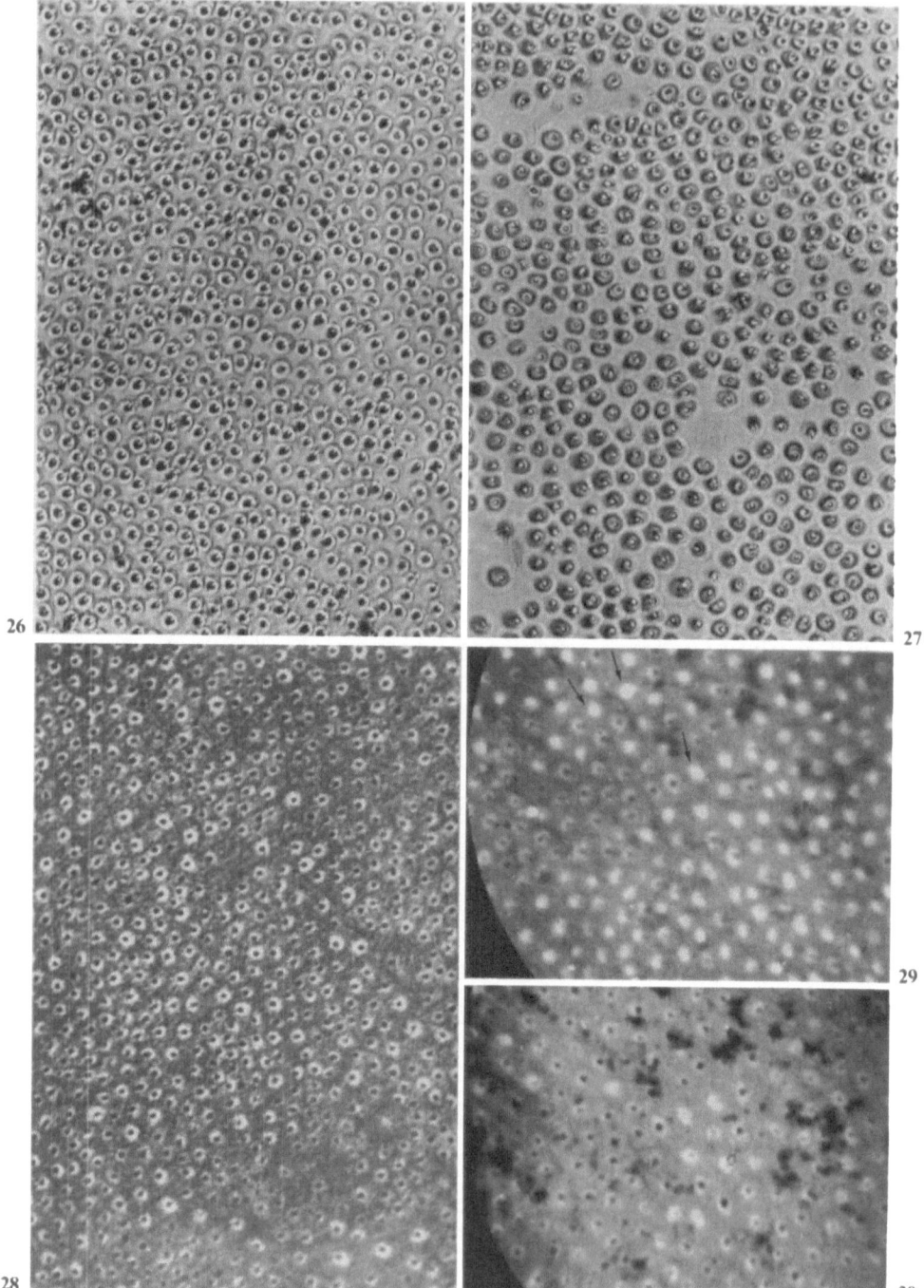

Fig. 26. Light micrograph of a transverse section of adult human coronal dentine with a homogenous distribution of dentinal tubule. The clear ring surrounding the dark central lumen corresponds to the peritubular dentine. × 645

GUSTAFSON 1951; BERNICK et al. 1952; FRANK 1952; HELMCKE and JAHN 1952; KENNEDY et al. 1953; MATSUMIYA and TAKUMA 1954; SCOTT 1952). Further progress was achieved with the use of buffered glutaraldehyde fixatives, epoxy resin embedding and diamond-knife sectioning, and more recently with SEM as well as special techniques of immunocytochemistry. Nevertheless, as noted by HOLLAND (1985), the problem of fixation still limits our understanding of the odontoblastic process, especially with respect to its length and composition in the more peripheral dentine.

The odontoblastic process, in contrast to the cell body, contains very little in the way of organelles for synthetic activity. Only a few cisternae of rough-surfaced or smooth endoplasmic reticulum, occasional polyribosomes and very sparsely occurring mitochondria are observed, mostly at the level of the predentine (Fig. 8). The process has many side branches that leave the main stem at right angles or that make an acute angle with the distal portion. These offshoots themselves frequently branch and may connect with comparable components of other odontoblasts. Cell junctions probably exist at these sites but none have been identified to date.

The odontoblastic process has a well developed cytoskeleton made up primarily of microfilaments and microtubules oriented parallel to the long axis of the process. The localization of intermediate filaments (vimentin) needs to be clarified. Microtubules and microfilaments have been observed in human, rodent, cat and even fish odontoblastic processes (FRANK 1966a, b; JESSEN 1967; FRANK 1968a; GARANT et al. 1968; REITH 1968; HARRIS and GRIFFIN 1969; FRANK et al. 1972; GARANT 1972; HOLLAND 1975b, 1976a, b).

The microtubules are proteinaceous organelles which, according to REITH (1968), are restricted to the main stem of the process and are not found in the side branches. They are made up of subunits assembled into elongated tubular structures with an average exterior diameter of 24 nm and an indefinite length, capable of rapid change in length by assembly or dissassembly of their subunit protein molecules or tubulins. These microtubules appear as elongated thin profiles in longitudinal or oblique section (Fig. 31) and as small annular structures in transverse section (Figs. 14, 35).

Fig. 27. Light micrograph of a transverse section of adult human coronal dentine. Some areas are devoid of dentinal tubules. × 570

Fig. 28. Microradiograph of a transverse section of human adult coronal dentine. The peritubular dentine, appearing as light rings, is more calcified than the intertubular dentine. × 530

Fig. 29. Secondary ion microscopy for calcium in a transverse section of adult human coronal dentine. Higher calcium content is observed in the peritubular dentine of some tubules. Sclerosed tubules can also be observed (*arrows*). In the intertubular dentine, there are areas deficient in calcium, appearing as dark spots. Primary beam $= O_2^+$. × 570

Fig. 30. Secondary ion microscopy for magnesium in a transverse section of adult human coronal dentine. The peritubular dentine and some sclerosed tubules show a higher content of magnesium. The dark spots indicate areas in the intertubular dentine where magnesium is absent. Primary beam $= O_2^+$. × 570

Fig. 31. Odontoblastic process (*Op*) of adult human coronal dentine observed in transmission electron microscopy. The cytoplasm contains microfilaments and microtubules (*arrows*). *CV* Coated vesicle; *S* Periodontoblastic space. *ID* Intertubular dentine. (From FRANK and VOEGEL 1980, courtesy of Caries Research)

Actin microfilaments, 5 to 8 nm in diameter, are found oriented longitudinally in the odontoblastic process and in its branches, as well. The actin filaments form a framework on which, by which, and with which the cytoplasm can undergo complex movements. The question can be raised as to whether the ends of these longitudinal microfilaments are attached to the cell membrane of the odontoblastic process and if so, how. Essentially no information is available on this. Perhaps, as in microvilli of intestinal epithelial cells, there is an attachment plate of some sort between the microfilaments and the plasmalemma where α-actin is involved (TILNEY 1983). As in the intestinal microvilli, the lateral surfaces of the microfilaments in the odontoblastic process are connected with each other and to the cell plasma membrane by short, thin (2–8 nm in diameter) cross connections (Figs. 8, 31).

As has been shown for certain detergent-extracted cells (PENMAN et al. 1983) a three-dimensional anastomosing network probably exists in the odontoblastic process. Such a microtrabecular lattice, different from the cytoskeleton and forming a space-filling lattice (PORTER et al. 1983) has, as yet, not been demonstrated in the odontoblast process.

In addition to the cytoskeleton, the odontoblastic process in adult dentine contains a number of small coated and smooth-surfaced vesicles, as well as large vacuoles or droplets of heterogeneous morphology. These will be described in detail. However, the characteristic small elongated secretory vesicles typically found in the odontoblastic process during dentinogenesis (see previous chapter) have never been observed in human adult dentine. Typical coated vesicles with an external fuzzy coat (Fig. 31) have repeatedly been described in the odontoblastic process (REITH 1968; FRANK 1968a; NAGAI 1970; SASAKI et al. 1984). Smooth-surfaced vesicles, or vesicles with dense cores similar in size to the coated vesicles, can also be identified in the process (Figs. 8, 14, 48). Through freeze-fracture replication, GOLDBERG and ESCAIG (1981) as well as SASAKI et al. (1984) studied the surface of the odontoblastic process at various levels and confirmed the reality of exchange phenomenona by the observation of a large number of particles in the plasmalemma. SASAKI et al. (1984) comparing the replicas with thin sections concluded that, for the proximal part of the odontoblastic process, depressions on the plasma membrane with particles represented sites of endocytosis associated with coated vesicles and indicated an absorptive activity of the odontoblast.

Multivesicular bodies have also been located in the odontoblastic process (NAGAI 1970; FRANK and VOEGEL 1980). They are quite large (about 1.5 μm

Fig. 32. Presence of uncalcified collagen fibrils in the periodontoblastic space (*S*) located between intertubular dentine (*ID*) and the odontoblastic process (*Op*)

Fig. 33. Longitudinal section of adult human coronal dentine. Uncalcified collagen fibrils in the periodontoblastic space (*S*) of the upper tubule; *ID* Intertubular dentine; *Op* Odontoblastic process. (From FRANK 1968a, courtesy of LIVINGSTONE)

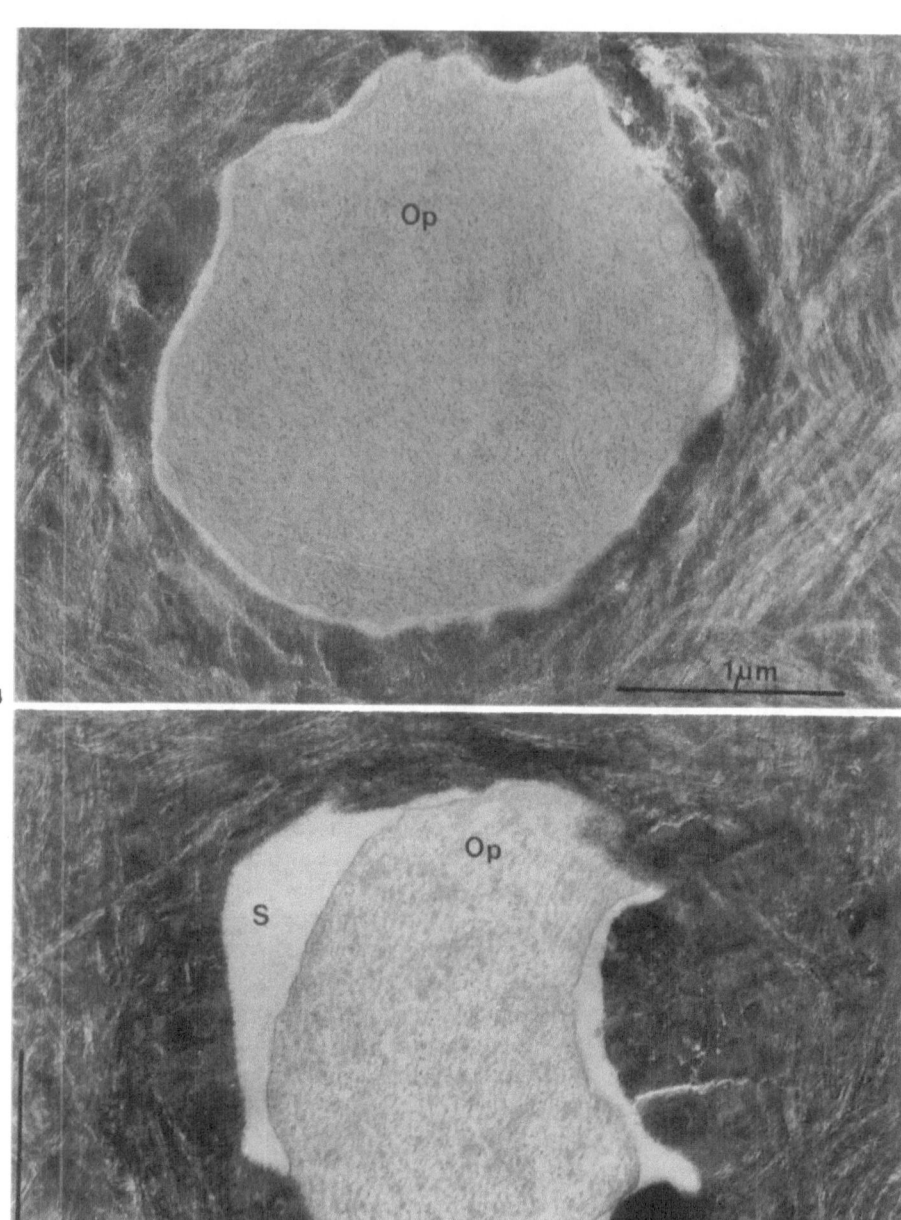

Fig. 34. Transverse section of a dentinal tubule in adult human coronal dentine. The odontoblastic process (*Op*) almost completely fills the dentinal tubule. The wall of the tubule consists of intertubular dentine

Fig. 35. Transverse section of a dentinal tubule in adult human coronal dentine. The odontoblastic process (*Op*) does not completely fill the tubular lumen. A periodontoblastic space (*S*) containing a finely granular material is present. (From FRANK 1968, courtesy of LIVINGSTONE)

in diameter) and contain a great number of smaller vesicles, each bound by a trilaminar membrane (Fig. 44).

Large vacuoles limited by a folded or distended trilaminar membrane were also observed in the odontoblastic process (Figs. 36, 41, 43). These had a diameter of 1 μm or more and were found at the level of predentine (Fig. 36) and the dentine (Figs. 41, 43). They often occupied the center of the odontoblastic process, seemingly producing a condensation of the cytoplasm at the periphery (Fig. 41). Accordingly, it was thought that the occurrence of large vacuoles in the extremities of the odontoblastic process (Fig. 45) might lead to a hyaline appearance of the cytoplasm, giving rise to peripheral degeneration of the process with concomittant reduction in length (FRANK 1966a, b). In fact, as will be discussed later, a hyaline appearance can also arise from incomplete fixation of the odontoblastic process.

The large vacuoles contain a fine-granular material, vesicles or membranous material (Figs. 36, 41, 43). Sometimes apparent fusion between two of these vacuoles was observed (Fig. 42). Accumulation of large droplets in the center of the odontoblastic process was also noted (Fig. 39). The content of these droplets was relatively electron dense and homogeneous. The significance and nature of these large vacuoles is a matter of speculation, but it is reasonable to assume that the multivesicular bodies as well as the large vacuoles containing cell remnants may be associated with lysosomal resorptive functions analogous to those in the odontoblastic cell body (Fig. 13). Other of these features are more difficult to interpret and could be related either to endocytic or exocytic activities (Fig. 37). The appearance of Fig. 37 could be explained either by formation of a large endocytic vacuole through invagination into the process or, alternatively, by fusion of the limiting membrane of a peripherally located vacuole with the plasma membrane and excretion of its contents into the extracellular space in the tubule. The presence of cross-cut collagen fibrils in the extracellular space, located in a deep invagination of a human odontoblastic process, raises the obvious unanswered question of a secretory vs. a resorptive phenomenon (Fig. 38). Similarly, on the lower side of the odontoblastic process illustrated in Fig. 40, it is possible that several vacuoles were in the process of formation or, alternatively, had been discharged into extracellular position. Further work is needed to clarify the function of the various vacuolar structures observed in the adult human odontoblastic process.

The extent or length of the odontoblastic process in the dentinal tubule is a matter of debate. Based on light-microscopic observations it has long been assumed that these cytoplasmic processes extend throughout the entire thickness of dentine, in concurrence with Sir John Tomes (1858) who described the presence of fibrils of soft tissue within dentinal tubules. With refinements of technique and the use of TEM as well as SEM, the odontoblastic process has been described as being limited to the pulpal third of the tissue in man and in a variety of animal species of different types (FRANK 1968a; BRÄNNSTRÖM and GARBEROGLIO 1972; FRANK et al. 1972; GARANT 1972; TSATSAS and FRANK 1972; HOLLAND 1976a, b; THOMAS 1979, 1983; TIDMARSH 1981; THOMAS and PAYNE 1983; THOMAS 1985).

Conversely other investigators using scanning electron microscopy have described the odontoblastic process in the peripheral region of dentine, near the

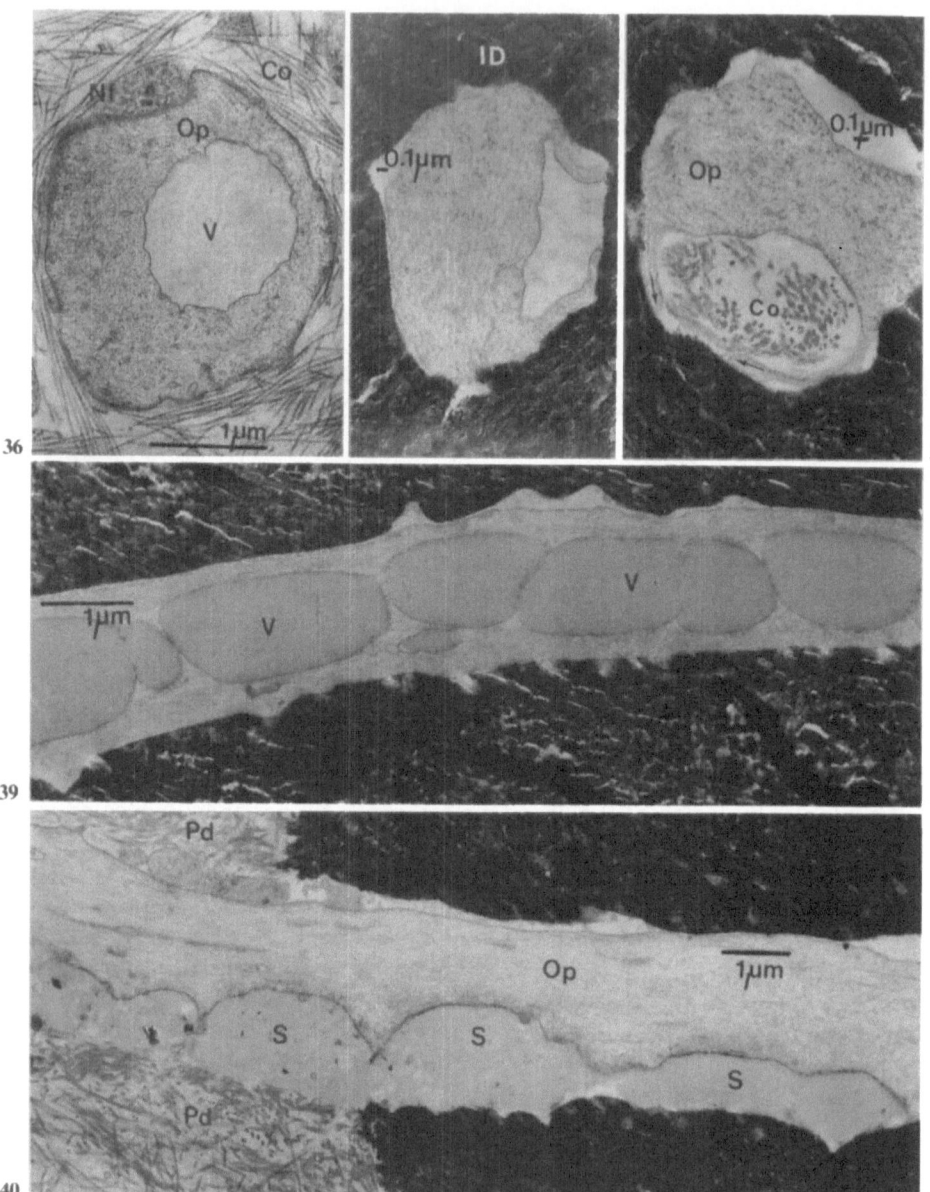

Fig. 36. Presence of a large vacuole (*V*) with a serrated membrane and containing a fine-granular material in a crosscut odontoblastic process (*Op*) in human adult predentine. A nerve fiber (*Nf*) is located in a depression at the surface of the odontoblastic process. *Co* Collagen fibrils. (From FRANK 1966a, courtesy of Archives Oral Biology)

Fig. 37. Transverse section of a dentinal tubule in adult human coronal dentine. The vacuolar content is now in extracellular position. This appearance is suggestive of an exocytosis through a mechanism of membrane fusion. *ID* Intertubular dentine. (From FRANK 1966a, courtesy of Archives Oral Biology)

dentinoenamel junction (KELLEY et al. 1981; GROSSMAN and AUSTIN 1983; CROOKS et al. 1983; GUNJI 1983; GUNJI and KOYABASHI 1983; MANIATOPULOS and SMITH 1983; YAMADA et al. 1983; SIGAL et al. 1984b). Using TEM, HAWKINSON and EISENMANN (1983) noted that in rat molars the odontoblastic process was frequently present throughout the full thickness of dentine. THOMAS and CARELLA (1983, 1984) and THOMAS (1985) have pointed out that great care must be taken in interpreting SEM observations of decalcified dentine due to the presence of an organic sheet-like structure lining the peritubular dentine throughout the length of the tubule. This sheet-like structure, referred to as the *lamina limitans* or limiting membrane (THOMAS 1983, 1984; THOMAS and CARELLA 1983, 1984; THOMAS and PAYNE 1983), in order to conform with terminology established for a similar structure seen in bone canaliculi (SCHERFT 1972), appears in SEM as a tube-like structure which has been improperly interpreted as the odontoblastic process. The limiting membrane, which assumes the same outer shape as the odontoblastic process, is set free from the tubular walls and can simulate the cellular process following demineralization and dissolution of the peritubular dentine (Fig. 51).

Using specific anti-actin and anti-tubulin antibodies with an indirect light-microscopic immunofluorescence technique, SIGAL et al. (1984a) and AUBIN (1985) localized intratubular labeling for tubulin extending up to the dentinoenamel junction in rat molars, whereas labeling for actin, although extending to the dentinoenamel junction, was more prominent in the pulpal third of rat dentine. Similar observations were made when the same techniques were applied to human teeth (SIGAL et al. 1984b; AUBIN 1985) and it was concluded that the presence of tubulin-containing structures extending to the dentinoenamel junction supports the hypothesis that the structures observed with SEM are odontoblastic processes that extend to the dentinoenamel junction.

If in fact the structures observed in SEM and reacting positively to antitubulin antibodies at the light-microscopic level were indeed odontoblastic processes, they should appear as such with TEM which permits identification of cellular element in any part of the organism. Taking into account that with customary glutaraldehyde-paraformaldehyde fixation of fragmented dentine specimens, the odontoblastic processes have always been found in the pulpal third of human

Fig. 38. Presence of uncalcified cross-cut collagen fibrils in an extracellular position. These collagen fibrils (*Co*) are surrounded by cytoplasmic extensions (*arrows*) of the odontoblastic process. This appearance is highly suggestive of exocytosis of an intracellular vacuole. *Op* Odontoblastic process. (From FRANK 1968a, courtesy of LIVINGSTONE)

Fig. 39. Longitudinal section of a human dentinal tubule. Numerous large vacuoles (*V*) with a dense content are located in the center of the odontoblastic process. (From FRANK 1968a, courtesy of LIVINGSTONE)

Fig. 40. Longitudinal section of an adult human dentinal tubule near the predentine (*Pd*). Along the lower side of the odontoblastic process (*Op*), the content of several vacuoles seems to be in the process of formation or to have been discharged into the extracellular position (*S*). (From FRANK 1968a, courtesy of LIVINGSTONE)

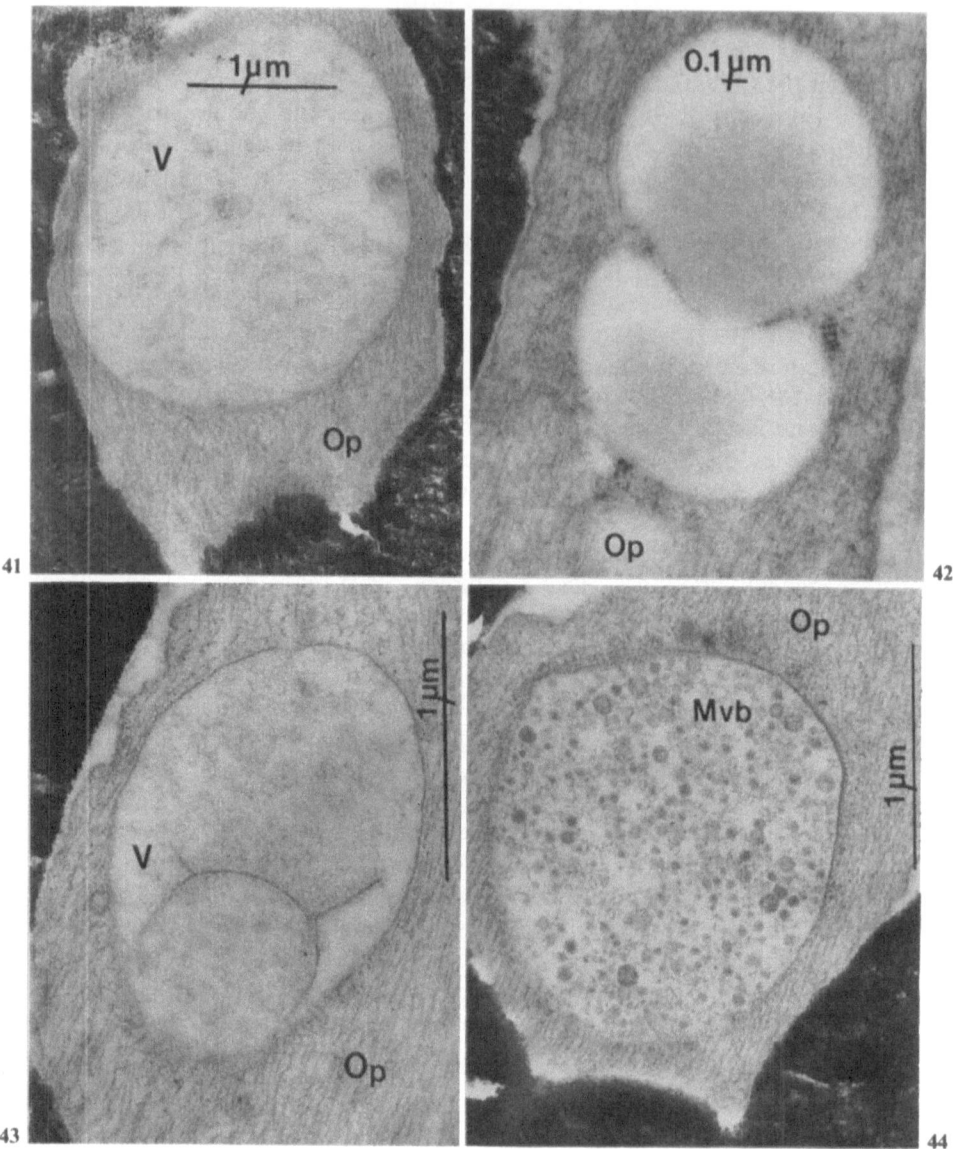

Fig. 41. Transverse section of tubule in human adult dentine. Presence of a large vacuole (*V*) in the center of an odontoblastic process (*Op*). (From Frank and Voegel 1980, courtesy of Caries Research)

Fig. 42. Longitudinal section of an odontoblastic process (*Op*). Presence of two confluent vacuoles within the cytoplasm of the process. (From Frank and Vofgel 1980, courtesy of Caries Research)

Fig. 43. Large membrane-limited vacuole (*V*) in the human odontoblastic process (*Op*) located in a dentinal tubule of the inner pulpal dentine. Membranous remnants are visible in the vacuole. (From Frank and Voegel 1980, courtesy of Caries Research)

Fig. 44. Presence of a large multivesicular body (*Mvb*) in the odontoblastic process (*Op*) of inner human pulpal dentine. (From Frank and Voegel 1980, courtesy of Caries Research)

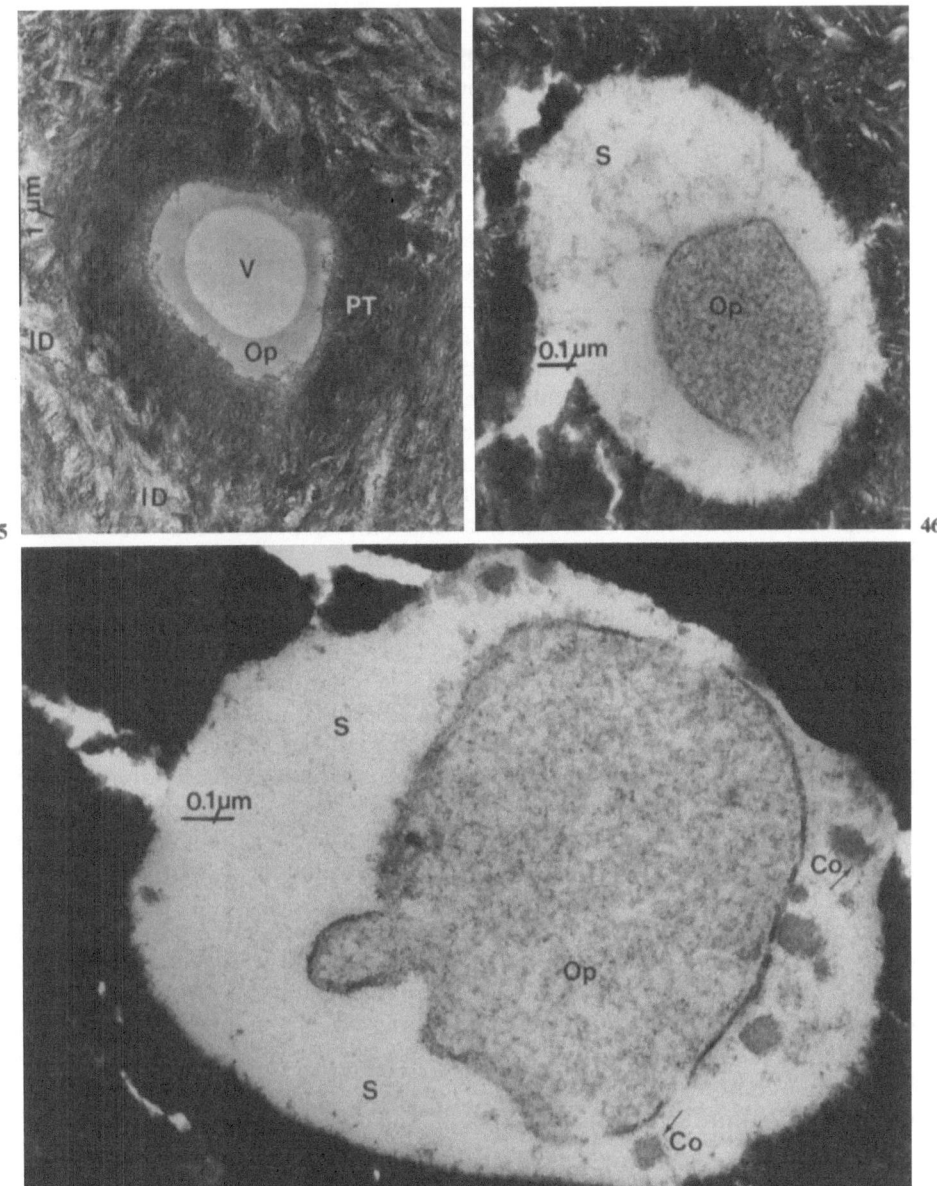

Fig. 45. Transverse section of a tubule in adult human dentine showing a prominent zone of peritubular dentine (*PT*). It is more highly calcified than the intertubular dentine (*ID*). Presence of a vacuolar space (*V*) in the center of the tubule with hyaline peripheral material (*Op*). (From FRANK 1966b, courtesy of Springer)

Fig. 46. Transverse section of a dentinal tubule in adult human coronal dentine, just under the dentino-enamel junction. Presence of an odontoblastic process (*Op*) limited by a trilaminar cell membrane. *S* periodontc

Fig. 47. Transverse section of a dentinal tubule in human coronal dentine, near the dentino-enamel junction. The odontoblastic process (*Op*), limited by a trilaminar cell membrane, does not fill the dentinal tubule. A few cross-sectioned collagen fibrils (*Co*) are observed in the periodontoblastic space (*S*). (From LAFLÈCHE et al. 1985, courtesy of Journal de Biologie Buccale)

dentine in transmission electron microscopy (Figs. 31–35, 37, 38). One of us, with La Flèche and Steuer, raised the hypothesis that the odontoblastic cell and its process may be similar to a retractable suspensor system (La Flèche et al. 1985) whereby in the living state the process extends throughout the entire dentinal tubule to its extremity. The process possesses compounds compatible with a stretched system, i.e. in its cytoskeleton and especially in its microtubules and actin filaments. The lateral surfaces of the actin microfilaments, as already described, are bound by cross connections to the plasma membrane of the odontoblastic process and it could very well be that the longitudinally oriented actin filaments are similarly attached by their tips at the most distal part of the plasmalemma of the odontoblastic process. If, in addition, we take into account the fact that this stretched cellular process is suspended in a liquid gel phase, filling the periodontoblastic space (which will be described later) it is perfectly conceivable that when this complex system is opened or disrupted by the various standard techniques of fixation or by any operative procedure, an immediate and sudden retraction of the cellular process may occur, with concomittant shrinkage and shortening in length.

La Flèche et al. (1985) proceeded to test this hypothesis. Intact human premolars and adult molars were frozen in liquid nitrogen for 30 mins or 1 h. They were then immediately fragmented and fixed in cacodylate-buffered glutaraldehyde and paraformaldehyde solutions. After post-fixation in similarly buffered osmic acid solution and embedding in Epon, without decalcification, sections were cut with a diamond knife, double stained with lead citrate and uranyl acetate, and studied with TEM. It was thus possible to locate odontoblastic processes in the outer parts of coronal dentine immediately adjacent to the dentinoenamel junction (Figs. 46, 47, 48). These processes, limited by a typical trilaminar plasma membrane, did not completely fill the lumen of cross-cut dentinal tubules. A periodontoblastic space generally filled with a finely granular material (Fig. 135) or a filamentous material (Fig. 46) sometimes contained cross-cut collagen fibrils. In transverse sections, the odontoblastic process contained a dense stippled material, occasionally having an electron-dense annular configuration. Some of these objects could very well correspond to cross-cut microfilaments. In any case, the detailed cytoplasmic ultrastructure of these peripheral odontoblastic processes was similar to that of the odontoblastic processes prepared in the standard fashion (Figs. 33, 34). The cytoplasmic nature

Fig. 48. Cross-sectioned dentinal tubule in peripheral human coronal dentine near the dentine-enamel junction. Presence of an odontoblastic process limited by a trilaminar cell membrane (*arrows*) and containing, dense granules (*1*) limited by a membrane as well as light-core vesicles (*2*). *S* periodontoblastic space. (From Laflèche et al. 1985; by courtesy of Journal de Biologie Buccale)

Fig. 49. Transverse section of a dentinal tubule in coronal human dentine near the dentine-enamel junction. Presence of two round-shaped cytoplasmic masses limited by a trilaminar membrane (*arrows*). Presence of a myelin figure in the upper structure. The periodontoblastic space (*S*) is filled with a fine granular material. (From Laflèche et al. 1985; by courtesy of Journal de Biologie Buccale)

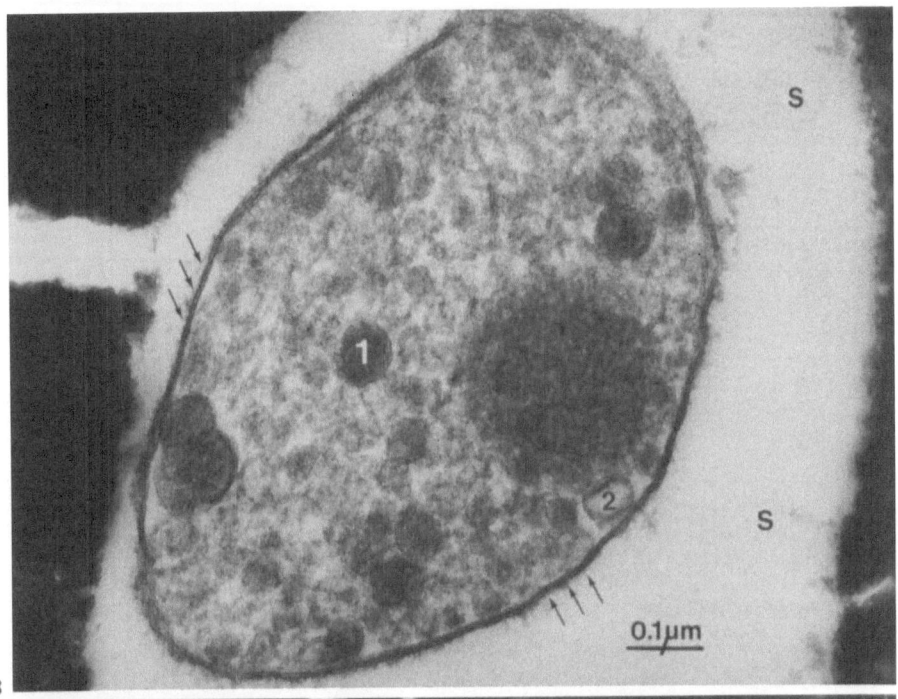

48

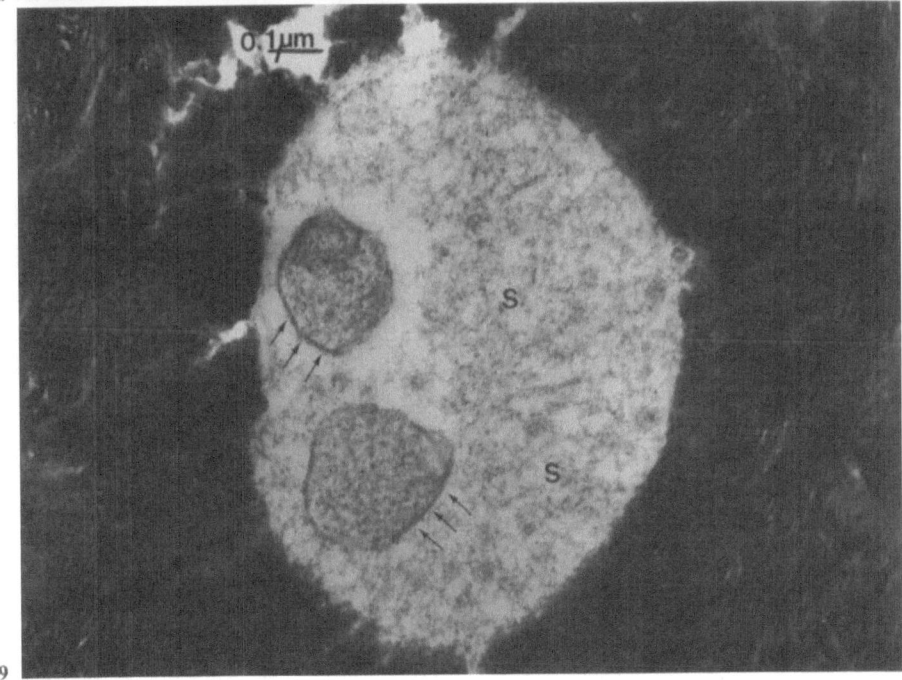

49

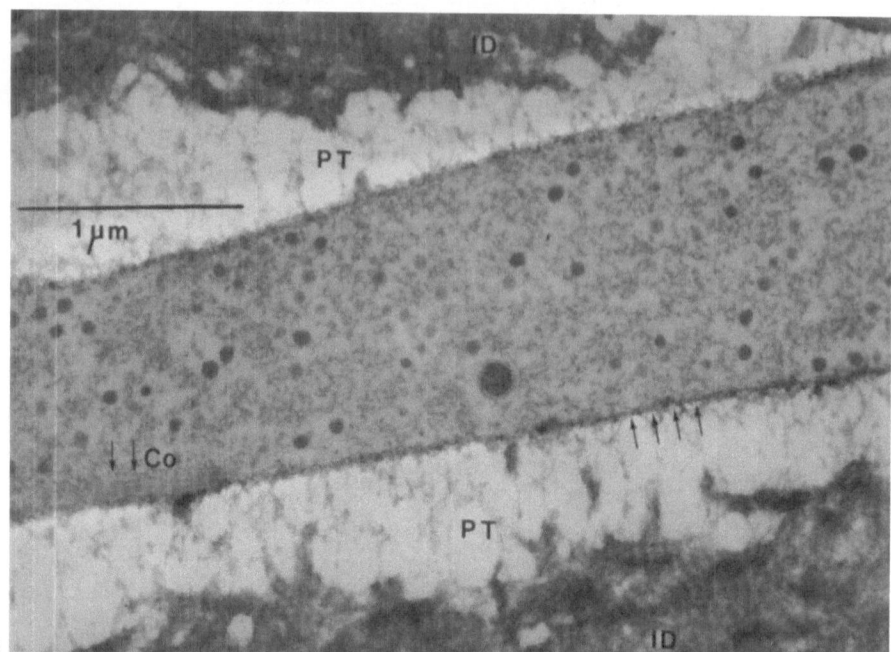

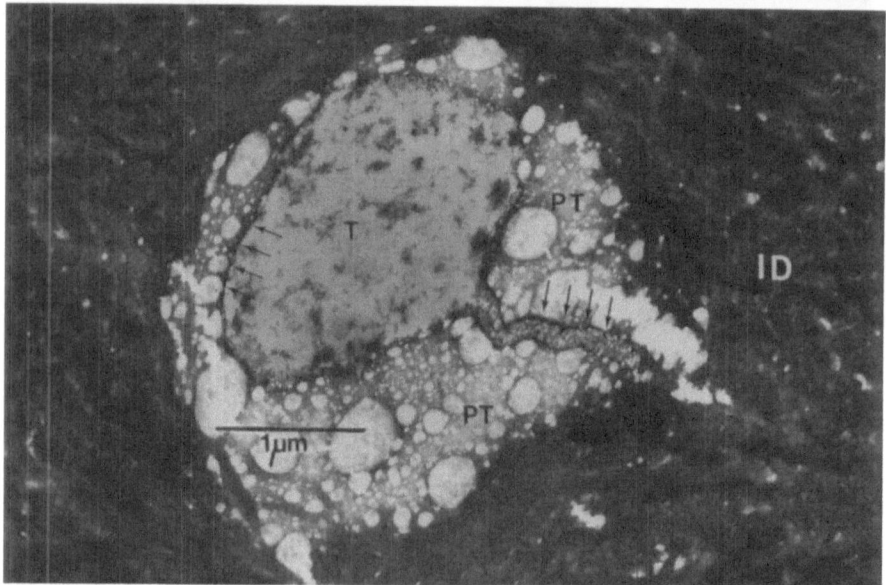

Fig. 50. Human peripheral dentinal tubule near the dentinoenamel junction in longitudinal section, following decalcification for 36 h in a 0.4-M EDTA solution at pH 7.4. Note the presence of a dense line (*arrows*) along the dentinal tubule walls. In the tubular lumen, a fine-granular material is dispersed between dense granules limited by a trilaminar membrane. A collagen fibril (*Co*) can be seen along the lamina limitans. *PT* Organic remnants in the peritubular dentine; *ID* Intertubular dentine. (From LAFLÈCHE et al. 1985, courtesy of Journal de Biologie Buccale)

Fig. 51. Transverse section of the peripheral part of a human dentinal tubule near the dentinoenamel junction, following decalcification for 36 h in a 0.4 M EDTA solution at pH = 7.4. The lamina limitans (*arrows*) circumscribing the tubule (*T*) is continuous around the lateral secondary branch (*arrows to the right*). *PT* Remnants of peritubular dentine; *ID* Intertubular dentine. (From LAFLÈCHE et al. 1985, courtesy of Journal de Biologie buccale)

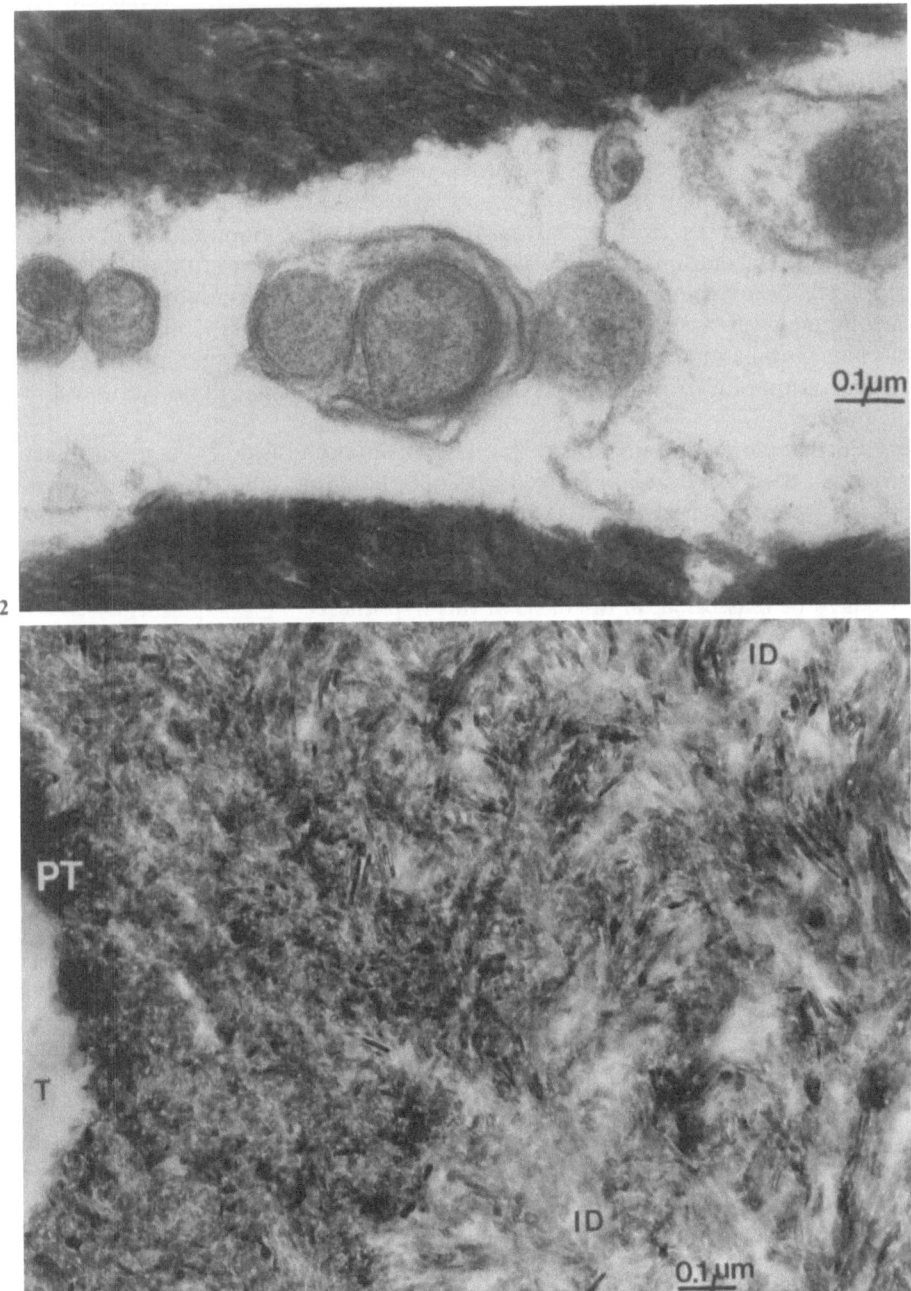

Fig. 52. Longitudinal section through the periphery of a human dentinal tubule, near the dentinoenamel junction. The tubular lumen contains round bodies with a complex array of folded membrane resembling myelin figures. (From LAFLÈCHE et al. 1985, courtesy of Journal de Biologie Buccale)

Fig. 53. Enlargement of a transverse section through a human dentinal tubule (*T*) showing the denser arrangement of inorganic crystals comprising the peritubular dentine (*PT*) as compared with the intertubular dentine (*ID*)

of these processes is further supported by the presence of granules (Fig. 48). These observations indicate that the odontoblastic process extends throughout the entire dentinal tubule to the dentinoenamel junction and that with proper care retraction of the cell extension can be prevented.

Moreover, with the use of liquid nitrogen we also observed in the peripheral parts of the human dentinal tubules, in longitudinal as well as in transverse sections, the presence of globular cytoplasmic masses limited by a trilaminar membrane (Figs. 49, 52). The size of these cytoplasmic globules was quite variable and it was not unusual to observe typical myelin figures in these structures (Fig. 52). On occasion numerous small granules with dense cores and trilaminar membranes were seen in the peripheral dentinal tubules (Fig. 50). These cytoplasmic globules and granules could represent either a type of peripheral cellular degeneration or a fixation artefact induced by liquid nitrogen in the odontoblastic process.

Further work is needed to substantiate our hypothesis that the odontoblastic process acts as a retractable system. It is possible that, apart from the odontoblastic process, nerve fibers extend into the peripheral dentine and that all fixation and operative or experimental procedures used for their study may induce their shrinkage as well. Our basic concept of the pulpo-dentinal complex needs to be reassessed on the basis of this hypothesis.

E. Predentine

A thin rim of predentine can be observed between the odontoblasts and the calcified dentine with light microscopy (Fig. 5) as well as with electron microscopy (Figs. 6, 7, 15, 16). It has a width varying from 10 to 40 μm in human teeth, and is traversed by the odontoblastic processes, the latter being narrower than the odontoblast cell bodies from which they originate (Fig. 7). The extracellular regions between the processes are occupied by the organic matrix of the predentine.

The predentine matrix for the most part consists of fibrils which are almost exclusively composed of type I collagen. The presence of minor amounts of type I-trimer is recognized but not unequivocally established (Butler 1984a). As discussed in the chapter on dentine development, the collagen precursors are elaborated by the odontoblasts and are secreted within the predentine as type I procollagen (Cournil et al. 1979; Karim et al. 1979). Type I procollagen is more concentrated in the inner half of the predentine and is progressively transformed into type I collagen as the outer half of the predentine is approached. A gradual increase in the diameter of the collagen fibrils has also been noted from the inner half to the outer half of rat incisor predentine in freeze-fracture replication studies (Goldberg and Escaig 1981). In the inner half, diameters varied from 35–60 nm, whereas in the outer half adjacent to the dentine, diameters ranged from 70 to 100 nm. Corresponding values of 100–120 nm were found in the adjacent calcified dentine. The 640 Å cross-striated collagen fibrils are primarily oriented perpendicular to the odontoblastic processes and parallel to the calcification front between dentine and predentine (Figs. 8, 14). At closer examination, however, they prove to weave and interlace in a horizontal plane between the odontoblastic processes (Fig. 14), and can even be found oriented circumferentially around them (LESTER and BOYDE 1967).

Besides type I collagen, predentine contains proteoglycans (PG) and glycosaminoglycans (GAG), glycoproteins, γ-carboxyglutamic (Gla)-containing proteins, phosphoproteins, and serum proteins (BUTLER 1984b). Using rutheniumred staining, PG and GAG were apparent as fine granules under electron microscopy, either associated with or not associated with the collagen fibrils (NAGAI et al. 1974; NYGREN et al. 1976). The proteoglycans stain with cuprolinic blue, taking the form of ribbon-like structures and appearing as radiating filaments (GOLDBERG 1983; GOLDBERG and SEPTIER 1983). These cytochemical observations are in agreement with biochemical analyses which have demonstrated high concentrations of PG and GAG in predentine (LINDE 1973, 1984). It is interesting to note, for comparison, that in dentine the amount in PG and GAG is significantly reduced.

Glycoproteins seemed to be more concentrated along the dentine-predentine front (MARTENS 1968; GOLDBERG et al. 1978; TAKAGI et al. 1981). The distribution of fibronectin has also been studied in predentine but its distribution there is somewhat controversical (LINDE et al. 1982; THESLEFF et al. 1979, 1981; CONNOR et al. 1984). LINDE et al. (1982) reported absence of fibronectin in circumpulpal predentine, whereas positive fibronectin labeling in predentine was observed by THESLEFF et al. (1979, 1981) and CONNOR et al. (1984).

Phosphoproteins occur primarily in the predentine near the calcification front. The levels of lipids are higher in predentine than in dentine (LINDE 1985).

Using SEM, LESTER and BOYDE (1967) and other investigators have studied the surface of the predentine-dentine junction after stripping away the pulp and predentine. This surface turns out not to be flat but constituted of the juxtaposition of globular calcospherites. WHITTAKER and KNEALE (1979) noted that in human teeth the calcospherites were very prominent in younger teeth and less conspicuous in older teeth.

F. The Dentinal Tubules and Their Contents

Dentine is permeated by a great number of dentinal tubules. In the crown, they extend from the predentine to the dentinoenamel junction and in the root, from the predentine to the cementodentinal junction. The dentinal tubules follow a primary S-shaped curvature in coronal dentine (Fig. 4), most apparent in longitudinal ground sections of the cervical region. Each dentinal tubule describes a slight reverse curve such that its outermost part near the tooth surface presents a convexity facing the occlusal or incisal edge of the tooth whereas the innermost part near the pulp presents a concavity towards the incisal or occlusal surface. Secondary curvatures of the dentinal tubules, consisting of smaller oscillations often described as "corkscrew forms", occur within the primary curvature (Bradford 1967). A mechanistic explanation of the primary and secondary curvatures has been presented by Osborn (1967). Accordingly, as the odontoblasts are gradually pushed back into the pulp chamber by accumulation of matrix, crowding of cells occurs, mechanically inducing an S-shaped course for the tubules. The secondary curves are produced by a buckling of the odontoblastic processes in the soft predentine because a greater length of process is formed as compared to the distance the odontoblasts recede.

The tubules which arise from the tips of the pulp horns are straight or nearly so (Fig. 3) as are the tubules in the apical third of the root. From the cervix towards the apex, the tubules become less and less curved (Bradford 1967).

While extending to the periphery, dentinal tubules may communicate with each other by lateral branches as can be convincingly demonstrated in light-microscopic preparations following silver nitrate impregnation (Fig. 23). Using a special cast technique, Weber (1983) showed that branching was extensive at the dentinoenamel junction; only minimal branching was seen in the middle third of dentine and no branching was observed in the pulpal third.

Usually the dentinal tubules terminate on the pulpal side of the dentinoenamel junction with a Y-shaped configuration (Figs. 22, 25). With SEM, the terminal portions of the tubules are seen to have secondary and even tertiary branches (Fig. 20).

The number of dentinal tubules per mm² decreases from the inner dentine towards the peripheral dentine as seen on cross-sections. According to Ketterl (1961) about 80% of the total volume of the dentine close to the pulp is composed of the lumens of the tubules, whereas only 4% of the volume consists of lumens in peripheral dentine. The number of dentinal tubules in human coronal dentine ranges from 45,000 to 65,000 tubules per mm² at the pulpal surface; from 29,500 to 35,000 tubules per mm² in mid-dentine and from 15,000–20,000 per mm² near the dentinoenamel junction (Frank 1966a; Garberoglio and Brännström 1976; Mjör 1979).

After removal of predentine, Whittaker and Kneale (1979) studied the entire pulpal surface of dentine with SEM and found the highest number of tubules to be along the coronal portion of the pulp chamber (42,000 tubules/mm²). The number diminished slowly along the length of the pulp canal when moving in the apical direction (radicular upper third = 43,000 tubules per mm²; radicular middle third = 37,000 tubules per mm²; radicular lower third = 32,000 tubules per mm²). In the apical region only 9,000 tubules per mm² were found.

Variations in the number of tubules per mm² between the pulpal and the outer dentinal surface may be explained by the fact that the pulpal surface of dentine is considerably smaller than the corresponding surface of the dentinoenamel junction. The gradual decrease in diameter of dentinal tubules must be taken in account in comparisons of the proportions of dentine occupied by tubules. Tubular diameters ranging from 2 to 3 μm at the pulpal end, from 1 to 2 μm in the midportion, and 0.5 to 0.9 μm at the dentinoenamel junction have been reported (Tronstad 1973a; Garberoglio and Brännström 1976;

TEN CATE 1980). On transverse non-decalcified sections of dentinal tubules we measured the following variations in tubule diameter in human dentine with TEM: 2–2.6 µm in the pulpal dentine; 1.5–2.0 µm in middentine; and 1–1.5 µm in the peripheral dentine. Taking into account all the tubule diameters measured for numerous deciduous and permanent teeth, FROMME and RIEDEL (1970) found the largest tubule to have a diameter of 2 µm and the smallest one, 0.74 µm. The foregoing observations underline the wide spectrum that exists for tubule diameters.

When human dentine is decalcified for light-microscopic study (Fig. 24), the peritubular dentine dissolves and disintegrates. Consequently, on transverse section, wide variation can occur in the resultant diameters of neighboring tubules observed at the same level.

Giant tubules 5–40 µm wide were recently described in human permanent and primary anterior teeth (HALS 1983). The number of such giant tubules varied from 0 to 30. They were located in the mesio-distal axes of the teeth and extended from the pulpal cavity to the incisal dentinoenamel junction. This finding supports the assumption made by TRONSTAD (1972) and MILLER (1981) that pulpal extensions can be found far out into the primary dentine.

As has been noted above, the odontoblastic process in the normal dentinal tubule can be shown to extend to the dentinoenamel junction if special care is exercised in fixation. This cytoplasmic process is separated from the calcified wall of the tubules by a periodontoblastic "space" which is quite narrow near the predentine and becomes wider towards the periphery of the dentine. Between the periodontoblastic space and the tubular wall, a thin sheet-like structure, the *lamina limitans* or limiting membrane is found on the surface of the inorganic material. These structures will now be described, along with the so-called enamel spindles which are dilated extensions of the dentinal tubules ending in the inner enamel. The sensory nerve fibers which are also found in the dentinal tubules will be covered later in the chapter on pulp.

I. The Periodontoblastic Space

The periodontoblastic space, a region between the odontoblastic process and the calcified walls of the dentinal tubule was initially described by FRANK (1966a, b). It contains extracellular material sometimes described as dentinal fluid and some non-calcified collagen fibrils, as well.

The features of this periodontoblastic space, as described with TEM are, in fact, reconcilable with the juxtacytoplasmic sheath or layer rich in acid mucopolysaccharides described in light microscopy by WEILL (1959, 1963), SYMONS (1961), and FIORE-DONNO and BAUME (1966). The width of the periodontoblastic space is quite variable, but at the pulpal end of the dentinal tubule it may be very narrow, the cell membrane of the odontoblastic process approaching contact with the tubular wall (Figs. 6, 7, 31, 33, 37, 39). With increasing distance from the predentine-dentine junction, the space increases in width (Figs. 32, 35), although in cross-sections it can be seen that the periodontoblastic space displays considerable variation in width around the cellular structure (Figs. 35, 46, 47). In peripheral dentine, the periodontoblastic space may have a substantial width (Figs. 46, 47). The entire region between the cell process and the tubule wall is occupied by extracellular material. In the longitudinal dimension the periodontoblastic space extends the full length of the tubule, i.e. to the Y-shaped endings.

Dense bundles of longitudinally oriented collagen fibrils with typical 640 Å cross-striations are found in the periodontoblastic space (Figs. 32, 33). LA FLÈCHE et al. (1985) even found a few collagen fibrils in the periodontoblastic space of a human dentinal tubule, close to the dentinoenamel junction (Fig. 47). Collagen fibrils have been described in the periodontoblastic space repeatedly by electron microscopy (HELWIG and MENKE 1949; ROUILLER 1951; FRANK 1966a, b, 1968a; LESTER and BOYDE 1968; SHIMAUCHI et al. 1973; TRONSTAD 1973a; THOMAS 1979; TIDMARSH 1981). The finding of these uncalcified collagen fibrils in adult human dentine is intriguing since, as will be clear later, only very few collagen fibrils are found in the peritubular dentine. As already mentioned, it is difficult to state whether the finding illustrated in Fig. 38 is related to secretion of collagen or phagocytosis. But the visualization of dumbbell-shaped organelles associated with intracellular collagen phagocytosis in the adult odontoblast (Fig. 12B) makes it tempting to assume that these images are related to endocytosis.

The collagen fibrils of the periodontoblastic space are invested in a gel-like extracellular material which is found not only between the cell process and the calcified wall of the tubule but also in the very terminal parts of the dentinal tubule where the process may be absent (Figs. 49, 50, 52). Ultrastructurally, this extracellular material is finely granular (Figs. 35, 47, 48) and/or fuzzy and filamentous in appearance (Figs. 46, 49, 50). In the periphery of the dentinal tubules near the dentinoenamel junction, small dense granules sometimes having a heterogenous content (Fig. 50), or globular cytoplasmic masses with trilaminar membranes (Fig. 49) or myelin figures (Fig. 52) can be observed in the absence of the odontoblastic process found more pulpally.

The extracellular material described above surrounds the cytoplasmic process which is suspended in the so-called dentinal fluid, described over the years. FISH (1927) and BODECKER and APPLEBAUM (1931) considered this fluid as dentinal lymph. SPRETER VON KREUDENSTEIN and VON STUBEN (1956) recovered dentinal fluid as a light yellow transparent fluid. HALDI et al. (1961) were the first to measure the protein concentration of the dentinal fluid, relative to the plasma concentration, and reported it to be about 1/5th that of plasma. As similar proteins were found in the two fluids (HALDI and WYNN 1963), the dentinal fluid was suggested to be a derivative of a capillary transudate or filtrate. The extracellular nature of the dentinal fluid was assessed by COFFEY et al. (1970) on the basis of a high Na^+/K^+ ratio and PASLEY et al. (1981) concluded that dentinal fluid may be used to determine the composition of pulpal interstitial fluid. The dentinal fluid also seems to contain glycoproteins. CONNOR et al. (1984) observed fibronectin labeling in the lumen of dentinal tubules in rat molars and incisors, using a monoclonal antibody and indirect immunofluorescence.

Using an intravascular lanthanum nitrate injection as a tracer, TANAKA (1980) observed passage of this electrondense substance into the periodontoblastic spaces of rat molar teeth, concluding that the dentinal fluid was a transudate from the terminal pulpal capillaries, circulating within the dentinal tubules. According to TANAKA (1980), the periodontoblastic space is the main route by which various substances reach the dentinoenamel junction in both mature and immature teeth. Similar findings were reported by SASAKI et al. (1982) who used intravenous injections of horseradish peroxidase (HRP), supporting the concept of a free exchange of dentinal fluid between pulp and predentine, through the odontoblast layer. It should be mentioned, however, that BISHOP (1985), using lanthanum nitrate and SATTELBERG and TURNER (1984), using HRP, obtained opposite results, i.e., no passage of the tracers through the distal junctional complexes of the odontoblasts.

Based on these recent detailed observations on the structure of the dentinal tubule, it is difficult to reconcile the existence of the so-called dead tracts described by FISH (1932) as empty dentinal tubules. Even in the case of a total or partial local degeneration of the odontoblast and its process, dentinal fluid would be expected to fill the tubular lumen.

II. The Lamina Limitans

On thin sections of dentine decalcified with acid or EDTA solutions and observed in transmission electron microscopy, an electron-dense limiting membrane can be identified as constituting a tubular coat on the dentinal tubule wall and remaining after dissolution of the peritubular matrix (Figs. 50, 51). This thin structure, described by THOMAS (1979, 1984, 1985) and THOMAS and CARELLA (1984), was termed *lamina limitans* after a similar structure found by SCHERFT (1972) lining bone canaliculi. It must be noted however that the *lamina limitans* cannot be identified in non-decalcified sections of dentine (Figs. 31–34, 35, 37, 46, 47) and is probably strongly adherent to the inorganic components of the dentine. Interestingly, a hypomineralized layer had already been described along the inner wall of the peritubular dentine by SHROFF et al. (1954, 1956), SCOTT (1955), JOHANSEN and PARKS (1962), TAKUMA and EDA (1966) and ISOKAWA et al. (1970).

On longitudinal decalcified sections, the *lamina limitans* appears as a finely granular electron-dense line (Fig. 50) and on transverse sections it can be seen that the limiting membrane not only delineates the tubular walls but also circumscribes the secondary branches of the tubules (Fig. 51). The *lamina limitans*, according to THOMAS (1984), is found throughout the length of the dentinal tubules. It is not affected by collagenase but its susceptibility to hyaluronidase digestion indicates a high content of glycosaminoglycans (THOMAS 1984; THOMAS and CARELLA 1984). According to these authors, the odontoblastic processes described in SEM in the peripheral parts of decalcified dentine are, in fact, the *lamina limitans* which assume an external tube-like shape including even the secondary branches (Fig. 51).

III. The Enamel Spindles

Occasionally dentinal tubules can cross the dentineoenamel junction and can end in the inner enamel layer (Figs. 64, 65). These so-called enamel spindles are most commonly seen on longitudinal ground sections near the tip or even at the tip of the dentinoenamel junction underlying the cusps or the incisal edges. They appear as a series of dilated ampulla-shaped extensions implanted on the dentinoenamel junction, and when a ground section is thin enough, the enamel spindles are seen to be direct continuations of the dentinal tubules across the junction (Fig. 65). The course of the enamel spindles is not straight and does not seem to follow that of the enamel prisms.

When viewed in transmission electron microscopy, it appears that the walls of the enamel spindles are composed of large enamel apatite crystals (Fig. 66). An annular amorphous or hyaline appearance very similar to that observed with cross-sectioned dentinal tubules (Fig. 45) is observed in a transverse section of an enamel spindle (Fig. 66), confirming the dentinal nature of the spindle content.

Enamel spindles are formed during tooth development. During early dentinogenesis, some odontoblastic processes are thought to extend between the ameloblasts, so that in the course of early enamel matrix formation, they become trapped to form the enamel spindles.

G. Peritubular Dentine

The term "peritubular dentine" was proposed by FEARNHEAD (1957) to distinguish the hypercalcified walls of the dentinal tubules. In fact, peritubular dentine should probably be considered as intratubular, since during dentinogenesis the dentinal tubule is initially lined by walls of intertubular dentine and the peritubular dentine is formed within it, thus reducing the initial diameter of the tubule.

In ground sections each dentinal tubule appears to be surrounded by a cuff or halo when cut in cross section (Figs. 26 and 27). This unique structure was first investigated by NEUMANN (1863) and came to be known as the sheath of Neumann. For almost a century highly controversial explanations were put forth until the true nature of this structure was established. WALKHOFF (1924) noted that the translucent layer, known today as peritubular dentine, decalcified more easily than the remainder of the dentine. It was confirmed that the peritubular dentine was more easily decalcified than the intertubular dentine during the etching process used for preparing replicas of dentine for electron microscopy (SCOTT and WYCKOFF 1947; SYRRIST 1949; ROUILLER 1951; TAKUMA 1960; SELVIG 1968).

The true hypercalcified nature of peritubular dentine was predicted by means of the light microscope (BRADFORD 1955) but clearly demonstrated by microradiography (MILLER 1954; BAUD and HELD 1956; BLAKE 1958; DREYFUSS et al.

1964; EDA and TAKUMA 1965; TRONSTAD 1973b; WEBER 1974). When transverse sections of dentinal tubules are observed in microradiographs (Fig. 28), the peritubular dentine appears as a white ring around the tubular lumen, with a higher radiopacity than the intertubular dentine.

FRANK (1966a) graphically demonstrated the hypercalcified character of peritubular dentine in transverse non-decalcified sections of human dentine examined with transmission electron microscopy. Further observations with the same technique were reported by PLACKOVA and STEPANEK (1960), TAKUMA (1960), JOHANSEN and PARKS (1962), EDA and TAKUMA (1965) and FRANK (1966a, b). The hypermineralized character of peritubular dentine was also documented with the electron probe, and TAKUMA et al. (1966) demonstrated a higher content of calcium and phosphorus in it. Using an electron microprobe combined with scanning electron microscopy, MILLER et al. (1971) estimated that coronal peritubular dentine was hypermineralized by a total of up to 9% by weight of Ca, P and Mg compared to intertubular dentine, although most tubules showed a difference of between 2 to 4%. Also in electron-microprobe investigations, HÖHLING et al. (1972) found the mean calcium content per unit volume to be about 40% higher in peritubular dentine than in intertubular dentine. Finally, a higher content of calcium (Fig. 29) as well as phosphorus and magnesium (Fig. 30) was demonstrated by secondary ion microscopy (LEFÈVRE et al. 1976).

Considerable species variation exists with respect to the occurrence of peritubular dentine (JONES and BOYDE 1984). In human permanent teeth from young individuals, peritubular dentine is not always present, and it is not found in interglobular dentine. Often it is absent from the pulpal part of the dentinal tubules near the predentine-dentine junction in young teeth (Figs. 31, 33, 34, 40).

Histochemically, the peritubular matrix stains intensely and metachromatically with methylene blue and toluidine blue, and deeply with alcian blue at pH 2.6 (SYMONS 1961). In recently erupted human teeth, SYMONS (1961, 1968) found that peritubular dentine starts a short distance from predentine-dentine junction and that in the innermost dentine, intertubular dentine walled the tubules. In older teeth, peritubular dentine can be recognized in microradiographs to extend to the most pulpal aspects of the dentinal tubules, i.e. to the predentine-dentine junction (WEBER 1974). Using a systematic point count histomorphometric method, WEBER (1969) demonstrated that for the inner half of human coronal dentine, the relative volumes of peritubular and intertubular dentine were almost equal. In the root, peritubular dentine is also present, but according to TAKUMA and EDA (1966) it is narrower there.

When observed in scanning electron microscopy (Fig. 21) or in transmission electron microscopy (Figs. 45, 53), peritubular dentine appears as a denser structure compared to intertubular dentine. In non-decalcified section (Fig. 45), periodic cross-striations indicative of collagen are easily visible in intertubular dentine but are not seen in peritubular dentine. A denser configuration of inorganic material is observed in the latter structure (Figs. 45, 53). Few studies have addressed the inorganic constituents of peritubular dentine. On carbon replicas submitted to stereophotogrammetric analysis, LESTER and BOYDE (1968) described spherical isodiametric crystals of approximately 25 nm in diameter, whereas Höhling (1966) and Frank (1966a, b) observed elongated needle-shaped crystals with their long axes parallel to those of the dentinal tubules and identified these particles as apatites by electron diffraction.

Using high-resolution transmission electron microscopy, Schroeder and Frank (1985) identified, in human peritubular dentine, hydroxyapatite crystal forms which had flattened hexagonal prismatic shapes, with length 1.5 times greater than their width which, in turn, was 2.6 times greater than the thickness. From non-decalcified longitudinal or transverse sections of these inorganic crystals (Figs. 54A, B, 55), various periodic fringe patterns were obtained, from which the exact shapes of the crystals were deduced. These crystals were found to have a mean length of 36.00 ± 1.87 nm, a mean width of 25.57 ± 0.37 nm and a mean thickness of 9.76 ± 0.69 nm. They consisted of platelets with a mean width-to-thickness ratio of 2.61, each being a flattened hexagonal prism of hydroxyapatite. The above description was based upon a) the electron-diffraction patterns obtained, and b) comparison of the values of equidistant periodic fringes observed directly along the various planes of section with the corresponding theoretical values of hydroxyapatite. It is thus evident that some crystals of peritubular dentine consist of hydroxyapatite platelets. The question remains as to whether other crystalline species are present there.

Along these lines, it is interesting to note that in peritubular dentine Lefèvre et al. (1976) found a higher content of Mg (Fig. 30) as well as a higher content of Ca and P (Fig. 29) using secondary ion microscopy. Subsequently, transverse non-decalcified sections of human premolar dentine were prepared from teeth of 12-year-old patients. No dentinal sclerosis was present and the dentinal tubule lumen was free from abnormal crystalline formation. Selective electron diffraction of the peritubular dentine confirmed the presence of whitlockite (Lefèvre et al. 1976). It can be estimated that approximately 1% of magnesium by weight is incorporated into the whitlockite structure of peritubular dentine.

The organic content of the peritubular dentine seems to be minimal. When submitted to acid decalcification without special care, peritubular dentine disintegrates completely (Fig. 24). However, with the use of EDTA or special techniques, the peritubular matrix can be preserved (Figs. 50, 51). Whereas Lester and Boyde (1968) and Sundström et al. (1970) did not describe collagen fibrils in peritubular dentine, a few fibrils were unveiled there by Frank (1966a, b), Takuma and Eda (1966), Johnson and Poole (1967), and Selvig (1968). Very recently, La Flèche et al. (1985), applying EDTA on non-decalcified thin ections of human dentine, demonstrated most convincingly the presence of a few collagen fibrils in peritubular dentine (Fig. 50). According to Sauk et al. (1976) peritubular dentine contains more glycosaminoglycans than intertubular dentine as judged from ruthenium-red staining. The presence of proteoglycans and glycoproteins was confirmed by electron-microscopic cytochemistry by Goldberg et al. (1978).

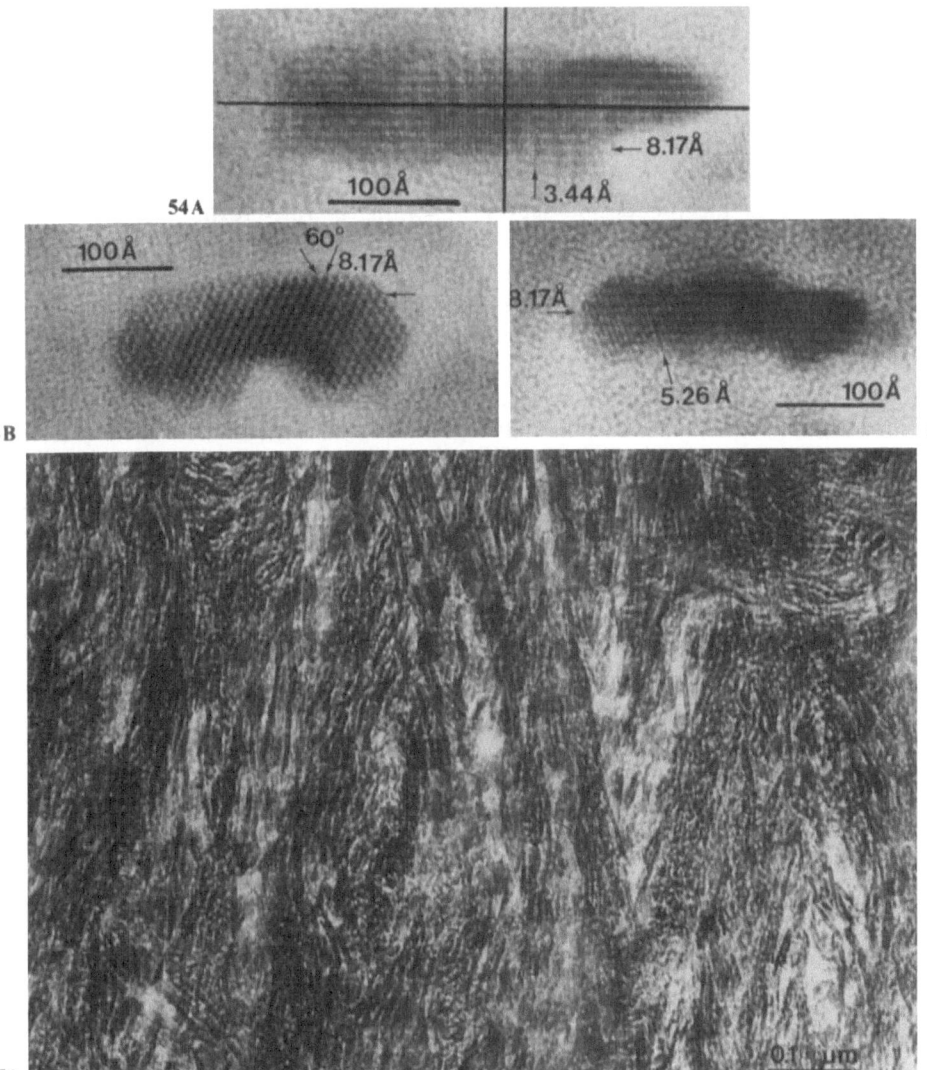

Fig. 54 A. Longitudinal section parallel to the *c*-axis of a human peritubular dentine crystal. Along direction [020], 2 series of perpendicular periodic fringes can be observed: 1) periodic fringes of (002) planes = 3.44 Å, and 2) periodic fringes of (100) plane = 8.17 Å. (From SCHROEDER and FRANK 1985, courtesy of Cell and Tissue Research)

Fig. 54 B. Transverse section perpendicular to the *c*-axis of a human peritubular dentine crystal. Along direction [001], three series of periodic fringes, forming 60° angles correspond to the (100) and (010) series of planes, with an equidistance of 8.17 Å. (From SCHROEDER and FRANK 1985, courtesy of Cell and Tissue Research)

Fig. 55. Periodic fringes of a human peritubular dentine crystal along direction [101]. Two series of periodic fringes can be seen: 1) periodic fringes of (010) planes = 8.17 Å, and 2) periodic fringes of (101) planes = 5.26 Å. (From SCHROEDER and FRANK 1985, courtesy of Cell and Tissue Research)

Fig. 56. Area of human intertubular dentine with longitudinally oriented and elongated inorganic crystals

H. Intertubular Dentine

Intertubular dentine is the first calcified dentine region formed, in both the peripheral mantle and the circumpulpal dentine. It occupies the region around and between the peritubular dentine regions (Figs. 21, 24, 26–28, 45) and extends to the predentine-dentine junction; in fully formed young teeth, it can even circumscribe the tubular lumens (Figs. 6, 7, 31, 33, 40). Intertubular dentine consists mainly of a calcified collagenous matrix. The collagen which constitutes 90–95% by weight of the organic material is almost exclusively type I collagen (Butler 1984a). Besides some lipids and citrate, the remainder consist of non-collagenous proteins such as proteoglycans, phosphoproteins, γ-carboxyglutamic acid (Gla)-containing proteins, serum proteins, glycoproteins, and fibronectin (Butler 1984b). The nature, distribution and function of these non-collagenous proteins has already been discussed. This collagenous organic matrix is intimately associated with minute hydroxyapatite crystals assuming a flattened hexagonal prismatic shape (Fig. 60).

Traditionally, along the dentinoenamel junction and the cementodentinal junction, a very thin layer of mantle dentine is described in polarizing light microscopy (Schmidt and Keil 1958) and electron microscopy (Jones and Boyde 1984). Along the human dentinoenamel junction, the mantle dentine, ranging in width from virtually nothing to a few tenths of a micron, contains coarse collagen fibrils oriented perpendicular to the junction. The root mantle dentine presents differences between species, but in primates it is apparently 15–30 μm thick (Lavelle et al. 1977) and the coarse collagen fibrils seem to be parallel to the cementodentinal junction (Ten Cate 1978, 1980).

The intertubular dentine of the circumpulpal region is continuous with that of the mantle dentine. Collagen fibrils of intertubular dentine are generally smaller than those in the mantle dentine and they are largely oriented perpendicular to the dentinal tubules. They are also interwoven in a transverse direction (Fig. 57). However, substantial bundles of parallel collagen fibrils can also be observed in circumpulpal dentine (Fig. 56). The crystallites of the mineral phase of intertubular dentine which are closely associated with the collagen fibrils, were initially shown by polarizing light microscopy, to be arranged in globular masses known as calcospherites (Schmidt and Keil 1958). The size, shape and distribution of calcospherites in intertubular dentine have been clearly described in anorganic dentine using scanning electron microscopy (Jones and Boyde 1984). The calcospherites are smallest and most numerous close to the peripheral limits of the dentine. They increase in size and decrease in incidence in a pulpal direction. The calcospherites in the inner half of dentine are again smaller, yet not so small as those of the initial layer. The presence of concentric bands in calcospherites was demonstrated by Quigley et al. (1965) on decalcified sections of human dentine incubated with papain and on metal-shadowed replicas of ground sections.

The globular configuration of intertubular dentine calcification is also clearly visible at the predentine-dentine junction when the pulp and predentine are removed and the surface of the junction is visualized with scanning electron

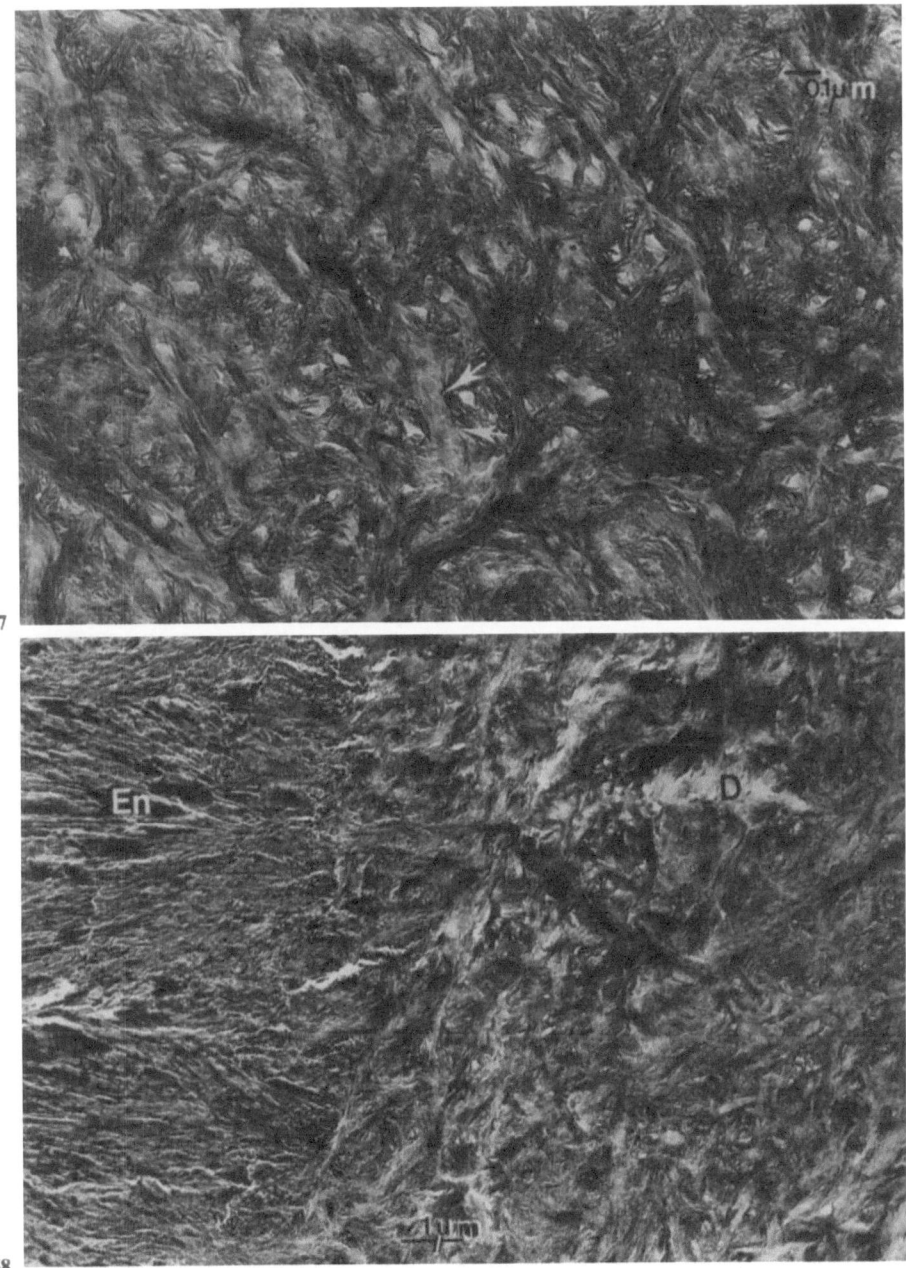

Fig. 57. Human intertubular dentine. Apatite crystals appearing as thin dense profiles or as transparent plates (*white arrows*) are oriented in various directions. Some crystals are deposited along the intermingled collagen fibrils

Fig. 58. The precise location of the dentino-enamel junction in an adult human tooth is difficult to follow. *D* Dentine; *En* Enamel. (From FRANK 1966, courtesy of Springer)

microscopy (LESTER and BOYDE 1967). In intertubular dentine, the apatite crystals are located not only between and at the surface of the collagen fibrils, but also within the collagen fibrils (GLIMCHER 1982). With low-angle X-ray diffraction combined with optical reconstruction and model building (ENGSTRÖM 1966), as well as with low-angle X-ray and neutron diffraction (WHITE et al. 1977; BERTHET-COLOMINAS et al. 1979), it was confirmed that the apatite crystals are deposited within the "holes" and pores, completely filling up all the intermolecular space within the collagen fibrils.

The apatite crystals of dentine give electron diffraction patterns typical of hydroxyapatite (Fig. 59). The c-axes of the apatite crystals are roughly parallel to the long axes of the collagen fibrils. Despite the close relationship the apatite crystals and the collagen fibrils, the crystals are deposited with respect to the transverse periodicity of the collagen fibrils (Figs. 56 and 57). Using stereoscopic techniques, JOHANSEN (1967) and JOHANSEN and PARKS (1962) observed individual crystals of intertubular dentine at different angles. Some crystals appeared as narrow, dense and elongated profiles in one view (Figs. 53, 56) and as broad, less dense profiles in another view. It was concluded that these crystals had a plate-like structure which, when viewed on edge, appeared as a narrow dense profile.

The plate-like structure of the apatite crystals of intertubular dentine has been confirmed by high-resolution transmission electron microscopy whereby it has been possible to demonstrate periodic lattice images in apatite crystals (SELVIG 1970; SPECTOR 1975; VOEGEL and FRANK 1977; FRANK and VOEGEL 1978; NAKAHARA 1982). When the represented crystal planes are nearly parallel to the electron beam, the periodic fringes obtained can be considered as the projection onto the screen of the various planes of the crystal. These periodic lattice fringes allow the determination of the precise size and shape of the crystals in relation to their crystallographic axes. The simultaneous visualization of three lattice plane sets equivalent to the (100) planes forming 60° angles between each other with an equidistance of 8.17 Å, allowed a precise determination of the plane of section of the crystal, since such hexagonal fringe patterns visible inside the dentine crystals (Figs. 59, 63) can only coincide to a section made strictly perpendicular to the c-axis of the crystal. On such sections (Figs. 61, 63) the width and the thickness of the apatite crystals (Fig. 62) can be determined precisely. With this method, VOEGEL and FRANK (1977) and FRANK and VOEGEL (1978) measured the width (W)-to-thickness (T) ratio of a great number of mature human apatite crystals in enamel, intertubular dentine and alveolar bone (Table 1).

Whereas the enamel crystals showed the highest mean width and mean thickness values, the variations in monocrystal sizes and W/T ratios were clearly more variable for intertubular dentine than for enamel and bone. Enamel crystals are slightly flattened hexagons with a W/T ratio of 1.99, while crystals of intertubular dentine and bone with a respective W/T ratio of 3.6 and 7.43 have a tendency to be flatter. A transition from the slightly flattened hexagonal prism toward a thin platelet shape is therefore apparent from enamel to dentine with a further transition from dentine to bone. The total length of the crystals is difficult to assess on thin sections. However, JOHANSEN (1967) reported

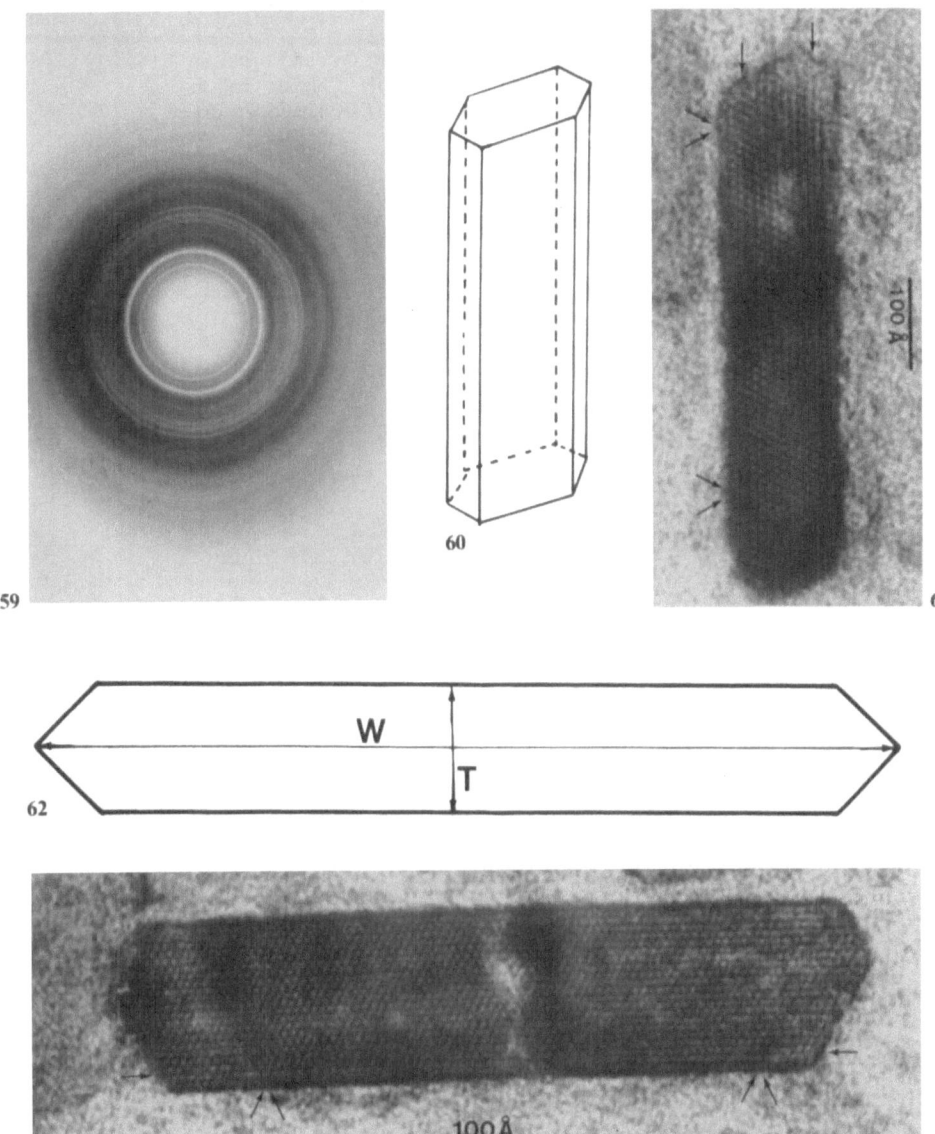

Fig. 59. Electron diffraction pattern of human intertubular dentine. Note the absence of arcing on the diffraction rings indicating absence of preferential orientation of the hydroxyapatite crystals

Fig. 60. Flattened hexagonal prismatic shape of a monocrystal of human intertubular dentine

Fig. 61. A crystal of human intertubular dentine at cross-section to the c-axis with a width: thickness ratio of 4.37. Three series of fringes (*arrows*) with a periodicity of 8.17 Å corresponding to (100) planes and equivalent planes are visible. (From FRANK and VOEGEL 1978, courtesy of Academic Press)

Fig. 62. Schematic cross-section perpendicular to the c-axis of a dentine apatite monocrystal with the designations of width (W) and thickness (T)

Fig. 63. Transverse section perpendicular to the c-axis of a monocrystal of human intertubular dentine with a width:thickness ratio of 4.62. Three series of fringes (*arrows*) with a periodicity of 8.17 Å corresponding to (100) planes and equivalent planes are visible

Table 1. Range and mean values of width (W) and thickness (T) as well as W/T ratios of apatite crystals of human enamel, intertubular dentine and alveolar bone. (Frank and Voegel 1978):

Width (W) in Å	Thickness (T) in Å	W/T
Human enamel		
Range: 539–1515 Å	216–1067	1.99 ± 0.10
Mean: 890.80 Å ± 25.80	503.60 ± 20.80	
Intertubular dentine		
Range: 100–900	40–170	3.60 ± 0.14
Mean: 364.50 ± 14.50	103.30 ± 2.70	
Alveolar bone		
Range: 270–900	40–130	7.43 ± 0.27
Mean: 562.10 ± 1900	79.10 ± 3.10	

lengths up to 1000 Å in intertubular dentine, whereas by high-resolution electron microscopy, Nakahara (1982) reported plate-shaped crystals measuring 50 to 700 Å in their longest direction.

In summary, the apatite crystals of intertubular dentine are flattened hexagonal prisms (Fig. 60) with a mean W/T ratio of 3.60. They are therefore very similar to some of the crystals observed in peritubular dentine by Schroeder and Frank (1985), although the latter had a mean W/T ratio of 2.61.

J. Dentinoenamel and Cementodentinal Junctions

The dentinoenamel junction is the first hard tissue interface to develop during tooth formation. It appears in ground sections either as a flat (Figs. 3, 70) or scalloped (Fig. 69) line. The shape of the junction in human teeth is usually said to confer strength and prevent the two tissues from separating during function. However, there is no evidence that an increase in surface area afforded by scalloping is functionally significant (Jones and Boyde 1984). When observed with transmission electron microscopy, the precise boundary between the two tissues is difficult to follow (Fig. 58) and intermingling of the tissue components at a finer scale would appear to provide more than adequate mechanical adhesion.

With respect to the cementodentinal junction, some controversial issues may be raised about the exact location, since considerable variation appears to exist in the structural organization and the degree of mineralization of dentine and cementum in this region. We agree with Jones and Boyde (1984) who have

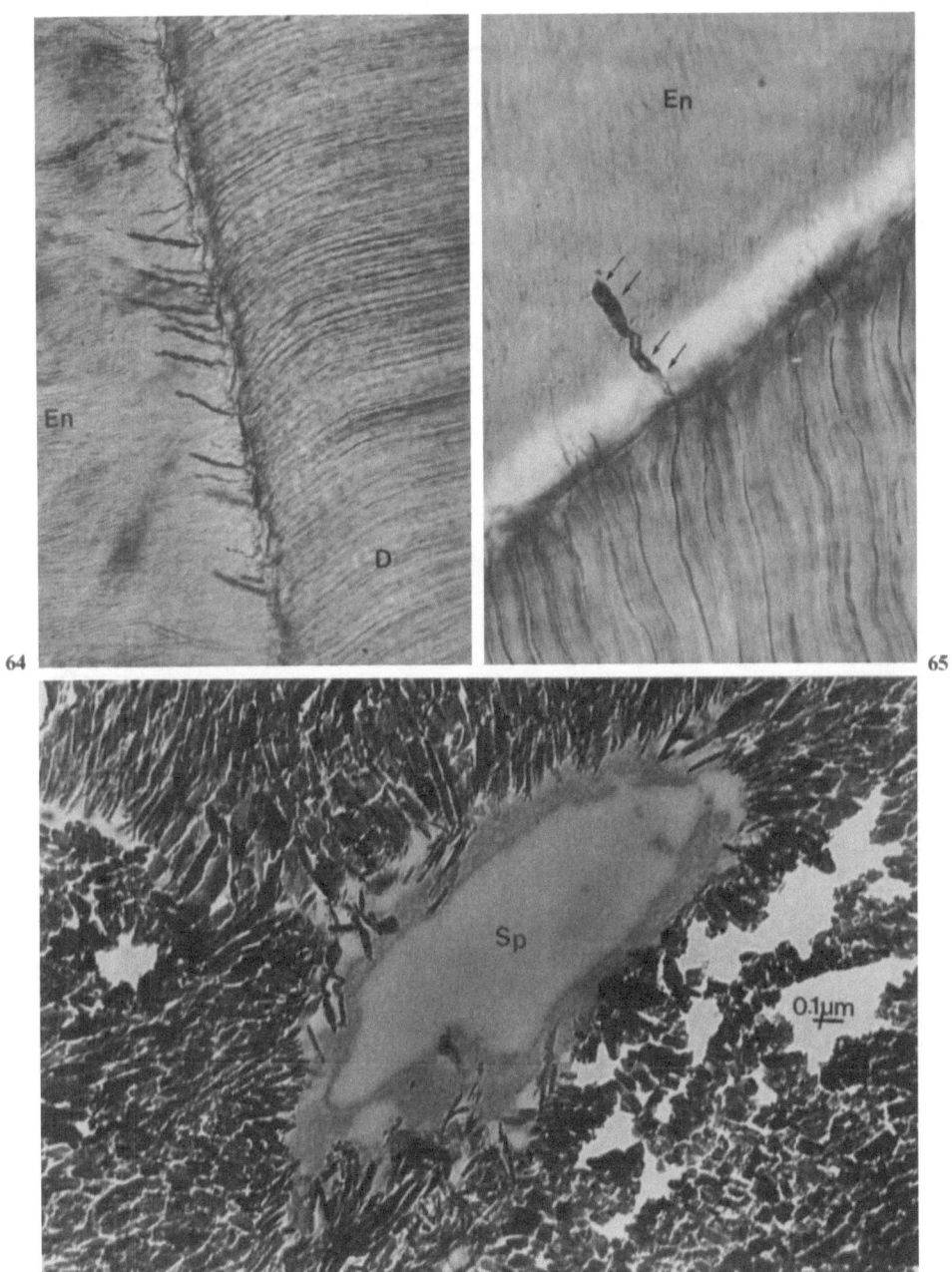

Fig. 64. Longitudinal ground section of a human dentinoenamel junction showing a great number of enamel spindles. *En* Enamel; *D* Dentine with dentinal tubules. × 110

Fig. 65. Higher enlargement of a ground section of a human dentinoenamel junction. The enamel spindle (*arrows*) clearly appears to be the termination of a dentinal tubule. *En* Enamel. × 185

Fig. 66. Cross-section in transmission electron microscopy of an enamel spindle (*Sp*) in a human tooth. (From FRANK 1966a, courtesy of Archives of Oral Biology)

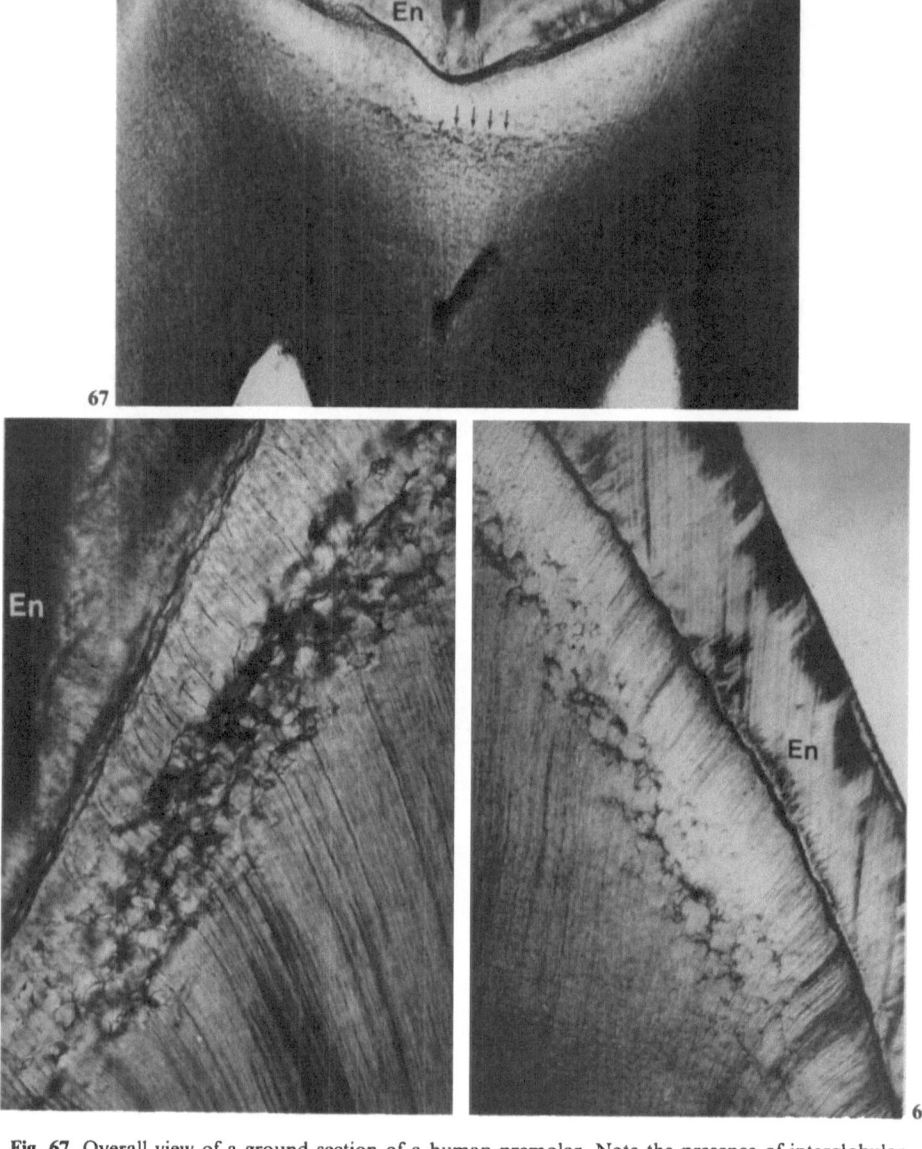

Fig. 67. Overall view of a ground section of a human premolar. Note the presence of interglobular dentine (*arrows*) in the peripheral dentine. *En* Enamel. × 15

Fig. 68. Presence of prominent layer of interglobular dentine in the peripheral part of vestibular human dentine. *En* Enamel. × 150

Fig. 69. Presence of a more discrete layer of interglobular dentine, parallel to the dentinoenamel junction. *En* Enamel. × 45

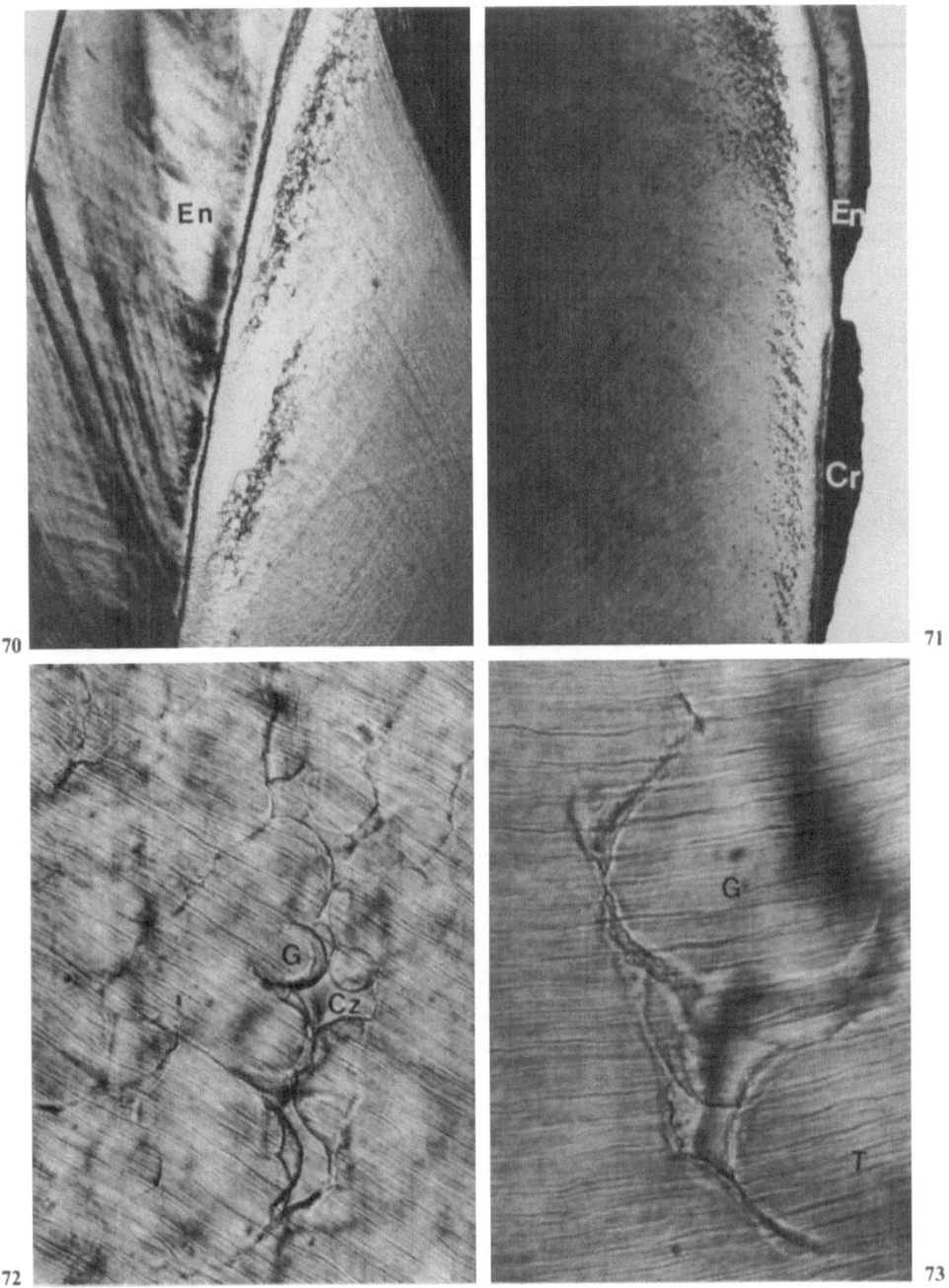

Fig. 70. Presence of 2 separate layers of interglobular dentine in the coronal part of a human canine. *En* Enamel. × 40

Fig. 71. Continuity of coronal and radicular interglobular dentine. *En* Enamel; *Cr* Radicular calculus. × 18

Fig. 72. Higher enlargement of human interglobular dentine with a typical lacelike appearance. *Cz* Interglobular space of Czermack; *G* Globular calcification. × 485

Fig. 73. Detail of human interglobular dentin. *T* Dentinal tubule; *G* Globular calcification. × 730

found that the granular layer of Tomes may extend right to the junction (Figs. 78, 80, 81) or may be separated from it by a more homogeneously mineralized dentine layer of varying width (Fig. 79) which could correspond to the so-called hyaline layer of Hopewell-Smith (1903). Jones and Boyde (1984) studied the most peripheral part of dentine by SEM backscattered electron imaging and found it to be more highly mineralized than dentine and cementum. This hypermineralized layer, which is an inconstant feature, belongs to dentine and is located just below and at the junction (Jones and Boyde 1984).

The hyaline layer described by Hopewell-Smith (1903) peripheral to the granular of Tomes has, however, been variously interpreted. It has been thought to be part of cementum (Hopewell-Smith 1903; Blackwood 1957) or alternatively of dentine (Bradford 1967; Owens 1972, 1975). At the ultrastructural level, not only has it been impossible to assign a clear definition to the hyaline layer of Hopewell-Smith but the precise boundary between dentine and cementum is difficult to follow. For clarification of these issues, a more precise understanding of the embryology of the cementodentinal junction may be required.

K. Interglobular Dentine

In the outer third of the coronal dentine, a zone of interglobular dentine is commonly found paralleling the dentinoenamel junction (Figs. 67–69, 71). Sometimes two nearly parallel zones are present (Fig. 70). These consist of a series of interglobular spaces, the density of which varies from tooth to tooth. They sometimes extend into the peripheral root dentine (Fig. 80) such that a continuous layer of interglobular dentine runs across the cervical region (Fig. 71). Interglobular dentine can also be absent, however.

Initially Czermak described such zones as being composed of empty spaces formed because of failure in the fusion of globular calcification centers. Kölliker (1852), a professor of Czermak, then recognized the empty appearance of these zones as due to artefactual drying of the preparations. There is currently universal agreement on the basis for these distinctive regions, i.e. incomplete fusion of the calcospherites associated with the normal mineralization pattern of dentine. Unmineralized dentinal matrix areas with lace-like outlines are clear-

Fig. 74. Ground section of a human permanent molar observed with transmitted light microscopy. Presence of incremental lines of von Ebner with a range of widths, running perpendicular to the direction of the dentinal tubules. *En* Enamel. × 60

Fig. 75. Higher enlargement of a human permanent molar observed in transmitted light microscopy showing incremental lines of von Ebner. Dentinal tubules (*arrows*) run almost perpendicular to the lines of von Ebner. × 60

Fig. 76. Human third molar examined with transmitted fluorescence microscopy showing fluorescent lines related to tetracycline intake. A thin line in the occlusal portion of the crown dentine, a

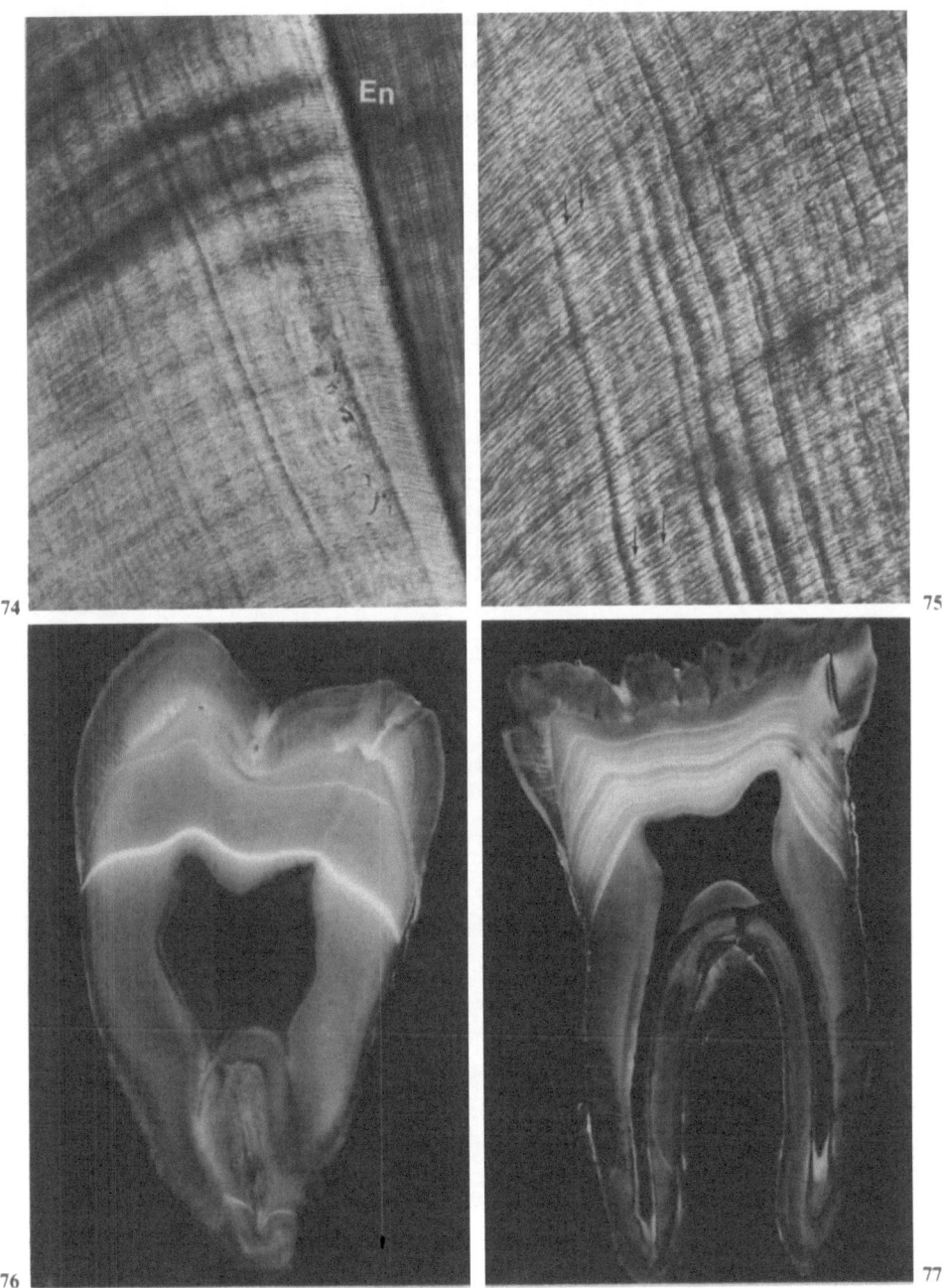

broad band and a thin line in the cervical region and another line in the apical dentine can be observed. (Nalbandian et al. 1982, courtesy of Journal de Biologie Buccale). × 3,5

Fig. 77. A human mandibular first molar examined with transmitted fluorescence microscopy displays 18 tetracycline lines in the crown region and one in the root. (Nalbandian et al. 1982, courtesy of Journal de Biologie Buccale). × 3,5

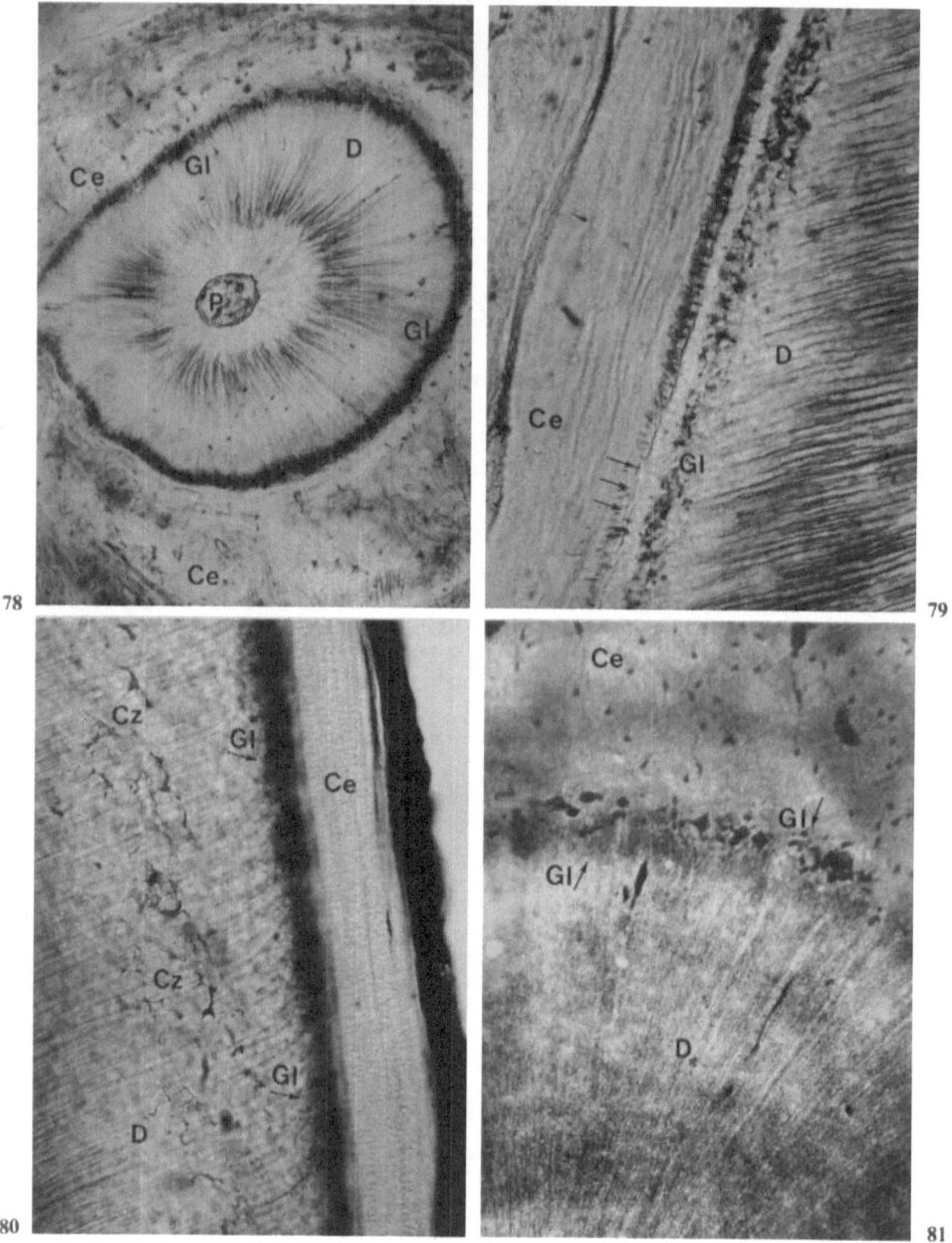

Fig. 78. Transverse ground section of a human root. *P* Pulp; *D* Dentine; *Gl* Tomes' granular layer; *Ce* Cementum. × 40

Fig. 79. Longitudinal ground section of the cementum (*Ce*), cementodentinal junction (*arrows*), Tomes' granular layer (*Gl*) and dentine (*D*). A thin hyaline layer is visible between the granular layer and the cementodentinal junction. × 110

Fig. 80. Longitudinal ground section of cementum (*Ce*), Tomes' granular layer (*Gl*), interglobular dentine (*Cz*) and dentine (*D*). × 100

Fig. 81. Microradiograph of a transverse section of radicular dentine (*D*) and cellular cementum (*Ce*) with Tomes' granular layer (*Gl*) in the peripheral zone of the root dentine. × 120

ly visible on relatively thin ground sections (Figs. 72, 73). Under the electron microscope, the absence of peritubular dentine, as well as large areas rich in uncalcified collagen fibrils were demonstrated in the interglobular dentine (HELMCKE 1968; TSTASAS and FRANK 1972; KOCKAPAN 1984).

Interglobular dentine, therefore, consists of regions of unmineralized dentine matrix and is considered a normal histological feature. However, with disturbances in mineralization such as in vitamin D deficiency, vitamin D-resistant rickets, or severe fluorosis (JONES and BOYDE 1984; NIKIFORUK and FRASER 1979), interglobular dentine becomes abnormally prominent. Interglobular dentine is also conspicuous in parathyroidectomized rats (SCHOUR et al. 1937; BERNICK 1969) and in rats with chronic calcitonin deficiency (KLINE and THOMAS 1977).

It is interesting to note that with secondary ion microscopy, we observed small localized unmineralized areas with irregular outlines in human intertubular dentine (Figs. 29, 30). These were deficient in calcium (Fig. 29) and magnesium (Fig. 30). Based on their configuration, they did not seem to be formed by an absence of fusion of calcospherites.

L. Granular Layer of Tomes

The granular layer of Tomes is found in the peripheral regions of the root dentine running parallel to the cementodentinal junction (Figs. 78–81). Its dark, finely granular appearance on ground sections registers as small irregular hypomineralized areas in microradiographs (Fig. 81). With respect to location, the layer has been described as extending to the cementodentinal junction (TEN CATE 1972b) or as separated from it by mantle dentine (MOSS 1974; SHACKLEFORD 1971) or by the hyaline layer of HOPEWELL-SMITH (LAVELLE et al. 1977). TEN CATE (1972b) proposed that the granular layer was due to an increase in the space occupied by the odontoblast processes because their development was out of phase with the rate of matrix production. This would lead to looping and buckling of the processes. A second explanation was based on the notion of deficient mineralization of the dentine matrix (SHACKLEFORD 1971). The possible inclusion of epithelial cells of Hertwig's sheath has also been advanced (LINDSKOG and HAMMARSTRÖM 1982). Yet another explanation is the presence of small calcospherites which fail to fuse near the cementodentinal junction, leaving small interglobular spaces (FURSETH 1974; HUGUES et al. 1982). In fact FURSETH (1974) described non-calcified collagen fibrils in these areas.

Backscattered electron imaging in scanning electron microscopy seems to indicate that the unmineralized regions in the granular layer are interglobular and that they often mark a change in sizes from small to larger calcospherites (BOYDE and JONES 1983). These authors present the following hypothesis to explain the occurrence of the unmineralized interglobular spaces. Although dentine formation starts slowly in the crown of human teeth and then accelerates, the reverse is true in the root where the rate of apposition is relatively high during formation of the granular layer (KAWASAKI et al. 1977). A reduction in the number of sites of initiation of mineralization while dentine development is rapid, would be conducive to the retention of unmineralized areas.

M. Incremental Lines and Neonatal Lines

Formation of dentine is a rhythmical process which proceeds with alternating periods of activity and quiescence. The periods of inactivity are indicated by incremental lines which can be seen in non-decalcified ground sections of normal teeth (Figs. 3, 74, 75) or can be demonstrated experimentally with various markers including lead acetate, various fluorochrome dyes, or the tetracycline antibiotics which are incorporated into the dentine and revealed as bright yellow fluorescent lines under ultraviolet light (Figs. 76, 77). Tetracycline fluorescence in human teeth is very common; in a random sample, the incidence of affected teeth was estimated to be 45.4% (NALBANDIAN et al. 1982). Tetracycline fluorescence in dentine occurs as sharply defined lines or bands corresponding to those anatomic sites mineralizing at the time of drug administration. Narrow fluorescent lines reflect short regimens of drug administration whereas broad bands correspond to long-term administration. Because of the incremental pattern of dentine mineralization and the permanence of the fluorescence, these markings constitute a vivid and accurate record of exposure to tetracycline in relation to the chronology of tooth development. HARCOURT et al. (1962), BREARLEY et al. (1968), BAKER and STORREY (1970) and EGAN et al. (1972) were among the first to describe the microscopic effects of tetracycline on ground sections. WEYMAN (1968) noted that the side of a tetracycline band toward the enamel-dentine junction was always more sharply demarcated that that on the pulpal side, a feature probably related to a high serum level at the commencement of therapy and a gradual decrease after intake has ceased.

Classically two types of incremental lines are described, the so-called contour lines of Owen and the incremental lines of VON EBNER. In fact, these two catagories are descriptively inadequate to distinguish between the numerous striae which can be revealed by various techniques. Confusion arises because the incremental lines arise due to a number of different reasons, such as variations in the rate at which preodontoblasts start their secretory activity, variations in the rate of organic matrix secretion during the life cycle of the cells and finally variations in the rate at which mineral is deposited in the matrix.

As originally described by OWEN (1845), the contour lines of Owen are produced by the coincidence of the small secondary curvatures between neighboring dentinal tubules. Alignment of these curvatures gives rise to the contour lines which run at right angles to the dentinal tubules but not necessarily parallel to the outer surface of dentine. However, other lines having the same disposition but caused by deficiencies in mineralization are now generally known as contour lines of Owen.

The incremental lines of VON EBNER are parallel markings which are much closer to each other than the contour lines of Owen's. In fact, two types of lines were independently described by ANDRESEN (1898) and VON EBNER (1906). ANDRESEN (1898) reported on the presence of regularly arranged hematoxylin-stained lines with a rhythm of 20 μm in human coronal dentine viewed in longitudinal decalcified sections, whereas VON EBNER (1906) reported on similar lines which he described as "Dentinlamellen", found not only in coronal but also in radicular dentine. According to KAWASAKI et al. (1980) the 2 types of incremental lines are the same. In young pigs injected at bi-weekly intervals with multiple fluorochrome markers, YILMAZ et al. (1977) confirmed the belief that VON EBNER lines are a visible manifestation of the daily rhythm of dentine formation.

In human dentine, KAWASAKI et al. (1980) found that the incremental lines are approximately 20 μm apart. Within these 20-μm bands, 5 further subdivisions about 4 μm wide were found. They corresponded reasonably well with the daily increments of dentine growth reported by SCHOUR and·PONCHER (1937) for man. KAWASAKI et al. (1980) differentiated at 24 h rhythm with about 4 μm of organic dentine matrix formation from the mineral deposition in the dentine calcospherites which occurred at the rate of approximately 2 μm every 12 h.

Besides the circadian rhythm association with VON EBNER'S incremental growth lines, ROSENBERG and SIMMONS (1980) have identified ultradian and infradian periods of dentinogenesis. They injected rabbits during the day or at night at 2-week intervals and examined transverse sections of lower incisors with electron-microprobe analysis for calcium and sulfur. They found that the mineralization rhythm, as followed by calcium determination, was independent of the rhythm of matrix deposition, as followed by the sulfur contained in dentinal proteoglycans. Further analysis of the incremental pattern in sections of rabbit dentine as thick as 10–20 μm revealed a circadian pattern of 14–30 μm per day, while analysis of 5 μm-thick sections revealed that the daily increments were composed of 2–3 narrower increments which were, on the average, 10 μm wide and which were deposited over a period from 8–12 h.

Apparently, many factors can influence the production rate of dentine. SCHOUR and DYKE (1932) observed an increased growth of maxillary rat incisors after injection of growth hormone to hypophysectomized rats. Similarly, bovine growth hormone was found to increase the dentine production slightly in normal rats and significantly in hypophysectomized rats (HANSSON et al. 1978).

Rhythmicity in daily dentine deposition was influenced in rats by lesion of the hypothalamic nuclei, produced by proteolysis or electrolysis with the use of stereotaxic apparatus (ASHIDA 1983). From this and other observations it is evident that dentine deposition is a complex and highly regulated phenomenon.

Finally neonatal lines have been described in teeth developing at the time of birth. In those teeth, i.e. the deciduous teeth and frequently the first permanent molar, a particularly accentuated incremental line is found demarcating the dentine formed before and after birth. The period of arrest of dentine formation in the paranatal period has been estimated to be of the order of 15 days (MASSLER and SCHOUR 1946).

N. Translucent Dentine

Translucent dentine represents a special tissue change occurring either under normal physiological conditions related to aging, especially in root dentine, or under various pathological conditions such as dental caries, attrition or dental erosion, generally affecting coronal or cervical root dentine. The translucency is due to dentinal sclerosis, an obliteration of the tubular lumens by calcified material which has a refractive index similar to that of the rest of the dentine. Microscopically, on a ground section the sclerosed or translucent dentine appears transparent.

Miller (1980) appears to be among the first to attract attention to the probable significance of certain histological changes in the dentine in response to aging, wear and tear. Previously Fish (1932), Beust (1934), and Bödecker and Lefkovitz (1946) studied the nature of the changes evidenced by the relative degree of opacity versus transparency of dentine. They demonstrated a decrease in dentine permeability to dyes in connection with aging and pathology, a finding suggesting changes within the dentinal tubules.

Nalbandian and Sognnaes (1960) and Nalbandian et al. (1960), observed that coincident with aging, root dentine becomes transparent, giving rise to a glass-like appearance in transmitted light, commencing at the root tip and creeping towards the crown. Micrographic observations indicated that the distribution of root transparency of old teeth is correlated with the degree of tubular sclerosis. Through microhardness test, Grajower et al. (1977) found sclerotic root dentine to be harder than normal opaque root dentine. Vasiliadis et al. (1983a) confirmed that the amount of root sclerosis increased linearly with age and was not markedly affected by the function of the tooth or external stimuli encountered during life. Sclerosis started at the apical dentine adjacent to cementum and extended coronally and towards the root canal with increasing age. In planes transverse to the long axis of the root, sclerosis appeared first at the mesial and distal sides so that the sclerotic zones had a butterfly shape. Vasiliadis et al. (1983b) noted that translucence of root dentine appeared before all tubules are occluded.

Sclerosis of root dentine was one of the six criteria used by Gustafson (1950) for forensic age determination of a human individual. He appraised six dental age changes observed in routine longitudinal ground sections and combined semi-quantitative estimates of the extent of these changes, namely:

1) the increasing degree to which enamel and subsequently the dentine was worn at the chewing surfaces of the teeth (attrition);

2) the increasing amount of internal dental apposition which results in decrease in the size of the coronal pulp chamber (secondary dentine);

3) apical migration of the periodontal attachment;

4) increasing amounts of cementum apposition;

5) increasing number of Howship's lacunae on the root surface (root resorption);

6) increased area of root dentine sclerosis as evidenced by its glassy appearance in transmitted light (root transparencies).

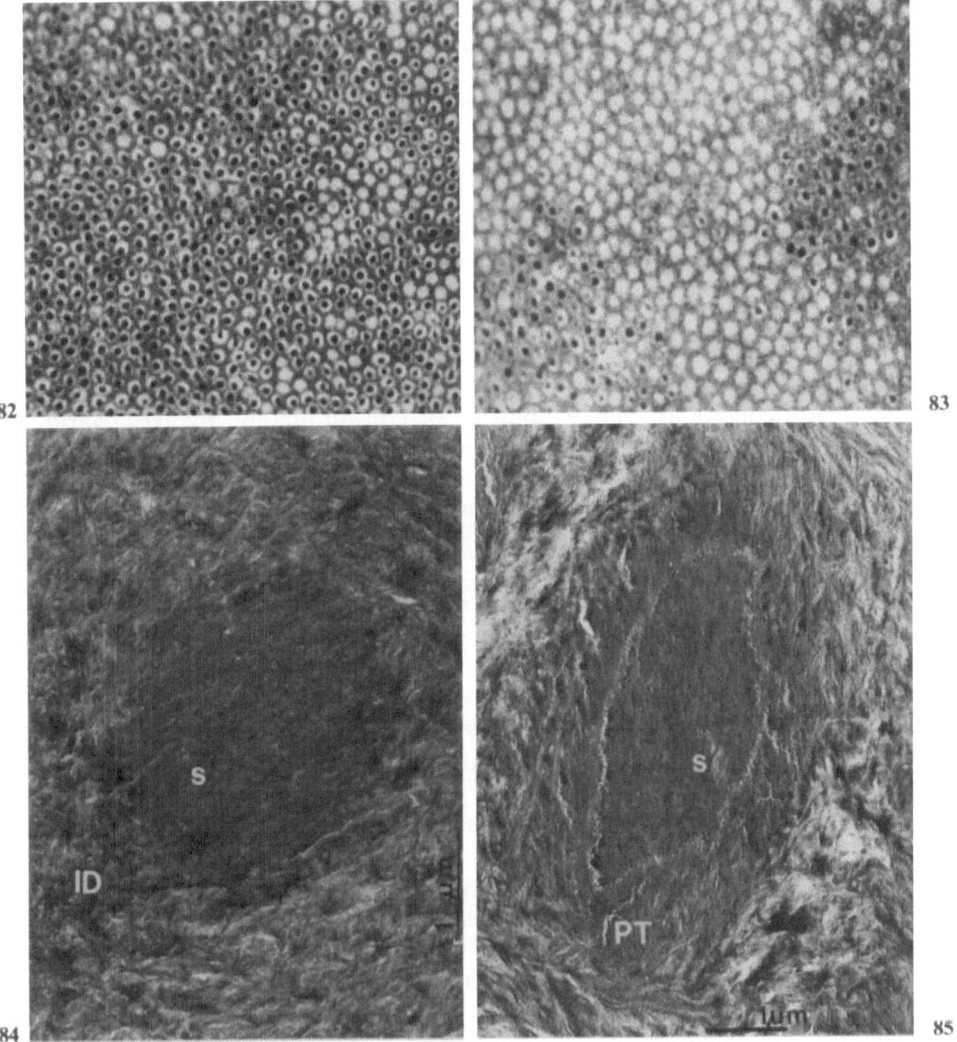

Fig. 82. Microradiograph of a cross section of human noncarious dentine in a molar of a 55-year-old man. Most of the dentinal tubules are surrounded by a calcified white ring of peritubular dentine and show a open lumen (black). However, a few sclerosed tubules are seen, completely calcified (white). × 530

Fig. 83. Microradiography of a cross section of human noncarious coronal dentine in a premolar of a 61-year-old woman with signs of occlusal abrasion. Note the numerous sclerosed tubules. × 530

Fig. 84. Appearance in transmission electron microscopy of a sclerosed tubule (*s*) in cross-section. Note the homogeneous nature of the intratubular calcification, *ID* Intertubular dentine

Fig. 85. Transverse section of a sclerosed human tubule in transmission electron microscopy. The boundary between the calcified lumen (*s*) and the peritubular dentine (*PT*) is visible

By this combination of six dental landmarks, Gustafson (1950) was able to arrive at a good prediction of chronological age within a sample of Swedish individuals. The six dental age criteria were quantitated by histological examination, each factor given a point value of 0–3 depending on the extent of the age change. The points for a given tooth were added and a plot was made of total point values against true chronological ages. Nalbandian and Sognnaes (1960) tested this method on an American population. The multifactorial approach previously tested on a Swedish group was applied to a sample of Boston teeth of comparable size with comparable results. Average error of estimation was ±7.9 years in the Boston sample and ±3.7 years in the Swedish study.

Dentinal sclerosis related to aging can also be observed in coronal dentine (Fig. 82) but it is more difficult here to be sure that other environmental or pathological conditions such as abrasion or erosion are not involved. Various methods have been used to study dentinal sclerosis. Keil (1939) detected crystal deposits in the tubules of carious dentine using polarizing light microscopy. Microradiography was widely used (Bergman and Engfeldt 1955; Nalbandian et al. 1960; Dreyfuss et al. 1964; Mendis and Darling 1979; Vasiliadis 1983 a, b). With the latter technique the calcified tubular lumen cannot be distinguished from the peritubular dentine on transverse sections of sclerosed dentinal tubules (Figs. 82, 83). Examination of sclerosed tubules with secondary ion microscopy demonstrates a higher content of calcium (Fig. 29) and phosphorus, as well as a higher content of magnesium (Fig. 30), potassium and sodium than in the intertubular dentine (Lefèvre et al. 1976). The substance obliterating the dentinal tubule lumen likewise cannot be differentiated from peritubular dentine with this method.

Dentinal sclerosis has been studied extensively with scanning and transmission electron microscopy. With scanning electron microscopy of sclerotic root dentine, the occluding material and peritubular dentine are almost indistinguishable (Vasiliadis et al. 1983 b). With transmission electron microscopy, the ultrastructural aspects of dentinal sclerosis seemed to be highly variable. According to Takuma and Eda (1966) the development of peritubular dentine is different from that of sclerosed root dentine. The peritubular dentine grows in an appositional fashion with a clear forming front, whereas in sclerosed tubules the occlusion occurs rather diffusely.

Sclerotic tubules observed in transverse non-decalcified sections with transmission electron microscopy seem to be obliterated by a calcified material having the same ultrastructure as the peritubular dentine (Fig. 84). The limit between the calcified lumen and the peritubular dentine may (Fig. 85) or may not (Fig. 84) be differentiated. In carious translucent dentine, large rhombohedral or cuboid crystals, 2000 to 6000 Å in width were identified in sclerosed tubules by Helmcke (1955), Lenz (1955), Frank (1957), and Frank et al. (1964). Similar crystals were also found in sclerosed tubules located under occlusal attrition (Tronstad and Langeland 1971; Mendis and Darling 1979). These large rhombohedral crystals were identified as $\beta - Ca_3(PO_4)_2$ (Whitlockite) by electron diffraction (Vahl et al. 1964; Tronstad and Langeland 1971).

Studying the development of dentinal sclerosis by transmission electron microscopy and electron diffraction in transparent carious dentine, Frank and

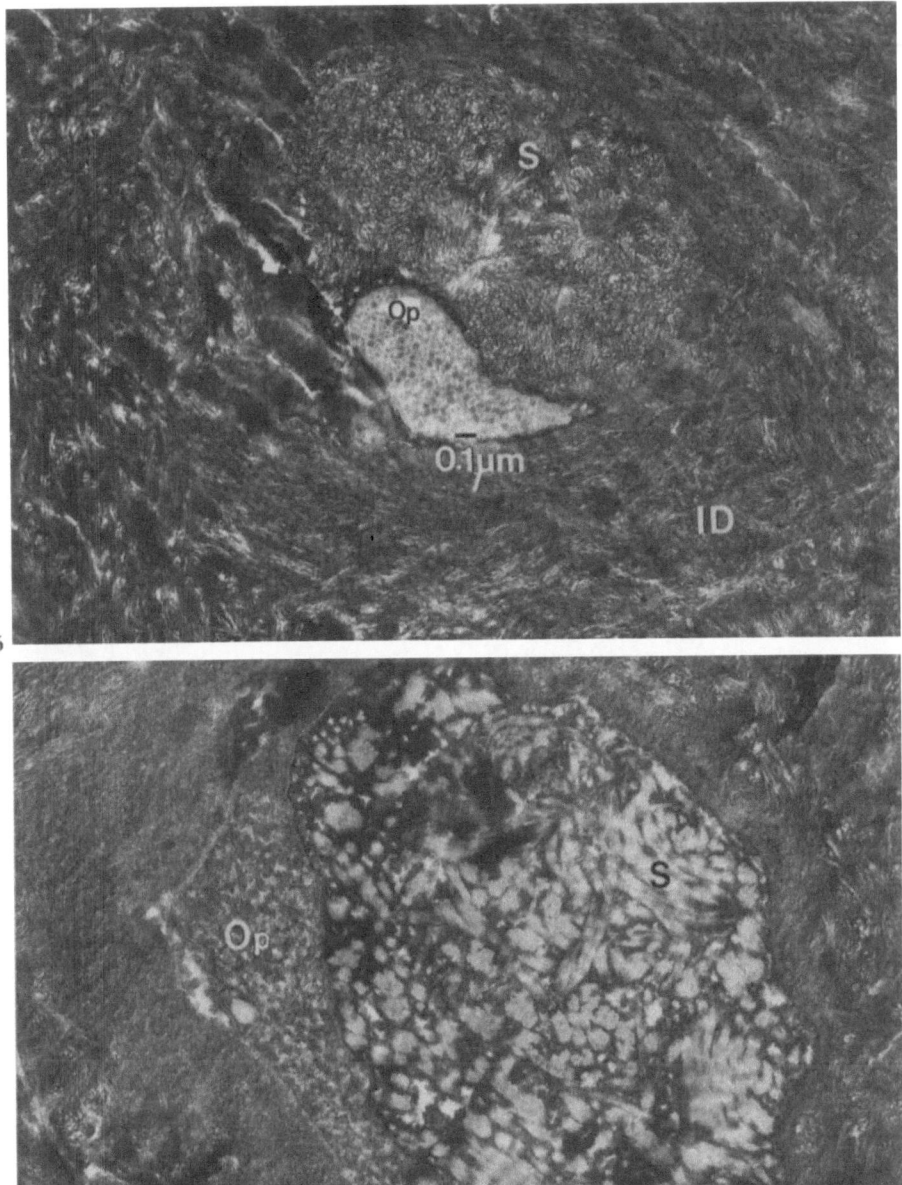

Fig. 86. Transverse section of a dentinal tubule in the process of sclerosis. Advanced calcification of the periodontoblastic space (*S*). A small portion of the odontoblastic process (*Op*) with typical microtubules and microfilaments is completely surrounded by calcified dentine. *ID* Intertubular dentine. (From FRANK and VOEGEL 1980, courtesy of Caries Research)

Fig. 87. Transverse section of a dentinal tubule in the process of sclerosis. The calcified odontoblast process (*Op*) is adjacent to the calcifying periodontoblastic space (*S*) where dense granular material is laid down in a honeycombed pattern between collagen fibrils that show typical cross-striations. (From FRANK and VOEGEL 1980, courtesy of Caries Research)

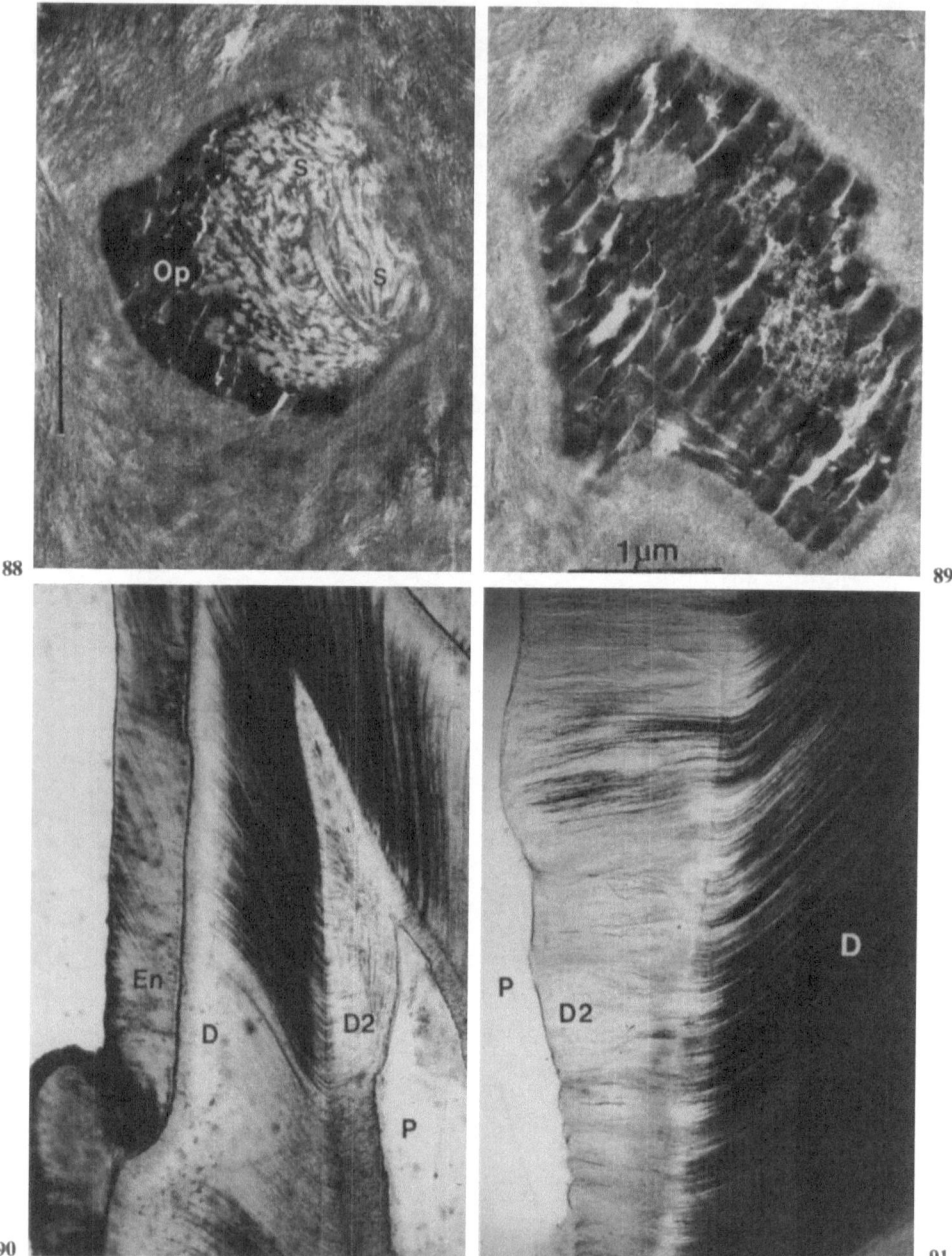

Fig. 88. Transverse section of a dentinal tubule in the process of sclerosis. Homogeneous and dense calcification of the odontoblastic process (*Op*). The periodontoblastic space (*S*) shows deposition of a fine, dense granular material between collagen fibrils with periodic cross-striations

Fig. 89. Cross section of a sclerosed dentinal tubule. Filling the tubule, a large mass of dense homogeneous mineral is seen, with shatter lines due to sectioning. (From Frank and Voegel 1980, courtesy of Caries Research)

VOEGEL (1980) observed one type of intratubular calcification which started by an initial mineralization of the periodontoblastic space, followed by calcification of the odontoblastic process (Fig. 86). A second type of dentinal sclerosis consisted of an initial intracytoplasmic calcification of the odontoblastic process followed by secondary mineralization of the periodontoblastic space (Figs. 87, 88). Sometimes large masses of mineral, exhibiting shatter due to sectioning, filled the lumen of the tubule (Fig. 89). By electron diffraction, the mineral contained there in was identified as hydroxyapatite.

O. Regular and Irregular Secondary Dentine

As has already been pointed out in the chapter on classification, considerable confusion exists regarding the nomenclature for the various types of dentine, but the classification of KUTTLER (1959) into primary, secondary, and tertiary dentine seems to be generally accepted. A precise definition of secondary dentine as given by BAUME (1980) deliniates that part of the circumpulpal dentine consisting of regular dentine (orthodentine) in continuity with the primary dentine, but produced circumpulpally throughout the later periods of vitality. The term "physiological secondary dentine" has been used as a synonym for the term "regular secondary dentine".

The list of names used for tertiary dentine formed as a response to various irritation factors is confusingly long: reparative dentine, irregular secondary dentine, irritation dentine, replacement dentine, adventitious dentine, defense dentine, and atypical dentine (TAINTOR et al. 1981; KARJALAINEN 1984). Tertiary dentine was defined as follows by BAUME (1980): tertiary dentine is the dentine, more or less regular in structure, deposited at sites on the pulpal aspects of primary or secondary dentine and corresponding to areas of external irritation. In other words, the differentiation between secondary and tertiary dentine is not based on histological grounds but is based on the causative factors responsible for the tissue elaboration and which are either physiological (for secondary dentine) or are pathological or iatrogenic (for tertiary dentine).

Since we are dealing here with structure and ultrastructure of dentine, we will use histological criteria exclusively to describe regular and irregular secondary dentine. Therefore, under regular secondary dentine we will include the secondary dentine (orthodentine) in continuity with primary dentine, which pres-

Fig. 90. Longitudinal ground section of an upper incisor in a 45-year-old man. Presence of irregular secondary dentine (*D2*) obliterating the upper part of the pulp chamber. *P* Pulp chamber; *D* Dentine; *En* Enamel. × 7,5

Fig. 91. Lateral dentinal wall of a pulp chamber in an upper premolar in a 35-year-old man. Regular configuration of the secondary dentine with respect to the dentinal tubule orientation. *D* Dentine; *D2* Secondary dentine; *P* Pulp chamber. × 38

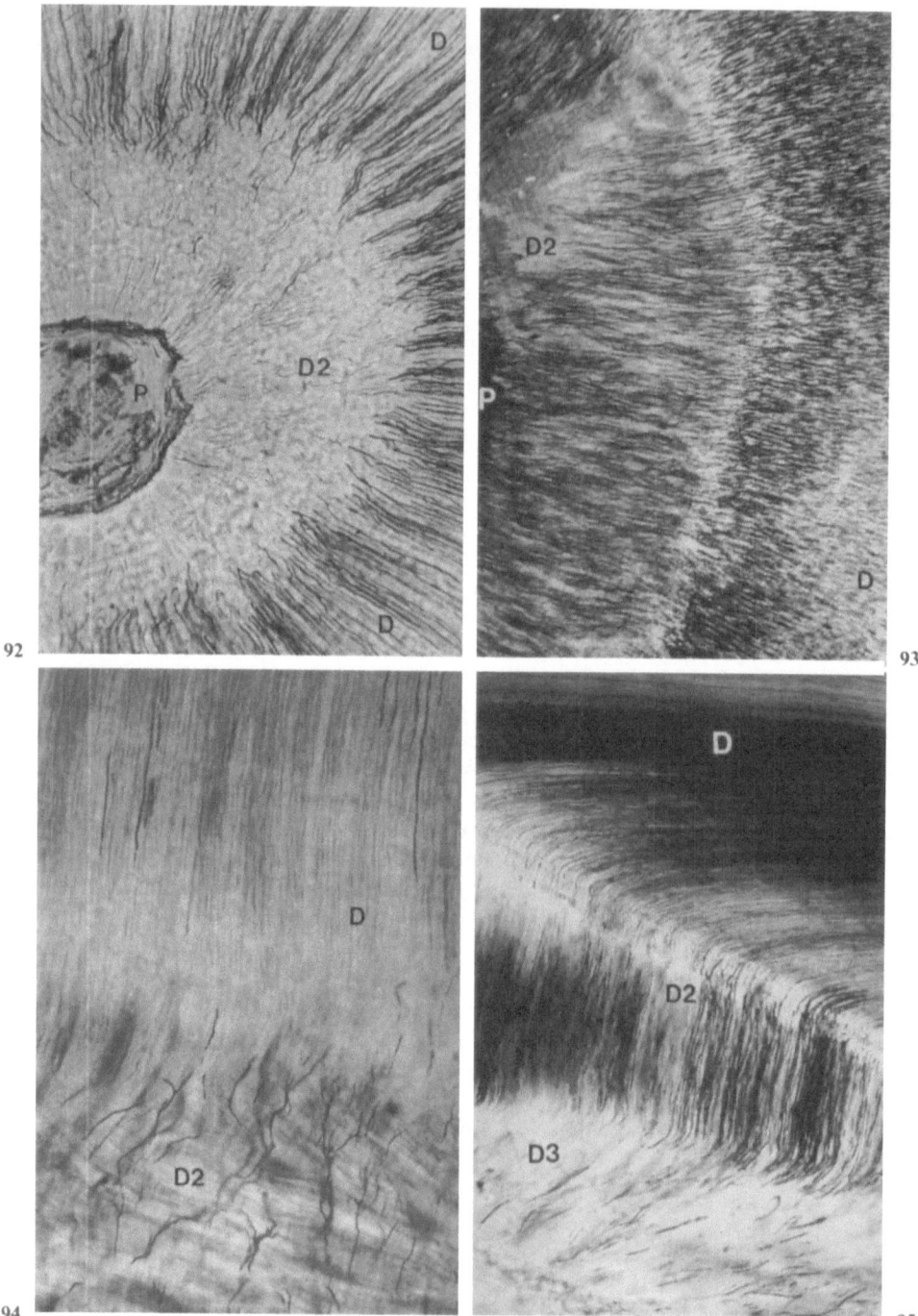

Fig. 92. Cross-section of a lower human canine from a 42-year-old man. Presence of irregular secondary dentine (*D2*) exhibiting a few dentinal tubules. *P* Pulp area; *D* Dentine with dentinal tubules.
× 110

ents a regular and parallel distribution of dentinal tubules and is normally calcified (Figs. 91, 93 and Fig. 95D2). The irregular secondary dentine is that part of circumpulpal dentine in continuity either with primary dentine (Figs. 90, 92, 94) or with regular secondary dentine (Fig. 95D3) and which shows a more or less irregular disposition of dentinal tubules, a complete absence of tubules, or even a variety of cellular inclusions. Regular and irregular secondary dentine are both formed after eruption of the teeth, i.e. after completion of primary dentine formation.

Regular secondary dentine is often found on the lateral sides of the pulp chamber (Fig. 91). In multirooted teeth, regular secondary dentine is usually much thicker on the floor of the pulp chamber than on the side walls. The transition between this type of secondary dentine and the primary dentine may occur as distinct demarcation line, without change in the direction of the tubules (Fig. 91). Such a demarcation is even visible on microradiographs (Fig. 93). In other cases a marked S-shaped change can be observed in the direction of the dentinal tubules (Fig. 95). Studying the junctional region between primary and regular secondary dentine, SCOTT and WEBER (1977) claimed that on thin section a dramatic reduction is observed in the number of tubules on the secondary dentine side. However, the intertubular radiodensity seems to be comparable. The latter observation was confirmed with our own microradiographs (Fig. 93).

Using casts of human dentinal tubules examined with SEM, WANG et al. (1985) described continuity between tubules of regular secondary dentine and primary dentine. A downward slope of Ca, P, and Mg profiles were observed by electron-microprobe analysis of secondary dentine, with marked elevations of Zn profiles and modest elevations of F and sometimes S profiles (TÖTDAL and HALS 1985).

Irregular secondary dentine can be found everywhere along the pulpal walls. However, it is frequently conspicuous in the regions formerly occupied by the pulpal horn (Fig. 90). This type of secondary dentine has a highly variable structure. If tubules are present, they are disposed somewhat irregularly, with various types of configurations (Figs. 92, 94, 95D3). Dentinal tubules can also be totally absent. The tissue then has an homogeneous appearance in light microscopy. In the latter instance, various types of irregular spaces or cellular inclusions can be found. Between primary and irregular secondary dentine, a small intermediary region devoid of dentinal tubules and consisting of an

Fig. 93. Microradiograph of a transverse coronal section of a premolar in a 38-year-old man. The regular secondary dentine (*D2*) presents numerous well-oriented dentinal tubules. *P* Pulp area; *D* Dentine. × 85

Fig. 94. Longitudinal ground section of coronal dentine in a lower molar of a 41-year-old man. The irregular secondary dentine (*D2*) shows a tortuous arrangement of dentinal tubules. *D* Midway dentine with parallel dentinal tubules. × 95

Fig. 95. Longitudintal ground section of coronal dentine in an upper central incisor. Regular secondary dentine consisting of an outer layer with regular and almost parallel dentinal tubules (*D2*) and irregular secondary dentine (*D3*) located pulpwards. *D* Primary dentine. × 80

amorphous calcified layer has been described as a hyaline zone by Fish (1932) and as interface dentine by Mjör (1983, 1985). On casts of human dentinal tubules examined with SEM, Wang et al. (1985) found no continuity between tubules in primary dentine and those of irregular secondary dentine.

At the end of this chapter on the structure and ultrastructure of dentine, it can be stated that dentine is a remarkable calcified tissue. It has a highly complex organization characterized by the presence of a very unique cell, the odontoblast, with its long cell extension into the tissue. The dentinal tubules which contain these cell extensions also house sensory nerve fibers. The innervation of the dentine will be described in the next chapter, dealing with the structure and ultrastructure of the pulp.

References

Andresen V (1898) Die Querstreifung des Dentins. Dtsch Monatschr Zahnheilk 16:386–389

Armstrong WD, Brekhus PJ (1937) Chemical constitution of enamel and dentine. I. Principal components. J Biol Chem 120:677–687

Ashida O (1983) Hypothalamic nuclei related to circadian rhythmicity in dentinogenesis of the rat incisor. Bull Kanagawa Dent Coll 11:15–27

Athenstaedt H (1971) Pyroelectric and piezo-electric behavior of human dental hard tissues. Arch Oral Biol 16:495–501

Aubin JE (1985) New immunocytochemical approaches to studying the odontoblast. J Dent Res 61:515–522 (special issue)

Baker KL, Storey E (1970) Tetracycline-induced tooth changes. 3. Incidence in extracted first permanent molar teeth. Med J Aust 1:109–113

Baud CA, Held AJ (1956) Silberfärbung, Röntgenmikrographie und Mineralgehalt der Zahnhartgewebe. Dtsch Zahnärztl Z 11:309–314

Baume LJ (1980) The biology of pulp and dentine: a historic, terminologic, taxonomic, histological, biochemical, embryonic and clinical survey. Monographs in oral science. Karger, Basel

Beersten W, Everts V, Hoff van den A (1974) Fine structure of fibroblasts in the periodontal ligament of the rat incisor and their role in tooth eruption. Arch Oral Biol 19:1087–1098

Bergman G, Engfeldt B (1955) Studies on mineralized dental tissues. Acta Odontol Scand 13:1–7

Bernick S (1969) Histochemical study of dentin in parathyroidectomized rats. J Dent Res 48:1251–1257

Bernick S, Baker RF, Rutherford RL, Warren O (1952) Electron microscopy of enamel and dentin. J Am Dent Assoc 45:689–696

Berthet-Colominas C, Miller A, White SW (1979) Structural study of the calcifying collagen in turkey leg tendons. J Mol Biol 134:431–445

Beust TB (1934) Demonstration of sclerosis of dentin in tooth maturation and caries. Dental Cosmos 76:305–317

Bishop MA (1985) Evidence for tight junctions between odontoblasts in the rat incisor. Cell Tissue Res 239:137–140

Blackwood HJJ (1957) Intermediate cementum. Br Dent J 102:345–350

Blake GC (1958) The peritubular translucent zones in human dentine. Br Dent J 104:57–64

Bödecker CF, Applebaum E (1931) Metabolism of the dentin. Its relation to dental caries and to the treatment of sensitive teeth. Dent Cosmos 73:995–1009

Bödecker CF, Lefkowitz W (1946) Further observations on the vital staining of dentin and enamel. J Dent Res 25:387–396

Boyde A, Jones SJ (1983) Backscattered electron imaging of dental tissues. Anat Embryol (Berl) 168:211–226

Braden M (1976) Biophysics of the tooth. In: Kawamura Y (ed) Frontiers of oral physiology, vol II. Karger, Basel, pp 1–37

Bradford EW (1950) An investigation in the structure of the pulpo-dentinal junction. Br Dent J 88:55–58

Bradford EW (1955) The interpretation of calcified sections of human dentine. Br Dent J 98:153–159

Bradford EW (1967) Microanatomy and histochemistry of dentine. In: Miles AEW (ed) Structural and chemical organization of teeth, vol II. Academic, New York, pp 3–34

Brännström M, Garberoglio AR (1972) The dentinal tubules and the odontoblast process. A scanning electron microscopic study, Acta Odontal Scand 30:291–311

Brearley LJ, Stragis AA, Storey E (1968) Tetracycline-induced tooth changes. 1. Prevalence in preschool children. Med J Aust 2:653–658

Butler WT (1984a) Dentin collagen: chemical structure and role in mineralization. In: Linde A (ed) Dentin and dentinogenesis, vol II. CRC Press, Boca Raton, pp 37–53

Butler WT (1984b) Matrix molecules of bone and dentine. Coll Relat Res 4:297–307

Calle A (1985) Intercellular junctions between human odontoblasts. A freeze-fracture study after demineralization. Acta Anat (Basel) 122:138–144

Coffey CT, Ingram MJ, Björndal AM (1970) Analysis of human dentinal fluid. Oral Surg 30:835–837

Connor NS, Aubin JE, Melcher AH (1984) The distribution of fibronectin in rat tooth and periodontal tissues: an immunofluorescence study using a monoclonal antibody. J Histochem Cytochem 32:565–572

Cournil I, Leblond CP, Pomponio J, Hand AR, Sederlof L, Martin GR (1979) Immunohistochemical localization of procollagens. I. Light microscopic distribution of procollagen I, III and IV antigenicity in the rat incisor tooth by the indirect peroxidase-antiperoxidase method. J Histochem Cytochem 27:1059–1069

Crooks PV, O'Reilly CB, Owens PDA (1983) Microscopy of the dentin of enamel-free areas of rat molar teeth. Arch Oral Biol 28:167–176

Cuvier G (1805) Lecons d'anatomie comparée. Dix septième leçon des dents. Crochars et Fantin, Paris 3:103–200

Czermak J (1850) Beiträge zur mikroskopischen Anatomie der Zähne. Z Zool 2:297–312

Dreyfuss F, Frank RM, Gutmann B (1964) La sclérose dentinaire. Bull Group Int Rech Sci Stomatol Odontol 7:207–229

Driessens FCM (1982) Mineral aspects of dentistry. In: Myers HM (ed) Monographs in oral science, vol 10. Karger, Basel, pp 1–215

Ebner V von (1906) Über die Entwicklung der leimgebenden Fibrillen insbesondere im Zahnbein. Sitzung der Akad Wiss Wien Math-naturwiss 115:281–346

Eda S, Takuma S (1965) Microstructure of the peritubular matrix in horse dentin. Bull Tokyo Dent Coll 6:1–14

Egan J, Tring FC, Prout RES (1972) Tetracycline deposits in children's teeth in relation to atmospheric pollution. Brit J Prev Soc Med 26:259–262

Engström A (1966) Apatite-collagen organization in calcified tendon. Exp Cell Res 43:241–245

Farquhar MG, Palade GE (1963) Junctional complexes of various epithelia. J Cell Biol 17:375–412

Fearnhead RW (1957) Histological evidence for the innervation of human dentine. J Anat 91:267–277

Fiore-Donno G, Baume LJ (1966) Etude histochimique de la dentinogenèse humaine. Helv Odont Acta [suppl 4] 10:141–195

Fish WE (1927) The circulation of lymph in dentin and enamel. J Am Dent Assoc 14:804–817

Fish WE (1932) An experimental investigation of enamel, dentine and the dental pulp. 1st edn. Bale and Danielson, London

Frank RM (1952) Données recentes sur l'infrastructure de la dent fournies par les techniques de microscopie électronique. Arch Stomatol 7:127–140

Frank RM (1957) Microscopie et diffraction électronique des éléments cristallins de la dent. Odont Rev 8:134–142

Frank RM (1959) Electron microscopy of undecalcified sections of human adult dentine. Arch Oral Biol 1:29–32

Frank RM (1966a) Etude au microscope électronique de l'odontoblaste et du canalicule dentinaire humain. Arch Oral Biol 11:179–199

Frank RM (1966b) Ultrastructure of human dentine. In: Fleisch H, Blackwood JHH, Owen M (eds) Third Europ Symp Calcified Tissues. Springer, Berlin, pp 259–271

Frank RM (1968a) Ultrastructural relationship between the odontoblast, its process and the nerve fibre. In: Symons NBB (ed) Dentine and pulp: their structure and reactions. Livingstone, London, pp 115–145

Frank RM (1968b) Attachment sites between the odontoblast process and the intradentinal nerve fibre. Arch Oral Biol 13:833–834

Frank RM (1970) Etude autoradiographique de la dentinogenèse en microscopie électronique à l'aide de la proline tritiée chez le chat. Arch Oral Biol 15:583–596

Frank RM (1979) Electron microscope autoradiography of calcified tissues. Int Rev Cytol 56:183–253

Frank RM, Voegel JC (1978) Dissolution mechanisms of the apatite crystals during dental caries and bone resorption. In: Berlin RD, Herrmann H, Lepow IH, Tanzer JM (eds) Molecular basis of biological degrative processes. Academic, New York, pp 277–311

Frank RM, Voegel JC (1980) Ultrastructure of the human odontoblast process and is mineralization during dental caries. Caries Res 14:367–380

Frank RM, Wolff F, Gutmann B (1964) Microscopie électronique de la carie au niveau de la dentine humaine. Arch Oral Biol 9:163–179

Frank RM, Sauvage C, Frank P (1972) Morphological basis of dental sensitivity. Int Dent J 22:1–19

Frank RM, Cimasoni G, Tsamouranis A, Matter J, Fiore-Donno G (1977) Collagen resorption by fibroblasts in human gingiva. J Biol Buccale 5:343–351

Fromme HG, Riedel H (1970) Messungen über die Weite der Dentinkanälchen an nichtentminera-lisierten bleibenden Zähnen und Milchzähnen. Dtsch Zahnärztl Z 25:401–405

Furseth R (1974) The structure of peripheral root dentin in young human premolars. Scand J Dent Res 82:557–561

Garant PR (1972) The organization of microtubules within rat odontoblast processes revealed by perfusion fixation with glutaraldehyde. Arch Oral Biol 17:1047–1058

Garant PR (1976) Collagen resorption by fibroblasts. A theory of fibroblastic maintenance of perio-dontal ligament. J Periodont Res 47:380–390

Garant PR (1978) Microanatomy of the oral mineralized tissues. In: Shaw JH, Sweeney EA, Cappuc-cino CC, Meller SB (eds) Textbook of oral biology, vol 5, 1st edn. Saunders, Philadelphia, pp 181–225

Garant PR, Szabo G, Nalbandian J (1968) The fine structure of mouse odontoblasts. Arch Oral Biol 13:857–876

Garberoglio R, Brännström H (1976) Scanning electron microscopic investigations of human dentinal tubules. Arch Oral Biol 21:355–362

Gerould CH (1944) Ultramicrostructure of human tooth as revealed by the electron microscope. J Dental Res 23:239–245

Glimcher MJ (1982) On the form and function of bone: from molecules to organs. Wolff's law revisited 1981. In: Veis A (ed) The chemistry and biology of mineralized connective tissue. Elsevier, North Holland, New York, pp 617–613

Goldberg M (1983) "Protéoglycanes de la dentine et de l'émail: interrelations avec les composants matriciels. Etudes histochimiques et ultrastructurales"; Thèse Dr d'état Sciences Nat, Université Paris 6

Goldberg M, Escaig F (1981) Odontoblastes: collagène dans la prédentine et la dentine de l'incisive de rat: étude par cryofracture. Biol Cell 40:203–216

Goldberg M, Septier D (1983) Electron microscopic visualization of proteoglycans in rat incisor predentine and dentine with cuprolinic blue. Arch Oral Biol 38:79–83

Goldberg M, Genotelle-Septier D, Weill R (1978) Glycoprotéines et protéoglycanes dans la matrice prédentinaire et dentinaire chez le rat: étude ultrastructurale. J Biol Buccale 6:75–90

Grajower R, Azaz B, Bron-Levi M (1977) Microhardness of sclerotic dentin. J Dent Res 56:446 (short paper)

Granström G, Linde A, Nygren H (1978) Ultrastructural localization of alkaline phosphatases in rat incisor odontoblasts. J Histochem Cytochem 26:359–368

Grön P (1978) Inorganic chemical and structural aspects of oral mineralized tissues. In: Shaw JH, Sweeney FA, Capuccino CC, Meller SM (eds) Textbook of oral biology, vol 15. Saunders, Phila-delphia, pp 484–507

Grossman ES, Austin JC (1983) Scanning electron microscope observations on the tubule content of freeze fractured peripheral vervet monkey dentine (*Cercopithecus pygerythrus*). Arch Oral Biol 28:279–281

Gunji T (1983) Morphological research on the sensitivity of dentin. Arch Histol Jpn 45:45–67

Gunji T, Koyabashi S (1983) Distribution and organization of odontoblast processes in human dentine. Arch Histol Jpn 46:213–219

Gustafson G (1950) Age determinations on teeth. J Am Dent Assoc 431:45–54

Haldi J, Wynn W (1963) Protein fractions of the blood plasma and dental pulp fluid in the dog. J Dent Res 42:1217–1221

Haldi J, Wynn W, Culpepper WD (1961) Dental pulp fluid. I. Relationship between dental pulp fluid and blood plasma in protein, glucose and inorganic content. Arch Oral Biol 3:201–206

Hals E (1983) Observations on giant tubules in human coronal dentin by light microscopy and microradiography. Scand J Dent Res 91:1–7

Hansson LI, Stenström A, Thorngren KG (1978) Effect of pituitary hormones on dentine production in maxillary incisors in the rat. Scand J Dent Res 86:80–86

Harcourt JK, Johnson NW, Storey E (1962) *In vivo* incorporation of tetracycline in the teeth of man. Arch Oral Biol 7:431–437

Harris R, Griffin CJ (1969) The fine structure of the mature odontoblasts and cell rich zone of the human dental pulp. Aust Dent J 14:168–177

Harris R, Griffin CJ (1978) Fine structure of nerve endings in the human dental pulp. Arch Oral Biol 13:773–778

Hawkinson RW, Eisenmann DR (1983) Electron microscopy of dentinal tubule sclerosis in the enamel free region of the rat molar. Arch Oral Biol 28:409–414

Helmcke GH (1968) Formen und Strukturen der Interglobulärräume im menschlichen Dentin. Bull Group Int Rech Sci Stomatol Odontol 11:317–328

Helmcke JG (1955) Elektronenmikroskopische Strukturuntersuchungen an gesunden und kranken Zähnen. Dtsch Zahnärztl Ztschrft 15:1461–1478

Helmcke JG, Jahn B (1952) Elektronenmikroskopische Untersuchungen über das Dentin im menschlichen Zahn. Naturwissenschaften 39:492–493

Helwig G, Menke E (1949) Elektronenmikroskopie an Zellfortsätzen im menschlichen Zahnbein. Naturwissenschaften 36:281–283

Höhling HJ (1966) Die Bauelemente von Zahnschmelz und Dentin aus morphologischer, chemischer und strukureller Sicht. 1st edn. Hanser, München

Höhling HJ, Steffens H, Heuck F (1972) Untersuchung zur Mineralisationsdichte im Hartgewebe mit Protein-Polysaccharid bzw. mit Kollagen als Hauptbestandteil der Matrix. Z Zellforsch 134:283–296

Holland GR (1975a) Membrane junctions on cat odontoblasts. Arch Oral Biol 20:551–552

Holland GR (1975b) The dentinal tubule and odontoblast process in the cat. J Anat 120:169–177

Holland GR (1976a) Lanthanum hydroxide labelling of gap junctions in the odontoblast layer. Anat Rec 186:121–126

Holland GR (1976b) The extent of the odontoblast process in the cat. J Anat 121:133–149

Holland GR (1985) The odontoblast process: form and function. J Dent Res 64:499–514 (special issue)

Hopewell-Smith A (1903) The histology and pathohistology of the teeth and associated parts. 1st edn. The Dental Manufacturing Co, London

Hugues K, Mead R, Adams D (1982) A scanning electron microscope study of root development in human unerupted wisdom teeth. J Dent Res 61:557 (abstract)

Hunter J (1778) The natural history of human teeth, 2nd edn. Johnson, London

Iguchi Y, Yamamura T, Ichikawa T, Hashimoto S, Horiuchi T, Shimono M (1984) Intercellular junctions in odontoblasts of the rat incisors studied with freeze fracture. Arch Oral Biol 29:487–497

Isokawa S, Toda Y, Kubota K (1970) A scanning electron microscopic observation of etched human peritubular dentine. Arch Oral Biol 15:1303–1306

Ivanyi D (1972) Nucleoli of human odontoblasts. Arch Oral Biol 17:931–936

Jessen H (1967) The ultrastructure of odontoblasts in perfusion fixed demineralized incisors of adult rats. Acta Odontol Scand 25:491–523

242   R.M. Frank and J. Nalbandian

Johansen E (1964) Microstructure of enamel and dentin. J Dent Res 43:1007–1020
Johansen E (1967) Ultrastructure of dentine. In: Miles AEW (ed) Structural and chemical organiza-
    tion of teeth, vol II. Academic, New York, pp 35–74
Johansen E, Parks HF (1962) Electron microscopic observations of sound human dentin. Arch
    Oral Biol 7:185–193
Johnson NW, Poole DFG (1967) Orientation of collagen fibers in dentine. Nature 213:695–696
Jones SJ, Boyde A (1984) Ultrastructure of dentin and dentinogenesis. In: Linde A (ed) Dentin
    and dentinogenesis, vol I. CRC Press, Boca Raton, pp 81–134
Karim A, Cournil I, Leblond CP (1979) Immunohistochemical localization of procollagens. II. Elec-
    tron microscopic distribution of procollagen I antigenicity in the odontoblasts and predentin
    of rat incisor teeth by a direct method using peroxidase linked antibodies. J Histochem Cytochem
    27:1070–1083
Karjalainen S (1984) Secondary and reparative dentin formation. In: Linde A (ed) Dentin and
    dentinogenesis, vol II. CRC Press, Boca Raton, pp 7–420
Katchburian E, Holt SJ (1968) Ultrastructural studies on lysosomes and acid phosphatase in odonto-
    blasts. In: Symons NBB (ed) Dentine and pulp: their structure and reactions. Livingstone, Edin-
    burgh, pp 43–57
Kawasaki K, Tanaka S, Ishikawa T (1977) On the increment lines in human dentine as revealed
    by tetracycline labelling. J Anat 123:427–436
Kawasaki K, Tanaka S, Ishikawa T (1980) On the daily incremental lines in human dentin. Arch
    Oral Biol 24:939–943
Keil A (1939) Über den Feinbau des normalen und krankhaften Zahnbeins nach Untersuchung
    im polarisierten Licht. Dtsch Zahn Mund Kieferheilk 6:347–364
Kelley KW, Bergenholtz G, Cox CF (1981) The extent of the odontoblast process in rhesus monkey
    (Macaca mulatta) as observed by scanning electron microscopy. Arch Oral Biol 26:893–897
Kennedy JJ, Teuscher GW, Fosdick LS (1953) The ultramicroscopic structure of enamel and dentin.
    J Am Dent Assoc 46:423–431
Ketterl E (1961) Studie über das Dentin der permanenten Zähne des Menschen. Stoma 14:148–163
Kline LW, Thomas NR (1977) The role of calcitonin in the calcification of dental matrix. J Dent
    Res 56:862–865
Kockapan C (1984) Elektronenmikroskopische Untersuchungen an menschlichen Zähnen über die
    Struktur des Interglobulardentins unter besonderer Berücksichtigung von Karies und Abrasion.
    Habilitationsschrift, Justus Liebig Universität, Giessen/Lahn
Köling A (1983) "Membrane structures in the human pulpdentin region. An electron microscopic
    investigation of permanent teeth using the freeze fracture technique"; Doctorat thesis, Universitet
    Centraltryckeriet, Uppsala
Kölliker A (1852) Handbuch der Gewebelehre des Menschen. 1st edn. Engelmann, Leipzig
Kreudenstein T von Spreter, Stuben J von (1956) Dentinstoffwechselstudien. IV. Über die Gewinnung
    des Dentinliquors durch Elution zum Zweck der Bestimmung seiner Bestandteile sowie Untersu-
    chungen über den Wassergehalt des Dentins und über das Dentin-Liquorspatium. Dtsch Zahnärztl
    Z 11:1214–1220
Kuttler Y (1959) Classification of dentin into primary, secondary and tertiary. Oral Surg 12:906–1001
La Flèche R, Frank RM, Steuer P (1985) The extent of the human odontoblast process as determined
    by transmission electron microscopy: the hypothesis of a retractable suspensor system. J Biol
    Buccale 13:293–305
Laikkö I, Larmas M (1979) Changes of dentinal inorganic phosphate in different areas of sound
    and carious human teeth. Caries Res 13:32–38
Larsson A, Bloom GD (1973) Studies on dentinogenesis in the rat. Fine structure of developing
    odontoblasts and predentin in relation to the mineralization process. Z Anat Entwickl Gesch
    139:227–246
Lavelle CLB, Shellis RP, Poole DFG (1977) Evolutionary changes to the primate skull and dentition.
    Thomas, Springfield/Illinois
Lefèvre ML, Manly RS (1938) Moisture, inorganic and organic constituents of enamel and dentin
    from carious teeth. J Am Dent Assoc 25:233–242
Lefèvre R, Frank RM, Voegel JC (1976) The study of human dentine with secondary ion microscopy
    and electron diffraction. Calcif Tissue Res 19:251–261

Lenz H (1955) Elektronenmikroskopischer Nachweis der Dentinveränderungen durch Karies. Dtsch Zahn Mund Kieferheilk 22:24–33

Lester KS, Boyde A (1967) Electron microscopy of predentinal surfaces. Calcif Tissue Res 1:44–54

Lester KS, Boyde A (1968) The surface morphology of some crystalline components of dentine. In: Symons NBB (ed) Dentine and pulp: their structure and reactions. Livingstone, London, pp 197–219

Leuwenhoek A van (1675) Microscopical observations on the structure of teeth and other bones. Phil Trans Martyn (London) 10:1002–1003

Linde A (1973) A study of the dental pulp glycosaminoglycans from permanent human teeth and rat and rabbit incisors. Arch Oral Biol 18:49–59

Linde A (1984) Non-collagenous proteins and proteoglycans in dentinogenesis. In: Linde A (ed) Dentin and dentinogenesis, vol II. CRC Press, Boca Raton, pp 55–92

Linde A (1985) The extracellular matrix of the dental pulp and dentine. J Dental Res 64:523–529 (Special Issue)

Linde A, Johansson S, Jonsson R, Jontell M (1982) Localization of fibronectin during dentinogenesis in rat incisor. Arch Oral Biol 27:1069–1073

Lindskog S, Hammarström L (1982) The possible origin and nature of Tomes' granular layer. J Dent Res 61:589 (abstract)

Magloire H, Dumont J (1976) Etude ultrastructurale de cellules pulpaires humaines cultivées "in vivo". J Biol Buccale 4:3–20

Maniatopoulos C, Smith DC (1983) A scanning electron microscopic study of the odontoblast process in human coronal dentine. Arch Oral Biol 28:701–710

Martens P (1968) Human dentinogenesis with special regard to the formation of peritubular crown dentine and zones in fetal deciduous and unabraded permanent teeth. Scand J Dent Res (Suppl) 76:5–169

Massler M, Schour I (1946) The appositional life span of the enamel and dentin-forming cells. I. Human deciduous teeth and first permanent molars. J Dent Res 25:145–150

Matsumiya S, Takuma S (1954) Atlas of electron micrographs of the human dental tissues, 1st edn. Dental College Press, Tokyo

Mendis BRR, Darling AI (1979) A scanning electron microscope and microradiographic study of closure of human coronal dentinal tubules related to occlusal attrition and caries. Arch Oral Biol 24:725–733

Menke E (1950) Elektronenmikroskopie der menschlichen Zahnhartsubstanz. Z Anat Dtsch 115:1–18

Mennehöh S, Fahnenbrock M (1951) Über elektronenmikroskopische Abbildung lebender Zähne mit Hilfe des Polymerisationsabdruckverfahrens. Kolloid Zeitschr 121:155–157

Miller J (1954) The microradiographic appearance of dentine. Br Dent J 97:7–9

Miller J (1981) Large tubules in dentine. J Dent Child 48:296–271

Miller WA, Eick JD, Neiders ME (1971) Inorganic components of the peritubular dentin in young permanent teeth. Caries Res 5:264–278

Miller WD (1980) Micro-organisms of the human mouth, 1st edn. SS White Dent Mfg Co, Philadelphia, p 156

Mjör IA (1979) Dentin and pulp. In: Mjör IA, Feyerskov O (eds) Histology of the human tooth, 1st edn. Munksgaard, Copenhagen, pp 43–73

Mjör IA (1983) Dentin and pulp. In: Mjör IA (Ed) Reaction patterns in human teeth. CRC Press, Boca Raton, pp 63–156

Mjör IA (1985) Dentin-predentin complex and its permeability: pathology and treatment overview. J Dent Res 64:622–627 (special issue)

Moss ML (1974) Studies on dentin. I. Mantle dentin. Acta Anat (Basel) 87:481–490

Nagai N (1970) Ultrastructural localization of acid phosphatase in odontoblasts of young rat incisors. Bull Tokyo Dental Coll 11:85–120

Nagai N, Takuma S, Goto Y, Ogiwara H (1974) Electron microscopy of dentine and predentine of developing rat molars stained with ruthenium red. J Biol Buccale 2:73–83

Nakahara H (1982) Electron microscopic studies of the lattice image and central dark line of crystallites in sound and carious human dentin. Bull Josai Dent Univ 11:209–215

Nalbandian J, Sognnaes RF (1960) Structural changes in human teeth. In: Shock NW (ed) Aging – some social and biological aspects. Am Assoc Adv Sci, Washington, pp 367–382

Nalbandian J, Gonzales F, Sognnaes RF (1960) Sclerotic age changes in root dentin of human teeth as observed by optical, electron and X-ray microscopy. J Dent Res 39:598–607

Nalbandian J, Hagopian M, Patters M (1982) The microscopic distribution of tetracycline in human teeth. J Biol Buccale 10:271–279

Neumann E (1863) Beitrag zur Kenntniss des normalen Zahnbein- und Knochengewebes. 1st edn. Voegel, Leipzig

Nikiforuk G, Fraser D (1979) Etiology of enamel hypoplasia and interglobular dentin: the roles of hypocalcemia and hypophosphatemia. Met Bone Dis Rel Res 2:17–23

Nygren H, Hansson HA, Linde A (1976) Ultrastructural localization of proteoglycans in the odonto-blast-predentin region of rat incisor. Cell Tissue Res 168:277–287

Osborn JW (1967) A mechanistic view of dentinogenesis and its relation to the curvatures of the processes of the odontoblast. Arch Oral Biol 12:275–280

Owen R (1845) Odontography or a treatise on comparative anatomy of the teeth: their physiological relations, mode of development and microscopic structure in vertebrate animals, 1st edn. Baillière, London

Owens PDA (1972) Light microscopic observations on the formation of the layer of Hopewell-Smith in human teeth. Arch Oral Biol 17:1785–1789

Owens PDA (1975) The fine structure of coronal root region of premolar teeth in dogs. Arch Oral Biol 20:705–712

Pasley DH, Nelson QR, Williams EC, Kepler EE (1981) Use of dentine fluid protein concentrations to measure pulp capillary reflection coefficient in dogs. Arch Oral Biol 26:703–706

Pease DC (1951) Electron microscopy of sectioned teeth. Anat Rec 110:539–547

Penman S, Capco DG, Fey EG, Chatterjee P, Reiter T, Ermish S, Wan K (1983) The three-dimensional structural networks of cytoplasm and nucleus: function in cells and tissue. Modern cell biology, vol 2. Liss, New York, pp 385–445

Plackova A, Stepanek J (1960) Zur Kenntnis der peritubulären Zone des Dentins. Z Zellforsch 52:730–738

Porter KR, Beckerle M, McNiven M (1983) The cytoplasmic matrix. Modern cell biology, vol 2. Liss, New York, 259–302

Posner AS, Tannenbaum PJ (1984) The mineral phase of dentin. In: Linde A (ed) Dentin and dentinogenesis, vol II. CRC Press, Boca Raton, pp 17–36

Quigley MB, Starrs JW, Zwarych PD (1965) Demonstration of calcospherites in mature human dentin. J Dent Res 44:794–800

Reith EJ (1968) Ultrastructural aspects of dentinogenesis. In: Symons NBB (ed) Dentin and pulp: their structure and reaction. Livingstone, Edinburgh, pp 15–57

Retzius G (1894) Zur Kenntniss der Endungsweise der Nerven in den Zähnen. Biol Untersuch 6:64–69

Rosenberg GD, Simmons DJ (1980) Rhythmic dentinogenesis in the rabbit incisor: circadian, ultra-dian and infradian periods. Calcif Tissue Int 32:29–44

Rothman JE (1981) The Golgi apparatus: two organelles in tandem. Science 213:1212–1219

Rouiller CL (1951) La gaine de Neumann; Thèse Dr Med n° 2064. Méd Hygiène ed, Geneve

Sasaki T, Eshida I, Higashi S (1982) Ultrastructure and cytochemistry on old odontoblasts in rat incisors. J Electron Microsc (Tokyo) 31:378–388

Sasaki T, Tominaga H, Higashi S (1984) Endocytic activity of kitten odontoblasts in early dentinogen-esis. 1. Thin section and freeze fracture study. J Anat 138:485–492

Sattelberg C, Turner DF (1984) Anatomical evidence for existence of zonula occludens between pulpal odontoblasts. J Dental Res 63:225 (abstract) (special issue)

Sauk JJ Jr, Brown DM, Corbin KW, Witkop CJ Jr (1976) Glycosaminoglycans of predentin, peritubu-lar dentin and dentin: A biochemical and electron microscopic study. Oral Surg 11:623–630

Scherft JP (1972) The lamina limitans of the organic matrix of calcified cartilage and bone. J Ultrastr Res 38:318–331

Schmidt WJ, Keil A (1958) Die gesunden und erkrankten Zahngewebe des Menschen und der Wirbel-tiere im Polarisationsmikroskop, 1st edn. Hanser, München

Schour I, Dyke HB (1932) The effect of replacement therapy on the eruption of the incisor of the hypophysectomized rat. Proc Soc Exp Biol Med 29:378–382

Schour I, Poncher HG (1937) The rate of apposition of human enamel and dentin as measured by the effects of acute fluorosis. Am J Dis Child 54:765–776

Schour I, Chandler SB, Tweedy WR (1937) Changes in the teeth following parathyroidectomy. Am Pathol 13:945–970

Schroeder L, Frank RF (1985) High resolution transmission electron microscopy of adult human peritubular dentine. Cell Tissue Res 242:449–451

Scott DB (1952) Electron microscopy of tooth structure. Oral Surg Oral Med Oral Pathol 5:527–535

Scott DB (1955) The electron microscopy of enamel and dentine. Ann NY Acad Sci 60:575–585

Scott DB, Kennedy JJ (1950) Electron microscopy of human dentine. J Dent Res 29:556–560

Scott DB, Wyckoff RWG (1947) Electron microscopy of tooth structure by the shadowed collodion replica method. Public Health Rep 62:1513–1516

Scott J, Weber DF (1977) Microscopy of the junctional region between human coronal primary and secondary dentine. J Morphol 154:133–145

Scott JH, Symons NBB (1982) Introduction to dental anatomy, 9th edn. Churchill, Livingstone Edinburgh, p 419

Selvig KA (1968) Ultrastructural changes in human dentine exposed to a weak acid. Arch Oral Biol 13:719–734

Selvig KA (1970) Periodic lattice images of hydroxyapatite crystals in human bone and dental hard tissues. Calcif Tissue Res 6:227–238

Shackleford JH (1971) The structure of Tomes granular layer in dog premolar teeth. Anat Rec 170:357–368

Shimauchi K, Yoshie T, Hasegaawa S, Fujioka S, Kizu T, Ideda K (1973) A scanning electron microscope study of intratubular dentin fibers. J Nihon Univ Sch Dent 15:113–117

Shroff FR, Williamson KI, Bertaud WS (1954) Electron microscope studies of dentine: true nature of the dentine canals. Oral Surg 7:662–670

Shroff FR, Williamson KI, Bertaud WS, Hall DM (1956) Further electron microscope studies of dentine. Oral Surg 9:432–443

Sigal MJ, Aubin JE, Ten Cate AR, Pitaru S (1984a) The odontoblast process extends to the dentino-enamel junction: an immunocytochemical study of rat dentine. J Histochem Cytochem 32:872–877

Sigal MJ, Pitaru S, Aubin JE, Ten Cate AR (1984b) A combined scanning electron microscopy and immunofluorescence study demonstrating that the odontoblast process extends to the dentino-enamel junction in human teeth. Anat Rec 210:453–462

Spector M (1975) High resolution electron microscope study of lattice images in biological apatites. J Microc 103:55–62

Stanley HR, White CL (1966) The rate of tertiary (reparative) dentine formation in the human tooth. Oral Surg Oral Med Oral Pathol 21:180–189

Stewart DJ (1968) Tetracyclines: their prevalence in children's teeth. Br Dent J 124:318–320

Sundström B, Takuma S, Nagai N (1970) Ultrastructural aspects of human dentine decalcified with chromium sulphate. Calcif Tissue Res 4:305–313

Symons NBB (1961) A histochemical study of the intertubular and peritubular matrices in normal human dentine. Arch Oral Biol 5:241–250

Symons NBB (1968) The formation of primary and secondary dentine. In: Symons NBB (ed) Dentine and pulp: their structure and reactions. Livingstone, London, pp 67–76

Syrrist A (1949) An introduction in electron microscopy with some results from histological investigations of enamel and dentine. Odont Tidskr 57:79–105

Syrrist A, Gustafson G (1951) A contribution to the technique of the electron microscopy of dentine. Odont Tidskr 59:500–513

Taintor JF, Biesterfeld RC, Langeland K (1981) Irritational or reparative dentin. A challenge of nomenclature. Oral Surg 51:443–449

Takagi M, Parmley RT, Denys FR (1981) Ultrastructural localization of complex carbohydrates in odontoblasts, predentin and dentin. J Histochem Cytochem 29:747–758

Takuma S (1960) Electron microscopy of the structure around the dentinal tubule. J Dent Res 39:973–981

Takuma S, Eda S (1966) Structure and development of the peritubular matrix in dentin. J Dent Res 45:683–692

Takuma S, Nagai N (1971) Ultrastructure of rat odontoblasts in various stages of their development and maturation. Arch Oral Biol 16:993–1011

Takuma S, Katagiri S, Ozasa S (1966) Electron probe microanalysis of horse dentin. J Electron Microsc (Tokyo) 15:86–89

Tanaka T (1980) The origin and localization of dentinal fluid in developing rat molar teeth studied with lanthanum as a tracer. Arch Oral Biol 25:153–162

Ten Cate AR (1967) A histochemical study of the human odontoblast. Arch Oral Biol 12:963–969

Ten Cate AR (1972a) Morphological studies of fibrocytes in connective tissue undergoing rapid remodelling. J Anat 112:401–404

Ten Cate AR (1972b) An analysis of Tomes' granular layer. Anat Rec 172:137–148

Ten Cate AR (1978) A fine structural study of coronal and root dentinogenesis in the mouse: observations on the so-called von Korff fibres and their contribution to mantle dentine. J Anat 125:183–197

Ten Cate AR (1980) Oral histology, development structure and function, 1st edn. Mosby, St Louis, p 449

Ten Cate AR, Deporter DA (1974) The role of the fibroblast in collagen turnover in the functioning periodontal ligament of the mouse. Arch Oral Biol 19:339–340

Ten Cate AR, Deporter DA, Freeman E (1976) The role of fibroblasts in the remodelling of periodontal ligament during physiologic tooth movement. Am J Orthod 69:155–168

Thesleff I, Stenman S, Vaheri A, Timpl R (1979) Changes in the matrix proteins, fibronectin and collagen during differentiation of mouse tooth germ. Dev Biol 70:116–126

Thesleff I, Barrach JH, Foldart JM, Vaheri A, Pratt RM, Martin GR (1981) Changes in the distribution of type IV collagen, laminin, proteoglycan and fibronectin during mouse tooth development. Dev Biol 81:182–192

Thomas HF (1979) The extent of the odontoblast process in human dentine. J Dent Res 58/D:2207–2218

Thomas HF (1983) The effect of various fixatives on the extent of the odontoblast process in human dentine. Arch Oral Biol 28:465–469

Thomas HF (1984) The lamina limitans of human dentinal tubules. J Dent Res 63:1064–1066

Thomas HF (1985) The dentin-predentin complex and its permeability: anatomical overview. J Dent Res 64:607–612 (special issue)

Thomas HF, Carella P (1983) A scanning electron microscope study of dentinal tubules from unerupted human teeth. Arch Oral Biol 28:1125–1130

Thomas HF, Carella P (1984) Correlation of scanning and transmission electron microscopy of human dentin tubules. Arch Oral Biol 29:641–646

Thomas HF, Payne RC (1983) The ultrastructure of dentinal tubules from erupted human premolar teeth. J Dent Res 62:532–536

Tidmarsh BG (1981) Contents of human dentinal tubules. Int Endo J 14:191–196

Tilney LG (1983) Interactions between actin filaments and membranes give spatial organization to cells. Modern Cell Biology, vol 2. Liss, New York, pp 163–199

Tomes CS (1898) Upon the structure and development of the enamel of the elasmobranch fishes. Phil Trans B190:443–464

Tomes J (1856) The presence of fibrils of soft tissue in the dentinal tubules. Proc R Soc Lond [Biol] 146:515–522

Tominaga H, Sasaki T, Higashi S (1984) Ultrastructural changes in odontoblasts during early development. Bull Tokyo Dental Coll 25:9–26

Torneck CD (1978) Intracellular destruction of collagen in the human dental pulp. Arch Oral Biol 23:745–747

Tötdal B, Hals E (1985) Electron probe study of human and red deer cementum and root dentin. Scand J Dent Res 93:4–12

Tovborg Jensen A, Möller A (1948) Determination of size and shape of the apatite particles in different dental enamels and dentin by the X-ray powder method. J Dent Res 27:524–531

Trautz OR, Klein E, Fessenden E, Addleston HK (1953) The interpretation of the X-ray diffractograms obtained from human dental enamel. J Dent Res 32:420–431

Tronstad L (1972) Optical and microradiographic appearance of intact and worn human coronal dentine. Arch Oral Biol 17:847–858

Tronstad L (1973a) Ultrastructural observations on human coronal dentin. Scand J Dent Res 81:101–111

Tronstad L (1973b) Quantitative microradiography of intact and worn human coronal dentine. Arch Oral Biol 18:533–542

Tronstad L, Langeland K (1971) Electron microscopy of human dentin exposed by attrition. Scand J Dent Res 79:160–171

Tsatsas B, Frank RM (1972) Ultrastructure of the dentinal tubular substances near the dentino-enamel junction. Calcif Tissue Res 9:238–242

Vahl J, Höhling HJ, Frank RM (1964) Elektronenbeugungsuntersuchungen an rhomboedrisch aussehenden Mineralbildungen an kariösem Dentin. Arch Oral Biol 9:315–320

Vasiliadis L, Darling AI, Levers BGH (1983a) The amount and distribution of sclerotic human root dentine. Arch Oral Biol 28:645–649

Vasiliadis L, Darling AI, Levers BGH (1983b) The histology of sclerotic human root dentine. Arch Oral Biol 28:693–700

Vermot-Gaud M (1967) Mise en évidence et recherches statistiques sur la fréquence des canaux pulpo-parodontaux sur les molaires de lait et leur incidence sur l'infection du septum interradiculaire. Rev Franc Odont Stomat 14:1487–1504

Voegel JC, Frank RM (1977) Ultrastructural study of apatite crystal dissolution in human dentine and bone. J Biol Buccale 5:181–194

Walkhoff O (1924) Neue Untersuchungen über den feineren Bau der Dentinkanälchen. Dtsch Monatschr Zahnheilk 42:521–540

Wang YN, Ashrafi SH, Weber DF (1985) Scanning electron microscopic observations of casts of human dentin tubules along the interface between primary and secondary dentine. Anat Rec 211:149–155

Watson ML, Avery JK (1954) The development of the hamster lower incisor as observed by electron microscopy. Am J Anat 95:109–162

Weatherell JA, Robinson C (1973) The inorganic composition of teeth. In: Zipkin I (ed) Biological mineralization, vol 3. Wiley and Sons, New York, pp 43–74

Weber DF (1969) Volume fractions analysis of human coronal dentin. Calcif Tissue Res 4:257–259

Weber DF (1974) Human dentine sclerosis: a microradiographic survey. Arch Oral Biol 19:163–169

Weber DF (1983) An improved technique for producing casts of the internal structure of hard tissues including some observations on human dentine. Arch Oral Biol 28:885–891

Weidenreich F (1925) Über den Bau und die Entwicklung des Zahnbeins in der Reihe der Wirbeltiere. Knochenstudien IV Teil. Z Anat Entwickl Gesch 76:218–260

Weill R (1959) Etude histochimique de la dentine. La zone translucide de Bradford. La gaine de Neumann. Ann Histochim 4:59–71

Weill R (1963) Histochimie et autoradiographie de la dent normale et pathologique. Arch Oral Biol [suppl] 7:111–123

Weinstock M, Leblond P (1974) Synthesis, migration and release of precursor collagen by odontoblasts as visualized by radioautography after (^{3}H-) proline administration. J Cell Biol 60:92–127

Weyman J (1968) Microscopic appearances of tetracycline deposition of human dentin. J Dent Res 47:742–745

White SW, Hulmes DJS, Miller A, Timmins PA (1977) Collagen-mineral axial relationship in calcified turkey leg tendon by X-ray and neutron diffraction. Nature 266:421–425

Whittaker D, Kneale M (1979) The predentine-dentine interface in human teeth. A scanning electron microscope study. Br Dent J 146:43–46

Winter GB (1962) Abscess formation in connection with deciduous molar teeth. Arch Oral Biol 7:373–380

Yamada T, Nakamura K, Iwaku M, Fusayama T (1983) The extent of the odontoblast process in normal and carious human dentin. J Dent Res 62:798–802

Yilmaz S, Newman HN, Poole DFG (1977) Diurnal periodicity of von Ebner growth lines in pig dentine. Arch Oral Biol 22:511–513

Young RA, Brown (1982) Structures of biological minerals. In: Nancollas GH (ed) Biological mineralization and demineralization. Dahlem Konferenzen. Springer, Berlin, pp 101–141

Structure and Ultrastructure of the Dental Pulp

R.M. FRANK and J. NALBANDIAN

A. Basic Anatomy

The dental pulp is a richly innervated and vascularized, loose connective tissue. As previously mentioned, the study of the dental pulp cannot be disassociated from that of dentine, since one of the principal functions of the pulpal tissue is, in fact, the elaboration of dentine. This is accomplished through the activity of the peripheral pulpal cells, the odontoblasts, which are an integral part of both tissues. From a histological and anatomical point of view, the dental pulp has classically been described as having three main regions, the peripheral odontoblastic layer, the sub-odontoblastic layer, and the central pulp. These zones form a continuum however (Figs. 1, 2), and in the radicular pulp, near the root apex, such zones are less apparent. According to BAUME (1980), the important relationship between dentine and pulp was first recognized in the mid 19th century by, among others, KÖLLIKER (1852) and TOMES (1856).

On an average wet-weight basis, the pulpal tissue is composed of 25% organic material and 75% water. Fibroblasts are the most common cell type, followed probably by the peripheral odontoblasts, whereas collagen (types I and III) is the principal protein component of the extracellular pulpal matrix. With aging, the dental pulp tissue is diminished in volume by secondary or tertiary dentine apposition. In addition, the tissue becomes less cellular, with proportionately more collagen.

The pulp is located in the pulp cavity, which includes the pulp chamber and the root canal or canals. The shape of the pulp cavity gives the approximate outer contour of the tooth. Under each cusp, narrow projections, the pulp horns, extend from the ceiling of the pulp chamber. These are particularly marked in young teeth. At the apex of each root is a foramen or foramina through which the blood vessels, the lymphatics, and the nerves enter the pulp cavity. Communication of pulp and periodontal ligament is not limited to the apical region. Accessory canals can be found at various levels; however, the majority appear to be encountered in the apical half of the root.

The pulps of primary human teeth do not seem to present significant structural differences from the pulps of young, permanent, human teeth. A subodontoblastic layer is present in the coronal third of the normal primary pulp but absent in primary teeth with occlusal attrition (FOX and HEELEY 1981). A unique characteristic of the pulp of primary human teeth seems to be its relatively short lifespan which is 8.3 years on average, according to AVERY (1976). This includes 1 year of development, 3.9 years of maturation and 3.6 years of remodelling and regression.

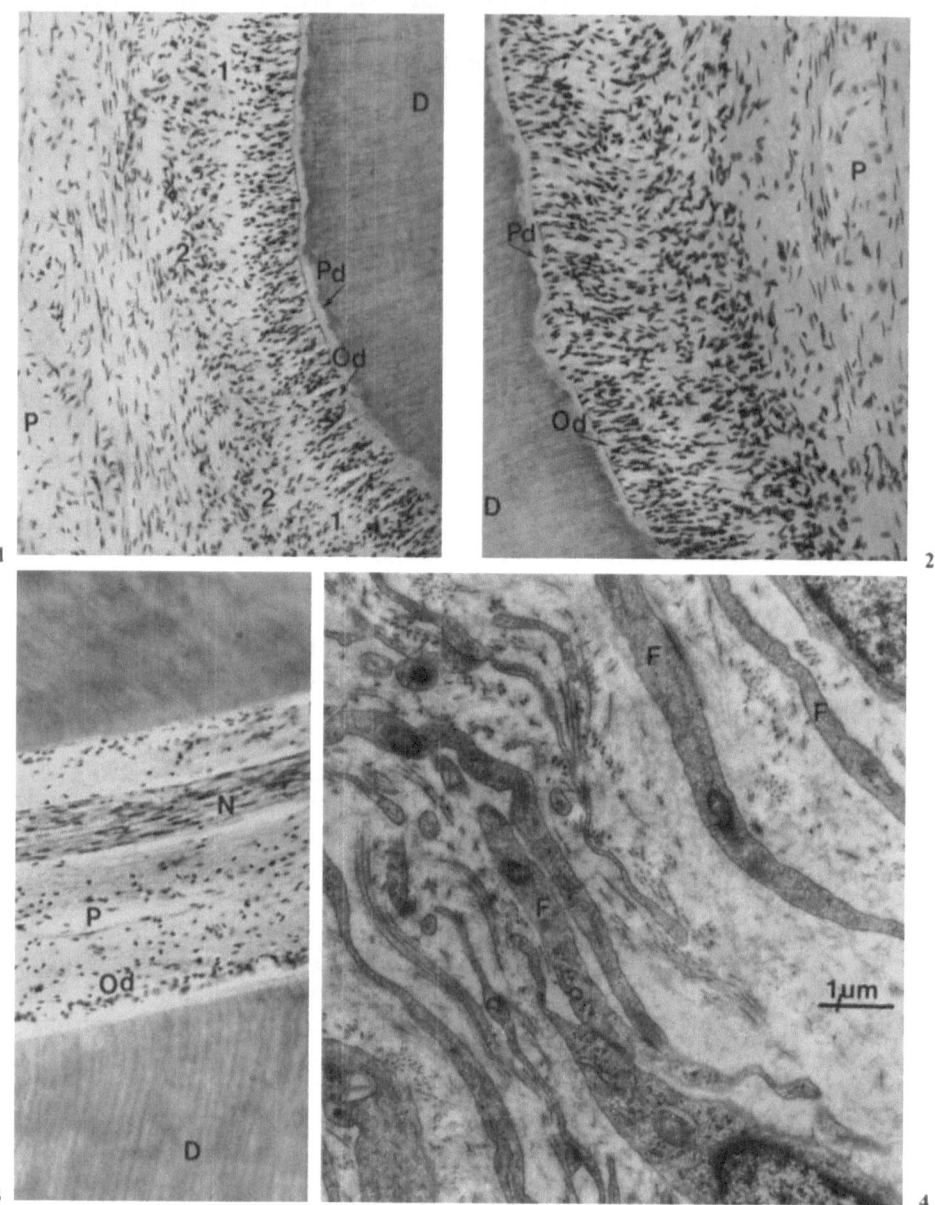

Fig. 1. Human pulp chamber. The subodontoblastic layer is composed of a cell-free zone (*1*) adjacent to the odontoblast cell layer (*Od*), and a cell-rich zone (*2*) continuous with the central pulpal tissue (*P*). *D* Dentine; *Pd* Predentine. × 150

Fig. 2. Human pulp chamber. The subodontoblastic layer consists of a cell rich zone without a cell-free layer. *P* Pulp core; *Od* Odontoblast cell layer; *D* Dentine; *Pd* Predentine. × 170

Fig. 3. Human pulp in the root. Absence of a distinct subodontoblastic layer. *P* Pulp core; *N* Nerve fibrils; *Od* Odontoblast layer; *D* Dentine. × 140

Fig. 4. Transmission electron microscopic view of the pulpal tissue. Section through slender processes of fibroblasts (*F*). Some bundles of collagen fibrils are visible in the extracellular matrix

The physiology of the human dental pulp is determined, to a large extent, by the fact that this elongated, highly vascular and innervated unit of loose connective tissue is enclosed in an inexpandible calcified shell of dentine. Dental pulp seems to have three principal functions. Its first task is to elaborate dentine, not only during the developmental phase of dentinogenesis but also throughout the later life of the tooth during which the pulp continues to elaborate and also repair the calcified tissue in which it is enclosed. This activity is characterized by the production of secondary and tertiary dentine as well as pulp stones.

A second function is related to innervation. The pulp responds to irritation by neural activity which gives rise to painful sensation. As will be related in the pages to follow, new developments have improved our understanding of dentine sensitivity. The presence of non-myelinated sensory nerve endings in the dentinal tubules has now been firmly established.

A third function of the pulp is to maintain the vitality of the dental cellular constituents by providing oxygen and nutrients for their metabolism. An environment conducive to cellular function is provided by the continual removal of the pulpal interstitial fluid. This exchange is accomplished as water and soluble metabolic substrates enter the tissue and filtration of plasma occurs across the capillary wall. The forces promoting filtration are the capillary hydrostatic pressure and tissue osmotic pressure. The average tissue pressure in the healthy human dental pulp was found to be about 25 mmHg (Beveridge and Brown 1965; Von Hassel 1971), a pressure unusually high as compared to that in other organs.

Continuous progress is being made in the area of pulp physiology. In recent years, for example, a prostaglandin system has been demonstrated in the pulp of rodents (Hirafuji et al. 1980; Antila and Pohto 1984).

B. Pulpal Cells

I. The Odontoblasts

The odontoblasts are an integral part of the dental pulp of which they constitute the peripheral cell layer. For the details, see chapter on dentine.

II. The Fibroblasts

The fibroblasts are the most numerous cells found in the human pulp. They have a long and slender configuration (Figs. 5–8) and the nucleus is found at one end of the cell (Fig. 6, upper cell to the right). The Golgi apparatus, surrounded by the endoplasmic reticulum, is located in the center (Fig. 6, upper cell, in the middle) and at the other end, a region with long processes is found (Figs. 4–7, 9). Pulpal fibroblasts have the same ultrastructural characteristics as those of other connective tissues and seem to be responsible not only for

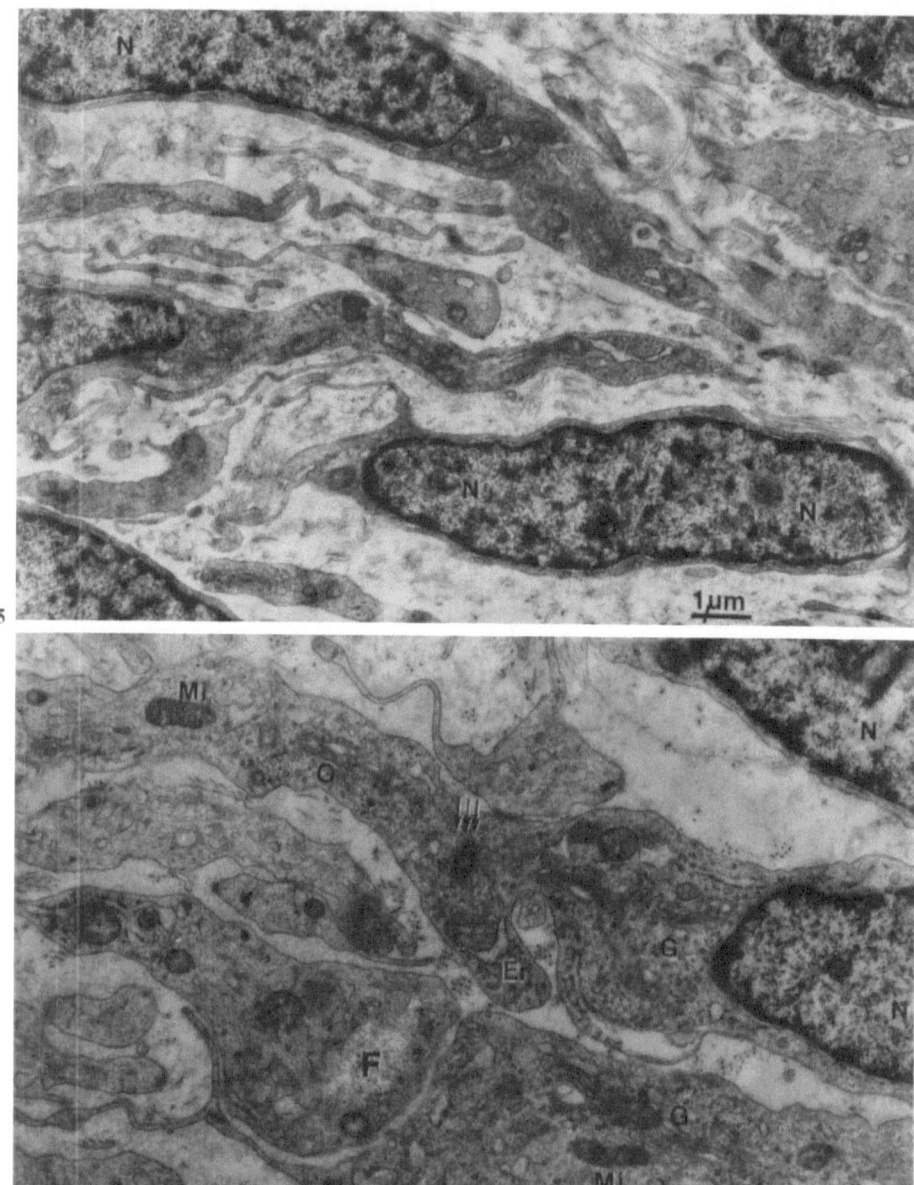

Fig. 5. Ultrastructural view of the pulp chamber. Fibroblasts have large oval nuclei (*N*) and their relatively scant cytoplasm contains all the typical organelles and cytoskeletal elements. Presence of collagen fibrils in the extracellular matrix. (From FRANK 1968a, courtesy of Livingstone)

Fig. 6. Fibroblasts (*F*) in the human pulp core showing a centriole (*arrows*). *G* Golgi apparatus; *Er* Endoplasmic reticulum; *Mi* Mitochondria; *N* Nucleus

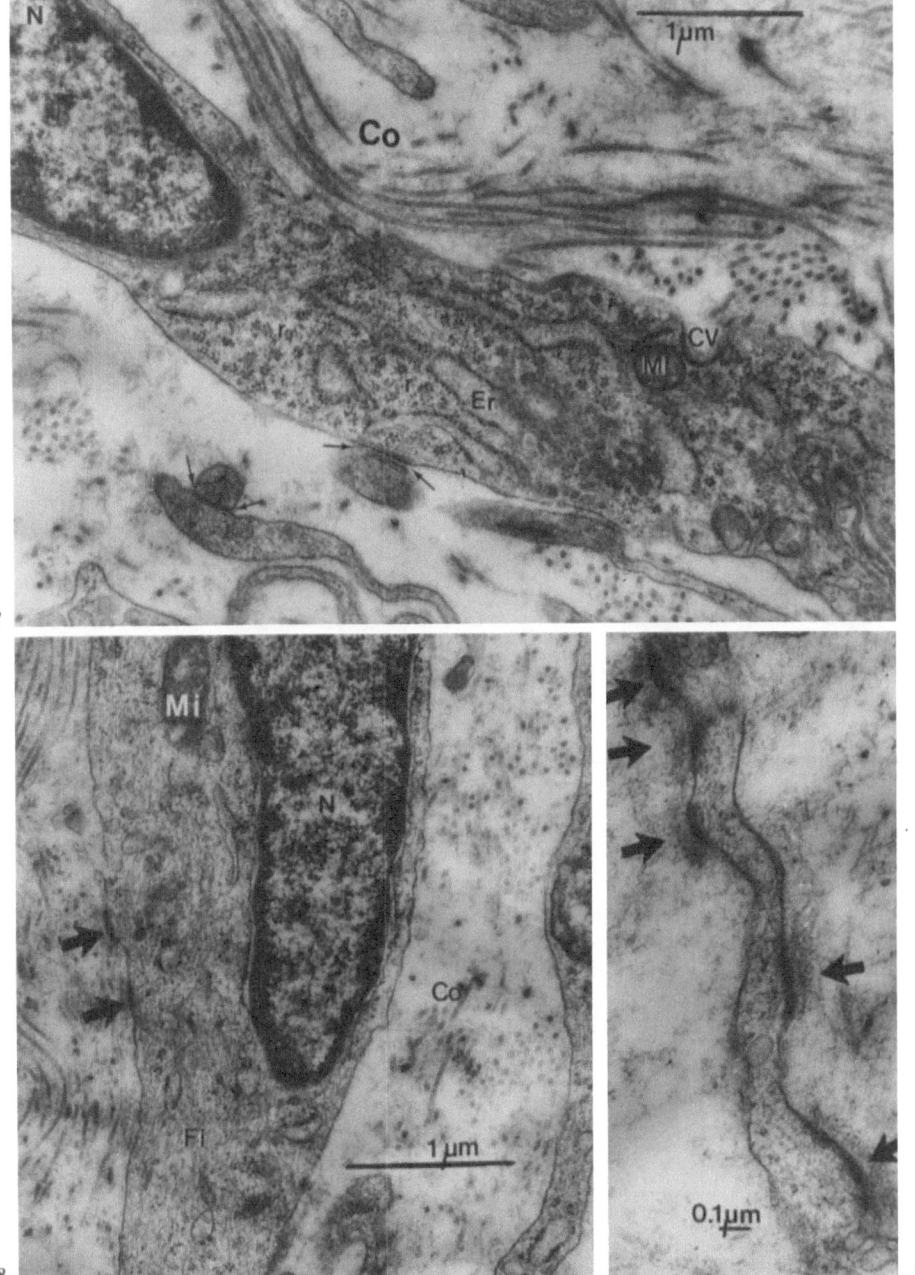

Fig. 7. Higher enlargement of pulpal fibroblasts and their processes connected by desmosome-like junctions (*arrows*). *N* Nucleus; *Er* Endoplasmic reticulum; *CV* Coated vesicle in formation; *Mi* Mitochondria; *r* Free ribonucleoprotein particles; *Co* Collagen fibrils

Fig. 8. Presence of dense thickenings on the cytoplasmic side of the fibroblast plasma membrane (*large arrows*). *N* Nucleus; *Mi* Mitochondria; *Fi* Microfilaments; *Co* Collagen fibrils. (From CAHEN and FRANK 1970, courtesy of Bull Group Rech Sci Stomatol)

Fig. 9. Human pulpal fibroblast. Presence of dense thickenings on the cytoplasmic side of the plasma membrane of a slender process of a fibroblast. Note the dense material concentrated in the extracellular space facing these areas (*large arrows*). (From CAHEN and FRANK 1970, by courtesy of Bull Group Rech Sci Stomatol)

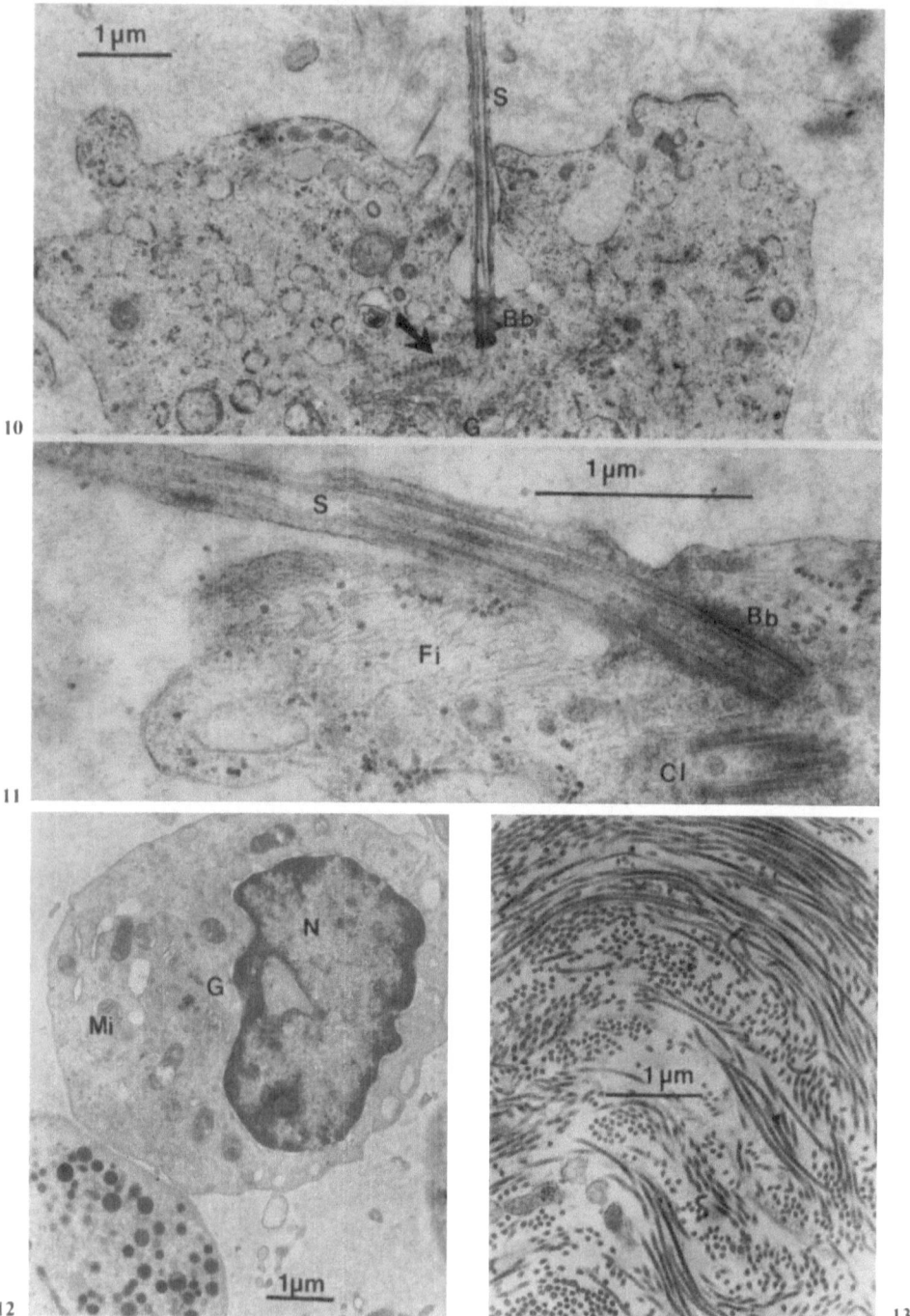

Fig. 10. Human pulpal fibroblast. View of a cilium (*S*) with a basal body (*Bb*) located next a centriole (*large arrow*). *G* Golgi apparatus

the synthesis and maintenance of the extracellular matrix, but also for the remo-delling and possible phagocytosis of matrix components such as collagen. Inter-cellular junctions between fibroblasts and also between fibroblasts and odonto-blasts, together with the unique cytoskeleton of these cells, may provide for important mechanical functions, both intra- and intercellularly.

The pulpal fibroblast has a range of diameters from 1 to 4 μm (CAHEN and FRANK 1970). Length is difficult to evaluate but can be estimated to be about 20 μm or more. The nucleus, located at one end of the cell, shows one or two nucleoli and is surrounded by a flattened envelope consisting of an inner nuclear membrane and an outer nuclear membrane continuous with the rough-surfaced endoplasmic reticulum. The latter having ribonucleoprotein granules attached to its external surface. Nuclear pores occur at intervals along the circumference of the envelope.

In young fibroblasts, an extensive, rough-surfaced endoplasmic reticulum (RER) with closely associated mitochondria is found around the nucleus and around the well-developed Golgi apparatus which occupies the center of the cell. Near the nucleus and adjacent to the Golgi, a pair of centrioles are seen (Figs. 6, 10, 11). These are cylindrical structures which often lie with their long axes perpendicular to each other. Each cylinder consists of nine groups of micro-tubules in triplets, arranged circumferentially and embedded in an amorphous matrix. One of the centrioles may form a cilium associated with a basal body (Figs. 10, 11). Cilia may be quite long and extend into the extracellular matrix for a distance of several microns.

A prominent Golgi apparatus occupies the center of young pulpal fibroblast (Fig. 6). The walls of the Golgi apparatus consist of four stacks of approximately four dictyosomes made up of flattened cisternae with dilated ends and associated with saccules, vesicles, and granules. The convex outer face of the stacks of dictyosomes, the so-called cis-face, is adjacent to the transitional endoplasmic reticulum that lacks bound ribosomes. Intermediate vesicles are observed there. The internal face of the dictyosomes or trans-face is concave. Various saccules, elongated secretory granules, coated vesicles, vacuoles of various sizes, and mul-tivesicular bodies occupy the central part of the Golgi apparatus. According to ROTHMAN (1981), the Golgi apparatus consists of two attached but different organelles (the cis and trans face) through which exported proteins are sequen-tially transported and purified.

Fig. 11. Human pulpal fibroblast. A cilium (S) with a basal body (Bb) is seen adjacent to an obliquely cut centriole (Ct). Fi Microfilament. (From CAHEN and FRANK 1970, courtesy of Bull Group Int Rech Sci Stomatol)

Fig. 12. Lymphocyte in the lumen of a pulpal vessel. N Nucleus; G Golgi apparatus; Mi Mitochondria.

Fig. 13. Extracellular matrix in human dental pulp. Presence of oriented collagen fibrils arranged in alternating bundles

A substantial cytoskeleton consisting of microtubules, microfilaments, and intermediary filaments is seen in the fibroblast. Microtubules composed of tubulin have an average diameter of 24 nm. Microfilaments of actin, with diameters ranging from 5 to 8 nm, are generally oriented parallel with the long axis of the cell. Intermediary filaments of vimentin are also found in the fibroblast. Lipid and glycogen granules are disseminated throughout the cytoplasmic matrix. The cells have long and slender cytoplasmic processes, often running parallel with each other (Figs. 3, 5).

The presence of a well-developed Golgi apparatus surrounded by rough-surfaced endoplasmic reticulum and closely associated mitochondria, is typical of young, actively synthesizing fibroblasts. In pulps of permanent teeth, fibroblasts with a reduction in Golgi apparatus, endoplasmic reticulum and mitochondria are also seen. Such resting fibroblasts have prominent nuclei and a reduced amount of cytoplasm (Fig. 5).

Various types of intercellular junctions have been identified between fibroblasts as well as between odontoblasts and fibroblasts in the sub-odontoblastic layer. Desmosome-like junctions have been described by several workers (Han 1968; Pischinger and Stockinger 1968; Byers et al. 1982; Koling 1983; Iguchi et al. 1984). In thin sections, these junctions seemed to be formed by the apposition of two unit membranes separated by a space of 30 to 70 nm. On the cytoplasmic surface of the opposing plasma membranes, a condensation of electron-opaque material, resembling attachment plaques, was described. The intercellular space in this region was occupied by an electron-opaque, amorphous material. However, the typical intercellular substructure seen in epithelial desmosome was not visible. In freeze-fracture replicas, no cluster of intramembranous particles, typical of desmosomes, was found on either the PF (protoplasmic) or the EF (extracellular) face.

With the use of freeze-fracture replicas, gap junctions have been identified between fibroblasts and between fibroblasts and odontoblasts (Köling 1983; Iguchi et al. 1984). In pulp explant cultures, Calle et al. (1985) using freeze-fracture, lanthanum impregnation and filipin detection for cholesterol, identified three kinds of intercellular junctions. In addition to desmosome-like junctions and gap junctions, they localized tight junctions on freeze-fracture replicas, although these were not found in ultrathin sections. Their gap junctions on freeze-fracture replicas were of two types, one having a conventional arrangement of intramembranous particles, the other having a crystalline array corresponding to a formative stage of this type of junction.

For fibroblasts cultured in vitro, attachment of the cells to the extracellular matrix has been demonstrated and shown to be based on transmembrane association of extracellular fibronectin fibrils and intracellular bundles of 5 nm microfilaments via the plasma membrane of the fibroblast. The junction occurs at the level of the plasmalemma as a dense patch. This tendency of extracellular fibrils to coalign with the intracellular fibrous components of the fibroblast was demonstrated by Hynes and Destree (1978) by double-labeling immunofluorescence which revealed a structural connection between extracellular fibronectin fibrils and intracellular bundles of actin filaments. Such an attachment, referred to as the "fibronexus" by Singer (1979), was initially only observed under tissue-culture conditions. In their study of the periodontal ligament of the squirrel monkey, GARANT et al. (1982) were the first to describe, in vivo, the presence of fibronexus junctions between fibroblasts and the extracellular matrix. The fibronexus has not been identified as yet in the fibroblasts of the pulp. However, CAHEN and FRANK (1970) observed dense patches along the plasma membranes of human pulpal fibroblasts (Figs. 8, 9). An electron-dense thickening was noted on the cytoplasmic side of the cell membrane. Intracellular microfilaments seemed to be related to these dense patches, and in the adjacent extracellular matrix, a condensation of electron-dense material as well as fine filaments was observed. CAHEN and FRANK (1970) speculated on the anchoring possibilities of these cell surface specializations. It would be interesting to perform immunocytochemical studies for the possible identification of fibronectin in the condensed extracellular material. If this attachment protein were present, these structures could represent pulpal fibronexuses.

Collagen, the main protein component of the extracellular pulpal matrix, is mainly of type I and type III (LINDE 1985). About 60 per cent of the collagen in pulp was found to be of type I as assessed by quantitative biochemical methods (VAN AMERONGEN 1984). Biosynthesis of type I collagen has been extensively studied in the fibroblast with the use of ultrastructural radioautographic methods, in locations other than the pulp organ, and it now appears that these cells secrete a collagen precursor which is assembled into fibrils in the extracellular compartment (MARCHI and LEBLOND 1983, 1984). As already described in detail in the chapter on dentine development, examination of odontoblasts during dentinogenesis with ^{3}H-proline radioautography (FRANK 1970; WEINSTOCK and LEBLOND 1974) and procollagen immunotyping (KARIM et al. 1979) led to the conclusion that these cells first elaborate procollagen, package it into secretory granules in the Golgi apparatus, and release it by exocytosis into the predentine matrix where the type I procollagen is transformed into collagen and assembled into fibrils. Hence, the formation of the collagen fibrils of dentine occurs extracellularly.

Somewhat different observations were reported in electron-microscopic radioautographic studies of fibroblasts. Studying collagen biosynthesis during wound healing, ROSS and BENDITT (1965) proposed that ^{3}H-proline first appearing in the rough-surfaced endoplasmic reticulum of the fibroblast, by-passed the Golgi apparatus and was transferred directly to the extracellular matrix. Similar observations were made by SALPETER (1968), MARTINEZ-HERNANDEZ et al. (1974), ASHHURST and COSTIN (1976). A different mechanism for collagen biosynthesis was proposed for fibroblasts of the periodontal ligament by CHO and GARANT (1981a, b, c) who described an alignment along a microtubule of two to twelve secretory granules arising in the Golgi apparatus, after collagen synthesis in the RER. The secretory granules fuse at their extremities within a long intracellular compartment in which the collagen precursors polymerize. Thus intracellular collagen is formed. Moreover, maximum labeling of individual and fused secretory granules as well as of intracellular collagen fibrils was said to occur at 30 min after ^{3}H-proline injection, i.e., about the time when the label was known to be released from cells as collagen. The authors concluded that this sequence depicts the intracellular formation of collagen fibrils followed by a release of the formed fibrils outside the cells. Yet a different mechanism was proposed by Trelstad and Hayashi (1979) working mainly on chick tendon fibroblasts. Apparently, secretory granules discharge their content into deep recesses at the cell surface where the assembly of collagen fibrils occurs. Thus fibril formation takes place within the confines of the cell but actually is extracellular.

In fact, Marchi and Leblond (1983, 1984) demonstrated that biosynthesis of type I collagen by rat foot pad fibroblasts is similar to that by odontoblasts during dentinogenesis. In fibroblasts, collagen precursors arise in the RER and ultimately reach the spherical Golgi distensions by way of intermediate tubules or vesicles. These seem to migrate from a cis to a trans direction while they become cylindrical, their looping threads straightening out into rods identifiable as procollagen. The cylindrical distensions are then freed from their saccules to become secretory granules. The procollagen content of these granules is released outside the cell and presumably transforms into collagen which subsequently polymerizes into fibrils. Thus, except for a minor fraction of the ^{3}H-proline-labeled precursors directed towards the lysosomal pathway, the bulk gives rise to collagen. Marchi and Leblond (1983, 1984) observed no labeling of the intracellular collagen contained in elongated vacuoles, concluding that collagen there does not grow but undergoes degeneration. These authors therefore concluded that the assembly of collagen into fibrils does not occur within the cells but is entirely extracellular. The reader interested in the problem of collagen biosynthesis is referred to the section on metabolism of the odontoblasts where the topic is discussed in greater detail.

Some of the intracellular transport processes for procollagen can be identified experimentally by certain substances. Thus Han et al. (1967) studied the effect of actinomycin D (which inhibits DNA-dependent RNA polymerase activities) on the fibroblast of the rat incisor pulp, using electron-microscopic radioautography following injection of ^{3}H-leucine. They noted a progressive incorporation of the isotope by the fibroblasts, reaching a maximum on day 7. The most intensively affected structures were the rough-surfaced endoplasmic reticulum, the Golgi apparatus, and the polyribosomes, considered to be directly involved in the elaboration and secretions of proteins.

Besides their capacity to participate in the elaboration of extracellular matrix, the fibroblasts are able to remodel the matrix. This was demonstrated initially for periodontal fibroblasts, with the observation of intracellular phagocytosis of collagen (Ten Cate 1972; Listgarten 1973; Beersten et al. 1974; Garant 1976; Ten Cate et al. 1976; Frank et al. 1977; Cho and Garant 1981 a, b, c). The breakdown of extracellular collagen fibrils requires the participation of collagenase produced and secreted by fibroblasts (Ten Cate 1980). Fragments of collagen fibrils initially attacked by collagenase enter the fibroblast by phagocytosis, one or several collagen fibrils becoming situated within elongated membrane-bounded vacuoles or phagosomes. Lysosomes of the fibroblast fuse with the phagosomes, adding hydrolytic enzymes. In these typical phagolysosomes, the collagen fibrils are destroyed. Such phagolysosomes have also been described in human pulpal fibroblasts undergoing dental caries (Torneck 1978; Furseth et al. 1980).

In addition, MAGLOIRE and DUMONT (1976) observed in differentiating odontoblast-like cells of explanted primary cultures of human dental pulp, a dumbbell-shaped intracytoplasmic body associated with intracellular collagen degradation, previously described in fibroblasts of periodontal ligament (TEN CATE and DEPORTER 1974). Similar dumbbell-shaped phagolysosomes were also observed by us in mature human odontoblasts (p. 181, Fig. 12 B) from a normal human permanent tooth.

Apart from their function of elaborating and remodelling extracellular matrix, the fibroblasts with their well-structured cytoskeleton and intercellular junctions constitute an organized framework throughout the pulp, probably capable of providing for certain mechanical functions. Along these lines MONTANDON et al. (1973) described a contractile fibroblast in granulation tissues during wound healing. This myofibroblast containing massive bundles of microfilaments, a crenated nucleus, and intercellular junctions was similar to a smooth muscle cell.

III. Other Pulpal Cells

In the normal human adult pulp there are special cells associated with the vascular and neural elements. Examples are pericyte and the SCHWANN cell, both of which will be described later together with the blood vessels and the neural elements.

Three other cell types can be considered as normal residents of human pulp. These are the undifferentiated mesenchymal cells, the macrophages and the lymphocytes (Fig. 12). Undifferentiated mesenchymal cells and macrophages are often found around blood vessels and capillaries. Macrophages are rounded cells with an irregular cell surface showing infolding and finger-like processes. Their nuclei are often indented. Besides the traditional organelles such as endoplasmic reticulum and mitochondria, the Golgi apparatus presents well-developed endocytic vacuoles, lysosomes and phagolysosomes. Numerous vesicles and vacuoles are related to surface ruffling.

According to YAMAMURA (1985), fibroblasts, endothelial cells and pericytes, when properly stimulated, are able to be transformed into undifferentiated mesenchymal cells which can then differentiate into odontoblasts capable of producing dentine matrix. Using germ-free rats, YAMAMURA (1985) in an experiment on dentine bridge formation, demonstrated that the pulp tissue has intrinsic healing potentials. ZACH et al. (1969) presented experimental evidence with tritiated thymidine that odontoblasts can be replaced by undifferentiated mesenchymal cells.

C. Extracellular Matrix of the Pulp

The composition of the extracellular matrix of the mature pulp is quite different from that of the dental papilla from which it is derived. In spite of a common ancestry (dental papilla), pulp and dentine differ considerably with respect to extracellular composition. In the pulp matrix, type I collagen is the main fibrous protein but type III collagen constitutes a significant proportion. Fibronectin is present and there is a high content of proteoglycans as well. Anionic non-collagenous proteins such as phosphoproteins and γ-carboxygluta-mate (Gla)-containing proteins are absent from the pulp. By contrast, the collagen in dentine is exclusively type I. Proteoglycans and other non-collagenous proteins are present but type III collagen and fibronectin are not found.

The basic biochemical and genetic structure of collagen, the major component of dental pulp, has already been discussed in the chapter on development of dentine, to which the reader is referred. Based on hydroxyproline analysis, collagen was found to constitute 34% of the total protein in human pulp and 11% in rabbit pulp (UITTOA and ANTILA 1971). In animals less than one year old, the corresponding values were 12.7% for porcine pulp and 16.8% for bovine pulp (ORLOWSKI 1974). The concentration of collagen in bovine pulp reaches 25% of the dry weight (HAYAKAWA et al. 1981). On a wet-weight basis the collagen content in human pulp was found to be 3–5% (VAN AMERONGEN et al. 1983). These values are relatively low in comparison to other collagenous connective tissues. In human pulp the amount of collagen varied among different teeth and in relation to age. The amount of collagen in dried pulps was 25.7% in premolars (9 to 15 years) and 31.9% in third molars (20 to 40 years) (VAN AMERONGEN 1984). These percentages are much higher than those reported for other species. The collagen content was lowest in the coronal part. It was highly cross-linked and therefore cannot be considered as immature (VAN AMERONGEN 1984). Increased proportions of insoluble collagen, i.e. fibrous collagen with a higher degree of cross-linking, have been demonstrated in mature bovine pulp in comparison to the dental papilla (SHUTTLEWORTH et al. 1978a).

Pulpal collagen is mainly of type I and type III (LINDE 1985). VAN AMERON-GEN (1984) using quantitative biochemical methods, found that about 60% of the collagen was of type I in human pulp. The presence of type III collagen in the pulp was initially demonstrated by SHUTTLEWORTH et al. (1978b). During development, the ratio of type III to type I collagen increased so that in bovine pulp type III constituted about 30% of the collagen (SHUTTLEWORTH et al. 1978, HAYAKAWA et al. 1981). An even higher value of 45% was found in bovine pulp (LECHNER and KALNITSKY 1981). The proportion of type III collagen in human pulp was reported to be 43% (VAN AMERONGEN et al. 1983).

In transmission electron microscopy, type I collagen appears as cross-striated fibrils with 640 Å bandings and a mean diameter of 100 nm (Figs. 13, 14). These fibrils are found in varying numbers and density throughout the dental pulp and there appears to be an increase in density with age. The presence of both type I and III collagen in the pulp has been confirmed to be in the rat incisor (COURNIL et al. 1979) and human pulp (MAGLOIRE et al. 1982) by immunohistochemical methods. The presence of type IV collagen has not been reported.

The use of type III collagen antibodies demonstrates that this type of collagen appears in the pulp as fine, branched filaments 15 nm in diameter or as electron-dense material disseminated throughout the tissue in close association with the plasma membrane of fibroblast (MAGLOIRE et al. 1982). In explants

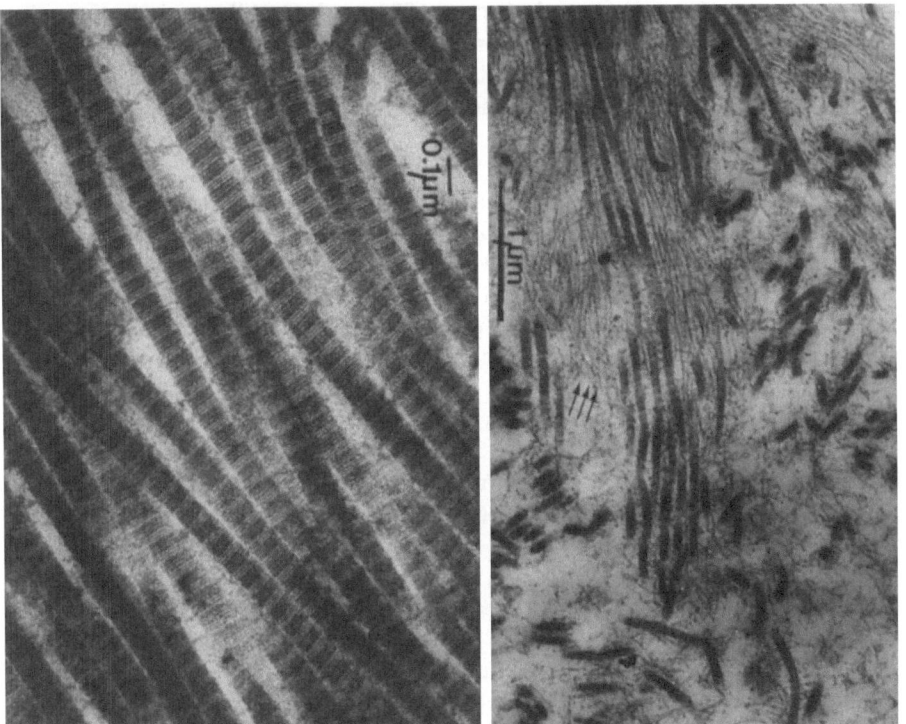

Fig. 14. Typical pulpal collagen fibrils with 640-Å primary periodic cross striations and a number of sub-bands

Fig. 15. Bundles of fine filaments (*arrows*) between some collagen fibrils in the extracellular space. Human pulp

of adult human pulp cultures, MAGLOIRE et al. (1981) noted that most of the pulpal cells gave rise only to type I collagen. However, peripheral odontoblast-like cells produced type I and III simultaneously.

Proteoglycans consist of a central protein core to which side chains of glyco-saminoglycans (GAG) and to a certain extent oligosaccharides are covalently linked. The GAGs (formerly described as acidic mucopolysaccharides) are long unbranched polysaccharide chains constructed from repeating disaccharides and containing many acidic, carboxylate and/or sulphate groups. The GAGs consti-tute from 50 to 90% of the proteoglycan molecular weight. Normally the GAG do not occur as free chains in vivo but as proteoglycans. The high molecular weight (from some 50000 to several millions) and the intense negative-charge density are the primary characteristics of proteoglycans. Earlier histochemical studies demonstrated high concentrations of polyanions, presumably GAGs, in the dental pulp (WISLOCKI and SOGNNAES 1950; MARTENS 1968).

As already mentioned in the chapter on dentine development, it is now known that the total amount of GAGs in the pulp decreases considerably at the completion of dentinogenesis (LINDE 1973a, 1985). All normally occurring connective-tissue GAGs were demonstrable in the pulp of adult teeth, although differences between species were evident. VAN AMERONGEN (1984) was the first to study the intact proteoglycans of the dental pulp. The presence of chondroitin-4-sulfate, keratan sulfate, dermatan sulfate, and hyaluronate were confirmed in bovine pulp; heparan sulfate, however, could not be detected.

BOSTRÖM and ODEBLAD (1953) and KENNEDY and KENNEDY (1957) were the first to demonstrate that the incorporation of $^{35}SO_4$ was much greater in the pulp than in other connective tissue. LINDE (1973b) studied the rate of sulfate GAG turnover in rat incisor pulp and found that the biological half-life of the chondroitin sulfate fractions was short, approximately 4 to 5 days, indicating that the extracellular matrix in rat incisor pulp constitutes a highly active metabolic pool.

Fibronectin is a high-molecular-weight glycoprotein present in the pulp as well as in blood plasma, connective tissues, cell surfaces, and basement membrane (RUOSLAHTI 1981). Fibronectin acts as a mediator of adhesion both of cells to each other and of cells and extracellular components such as collagen and GAGs.

Fibronectin is often found along collagen fibrils as small non-striated filaments 10 nm in diameter (HAY 1983). It seems that fibronectin through its binding to collagen and to the cell surface forms a bridge between the cell and the surrounding matrix. LINDE et al. (1982) demonstrated the presence of fibronectin in the dental pulp. Using indirect immunofluorescence microcopy, VAN AMERONGEN (1984) found that fibronectin is abundant in the odontoblastic layer, around blood vessels, and in the core of human pulp. A higher amount of fibronectin was extractable from the apical part as compared to the coronal and middle parts of human pulp (VAN AMERONGEN et al. 1984). In these explant cultures of human mature pulp, MAGLOIRE (1983) noted by immunostaining a close association of fibronectin and collagen fibrils. However, in an immunofluorescence study using a monoclonal antibody, no fibronectin labeling was observed in the pulp of rats (CONNOR et al. 1984).

Except for proteoglycans and fibronectin, reports on non-collagenous proteins in the dental pulp are scant. On the other hand, extensive studies have been performed on the lipid content of pulp (SHAPIRO and WUTHIER 1966; MANZOLI and GELLI 1968; ELLINGSON and SMITH 1975; VON MÜHLE and DORONIN 1976; RABINOWITZ and ROSSMAN 1979). Because of its cellularity, the dental pulp has quite a high lipid concentration, estimated to be 10–15% of the dry mass. However, the extent to which lipid constitutes a significant portion of the pulpal extracellular matrix is not clear.

Oxytalan fibers are fibers identified in light microscopy with aldehyde-fuchsin staining after peracetic acid oxidation (FULLMER et al. 1974). Oxytalan connective tissue fibers appear to be related to elastic fibers although they differ at the ultrastructural level from both elastic and prelastic fibers. Oxytalan fibrils for the most part, have been described in the periodontal ligament, but similar fibrils appearing as bundles of parallel and non-striated filaments have been ultrastructurally described in the pulp (Fig. 15) interspersed between and aligned parallel with collagen fibrils (PROVENZA et al. 1967; CAHEN and FRANK 1970;

BRADAMANTE et al. 1980). By electron-microscopic immunotyping, MAGLOIRE et al. (1982) estimated that these filaments represented type III collagen. In fact, it must be added that these extracellular non-striated fibrils have a similarity to fibronectin filaments and therefore further work is necessary to clarify the nature and distribution of the extracellular non-striated filaments of the pulp matrix.

D. The Cellular Organization of the Pulp

Classically, the dental pulp has been described as having three parts: the peripheral odontoblastic layer, the intermediary subodontoblastic layer and the central pulp.

I. The Odontoblastic Layer

Specialized post-mitotic cells, the odontoblasts, form a peripheral layer of pulpal cells adjacent to the predentine, sending long cytoplasmic processes into the dentinal tubules (Figs. 1, 2). These cells comprize a single layer but, due to their irregular lengths, there is an impression of a multilayer. The shape of the odontoblast cell body is generally cylindrical, but pear-shaped odontoblasts are often observed. In the radicular pulp, the odontoblasts are shorter and their cell bodies are more cuboid (Fig. 3). The structure and ultrastructure of this layer has been described in detail in the previous chapter on dentine to which the reader is referred.

II. The Subodontoblastic Layer

Initially the subodontoblastic layer was described as a cell-free zone by WEIL (1887). Actually the subodontoblastic region (Fig. 1) found in coronal pulp is composed of a cell-free zone located subjacent to the odontoblasts and a cell-rich inner zone (GOTJAMANOS 1969a, b; HARRIS and GRIFFIN 1969; CAHEN and FRANK 1970). The cell-free zone and the cell-rich zone appear rather late in development when the tooth erupts into the oral cavity. However, it was noted that the degree of organization in the subodontoblastic layer can be highly variable in human coronal pulp. The cell-free zone may be absent, in which case, a cell-rich zone is located directly beneath the odontoblast layer (Fig. 2). Sometimes only a cell-free zone is found and the cell-rich layer is lacking. No subodontoblastic layer is found in the radicular pulp (Fig. 3).

The principal components of the cell-free zone seems to be the rich network of nerve fibers constituting the plexus of RASCHKOW and a matrix containing collagen. The nerve fibrils found there are, for the most part, non-myelinated. Blood capillaries are also conspicuous in this area.

The cell-rich zone is composed of fibroblasts and undifferentiated mesenchymal cells, as well as occasional lymphocytes (Gotjamanos 1969a, b; Harris and Griffin 1969; Gvozadenovic et al. 1973). Zach et al. (1969), in experiments with tritiated thymidine, demonstrated that altered odontoblasts can be replaced by undifferentiated mesenchymal cells in this zone. Newly differentiated odontoblasts are capable of elaborating a dentine bridge. The cell-rich layer contains blood capillaries as well as myelinated and non-myelinated nerves enveloped by Schwann cells.

III. The Central Pulp

The mass of loose connective tissue filling the rest of the pulp cavity is referred to as the central pulp. It is subjacent to the inner part of the subodontoblastic layer in the coronal region (Figs. 1, 2) or subjacent to the odontoblasts in the radicular region (Fig. 3). It is abundantly laden with fibroblasts with long slender processes (Figs. 4, 5), separated by large intercellular spaces. As will be described in detail in subsequent sections, blood vessels, lymphatics, and nerve fibrils are found in the central part of the pulp.

E. Innervation and Dentine-Pulp Sensitivity

The dental pulp has an exceptionally rich innervation, provided primarily by bipolar sensory neurons whose cell bodies are located in the Gasserian (trigeminal) ganglion and whose central axons innervate the main sensory nucleus and spinal subnuclei of the trigeminal system. Besides the trigeminal sensory innervation, non-myelinated sympathetic axons containing noradrenalin or vasoactive intestinal peptides are routed principally from the superior cervical ganglion (Anneroth and Norberg 1968; Pohto and Antila 1972). Parasympathetic innervation has been suggested but not demonstrated conclusively.

Historically, tooth sensitivity as well as the presence of nerves in the pulp was recognized in the 19th century (Bell 1829) but the question of dentine innervation remained unanswered and controversial for over a hundred years. Raschkow (1835) studied mammalian teeth with methylene blue and described nerves entering the pulp and forming a plexus beneath the odontoblasts. Tomes (1856) described fibers in the dentine and suggested initially that they might be responsible for the sensitivity of dentine. Subsequently, Retzius (1894) described free, beaded terminations of nerve fibers between the odontoblasts in teeth of several vertebrate classes. However, the presence of sensory nerve fibrils was first established only in the 1960's when electron-microscopic studies clearly revealed nerve-like cells in the dentinal tubules (Frank 1966a, b, 1968a, b; Arwill 1967, 1968; Harris and Griffin 1968; Avery 1971; Frank et al. 1972; Corpron and Avery 1973; Dahl and Mjör 1973a; etc.). Further clarification came when Corpron et al. (1972) and Arwill et al. (1973) demonstrated with electron microscopy a degeneration of these dentinal nerves after transection of the inferior alveolar nerve but not after extirpation of the superior cervical sympathetic ganglion. The sensory nature and trigeminal origin of the nerve fibrils located in the dentinal tubules was further assessed by axonal-transport mapping techniques (for review see BYERS 1984a). Radioactive proline or leucine injected into the GASSERIAN ganglion were identified in the intratubular dentinal nerves with light and electron microscopic radioautography, indicating anterograde axonal transport. However, despite the fact that the presence of trigeminal sensory nerve fibrils in calcified dentine has been clearly established, the mechanism for dentine sensitivity is still not fully understood.

I. Histological Nerve Distribution in the Pulp-Dentine Complex

Nerves enter the pulp in company with afferent and efferent blood vessels and afferent lymphatics, generally following a similar straight course in the direction of the pulp chamber occupying the central part of the pulp (Fig. 3). Sensory nerve fibers enter the tooth as one or more dental nerves, similar in configuration to other peripheral sensory nerves in that relatively large numbers of myelinated and non-myelinated fibers are mixed and aligned in parallel (Figs. 16–18). According to OBST (1971) and BISHOP (1982), all nerves seen in the periapical region of the ligament are invested in a typical perineurium. In the apical region of the pulp a less typical perineurium is observed and ensheathment of nerve groupings by extensions of fibroblast-like cells has been demonstrated (JOHNSEN and KARLSSON 1974; FRIED and HILDEBRAND 1981).

The number of nerve fibers entering the human pulp is quite variable. Human incisors and canines have been shown to contain several hundred axons. The fiber counts made on cross sections of pulp in the apical region varied from 151 to 1296 for myelinated axons and from 40 to 650 for nonmyelinated ones (GRAF and BJÖRLIN 1951). Human premolars may have from 1000 to 2500 non-myelinated axons and from 350 to 700 myelinated axons which enter into their respective root pulp at the apex (JOHNSEN and JOHN 1978). A mean number of 926 nerve fibers was found by READER and FOREMAN (1981) on cross sections of human premolars made at the apical level.

Individual intradental axons branch extensively during their course from the root pulp to the pulp-dentine border area. Three times as many nerve fibers can be found in the middle third of the canine pulp of the cat as in the apical region (BEASLEY and HOLLAND 1978; HOLLAND and ROBINSON 1983). Each nerve fiber has been estimated to send about eight terminal axons to the plexus of Raschkow and, accordingly, the overlap of the receptive fields of individual fibers seems to be enormous (HARRIS and GRIFFIN 1968; MUMFORD and BOWSHER 1976).

ERLANGER and GASSER (1938) classified all nerve fibers into three groups. In group A, motor and sensory myelinated cerebrospinal fibers with diameters varying from 1–22 μm and a conduction speed of 6–120 m/sec are found. Group B includes myelinated autonomous vegetative system fibers with 1–3 μm diameters and a conduction speed of 3–5 m/sec. Non-myelinated small fibers (diame-

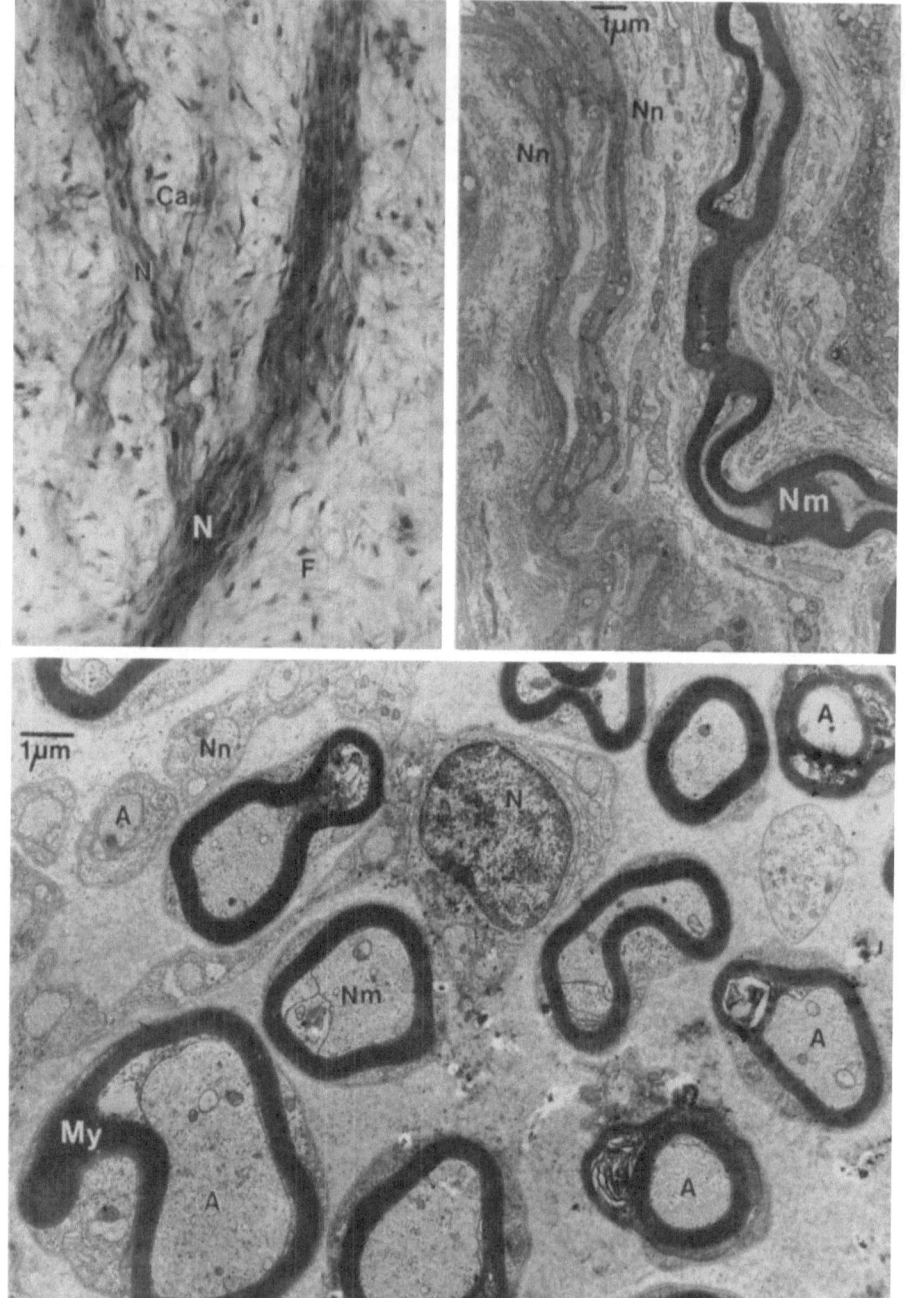

Fig. 16. Transmitted light microscopy of human dental pulp. Presence of nerve fibrils (*N*) with ramifications. *Ca* Blood capillary; *F* Fibroblast. × 385

Fig. 17. Longitudinal section of a myelinated nerve fibril (*Nm*) and two non-myelinated nerve fibrils (*Nn*) in transmission electron microscopy

Fig. 18. Transverse section in the central part of a human pulp, showing a great number of myelinated nerve fibrils (*Nm*) and non-myelinated nerve fibrils (*Nn*). *A* Axons; *My* Myelin sheath; *N* Nucleus of a Schwann cell. (From CAHEN and FRANK 1970, courtesy of Bull Group Int Rech Sci Stomatol)

ters less than 1 μm) with a conduction speed of 0.5–2 m/sec are found in group C. Nerve fibers entering the tooth apex varied considerably in diameter: the upper limit fell between 5–13 μm but 90% of the diameters were equal to or less than 4 μm (FEARNHEAD 1967). According to ERLANGER'S and GASSER'S (1938) classification, nerve fibers entering the teeth have been designated as myelinated A fibers and non-myelinated C fibers (TROWBRIDGE 1983). Most of the myelinated axons have diameters (1–6 μm) corresponding to the A-delta group. However, some axons with considerably greater diameters also seem to be present, probably corresponding to the A-beta group. It has been suggested that these rapidly conducting pulpal nerves may have some function other than conduction of pain impulses. The non-myelinated C fibers (0.25–1.5 μm in diameter) seem to be more numerous than myelinated axons. In various studies in man, monkey, dog, and cat, about 50–80% of the fibers in the apical region of the tooth are reported as non-myelinated (NARHI 1985). However, classification of these fibers is complicated because some myelinated axons may actually loose their myelin sheaths just before entering the apical area of the tooth (BYERS 1984b).

Nerve bundles pass through the radicular pulp while still in groupings (BERNICK 1948; RAPP et al. 1957). The association of nerve bundles and blood vessels may be described with the term neurovascular bundle. When the main nerve fasciculi reach the coronal pulp they disperse towards the walls and roof of the coronal pulp chamber. On approaching the cell-free zone of Weil in the subodontoblastic region, these fibers change direction frequently and branch repeatedly to give rise to an overlapping network of nerves that has come to be known as the plexus of RASCHKOW (1835). The plexus of Raschkow is not prominent until root formation is complete (FEARNHEAD 1957, 1961, 1963, 1967). Once it is established, many nerve fibers (mostly non-myelinated and without Schwann cell coverings) apparently turn from this plexus towards the predentine and extend between the odontoblast cell bodies. The majority of these nerve fibers are less than 2 μm in diameter. Many of them pass into the predentine where they form a marginal predentinal plexus, branching frequently in their course. They often run perpendicular to the odontoblastic processes in the predentine. Some of the branches of the marginal plexus are confined to the predentine while others turn into the dentinal tubules. Other nerve fibrils without Schwann cell coverings pass between the odontoblastic cell bodies and strictly follow the odontoblastic process in a straight or spiraling course.

GUNJI (1982) described 4 types of nerve endings in the peripheral coronal pulp: 1) marginal pulpal nerve fibrils ending in the cell-rich or cell-free zone or near the odontoblast cell bodies, 2) simple predentinal nerve fibers with a straight course ending in the predentine, 3) complex predentinal nerve fibers showing complicated terminal ramifications in the predentine like the root of a tree, and 4) dentinal nerve fibers penetrating dentinal tubules up to 100 μm.

Summing up a large number of studies conducted with axonal mapping techniques in the rat, cat, and monkey, BYERS (1984a) concluded that: 1) sensory nerves innervate dentine and can extend up to 0.2 mm into dentinal tubules at the tip of the crown but usually penetrate a shorter distance in other coronal regions, 2) certain kinds of dentine are not innervated (interradicular dentine) or are poorly innervated (reparative dentine), 3) the innervation of tubular dentine (orthodentine) is graded, with the greatest innervation adjacent to the tip of the pulp horn. Fewer nerves are present in mid-crown dentine, even fewer in the intercuspal and cervical dentine, and fewest in the root dentine, 4) near the tip of the crown more than 50% of the dentinal tubules can be innervated, 5) the nerves in dentine are oriented parallel to the dentinal tubules and do not form recurrent loops or perpendicular branches.

Root dentine can be innervated but innervation is less apparent than in the crown. Autoradiographic studies of dental nerves in the cat demonstrated the existence of occasional contralateral innervation in maxillary central incisors but not in other contralateral teeth (BYERS and MATTHEWS 1981).

Histochemical and degeneration studies have illustrated functional hetero-geneity of pulpal nerve fibers. Dental innervation has distinct sets of 1) large and small myelinated and non-myelinated sensory axons, 2) small substance P-sensory axons, and 3) sympathetic non-myelinated axons containing non-adrenaline or vasoactive intestinal peptide or both.

Sensory myelinated and non-myelinated axons are very numerous and con-tain acetylcholinesterase activity (KUKLETOVA et al. 1968; POHTO and ANTILA 1968a, b; AVERY et al. 1980) suggesting that they contain acetylcholine. The numerous cholinergic endings have a fine beaded structure forming a plexus along the blood vessels or serve as terminal axons of various sizes in the peripher-al pulpal plexus (KUKLETOVA et al. 1968). The myelinated and non-myelinated nerve fibers have been shown to disappear after sectioning of the inferior alveo-lar nerve (ARWILL et al. 1973). In this case the nerve terminals in the peripheral pulp, predentine and dentine are also lost. Functional studies indicate that mye-linated nerve fibers with their endings in the pulp-dentine border area are respon-sible for the sensitivity of dentine (NÄRHI et al. 1982a, b; NÄRHI and HAEGER-STAM 1983).

Another group of non-myelinated nerve fibers contains substance P and probably some other peptides (OLGART et al. 1977a, b; GRÖNBLAD et al. 1984; WASIKOWA et al. 1984). Numerous small axons in cat canines (OLGART et al. 1977a, b) were shown to be reactive to substance P. In the human tooth, sub-stance P-like reactive fibers predominated in the pulp center, while a thicker type of enkephalin-like reactive nerve fibers were seen at the periphery of the pulp (GRÖNBLAD et al. 1984). Substance P is a polypeptide which may act as a neurotransmitter possibly involved in vasodilation and in the pain mechanism. Substance P is present in various neuronal pathways including primary afferent sensory neurons. However, the functional role of substance P in the pulp remains unknown. By indirect immunofluorescence, WASIKAWA et al. (1984) demon-strated substance P-like immunoreactivity in pulp and in dentine. Some sub-stance P-containing fibers ended at the odontoblast layer and did not reach predentine. Other fibers terminated at the predentine surface or penetrated into the predentine. In the predentine, some of the substance P-fibers accompanied the odontoblast processes and ended near the mineralized dentine; others changed course transversely at various levels (WASIKAWA et al. 1984). The pres-ence of substance P fibers in predentine and dentine suggested that these fibers may be involved in the pain transmission mechanism. A role in vasodilatory responses to various external irritants and in local regulation of blood flow in the pulp has also been suggested (GAZELIUS et al. 1981; OLGART 1979). Since these fibers have been shown to resist the disruption of the superior cervical ganglion but to disappear when the inferior alveolar nerve is cut (OLGART et al. 1977b), they are probably afferent sensory C-fibers. Substance P-containing sensory fibers are present at fairly early stages in both dental papilla and dental follicle (MOHAMED and ATKINSON 1982).

Sympathetic innervation is routed principally from the superior cervical ganglion (ANNEROTH and NORBERG 1968; POHTO and ANTILA 1972). Using tracers and fluorescent-dye techniques, sympathetic fibers forming plexuses can be identified, usually around arterioles. One of the methods of choice for visualization of sympathetic terminals was that of intraperitoneal injection of 5-hydroxydopamine, used by AVERY et al. (1980) to study distribution of adrenergic endings in mouse molars. The greatest percentage of endings was in the pulp horn (49%) and in the central part of coronal pulp near the blood vessel walls (36%). The bifurcation area contained far fewer endings than did the radicular pulp with 8%. Surprisingly, adrenergic endings were found in the odontoblast layer as free endings and as vessel related endings. In addition to blood flow regulation, the pulpal sympathetic fibers may play a role in regulation of dentinogenesis as has been shown following resection experiments (AVERY et al. 1971).

· BYERS (1984a) described the development of innervation as having phases. In the early developmental period, a few pioneer axons enter the dental papilla during crown formation and most axons are associated with blood vessels. During the eruptive period in which the roots develop, the plexus of Raschkow is formed as are the terminals in the odontoblast layer, the predentine, and dentine. The first pioneer axons enter the tubules just prior to eruption (BYERS 1980). The nerve fibers appear first at the tip of the pulp horn and gradually spread to include most of the crown. During the adult period, a gradual increase in axon numbers, density of the pulpal plexus and dentinal innervation is observed.

A concept of continuous dynamic activity and readjustment of the pulpal nerve system, once it is established in the adult teeth, has been suggested by HATTYASY (1961). He proposed that axonal changes may be brought about by changes in local environment. Retraction of terminal parts of nerve fibrils would then be possible, followed by growth and outgrowth of new terminal nerve fibrils in new locations.

II. Microscopic and Fine Structure of Pulpal
and Dentinal Nerve Fibrils

In light microscopy, the structure and distribution of pulpal nerve fibrils has been studied mainly with the use of silver impregnation techniques. Since argyrophilic non-neural protein fibers are often stained with this procedure, BERNICK (1968) used, in addition, proteolytic enzymes to digest the argyrophilic fibers. One of the most convincing light-microscopic demonstrations of the presence of nerve fibrils is that of FEARNHEAD (1957, 1961). Small, beaded intratubu-

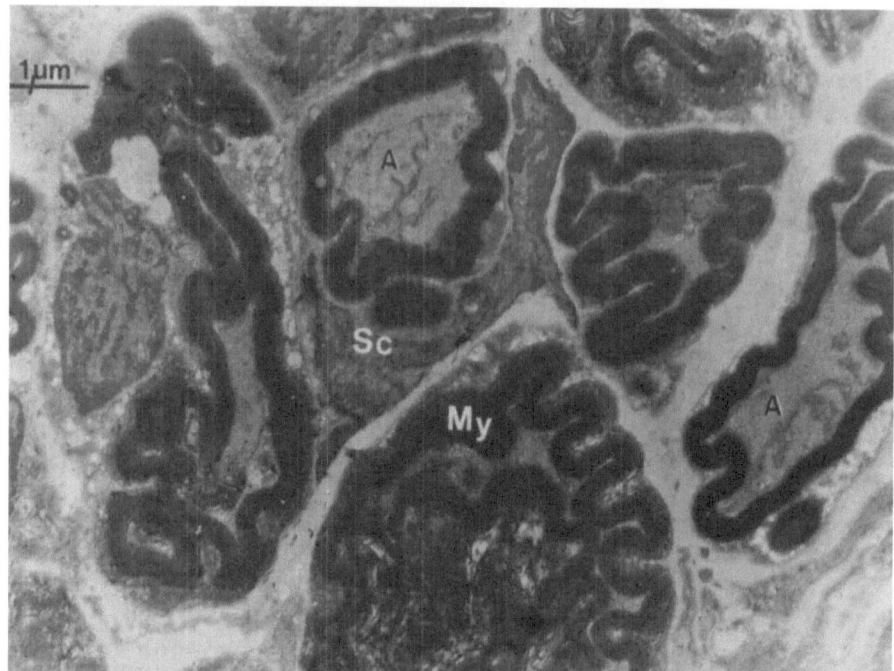

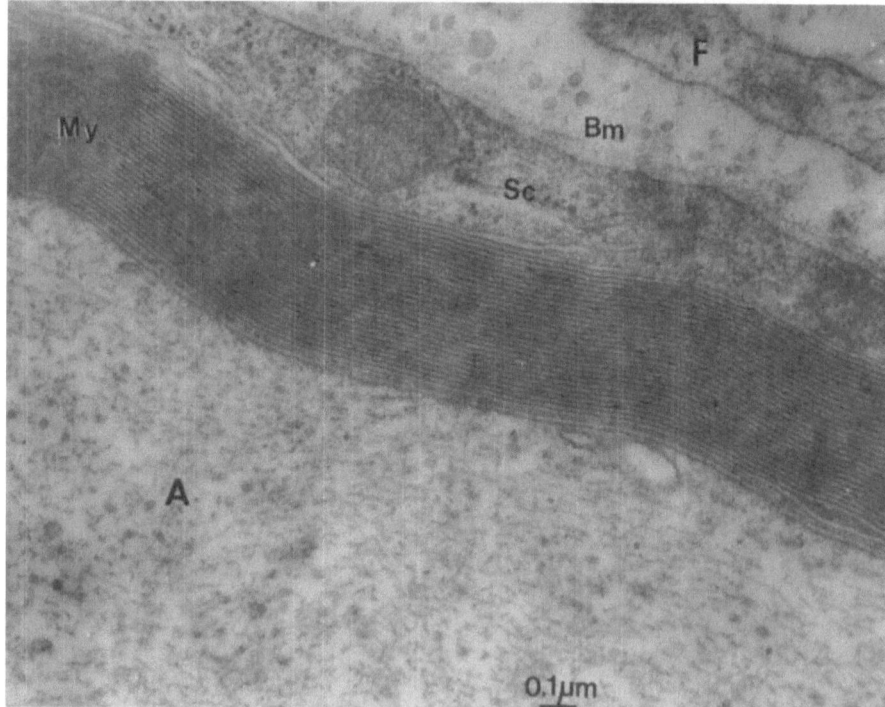

Fig. 19. Human dental pulp showing several myelinated nerve fibrils. Conspicuous folds are observed in the myelin sheaths (*My*). *Sc* Schwann cell; *A* axon. (From CAHEN and FRANK 1970, courtesy of Bull Group Int Rec Sci Stomatol)

Fig. 20. Part of a myelinated nerve fibril. *F* Fibroblast; *Bm* Basement membrane; *Sc* Schwann cell; *My* Myelin sheath; *A* Axon. (From FRANK et al. 1972, courtesy of Int Dental J)

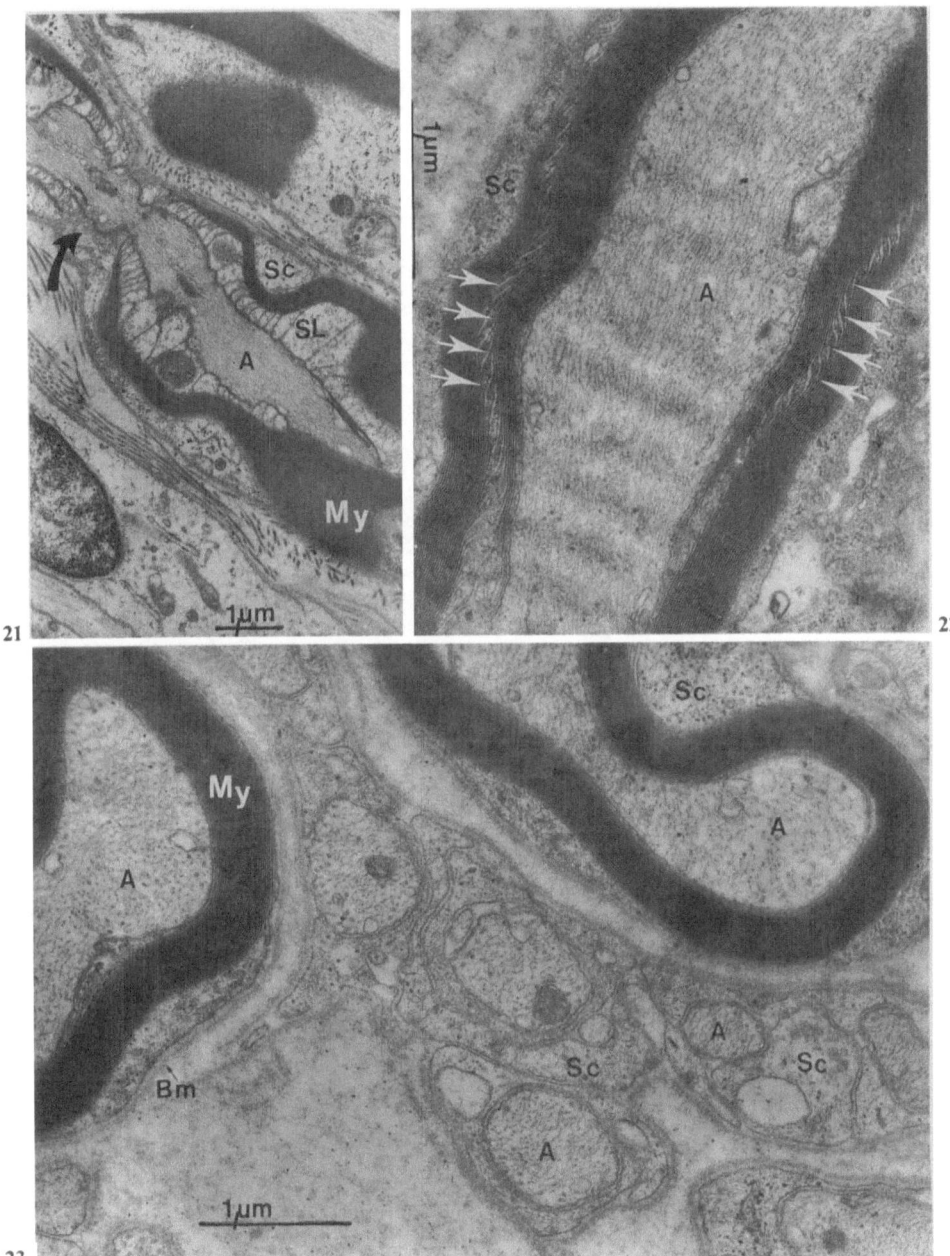

Fig. 21. Nerve fibril in human pulp showing a node of Ranvier (*large arrow*). *My* Myelin sheath with Schmidt-Lanterman clefts (*SL*); *Sc* Schwann cell; *A* Axon

Fig. 22. Presence of typical Schmidt-Lanterman clefts (*white arrows*) in the myelin sheath. *A* axon; *Sc* Schwann cell. (From Cahen and Frank 1970, courtesy of Bull Group Int Rech Sci Stomatol)

Fig. 23. Cross sections of myelinated and non-myelinated human pulpal nerve fibrils. *A* Axon; *Sc* Schwann cell; *Bm* Basement membrane; *My* Myelin sheath

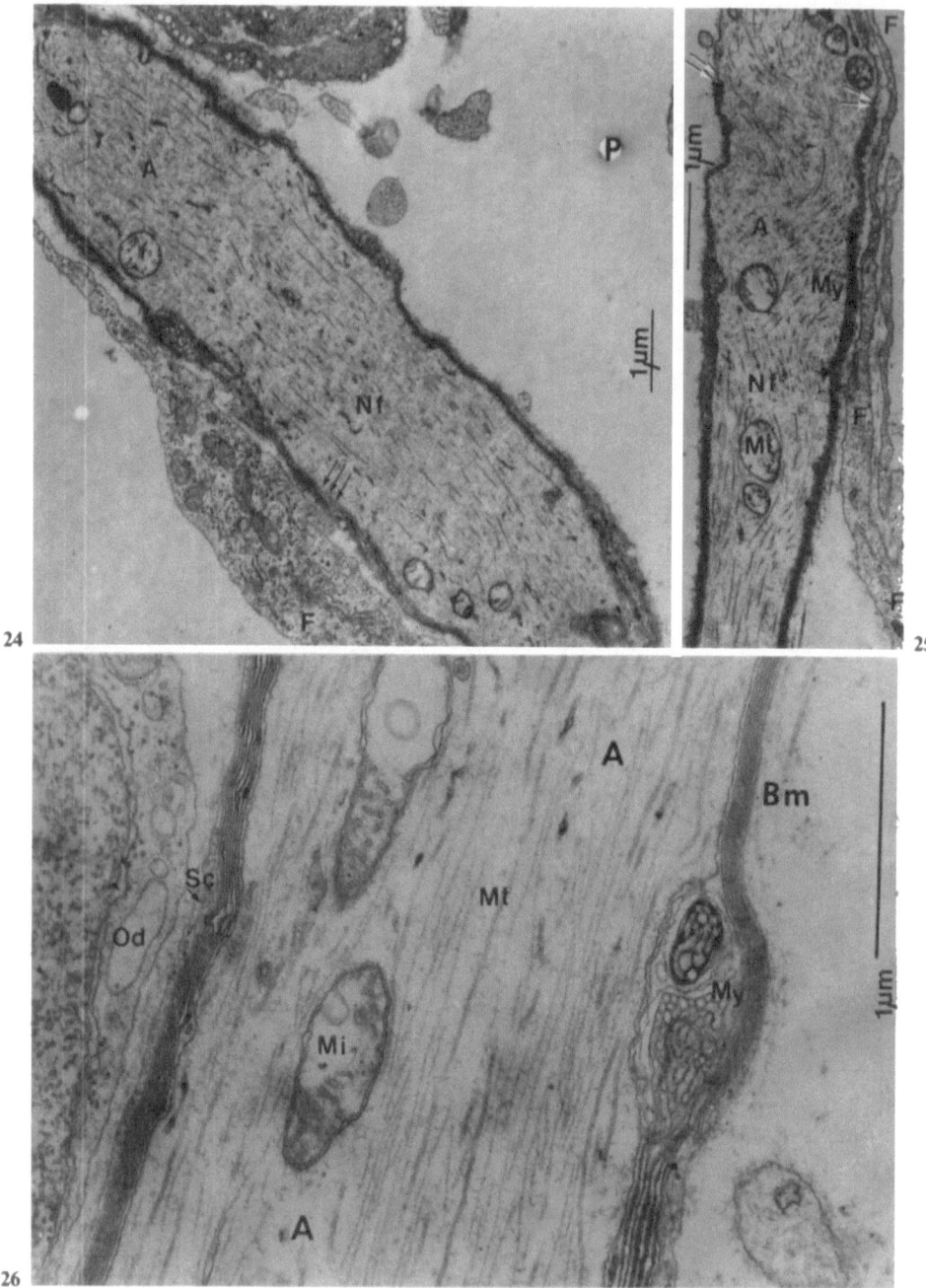

Fig. 24. Dental pulp of *Carassius auratus*. A nerve fibril (*Nf*) with numerous neurotubules and neurofilaments in the axon (*A*) exhibits a very thin myelin sheath (*arrows*). Note the presence of a fibroblast (*F*) in close association to the nerve fibril. *P* Pulp extracellular matrix. (From FRANK et al. 1972, courtesy Int Dent J)

lar fibrils, about 0.2 µm in diameter, were visible in human dentine, lying between the odontoblastic process and the tubule wall. They were traced about 0.4 mm into the coronal dentine. FAERNHEAD (1961) was most stringent in applying criteria to be satisfied when using silver impregnation before accepting a fiber as a nerve fibril. These criteria were as follows:

1. the diameter of the fiber should fall in the size range of terminal fibers, i.e. 1.0 µm or less,

2. the silver must be deposited with a very fine grain size,

3. beads or vesicular swellings along the fibers are considered to be an important morphological feature of terminal axons,

4. the fibers should branch,

5. fibers showing the criteria defined under 1–4 above should be in continuity with large fibers,

6. sectioning of the main nerve trunk should result in disappearance of these fibers in experimental animals.

In transmission electron microscopy, the dental nerve located in the central part of the radicular and coronal pulp is constituted by juxtaposition of numerous myelinated and non-myelinated fibrils, running parallel to the long axis of the root pulp (Figs. 17, 18). Between the nerve fibrils, a collagenous matrix and some fibroblasts can be observed. The pulp nerves in close proximity to blood vessels are surrounded peripherally by fibroblast-like cells with slender extensions, but a typical perineurium is not present (HAIM 1965; PISCHINGER and STOCKINGER 1968; JOHNSON and KARLSSON 1974; FRIED and HILDEBRAND 1981).

In the apical region, myelinated axons are quantitatively less numerous than non-myelinated fibers in man, monkey, dog, and cat (NÄRHI 1985). The myelinated nerve fibrils consists of a centrally located axon surrounded by a lamellar myelin sheath, itself circumscribed by a Schwann cell covering (Figs. 18–20, 23). A continuous basement membrane (Figs. 20, 23) surrounds the myelinated fibril, closely paralleling the outer plasma membrane of the Schwann cells. The centrally located axons are delimited by a continuous axolemma. The axoplasm is filled with neurofilaments and neurotubules and, in addition, it contains some mitochondria, lysosome-like bodies, and light-core and dense-core vesicles (Figs. 18–20, 22, 23, 26).

The myelin sheath appears as an elongated tube consisting of concentric lamellar layers of major dense lines exhibiting a periodicity of 128 Å. The thick-

Fig. 25. Dental pulp of *Carassius auratus*, illustrating a nerve fibril (*Nf*) with a thin myelin sheath (*My*). Note the end (*arrows*) of the myelin sheath in the upper part of the axon (*A*). The slender fibroblast process (*F*) comes into contact with the surface of the nerve fibril. *Mi* Mitochondria. (From FRANK et al. 1972, courtesy of Int Dent J)

Fig. 26. Dental pulp of *Carassius auratus*. Presence of a myelinated nerve fibril in contact with an odontoblast (*Od*). *A* Axon; *My* Myelin sheath; *Mi* Mitochondria; *Mt* Microtubule; *Sc* Schwann cell; *Bm* Basement membrane. (From FRANK et al. 1972, courtesy of Int Dent J)

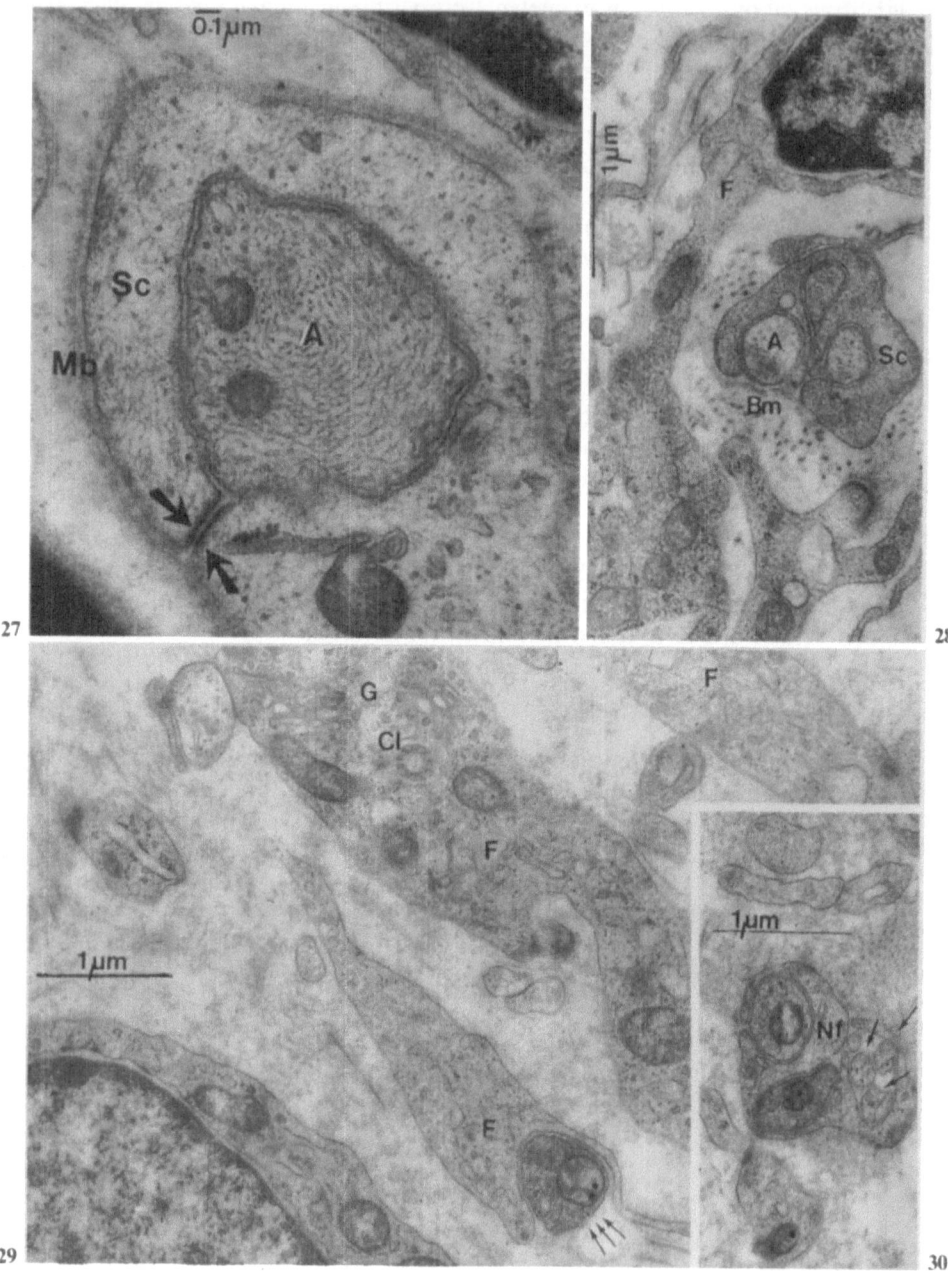

Fig. 27. Non-myelinated nerve fibril in the human dental pulp. *A* Axon with cross-sectioned neurotubules, neurofilaments and mitochondria; *Sc* Schwann cell with a mesaxon (*large arrows*); *Mb* Basement membrane. (From CAHEN and FRANK 1970, courtesy of Bull Group Int Rech Sci Stomatol)

Fig. 28. Human sub-odontoblastic layer showing a non-myelinated nerve fibril with three axons (*A*). One of these axons not completely enveloped by the Schwann cell (*Sc*) is observed adjacent to the basement membrane (*Bm*). *F* Fibroblast. (From FRANK 1968a, by courtesy of LIVINGSTONE)

ness of the myelin sheath can vary from 0.2 µm to 1 µm. It may be smooth and regular (Figs. 18, 20, 23), but in some instances, myelin sheathes can be extremely folded (Fig. 19). This folded appearance, noted by VACEK et al. (1969) and CAHEN and FRANK (1970), is typical of pulpal nerves. The myelin sheath is interrupted at regular intervals along its length at nodes of RANVIER (Fig. 21). Myelin internodes show a series of small clefts or splits running obliquely for some distance across the thickness of the myelin sheath. These are known as Schmidt-Lanterman incisures or clefts (Figs. 21, 22). They represent regions in which the major dense lines are separated over a short distance so that a small amount of Schwann cell cytoplasm is inserted into the myelin wrapping. The cleavage can affect all lamellae (Fig. 22). KÖLING (1985) studied freeze-fracture replicas of myelinated nerve fibers in human pulp and found membrane specializations of the tight-junction type at the outer and inner mesaxons of the myelin sheath, as well as at the nodes of Ranvier and Schmidt-Lanterman incisures. Around the myelin sheath, a Schwann cell covering is found and a continuous basement membrane delineates the myelinated nerve fibril from the collagenous interfibrillar matrix (Figs. 18, 20, 23).

The non-myelinated nerve fibrils alternate with myelinated fibrils (Figs. 17, 18, 23). Two to five non-myelinated axons with similar ultrastructure as those found in myelinated fibrils are ensheathed in one Schwann cell (Fig. 23). These cells present a nucleus (Fig. 18) surrounded by an endoplasmic reticulum and a Golgi apparatus, with small mitochondria, coated vesicles and some myelin figures. Coated vesicles are occasionally seen in continuity with the cell membrane facing the axons. The Schwann cells have secretory characteristics. The outer cell membrane of the Schwann cell is invaginated towards the cytoplasm around the non-myelinated axons, constituting elongated mesaxons (Figs. 18, 23, 27). The biggest axons found in the center of non-myelinated fibrils have diameters of 1–1.5 µm, whereas smaller axons 0.10–0.15 µm in diameter are found in the periphery. A continuous basement membrane is seen around the Schwann cell covering of the non-myelinated nerve fibrils (Figs. 23, 27).

When approaching the cell-free zone of the subodontoblastic layer, a number of myelinated axons lose their myelin sheath and the number of non-myelinated fibrils increases there (PISCHINGER and STOCKINGER 1968; HARRIS and GRIFFIN 1968; FRANK and CAHEN 1970). In addition, non-myelinated axons of the plexus of Raschkow have a tendency not to be invaginated into the Schwann cell covering and hence come into contact with the basement membrane (Figs. 28, 30). Some of these axons in human and cat pulp even become free from the Schwann cell coverings and basement membrane and are found in close contact with fibroblasts (Fig. 29) or odontoblasts (Fig. 31) and even in grooves or gutters

Fig. 29. Human sub-odontoblastic layer. Presence of a bare axon (*arrows*) in close contact with a fibroblast (*F*). *G* Golgi apparatus of a fibroblast; *C1* Centriole

Fig. 30. Human sub-odontoblastic layer. Non-myelinated nerve fibril (*Nf*). A few axons are in the process of being uncovered by the Schwann cell (*arrows*). (From CAHEN and FRANK 1970, courtesy of Bull Group Int Rech Sci Stomatol)

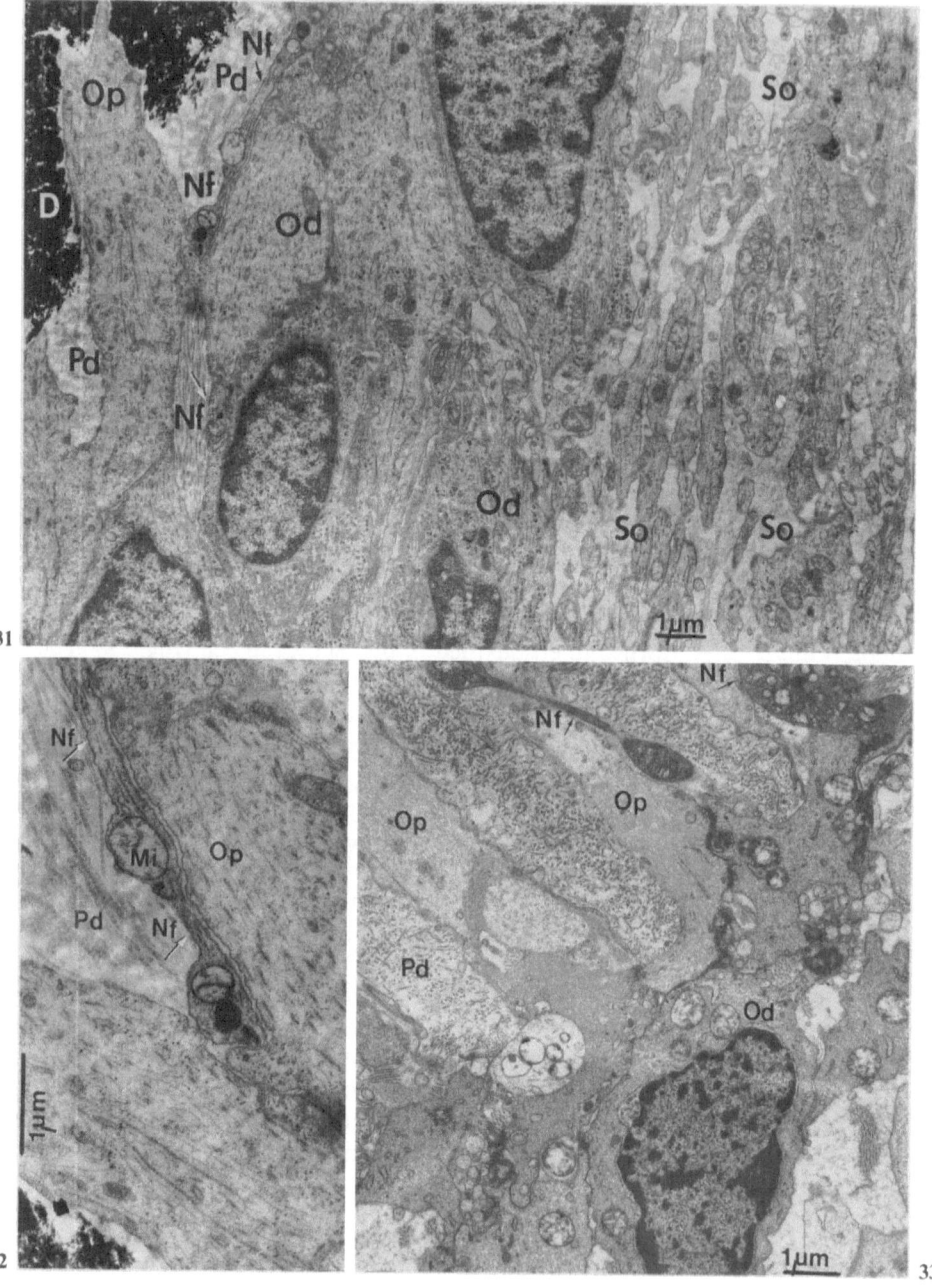

Fig. 31. Overview of the subodontoblastic layer (*So*), on the right, the odontoblast layer (*Od*), the predentine (*Pd*) and dentine (*D*). Non-myelinated nerve fibrils (*Nf*) can be followed between odontoblasts and their process (*Op*). Adult cat molar. (FRANK et al. 1972, courtesy of Int Dent J)

Fig. 32. A non-myelinated nerve fibril (*Nf*) along an odontoblastic process (*Op*) in predentine (*Pd*). *Mi* Mitochondria. Adult cat molar

Fig. 33. Odontoblast layer (*Od*) adjacent to predentine (*Pd*) in a human premolar. Note the presence of non-myelinated nerve fibrils (*Nf*). *Op* Odontoblastic process

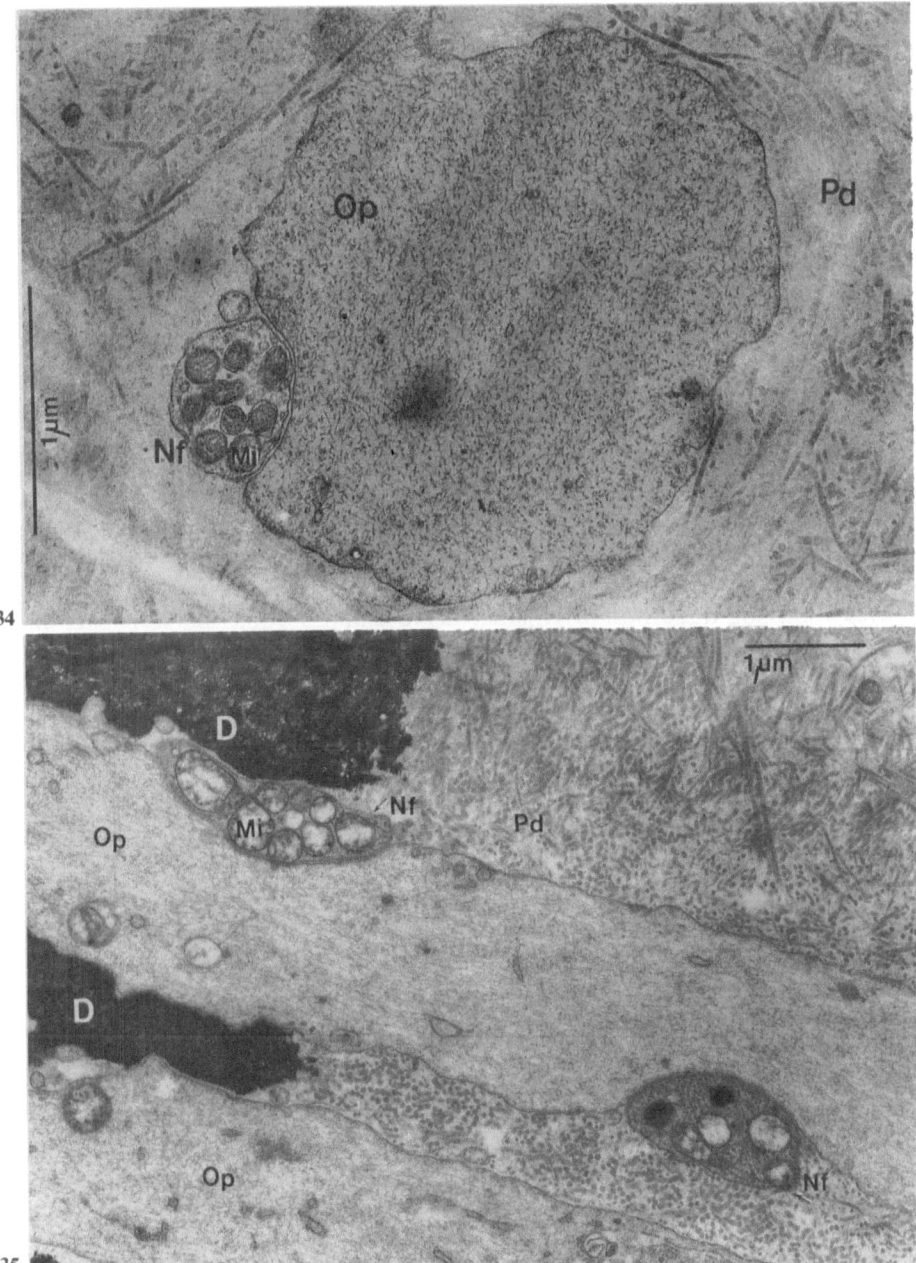

Fig. 34. Transverse section of an odontoblastic process (*Op*) of human predentine (*Pd*). A non-myelinated nerve fibril (*Nf*) containing numerous mitochondria (*Mi*) is closely associated with the odontoblastic process. (From Frank 1966a, courtesy of Arch Oral Biol)

Fig. 35. Longitudinal section of human predentine (*Pd*) and dentine (*D*). A nerve fibril (*Nf*), closely associated with an odontoblastic process (*Op*), is sectioned at 2 different levels. *Mi* Mitochondria

along these cells. The loss of the myelin sheath and Schwann cell covering was followed by Frank et al. (1972) in the subodontoblastic layer of pharyngeal teeth of one-year-old goldfish, *Carassius auratus*. Myelinated nerve fibrils without a Schwann cell covering and basement membrane were seen in close contact with subodontoblastic fibroblasts and their long slender processes (Figs. 24, 25). Such myelinated nerve fibrils were also seen in close approximation to odontoblasts (Fig. 26). It appears as though the nerve fibril losing its Schwann cell covering recovers another supporting cell (fibroblast or odontoblast) in the periphery of the pulp. Incompletely sheathed nerve fibers were also observed in the plexus of Raschkow in cat pulp (Holland 1980), and axons were also observed leaving the Schwann cell, sometimes ending in the subodontoblastic layer or among the odontoblasts.

In the coronal pulpodentinal region of both human and cat teeth, Frank (1966a, b, 1968) and Frank et al. (1972) observed a number of non-myelinated axons passing through the odontoblastic layer and penetrating the predentine and dentine (Figs. 31–33). In contrast to the observations of Roane et al. (1973), neither myelinated fibers nor fibers sheathed with Schwann cells were found in predentine and dentine.

With scanning electron microscopy (p. 187, Fig. 15), branching and complex ramifications of these non-myelinated nerve endings could be seen over the odontoblast layer and the predentine. Nerve fibers were observed running between the odontoblastic processes obliquely or at right angles. A similar appearance was clearly visualized with transmission electron microscopy (Figs. 33, 36, 37). Besides seeing such non-myelinated predentinal fibers branching between odontoblasts and their processes (also visualized by Hattyasy 1961; Fearnhead 1963, 1967; Langeland and Yagi 1972; and Gunji 1982), beaded non-myelinated fibrils were also observed closely associated with the odontoblastic cell body (Fig. 31) and its processes (Figs. 32, 34, 35, 38–41). In this case, the nerve fibril was often located in a groove or gutter made in the odontoblastic process (Figs. 34, 35). Sometimes one nerve fibril was found in close approximation to two odontoblastic processes (Fig. 38). These non-myelinated fibrils found in close association with the odontoblasts and their processes in predentine and dentine follow either a straight course or, more often, a spiraling or helicoidal course around the process (Figs. 34, 35, 39–41). On longitudinal sections, the corkcrew nerve fibrils are cut at different levels (Figs. 35, 41).

Ultrastructurally, the non-myelinated intradentinal nerve fibrils had a characteristic shape and content. They were regularly beaded (Figs. 32, 33, 39, 40) and the diameters of the those closely associated with the odontoblasts and their processes were much smaller than the non-branching predentinal fibrils (Figs. 36, 37). Both types of fibrils contained an axoplasm rich in mitochondria, neurofilaments, and neurotubules as well as some light-core and dense-core vesicles. Mitochondria were often found in the enlarged beaded parts of the fibrils (Figs. 32, 33, 39, 40). The numerous mitochondria contained in the nerve fibrils contrasted with the scarcity of organelles in the odontoblastic process (Figs. 34, 35, 38).

In the inner third of human coronal dentine, Frank (1968a, b) found complex invagination of the nerve fibrils into the odontoblastic process (Figs. 41,

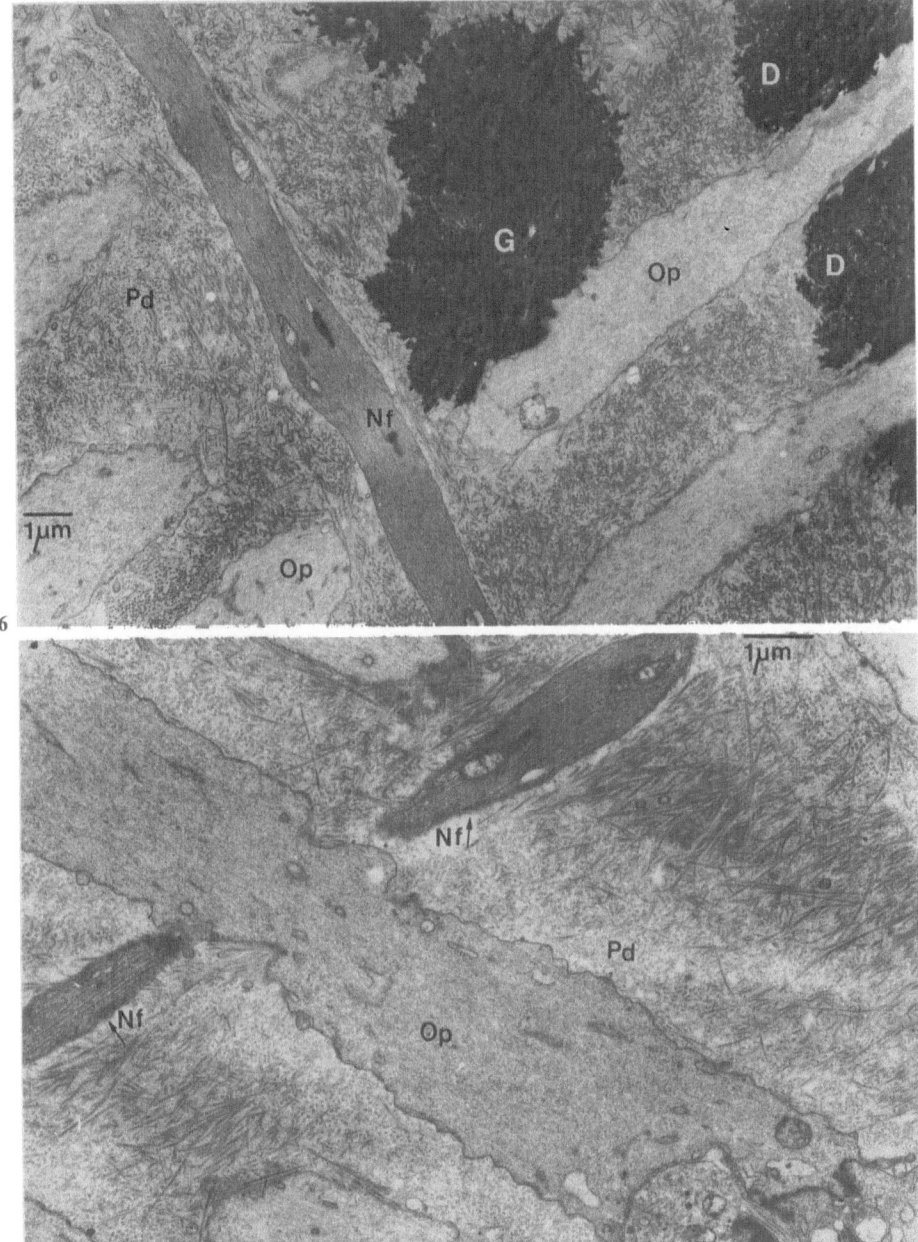

Fig. 36. Human predentine-dentine in longitudinal section. The odontoblastic processes (*Op*) can be followed from predentine (*Pd*) to dentine (*D*). A non-myelinated nerve fibril (*Nf*) extends perpendicular to the odontoblastic processes, almost parallel to the dentine-predentine junction. *G* Globular calcification

Fig. 37. Nonmyelinated nerve fibril (*Nf*) extending transversely around an odontoblastic process (*Op*) in human predentine (*Pd*)

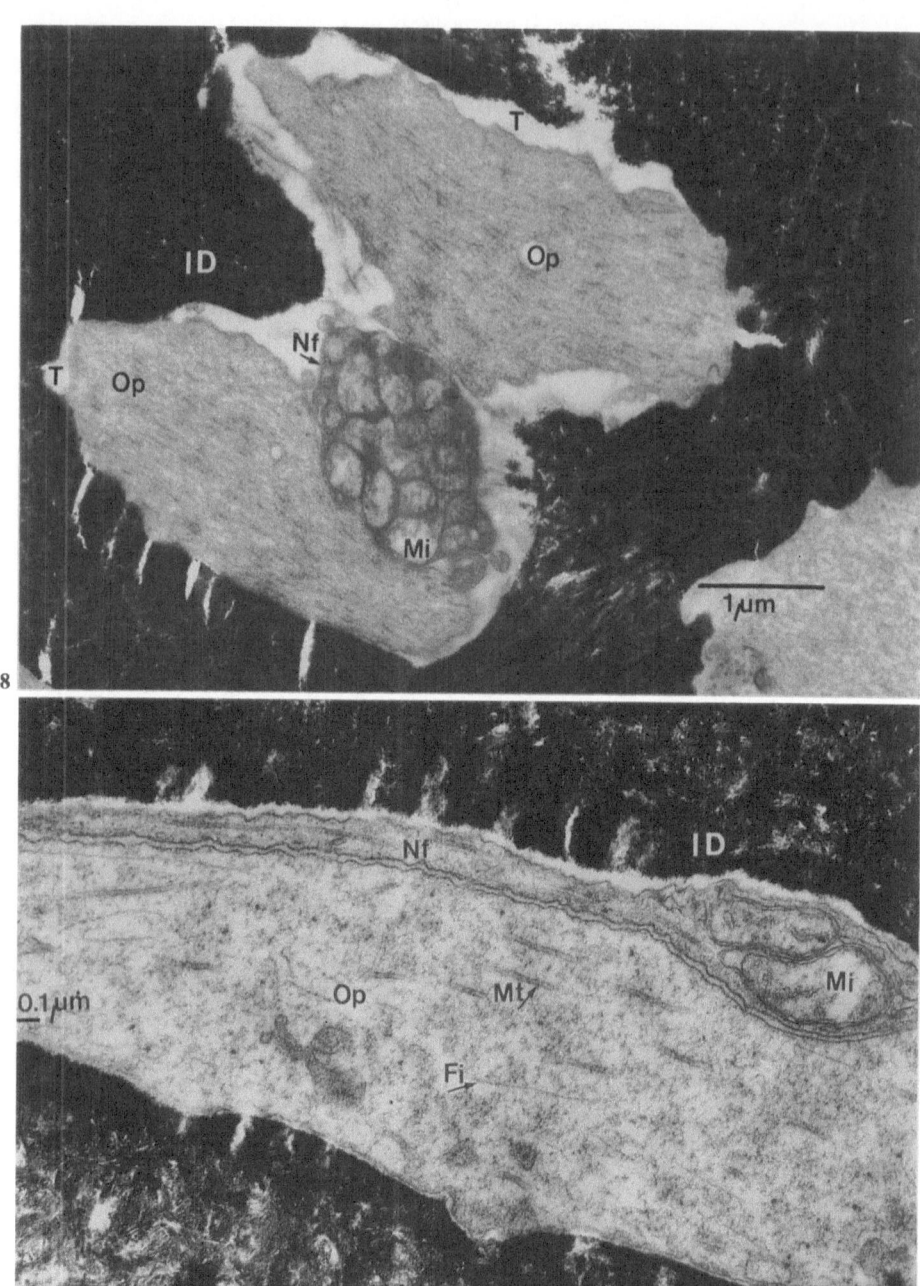

Fig. 38. A non-myelinated nerve fibril (*Nf*) containing numerous mitochondria (*Mi*) in contact with 2 human odontoblastic processes (*Op*) located in 2 incompletely closed dentinal tubules (*T*). *ID* Intertubular dentine

Fig. 39. Longitudinal section of a dentinal tubule in inner dentine of an adult cat. *Op* Odontoblastic process; *Nf* non-myelinated nerve fibril; *Mi* Mitochondria; *Mt* Microtubule; *Fi* Microfilament; *ID* Intertubular dentine. (From FRANK et al. 1972, courtesy of Int Dent J)

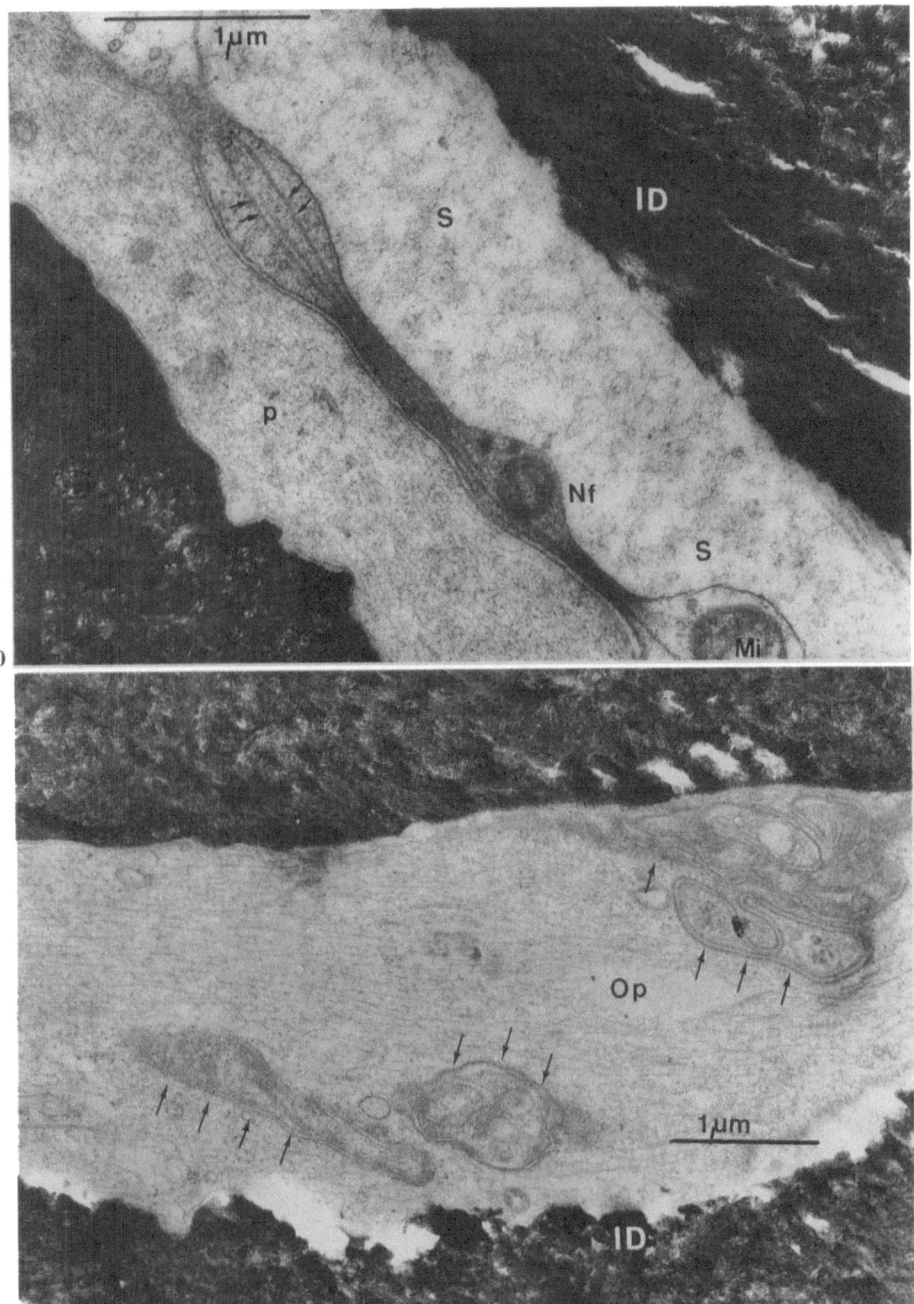

Fig. 40. Non-myelinated nerve fibril (*Nf*) shaped like a string of beads and containing neurotubules (*arrows*) and mitochondria (*Mi*), in close association with an odontoblastic process (*p*) in inner human dentine. *S* Periodontoblastic space; *ID* Intertubular dentine

Fig. 41. In addition to the complex infolding of the nerve ending in a human odontoblast process (to the right), the numerous sections of the axon (*arrows*) suggest a spiral course around the odontoblast process (*Op*). *ID* Intertubular dentine. (From FRANK 1968a, courtesy of Livingstone)

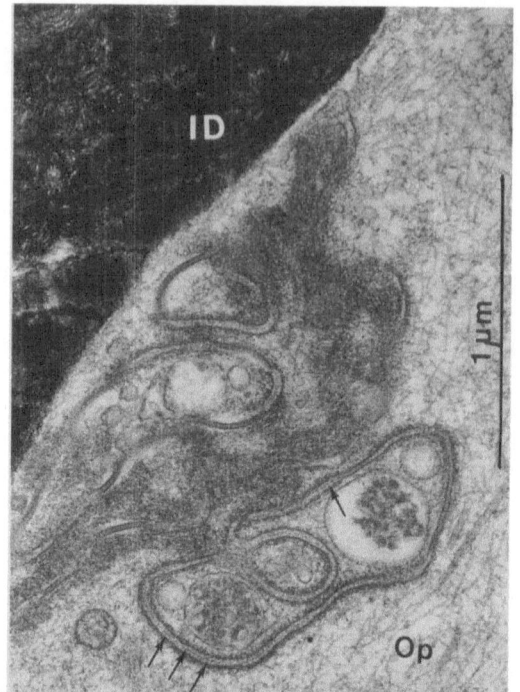

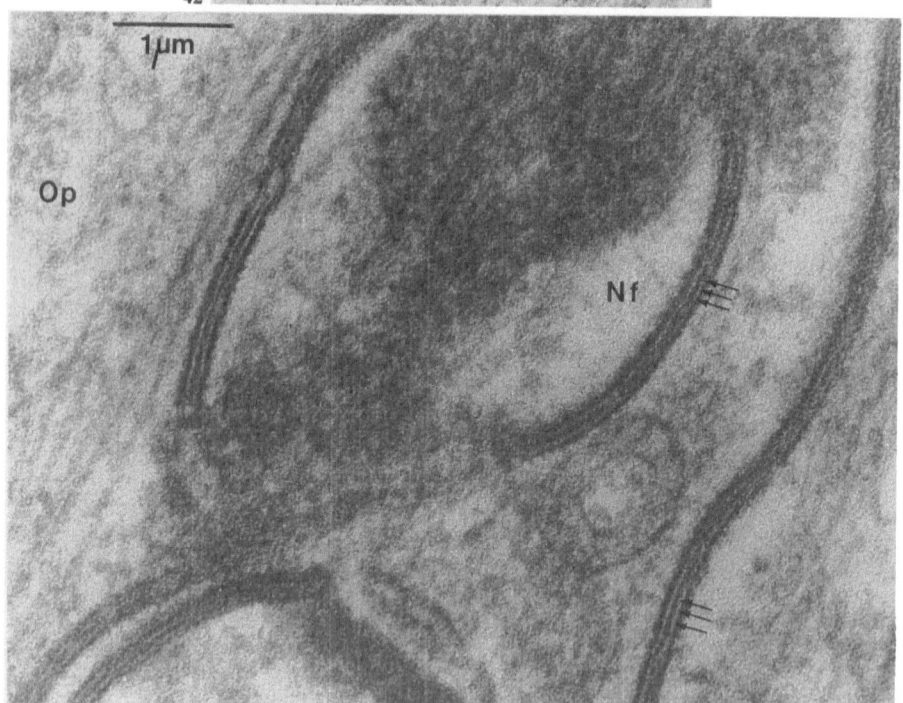

42). This configuration was observed in several human teeth and the high level of preservation of the tubular content excluded the possibility of cellular displacement. An electron-dense material was condensed along the inner leaflet of both the nerve fibril and the odontoblastic process (Fig. 42). The outer leaflets of both plasma membranes came into close contact (Fig. 43A), and FRANK (1968b) concluded that tight junctions existed at these sites. In fact, membrane fusion is not complete and the extracellular space is not totally occluded (Fig. 43A). Repetitive and punctual contacts between the outer leaflets of the plasma membranes are compatible with gap junctions, several of which can be found in these complex invaginated regions.

The above findings are in disagreement with BYERS (1984a) and HOLLAND (1985) who reviewed the issue of junctions between the nerve endings and the odontoblasts or their processes. They concluded that no synapses, tight junctions, or gap junctions occurred between these types of cells. However, KÖLING (1983) using the freeze-fracture technique found gap junctions between the nerve fibrils and both the subodontoblastic fibroblasts and the odontoblast cell bodies and their processes. Further investigations are necessary to define these interrelationships more precisely.

The trigeminal sensory origin of most of the beaded non-myelinated nerve fibrils was initially demonstrated by sectioning of the inferior alveolar nerve and the superior sympathetic cervical ganglion (CORPRON et al. 1972; ARWILL et al. 1973) and further substantiated by techniques mapping axonal transport (see review by BYERS 1984a). The anterograde axonal-transport method depends upon the fact that neuronal protein synthesis occurs almost exclusively in the nerve cell body or in the dendrites. Although a few small peptides can be synthesized in the axons and a small proportion of axonal proteins are made by the Schwann cells, almost all the proteins needed in the axons and nerve endings are supplied via anterograde axonal transport. LASEK et al. (1968) were the first to recognize that this neuronal phenomenon could be used to map nerve pathways by allowing sufficient time for axonal transport of radioactive proteins which are then detected by radioautography. DROZ and LEBLOND (1963) and DROZ (1967) reported a migration of radioactive protein molecules from the neurons towards the nerve endings, allowing mapping of the terminations. After injection of ^{3}H-proline into the Gasserian ganglion, the transported radioactive proteins remained within the sensory endings of the trigeminal axons, with survival times up to one week (FINK et al. 1975; BYERS and HOLLAND 1977). Thus when ^{3}H-proline or ^{3}H-leucine was injected into the trigeminal ganglion of rats, cats, or monkeys, it was incorporated into radioactive proteins that

Fig. 42. Numerous complex infoldings of an intradentinal nerve ending in the odontoblast process (*Op*) of inner human dentine. Note the parallelism and thickening of both plasmalemmata and the dense line in the small intercellular space (*arrows*). *ID* Intertubular dentine. (From FRANK 1968a, courtesy of Livingstone)

Fig. 43A. Attachment sites between the nerve fibril (*Nf*) and the odontoblastic process (*Op*) in the region of the complex infoldings. Note the punctual fusions between the outer leaflets of both plasma membranes (*arrows*)

moved rapidly by axonal transport into the predentinal and dentinal nerve endings previously described. This phenomenon has been demonstrated with both light- and electron-microscopic radioautography (FINK et al. 1975; WEILL et al. 1975; BYERS and KISH 1976; BENSADOUN 1976; BYERS and DONG 1983; BYERS 1984a, b).

FEARNHEAD (1963) estimated that the innervation of inner human coronal dentine can be highly variable with extremes estimated at 1 nerve fiber for every 200 tubules or 1 nerve fiber for every 2000 tubules. FRANK (1966a), based on an estimation of 60000 tubules per mm^2 in the inner dentine, calculated that 30 nerve endings per mm^2 would exist in the case of one nerve for every 2000 tubules, or 300 nerve endings per mm^2 in the case of one nerve for every 200 tubules. LILJN and FAGERBERG-MOHLIN (1984) studied impacted third molars in patients aged 20 to 51 years and estimated that 15% of the odontoblasts were accompanied by nerve fibrils in coronal and root predentine whereas only 2% of coronal odontoblasts were associated with nerve fibrils in mineralized dentine.

III. Dentine-Pulp Sensitivity

Dentine is one of the few tissues in the body from which only pain is perceived whether the stimulus is due to temperature changes, mechanical grinding or cutting, pressure, a direct stream of air, or electrical or chemical stimuli. All produce a nociceptive response. Moreover, this pain is often diffuse, making it difficult on occasion to localize the precise anatomic site of the stimulus. Dentine is not uniformly sensitive. It is well established from clinical impression that dentine is most sensitive at the dentinoenamel junction and also quite sensitive close to the pulp.

Three mechanisms might explain dentine-pulp sensitivity. These are 1) that the odontoblasts serve as receptor cells, 2) that rapid outward or inward movements of tubular contents induce a neural stimulus through hydrodynamic phenomena and 3) that the dentine contains nerve endings which respond directly to stimulation.

1. The Odontoblast as a Receptor Cell

It has been suggested that the odontoblasts with their processes extending into the dentinal tubules act as receptor cells, activated by the external stimuli and mediating their effects on the nerve fibers in the dentine-pulp border area. As early as 1856, John Tomes thought that the "dentinal fibers" might be responsible for the sensitivity of dentine. Describing the pulpal nerve supply beneath the odontoblastic layer, HOPEWELL-SMITH (1916) viewed the odontoblast as a transducer which relayed the external stimuli from the dentine to the nerve endings.

The transduction theory, initially based on the histochemical studies of AVERY and RAPP (1959), AVERY (1963), and RAPP et al. (1968) is based on the premise that the odontoblastic process acts as the primary nociceptor and is further capable of transmitting this impulse to the nerve endings in the dentinal tubules or in the peripheral pulpal layers. The neural-crest origin of the odontoblast is supportive of this view. AVERY and RAPP (1959) and RAPP et al. (1969) demonstrated the presence of acetylcholinesterase (AChE) throughout the length of the odontoblastic process. AChE was also found in the subodontoblastic layer and the pulpal nerves. Since acetylcholine and acetylcholinesterase may play an important role in the transmission of neural impulses (NACHMANSOHN 1948), AVERY and RAPP (1959) postulated that the stimuli passing along the odontoblastic process and the cell body of the odontoblast pass to the nerve ending via a synapse. The fact that the odontoblastic process can extend to the dentinoenamel junction, as shown recently (see chapter on structure and ultrastructure of dentine), supports the view that stimuli could affect the odontoblastic process right at the dentinoenamel junction.

However, other histochemical studies conducted by TEN CATE and SHELTON (1966) and POHTO and ANTILA (1968) with more sensitive methods did not reveal cholinergic activity in the odontoblasts or their processes. Moreover, if the odontoblast is a neural receptor it must make a synaptic connection with an axon and must produce an action potential. A synapse has not been described between odontoblasts and nerve endings to date. Membrane potentials of odontoblasts have been measured. They are 25–40 mV and not indicative of a receptor function as such (KROEGER et al. 1961; WINTER et al. 1963; YAMADA et al. 1971). Moreover, in these studies no change in the membrane potential was observed in response to stimulation. However, MAGLOIRE et al. (1979) recorded membrane potentials as high as −80 mV in human odontoblast-like cells maintained in peripheral pulp explant cultures, different from those of the inner dental pulp cells which had considerably low resting membrane potentials (−30 mV). They assumed that the peripheral cells in their explant cultures were odontoblasts and that their findings supported the idea that these cells can act as receptor cells.

According to AVERY et al. (1984) evidence of odontoblastic transduction is actually inconclusive but the concept remains viable based on the demonstration that the odontoblastic process extends all along the dentinal tubule.

2. The Hydrodynamic Hypothesis

Since the peripheral dentine is known to be sensitive and since nerve fibers have been shown to exist only in the deeper layers of the dentine, the notion that sensory responses from peripheral dentine are not induced by direct activation of nerve fibers gained strong support. The hydrodynamic hypothesis of dentine sensitivity is based on the finding that different pain-producing stimuli applied to dentine are able to induce movements and fluid flow of the tubular contents. These movements of tubular contents, if rapid enough, would be able to activate the nerve endings. This thesis was first presented by GYSI (1900) and later extensively supported by BRÄNNSTRÖM (1960a, b, 1962, 1963, 1966, 1981); BRÄNNSTRÖM and ASTRÖM (1964, 1972); and BRÄNNSTRÖM et al. (1967, 1969). According to these authors, capillary action is the main factor responsible for dentine sensitivity. With the exception of heat and electrical stimulation, all stimuli that produce dental pain, e.g., compressed air, application of dry paper, and hypertonic solutions (sugar, $CaCl_2$ solutions, etc.), mechanical stimulation (probing, drilling, etc.) result in an outward movement of fluid. In order to produce pain, the outward movement of the tubular content must be fast enough. In chemical stimulation with hyperosmotic solutions such as hypertonic $CaCl_2$ and saccharose, it has been shown that the pain-producing capacity of the solution is related to its osmotic pressure and to its capacity to induce fluid flow in dentinal tubules rather than to its chemical composition (Anderson and Ronning 1962; Anderson 1963; Anderson and Matthews 1967).

Changes in temperature applied to dentine may produce pain by different mechanisms, in the case of cold vs. heat. Whereas cold inducing contraction of the tubular content produces an outward movement of fluid, heat results in an inward movement by expansion of the tubular content.

With the hydrodynamic hypothesis, it is not exactly known how the energy induced by outward or inward movements of tubular contents is transformed into neural impulses. However, due to the fact that nerve fibers have not been described in peripheral dentine, this theory continues to be popular.

3. Direct Neural Stimulation

This theory presumes the presence of nerve endings in the dentinal tubule all along its lumen and extending to the dentinoenamel junction. The nociceptive stimuli are picked up at or below the dentinoenamel junction and carried to the pulp via activation of nerve endings and direct transmission to the trigeminal system. In a light-microscopic study, Stella and Fuentes (1963) claimed to have found nerve endings in the peripheral coronal dentine, but these findings have not been confirmed.

One of us, with La Flèche and Steuer, has recently studied the tubular content of human peripheral coronal dentine with transmission electron microscopy after freezing the tubular contents with the use of liquid nitrogen (La Flèche et al. 1985). Odontoblastic processes were thus demonstrated immediately within the dentinoenamel junction. La Flèche et al. (1985) considered the odontoblastic process as a retractable suspensor system. If proper care is not taken prior to fixation, the cell processes retract very quickly. Using the same technique of fixation after freezing in liquid nitrogen, La Flèche et al. (1985) were able to observe structures in close contact with the odontoblastic process near the dentinoenamel junction, structures limited by a cell membrane and which appeared to be non-myelinated nerve fibrils (Fig. 43B). The negative morphological and physiological findings concerning peripheral dentine innervation can be explained by the simultaneous rapid retraction of the odontoblastic process along with their closely related nerve fibrils upon exposure of the dentinal tubules.

If the above observations are substantiated, direct neural stimulation at the dentinoenamel junction will become a plausible explanation. In this case the odontoblast and its process would act as a supporting cell to the fine terminal trigeminal axons, as already suggested by Byers (1984a).

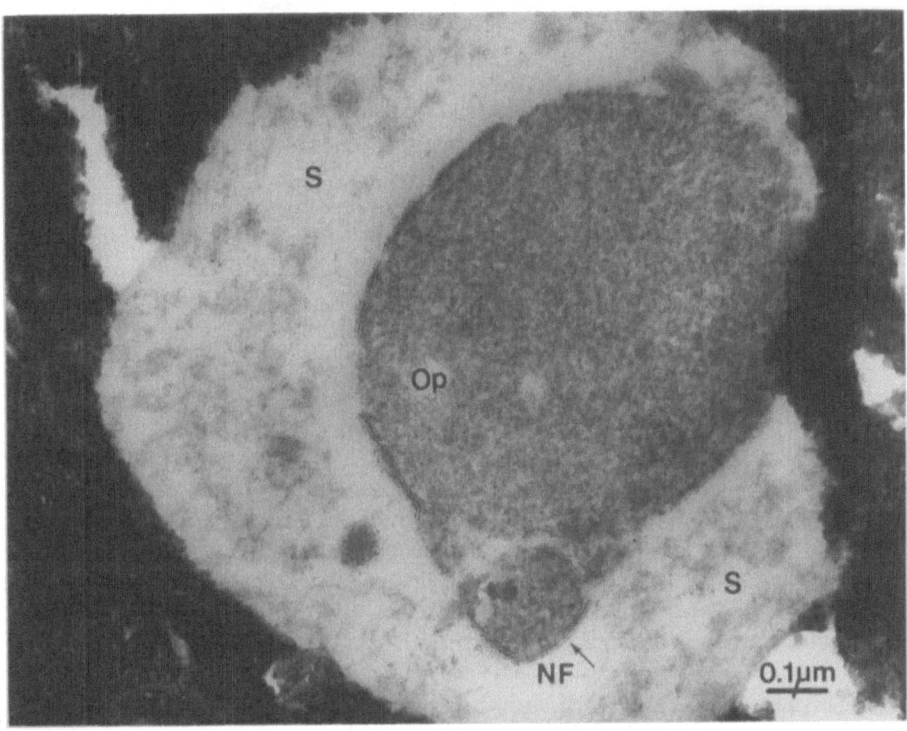

43B

Fig. 43B. After special fixation with liquid nitrogen, an odontoblastic process (*Op*) is present in a cross-section of a human coronal dentinal tubule located near the dentino-enamel junction. A non-myelinated nerve fibril (*NF*) is closely associated with the odontoblastic process. *S* Periodonto-blastic space. (From LAFLÈCHE et al. 1985, courtesy of J Biol Buccale)

F. Vascular Supply

I. Blood Vessels

Blood vessels enter and exit the pulp via the apical and accessory foramina. One or sometimes two or three vessels of arteriolar size (150 μm) enter the apical foramen with the nerve bundles. Smaller arterioles without accompanying nerve bundles can enter the pulp via the minor foramina. The blood vessels of the pulp form a complex network of afferent arterioles which branch and are continuous with a rich capillary system. The blood of the capillaries is drained by veins. Pulpal blood vascular architecture can be visualized by a number of different methods. KRAMER (1960, 1968) used India ink perfusion, whereas DE SAUNDERS (1952, 1967) and DE SAUNDERS and RÖCKERT (1967) studied the blood vessels with microradiography following retrograde aortic injection of radioopaque solutions. TAKAHASHI (1985) observed the blood vascular tree of the dog pulp with scanning electron microscopy, using corrosion resin casts after injection of colored synthetic resin. The latter method allows one to differentiate an arteriole from a venule by the surface appearance of the endothelial cells. The apical arteriole or arterioles on entering the apex and sides of the root tended to divide almost immediately, giving rise to central or principal pulpal arterioles that ascended the root canal and pulp chamber (DE SAUNDERS and RÖCKERT 1967).

KRAMER (1968) noted marked macroscopic differences between arterioles and venules. Veins are larger; their course is tortuous and their outlines are irregular. In contrast, the arterioles are slender and their outlines are smooth. They pursue a more direct course. The veins tend to lie near the inner and interradicular aspect of the root canal while the arterioles run on the outer aspect.

The pulpal arterioles range from 100–150 μm in diameter (PROVENZA 1958). As the arterioles travel longitudinally along the root canal, there is some diminution in their caliber. As they progress coronally, they also give rise to numerous branches which can anastomose with those of adjacent arterioles. There is a close relationship between blood vessels and nerves. In the basal region of the pulp chamber, the arterioles branch so that one of the vessels veers centralward and increases greatly in diameter while other vessels continue towards the ceiling of the pulp chamber (PROVENZA 1968). The terminal part of an arteriole, often called a metarteriole is smaller and continuous with the capillaries.

The pulpal arterioles are composed of 3 layers. The *intima* is formed by a single layer of endothelial cells, covered on the outer surface by a basement membrane. The *media* is made up by one or two layers of continuous smooth muscle cells arranged circumferentially. The outer adventitial layer is ill-defined and made up of loose connective tissue with fibroblasts and a collagenous matrix. With specific histochemical stains, HALS and TONDER (1981) failed to identify any true elastin in the middle layer of arterioles of human and dog pulp. However, they found pseudoelastic fibers in both animals.

In the endothelium of arterioles, EKBLOM and HASSON (1984) found tight junctions which they classified as relatively tight, but they did not observe any gap junctions there. A few of the latter were found among smooth muscle cells.

The pulpal capillaries are continuous with the metarterioles and constitute an important part of the pulpal circulatory network. Capillaries can be found throughout the pulp (CAHEN and FRANK 1970) but the peripheral subodontoblastic capillary plexus, located principally in the cell-free zone of the subodontoblastic layer of the pulp chamber, has been particularly well studied (DE SAUNDERS 1967; HAN 1968; CAHEN and FRANK 1970; HARRIS and GRIFFIN 1971; CORPRON et al. 1973; DAHL and MJÖR 1973b; RAPP et al. 1977; KÖLING 1983; BISHOP 1985). From this peripheral plexus, looping branches pass between the odontoblast cell bodies towards the predentine surface.

Two types of capillaries are found in the dental pulp: the continuous capillaries (Figs. 44–46) and the fenestrated capillaries (Fig. 47). The latters seem to constitute 4–5% of the whole capillary system (HARRIS and GRIFFIN 1971; KÖLING 1983). CAHEN and FRANK (1970) and RAPP et al. (1977) found fenestrated capillaries throughout the human pulp but especially in the subodontoblastic

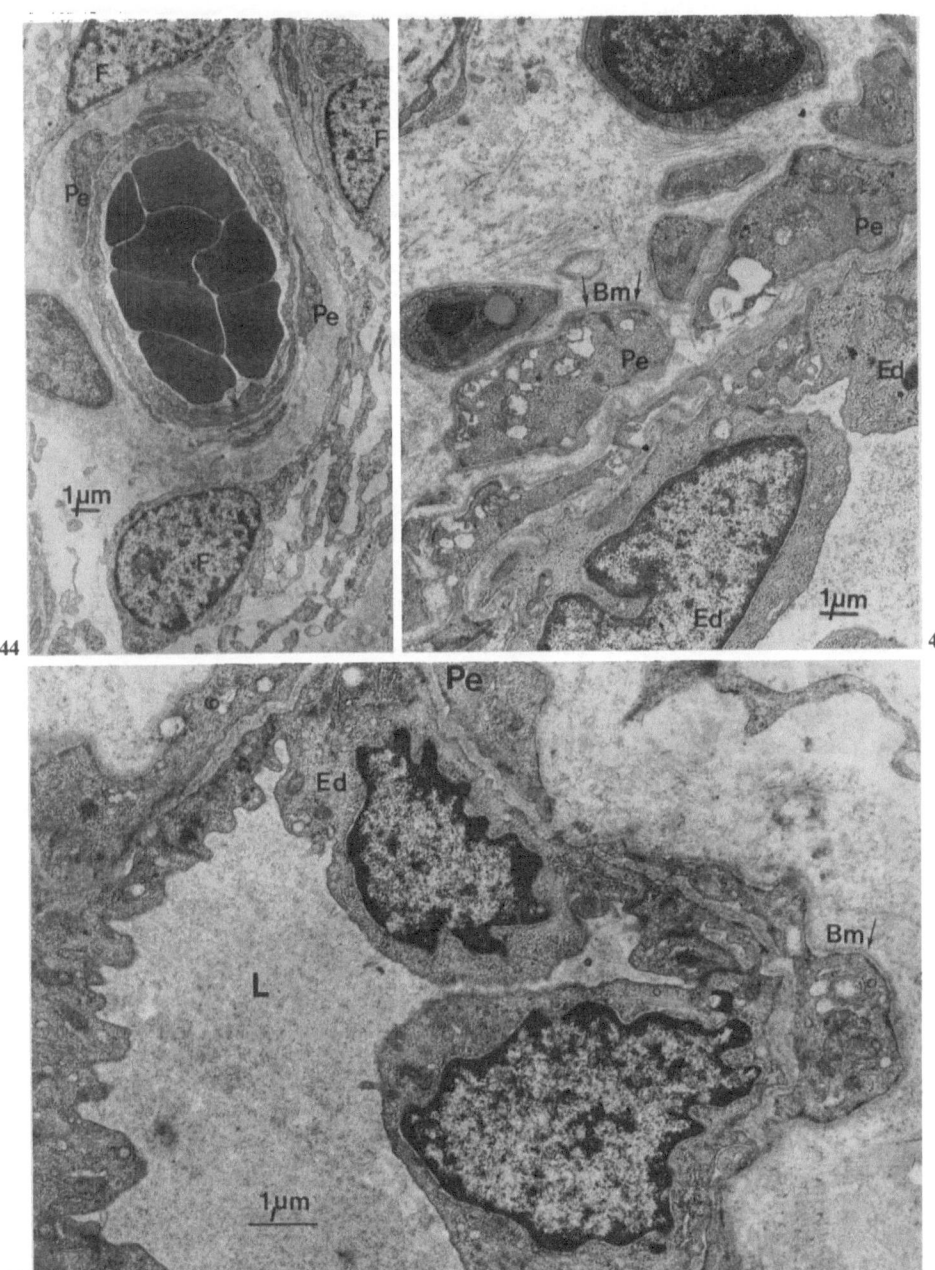

Fig. 44. Transverse section of a blood capillary filled with erythrocytes and limited by a continuous endothelium. Cat molar pulp. *Pe* Pericyte; *F* Fibroblast

Fig. 45. Blood capillary in the human pulp with a continuous endothelium (*Ed*). Pericytes (*Pe*) enveloped by a continuous basement membrane (*Bm*)

Fig. 46. Blood capillary with continuous endothelium (*Ed*) in the human dental pulp. The pericytes (*Pe*) are enveloped by a continuous basement membrane (*Bm*). *L* Lumen of the capillary

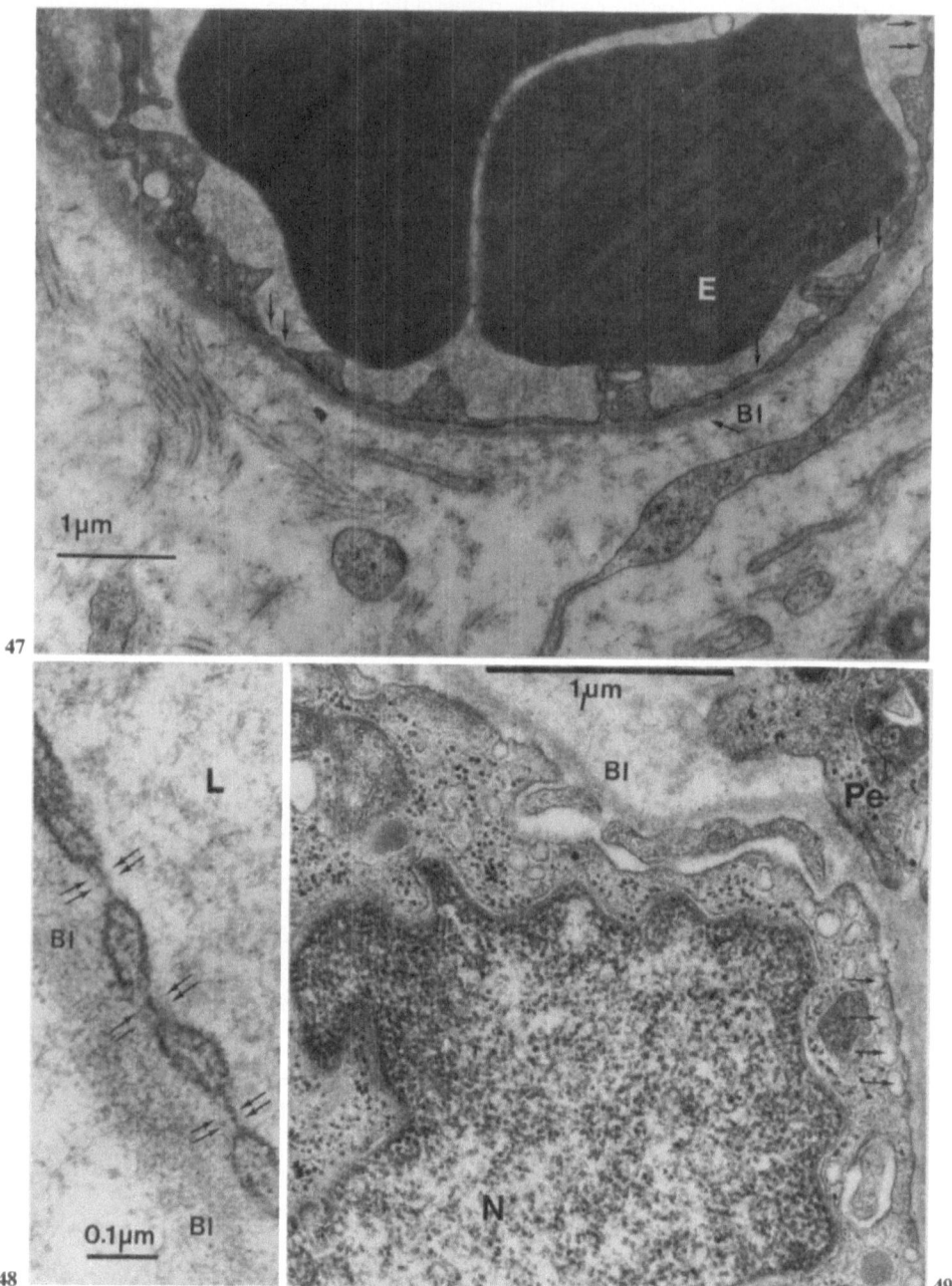

Fig. 47. Fenestrated.blood capillary in the human pulp. The endothelium is thin and contains pores covered by a diaphragm (*arrows*). Note the thick basal lamina (*Bl*). *E* Erythrocyte

Fig. 48. Higher enlargement of the wall of a human (fenestrated) blood capillary. The thin endothelium contains pores covered by a diaphragm (*arrows*). *Bl* Basal lamina; *L* Capillary lumen. (From CAHEN and FRANK 1970, courtesy of Bull Group Int Rech Sci Stomatol)

Fig. 49. Endothelial cell of a blood capillary wall in human pulp. Note the presence of numerous coated vesicles adjacent to the basal plasma membrane (*arrow*). *Bl* basal lamina; *N* Nucleus; *Pe* Pericytes

plexus and between the odontoblasts. BISHOP (1985) studied the pulp of rats perfused through the aorta with lanthanum nitrate. In the peripheral pulp, a few capillaries were permeable but the most permeable capillaries were found between the odontoblasts. The pulpal permeability was attributed to the fenestrated capillaries, contrasted with continuous capillaries lying among nerve fibers and impermeable to lanthanum.

The continuous capillaries (Figs. 44–46) are made up of a continuous endothelium, consisting of a monolayer of flat cells making contact with each other. The ovoid nucleus with its occasionally wrinkled contour (Fig. 46) produces a central bulge in the cytoplasm. The latter contains a small Golgi, a few short profiles of endoplasmic reticulum, small mitochondria and a cytoskeleton. In addition, numerous small micropinocytic vesicles 60 to 70 nm in diameter are observed in the cytoplasm as well as along the internal and external plasmalemmas (Fig. 49). These plasmalemmal vesicles open on both surfaces of the cell and are more numerous in the capillary endothelium than in the endothelial of arterioles or venules. EKBLOM and HANSON (1984) found an occasional solitary cilium in the endothelial cells, with a $9 + 0$ microtubular pattern.

No gap junctions were found between the endothelial cells (EKBLOM and HANSSON 1984), but tight junctions were described and classified as leaky (KÖLING 1983; EKBLOM and HANSSON 1984). The endothelial cells are covered on their outer surface by a continuous basement membrane of approximately 20 to 50 nm in thickness. This splits to enclose the pericytes or Rouget cells (Figs. 44–46). Pericytes are spindle-shaped cells with a well-developed Golgi apparatus and endoplasmic reticulum. BENNINGHOF (1923) was the first to state that the pericyte cytoplasm in mammals was similar to that of the fibroblast. When stimulated, this cell can proliferate and show histiocytic characteristics such as ingestion of vital dyes. It has been considered as a relatively undifferentiated mesenchymal cell.

The fenestrated capillaries present a thin endothelium layer (approximately 60–100 nm thick) exhibiting circular transcellular openings, the so-called fenestrae or pores with a diameter of approximately 60 to 80 nm (Figs. 47, 48). Around the periphery of the pores, the cell membrane of the blood front is continuous with the cell membrane of the tissue front. Each pore is usually closed by a single-layered diaphragm displaying a central knob; the basal lamina is continuous over the pores (Fig. 48). KÖLING (1983) studied the pores in fenestrated pulpal capillaries using freeze-fracture replicas. The pores appeared as circular depressions on the PF faces (protoplasmic faces) and as elevated or crater-like structures on the EF faces (extracellular faces). Most of the fenestrated capillaries are found in the subodontoblastic plexus, primarily between the cell bodies of the odontoblasts (FRANK and CAHEN 1970; CORPRON et al. 1973; RAPP et al. 1977; KÖLING 1983; BISHOP 1985).

The pulpal capillary plexus drains into large thin-walled venules which in turn are gathered together to form several veins escaping through each apical foramen. In the apical region, these veins can be one half of the caliber of their coronal counterparts (KRAMER 1968). This arrangement results in a sluggish venous circulation, thus requiring more veins than arteries to keep the afferent blood flow balanced. Venules leaving the dental pulp run in close association with arterioles and nerve fibers.

Transition from capillaries to venules occurs gradually. The immediate post-capillary venules are characterized by the presence of pericytes (pericytic venules). They are drained by venules with increasing diameters presenting a structure similar to arterioles. The walls are perhaps thinner and the media has a discontinuous layer of smooth muscle cells (DAHL and MJÖR 1973). EKBLOM and HANSSON (1984) did not find any gap junction in the endothelium of the venules and the tight junctions found there were classified as very leaky.

Finally it must be mentioned that an arterio-venous shunt system has been observed in the dental pulp by PROVENZA (1958, 1968), KRAMER (1960), BENNETT et al. (1965), and DAHL and MJÖR (1973). Arterio-venous anastomoses are vascular channels which directly connect arterioles with venules, thereby evading the constituents of the capillary plexus. Structurally the arterio-venous anastomosis is characterized by a relatively narrow lumen surrounded by a wall which, on the afferent side, is more arteriole-like and, on the distal or efferent side, is venule-like. The wall of this vessel consists of a well-developed endothelium lining, and a pronounced media which is externally bound by a loose network of collagenous elements.

II. Pulpal Lymphatics

Lymphatic vessels are difficult to identify with certainty in the pulp. For conventional light microscopy, the pulp lymphatics have been stained by sealing dyes in the pulp chamber (FISH 1927) or in cavities prepared in coronal dentine (ISOKAWA 1960). Various dyes such as Prussian blue, Berlin blue, trypan blue or patent blue have been used (SCHWEITZER 1909; DEWEY and NOYES 1917; FISH 1925; NOYES and LADD 1929; OEHMKE and HAGER 1983). Lead acetate (MACGREGOR 1936) and hydrogen peroxide (MAGNUS 1922) have also been used. By retrograde lymphography using a carbon suspension, RUBEN et al. (1971) demonstrated lymphatics in the pulp and periodontal tissues. BERNICK and PATTEK (1969) and BERNICK (1977) staining decalcified sections of human teeth with iron hematoxylin described lymphatic capillaries originating as blind sacs in the odontoblastic layer and in the subodontoblastic area. They drained into small thin-walled collecting vessels that were irregular in shape. Communications between these vessels were common. The large conducting vessels accompanied the blood vessels and the nerve fibers in their course through the pulp. The lymphatic vessels could be identified by their thin walls and their small sizes. The large-caliber lymphatic vessels contained, in addition, valves, structures not present in the veins of the same size. The conducting lymphatic vessels

Fig. 50. Lymphatic capillary in the human dental pulp lined by thin endothelial cells (*Ed*) at the right. At the left, note the presence of a cleft (*arrow*) adjacent to a thick endothelial cell. Absence of pericytes and basement membrane. Some carbon particles are present in the lumen (*L*). Upper canine of a 14-year-old girl. Occlusal cavity filled for 10 days with gutta percha. The tooth was extracted 2 h after intrapulpal injection of 0.2 ml sterile colloidal carbon. (From FRANK et al. 1977, courtesy of Cell Tissue Res)

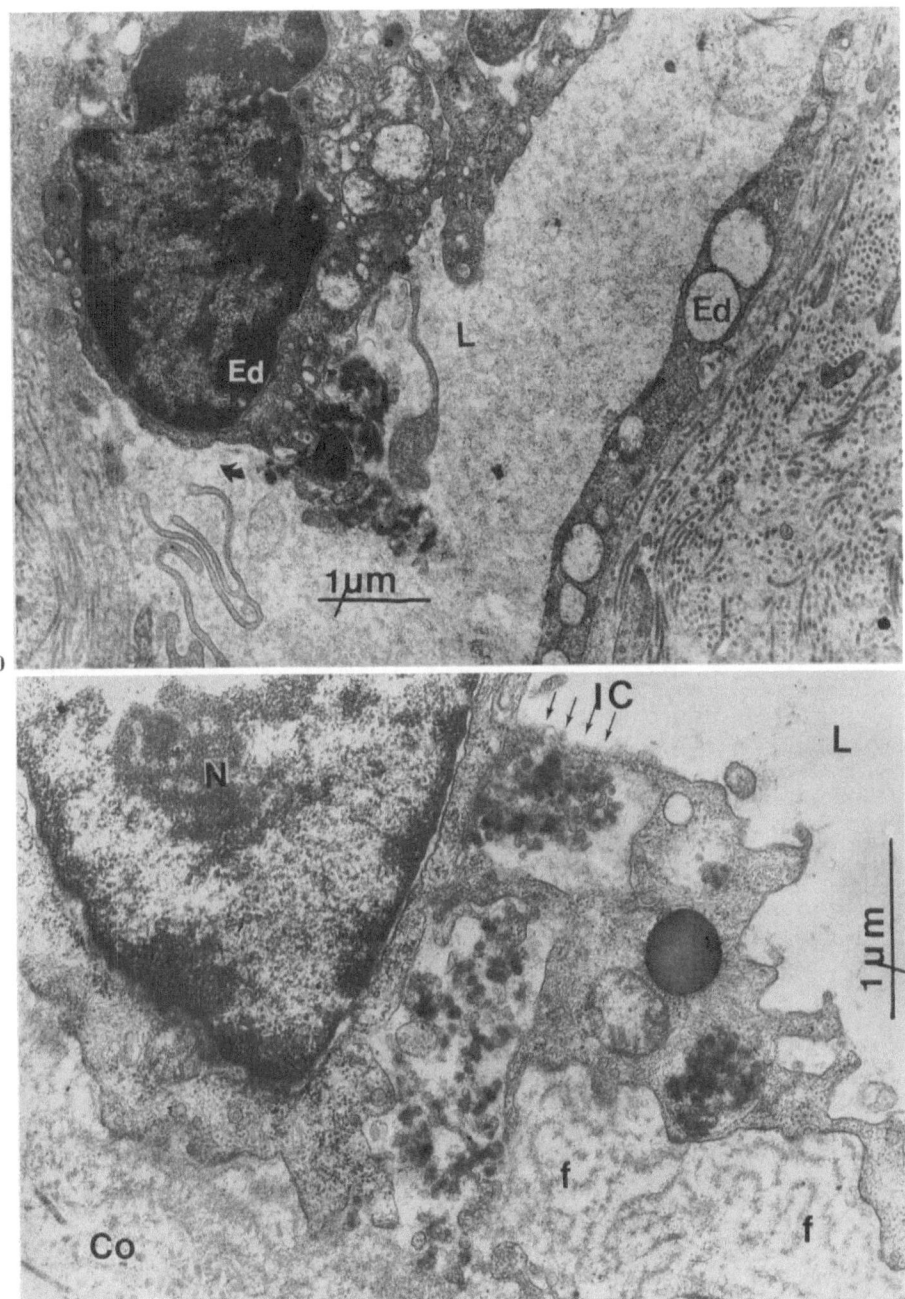

Fig. 51. Lymphatic capillary with an intercellular cleft (*IC*) between two endothelial cells. This intercellular gap (*IC*) is filled with carbon particles. A filamentous material (*f*) is visible under the endothelial cells. *N* Nucleolus of an endothelial nucleus; *Co* Collagen fibril; *L* Lumen. Upper canine of 14-year-old girl with an occlusal filling of gutta percha for 10 days. The tooth was extracted 2 h after intrapulpal injection of 0.2 ml sterile colloidal carbon. (From FRANK et al. 1977, courtesy of Cell Tissue Res)

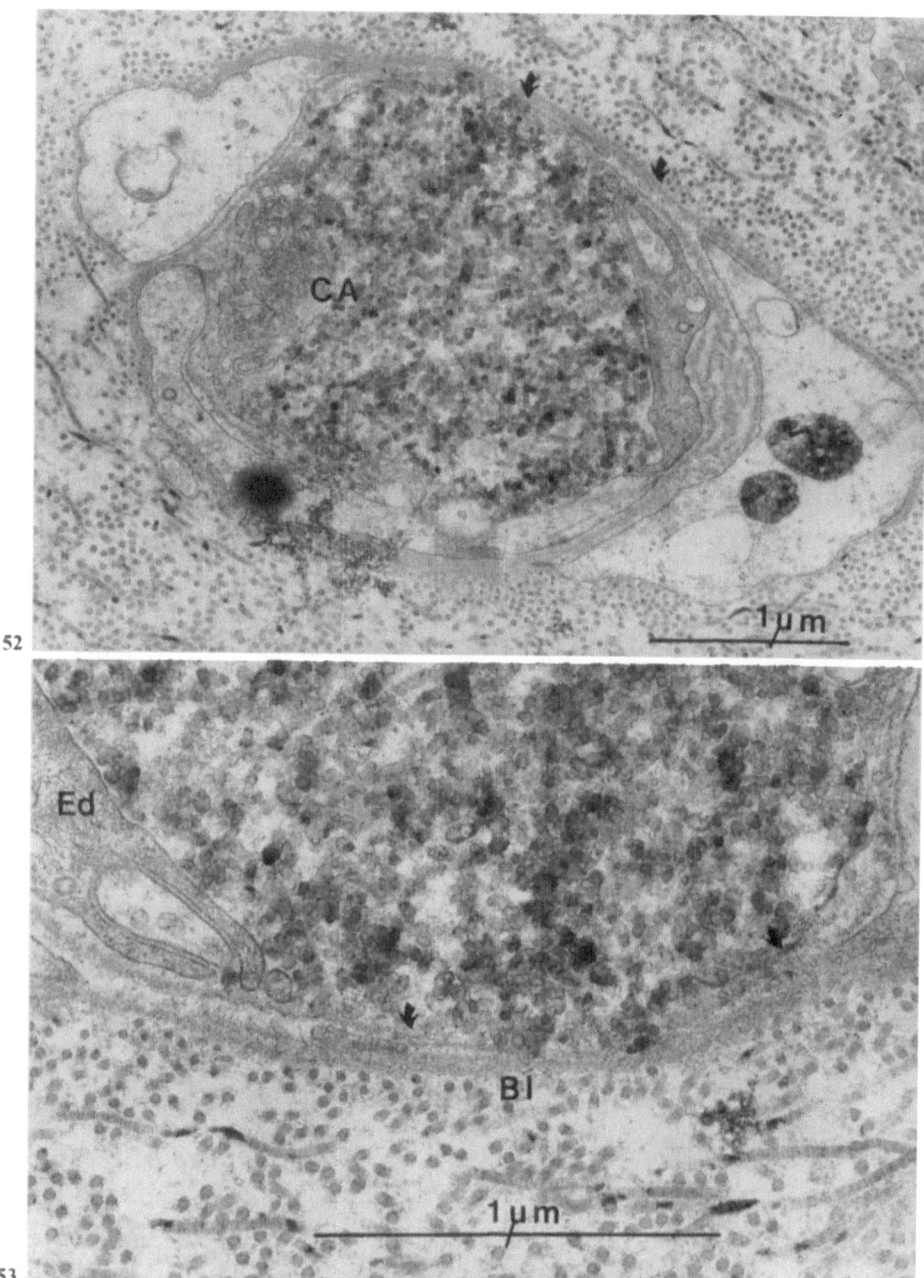

Fig. 52. Transverse section of a small lymphatic capillary with a discontinuous endothelial lining (*arrows*). The lumen is filled with carbon particles (*CA*). A basement membrane and some pericytes are present. Upper canine of a 14-year-old girl with a gutta percha filling for 10 days. The tooth was extracted 2 h after injection of 0.2 ml sterile colloidal carbon solution. (From Frank et al. 1977, courtesy of Cell Tissue Res)

passed through the roots as individual units without draining into a large single vessel.

The ultrastructural characteristics of lymphatic capillaries in general, located outside the pulp. are well documented (CASLEY-SMITH and FLOREY 1961; LEAK and BURKE 1966; WEISS 1977). They consist of a single layer of flattened endothelial cells and there is no basement membrane or only a very poorly developed one. Collagen fibrils and extracellular matrix surround the vessels, and fine anchoring filaments run from the collagen and attach to the outer surface of the endothelium. No pericytes are present.

In transmission electron microscopy, vessels presumed to be lymphatic capillaries have been described in the pulp by RIEDEL et al. (1966); KUKLETOVA (1970) and DAHL and MJÖR (1973 b) on the basis of their thin walls and large openings in the endothelial lining. OEHMKE and HAGER (1983) observed pulpal lymphatic capillaries to have a thin basement membrane with large interruptions, in addition to a thin endothelium.

In young permanent human teeth to be extracted for orthodontic or surgical reasons, FRANK et al. (1977) produced an experimentally localized pulpitis by cutting an occlusal cavity in dentine and placing a gutta percha filling there for 7 to 11 days (MJÖR 1972). Development of inflammation was planned for the induction of a lymphatic ingrowth (COLLIN 1971) as well as for dilatation of the vascular lumen (VIRAGH et al. 1971; WEISS 1973). An electron-microscopic study was undertaken after a local injection of sterile colloidal carbon into the superficial layer of the pulp. The teeth were extracted 1 to 3 h after injection.

FRANK et al. (1977) identified lymphatic capillaries either by their typical ultrastructure or by the presence of colloidal carbon in their lumen or both. A very thin endothelial covering associated with a discontinuous basement membrane and the absence of pericytes characterized such capillaries. They contained some lymphocytes or polymorphonuclear leukocytes. Another characteristic was the presence of clefts in the endothelium with intercellular space (Figs. 50, 51). Carbon particles were sometimes observed in the intercellular cleft (Fig. 51).

Other lymphatic capillaries were recognized by intraluminal condensation of carbon particles (Figs. 52, 53). Thin endothelial cells circumscribed the filled lumen and very large clefts were observed in the endothelial lining. These openings were more than 1 micron wide (Figs. 52, 53). In this case, however, a basement membrane was visible, as were some pericytes with intracellular vacuoles containing carbon. FRANK et al. (1977) concluded that lymphatic capillaries are indeed present in the pulp. Despite some ultrastructural variations, they can be identified by the following features: 1) thin endothelial lining with occasional large clefts, 2) absence of basement membrane, or incompletely developed basement membrane with the presence of filamentous material between the endothelial cells and collagen fibrils (Fig. 51), 3) absence of luminal red blood cells. Occasional pericytes can also be observed.

Fig. 53. Enlargement of the endothelial gap region (*arrows*) of Fig. 52. The gap is more than 1 μm large. *Ed* Endothelial cell; *Bl* Basal lamina. (From FRANK et al. 1977, courtesy of Cell Tissue Res)

G. Age Changes and Pulpal Calcifications

A number of changes occur in the pulpal connective tissue with increasing age. BHUSSRY (1968) performed a histological study of the modifications seen with aging in a total of 60 human teeth. The pulp showed a fairly constant cell population at different ages. However, the odontoblastic layer appeared reduced in size from columnar in younger teeth to cuboidal in older teeth. According to BHUSSRY (1968) there was a tendency towards increase in collagen fibers in the aging pulp. On the other hand, NIELSEN et al. (1983) and VAN AMERONGEN (1984) suggested that the concentration of collagen is independent of age, agreeing with an earlier report by STANLEY and RANNEY (1962).

BENNETT et al. (1965), using infusion with India ink, noted that the vascularity of the human pulp diminishes with age. The number of vascular structures decreases, with loss of the peripheral plexus. BERNICK (1967) reported an apparent decrease of coronal nerve branches as a result of diffuse calcifications occurring around nerve fibers in the apical portion of the root.

In fact, pulpal calcifications are among the most typical features associated with aging, although they occur in young teeth as well. According to BERNICK (1967), 90 per cent of the pulps of old human teeth exhibited pulpal calcifications. Various types of pulpal calcifications can be observed. The pulp chamber can be progressively reduced in size by deposition on the dentinal walls of regular and irregular secondary dentine.

True and false pulp stones or denticles (Figs. 55, 56) can be observed in various positions in the coronal or radicular pulp. Sometimes this type of nodular calcification is very extensive and the pulp cavity may be completely obliterated by a calcified mass (Fig. 54). True denticles consists of dentine exhibiting tubules, surrounded by a layer of typical odontoblasts. False denticles are composed of concentric layers of calcified tissue (Fig. 56) without dentinal tubules. An absence of odontoblasts is noted around false pulp stones, and remnants of necrotic cells have classically been described. Most pulp stones are false denticles and can be incorporated into the dentine as shown by microradiography (Fig. 57). AOBA et al. (1980) studied denticles by X-ray diffraction, and energy-dispersive- and chemical analysis, reporting that the mineral component of these pulpal calcifications are apatites and possibly carbonate-containing apatites with high concentration of iron.

Besides these nodular calcifications, diffuse pulpal calcifications have been described with aging. These start as granular deposits observed along the external walls of blood vessels or nerve fibers (Fig. 54, 58). BERNICK (1967) noted that this type of calcification first involved the connective tissue surrounding the nerve fibers and then the nerve fibers themselves. We studied these diffuse

Fig. 54. Prominent calcification (*Pc*) of the pulp chamber of a human noncarious lower third molar which in addition shows conspicuous secondary dentine reaction (*Sd*). Diffuse calcifications are observed in the remaining radicular pulp (*P*). × 23

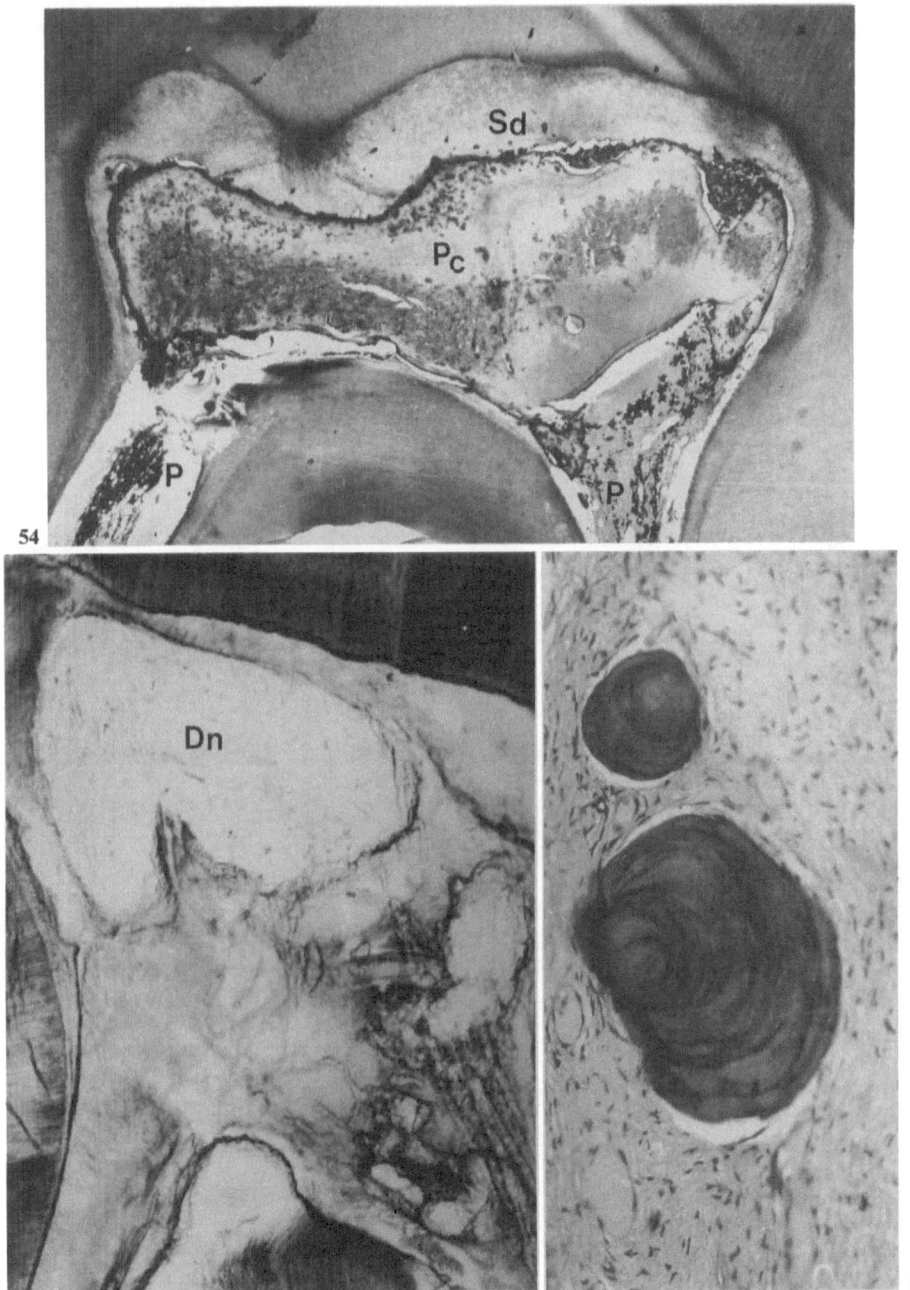

Fig. 55. Ground section of a molar of an Egyptian mummy from the Middle Empire (2160–1580 B.C.) Presence of large denticle (*Dn*) amidst fibrous tissue. (BOLENDER et al. 1964, courtesy of Actual Odontostomatol). × 30

Fig. 56. Presence of 2 false denticles in a human pulp chamber. Note their concentric lamellar structure with absence of peripheral odontoblasts. × 110

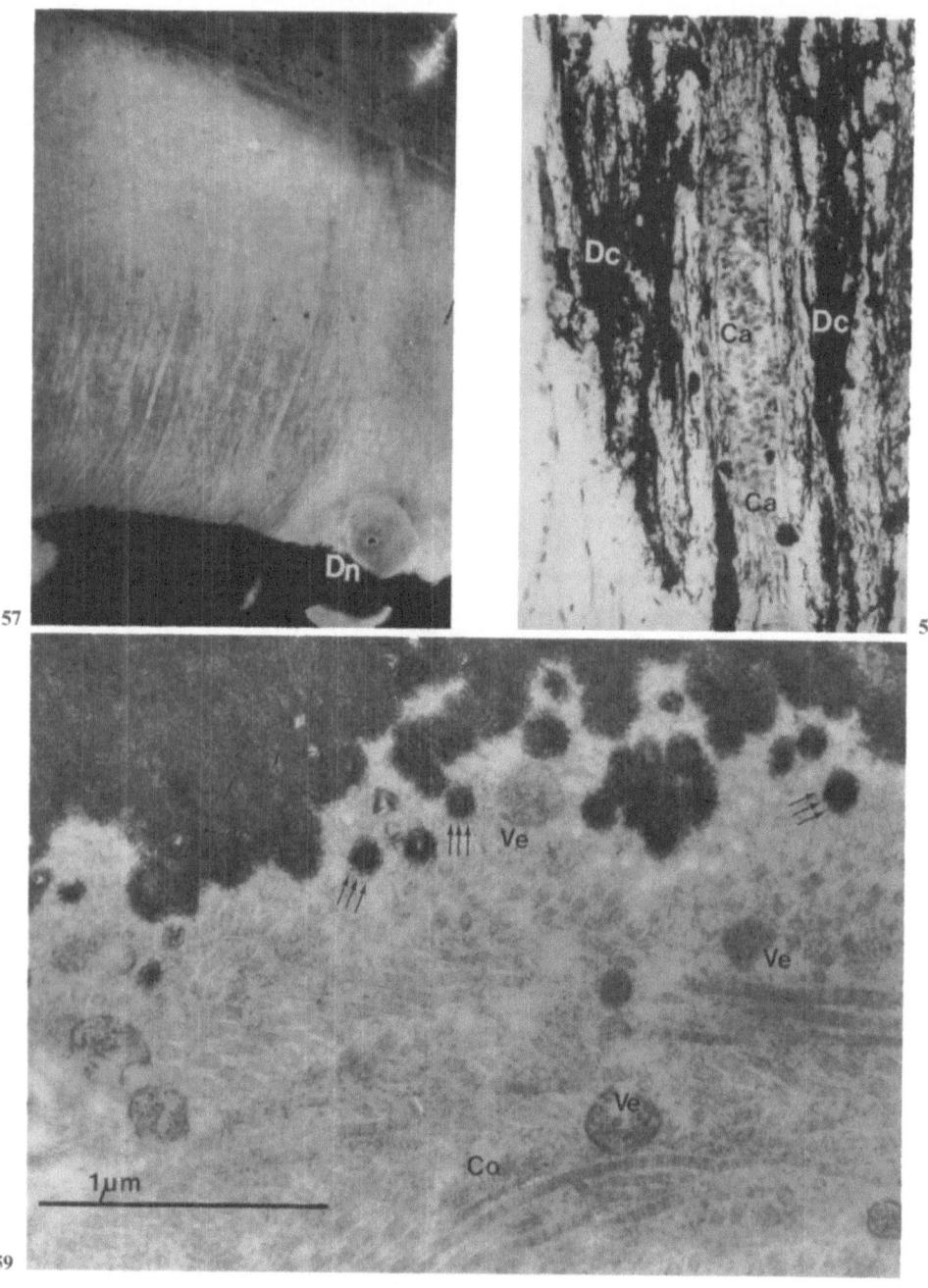

Fig. 57. Microradiography of a dentine wall in a human tooth showing tubules cut longitudinally and the inclusion of a denticle (*Dn*) in the dentinal wall. × 15

Fig. 58. Presence of diffuse calcifications (*Dc*) around a blood capillary (*Ca*) in the pulp of a noncarious human molar. × 550

Fig. 59. Transmission electron microscopy of the edge of diffuse calcification in a non-carious human pulp. Round vesicles (*Ve*) are observed among typical collagen fibrils (*Co*). The denticle seems to be formed by calcification of the vesicles (*arrows*) followed by their fusion

calcifications in human pulps with transmission electron microscopy and observed the presence of typical matrix vesicles which undergo calcification (Fig. 59). The diffuse calcifications increase by confluence of small calcified nodules composed of hydroxyapatite crystals. Matrix vesicle calcification was also observed by SELA et al. (1981) in dentine bridge formation induced in rat molar teeth by pulp exposure and capping with calcium hydroxide.

This brief review of age changes in the pulp concludes the account of the biology of the pulpal tissue, a loose connective tissue whose structure and function is highly integrated with that of the dentine which it forms, nourishes, innervates and repairs.

References

Amerongen JP van (1984) The extracellular matrix of dental pulp. A biochemical study. Doctoral Thesis, University Utrecht

Amerongen JP van, Lemmens IG, Tonino GJM (1983) The concentration, extractability and characterization of collagen in human dental pulp. Arch Oral Biol 28:339–345

Amerongen JP van, Lemmens IG, Tonino GJM (1984) Immunofluorescent localization and extractability of fibronectin in human dental pulp. Arch Oral Biol 29:93–99

Anderson DJ (1963) Chemical and osmotic excitants of pain in human dentin. In: Anderson DJ (ed) Sensory mechanisms in dentine. Pergamon, Oxford, p 88

Anderson DJ, Matthews B (1967) Osmotic stimulation of human dentin and the distribution of dental pain thresholds. Arch Oral Biol 12:417–426

Anderson DJ, Rönning GA (1962) Osmotic excitants of pain in human dentin. Arch Oral Biol 7:513–523

Anneroth G, Norberg KA (1968) Adrenergic vasoconstrictor innervation in the human pulp. Acta Odontal Scand 26:89–93

Antila R, Pohto P (1984) In vitro studies on the prostaglandin system in tooth pulp. Proc Finn Dent Soc 80:245–252

Aoba T, Ebisu S, Yagi T (1980) A study of the mineral phase of pulp calcification. J Oral Pathol 9:129–136

Arwill T (1967) Studies on the ultrastructure of dental tissues. II. The predentine-pulpal border zone. Odont Revy 18:191–208

Arwill T (1968) The ultrastructure of the pulpo-dentinal border zone. In: Symons NBB (ed) Dentine and pulp: their structure and reactions. Livingstone, London, pp 147–167

Arwill T, Edwall L, Lilja J, Olgart L, Svenson SE (1973) Ultrastructure of nerves in dentinal-pulp border zone after sensory and autonomic transection in the cat. Acta Odontol Scand 31:273–278

Ashhurst DE, Costin NM (1976) The secretion of collagen by insects: uptake of ^{3}H-proline by collagen-synthesizing cells in Locusta migratoria and Galleria mellonella. J Cell Sci 20:377–403

Avery JK (1963) A possible mechanism of pain conduction in teeth. Ann Histochem 8:59–64

Avery JK (1971) Structural elements of the young normal human pulp. Oral Surg 32:113–125

Avery JK (1976) Dentin. In: Bhaskar SN (ed) Orban's oral histology and embryology, 8th edn. Mosby, St Louis, pp 105–140

Avery JK, Rapp R (1959) An investigation of the mechanism of neural impulse transmission in human teeth. Oral Surg 12:190–198

Avery JK, Strachnan DS, Corpron RE, Cox CF (1971) Morphological studies of the altered pulp of the New Zealand white rabbit after resection of the inferior alveolar nerve and/or the superior cervical ganglion. Anat Rec 171:495–508

Avery JK, Cox CF, Chiego DJ (1980) Presence and location of adrenergic nerve endings in the dental pulps of mouse molars. Anat Rec 198:59–71

Avery JK, Cox CF, Chiego DJ (1984) Structural and physiological aspects of dentin innervation. In: Linde A (ed) Dentin and dentinogenesis, vol I. CRC Press, Boca Raton, pp 19–46

Baume LJ (1980) The biology of pulp and dentine: a historic terminologic, taxonomic, histological, biochemical and clinical survey. Monograph in Oral Sciences. Karger, Basel

Beasley WL, Holland GR (1978) A quantitative analysis of the innervation of the pulp of the cat's canine tooth. J Comp Neurol 178:487–494

Beersten W, Everts V, Hoff van den A (1974) Fine structure of fibrobasts in the periodontal ligament of the rat incisor and their role in tooth eruption. Arch Oral Biol 19:1087–1098

Bell C (1829) Anatomy and physiology of the human body, 7th edn, vol I. Longman, Rees, Orme, Brow and Green, London, p 236

Bennett CG, Kelln EE, Biddington WR (1965) Age changes of the vascular pattern of the human dental pulp. Arch Oral Biol 10:995–998

Benninghoff A (1923) Über die Formenreihe der glatten Muskulatur und die Bedeutung der Rouget-schen Zellen an den Capillaren. Z Zellforsch Mikrosk Anat 4:125–170

Bensadoun R (1976) "Contribution à l'étude de l'innervation de la dentine par la technique autoradio-graphique chez le chat et le rat"; Thèse Doct 3ème cycle Sciences Odont, Université Paris V

Bernick S (1948) Innervation of the human tooth. Anat Rec 101:81–107

Bernick S (1967) Effect of aging on the nerve supply to human teeth. J Dent Res 46:694–699

Bernick S (1968) Innervation of the teeth. In: Finn SB (ed) Biology of the dental pulp organ: a symposium. University of Alabama Press, Birmingham, pp 284–302

Bernick S (1977) Lymphatic vessels of the human dental pulp. J Dent Res 56:70–77

Bernick S, Pattek PR (1969) Lymphatic vessels of the dental pulp in dogs. J Dent Res 48:959–964

Beveridge EE, Brown AC (1965) The measurement of human dental intrapulpal pressure and its response to clinical variables. Oral Surg 19:655–668

Bhussry BR (1968) Modification of the dental pulp organ during development and aging. In: Finn SB (ed) Biology of the dental pulp organ: a symposium. University of Alabama Press, Birmingham, pp 144–165

Bishop MA (1982) A fine structural investigation on the extent of perineural investment of the nerve supply to the pulp in rat molar teeth. Arch Oral Biol 27:225–234

Bishop MA (1985) Vascular permeability to lanthanum in the rat incisor pulp. Comparison with endoneurial vessels in the inferior alveolar nerve. Cell Tissue Res 239:131–136

Boström H, Odeblad E (1953) The influence of cortisone upon the sulphate exchange of chondroitin sulfuric acid. Arkiv Demi 6:39–42

Bradamante Z, Pecina-Hrncevic A, Ciglar I (1980) Oxytalan fibres in human dental pulp. Experientia 36:1210–1211

Brännström M (1960a) Dentinal and pulpal response. I. Application of reduced pressure to exposed dentin. Acta Odontol Scand 18:1–15

Brännström M (1960b) Dentinal and pulpal response. II. Application of an air stream to exposed dentin. Short observation period. Acta Odontol Scand 18:17–28

Brännström M (1962) The elicitation of pain in the human dentin and pulp by chemical stimulation. Arch Oral Biol 7:59–62

Brännström M (1963) Dentine sensitivity and aspiration of odontoblasts. J Am Dent Assoc 66:366–370

Brännström M (1966) Sensitivity of dentine. Oral Surg 21:517–526

Brännström M (1981) Dentin and pulp in restorative dentistry, 1st edn. Dental Ther, Nacka

Brännström M, Aström A (1964) A study on the mechanism of pain elicited from the dentine. J Dent Res 43:619–625

Brännström M, Aström A (1972) The hydrodynamics of the dentine: its possible relationship to dentinal pain. Int Dent J 22:219–227

Brännström M, Linden KLA, Aström A (1967) The hydrodynamics of the dental tubule and of pulp fluid. A discussion of its significance in relation to dentinal sensitivity. Caries Res 1:310–317

Brännström M, Johnson G, Linden LA (1969) Fluid flow and pain response in the dentine produced by hydrostatic pressure. Odont Revy 20:15–30

Byers MR (1980) Development of sensory innervation in dentine. J Comp Neurol 191:413–427

Byers MR (1984a) Dental sensory receptors. In: Smythies JR, Bradley RJ (eds) Int Rev Neurobiol, vol 25. Academic, Orlando, pp 39–94

Byers MR (1984b) "Autoradiographic mapping of receptive fields of individual sensory axons in dentine and pulp". Abstract of the satellite symposium to the fourth world congress on pain: recent developments in oro-facial pain. Basic and Clinical Research. Port Ludlow, Washington, September

Byers MR, Dong W (1983) Autoradiographic location of sensory nerve endings in dentin of monkey teeth. Anat Rec 205:441–454

Byers MR, Holland GR (1977) Trigeminal endings in gingiva, junctional epithelium and periodontal ligament of rat molars as demonstrated by autoradiography. Anat Rec 188:509–524

Byers MR, Kish SJ (1976) Delineation of somatic nerve endings in rat teeth by radioautography of axon-transported protein. J Dent Res 55:419–425

Byers MR, Matthews B (1981) Autoradiographic demonstration of ipsilateral and controlateral sensory nerve endings in cat dentin, pulp and periodontium. Anat Rec 201:249–260

Byers MR, Neuhans SJ, Gehrig JD (1982) Dental sensory receptor structure in human teeth. Pain 13:231–235

Cahen PM, Frank RM (1970) Microscopie électronique de la pulpe dentaire humaine normale. Bull Group Int Rech Sci Stomatol Odontol 13:421–443

Calle A, Magloire H, Joffre A (1985) Intercellular junctions in human tooth-pulp cells in culture in vitro revealed by freeze fracture, lanthanum impregnation and filipin treatment. Arch Oral Biol 30:283–289

Casley-Smith J, Florey JW (1961) The structure of normal small lymphatics. QJ Exp Physiol 46:101–106

Cho MI, Garant PR (1981a) Sequential events in the formation of collagen secretion granules with special reference to the development of segment-long-spacing-like aggregates. Anat Rec 199:309–320

Cho MI, Garnat PR (1981b) Role of microtubules in the organization of the Golgi complex and the secretion of collagen secretory granules by periodontal ligament fibroblasts. Anat Rec 199:459–471

Cho MI, Garant PR (1981c) An electron microscopy radioautographic study of collagen secretion in periodontal ligament fibroblasts of the mouse: I. Normal fibroblasts. Anat Rec 201:577–586

Connor NS, Aubin JE, Melcher AH (1984) The distribution of fibronectin in rat tooth and periodontal tissues: an immunofluorescence study using a monoclonal antibody. J Histochem Cytochem 32:565–572

Corpron RE, Avery JK (1973) The ultrastructure of intradentinal nerves in developing mouse molars. Anat Rec 175:585–606

Corpron RE, Avery JK, Cox CF (1972) Ultrastructure of intradentinal nerves after resection of the inferior alveolar nerve in mice. J Dent Res 51:673

Corpron RE, Avery JK, Lee SD (1973) Ultrastructure of capillaries in the odontoblastic layer. J Dent Res 52:393

Cournil I, Leblond CP, Pomponio J, Hand AR, Sederlof L, Martin GR (1979) Immunohistochemical localization of procollagens. I. Light microscopic distribution of procollagen I, III and IV. Antigenicity in the rat incisor tooth by the indirect peroxidase-antiperoxidase method. J Histochem Cytochem 27:1059–1069

Dahl E, Mjör IA (1973a) The structure and distribution of nerves in the pulp-dentin organ. Acta Odontol Scand 31:349–354

Dahl E, Mjör IA (1973b) The fine structure of the vessels in the human dental pulp. Acta Odontol Scand 31:223–230

Dewey K, Noyes FB (1917) A study of the lymphatic vessels of the dental pulp. Dent Cosmos 59:436–444

Droz B (1967) Synthèse et transfert des protéines cellulaires dans les neurones ganglionnaires. Etude radioautographique quantitative en microscopie électronique. J Microsc 6:201–228

Droz B, Leblond CP (1963) Axonal migration of proteins in the central nervous system and peripheral nerves as shown by radioautography. J Comp Neurol 121:325–346

Ekblom A, Hansson P (1984) A thin and freeze fracture study of the pulp blood vessels in feline and human teeth. Arch Oral Biol 29:413–424

Ellingson JS, Smith M (1975) Phospholipid composition of rat, rabbit and bovine dental pulp. Arch Oral Biol 20:731–734

Erlanger J, Gasser HS (1938) Electrical signs and nervous activity, 1st edn. Pennsylvania Press, Philadelphia

Fearnhead RW (1957) Histological evidence for the innervation of human dentine. J Anat 91:267–277

Fearnhead RW (1961) The neurohistology of human dentine. Proc Roy Soc Lond [Biol] 54:884–887

Fearnhead RW (1963) The histological demonstration of nerve fibres in human dentine. In: Anderson DJ (ed) Sensory mechanisms in dentine. Pergamon, Oxford, p 15

Fearnhead RW (1967) Innervation of dental tissues. In: Miles AEW (ed) Structural and chemical organization of teeth, vol I. Academic, New York, pp 247–281

Fink R, Kish SJ, Byers MR (1975) Rapid axonal transport in trigeminal nerve of rat. Brain Res 90:85–95

Fish EW (1925) Circulation of lymph in the dentine. Proc Roy Soc Med 18:35–37

Fish EW (1927) The lymph supply of the dentine and enamel. Proc Roy Soc Med 20:225–236

Fox AG, Heeley JD (1981) Histological study of pulps of human primary teeth. Arch Oral Biol 25:103–110

Frank RM (1966a) Etude au microscope électronique de l'odontoblaste et du canalicule dentinaire humain. Arch Oral Biol 11:179–199

Frank RM (1966b) Ultrastructure of human dentine. In: Fleisch H, Blackwood HJJ, Owen M (eds) Third Eur Symp Calcified Tissues. Springer, Berlin, pp 259–271

Frank RM (1968a) Ultrastructural relationship between the odontoblast, its process and the nerve fibre. In: Symons NBB (ed) Dentine and pulp: their structure and reactions. Livingstone, London, pp 115–145

Frank RM (1968b) Attachment sites between the odontoblast process and the intradentinal nerve fibre. Arch Oral Biol 13:833–834

Frank RM (1970) Etude autoradiographique de la dentinogenèse en microscopie électronique à l'aide de la proline tritiée chez le chat. Arch Oral Biol 15:583–596

Frank RM, Sauvage C, Frank P (1972) Morphological basis of dental sensitivity. Int Dent J 22:1–19

Frank RM, Cimasoni J, Tsamouranis A, Matter J, Fiore-Donno G (1977) Collagen resorption by fibroblasts in human gingiva. J Biol Buccale 5:343–351

Frank RM, Wiedemann P, Fellinger E (1977) Ultrastructure of lymphatic capillaries in the human dental pulp. Cell Tissue Res 178:229–238

Fried K, Hildebrand C (1981) Pulpal axons in developing mature and aging feline permanent incisors. A study by electron microscopy. J Comp Neurol 203:23–26

Fullmer HN, Sheetz JH, Narkates AJ (1974) Oxytalan connective tissue fibers: a review. J Oral Pathol 3:291–316

Furseth R, Mjör IA, Skogedal O (1980) The fine structure of induced pulpitis in a monkey (*Cercaptithecus aethiops*). Arch Oral Biol 24:883–888

Garant PR (1976) Collagen resorption by fibroblasts. A theory of fibroblastic maintenance of periodontal ligament. J Periodontol 47:380–390

Garant PR, Cho MI, Cullen MR (1982) Attachment of periodontal ligament fibroblasts to the extracellular matrix in the squirrel monkey. J Periodont Res 17:70–79

Gazelius B, Brodin E, Olgart L (1981) Depletion of substance P-like immunoreactivity in the cat pulp by antidromic nerve stimulation. Acta Physiol Scand 111:319–327

Gotjamanos T (1969a) Cellular organization in the sub-odontoblastic zone of the dental pulp. I. A study of cell free and cell rich layers in pulps of adult rat and deciduous monkey teeth. Arch Oral Biol 14:1007–1010

Gotjamanos T (1969b) Cellular organization in the sub-odontoblastic zone of the dental pulp. II. Period and mode of development of the cell rich layer in rat molar pulp. Arch Oral Biol 14:1011–1019

Graf W, Björlin G (1951) Diameters of nerve fibers in human tooth pulp. J Am Dent Assoc 43:186–193

Grönblad M, Liesi P, Muck AM (1984) Peptidergic nerves in human tooth pulp. Scand J Dent Res 92:319–324

Gunji T (1982) Morphological research on the sensitivity of dentin. Arch Histol Jpn 45:45–67

Gvozdenociv-Sedlecki S, Olvist V, Hansen HP (1973) Histologic variations in the pulp of intact premolars from young individuals. Scand J Dent Res 81:433–440

Gysi A (1900) An attempt to explain the sensitiveness of dentin. Br J Dent Sci 43:865–868

Haim G (1965) Elektronenmikroskopische Untersuchungen der Zahnpulpa. Dtsche Zahnärztl Zeitschr 20:583–588

Hals E, Tonder KJ (1981) Elastic (pseudoelastic) tissue in arterioles of the human and dog dental pulp. Scand J Dent Res 89:218–227

Han SS (1968) The fine structure of cells and intercellular substances of the dental pulp. In: Finn SB (ed) Biology of the dental pulp organ: a symposium. University of Alabama Press, Birmingham, pp 103–140

Han SS, Avery JK, Bang JS (1967) The effect of actinomycin D on the fibroblast of the pulp of the rat incisor. Arch Oral Biol 12:503–512

Harris R, Griffin CJ (1968) Fine structure of nerve endings in the human dental pulp. Arch Oral Biol 13:773–778

Harris R, Griffin CJ (1969) The fine structure of the mature odontoblasts and cell rich zone of the human dental pulp. Aust Dent J 14:168–177

Harris R, Griffin CJ (1971) The ultrastructure of small blood vessels of the normal human pulp. Aus Dent J 16:220–226

Hassel JH von (1971) Physiology of the human dental pulp. Oral Surg 32:126–134

Hattyasy D (1961) Continuous regeneration of the dentinal nerve endings. Nature 189:72–74

Hay ED (1983) Cell and extracellular matrix: their organization and mutual dependence. Modern Cell Biology, vol 2. Liss, New York, pp 509–548

Hayakawa T, Iijima K, Hashimoto Y, Mioke Y, Takei T, Matsui T (1981) Developmental changes in the collagens and some collagenolytic activities in bovine dental pulp. Arch Oral Biol 26:1057–1062

Hirafuji M, Satoh S, Ogura Y (1980) Prostaglandins in rat pulp tissue. Dent Res 59:1535–1548

Holland Gr (1980) Non-myelinated nerve fibres and their terminals in the sub-odontoblastic plexus of the feline dental pulp. J Anat 130:457–467

Holland GR (1985) The odontoblast process: form and function. J Dent Res 64:499–514 (special issue)

Holland GR, Robinson PP (1983) The number and size of axons at the apex of the cat's canine tooth. Anat Rec 20:215–222

Hopewell-Smith A (1916) The so-called innervation of the dentin: an epicriticism. Dental Cosmos 58:421–427

Hynes RO, Destree AT (1978) Relationship between fibronectin (Lets protein) and actin. Cell 15:866–875

Iguchi Y, Yamamura T, Ichikawa T, Hashimoto S, Horiuchi T, Shimono M (1984) Intercellular junctions in odontoblasts of the rat incisor studied with freeze fracture. Arch Oral Biol 29:487–497

Isokawa S (1960) Über das Lymphsystem des Zahnes. Z Zellforsch 52:140–149

Johnsen D, John S (1978) Quantitations of nerve fibres in the primary and permanent canine and incisor teeth in man. Arch Oral Biol 23:825–829

Johnsen DC, Karlsson UL (1974) Electron microscopic quantitations of feline primary and permanent innervation. Arch Oral Biol 19:671–678

Karim A, Cournil I, Leblond CP (1979) Immunohistochemical localization of procollagens. II. Electron microscopic distribution of procollagen I antigenicity in the odontoblasts and predentin of rat incisor teeth by a direct method using peroxidase linked antibodies. J Histochem Cytochem 27:1070–1083

Kennedy JS, Kennedy GDC (1957) Sulphated mucopolysaccharides in rodent teeth. J Anat 91:398–408

Köling A (1983) Membrane structures in the human pulpdentin region. An electron microscopic investigation of permanent teeth using the freeze fracture technique; Doctoral Thesis, Centraltryckeriet, Uppsala Universitet

Köling A (1985) Membrane architecture of myelinated nerve fibres in the human dental pulp studied by freeze fracturing. Arch Oral Biol 30:121–128

Kölliker A (1852) Handbuch der Gewebelehre des Menschen, 1st edn. Engelmann, Leipzig

Kramer IRH (1960) The vasculature of the human dental pulp. Arch Oral Biol 12:177–189

Kramer IRH (1968) The distribution of blood vessels in the human dental pulp. In: Finn SB (ed) Biology of the dental pulp organ: a symposium. University of Alabama Press, Birmingham, pp 368–377

Kroeger DC, Gonzales F, Krivoy W (1961) Transmembrane potentials of cultured mouse dental pulpal cells. Proc Soc Exp Biol Med 108:134–136

Kukletova M (1970) An electron microscopic study of the lymphatic vessels in the dental pulp in the calf. Arch Oral Biol 15:1117–1124

Kukletova M, Zahradka J, Lukas Z (1968) Monoaminergic and cholinergic nerve fibers in the human dental pulp. Histochemie 16:154–158

LaFlèche R, Frank RM, Steuer P (1985) The extent of the human odontoblast process as determined by transmission electron microscopy: the hypothesis of a retractable suspensor system. J Biol Buccale 13:293–305

Langeland K, Yagi T (1972) Investigations on the innervation of teeth. Int Dent J 22:240–269

Lasek R, Joseph BS, Whitlock DG (1968) Evaluation of a radioautographic neuroanatomical tracing method. Brain Res 8:319–336

Leak LV, Burke JF (1966) Fine structure of the lymphatic capillary and the adjoining connective tissue area. Am J Anat 118:785–810

Lechner JH, Kalitsky G (1981) The presence of large amounts of type III collagen in bovine dental pulp and its significance with regard to the mechanism of dentinogenesis. Arch Oral Biol 26:265–273

Liljn J, Fayerberg-Mohlin B (1984) Dentinal innervation of impacted human third molars. Scand J Dent Res 92:485–488

Linde A (1973a) A study of the dental pulp glycosaminoglycans from permanent human teeth and rat and rabbit incisors. Arch Oral Biol 18:49–59

Linde A (1973b) Glycosaminoglycan turnover and synthesis in the rat incisor pulp. Scand J Dent Res 81:145–154

Linde A (1985) The extracellular matrix of the dental pulp and dentin. J Dent Res 64:523–529 (special issue)

Linde A, Johansson S, Jonsson R, Jontell M (1982) Localization of fibronectin during dentinogenesis in rat incisor. Arch Oral Biol 27:1069–1073

Listgarten MA (1973) Intracellular collagen fibrils in the periodontal ligament of the mouse, rat, hamster, guinea pig and rabbit. J Periodont Res 8:335–342

MacGregor A (1936) An experimental investigation of the lymphatic system of the teeth and jaws. Proc Roy Soc Med 29:1237–1272

Magloire H (1983) Elaboration de la trame organique prédentinaire: ultrastructure, cytochimie, immunochimie. Thèse Dr Scienc Odont Lyon

Magloire H, Dumont J (1976) Étude ultrastructurale de cellules pulpaires humaines cultivées "in vitro". J Biol Buccale 4:3–20

Magloire H, Vinard H, Joffre A (1979) Electrophysiological properties of human dental pulp cells. J Biol Buccale 7:251–262

Magloire H, Joffre A, Grimaud JA, Herbage D, Couble ML, Chavrier C, Dumont J (1981) Synthesis of type I collagen by human odontoblast-like cells in explant culture: light and electron microscope immunotyping. Cell Mol Biol 27:429–435

Magloire H, Joffre A, Grimaud JA, Herbage D, Couble MC, Chavrier C (1982) Distribution of type III collagen in the pulp parenchyma of human developing teeth. Light and electron microscope immunotyping. Histochemistry 74:319–328

Magnus G (1922) Über den Nachweis der Lymphgefässe in der Zahnpulpa. Dtsche Mschr Zahnheilk 40:661–670

Manzoli FA, Gelli M (1968) Quantitative determination of dental lipid in the dental pulp of Bos taurus during development. Arch Oral Biol 13:705–708

Marchi F, Leblond CP (1983) Collagen biogenesis and assembly into fibrils as shown by ultrastructural and ^{3}H-proline radioautographic studies on the fibroblasts of the rat foot pad. Am J Anat 168:167–197

Marchi F, Leblond CP (1984) Radioautographic characterization of successive compartments along the rough endoplasmic reticulum Golgi pathway of collagen precursors in foot pad fibroblasts of (^{3}H)proline-injected rats. J Cell Biol 98:1705–1709

Martens P (1968) Human dentinogenesis with special regard to the formation of peritubular crown dentine and zones in fetal deciduous and unabraded permanent teeth. Scand J Dent Res [suppl] 76:5–169

Martinez-Hernandez A, Nakane PK, Pierce GB (1974) Intracellular localization of basement membrane antigen in parietal yolk sac cells. Am J Pathol 76:549–555

Mjör IA (1972) Experimental pulpitis. Norske Tannl Tidsk 82:268–270

Mohamed SS, Atkinson ME (1982) The ontogeny of substance P containing nerve fibres in the mouse dentition. Anat Embryol (Berl) 164:153–159

Montandon D, Gabbiani G, Ryan GB, Majno G (1973) The contractile fibroblast. Plast Reconstr Surg 52:286–290

Muhle W von, Doronin PP (1976) Dünnschichtchromatographische Analyse der Phospholipide aus Pulpagewebe von Schneidezähnen der Ratte. Zahn Mund Kieferheilkd 64:561–568

Mumford JM, Bowsher D (1976) Pain and protopathic sensibility. A review with particular reference to teeth. Pain 2:223–243

Nachmansohn D (1948) Role of acetylcholinesterase in conduction. Bull John Hopkins Hosp 83:463–496

Närhi MVO (1985) Dentin sensitivity: a review. J Biol Buccale 13:75–96

Närhi MVO, Haegerstam G (1983) Intradental nerve activity induced by reduced pressure applied to exposed dentine in the cat. Acta Physiol Scand 119:381–386

Närhi MVO, Hirvonen TJ, Hakumäki MOK (1982a) Activation of intradental nerves in the dog to some stimuli applied to the dentine. Arch Oral Biol 27:1053–1058

Närhi MVO, Virtanen A, Huopaniemi T, Hirvonen TJ (1982b) Conduction velocities of single pulp nerve fibrils units in the cat. Acta Physiol Scand 116:209–213

Nielsen CJ, Bentley JP, Marshall FJ (1983) Age related changes in reducible cross-links of human dental pulp collagen. Arch Oral Biol 28:759–764

Noyes FB, Ladd RL (1929) The lymphatics of the dental region. Dent Cosmos 71:1041–1047

Obst T (1971) Über das Endgebiet des Perineurium an den Zahnnerven der Ratte. Z Zellforsch 114:515–531

Oehmke HJ, Hager C (1983) Zum Problem des Lymphabflusses in der Zahnpulpa. Dtsche Zahnärztl Z 38:959–962

Olgart L (1979) Local mechanisms in dental pain. In: Beer RF Jr, Bassette EG (eds) Mechanisms of pain and analgesia compounds. Raven, New York, pp 285–294

Olgart L, Gazelium B, Brodin E, Nilson G (1977a) Release of substance P-like immunoreactivity from the dental pulp. Acta Physiol Scand 101:510–512

Olgart L, Hökfelt T, Nilsson G, Pernow B (1977b) Localization of substances P-like immunoreactivity in the tooth pulp. Pain 4:153–159

Orlowski WA (1974) Analysis of collagen. Glycoproteins and acid mucopolysaccharides in the bovine and porcine dental pulp. Arch Oral Biol 19:255–258

Pimenidis MZ, Hinds JW (1977) An autoradiographic study of the sensory innervation of teeth. I. Dentin. J Dent Res 56:827–834

Pischinger A, Stockinger L (1968) Die Nerven der menschlichen Zahnpulpa. Z Zellforsch 89:44–61

Pohto P, Antila R (1968a) Demonstration of adrenergic nerve fibers in human dental pulp by histochemical fluorescence method. Acta Odontol Scand 26:137–144

Pohto P, Antila R (1968b) Acetylcholinesterase and noradrenalin in the nerves of dental pulp. Acta Odontol Scand 26:641–659

Pohto P, Antila R (1972) Innervation of blood vessels in the dental pulp. Int Dent J 22:228–239

Provenza DV (1958) The blood vascular supply of the dental pulp with emphasis on capillary circulation. Circ Res 6:213–218

Provenza DV (1968) Comparative morphology of the pulp vascular system. In: Finn SB (ed) Biology of the dental pulp organ: a symposium. University of Alabama Press, Birmingham, pp 353–363

Provenza DV, Fischlschweiger W, Sisca RF (1967) Fibres in human dental papillae. A preliminary report on the fine structure. Arch Oral Biol 12:1533–1539

Rabinowitz JL, Rossman S (1979) Lipid composition of human dental pulp. Arch Oral Biol 24:477–478

Rapp R, Avery JK, Rector RA (1957) A study of the distribution of nerves in human teeth. J Can Dent Assoc 23:447–453

Rapp R, Avery JK, Strachman DS (1968) Possible role of acetylcholinesterase in neural conduction within the dental pulp. In: Finn SB (ed) Biology of the dental pulp organ. A symposium. University of Alabama Press, Birmingham, pp 309–325

Rapp R, El-Labban NG, Kramer IRH, Wood D (1977) Ultrastructure of fenestrated capillaries in human dental pulp. Arch Oral Biol 22:317–319

Raschkow I (1835) Meletamata circa mammalium dentium evolutionem. Fridloender, Vratislaviae

Reader A, Foreman DW (1981) An ultrastructural quantitative investigation of human intradental innervation. J Endod 7:493–499

Retzius G (1894) Zur Kenntnisse der Endungsweise der Nerven in den Zähnen. Biol Untersuch 6:64–69

Riedel H, Fromme HG, Tallen B (1966) Elektronenmikroskopische Untersuchungen zur Frage der Kapillarmorphologie in der menschlichen Zahnpulpa. Arch Oral Biol 11:1049–1055

Roane JB, Foreman DW, Melfi RC, Marshall FJ (1973) An ultrastructural study of dentinal innervation in the adult human tooth. Oral Surg 35:94–104

Ross R, Benditt EP (1965) Wound healing and collagen formation. V Quantitative electron microscope radioautography observations of proline-H^3 utilization by fibroblasts. J Cell Biol 27:83–106

Rothman JE (1981) The Golgi apparatus: two organelles in tandem. Science 213:1212–1219

Ruben MP, Prieto-Hernandez JR, Gott FK, Kramer GM, Bloom AA (1971) Visualization of lymphatic microcirculation of oral tissue. J Periodontol 42:774–784

Ruoslahti E (1981) Fibronectin. J Oral Pathol 10:3–13

Salpeter MM (1968) ^{3}H-proline incorporation into cartilage. Electron microscope autoradiographic observations. J Morphol 124:387–391

Saunders RL de CH (1952) X-ray microscopy of human dental pulp vessels. Nature 180:1353–1354

Saunders RL de CH (1967) Microangiographic studies of periodontic and dental pulp vessels in monkey and man. J Canada Dent Assoc 33:245–252

Saunders RL de CH, Röckert HOE (1967) Vascular supply of dental tissues including lymphatics. In: Miles AEW (ed) Structural and chemical organization of teeth, vol I. Academic, New York, pp 199–245

Schweitzer G (1909) Über die Lymphgefäße des Zahnfleisches und der Zähne beim Menschen und Säugetieren. Arch Mikr Anat 75:927–999

Sela J, Tamari I, Hirchefeld Z, Bab I (1981) Transmission electron microscopy of reparative dentin in rat molar pulps. Primary mineralization via extracellular matrix vesicles. Acta Anat (Basel) 109:247–251

Shapiro IM, Wuthier RE (1966) A study of the phospholipids of bovine dental tissues. II. Developing bovine foetal dental pulp. Arch Oral Biol 11:513–519

Shuttleworth CA, Ward JL, Hirschmann PN (1978a) Extraction of collagen fractions from bovine and rabbit dental follicle, papilla and pulp. Arch Oral Biol 23:235–236

Shuttleworth CA, Ward JL, Hirschmann PN (1978b) The presence of type III collagen in the developing tooth. Biochim Biophys Acta 535:348–355

Singer II (1979) The fibronexus: a transmembrane association of fibronectin-containing fibers and bundles of 5 nm microfilaments in hamster and human fibroblasts. Cell 16:675–685

Stanley HR, Ranney AR (1962) Age changes in the human pulp. I. The quantity of collagen. Oral Surg Oral Med Oral Pathol 15:1396–1404

Stella A, Fuentes A (1963) Inervación dentinaria intracanalicular. Su demonstración por el metodo de la hematoxilina-ferrica de Heidenhain. An Fac Odont (Montevideo) 9:157–206

Takahashi K (1985) Vascular architecture of dog pulp using corrosion resin-cast examined under a scanning electron microscope. J Dent Res 64:579–584 (special issue)

Ten Cate AR (1972) Morphological studies of fibrocytes in connective tissue undergoing rapid remodelling. J Anat 112:401–404

Ten Cate AR (1980) Oral histology, development, structure and function. Mosby, St Louis

Ten Cate AR, Deporter DA (1974) The role of the fibroblast in collagen turnover in the functioning periodontal ligament of the mouse. Arch Oral Biol 19:330–340

Ten Cate AR, Shelton L (1966) Cholinesterase activity in human teeth. Arch Oral Biol 11:423–428

Ten Cate AR, Deporter DA, Freeman E (1976) The role of the fibroblast in the remodelling of periodontal ligament during physiologic tooth movement. Am J Orthod 69:115–168

Tomes J (1856) On the presence of fibrils of soft tissue in the dentinal tubes. Philos Trans Soc Lond [Biol] 146:515–522

Torneck CD (1978) Intracellular destruction of collagen in the human dental pulp. Arch Oral Biol 23:745–747

Trelstad RL, Hayashi K (1979) Tendon collagen fibrillogenesis: intracellular subassemblies and cell surface changes associated with fibril growth. Dev Biol 71:228–242

Trowbridge H (1983) Pulp histology and physiology. In: Cohen S, Burns RC (eds) Pathways of the pulp. Mosby, St Louis, pp 323–378

Uitto VJ, Antila R (1971) Characterization of collagen biosynthesis in rabbit dental pulp in vitro. Acta Odontol Scand 29:609–617

Vacek Z, Plackova A, Bures H (1969) Electron microscopy of innervation of human dental pulp. Folia Morphol (Praha) 17:97–109

Viragh SZ, Papp M, Rusznyak I (1971) The lymphatics in oedematous skin. Acta Morph Acad Sci Hung 19:203–213

Wasikawa S, Ichikawa H, Nishimoto T, Matsuo S, Yamamoto K, Nakata T, Akai M (1984) Substance P-like immunoreactivity in the pulp-dentine zone of human molar teeth demonstrated by indirect immunofluorescence. Arch Oral Biol 29:73–75

Weil A (1887) Zur Histologie der Zahnpulpa. Dtsche Monatsschr Zahnheilkd 5:335–356

Weill R, Bensadoun R, Touurniel F de (1975) Démonstration autoradiographique de l'innervation de la dent et du parodonte. CR Academie Sciences Série D 281:647–650

Weinstock M, Leblond CP (1974) Synthesis, migration and release of precursor collagen by odontoblasts as visualized by radioautography after (^{3}H)proline administration. J Cell Biol 60:92–127

Weiss L (1977) Lymphatic vessels and lymph nodes. In: Greep RO, Weiss L (eds) Histology, 4th edn. MacGraw Hill, New York, pp 523–544

Winter HF, Bishop JG, Dorman HL (1963) Transmembrane potentials of odontoblasts. J Dent Res 42:594–598

Wislock GB, Sognnaes RF (1950) Histochemical reactions of normal teeth. Am J Anat 87:239–266

Yamada MK, Sagi T, Sheds R (1971) Mechanisms of excitation of nerve and tooth by thermal stimulation. In: Dubner R, Kawamura Y (eds) Oral facial sensory and motor mechanisms. Appleton Centry Crofts, New York, pp 73–82

Yamamura T (1985) Differentiation of pulpal cells and inductive influences of various matrices with reference to pulpal wound healing. J Dent Res 64:534–540 (spec. issue)

Zach L, Topal R, Cohen G (1969) Pulpal repair following operative procedures: radioautographic demonstration with tritiated thymidine. Oral Surg 28:587–597

Enamel

A. Boyde

A. Introduction

The enamel which we consider in this chapter is the hard, white, external covering of human teeth. We shall also make considerable reference to the same tissue in the teeth of other mammals, which is distinguished from analogous coatings on teleost, chondrichthyan, reptilian and amphibian teeth by its division into microscopic units, roughly corresponding to the size of its secretory cells, called prisms. At least, that has been accepted as "a fact" until recently, but recent reports have described prisms in reptilian enamel (Cooper and Poole 1973; Sahni 1984; Dauphin 1987a, b; see also Poole 1956). Not all mammals have enamel. The order Edentata is distinguished by having none, and not all of mammalian enamel is white. Parts of the most superficial enamel are pigmented red by ferric iron in rodent incisors and shrew molars (Boyde et al. 1961).

The white colour of enamel in man (and other animals) is only relative and depends upon the light-scattering properties of the tissue, which reduces the translucency in the immature tissue. Thus, the translucency is less (and the whiteness greater) in deciduous teeth than in permanent teeth, and it is less in recently erupted permanent teeth than in teeth present in the mouth for a long time. The latter acquire a yellowish hue, as the yellow colour of the dentine may be seen through the enamel. If the dentine becomes discoloured – as, for example, on its darkening in teeth in which the pulp has died, and its acquisition of a grey or brown colour in individuals who have consumed one of the tetracycline antibiotics during dentine-formative stages – then the enamel appears to acquire the same coloration.

Enamel is hard because of its very high proportional content of a mineral component bearing a strong resemblance to hydroxyapatite, a calcium phosphate. In the mature tissue, the mineral component may reach 89–91% by volume (corresponding to 96–98% by weight), the remaining proportion of the tissue volume being occupied by the organic matrix and water. The proportion of mineral in the tissue increases steadily from the moment of the inception of mineralisation, which closely follows the initial secretion. When first completed, the tissue is still relatively soft and unsuitable to serve its normal masticatory function. The process of increase in the degree of mineralisation, known as maturation, extends over a long period. When the tooth has erupted into the oral cavity, a reversal of this positive uptake of mineral ions to increase the proportional volume of the mineral phase may, under certain circumstances, occur, and this is commonly found in modern human beings in the disease process called caries.

The function of enamel is to provide a hard, wear-resistant outer coating to the surface of the teeth, to damage food, yet itself to remain undamaged. The physical properties of enamel are reviewed by BRADEN (1976). We shall see how the structure of enamel at every scale of its organisation is suited to this purpose. Enamel, however, is never present without the mould of dentine upon which it is initially formed. It is intimately locked onto the surface of the dentine and cannot function without this support. A shell of unsupported enamel will crack and craze and be demolished under subnormal masticatory loading.

We shall explain the secretion of enamel from the inner layer of cells of the enamel organ of the embryonic tooth germ. The same layer of cells, before it begins to secrete enamel, is responsible for inductive interactions with the peripheral layer of cells of the (ecto)mesenchymal papilla of the tooth germ. Without the enamel epithelium, therefore, there would be no dentine for the enamel epithelium to deposit enamel upon later.

The highly evolved enamel structure in mammals is doubtless of considerable biological importance. This tissue has allowed mammalian teeth to have a longer functional lifespan, which, given the restriction of mammals to one deciduous and one permanent or successional dentition, must be of importance in determining the possibility of a longer lifespan for the individual mammal. The alternative strategy of replacing worn-out teeth in numerous sets (seen, for example, in reptiles) did not seem to give rise to a precision of form equivalent to that of the teeth found in modern mammals. The longer functional lifespan of the first, or deciduous, tooth set has made possible a much longer developmental life history for the individual successional teeth, which have, thereby, been able to be perfected. With notable exceptions, such as whales and anteaters, a mammal will die if its teeth wear out, fall out or are badly damaged. Modern man is an exception to this rule because he can arrange that his food is pre-pulped, but from a more general biological standpoint, it would not be unfair to say that no mammal will live longer than its enamel will last (with some exceptions).

B. Gross Anatomy

I. Distribution

In human teeth, the portions of the teeth covered with enamel are known as the anatomical crowns. A clinical crown is that part of the tooth which has erupted and is visible in the oral cavity. The enamel may be up to 2.3 mm thick over the tips of cusps of human permanent cheek teeth. It is more usually close to 1–1.3 mm thick over the lateral surfaces of human permanent teeth and is 1 mm thick, or less, in the deciduous teeth (detailed measurements in HUSZAR 1971). Enamel is thickest over biting edges and biting surface, and thinnest at the neck, or cervix, of the tooth, where it tapers to nothing and gives way to the coating of cementum – the tissue which attaches the tooth via the collagen fibre bundles of the periodontal ligament to the surrounding alveolar bone.

In the teeth of mammals which are subject to heavy and continuous wear, as in the majority of herbivorous mammals, an evolutionary strategy has been adopted of compensating for wear by increasing the potential height of the crown of the tooth. A high-crowned, or hypsodont, tooth may begin to function in mastication long before enamel formation ceases and an anatomical root is formed on that tooth. In such teeth, cementum is attached to the external surface of the enamel in order to attach the periodontal ligament to the bone.

Elevations on the functional surfaces of teeth are generally known as cusps. The valleys between the cusps on the biting, crushing or grinding surfaces of cheek teeth with two or more rows of cusps are known as fissures. Shallower depressions are known as grooves. The most anterior, or incisor, teeth are also equipped with cusp-like elements (known as mamelons) when first formed, and before these are worn away in function. Cusps are generally extended into ridges, which may extend like cols joining adjacent cusps. Details of the anatomy of human teeth and of the comparative anatomy of mammalian teeth are given in many standard texts.

In teeth of extensive or continuous growth, including the high-crowned, or hypsodont, teeth of herbivores, the features equivalent to fissures may be extremely deep. They would, if they remained in this condition, constitute a profound source of weakness to the total structure of the tooth. Fissures in such teeth (hypsodont, lophodont and selenodont teeth included) have the space equivalent to the fissure filled in with a bony tissue (called cementum because it is applied to a tooth), which packs out the space and binds the adjacent flattened cusp-like elements, or lophs, together.

The enamel coating in mammals other than man covers the extremes of ranges from a few μm to tens of μm thick, up to 3–4 mm thick in modern elephants. In the very thinnest enamels, the fundamental histological features which we shall describe may be missing. Such tissue may show neither prisms nor complex features of prisms, such as the Hunter-Schreger bands. Generally, however, the range of phenomena which we shall describe in detail for man will be found in some equivalent form in other mammals possessing thick enamel.

II. Naked-Eye and Magnifying-Glass-Range Features

Close examination of the crown of a tooth will show that it is very smooth over cusps and biting edges and may take on a high polish where it has been worn flat against a tooth in the opposing jaw. The boundary line between the tooth surface and a worn facet may present a very sharp edge, and that edge may be chipped. On the lateral surfaces of teeth which do not wear against teeth in the same or the other jaw, a series of circular wrinkles will be seen, first described by ANTHONY VAN LEEUWENHOEK in 1689. These features are today known as the perikymata, meaning "waves around the tooth", or imbrication lines, i.e. overlapping layer lines. These features become increasingly common in progressing from the crown towards the cervix of the tooth. They will be seen to be missing in extracted teeth where these teeth have worn against the adjacent tooth in the same jaw at the approximal contact facet, sometimes known as the interproximal wear facet. Some of the wave troughs of the perikymata pattern may be unusually deep, and these are known as hypoplastic (underdeveloped) grooves. Isolated areas of depressed enamel thickness are known as hypoplastic pits.

The edge of the enamel at the cervix of the tooth is generally neat, representing a smoothly curved line with only minor excrescences. This represents the fact that the cells of the internal enamel epithelium synchronously cease to be able to produce enamel but continue to be able to induce the differentiation of dentine-forming cells, odontoblasts, from the adjacent peripheral layer of cells of the mesenchymal papilla. Occasionally, the internal enamel epithelium is able to restart enamel formation in a localised patch, giving rise to enamel pearls on an area of the tooth which is otherwise clearly defined as root.

Close examination of the internal enamel structure through the tooth surface in carnivorous mammals may reveal another feature: "striping of the enamel", alternate light and dark striping altering with the direction of illumination of the tooth. These bands are the Hunter-Schreger bands (discussed later), which may be seen if they approach the tooth surface closely. The equivalent features may always be seen in a tooth which is fractured in the longitudinal direction.

Longitudinally fractured teeth show a crystalline, or fibrous, appearance, with bands about one-tenth of a millimetre in width radiating from the dentine, where they may be clearly distinguished, and becoming less clear towards the surface of the enamel. These fibres, visible with the naked eye or with a low-powered magnifying glass, were first recognised by DE LA HIRE (1699) and later also described by HUNTER (1770) and SCHREGER (1800), the names of the latter two authors being generally attached to these features today. If a tooth is cut longitudinally, rather than fractured, the same bands may be recognised using appropriate illumination conditions. Here, it would be found that the position of lighter and darker bands changes with changes in lighting direction. Another class of feature becomes visible by reflected light on a longitudinally cut surface; namely, a set of nearly parallel lines which appear white, or blueish-white, extend from the enamel-dentine junction all the way over the tips of cusps to joint the enamel-dentine junction again on the other side of the tooth in the occlusal regions. In the lateral regions, these lines radiate obliquely outwards and occlusally from the dentine surface to the enamel surface, where they equate with the grooves of the perikymata pattern.

If the tooth is sawn twice longitudinally to produce a thin, "ground" section and this is illuminated with transmitted light, the same incremental, or growth, lines may appear to be brown or orange, rather than the blueish-white seen in the reflected light. They were first clearly described by RETZIUS (1836) and are known as the brown striae of RETZIUS.

In rather more than 50% of human teeth, it is possible to find a zone of the enamel surface at the ultimate cervix in which the enamel thickness is somewhat reduced generally, but with small local projections of the enamel surface (BOYDE 1970). It is nearly always necessary to use a low-magnification loupe to see such features distinctly. All other features and levels of organisation of enamel structure are so small as to require really significant magnification for study; in other words, we have to resort to microscopy.

C. Microscopy and Technical Methodology for the Study of Enamel

I. Light Microscopy

1. Sectioning Enamel for Light Microscopy

As the hardest tissue, enamel is in many respects the hardest to prepare for microscopic study. Not only does enamel present problems in its mature and very hard condition, but it is also difficult to study in its most immature and developing condition. Young enamel is a mixture of crystals and an organic matrix which is inherently unstable, and most of which, indeed, can be dissolved in water or buffer at laboratory temperatures. Although soft enough to be cut with conventional microtome knives, immature enamel is inevitably attached to the surface of dentine, which is as hard and mature as it will ever be. To soften the underlying dentine and make it cuttable, it is conventional to demineralise the tooth either by acid dissolution of the calcium phosphate component of both the dentine and the enamel or by the use of chelating agents which may work at neutral or alkaline pH. The unstable enamel matrix may be stabilised by appropriate fixation with, for example, glutaraldehyde, formaldehyde or osmic acid.

Efficient demineralisation of mature enamel leads to the total dissolution of the tissue because a very large proportion of its volume is composed of the calcium phosphate crystals, and the intervening, extremely thin, ensheathing organic matrix is so delicate that it cannot exist without the support of the crystal. Methods for demineralising mature enamel using an organic acid buffer system at about pH 4.5 selectively dissolve only a portion of the enamel mineral and lead to a regrowth and reprecipitation phenomenon within the enamel. The so-called demineralised enamel is not such: it is composed of large single crystals differing in shape and size from the original enamel crystals, and the space in between can be filled with embedding medium to make the whole substance cuttable. The residual material is, however, very largely inorganic, and that inorganic component is not the original, unaltered mineral component of the enamel.

Owing to its unique properties, it is not possible to cut sections of mature enamel except as ultrathin sections for transmission electron microscopy (TEM), the latter sections containing severely deformed tissue: it is not possible to microtome sections for light microscopy. The usual approach is to saw sections from the intact tooth, making use of the hardness and rigidity of both the enamel and the dentine. Teeth may be cut with thin, stiff metal wheels carrying hard abrasive material at their periphery. This abrasive may be bonded to the wheel, as in diamond- or silicon-carbide-impregnated discs, or it may be a slurry carried with a soft metal wheel, as in the use of aluminium oxide with copper cutting wheels. Continuous wire saws, either impregnated with diamond or carrying an abrasive slurry, may also be used to cut teeth. Cutting-wheels or discs may be constrained to cut in a much flatter plane if the inside of an internal hole is used. Annular saws therefore may cut thinner ground sections.

Ground sections may also be prepared literally by grinding. One side of the tooth is reduced by rubbing against a flat surface of a hard abrasive material using first rough grades and then progressively finer grades until one side of a section through the tooth is polished. This is then stuck to a glass slide with a suitable adhesive, such as the thermoplastic Lakeside 70 cement or a cyanocrylate ("super-glue") cement. The other side of the tooth is then removed on the same abrasive materials, ultimately using very fine abrasives as the section thickness approaches its final desired value.

One may be able to expect to saw sections with an outside cutting-wheel at 150–100 μm in thickness, and with an inside cutting-wheel from 100 to 75 μm in thickness. Further reduction in section thickness must be done by grinding the section; this is best done, as already intimated, by sticking the section on one side to another support (a glass light-microscope slide). Very careful hand or automatic finishing procedures can be used to grind and polish sections as thin as 10 μm (and in some small areas even thinner) without causing total disruption of the section. Most ground-section histology, however, is based upon the examination of sections at least 100 μm thick, and this leads to a very different set of problems in interpretation than is normally encountered in any other branch of histology. We would normally cut sections to be not more than one cell thick, and usually a small fraction of the cell thickness characteristic of the tissue type under examination. In enamel, the fundamental light-microscopically visible units which we shall study, the enamel prisms, are of the order of 6–7 μm thick, so that we examine sections which are 13 times thicker than this value. Since most light-microscopic methods lead to an addition, summation or averaging of information through the entire section thickness, it would be obvious at the outset that we should have to expect false impressions, artefacts or misinterpretations as a consequence.

2. Examination of Ground Sections in Light Microscopy

Permanent mounts of ground sections for light microscopy are normally mounted in a medium with a high refractive index, such as canada balsam, whose refractive index is similar to that of the glass slide and cover slip but lower than that of the mineral component of enamel (balsam 1.54, hydroxyapatite 1.62). A well-cleared section, i.e. one in which the embedding material has run into the finest interstices of the tissue, may well, therefore, show very little detail in light microscopy. Part of the art of preparing good ground sections, or obtaining good results from ground sections, is to use a thick mounting medium which does not penetrate the finer spaces, and to make histological observations shortly after mounting, not waiting until the section is entirely cleared. A more sensible approach, however, is not to mount the section in the high-refractive-index medium in the first place, but to leave it wet in water and to examine it as a temporary mount. The refractive index difference between water and the majority component of enamel is much greater, and it will normally be possible in such a section to see all the histological features to be mentioned in this chapter. In order to improve the contrast of features visible in ordinary bright-field, transmitted-light microscopy, it is possible to restrict the aperture of illumination by stopping down the substage condenser diaphragm. While this increases contrast, it also reduces lateral and vertical resolution leading to interfering diffraction phenomena, and making it impossible to interpret the origin of contrasts from the multiple layers of the specimen. As regards interpretation of histological information, it is fortunate that we have many other methods to resort to in enamel, including polarisation (CARLSTROM 1963; DARLING 1961; G. GUSTAFSON 1945; AG GUSTAFSON 1959; GUSTAFSON and GUSTAFSON 1961; POOLE 1966; OSBORN and ROBERTS 1971), phase-contrast and interference methods, as well as dark-field and confocal scanning optical microscopy (PETRAN et al. 1985).

Much of the contrast originating from ordinary illumination conditions in transmitted light is dependent upon the reflection, or scattering, of light at interfaces (notably the prism boundary junctions, or sheaths), which subtract light from the original illuminating beam (OSBORN 1971; OSBORN and ROBERTS 1971). Examining the section in reflected light, we see a reversal of contrasts due to this origin. At low magnification, it is possible to use to some effect any conventional light-microscope objective with an oblique source of illumination. It is much better, however, to use dark-field illumination to witness this scattering or reflection from discontinuities in the enamel structure. In reflected light, also, the light collected by the objective comes from all layers within the sample. Under dark-field illuminating conditions, all the features seen are essentially self-luminous but lie within the entire thickness range of the section. Although the layer in focus will be more sharply imaged, details will be clouded out by reflected light from other planes in the section. This difficulty may be overcome by using one of the new genus of microscopes known alternatively as tandem scanning microscopes (TSM; PETRAN et al. 1985) or confocal scanning light microscopes (CSLM), both names denoting microscopes which yield focal-plane-specific information. In these microscopes, an image is built up sequentially, though it may be at such a high frame speed that it is not possible to detect the difference from a normal microscope in normal use. At any one instant, only one point or series of points in the focal plane is intensely illuminated, and only light from those brightly illuminated points is collected through conjugate apertures placed in the intermediate image plane of the observing objective lens. The most convenient configuration of such a microscope for study of enamel has proved to be the TSM (PETRAN et al. 1985). It is possible to configure a light microscope so that all the normal modes of operation may be conducted at the same time as the confocal mode, and this approach has allowed us to interpret optical appearances in enamel histology with increased confidence.

3. Reflected-Light Microscopy of Well-Polished Section Surfaces

This is a method with considerable potential, which has hardly been exploited. The top surface of the section to be examined is dry and examined with a dry objective, using a conventional epi-illumination reflection microscope or (better still) a TSM. The extremely low level of relief generated by the polishing process is sufficient to reveal the distribution of the discontinuities in the enamel structure, and thus the appearance of the prisms in any plane of section revealed at the surface. Softer polishing produces a greater degree of surface relief, so that at low magnification the Hunter-Schreger bands – zones of prisms having common orientation properties – are revealed as topographical relief on the polished surface.

A procedure which is often used to examine prism packing patterns in enamel is lightly to etch a polished surface and stain the prism boundaries just subjacent to the surface. Haematoxylin has been used for this purpose (SHOBUSAWA 1952), but better still is to stain by immersion in a dilute silver nitrate solution, followed by exposure to ultraviolet light and slight repolishing to remove any silver from the surface of the section. Visualisation of enamel prisms in polished, polished and etched, or polished, etched and stained section surfaces can also be well conducted in the SEM (see below).

II. Transmission Electron Microscopy of Enamel

1. Sectioning Enamel for Transmission Electron Microscopy

Transmission electron microscopy (TEM) of enamel presents unusual difficulties owing to the extreme hardness of the tissue. Enamel was one of the first problem areas for which diamond knives were invented and used in electron microscopy. The advantage of the use of the diamond knife is now widely recognised in soft-tissue histology. Young, immature enamel embedded in any conventional embedding resin can be cut even with glass knives, though these do not last too long. Sections are more easily cut in the transverse to the crystallite axis, or so that the crystals roll over the edge of the knife. The least favourable direction is when crystals are lying in the plane of the block face and approaching the knife edge head on. This tends to lead to the less regular fracturing of enamel crystals. Whether the tissue is immature or mature, ultramicrotomy of enamel inevitably leads to the breaking-up of the enamel crystallites, and too much confidence in the interpretation of appearances of sections has led to the widespread propagation of an error in the literature concerning the length of enamel crystallites. Fortunately, there are two other approaches to the electron microscopy of enamel which can be usefully applied.

2. Transmission Electron Microscopy of Replicas

The older approach, which has not been referred to in recent years but is nevertheless very reliable, is the simple replication of fractured surfaces of enamel (WOLF 1940). This was the first method applied to the electron-micro-

scopic study of enamel by SCOTT and WYCKOFF (1947), SCOTT (1952) and
HELMCKE (1953, 1955). It exploits the fact that enamel, as we shall explain
later, tends to break parallel with its crystals and parallel with the long axis
of the crystal orientation discontinuity boundaries, the prism junctions or
sheaths. An ultramicroscopic relief is developed on the fractured surface, showing the extent of the individual enamel crystallites. The fractured surface may
be replicated in a one- or two-step process. In the one-step process, a coating
of silicon monoxide, carbon or a stripping plastic replica material like Formvar
is applied to the surface, with or without oblique metal shadowing to improve
the contrast in the TEM. The specimen may then be dissolved away in a sequence
of acid and alkali reagents, leaving the replica floating in solution to be picked
up on a grid. In the two-step replication procedure, a flexible plastic replicating
material is applied to the fracture surface. This is then coated with, for example,
carbon – which again may be shadowed with a heavy metal – and the original
replicating plastic is dissolved. The advantages of the TEM replica approach
are very considerable. The resolution is as good as the resolution of the replicating medium, and at about twenty angstroms (2 nm) is far better than the repeating width of the enamel crystallites at 50–100 nm. Since the replica is of a
3-D surface, it is necessary to record stereopairs in the TEM and to evaluate
the surface morphology by stereoscopic observation and/or measurement.
Investment of time and in apparatus is minimal with this technique, and it
is very surprising that it is not more widely used today. This is probably because
it has been displaced by scanning electron microscopy.

3. Ion-Beam Thinning

For mature dental enamel, it is possible to produce ultrathin sections from
thin ground sections by further machining with ion beams, approaching the
surface of the section at oblique incidence (BOYDE and PAWLEY 1975). It is
normal to use a noble gas to avoid chemical-reactive effects, and argon is the
most commonly employed. Once the argon ions have been accelerated to the
voltage of the ion gun, it is possible to neutralise them on their flight path
to the specimen. The effects are similar whether ions or neutrals are used.
For ion-beam thinning it is usual to arrange for the incoming beam to strike
the specimen at a low angle of incidence, say 10–15°. One or two beams may
be used on one or both sides of the section. The section is rotated continuously
and is worn away gradually by the sputtering process until a hole with a thin
edge develops through the section. Only the small region at the edge of the
hole can be examined in the TEM. ORAMS (1976), PALAMARA et al. (1981) and
PHAKEY et al. (1984) have developed this technique and are able to choose
the location at which the thinning process will occur maximally by pre-dimpling
the ground section with air-abrasive technology. The rates of sputtering under
ion or neutral beam bombardment depend upon density and crystallographic
orientation. The method is only suitable for relatively homogeneously dense
areas and could never be used, for example, to examine the interface between
cells and the formative tissue. It is, however, the only way of producing sections
of crystals which have not been bent and broken in the process. The untoward

effects of this procedure are ion-implantation damage and heating, which gives rise to a so-called frothy appearance of the enamel crystals. Nevertheless, it is clear that the enamel crystals are essentially undamaged since Bragg extinction contours (see below) can be followed along individual crystals as they are tilted in the electron beam.

For very high resolution work, it does not matter if only small fragments of enamel crystallites are recovered, and these may be obtained even from micro-tomed sections of enamel crystallites (FRAZIER 1968; SELVIG 1973). Thus, at the resolution required to image the crystal lattice of hydroxyapatite, either microtomy or ion-beam thinning may be used. In order to have a true image of the extent of crystals and their relative orientations, ion-beam thinning is the preferable process.

III. Electron Diffraction of Enamel Crystals

1. Isolation of Enamel Crystallites

Preparations of single crystallites from less mature tissue are simply made by dissociating the crystals in water by mechanical means. A suspension of extremely elongated crystal fragments is obtained and dried down upon a holey electron microscope support film. These isolated crystals provide ideal test specimens for evaluation of the crystal lattice structure (FEARNHEAD 1961; DACULSI and BERTRAND 1978).

2. Electron Diffraction

Enamel crystallites are sufficiently robust to withstand the irradiation of the electron beam and yet not change their basic lattice structure to a significant degree. Thus, it is possible to record electron diffraction patterns from enamel crystallites, at least for those spacings which cross the long axis of the extraordinarily elongated crystals. Measurements of these crystal spacings can then be obtained to very high precision, and for the spacings which can be measured the data confirm the likely hydroxyapatitic nature of the calcium phosphate component.

3. Origin of Contrasts

The origin of contrasts from crystalline materials irradiated with electron beams is not as simple as that of amorphous solids, which constitute the vast proportion of the biological samples examined by TEM. A less regular arrangement of atoms in a crystal lattice – a more amorphous solid – may scatter electrons more efficiently than a more regular arrangement. The angular deflection of electrons passing through a crystalline solid will be greatly increased if the conditions for diffraction are satisfied (i.e., if Bragg's law, $n\lambda = 2D \sin\theta$, is satisfied). Since enamel crystallites may be strained (i.e., bent), it is possible for closely neighbouring regions within the same crystal to show dramatic differences in electron contrast. Where the conditions for diffraction are exactly satisfied, a Bragg extinction contour may be seen passing across the crystal. This feature will move as a strained crystal is tilted and different parts of the crystal satisfy the conditions for the lattice plane to be oriented exactly in the necessary direction with respect to the electron beam. The movement of these Bragg extinction contours is an important element in demonstrating the single-crystal nature of the very elongated enamel crystallites.

IV. Scanning Electron Microscopy

1. Scanning Electron Microscopy of Enamel

Scanning electron microscopy (SEM) has contributed greatly to the development of the understanding of 3-D relationships in enamel (BOYDE 1964, 1965, 1967, 1968, 1969, 1970). No other single method since polarised-light microscopy has been as important. In conventional SEM, the specimen is surrounded by a vacuum and irradiated with a very fine electron probe, which is scanned across the surface of the specimen. One of many interactions is measured, and the intensity of that interaction is used to modulate the brightness of a cathode-ray tube, which is scanned in synchronism with the spot scanning the specimen in the microscope. In the conventional mode, therefore, the specimen must be dry. In this respect, it is a great advantage that enamel is the least hydrated of all the tissues. It is not possible to remove water from any normal soft tissue without an inevitable shrinkage of 6% linear, or 21% by volume. The proportional shrinkage in mature enamel is very much less than this, immeasurably small, but nevertheless significant and leading to the development of cracks in samples which are placed in the scanning electron microscope and become super-dry over a long period.

It will be necessary to consider the different signal modes in the SEM which make a contribution to understanding enamel.

a) Secondary Electrons

Secondary electrons are the low-energy electrons able to escape from within a few Å of the specimen surface, and which can be collected by a field of a few hundred volts applied to a grid in front of a detector. Because the low-energy electrons leaving the sample can be bent through a large angle and swept into a large-angle detector, electrons leaving the sample in many different directions can be collected and used to produce an image. Image brightness depends upon, amongst other things, local variations in the secondary-electron emission coefficient and local variations in the angle of the specimen surface with respect to the collector system, as well as more complicating geometrical factors, such as whether the part of the specimen being looked at is inside a deep (dark) hole. The specimen irradiated by an electron beam may suffer charging, and the development of a voltage in the surface zone of the specimen will lead to an enhanced emission of electrons from that region. In order to allow the injected electron beam to leak away to earth, and not to make the specimen into an electron mirror, it is normal to coat the sample with a thin layer of a suitable material. For secondary-electron emission, this would normally be gold, gold-palladium alloy, platinum, silver or aluminium. At very high resolution, a metal which condenses with a reduced tendency to form crystals is preferred, and for this purpose tungsten has been recommended.

The depth of focus of the SEM depends upon the distance over which the electron beam is narrower than a chosen diameter. This is, relatively speaking a much greater distance than in other methods of microscopy, and the depth of field is great. Because of this and the complications of the different origins of contrasts in a secondary-electron image, these should always be recorded as stereoscopic pairs and evaluated by 3-D techniques (BOYDE 1973). The fact that this is not regularly done has led to the propagation of many conceptual errors, even to the present date, and the field of investigation of enamel structure is no exception to this rule.

b) Back-Scattered Electrons

Back-scattered electrons (BSE) are those electrons interacting with the specimen atoms which suffer collisions, or multiple collisions, leading to their exiting from the specimen surface again. An arbitrary definition of the distinction between secondary electrons and BSE is that the former have energies less than 50 eV and the latter energies greater than that. The majority of the BSE have energies much greater than 50 eV and up to the limiting energy of that of the incoming electron beam. Electron back-scattering happens at depths within the sample surface dependent upon the incoming beam voltage, but in the typical case with a 20- to 30-kV beam and mature enamel, we will be dealing with the majority of the electrons coming from a half μm to one μm surface layer. BSE cannot be deflected by the electric field which could be applied to a grid in front of a realistic collector, so that they are collected by line-of-sight devices. Much smaller solid angles of electrons can be collected than with SE. The contrasts due to topographic relief on the surface are consequently also much sharper. If the SEM is equipped with only one detector, this will be the conventional Everhart-Thornley detector, with an optional biased grid in front of a scintillator. To collect BSE, a negative voltage is applied to that grid to prevent the acceptance of low-energy electrons: from the position at which the Everhart-Thornley collector can see the specimen surface, it is inevitable that contrasts will be due largely to topography.

To minimise topographic relief, it is necessary to have a sample which has minimal topography, acquired by suitable polishing or micromilling procedures, and to have a collector which collects a large solid angle of BSE and views at near-normal mean incidence (BOYDE and JONES 1983). This means that the collector must be an annular detector surrounding the electron beam which views the specimen surface at normal incidence. If a flat-surfaced specimen (which shows no charging effects at the voltages concerned) is viewed at normal incidence, then it is possible to interpret the resulting contrasts as a measure of mean atomic number, or effectively the density of the sample surface layer.

This has been of great value in investigating enamel structure in recent years (BOYDE and JONES 1983). However, it is necessary to prove that the sample surface has no topographic relief. To do this, the signal entering opposing collection solid angles should be examined to see if there are differences. If more BSE enter one solid angle than another equal and opposite angle, then there is topographic relief. This simple test should be applied, and what is called the A-B signal compared and recorded, if the A + B signal is to be used as a measure of density. A two- or four-sectored solid-state BSE detector is preferred for this purpose.

The first SEM images of enamel were recorded using BSE collected with a scintillator collector which surrounded the electron beam. Such devices are commonly available today. A general advantage of the use of BSE for topographic imaging in the SEM is that the problem of charging is very greatly reduced. Specimens have to charge to hundreds or thousands of volts to disturb the image formation, whereas they only have to charge to a few volts, or tens of volts, to cause serious disturbances in secondary-electron imaging.

c) Cathodoluminescence

Cathodoluminescence (CL) is the emission of light under electron bombardment. The natural cathodoluminescence of enamel is of no value in displaying morphological features of this tissue. Cathodoluminescent materials may be introduced into enamel to secure some advantage. Natural cathodoluminescence may develop in some prism boundaries during the process of fossilisation of enamel. Dental restorative materials may contain two kinds of cathodoluminescent components: first, the organic resin phase is almost certain to be cathodoluminescent, and second, the inorganic "filler" particles may also be cathodoluminescent. The tetracycline antibiotics are also cathodoluminescent and may be used as markers of physiological uptake of a substance in the bloodstream into immature enamel (BOYDE and REID 1983). Tetracycline is lost from mature enamel, unlike the mineralising connective tissues – cement, dentine and bone.

2. Sample Preparation

a) Sample Preparation Procedures for Adult Enamel

Enamel is a brittle tissue which fractures to produce a very significant surface relief, which can be interpreted to understand most of the important basic features of the tissue structure. Nevertheless, it is sometimes necessary to apply more refined preparative procedures to adult enamel structure in order to show details in a defined plane or at a chosen level within the tissue. Such preparative procedures may be grouped together under the general heading of etching.

i) Polishing is the simplest means of etching enamel because the hardness of the tissue depends upon the orientation of its structural components. In general, structural elements which are parallel with the surface will be removed from the surface more rapidly than those which are perpendicular to it. Furthermore, elements which are close to a structural discontinuity boundary will be removed more rapidly than those farther away from such a boundary. Diamond-polishing of enamel is a means of producing a fine surface relief on the enamel. Polishing on soft laps produces a higher surface relief (BOYDE 1983; BOYDE and FORTELIUS 1986).

ii) Air-polishing involves treatment with an impinging high-velocity beam of soft abrasive particles in an air stream, surrounded by a water stream. Fine sodium bicarbonate powder is most commonly used for this purpose. This technique produces a good surface relief on enamel, which can be interpreted to indicate the distribution of prisms, prism boundary cross sections and zones of prisms having common orientation properties (BOYDE 1983; BOYDE and FORTELIUS 1986).

iii) Ion-beam etching is a second physical etching technique. Differential rates of removal of enamel crystals relate to the orientation of the enamel crystals because sputtering of oriented crystals depends upon the depth to which the incoming ions or neutrals can enter the fundamental crystallographic lattice structure. Atomic collision sequences initiated nearer to a crystal surface where the incoming ions do not intercept open crystallographic directions lead to more rapid rates of removal of material. If the incoming ions find relatively open directions and initiate collision sequences deep to the surface, the rate of removal of the surface is reduced. These effects are combined with the possibility for material to escape from internal surfaces, such as the prism boundary discontinuities, where these proceed at a large angle to the surface of the tissue under attack. Thus, bombardment with an ion beam produces a low surface relief which accentuates the prismatic structure of enamel without removing material in depth (BOYDE and STEWART 1962). This advantage for SEM is a considerable disadvantage in TEM, where we try to use ion beams to thin and produce plane parallel sections of the tissue.

iv) Acid-etching removes enamel structure in depth. The surface topography develops because of differential solubilities in different directions of the enamel crystals, but more importantly because less soluble, more acid, calcium phosphate species develop in the depths of the enamel structure close to the more porous prism boundary discontinuity regions. A zone of considerable porosity develops ahead of an acid-etched enamel surface, and this more porous enamel is severely subject to a mud-cracking type of drying distortion upon the desiccation of that tissue layer. In bad cases, a large part of the etched structure develops due to air-drying cracking and is hardly interpretable (BOYDE et al. 1978).

A general objection to all etching techniques applies with special force to acid-etching of enamel. The relief developed depends upon the differential solubility properties of the layers which have been removed from the sample surface, and which are no longer present and not available for inspection. We have available a surface whose topographical relief depends upon the complicated sequence of events which happened in the overlying, now missing, layers. This must make it difficult to arrive at a proper interpretation in many instances. Nevertheless, acid-etching of enamel is widely used in the investigation of its structure and with little care given to this general, philosophical point. The reason why it is so popular is that it is so simple and so easy to apply. If the acid-etching is superficial and only removes a few µm from the polished surface at which it begins, and if the aim is to examine such simple information as the prism cross-sectional pattern, then there is little objection to its use.

v) Etching with chelating agents may have substantial advantages in bringing out significant features of enamel structure. The most commonly used reagent is EDTA, which may be used over a wide range of pHs adjusted by NaOH or KOH. EDTA-etching depends upon the chelation of calcium ions from the hydroxy-apatite lattice structure and does not apparently lead to the formation of different crystal species with different solubilities deep to the general level of the demineralisation front, as is the case with acid-etching. Because of the large size of this molecule, it tends to penetrate more effectively down the prism boundaries and thus thins down the prisms, attacking from the more porous prism boundary spaces. It therefore leads to the production of long, slender fragments of prisms standing out from an etched surface which has great 3-D relief. Substantial portions of the long axis of the prismatic crystallite groups can then be seen, and their relative orientations in space determined. Again, it is necessary to pay attention to the drying technology used since this enamel will be somewhat porous and subject to drying-down artefacts. However, it is likely that the crystals will shrink closer together within their prismatic groupings and the tissue will not develop artefactual splits which do not relate to the original morphology of the etching process.

The tendency of EDTA to exploit the prism boundary discontinuities (POOLE and JOHNSON 1967; BOYDE 1969; SWANCAR et al. 1970) can be enhanced by pre-treatment with a regime designed to dissolve the organic matrix of the enamel which is concentrated in the prism boundary spaces. This can be done by prolonged extraction with hot 1,2-ethane diamine, by treatment with a hot alkali, with Na_2O_2 at 50° C or prolonged treatment with cold NaOCl solution prior to the use of EDTA. Such a regime will also tend to remove dentine by removing its organic matrix.

vi) Removing dentine from enamel can thus be accelerated by making the dentine anorganic or deproteinised by prolonged treatment with one of the above reagents. The mechanical strength of the dentine now depends only upon the degree of interlocking of its mineral component. Wet anorganic dentine is extremely weak and can be washed away with a water jet or with an air-polishing stream to reveal the enamel-dentine junction. Enamel may be completely removed from dentine to show the dentine surface of the enamel-dentine junction by dissolving it in an acid or a chelating agent. Such a procedure inevitably leads to demineralisation of the surface layer of the dentine, which will then suffer some shrinkage distortion during the subsequent drying process.

b) Preparative Procedures for Immature Enamel

Preparative procedures for immature enamel are more difficult than those for the majority of soft tissues because of the very high degree of hydration of newly formed enamel. Even in the most careful freeze-drying and critical-point drying regimes, the pre-existing spaces in enamel are exaggerated by the shrinkage of the prismatic bundles of crystallites on to their own centres of gravity. We can, however, exploit the fact that the most interesting structural component in the immature enamel is the hydroxyapatite mineral component. As will be seen shortly, we must be particularly interested in the details of 3-D morphology of the mineralising front in forming dental enamel. This can be prepared with minimum tissue distortion by first embedding the tissue in polymethyl methacrylate (PMMA). The methyl methacrylate monomer from which the infiltration process begins will penetrate the finest pore spaces in developing enamel. When polymerised, the solid PMMA effectively replaces the original tissue water. To remove the PMMA, it is only necessary to oxidise it by exposure to an oxygen plasma at ambient temperature. Plasma-ashing removes all the organic components of the enamel matrix and the adjacent ameloblasts, leaving the delicate enamel crystallites in their original locations. The best preparations of developing enamel have been made by this procedure (BOYDE and MARTIN 1982; BOYDE and FORTELIUS 1986).

The shrinkage due to freeze-drying or critical-point drying can be exploited (if it is not too great and if it is under control) to accentuate the enamel prism boundaries and make them more visible.

However a developing enamel surface is prepared, if it is clean and unde-formed by the preparative procedure, its examination constitutes absolutely the best means of determining the 3-D relationships of enamel crystallites, of their groupings and of the orientation of the prism boundary discontinuities. It is possible to orient the SEM specimen so that one looks along the prism boundary discontinuity at an exact cross section of the enamel prism (BOYDE and MARTIN 1982). Only in this way can this profile be reliably determined. In the same material, it is possible to examine the mean plane of the developing enamel surface at normal incidence, and to determine the exact shape of the Tomes' process pits in the surface of the tissue. It is also possible to record the size of these "secretory territories" of the ameloblasts.

3. X-ray Emission Microanalysis

a) Characteristic X-ray Emission

X-rays are emitted from the specimen under electron bombardment in an SEM. The X-ray spectrum shows a continuous portion due to the bremsstrah-lung, with superimposed characteristic peaks at specific wavelengths or energies which prove the atomic composition of the sample under electron bombardment.

b) Wavelength-Dispersive X-ray Analysis

Wavelength-dispersive X-ray analysis (WDX) produces the more sensitive separation of the characteristic X-ray peaks and is the basis of the more accurate quantitative techniques for determining the relative mix in the surface layers of the sample under analysis. WDX has been used to record the increase in mineral concentration during enamel maturation, and to characterise the differ-ences in the degree of mineralisation across the thickness of enamel in the mature tissue (BOYDE et al. 1961; ROSSER et al. 1967).

c) Energy-Dispersive X-ray Analysis

Energy-dispersive X-ray analysis (EDX) is much simpler to institute but does not have equal resolution to WDX. It has, however, the greater convenience that many characteristic peaks can be recorded and measured simultaneously, and thus it gives a better "fingerprint" of the concentration of elements in the sample surface.

All X-ray microanalytical procedures on bulk samples suffer from the same disadvantage that we do not know the volume from which X-rays are collected. This volume is always a great deal larger than that from which we collect any other signal used in forming an image in an SEM. It is unfortunately not the same for different elements since different characteristic radiations will have different abilities to escape from different depths within the sample. The only way round these difficulties is to use thin, plane-parallel sections through which the greater portion of the electron beam can penetrate. It is very difficult to produce such sections of enamel.

V. X-ray Microscopy and Microradiography

1. Contact and Projection Microradiography

X-ray microscopy may be conducted with the specimen in contact with the recording photographic emulsion (contact microradiography) or using a projected image of the sample held at some distance from the recording emulsion (projection microradiography) (GUSTAFSON and GUSTAFSON 1961; HAMMAR-LUND-ESSLER 1958; SILNESS 1969; SUGA et al. 1970; SUNDSTROM 1966).

X-ray absorption depends upon the density (mass distribution) of the sample. It is possible to exploit interesting effects due to the differential absorption of characteristic X-rays on either side of the absorption peak. Thus, an element becomes much more transparent to a radiation which has an energy just below that of the characteristic absorption maximum.

Specimens for X-ray microscopy should ideally be plane-parallel, very thin sections. Again, these are very difficult to produce from enamel. The process of preparation of the section will involve polishing on both sides of the section, and the rates of removal of tissue will depend upon the orientation of its fundamental structural elements. Thus, in those Hunter-Schreger bands where the prisms are parallel with the surface of the section (parazones), the rate of tissue removal will be greater, and the section thickness will differ significantly, if subtly. These differences, however, will be exaggerated in an X-ray microscopic image, and this is the reason for the appearance of Hunter-Schreger bands in X-ray microscopy (GLAS and NYLEN 1965).

2. Scanning X-ray Microscopy

By measuring the transmission of X-rays through a specimen which is mechanically scanned past the X-ray beam and the X-ray detection device, it is possible to produce an image of a strictly quantitative measure of the passing X-ray flux. This procedure can be of considerable value in dynamic in vitro experimentation upon mineralisation and demineralisation effects in enamel because the problems due to the variations in the starting thickness of the enamel are compensated for by the shift to observing changes with time during an experimental procedure (LANGDON et al. 1980).

3. X-ray Diffraction

Since there is so little water in enamel, this is also the tissue subject to the least doubt in the interpretation of X-ray diffraction images of even the dried tissue (GLAS 1962). X-ray diffraction proves the apatitic nature of the calcium phosphate component of enamel and allows the determination of the statistical pattern of preference of orientation of the enamel crystals in lateral enamel at about 17° inclined cervically away from the long axis of the enamel prisms (GLAS 1962; HAMMARLUND-ESSLER 1958; LYON and DARLING 1957).

D. Development of the Tooth Germ Prior to Enamel Formation

I. The Tooth Germ

Excellent accounts of the stages of differentiation of the tooth germ from downgrowths of the oral ectoderm, and the associated mutual inductive processes between ectodermal or epithelial components and the adjacent ectomesenchymal tissue, are available in many standard texts.

The first morphological evidence for the future development of the dentition occurs as a thickening of the embryonic oral ectoderm known as the primary epithelial band. The greater mitotic rate in this band leads to its relative expansion into a dental lamina, upon which a series of thickened elements, or tooth buds, may eventually be identified. At the time that the epithelial buds can be distinguished, there is a concentration of cells in the ectomesenchyme immediately surrounding these buds, and particularly at their aboral poles. Further growth of the epithelial tooth buds is associated with the relative spacing of the cells in the centre of the epithelial downgrowth, and with an invagination of the epithelial downgrowth on its aboral pole so that it partly surrounds the adjacent mesenchymal condensation. When the latter projects within the epithelium, it is referred to as the mesenchymal papilla. From the peripheral layer of cells of the mesenchymal papilla will differentiate the odontoblasts, which give rise to dentine. The tooth germ is referred to as being at the "cap" stage when the epithelium is invaginated by the mesenchyme.

The continued expansion of the epithelial component is associated with deepening of the invagination by the mesenchymal papilla and the distinction of layers in the epithelial component. The mass of cells in the central thickness of the epithelium develops a relatively greater spacing than is found in other epithelia, even in the embryonic condition. The concentration of extracellular material leads to the stretching of the individual cells into a stellate form, and the bulk tissue mass is known as the stellate reticulum or enamel pulp. This is surrounded on the outside of the tooth germ by a single layer of cells called the outer, or external, enamel epithelium (from its location) and continuous with the single layer of low columnar cells on the invaginated side called the inner dental or enamel epithelium; from this epithelium the future enamel-forming cells, or ameloblasts, will eventually differentiate. The inner enamel epithelium and the stellate reticulum are separated by a layer of cells a few cells thick, polygonal cells more extensive in the plane of the sheet, called the stratum intermedium.

Experimental embryology shows that the differentiation of the mesenchymal papilla depends upon the presence and position of the epithelial downgrowth (LUMSDEN 1979; SLAVKIN 1970; THESLEFF et al. 1979). Inductive influences from the internal epithelial cells result in the differentiation of the future odontoblasts. Inductive influences from this peripheral layer of ectomesenchymal cells result in the differentiation of the internal enamel epithelial cells into pre-ameloblasts, and eventually ameloblasts. Elaborate recombination experiments of epithelia and mesenchyme from different sites leave no room for doubt as to the specific nature of these inductive interactions (KOLLAR and FISHER 1980; RUCH and KARCHER-DJURICIC 1971; SLAVKIN et al. 1983; LUMSDEN 1979).

The only vertebrates not to possess teeth are the Aves. Nevertheless, KOLLAR and FISHER (1980) have shown that chick epithelium may be induced to develop enamel organs by mouse tooth germ mesenchymal elements. These experiments indicate not only that the genetic mechanisms necessary for tooth germ production and enamel synthesis are still retained by birds but that the morphologically apparent primary influence of the epithelium on the mesenchyme is incorrectly deduced. It must be that the message from the ectomesenchymal growth element, itself centred upon the distribution of nervous elements in the jaw region, is the more important factor.

II. Ameloblast Differentiation and the Pre-ameloblast

Prior to the development of dentine, which is secreted by the peripheral layer of cells of the mesenchymal papilla, the inner enamel epithelium cells undergo considerable morphological differentiation. At the lowest magnification level of analysis, perhaps the most prominent feature is that known as the reversal of polarity. Columnar epithelial cells normally have their nuclei at the basal end, that is, at the end facing the adjacent embryonic connective tissue layer. This is true of internal enamel epithelium cells when they are first recognisably columnar. Just prior to the secretion of dentine on the other side of the basement membrane, the nuclei of the enamel epithelium cells migrate to occupy a position towards the stratum intermedium end of these cells, i.e. proximal. The Golgi apparatus, previously away from the basement membrane, now moves towards the basement membrane side of the nucleus. This reversal of polarity, then, reflects not only a reversal of the situation with respect to other epithelia but a real change happening within these cells at a defined phase of differentiation. A terminal-bar apparatus connects the basement membrane end of the ameloblasts at an early stage. Following the reversal of polarity, a second terminal-bar apparatus develops at the stratum intermedium end of the pre-ameloblasts.

Prior to the demonstrable secretion of enamel matrix, the ameloblasts develop a considerable rough-surface, endoplasmic reticulum component and an expanded Golgi apparatus, and secretory granules may begin to be seen. Immunocytochemical labelling techniques show demonstrable mRNA for amelogenins (Young et al. 1988) and synthesis of enamel matrix protein (Graver et al. 1978; Nanci and Warshawsky 1984a, b; Nanci et al. 1984, 1988) prior to the morphological recognition of extracellular enamel matrix production. Mitochondria become concentrated in the pole of the ameloblasts towards the stratum intermedium.

E. Enamel Formation

I. The Fully Differentiated Ameloblast

The fully differentiated ameloblast is recognised by its juxtaposition to the enamel matrix which it has secreted on the surface of the dentine (Figs. 1, 2). There is evidence in some reports that, prior to the deposition of enamel, the ameloblast may participate in the formation of the enamel-dentine junction region in another way. Dentine is secreted against the basal lamina/basement membrane running between the odontoblasts and the enamel epithelium (Slavkin et al. 1983; Thesleff et al. 1979). This material is removed prior to the deposition of enamel matrix. It is the view of Slavkin and others that the future ameloblast removes this material by some form of extracellular proteolytic activity. That the ameloblast may engage in a resorptive activity and that this may account for an aspect of the morphology of the enamel-dentine junction was mooted by Abbott as early as 1889.

The discovery that enamel is an extracellular secretory product was made separately by Fearnhead (1960) and Watson (1960) and depended upon the demonstration of an intact plasma membrane at the distal Tomes' process pole of the ameloblast, separating it from the extracellular secretory product. Enamel matrix is recognised in most TEM preparations as "stippled material" (War-

Fig. 1. **a** Surface of dentine in developing human third permanent molar (prepared by dissecting away enamel organ and freeze-drying), showing the texture of peripheral collagen of dentine, upon which initial enamel matrix can be seen in top left corner. Field width 59 µm. **b** Inset, higher magnification. Field width 8 µm. **c** Surface of developing enamel on dentine, showing initial secretion in interameloblastic (interpit) location. Note presence of several fine pits within floors of pits. These are sites of development of the enamel tubules. Field width 90 µm

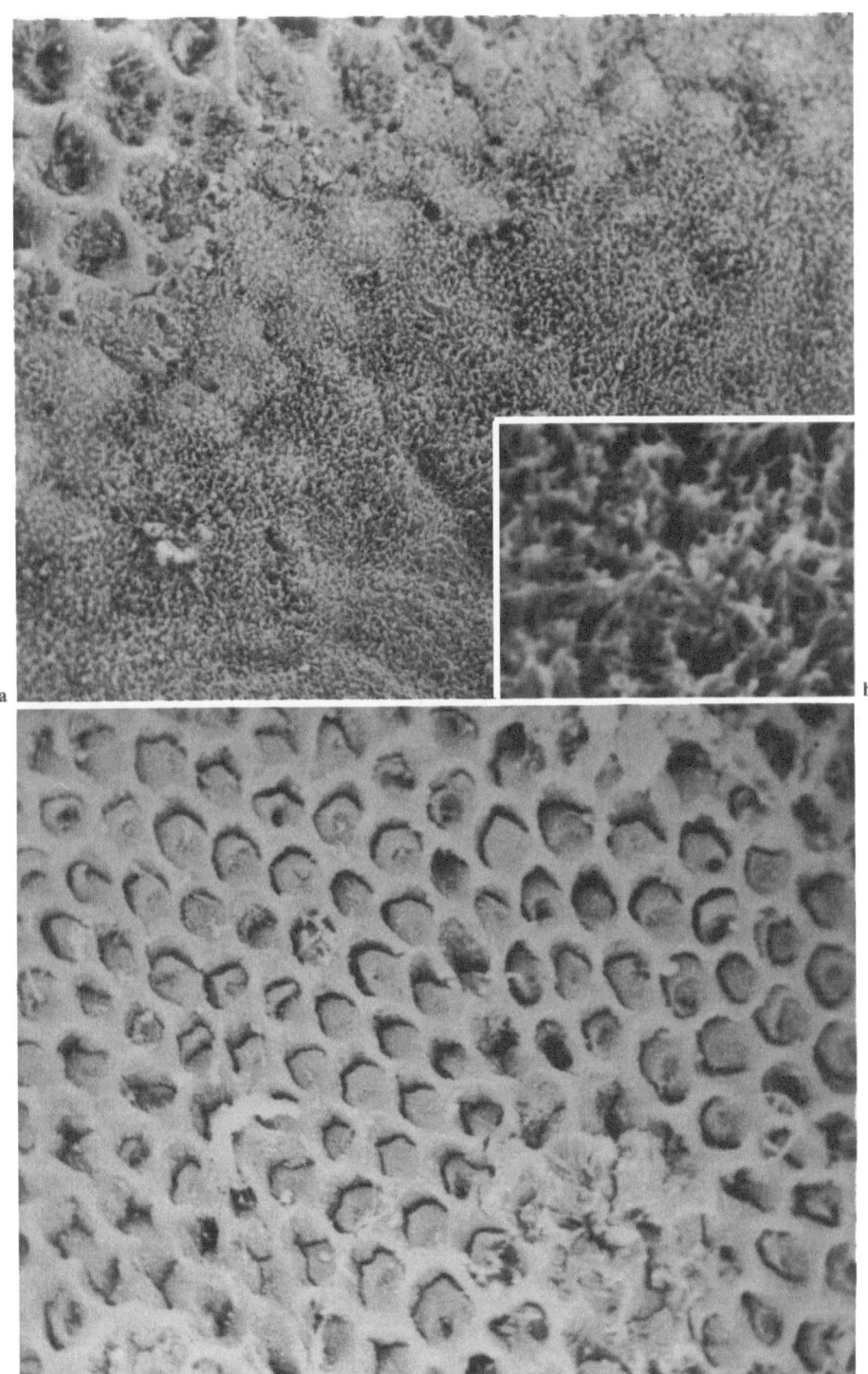

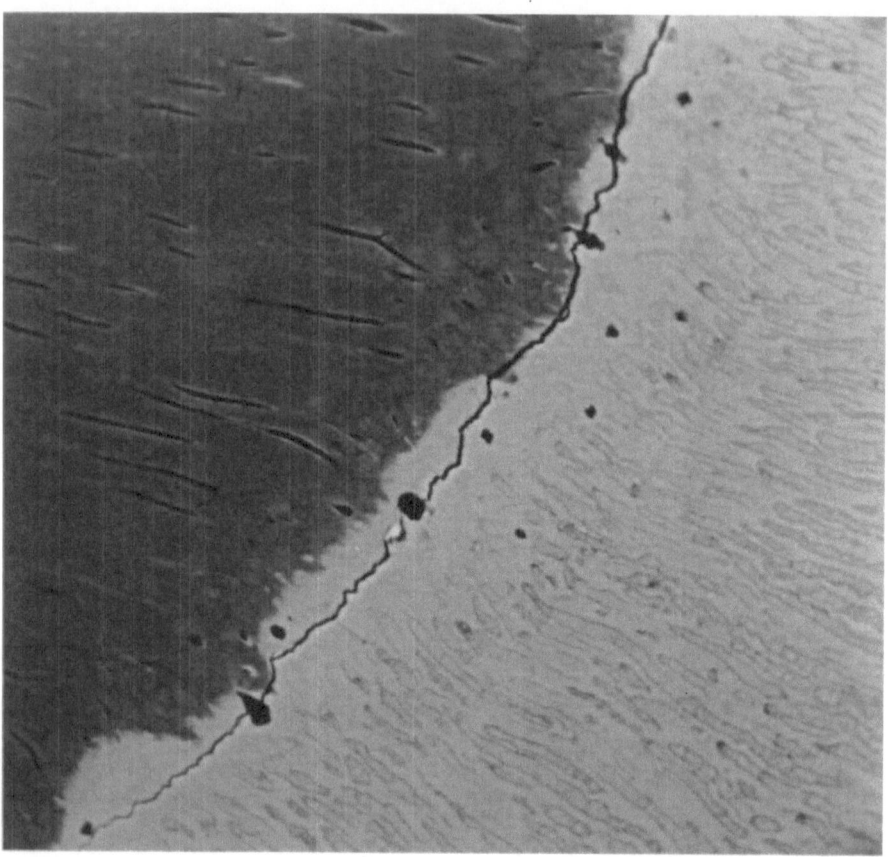

Fig. 2. Longitudinal section surface of human tooth, flat-polished, carbon-coated and imaged with symmetrical collection of back-scattered electrons (BSE) in an SEM. The enamel-dentine junction shows the characteristic embayed profile junction and the very fine scale of interdigitation between enamel and dentine, and demonstrates the continuity of enamel and dentine tubules. The enamel prism boundaries show as black in this image, as do dentine tubules. The cross-striations of the enamel prisms appear as alternate dark-grey and white bands across the prism axis. Note the prism-free layer close to the dentine. Field width 119 μm

SHAWSKY et al. 1984b). It acquires this non-uniform electron density in the process of fixation and electron staining using heavy metals, and some methods of preparation do not give rise to this clustering of density in the matrix (NANCI and WARSHAWSKY 1984a). At the time of writing, there are no data (from, for example, STEM of unstained enamel matrix) to suggest that there is a recognisable non-uniform distribution of the protein matrix gel. There is probably, therefore, no significance in this morphological feature of the enamel matrix, other than that it enables it to be recognised clearly and distinctly from any other extracellular matrix product, including dentine matrix collagen and basal lamina/basement membrane collagen, in the region of the future enamel-dentine junction.

The most important difference between the elaboration and secretion of enamel matrix protein and any other extracellular secretory product liberated by columnar epithelial cells is that the enamel matrix acquires a permanent form through the replacement of its majority water phase by a majority calcium-phosphate-crystal phase. The inception of mineralisation occurs extremely close to the secretory front, so that there is scarcely a delay between secretion and the progress of mineralisation into newly produced matrix. This is in strong contrast to the situation in dentine and bone, where the delay between matrix production and mineralisation in the adult tissues is of the order of several days. We can infer that this delay is only a matter of minutes, maximally tens of minutes, in the case of enamel mineralisation.

The mineralisation of enamel is clearly strongly dependent upon the blood supply to the enamel organ, which begins as enamel mineralisation begins (TOBIN 1972) and continues throughout maturation (HODDE et al. 1983; IWAKU and OZAWA 1979).

The degree of rigidity conferred by the mineral component of the enamel matrix makes it possible to rescue information about the shape of the interface between the ameloblasts and their secretory product from the mineral phase alone, and it proves to be more sensible to adopt this approach in making 3-D analyses (see below).

The fully differentiated ameloblast shows the following important features (Figs. 3–7). It is an extremely elongated cell, having an elongation ratio of >10:1. The elongated nuclei are polarised at the end away from the enamel. Between this now proximal or basal end of the nucleus and the stratum intermedium end of the cell lies the greater concentration of mitochondria. Distal to the nucleus (towards the enamel) lies the Golgi apparatus and the long cell body full of flat stacks of rough-surfaced, endoplasmic reticulum. The secretory product is packaged in the Golgi to form secretory granules, which migrate towards the distal secretory pole of the cell to fuse with the distal plasma membrane, so that the secretory product is exocytosed. Actin filament bundles are a prominent feature in ameloblasts throughout the secretory and maturation zones (REITH and ROSS 1973; NISHIKAWA and KITAMURA 1983).

Much of the enamel structure depends upon the fact that the secretory product is released not in a haphazard fashion from the distal pole of the cell, but in a very special relationship to certain parts of the cell distal to the distal terminal web. Thus, in the most common form of ameloblast, which has a flattish face on one side of Tomes' process (in human enamel roughly perpendicular to the prism direction within the forming tissue), this is a major secretory site. Remaining secretory locations are closer to the main cell body at the intercellular location closest to the neighbouring ameloblasts. The secretory pole of the cell projects into the enamel matrix, causing and occupying a pit. The projecting portion is called the Tomes' process, and the pit we may call the Tomes' process pit. It is commonly as deep as it is wide in human enamel, but it may be much deeper than it is wide in many other mammalian species, the much-studied rat incisor enamel being a good case in point (REITH 1961, 1963, 1970; FRANK 1968; KATCHBURIAN and HOLT 1972; KURAHASHI and MOE 1969; RONNHOLM 1962; SASASKI and HIGASHI 1983; SASAKI 1983,

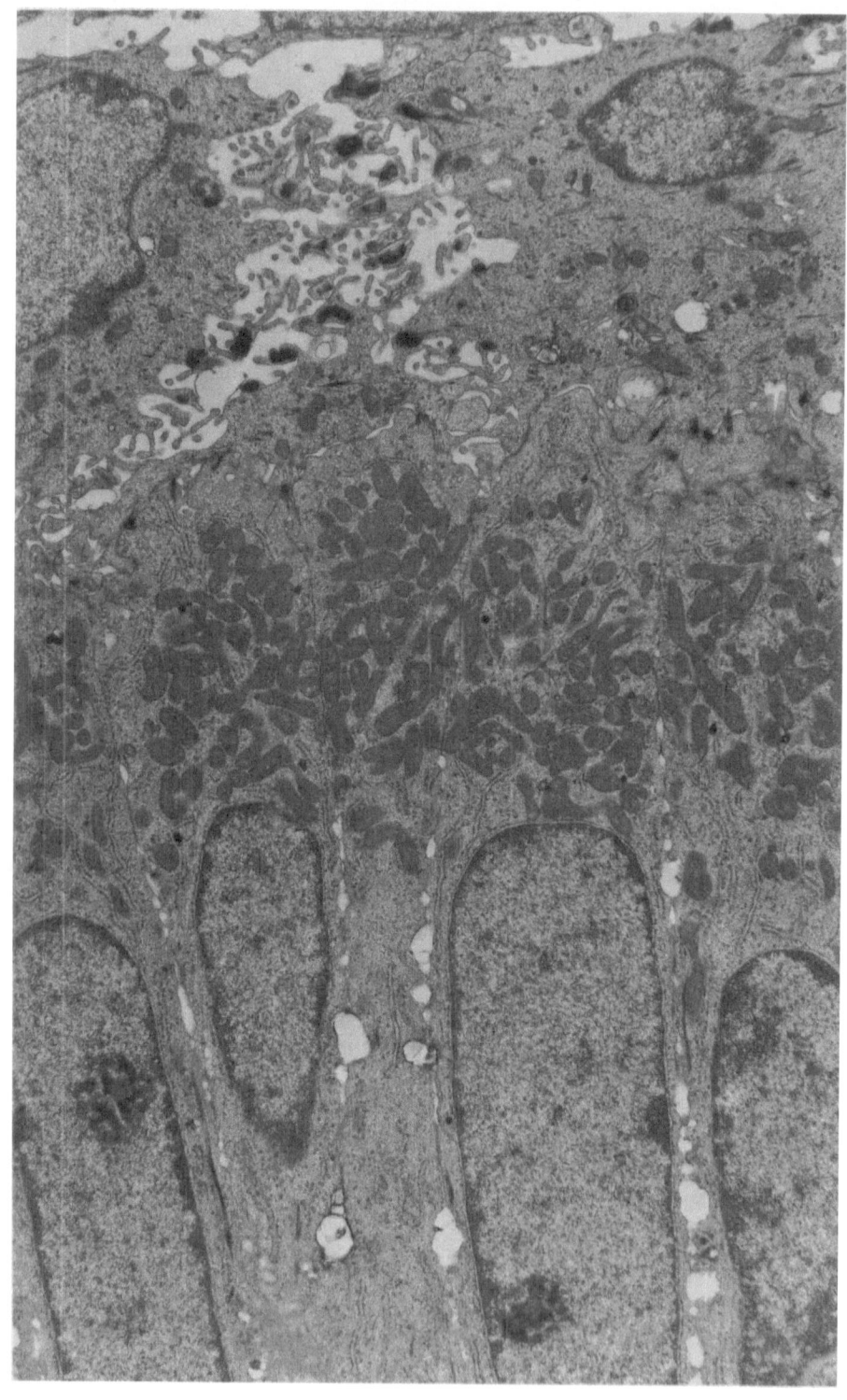

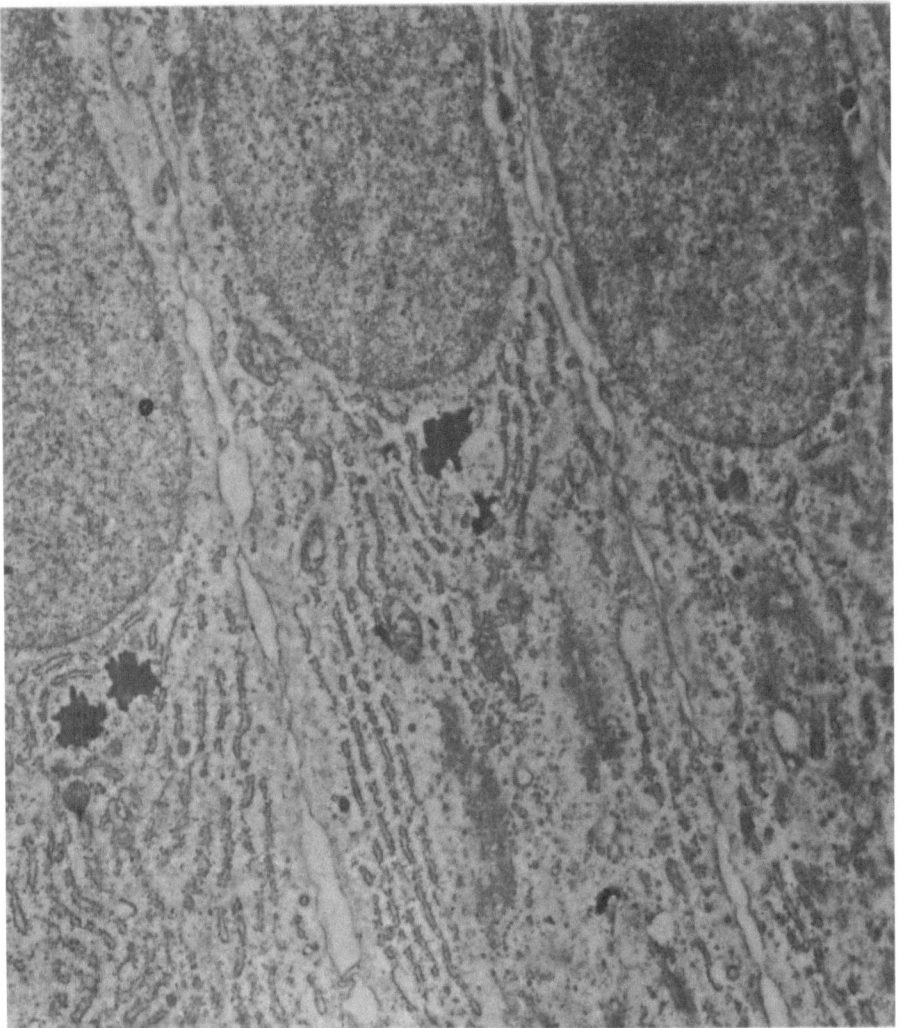

Fig. 4a. TEM of longitudinal section of monkey premolar secretory ameloblasts, showing the Golgi regions in the cytoplasm just distal to the nuclei; black patches are lipid droplets. Field width 12 μm

1984b; SASAKI et al. 1981, 1988a; WAKITA et al. 1981; WAKITA and HINRICHSEN 1980; WAKITA and KOBAYASHI 1983; WEINSTOCK and LEBLOND 1971; SKOBE et al. 1981; WARSHAWSKY 1968; WARSHAWSKY et al. 1981).

The secretory ameloblasts are closely apposed to each other, with little intercellular space which can be recognised by TEM. In suitable SEM specimen

◁ **Fig. 3.** TEM of longitudinal section through rat incisor inner-enamel secretory ameloblasts and their junction with the stratum intermedium (papillary layer), showing the concentration of mitochondria proximal to the nucleus. There are gap junctions between ameloblasts and papillary layer cells. Field width 15 μm

Fig. 4b. TEM of longitudinal section of rhesus monkey premolar secretory ameloblasts, showing Tomes' process pits in the surface of the enamel. Field width 11 μm

preparations in which the cells are encouraged to part from one another, it can be demonstrated that the lateral intercellular surface is basically smooth. There are, however, bulbous projections of one cell into the territory of another at least in the case of rat molar ameloblasts (REITH 1970).

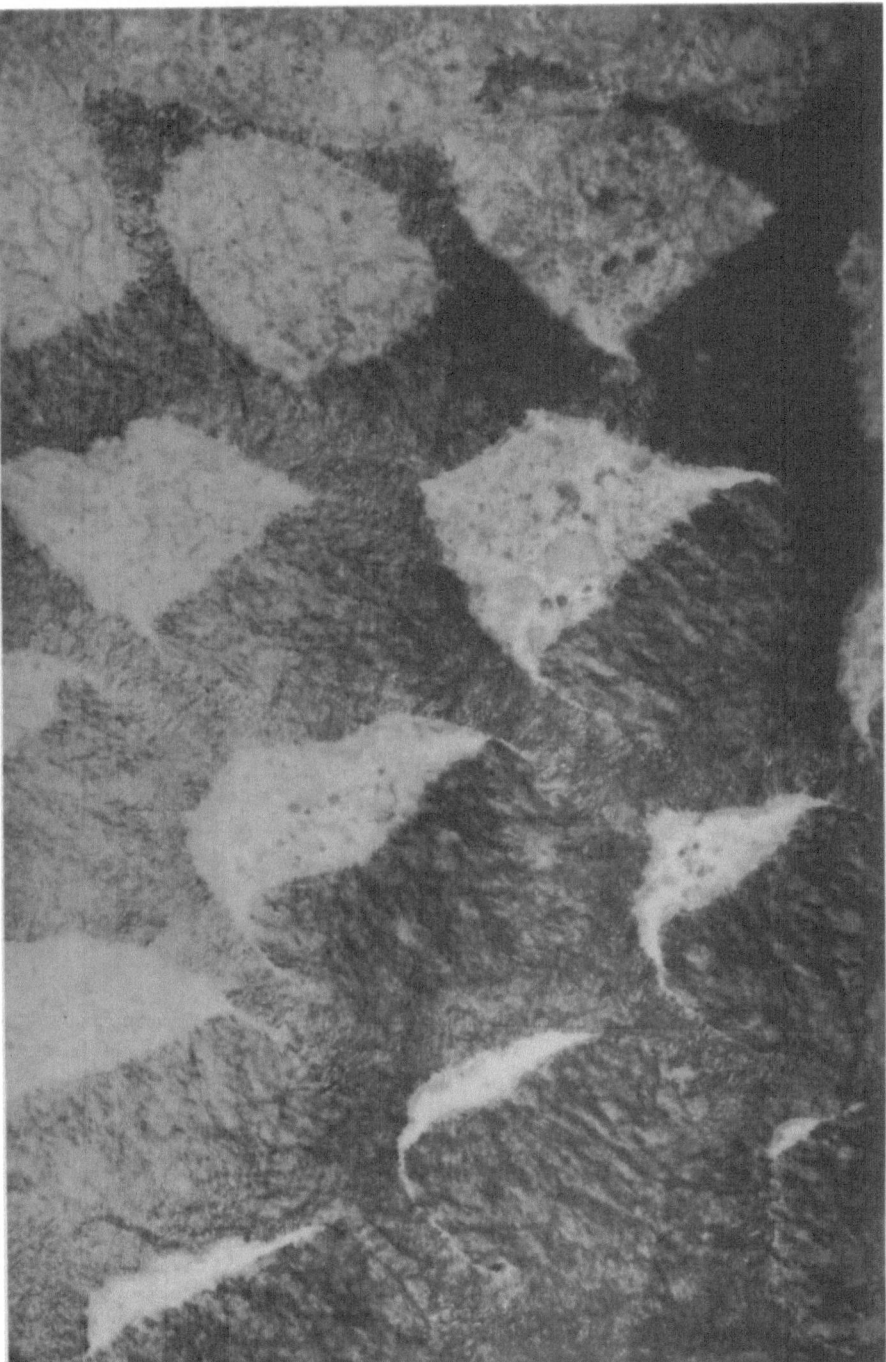

Fig. 5a. TEM of transverse section of Tomes' processes in tangential section of developing enamel surface in cat deciduous molar, showing the eccentric secretion resulting in the filling-in of the pits in the developing enamel surface from one side, generating Pattern 3 prisms. Field width 5 μm

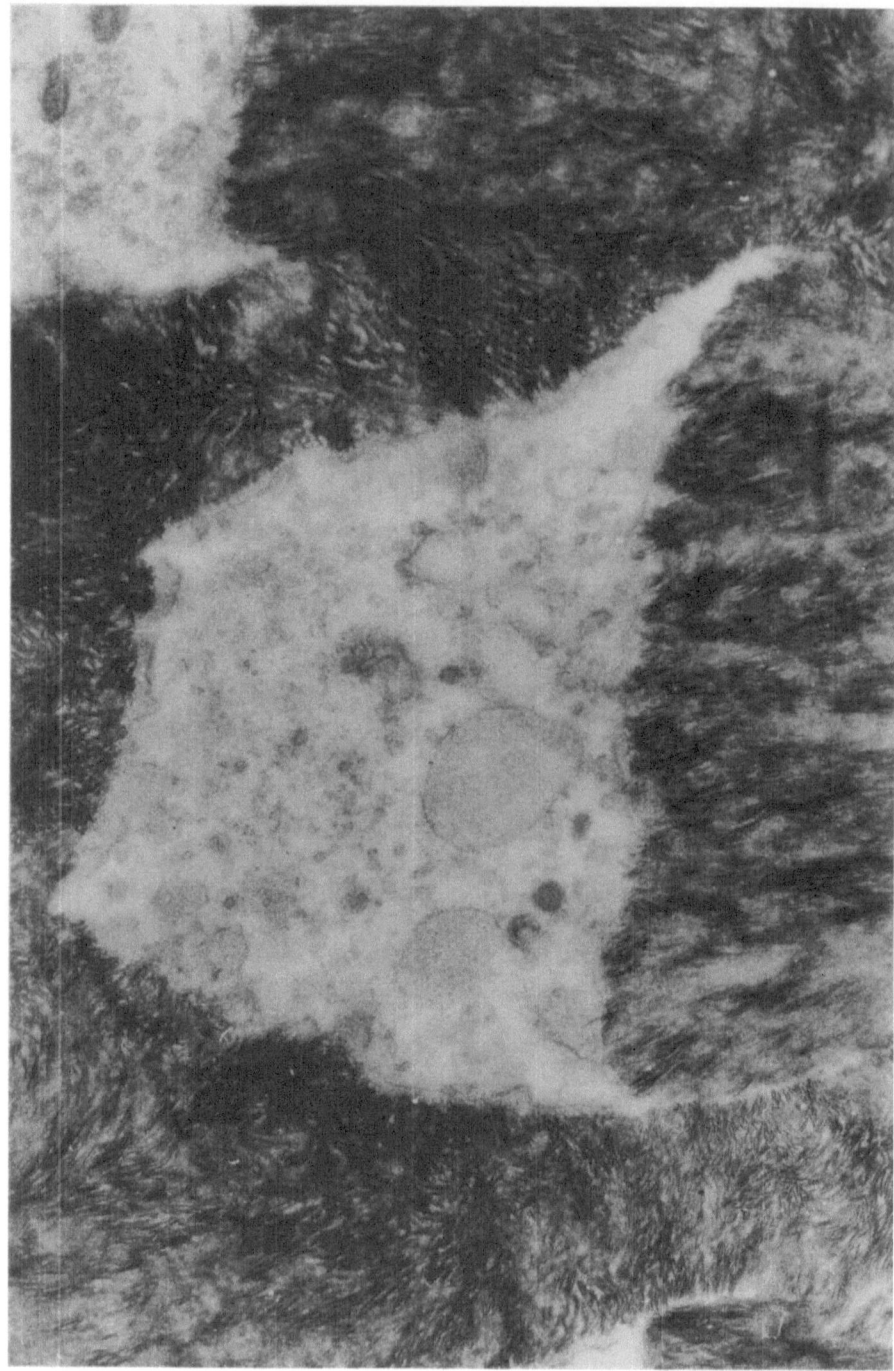

Fig. 5b. Same as **a** showing high magnification of one Tomes' process pit. Field width 1 μm

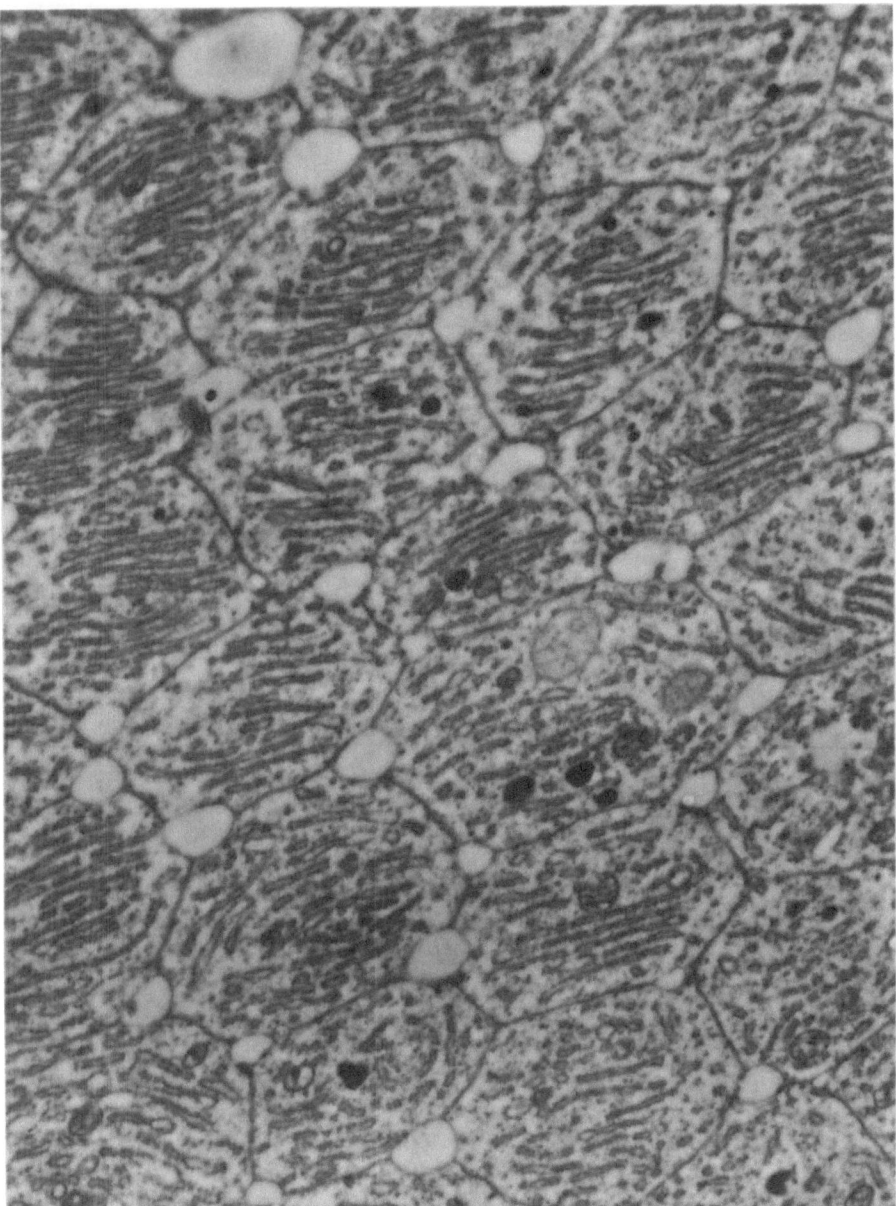

Fig. 6. TEM of transversely sectioned secretory ameloblasts of pig deciduous molar, showing alignment of rough endoplasmic-reticulum nearly parallel with the longituduinal-row axis of the ameloblasts, which are making Pattern 2 enamel. Field width 12 μm

Transverse section images of secretory ameloblasts confirm the limited extent of the lateral intercellular space compartment, contrasting with the extensive space between ameloblasts and adjacent stratum intermedium cells, and with the extensive extracellular space occupied by the enamel matrix at the secretory

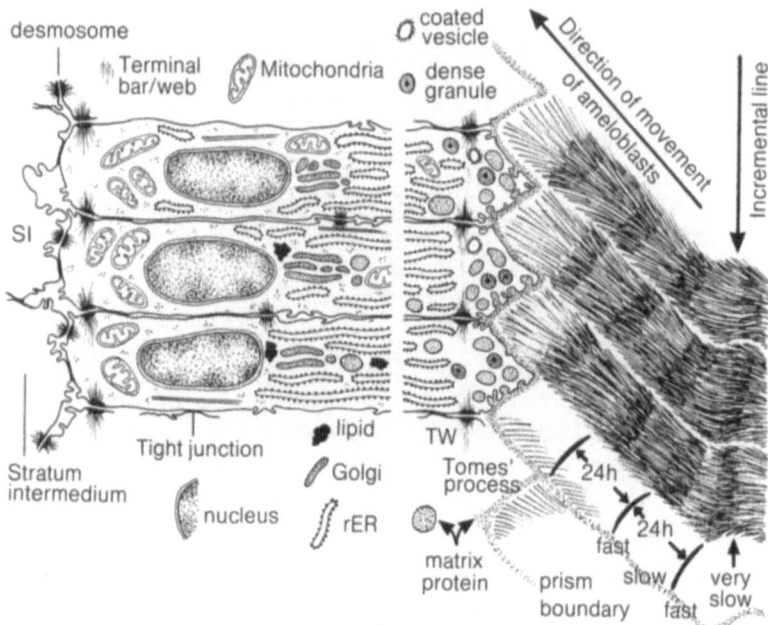

Fig. 7. Amelogenesis. Tall columnar ameloblasts secrete the protein matrix in which hydroxypatite crystals grow. Bundles of about 10000 crystallites constitute prisms, which are demarcated from one another by a sharp change in crystal orientation generated by a corresponding change in the direction of the surface of the developing enamel. The cross-striations of the enamel prisms are undulations of the enamel-prism boundary and correspond to 24-hour increments of formation. Unknown cyclical growth disturbances cause more severe changes in prism direction corresponding to the icremental lines or brown striae of Retzius

pole. One interesting aspect of enamel development in mammals is the tendency to behave in groups organised in a row-like fashion. In the case of ameloblasts making Pattern 2 enamel (see below), in which the cells are organised as longitudinal rows, it is common to find the rER profiles elongated in the direction of the longitudinal row of cells responsible for making the longitudinal rows of prisms (Fig. 6).

II. Ultrastructural Detail Within Secretory-Phase Enamel

As already described, a thin layer of extracellular stippled material, the EM equivalent of the enamel protein matrix gel, can be found between the plasma membrane of Tomes' process of the ameloblast and the layer of the same matrix containing recognisable mineral crystallites. The crystallites are recognised as electron-dense "fibres", most usually oriented perpendicular or at a large angle to the nearest part of the surface – the interface with the ameloblasts. These fibrous elements are generally accepted as being single hydroxyapatite-like crystals.

By mixing some of the unfixed secretory-phase enamel with water and making a suspension, which can be spread on an electron-microscope grid, it is possible to demonstrate that these single crystals can be isolated as tape or lath-like features of very considerable length. These crystals demonstrate moving Bragg extinction contours when carefully tilted under the electron beam, demonstrating their single-crystal nature. In tangential sections of the developing enamel surface, it can be seen that these crystallites have a flattened hexagonal section, with a dark zone at the centre of the crystal running in the longest diagonal of the hexagon. Such sections frequently show groups of crystallites in which the longer sides of the hexagons are aligned parallel, so that the orientation in this sense is not entirely random (BOYDE 1964).

The space between the adjacent enamel crystal profiles rapidly reduces with distance from the secretory front as the individual crystallites thicken. There is no evidence that any new crystals are deposited in the enamel. All new crystals in enamel grow in the related matrix region very close to the level of the mineralising front. The increase in the degree of mineralisation of enamel happens by a process of growth of these original colonising crystals.

III. Discontinuities in the Crystal Orientation Pattern

Discontinuities in the crystal orientation pattern can be recognised in any plane of section of the developing enamel surface. Several characteristic profiles of the developing enamel surface can be recognised in sections which cut this surface nearly perpendicularly. These characteristic profiles are shown in Fig. 8. The essential features to be abstracted from a TEM study are the following: if the Tomes' process is represented as a pointed feature (as in the so-called saw-tooth or picket-fence profile planes of section, in which a single crystallite orientation discontinuity is intercepted per Tomes' process pit), it will be seen that there is a sudden change in orientation of the crystals at either side of the pointed, most distal portion of Tomes' process. The sharp concavity in the surface of the developing enamel induces a sharp change in the orientation of the crystallites, which are otherwise more nearly perpendicular to the surface.

Perpendicularity to the surface may be nearly exact on one side of this discontinuity line, the side occupied by the face of the ameloblast which is actively secreting new matrix product. On the other side, which happens to correspond to the continuous curved lateral and cuspal walls of the pits in the case of human enamel, the crystals may even be nearly parallel with the surface. This is a non-growth surface of the pits (BOYDE 1964; FRANK 1968; KALLENBACH 1977; WAKITA et al. 1981; WAKITA and KOBAYASHI 1983).

In those planes of section in which two prism boundary discontinuities are present per Tomes' process pit, the secretory face of the ameloblast is represented as the floor of the pit, the most distal portion of the Tomes' process. The walls of the pits will contain crystallites whose fragment orientation is more

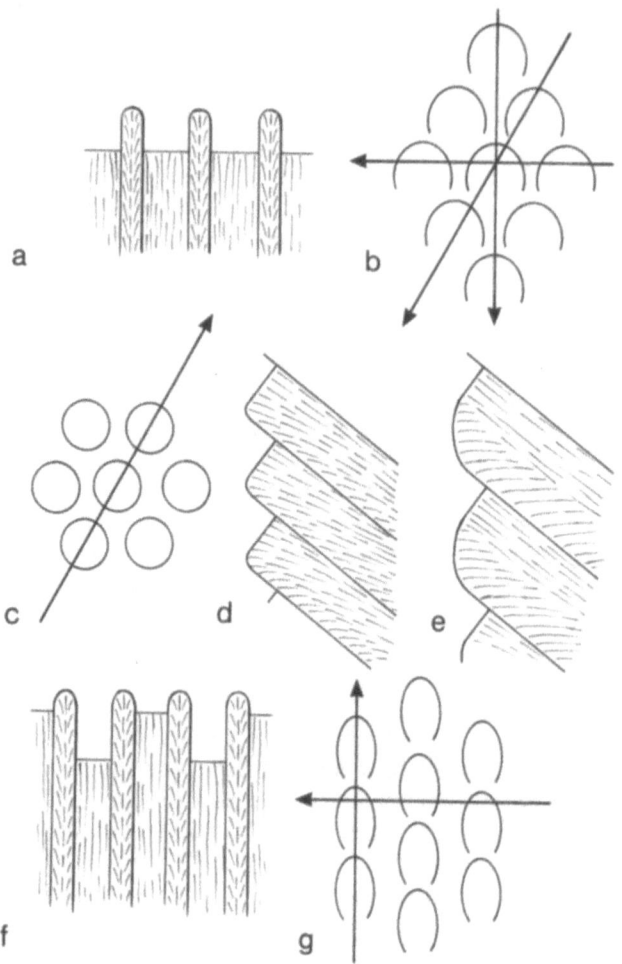

Fig. 8 a–g. The profile of the developing enamel surface and the underlying "grain" direction of orientation of the crystal fragments which would be seen in the corresponding plane of section for three different section directions through Patterns 1, 2 and 3 enamel. The battlements profile is seen for all section directions through Pattern 1 enamel, the apparent depths of the pits being the same for all directions (**a**). The battlements profile is only found in the "transverse" section direction in Pattern 3 enamel. In Pattern 2 enamel, the pits appear at alternate depths (**f**). The picket-fence profile, one prism boundary per pit, occurs in both Patterns 2 and 3 enamel in the directions indicated. The very long interval between prism boundaries is only found in the ideal "longitudinal", head-to-tail section direction in Pattern 3 enamel. Sections in the same direction in Pattern 2 enamel show a parallel field of crystals perpendicular to the surface since this section direction does not cross a prism boundary

nearly parallel with the local surface (again, the non-growth surface of the enamel). The crystal orientation discontinuity can also be recognised in tangential sections of the developing enamel surface, though in these planes of section all the crystals are obliquely or transversely sectioned, and it is not easy to recognise the discontinuity from the lengths of the crystal fragments so much as from the small gap which may exist in this location. Tangential sections of the surface show that this discontinuity is horseshoe or arch shaped, and that the individual horseshoes or arches are not continuous with one another. They also clearly show that the flat side of the filling-in pit in the developing enamel surface is the growth side, at which it is easy to recognise the process of secretion of enamel matrix (BOYDE 1964; FRANK 1968; KALLENBACH 1977).

The 3-D interpretation of the various appearances of the planes of section of the developing enamel surface seen by TEM is best correlated with a 3-D view of the developing enamel surface such can be obtained from SEM preparations of the enamel surface from which the ameloblasts have been removed, or of the mineralising front of the enamel (from which both cells and matrix have been removed) (see Section F.III.).

IV. Amelogenins and Other Matrix Components (Biochemistry, Histochemistry, Histology and Retention During Maturation)

Two principal protein groups are found within the newly secreted enamel matrix. These are known as amelogenins (the proteins found during the formative stages of enamel) and enamelins (those proteins which are retained into the mature tissue). The amelogenins are largely lost during the maturation process in this tissue, which begins as soon as any one volume is secreted and continues for a long time after the full thickness of the enamel has been elaborated (DEUTSCH et al. 1984; EASTOE 1982; ROBINSON et al. 1971, 1975, 1977, 1978, 1979, 1981 a, b, 1984; ROBINSON and KIRKHAM 1984).

Amelogenins are inherently unstable proteins. The principal component has a molecular weight of about 25 kilodaltons. Bovine and porcine amelogenins have been sequenced. The enamelins are inherently stable proteins with a much higher molecular weight. Several estimates of the principal component show the molecular weight to be about 56 kilodaltons. Amelogenins are the majority phase in young enamel and distinctly the minority phase in mature enamel. The amino acid sequence has been determined for bovine and porcine amelogenins (SASAKI et al. 1988 b).

Amelogenins are very pH sensitive and will be rendered unstable by a pH change from neutral to the 6.8 which can be inferred to exist close to the vicinity of the growing crystals. The amelogenin matrix gel is thixotropic. The gel will turn to a sol upon the application of pressure and then re-gel. Although of limited physiological significance, it is important to note that the enamel protein matrix gel will turn to a sol on a reduction of temperature from 37° C to room temperature (20° C) and re-gel on warming to physiological temperature. Various workers have suggested that these interesting physicochemical properties of the enamel matrix proteins provide explanations for the removal of the amelogenin matrix during the mineralisation process of the enamel independent of the involvement of any enzymatic mechanisms. Nevertheless, there are several workers who are convinced that proteases are secreted by the ameloblasts with their matrix and after the cessation of mineralisable matrix secretion, and that these are the important factors in the removal of surplus protein from the mineralising enamel (CRENSHAW and BAWDEN 1984; FINCHAM 1984).

KAKEI and NAKAHARA (1983 a, b) have pointed to the immunological relationships between immature enamel matrix proteins (amelogenins) and carbonic anhydrase isoenzyme C, and show that the distribution of carbonic anhydrase activity corresponds to the regions known to be rich in amelogenins. They speculate that the matrix proteins may play a direct role in the mineralisation process, carbonic anhydrase activity supplying carbonate ions, which become incorporated during the initial phase of crystal formation. AOBA et al. (1988), BELCOURT (1984), GLIMCHER et al. (1965, 1977), GRAVER et al. (1978) and others have shown amelogenins in pre-secretory ameloblasts.

V. Histology of Demineralised Enamel Matrix

If a tooth germ is fixed prior to decalcification, dehydration and embedding for routine histology, it is routinely possible to retain the organic matrix of the enamel from the secretory-phase tissue. Depending upon the rate of maturation of the tooth, which in turn depends upon the rate of secretion, the tooth type (deciduous versus permanent) and the species, some of the organic matrix is retained even after the secretory phase is completed. The organic matrix of enamel retained after fixation and demineralisation is basophilic. In routine haematoxylin-and-eosin-stained preparations, it is deep purple. In some preparations, it is possible to distinguish histological features in the demineralised enamel matrix. Thus, enamel prism boundaries or sheaths and cross-striation features, as well as longer periodic incremental line features, can be recognised. However, no histological detail is present in the matrix in most demineralised enamel preparations. It is not known what determines whether or not such features may be seen. That they are more generally absent, however, does suggest that the major feature or factor determining the recognisable features of enamel histology is the mineral component. This interpretation is strongly backed by the information obtained from TEM and SEM of developing enamel.

Weill and Goldberg and collaborators have published extensively on the subject of the histochemistry of the enamel matrix and conclude that some form of protein-polysaccharide complex or proteoglycan can be recognised by this means (refs. in GOLDBERG et al. 1987). The histochemical staining reaction with Sudan Black might also be taken to indicate the presence of a lipid component in the enamel matrix. However, this reaction specifically demonstrates hydrophobic compartments, and it is known that the enamel matrix proteins are hydrophobic. In the maturation-stage enamel, cyclical changes in Sudan Black staining – hydrophobicity – can be demonstrated in association with the cyclical changes in morphology of these cells (BOYDE and REITH 1982). Lipids can be extracted from enamel, but the amount increases in erupted teeth. GOLDBERG et al. (1987) conclude that the lipid component in unerupted teeth equates with residual ameloblastic debris: the increase amounts to inclusions derived from bacteria and saliva.

ROBINSON et al. (1977, 1978, 1979, 1981a, b), ROBINSON and KIRKHAM (1984), COOPER (1967, 1968), DEUTSCH and PEER (1982) and DEUTSCH et al. (1984) have studied the phenomenon of maturation by dissecting enamel into small fragments, which have been analysed by chemical and biochemical techniques to assess the organic matrix and mineral composition and fraction.

VI. Histological and Histochemical Changes with Maturation

Histological study of demineralised enamel preparations shows that there is a general reduction in staining intensity in enamel formed at earlier periods (EASTOE 1982; GUSTAFSON and GUSTAFSON 1967). Deeper enamel, formed earlier at any one anatomical level, is less intensely stained than the more recently formed peripheral enamel. Furthermore, the enamel formed at earlier times, i.e. at more cuspal or incisal locations, is also less intensely stained. In a tooth germ in which secretory enamel is present at the same time as very mature stages, it will be found that there is one roughly horizontal level (corresponding with what would be expected to be the prism direction at that location) at which the organic matrix rather suddenly disappears. With haematoxylin-and-eosin staining, the reaction changes from basophilic in the newly secreted matrix to eosinophilic just before it is no longer present. Several histochemical staining reactions show a matching colour change prior to the "disappearance" of the enamel matrix. Two conclusions can be drawn from such appearances. Firstly, a point is reached at which the organic matrix is so tenuous that it can no longer be retained in the absence of the support of the majority mineral phase; secondly, the material which is retained up to this given limit is of a different histochemical nature or balance than the material present at the early secretory phase. The biochemical data would suggest that one of the main reasons for this shift is the shift in the balance of amelogenin, the majority secretory-phase protein, to enamelin, the residual protein in the mature tissue. Histochemistry is not a sufficiently precise science to make further detailed conclusions justifiable (WEILL and TASSIN 1965).

F. Enamel Crystals and Prisms

I. Crystal Chemistry of the Inorganic Component of Enamel

The similarity of the calcium phosphate component of enamel to the naturally occurring mineral hydroxyapatite is indicated by several lines of evidence. It has a similar density, refractive index and birefringence (difference in the refractive index for light vibrating parallel and perpendicular to the major crystal axis). It has a similar, but not identical, ratio of calcium to phosphate on chemical analysis. X-ray and electron diffraction show identical lattice spacings for those lattice planes which are extensive, i.e. those perpendicular to the long axis of the elongated crystals. High-resolution TEM lattice images of enamel crystallites show details of the lattice which are indistinguishable from pure hydroxyapatite (SELVIG 1973). However, it should be noted that whereas all these and other lines of evidence indicate similarity, they all also show significant differences between enamel mineral and pure hydroxyapatite. Not the least of these differences concerns the natural crystal habit of apatite, which forms crystals more extensive in the b-axis direction than the c-axis direction – the opposite of enamel crystallites. The birefringence data are not identical, and the chemical data show a non-stoichiometric ratio of Ca:P. The ideal formula for a unit cell of hydroxyapatite is $Ca_{10}(PO_4)_6 \cdot 2H_2O$, giving a Ca:P ratio of 1.66. Measured ratios are ~ 1.61.

Enamel mineral contains substantial amounts of carbonate and magnesium, in addition to potassium, sodium and fluoride ions, and many other minority-phase ions. For more information see BROWN et al. (1984) and ELLIOTT (1965, 1969). The proportion of magnesium and carbonate in enamel crystals is considerably greater in the newly secreted tissue (ROBINSON et al. 1979, 1981, 1984).

These crystals are distinguished by being the central parts of the crystals which later grow in width to occupy a greater volume. Thus, the developmental data indicate that these ions are concentrated in the crystal centres. Other lines of evidence confirm this viewpoint: in natural enamel caries, it can be shown that the centres of the mature enamel crystallites are selectively dissolved (JOHANSEN 1964; JOHNSON 1967). Measurements of magnesium and carbonate in such carious enamel show a substantial reduction of these two components (HALLSWORTH et al. 1972, 1973). This selective removal of magnesium and carbonate can be imitated in artificial enamel carious lesions generated by treatment of the enamel with appropriate organic acid buffer systems.

A different composition for the cores of the enamel crystals is also indicated by their appearance in TEM, where a dark central line is seen (NYLEN et al. 1963; DACULSI and BERTRAND 1978; FRAZIER 1968). This is explained as a concentration of screw dislocations by ARENDS and JONGEBLOED (1979), who indicate that such imperfections would explain the selective solubility of the crystal cores. A magnesium- and carbonate-rich central phase would be more soluble in its own right.

The rate at which the concentration of the mineral phase of enamel increases during secretion and maturation differs markedly in different species. Owing to the difficulties of preparing plane-parallel sections for measurement by quantitative microradiography or X-ray microscopy, it is perhaps probable that the data obtained from plastic-embedded blocks examined by electron probe X-ray emission microanalysis are marginally more reliable (BOYDE 1964; ROSSER et al. 1968). Both X-ray microscopic and X-ray microanalytical data are certainly more reliable than the interpretation of ground sections imbibed with media having different refractive indices and different molecular diameters (CARLSTROM 1963; ALLAN 1959a, b, c, 1967; CRABB 1959, 1968; SUGA et al. 1970; ANGMAR et al. 1963; ANGMAR-MANSSON 1971; AVERY et al. 1961; BELANGER 1957; VON EBNER 1906; HAMMARLUND-ESSLER 1958).

In the rat molar, the enamel may reach nearly its full thickness with the earliest secreted enamel next to the enamel-dentine junction having reached only 25% of the calcium concentration in a pure apatite standard. In human third permanent molar enamel, the tissue may reach 85% of the concentration of apatite within 100 µm of the secretory enamel surface. In both cases, however, the rate of increase of mineralisation from the secretory front is initially rapid and then slows down deeper in the tissue (ROSSER et al. 1967).

Surveys of the increase in mineralisation in enamel during maturation using polarized-light microscopy, X-ray absorption or both show matching trends which do not relate simply to the sequence of formation and the relative age of tissue microvolumes (ALLAN 1959a, b, 1967). To begin with, the tissue next to the enamel-dentine junction is generally substantially more mature than that formed later. This effect may be explained by a putative access of mineral ions via the pore spaces of the dentine side of the enamel-dentine junction. Secondly, in most regions of the tooth where matrix secretion has stopped, the more superficial layers of the enamel are substantially denser than those lying deep. This would be explained by the preferred access to the mineral ions necessary for growth of the crystal surfaces closer to the enamel organ cells.

Thirdly, there is a trend for the more incisal or cuspal levels of the enamel to be more mature at any given relative thickness level within enamel than those formed later. All of these trends are superimposed upon the first trend mentioned, namely, that opposite any given ameloblast the enamel which is formed first tends to be denser than that which is formed last whilst secretion is still under way.

II. Size and Shape of Enamel Crystallites

Crystallites in the zone of interdigitation of enamel and dentine are larger than those of the dentine but smaller than those which typify the bulk of the enamel (ORAMS 1976). In the bulk of human enamel, crystallite diameters lie mostly in the range 50–100 nm (FRAZIER 1968; JONGEBLOED et al. 1975). The crystals are flattened hexagons in cross section but not nearly as unequal in the ratio of their two diameters as is the case in the developing enamel crystals (WATSON and AVERY 1954; FEARNHEAD 1960, 1961; FEARNHEAD and ELLIOTT 1962; RONNHOLM 1962; BOYDE 1964; JOHANSEN 1964). Furthermore, they are much more unequal and irregular in shape than in the developing tissue. Several studies have employed demineralised material to examine the shape of the enamel crystallites, as it were, in the negative created by the organic matrix component. Special fixation techniques are required to preserve the unstable enamel matrix (SUNDSTROM and ZELANDER 1968).

According to the studies of DACULSI and BERTRAND (1978) and DACULSI et al. (1984), not only do crystallites increase in thickness with depth from the formative surface but they reduce in number – data which could indicate the dissolution of a proportion of the crystallites, or their mutual fusion. DACULSI and KEREBEL (1978) reported a reduction in the number of crystallites per unit cross-sectional area, from 1240 per μm^2 at 25 µm deep to the formative surface, to 781 at 200 µm, to 581 in mature deciduous enamel. However, ABRIGO (1972) found no change in the numbers of enamel crystallites with this stage of maturation. His data indicate a constant number of about 850 crystallites per μm² in cross section.

DACULSI and BERTRAND (1978) also provided precise data on the cross-sectional dimensions of human deciduous enamel crystallites during their maturation by counting standard lattice repeats; as they indicate, this is a procedure which obviates any objections as to errors in the calibration of magnification by working to an atomic or crystal standard.

Many accounts indicate that the mature enamel crystallites are arranged as blocks of the order of 150 nm in length, arranged end to end (NYLEN et al. 1963; SELVIG and HALSE 1972; GROVE et al. 1972). This interpretation is most probably based upon a standard artefact of ultramicrotomy due to the shattering of the enamel crystals across their long axes. Replicas of fractured enamel, the data from the developing tissue (including preparations of isolated crystals) and the data obtained from ion-beam thinning studies of the mature tissue indicate that enamel crystals are extremely long when compared with their narrow cross-sectional diameter. Crystals as long as 100 μm but < 30 nm in diameter have been isolated from developing enamel (FEARNHEAD and ELLIOTT 1962; DACULSI et al. 1984). It would seem to be conceptually possible that a single enamel crystallite might extend all the way from the enamel-dentine junction to the surface of the enamel. Even if this remarkable circumstance were true, however, it would probably never be possible to prove the case. An enamel crystallite must travel in a curved bundle, the prism, and the substantial curvatures of that bundle make it extremely unlikely that such a small element could be traced for a significant distance. However, it is also unlikely to be of any importance to demonstrate that crystals have such a very extreme length since structures made of parallel bundles of fibrous elements depend for their strengths and properties on the relative length of overlap, not on the absolute length of such whisker-like, or fibrous, elements. It is clear that enamel crystallites are many tens of μm long. It is also certain that some of them end within the enamel since it is a property of some prism boundary discontinuities in enamel that a proportion of the crystallites end by butting against those in a neighbouring batch in a different orientation. Since this is so, and since the crystal diameters seem to be relatively uniform throughout the thickness of the enamel, it is also logically the case that new crystals begin to form somewhere within the centres of the enamel prisms in order to balance the numbers of those lost by collision with crystals in adjacent prisms at the prism boundaries.

The gap between adjacent crystallites within one bundle is extremely slender and can scarcely be measured accurately, though it is probably on the order of 5 nm or less. This gap is exaggerated at the prism boundary. There is some evidence that the thickness of the prism boundary increases during enamel maturation. The prism boundary in the mature tissue is also distinguished by the presence of larger diameter crystals with a reduced elongation ratio (JOHNSON 1967; SUNDSTROM and ZELANDER 1968; GUSTAVSEN and SILNESS 1969), but one must be careful to distinguish this as a normal phenomenon from the recognised formation of whitlockite crystals at prism borders as a "normal change" in early caries (VAHL et al. 1966; VAHL and PLACKOVA 1967; LENZ 1956; HELMCKE 1955). These should be considered as due to recrystallisation (JOHNSON 1967; for review see SHELLIS and HALLSWORTH 1987).

III. Crystal Orientation and the Existence of Prisms

1. The Enamel-Dentine Junction

The evidence from ion-beam thinning studies of mature tissues shows that there is a gradation of properties between the crystallites belonging properly to the enamel and distinct from those of the dentine. However, the evidence from TEM of ultramicrotomed sections of the developing enamel shows that there may be a layer in which the crystallite orientation, if not exactly perpendicular to the mean plane to the enamel-dentine junction, does not contain the regularly arranged discontinuities following the development of the Tomes' processes of the ameloblasts, which happens after a few microns of enamel matrix have been secreted (Fig. 2) (BOYDE 1964). There is a layer at the enamel-dentine junction which does not etch differentially when subjected to ion-beam erosion, indicating a uniformity of composition and crystallite orientation (BOYDE and STEWART 1962). Microradiography (ALLAN 1959; GUSTAFSON 1959) and electron probe X-ray emission microanalytical data (BOYDE et al. 1961; ROSSER et al. 1968) indicate that there is a layer at the enamel-dentine junction in mature tissue which is more substantially mineralised than the later deposited tissue. This is confirmed at a much coarser scale of resolution by microchemical analysis of small, dissected fragments of enamel (WEATHERELL et al. 1967, 1968; ROBINSON et al. 1971). Optical microscopy, particularly polarised-light and dark-field (GUSTAFSON 1959) and confocal microscopy of thin optical sections of enamel, shows that there is a relatively uniform layer with no discontinuities for a few microns outward from the junction. The high degree of negative birefringence of this junctional enamel may be explained both by its uniformity of crystal orientation and by a higher degree of mineralisation (VON EBNER 1906). BSE SEM (atomic number contrast) images of micromilled or polished section surfaces usually show an absence of low-density, prism boundary features in the enamel next to the junction (BOYDE and JONES 1983). Preparations of the junctional plane made by air-polishing away deproteinised dentine may show the prism boundary cross sections, but this may be supposed to be due to the removal of some enamel (albeit a very thin layer) due to the erosion caused by this procedure.

It would seem, therefore, that prior to the development of Tomes' processes on the ameloblasts, i.e. prior to the development of the substantive pattern of secretion from selective sites of the secretory pole of the ameloblast, the secretory front may be relatively flat, and the degree of divergence of crystallite orientation is relatively slight. A similar circumstance prevails under normal situations during the development of the most peripheral layer of mammalian enamel.

Fig. 9. Model of developing enamel surface made by constructing the surface using modelling clay to fit the stereoscopic optical model formed from a stereopair of scanning electron micrographs of developing human deciduous molar cuspal enamel. A similar field is seen in through-tilt series in Fig. 11. The lines drawn on the sides of the model indicate expected crystallite orientations and are derived from studies of transmission electron micrographs of matching planes of section and the data from scanning electron micrographs of fractured preparations of developing enamel

2. Majority-Phase Enamel Secretion

During the secretion of the bulk of enamel, crystallite orientations are confined by the fact that the enamel crystallites are extremely long and thin, and there is little chance of their being other than parallel with one another: over ranges of several tens of crystallite diameters, the majority are nearly parallel in the c-axis direction. Crystallite orientation patterns during the majority phase of enamel secretion are relatively simple (Fig. 8). Every enamel crystal is parallel with its neighbour except at the enamel prism boundary discontinuity, the origin

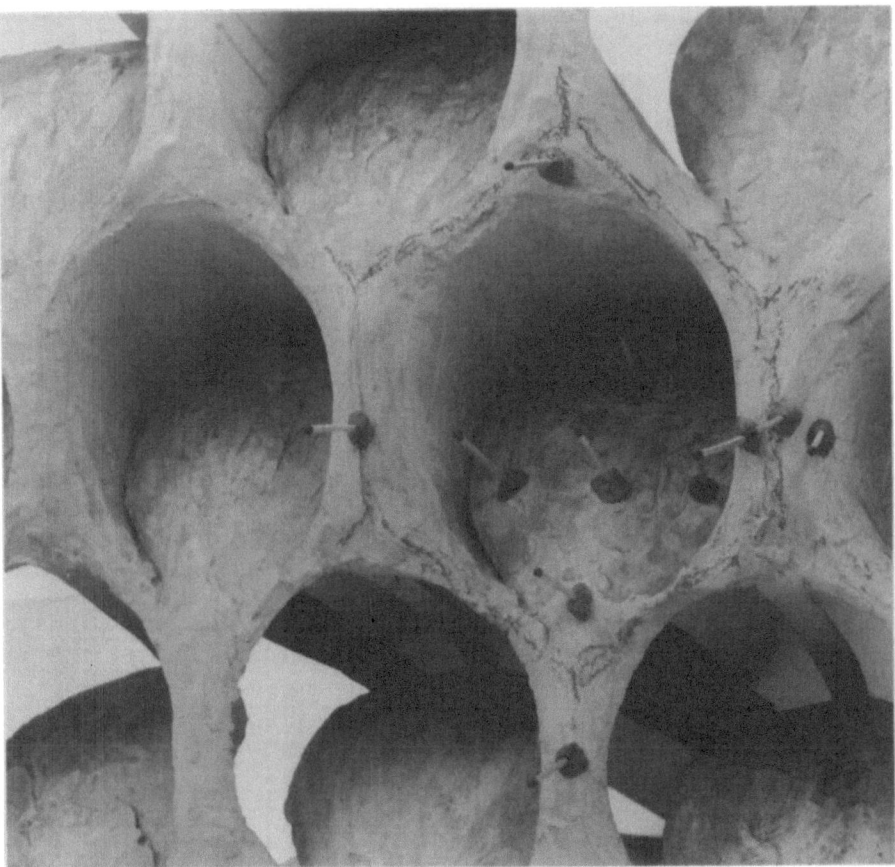

Fig. 10. Macrophotograph of a macromodel built at the scale of 1 inch to the µm, showing typical shape of pits in developing enamel surface of human Pattern 3 enamel. Centre-to-centre distance between the pits in the original model is 6 inches. Matches have been stuck to the model surface in several locations to indicate the orientation of the crystallites deep to the corresponding portion of the surface. In constructing the model, some of the parts of the surface have not been filled in: these missing parts are the relative non-growth surfaces of the interface between the ameloblasts and their secretory product

of which is clearly related to the presence of sharp concavities in the orientation of the interface between the ameloblasts and their secretory product (or the shape of the mineralising front, the two being almost synonymous).

As regards alignment and direction in the other sense of the crystallites (i.e. of their a,b-axes), there is no evidence of preferred orientation except over very short ranges and in the developing tissue, where groups of the flattened crystals may be seen to be stacked with their flat sides parallel (BOYDE 1964). Evidence for any such arrangement in the adult tissue is very scanty and has hardly been commented on in the literature. However, acid-etching enamel generates features with sizes closer to crystal groups than to single crystals.

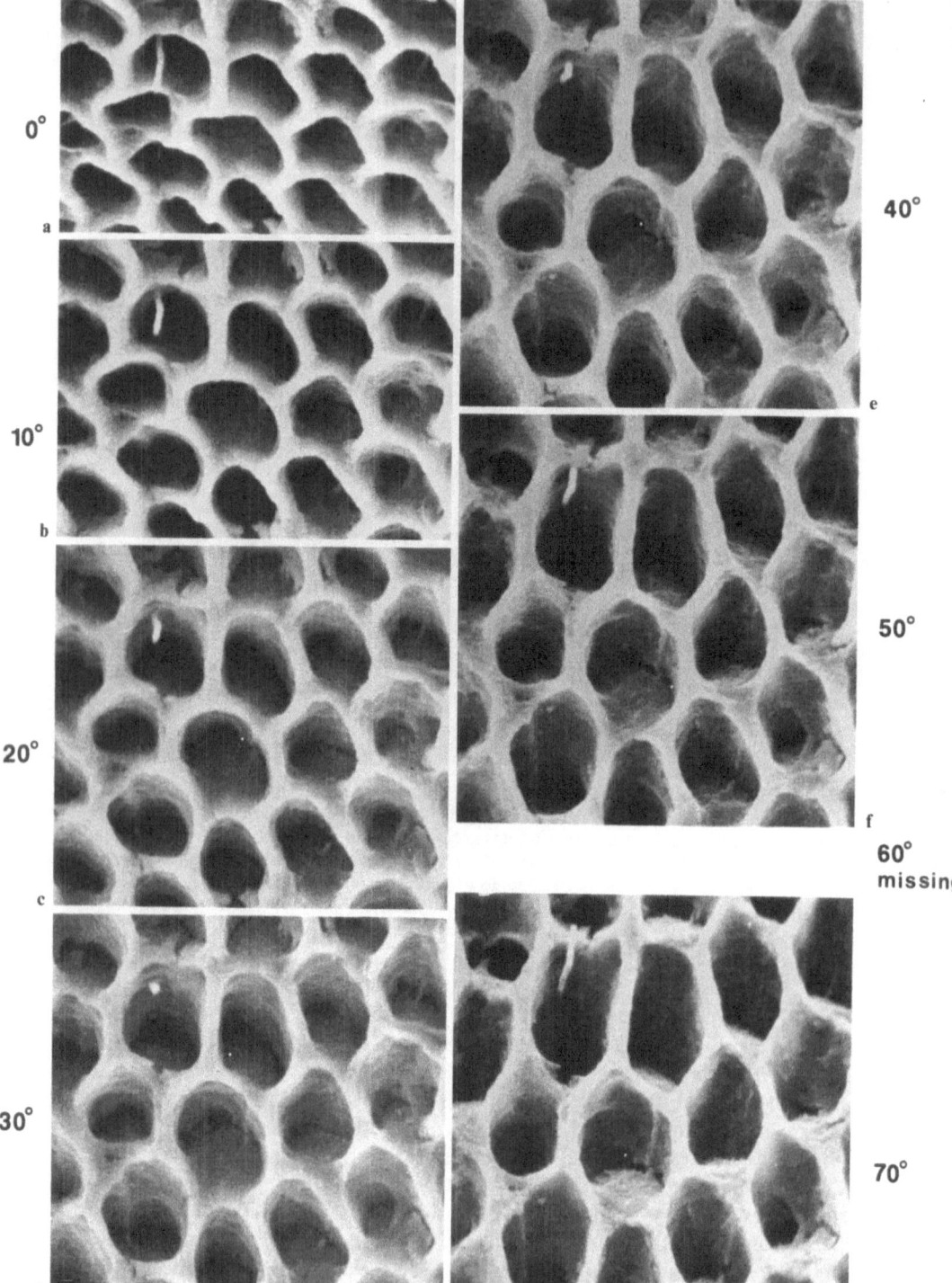

0° a

10° b

20° c

30° d

40° e

50° f

60°
missing

70° g

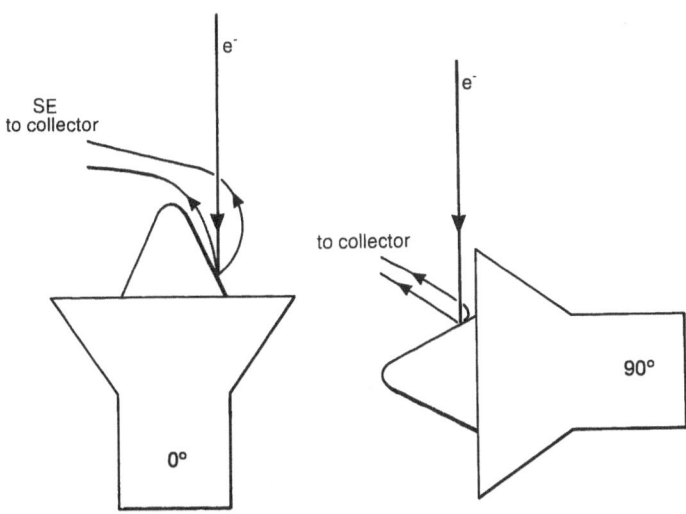

h

Fig. 11 a–h. Through-tilt series of SEM views of the same portion of developing enamel surface of human deciduous molar cusp. Ameloblasts removed by washing away enamel organ. Dehydration in ethanol and drying from absolute ethanol. The prepared surface is not perfectly clean, and debris obscures detail in some of the pits. This field was selected for the particular debris feature seen in the upper left-hand field of view as a spike projecting from a pit. As the specimen is tilted, the spike comes to be viewed end-on in the fourth member of the series. The second member of the series (**b**) views some of the pits nearly along the long axis of the prisms. The last member of the series views the developing enamel surface nearly perpendicularly. Turning the page sideways, each adjacent pair can be viewed as a stereoscopic pair. Note how the apparent details of the morphology change with the angle of view of the surface. The accompanying diagram (**h**) shows the orientation of the views with respect to the geometry of a scanning electron microscope. The tilt-angle difference between each view is 10°. The seventh member of the series, corresponding to a 60° tilt with respect to **a,** has been omitted from the series to save space. The relative lack of detail in **a** and **b** is due to the difficulties of secondary electron detection from a surface hidden from view from the electron collector system. The diagram (**h**) shows the orientation of the specimen.
Field width 20 μm

3. Development of a Prism-Free Surface Layer

On more than half of human permanent enamel (and in many other mammalian enamels), the depth of the Tomes' process pits in the secretory enamel surface undergoes an important reduction during the formation of the most peripheral layers of the enamel: this can be seen on appropriately staged developing teeth examined by SEM. TEM and SEM show that the interface becomes relatively flat, and the crystallites become predominantly parallel to each other and perpendicular to the surface. A layer of "prism-free" enamel develops, in which the prism boundary discontinuities may cease. Microradiography, electron probe X-ray emission microanalysis and wet chemical analysis show an increased degree of mineralisation in this layer (ALLAN 1959b; GUSTAFSON 1959; ROSSER et al. 1967; ROBINSON et al. 1971). Polarised-light microscopy shows a uniform extinction, indicating parallelism of the crystallites, and a higher degree of negative birefringence than the bulk enamel (ALLAN 1959a). Ion-beam erosion does not etch this layer as it does the prismatic bulk tissue (BOYDE

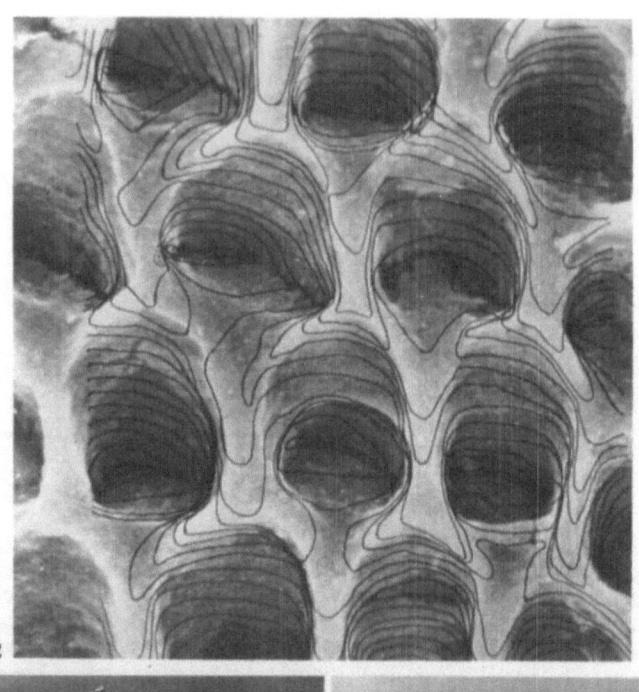

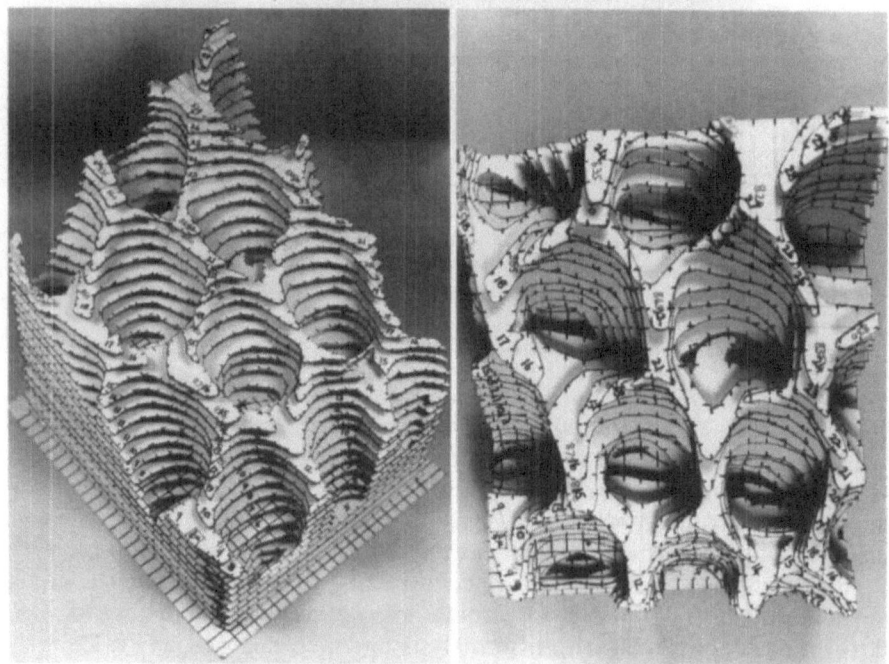

Fig. 12. SEM of developing enamel surface of human deciduous molar cusp, with contour lines reconstructed by stereophotogrammetry drawn back onto the surface. The contours represent 0.75-μm intervals

Figs. 13 and 14. Reconstruction of same field of view from same set of contour lines as in Fig. 12, with contour sections spaced appropriately. Square sections seen in background also represent 0.75-μm intervals

and STEWART 1962). Fractured preparations examined in the SEM also show a preferred orientation perpendicular to the surface.

The prism-free layer is generally more continuous and thicker and may be exceptionally thick in human deciduous enamel compared with the discontinuous layer in permanent teeth, where incremental lines reach the surface in the perikymata (RIPA et al. 1966, 1967; SHROFF and ROMANIUK 1964; SILNESS 1969).

In some mammals, a major part or the majority of the enamel thickness is composed of such a layer of prism-free enamel, strongly resembling the prism-free enamel found in many reptiles (BOYDE 1980; ISHIYAMA 1984).

It used to be thought that reptilian enamel is necessarily non-prismatic, or that if it is subdivided, then into pseudo-prismatic units, in which the crystallite orientation changes but the crystals do not abut at sharp boundary lines; this view has been reversed by the studies of COOPER and POOLE (1973), DAUPHIN (1987a, b) and SAHNI (1984). Dauphin believes that proper prisms may exist in representatives of all the reptilian taxa, but some of her data are unlikely to gain widespread acceptance because of the absence of distinct prism boundaries. The figures in the paper of COOPER and POOLE (1973) are, however, very convincing.

4. Modal Crystallite Orientation

Both polarised light and X-ray diffraction have been used to study the statistics of crystallite orientation with respect to prisms. LYON and DARLING (1957) found that the crystallites always deviated towards the cervical part of the tooth, the angle between the crystals and the prisms being about 17° in mid-lateral enamel; this result was essentially confirmed by GLAS (1962). The X-ray diffraction data give the statistics of crystallite orientation in great detail, but to put this information together as a three-dimensional model requires more information than can be obtained by this combination of low-resolution light microscopy and X-ray diffraction alone. The most important data about crystallite orientation in enamel have therefore been obtained by transmission electron microscopy and scanning electron microscopy. POOLE and BROOKES (1961) provided a model of the arrangement of crystallites in enamel prisms based largely on the polarised-light evidence interpreted through the information then available from electron microscopy.

The deviation of crystal from prism axes in mid-lateral central-thickness enamel of human permanent teeth is a remarkably constant value. This is shown not only by the data of LYON and DARLING (1957) and GLAS (1962) but by thousands of personal observations made by the author and his students and colleagues over many years. In this respect, there does not seem to be much variability in human enamel structure.

5. Explanation of Crystallite Orientation in 3-D as a Function of the Developmental Enamel Surface Morphology

Most of the facts gleaned about the relationship of the crystal orientation to the shape of the developing enamel surface in mammals are in agreement with the following simplified explanation (Figs. 8–20).

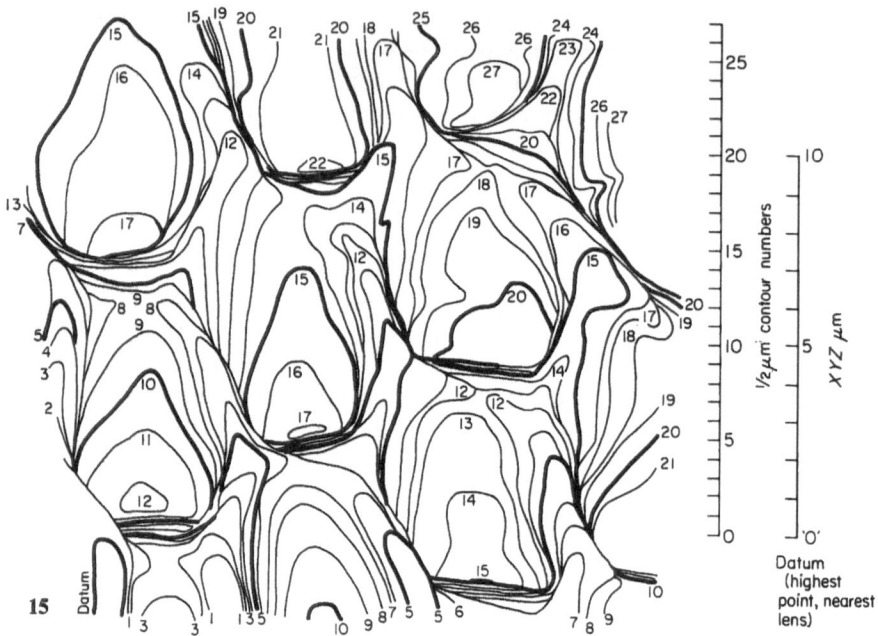

15

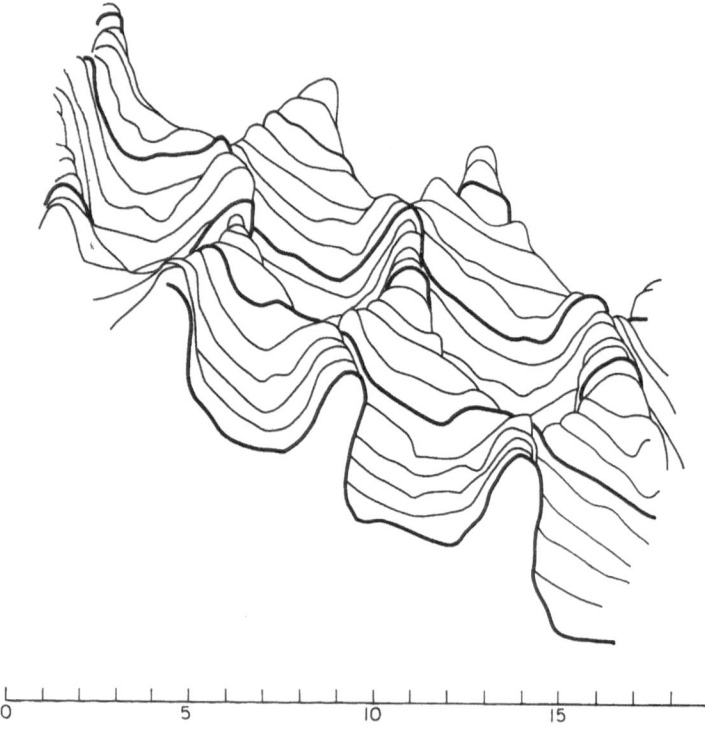

16

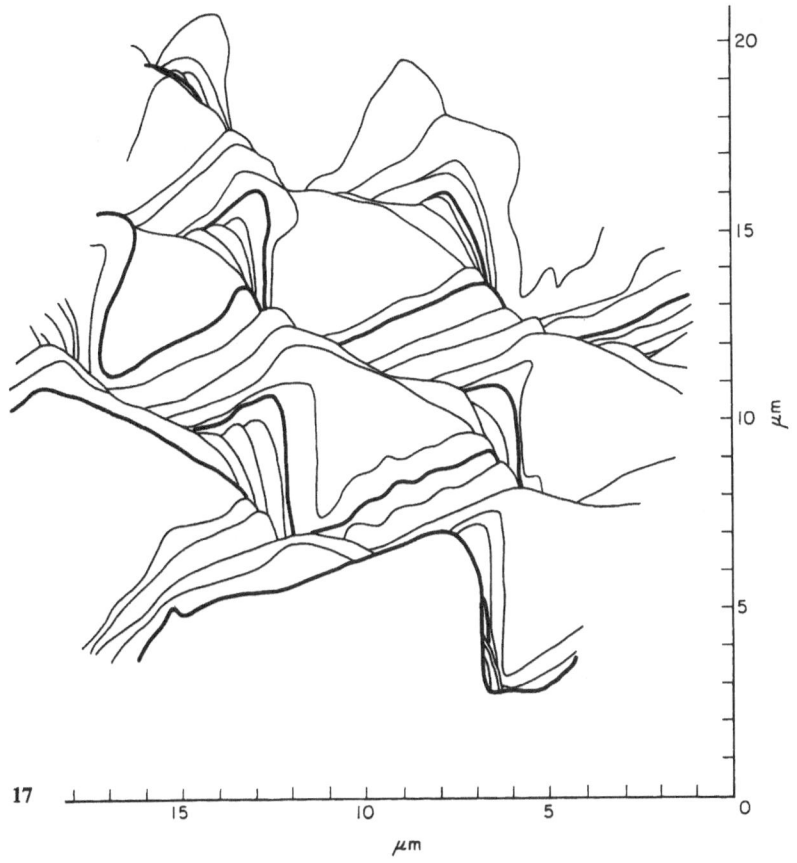

Figs. 15, 16 and 17. Contour map and profiles (Figs. 16 and 17) reconstructed from stereopair scanning electron micrographs of surface of human developing deciduous molar enamel. Figure 15 shows a contour section normal to the mean electron-optical axis. Figure 16 shows a set of profiles reconstructed from cuts parallel with the electron-beam direction (nearly perpendicular to the developing enamel surface) in the transverse section direction across the tooth, and showing the battlements profile of the developing enamel surface. Figure 17 shows a set of profiles reconstructed for cuts parallel with the electron-optical direction and in the longitudinal section direction of the tooth generating a saw-tooth or picket-fence profile. (From BOYDE 1973)

Crystals grow perpendicular to the growth portions of the interface between the ameloblasts and the developing enamel surface. If this interface is nearly flat, as may be the case during the development of the first layers of enamel at the enamel-dentine junction (Fig. 2) and during the formation of the last layers of enamel at the enamel surface in man and in some mammals (and for a high proportion of the period of enamel formation in odontocetes), then the crystallite orientation is uniformly perpendicular to the surface. During the period in which the surface is pitted by the existence of the Tomes' processes of the ameloblasts, different facets of the microsurface meet at sharp concavities in the surface. This situation causes the development of surfaces within the

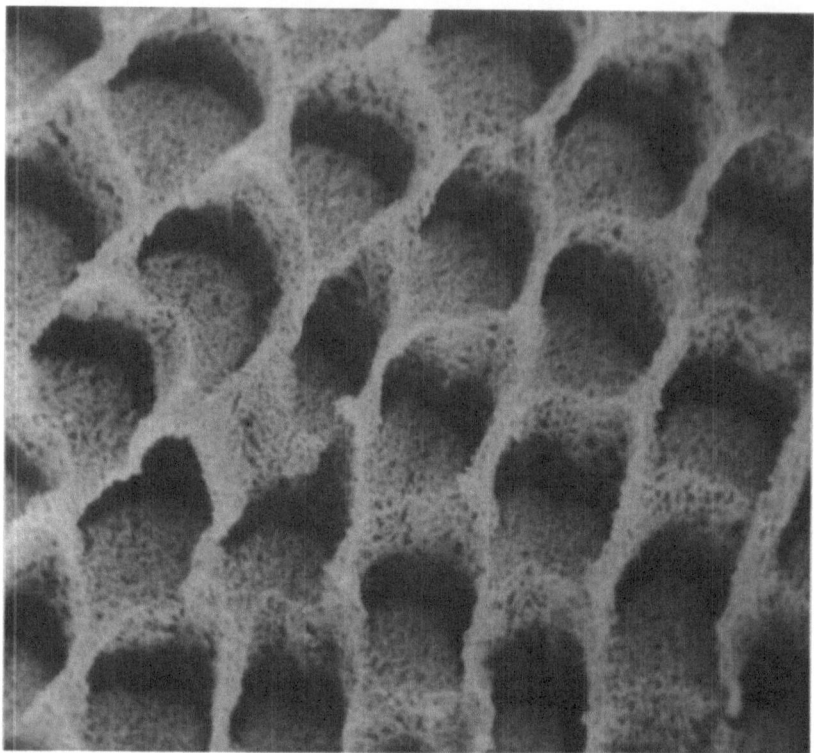

Fig. 18. SEM freeze-dried anorganic developing enamel surface of human deciduous incisor, showing the shape of the Tomes' process pits in the mineralising front. The small holes in the surface of this preparation are generated by ice-crystal formation during the freezing stage of preparation and have no morphological significance. Field width 31 μm

enamel – the prism boundaries or discontinuities – at which the crystal orientation changes through "collision" or crossing. The rates of matrix secretion and crystal growth on either side of such boundaries may be grossly unequal. In many instances, little or no growth is occurring on the side of the pit whose orientation is near the parallel with the general direction of progress of the ameloblast over the previous secretory history (Boyde 1964, 1965, 1967; Frank 1968; Kallenbach 1977; Sasaki and Higashi 1983; Wakita et al. 1981; Wakita and Kobayashi 1983; Warshawsky et al. 1981).

There are two main fields at which enamel grows, i.e. at which new matrix is released, and into which new crystal growth can occur. These are (Figs. 8–20):

1. At the most prominent parts of the enamel surface, where secretion occurs in a mutual interameloblastic location, and where the local surface has the same orientation as the general plane of the developing enamel surface. These locations we refer to as interpit. Crystals growing in these locations have a common orientation throughout the enamel, being perpendicular to the plane of the incremental surface. In most enamel, this constitutes a continuous phase, uninterrupted and joining all the enamel together.

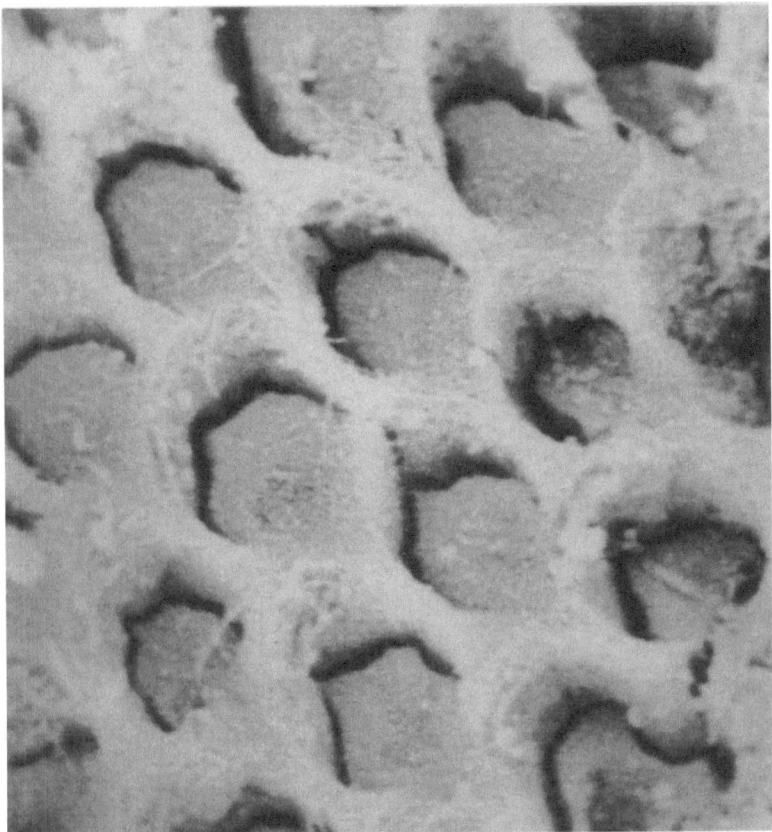

Fig. 19. SEM critical-point-dried developing enamel surface of human third permanent molar, low-lateral enamel showing Tomes' process pits. Field width 27 μm

2. The second type of region in which active new matrix secretion and crystal growth occurs is in the floors of the pits. These are continuous with the tops of the interpit walls on one side of the pit in the most common [Pattern 3 in man, Pattern 2 in many other mammals (see next section)] types of arrangement. This secretory site "belongs to" one cell: in tangential sections of the developing enamel surface seen in TEM, it appears that the cell is filling in the pit by secreting along one (flat) face only.

6. Prism Patterns

a) The Characteristic Profiles of the Developing Enamel Surface and the Three Packing Arrangements of Prisms

For descriptive purposes, we may divide the commonest morphologies of the developing enamel surface and of the resultant prism cross-sectional packing arrangements into three types, designated as *Patterns* (BOYDE 1964, 1965, 1968,

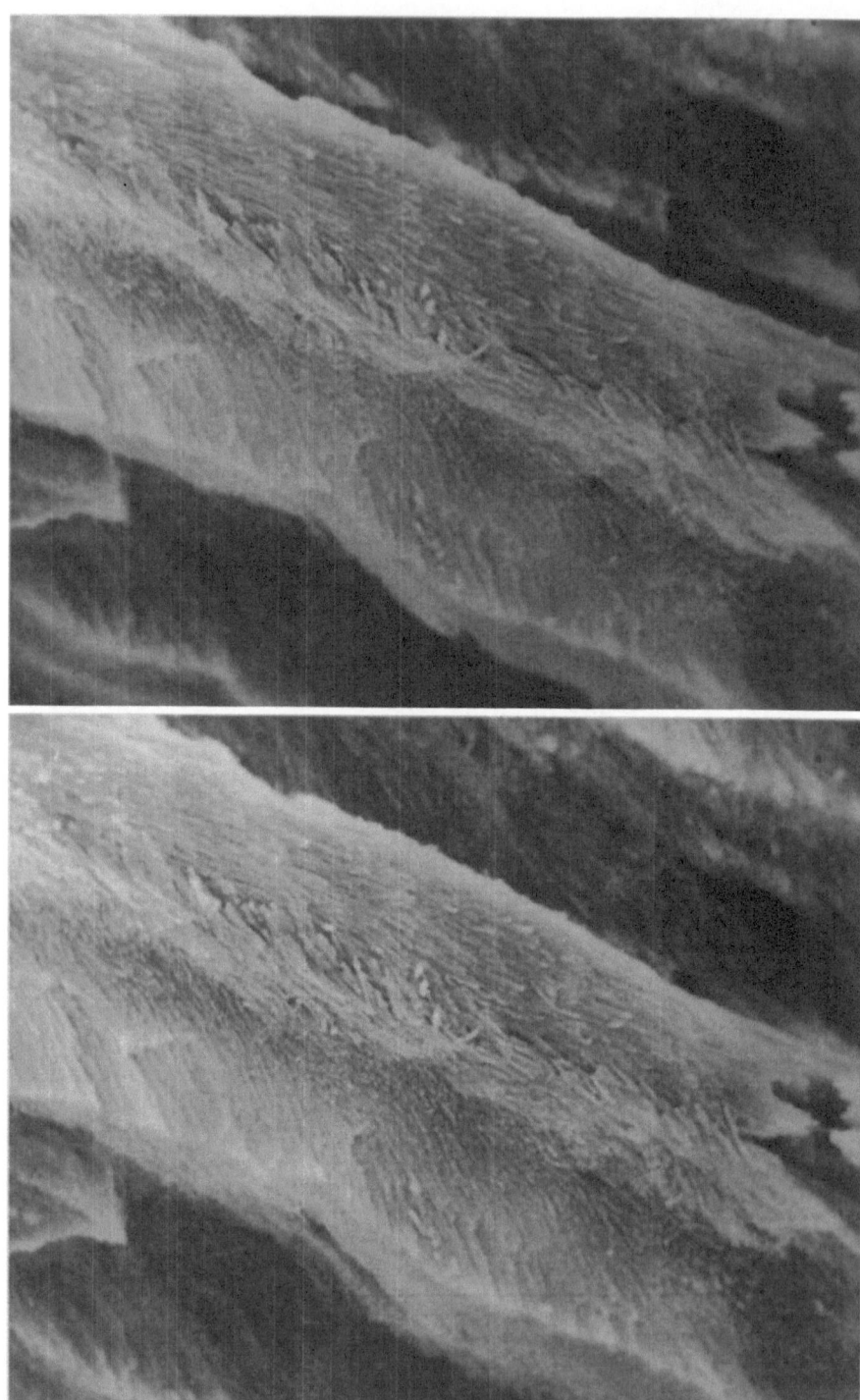

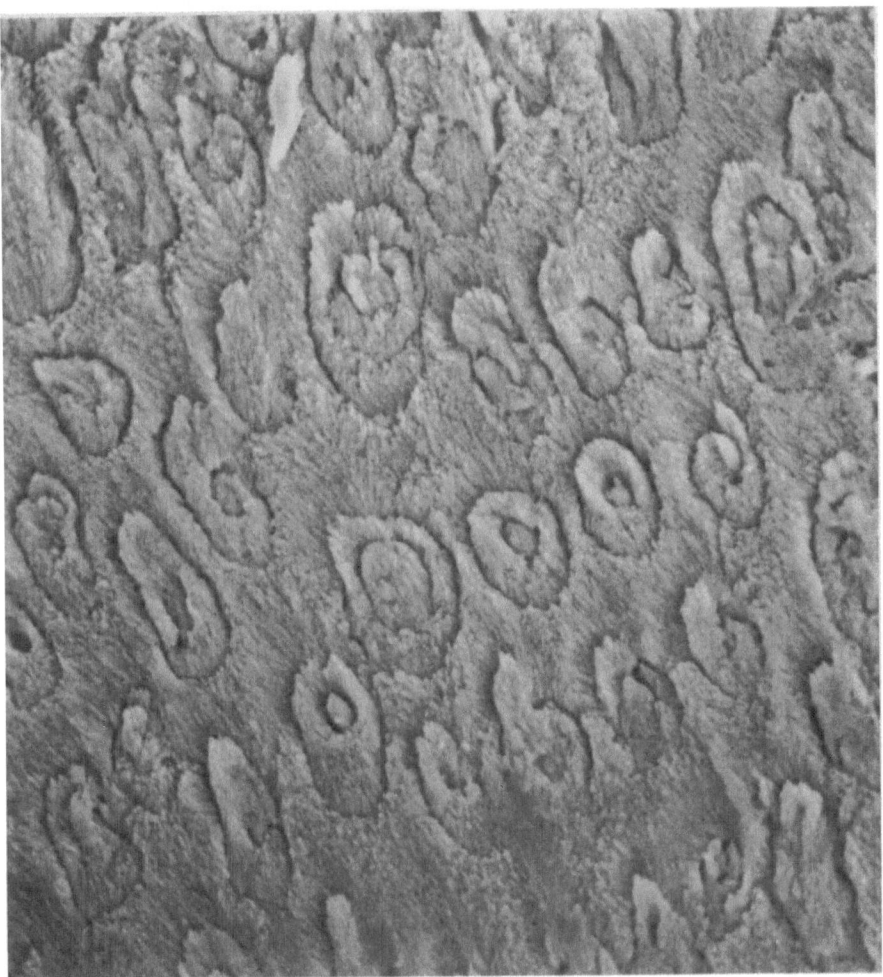

Fig. 21. SEM of *Dugong dugong* incisor enamel etched with 0.1 molar H_3PO_4 with constant stirring for 1 min, showing Pattern 1 enamel prisms, with prisms within prisms. This feature develops where pits fill in a two-stage process. It is common through mammals, including man. Field width 52 μm

1976a, b). There are also three classical (idealised) appearances of sections of the forming surface, as seen to best advantage in TEM, where the orientation of the crystallite fragments can also be determined – crudely, by inspection, and accurately, by 3-D imaging and measurement.

◁ **Fig. 20.** SEM stereopair showing one prism dissected from the same preparation as is shown in Fig. 19, viewed side-on. The crystallites in the head portions of the prisms are running in the long axis of the prism: this will be seen to project above the rest of the surface when the image is viewed stereoscopically. The crystallites in the tail portion correspond to the enamel formed in the interpit locations. Field width 19 μm

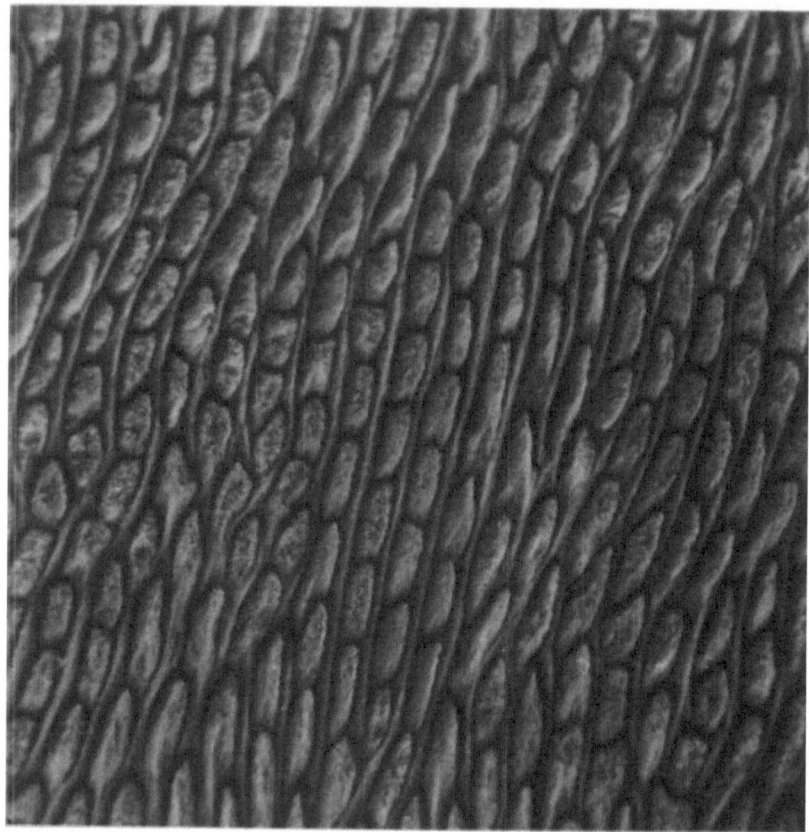

Fig. 22. Pattern 2 enamel in pig deciduous molar: facial surface, ground and etched, showing longitudinal rows of prisms separated by interrow sheets. Field width 55 µm

Tangential sections cut the Tomes' processes transversely and make the pits appear as circumscribed features: the pits appear round or oval in Pattern 1 (Figs. 8, 9, 21–29), and with one flat side in Patterns 2 and 3.

Longitudinal sections tilted to follow the "prism direction" below the surface generate a saw-tooth or "picket-fence" (American English) profile of the surface (Fig. 8d, e; BOYDE 1964, 1965, 1967, 1973). Deep to the cut surface, only one sharp change in the crystallite orientation is found per pit: this discontinuity is located opposite the most distal portion of Tomes' process (at the tip of the saw tooth), where there is a sharp concavity in the surface of the forming enamel. In the case of human (Pattern 3) enamel (Figs. 8b–19), such sections may intercept prism boundaries either at intervals of roughly 6 µm (Fig. 8d) or, less commonly, at roughly twice this interval (Fig. 8e). The origins of these differences may be understood from a diagram. The 6-µm repeat arises when the section runs at 30° to the (idealised, mean) longitudinal direction, the 12-µm repeat when it is exactly in that direction (Fig. 8d, e). The picket-fence profile does not arise in the case of Pattern 1 (Fig. 8e). It only occurs over a limited

Fig. 23. Stereopair, tilt-angle difference 10°, of fractured sheep molar enamel, Pattern 2. Stereoviewing will show the interrow sheets projecting upwards, the prisms lying more in the plane of the paper. Field width 25 µm

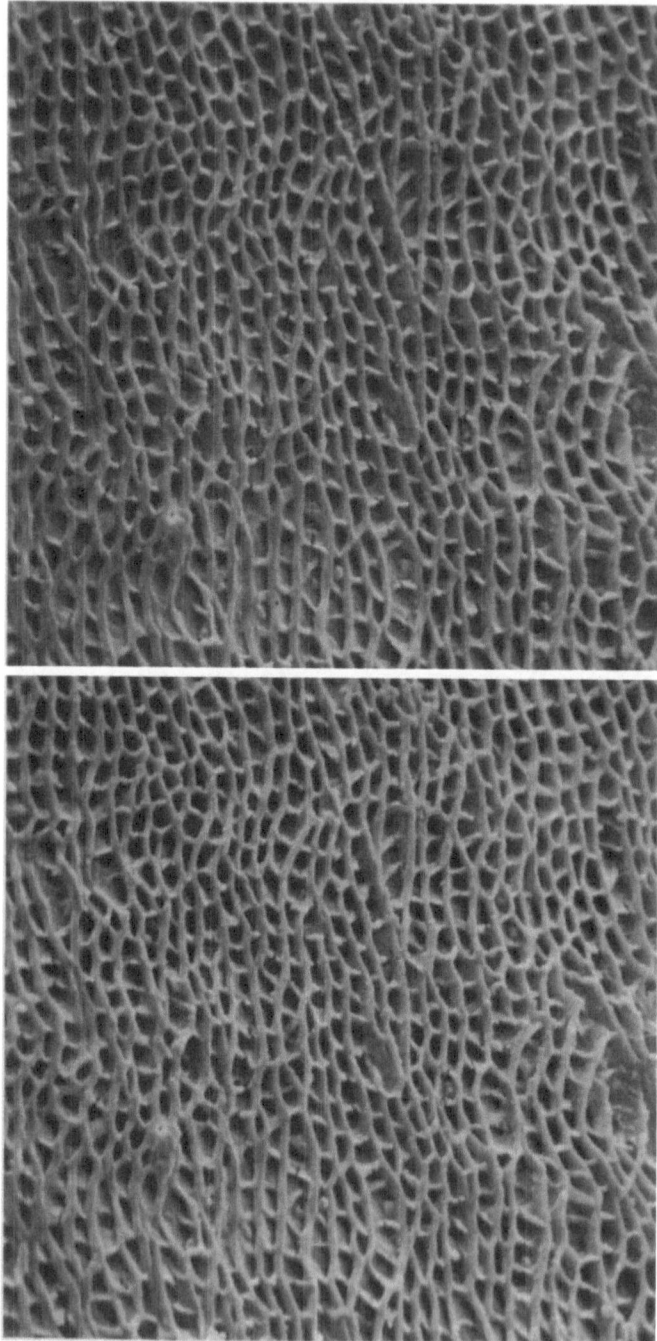

Fig. 24. Developing *Myocastor coypus* incisor inner enamel, dehydrated in ethanol and air-dried from ethanol. The pits form rows orientated in the longitudinal direction of the tooth, which are disturbed by their sideways inclination in alternate zones, resulting in the formation of decussating zones (Hunter-Schreger bands). Field width 30 μm

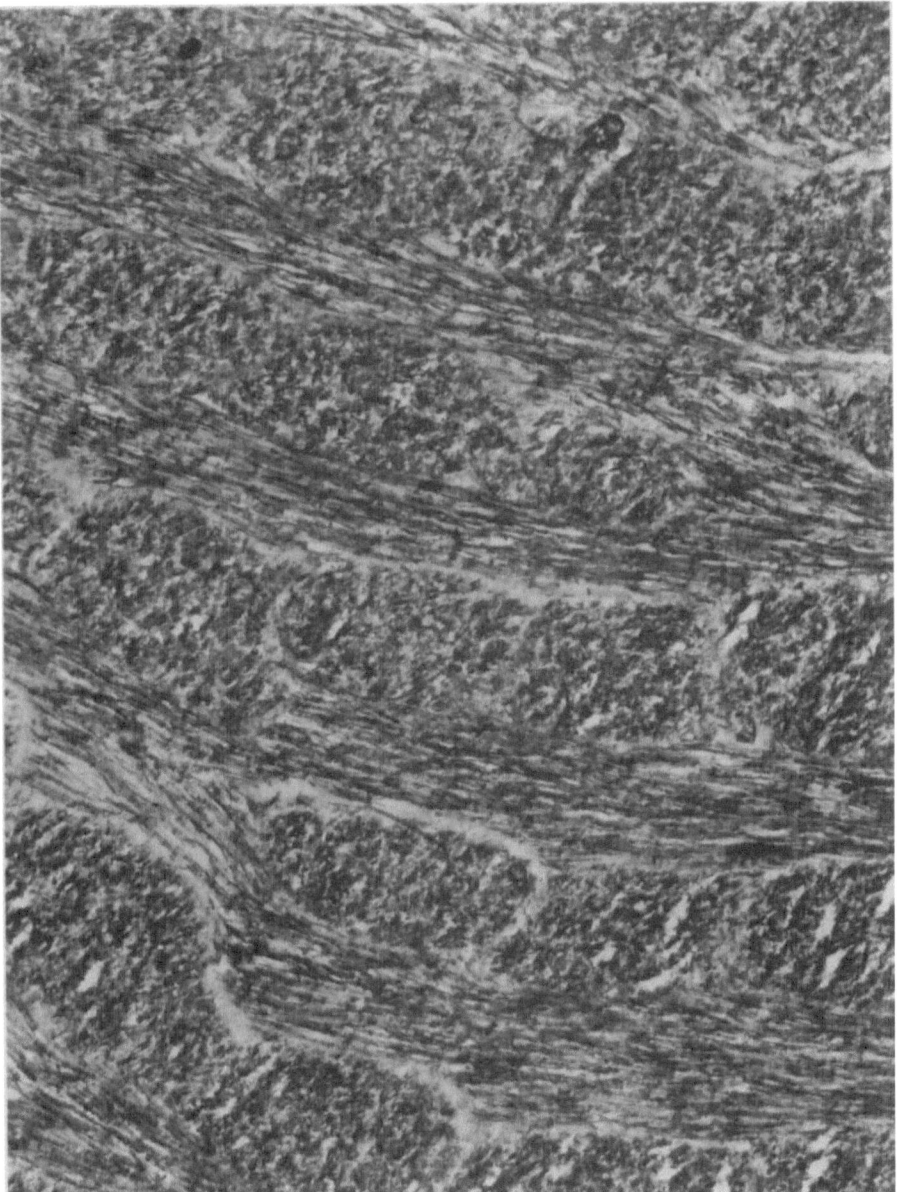

Fig. 25. Transmission electron micrograph of immature cow enamel (*Bos taurus*), cut more perpendicular to the crystallites in the prisms and more parallel with the crystallites in the interprismatic interrow sheets. Field width 12 µm

range of distance and over a limited range of directions about the idealised longitudinal section plane, and only at the abbreviated interval of about 4.5 µm in well-developed (ungulate type after SHOBUSAWA 1952) Pattern 2 enamel.

Obliquely tilted transverse sections which cut along the long axes of the prisms generate a boxy or battlements profile of the surface: two sharp changes

Fig. 26. TEM facial section of immature developing inner enamel in *Myocastor coypus* (Rodentia, Hystricomorpha) incisor, showing longitudinal interrow sheets of interprismatic material of interpit origin, separating rows of prisms. The crystallite orientation in each prism and in one row have been accentuated by a *straight-line overlay* mark showing how this changes. Field width 12 μm

Fig. 27. Section of developing *Myocastor coypus*, incisor inner enamel. Portions of prisms belonging to one zone are labelled *P*. Interrow sheet (*IRS*) is interprismatic material of interpit origin. Just to the left of label *IRS*, it can be seen that two layers of crystallites belonging to the prism and the interrow sheet cross each other within the thickness of the section. It is important to realise that many prism boundaries only represent changes in orientation between bundles of crystallites lying in different planes. Just above the label *H* over a hole in the section at the bottom of the field of view can be seen complexly curved crystals. Field width 6 µm

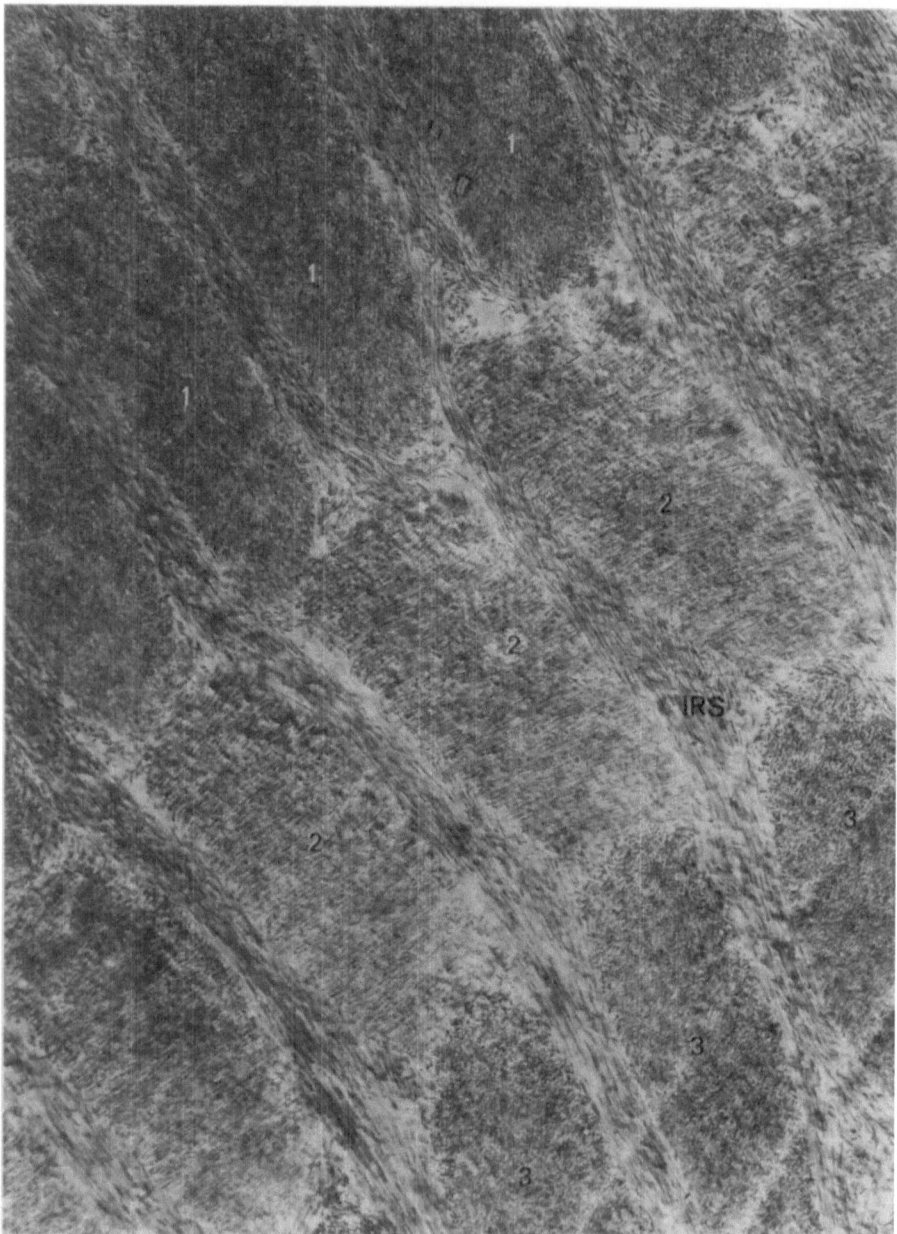

Fig. 28. TEM of façial section of immature inner enamel in rat incisor, showing alternate transverse rows of decussating prisms. *1* prisms of one row, *2* those of the next, *3* those of a row corresponding to row one. *IRS* interrow sheet. A small area *outlined* in a prism of row 1 at the top of the field of view shows a group of crystallites whose flat sides are parallel with each other: another group is marked by *two parallel lines*. Field width 12 µm

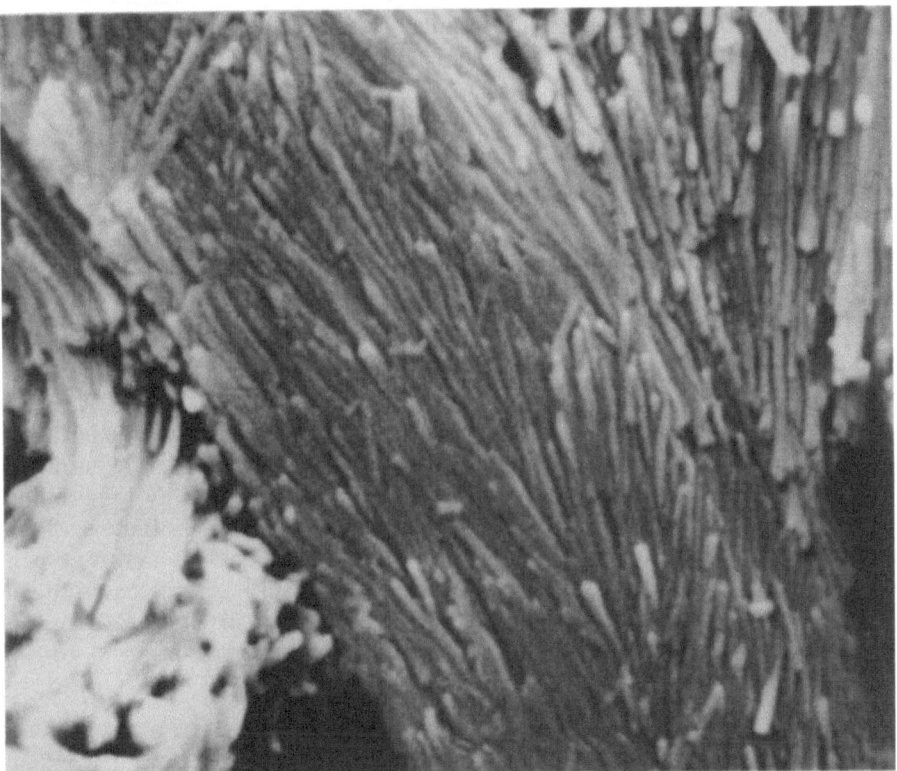

Fig. 29. SEM of fractured rat incisor enamel, showing details of crystallites at prism boundary plane. Field width 6 μm

in the orientation of the surface are intercepted per pit, associated with two discontinuities in the ,crystal packing arrangement [i.e. two prism boundaries per pit (Fig. 8a, f)]. The change in crystal orientation is mainly in the sense of the length of the crystal fragments to either side of the line of discontinuity and is therefore more difficult to discern by simple inspection of a single TEM image: crystals cross, rather than "collide", at such boundaries. The battlements profile appears in all planes of section along the long axis of Pattern 1 pits (Fig. 8a, c). The depth of suceeding pits alternates between shallow and deep in Pattern 2 enamel cut in this way (Fig. 8f, g). The depths of the pits are nearly the same for all pits in Pattern 3 enamel (Fig. 8a, b), but they may be narrower and shallow, wider and of intermediate depth, or deepest and again narrower, depending upon the latitude at which the row of pits is cut (BOYDE 1964).

b) Pattern 1 Enamel: Complete Prism Boundaries

The floors are separated from the interpit walls by a continuous declivity. In this case, a continuous circular prism boundary discontinuity will be formed within each pit. In the adult tissue, circular prisms will be found, and we refer to this type of enamel as Pattern 1 enamel (Fig. 8c).

c) Patterns 2 and 3: Incomplete Prism Boundaries

Ameloblasts and Tomes' process pits are most commonly hexagonally packed, and the floor of the pit is most commonly continuous with and indistinguishable from the wall of the pit over about one-sixth to just less than one-half of the circumference of the pit. Visualising the secretory territories as hexagons, two further main common types of enamel are distinguished by the detail of where the floor joins the pit wall (Fig. 8b, g).

d) Pattern 2 Enamel

In Pattern 2 enamel (Fig. 8g), the floor joins one of the sides of the hexagon, whereas in Pattern 3 (Fig. 8b), it joins one of the corners of the hexagon. This gives rise to the fact that the prisms are stacked in mainly longitudinal rows in Pattern 2 enamel, with the interpit phase forming continuous sheets separating these rows (Figs. 24, 25). The interpit-phase crystals still grow perpendicular to the general plane of developing enamel surface, whereas those in the pit floors grow perpendicular to the orientation of the formative surface in the pit floor, so that there are two predominant crystal orientations occurring in sheets.

e) Pattern 3 Enamel

In Pattern 3, the main part of the secretory product of each ameloblast is formed within the floor of the pit and contains crystals oriented essentially parallel with the prism "head". These are continuous with the interpit-phase crystals at the region where the floor joins the wall, where no discontinuity is formed because there is a gradual change in the slope of the surface and a gradual change in the crystallite orientation (Figs. 11–19). The same two sets of distinctive crystallite orientations occur – in the interpit phase, perpendicular to the general plane of the developing enamel surface, and in the pit floor, perpendicular to the pit floors – but now extensive sheets of crystals having one common orientation (Figs. 11–19) are no longer found. The prisms most commonly seem to be stacked as rows, which are identified in the transverse direction of the tooth: rows can also be identified 30° to each side of the longitudinal direction.

In man, Patterns 1, 2 and 3 enamel are all found in any search of a significant volume of tissue, and patches of all three may be found within one field of view. Pattern 3, however, is strongly predominant. Pattern 1 is commonest in the vicinity of the enamel-dentine junction and the just-sub-surface enamel, whereas Pattern 3 constitutes the great majority of the bulk of the enamel. Pattern 3 enamel is also found in other higher primates as the predominant type (SHOBUSAWA 1952; GANTT et al. 1977; GANTT 1979; VRBA and GRINE 1978; BOYDE and MARTIN 1982, 1984a, b, 1987; MARTIN and BOYDE 1984; MARTIN et al. 1988; SHELLIS and POOLE 1979; SHELLIS 1984a, b; DOSTAL 1987).

f) Distribution of Principal Prism Types

Regarding the distribution of the principal prism types amongst non-primate mammals, Pattern 3 predominates in the orders Carnivora, Pinnipedia and Proboscidea (SHOBUSAWA 1952; BOYDE 1964), and in hippopotamuses (Artiodactyla) and rhinoceroses (Perissodactyla) (FORTELIUS 1984, 1985; BOYDE and FORTELIUS 1986). Patterns 2 and 3 have been found in odontocetes in the deeper enamel, but Pattern 1 or prism-free enamel in relatively thick surface layers (BOYDE 1980; ISHIYAMA 1984, 1987). Pattern 2 enamel, on the other hand, is predominant in most members of both the perissodactyl and artiodactyl ungulate orders, as well as in the Marsupialia (CARTER 1922; BOYDE 1964; BOYDE and LESTER 1984; PHAKEY et al. 1984), Lagomorpha and Rodentia (in areas where the basic pattern is not distorted beyond recognition by prism decussation; BOYDE 1964). Pattern 1 is so commonly found in sub-surface enamel that it is probable that all reports in the literature (which mostly have dealt with subsurface layers in this respect) should be treated with circumspection. However, reports that it is predominant in Sirenia and Insectivora, as well as being common in some marsupials and Chiroptera, have not yet been gainsaid (SHOBUSAWA 1952; BOYDE 1964). Details of prism patterns in fossil mammals have been given by SAHNI (1984) and CARLSON and KRAUSE (1985), who also cite other relevant literature.

An interesting feature of unknown significance concerning the distribution of the prism packing types concerns the relationship to size. BOYDE (1968), in a study of developing enamel surfaces, noted that the numbers of ameloblastic pits per unit area of surface – the secretory territories of the ameloblast – differed in the different types, with Pattern 2 having the smaller size distribution, overlapping with Pattern 1, intermediate in size between Pattern 2 and Pattern 3, and the latter having the largest (Fig. 31). However, this relationship was only characterised for a small number of recent species, and even though more detailed study has confirmed the data for one taxon (Artiodactyla, Caprinae; GRINE et al. 1987) and abundant data are available from the studies of FOSSE (1968a, b, c) for human teeth, studies of earlier fossil mammals (e.g. CARLSON and KRAUSE 1985) have already strongly questioned the value of any such generalisation.

There are many fine details of enamel prism cross-sectional shape which differ between species within closely related groups. Good examples of this are found in the Anthropoidea and may conceivably prove to be of value in future studies of taxonomic affinities amongst this group (BOYDE and MARTIN, unpublished data and 1982, 1984a, b; MARTIN 1983; MARTIN and BOYDE 1984; BOYDE and GRINE 1988; GRINE, unpublished). Such details need to be studied more exhaustively before general conclusions can be drawn as to their significance. We mention them here so that the reader may know that this subject may have further complications. However, it is clear that as the only "born fossil", enamel must have a special significance in palaeoanthropological studies and will be of particular interest to us in working out our own evolutionary history (MARTIN 1983; DEAN 1987).

g) Arcades or Arches in Human Enamel Pattern 3 Prisms: A Question
of Depth or Decussation?

The majority of published illustrations of cross-sectional outlines of human
enamel prisms reflect the apparent reality of the situation: that prism boundary
discontinuities in man are not continuous (Fig. 32). High-resolution data are
necessary to decide on this point since the apparent extent of the "prism sheath"
is expanded in conventional optical images (OSBORN 1968a, b, 1971a, b). TEM
images may not lie in this respect, but one has to contend with the possible
artefacts introduced by the specimen preparation. Decalcification, etching and
sectioning may all extend the discontinuity (see images in SWANCAR et al. 1970;
HINRICHSEN and ENGEL 1966). Nevertheless, it would seem to be clear that
prism boundaries are continuous to a limited extent and in defined circumstances: they may be irregularly joined in the most superficial layers of the
enamel (SUNDSTROM and ZELANDER'S 1968 figures), they may be joined in the
plane of the incremental lines (see next section and Figures in HINRICHSEN
and ENGEL 1966) and they may be regularly joined to form arcades. GLAS
and NYLEN'S (1965) Figures 5 and 6 show a convincing example of a row
of prism boundaries joined at their cervical "tails" (term of MECKEL et al.
1965). Prisms arranged in this way are said to be *arkadenförmig* and to have
Flügelfortsätze in the German literature (VON EBNER 1903; SMREKER 1905). The
equivalent English term is "winged process" (MUMMERY 1916).

Winged processes and arcades are very well developed in elephant enamel,
and they are well illustrated in MUMMERY'S (1916) study. The developing surface
which gives rise to arcades is a Pattern 3 arrangement in which the prism
boundary approaches the surface at a shallow angle; i.e. there is a strong component of movement of the cell across the formative tissue surface. The geometry
of the surface (Fig. 33) then dictates that boundaries join to form arcades. In
human teeth, this occurs more frequently in deep than in superficial, and
in cuspal/occlusal/incisal (WEBER 1973) than in lateral enamel; in fact, it occurs
where prism decussation is the most marked. Teleologically speaking, this is
a fortunate circumstance since the joining of boundaries would tend to weaken
the tissue which fails by joining up boundaries. Deeper enamel is not exposed
to the risk of catastrophic failure, which is resisted by the complicated courses
of the prisms.

h) The Prism Boundary Discontinuity

The physical dimensions of the prism boundary discontinuity are unclear.
The boundary may consist of the line of junction between crystals in two different orientations, as, for example, longitudinal crystals in the prismatic bundle
being crossed by those forming an interrow sheet interprismatic material in
Pattern 2, the interprismatic phase in Pattern 1 or the tail portions of the
prisms in Pattern 3 enamel. The junctions may also consist of crystals in the
tail portions or in the interprismatic regions ending by colliding with the crystallites in the prismatic bundle. In early developing material, there is no distinguishing feature in this vicinity other than the change in crystal orientation in mature

tissue. However, a greater amount of a perhaps denser organic matrix can be recovered from this zone in stained and decalcified material (e.g. SUNDSTROM and ZELANDER 1968). The width of the discontinuity may vary from no more than the normal width of organic matrix between two adjacent crystallites, of the order of 5 nm (ORAMS 1976), to perhaps 150 nm. In either event, the prism boundary should be sub-microscopic, i.e. not visible as such by light microscopy. It is, however, clearly visible and appears much thicker than this by light microscopy. OSBORN and ROBERTS (1971) have provided an explanation for this based upon optical fringe effects. Whatever the explanation, it is clear that there is a large refractive index difference at the prism boundary, and that because of the complications of the shape of the prism boundary at a fine scale, it projects in the thickness of the thinnest optical section as an artefactually thick structure.

i) Seams, Pseudo-prisms and Evolution

A novel feature of the developing enamel surface in chiropterans was recently described by LESTER and BOYDE (1987). An additional crystallite orientation discontinuity line, or minor *seam*, was found in the central cervical region of each horseshow-shaped prism in what is essentially Pattern 2 enamel. The feature relates to a groove in the more superficial part of the developing floor wall of the Tomes' process depression, reflecting the absence of the most prominent parts of the developing enamel at the border between adjacent (Pattern 2) pits in the direction of their longitudinal alignment.

LESTER (1988) has recently incorporated this information into a conceptual scheme to explain enamel evolution. The scheme uses the concepts outlined here to explain: (a) aprismatic enamel as parallel crystallite groups oriented perpendicular to a nearly flat developing enamel surface, (b) pseudo-prisms demarcated by minor boundary planes within which prisms (later) reside, and (c) prisms demarcated by major boundary planes and constituting repetitive, recognisable domains.

IV. Incremental Phenomena

1. Cross-striations and/or Varicosities

Enamel prisms show a banding crossing their long axes at regular intervals, in human enamel most commonly being in the range 2–5 microns. Because the prisms in transmitted-light, polarised-light and reflected-light microscopy and in dark-field light microscopy appear to be striped, these features are known as cross-striations (Figs. 34, 35). There may also appear to be variations in the width contained between two parallel prism boundaries, an appearance frequently seen in fractured preparations examined in the SEM but which can also be seen in light microscopy and which gives rise to the name "varicosities" (Figs. 36–38). HELMCKE et al. (1963) provided an explanation of the cross-stria-

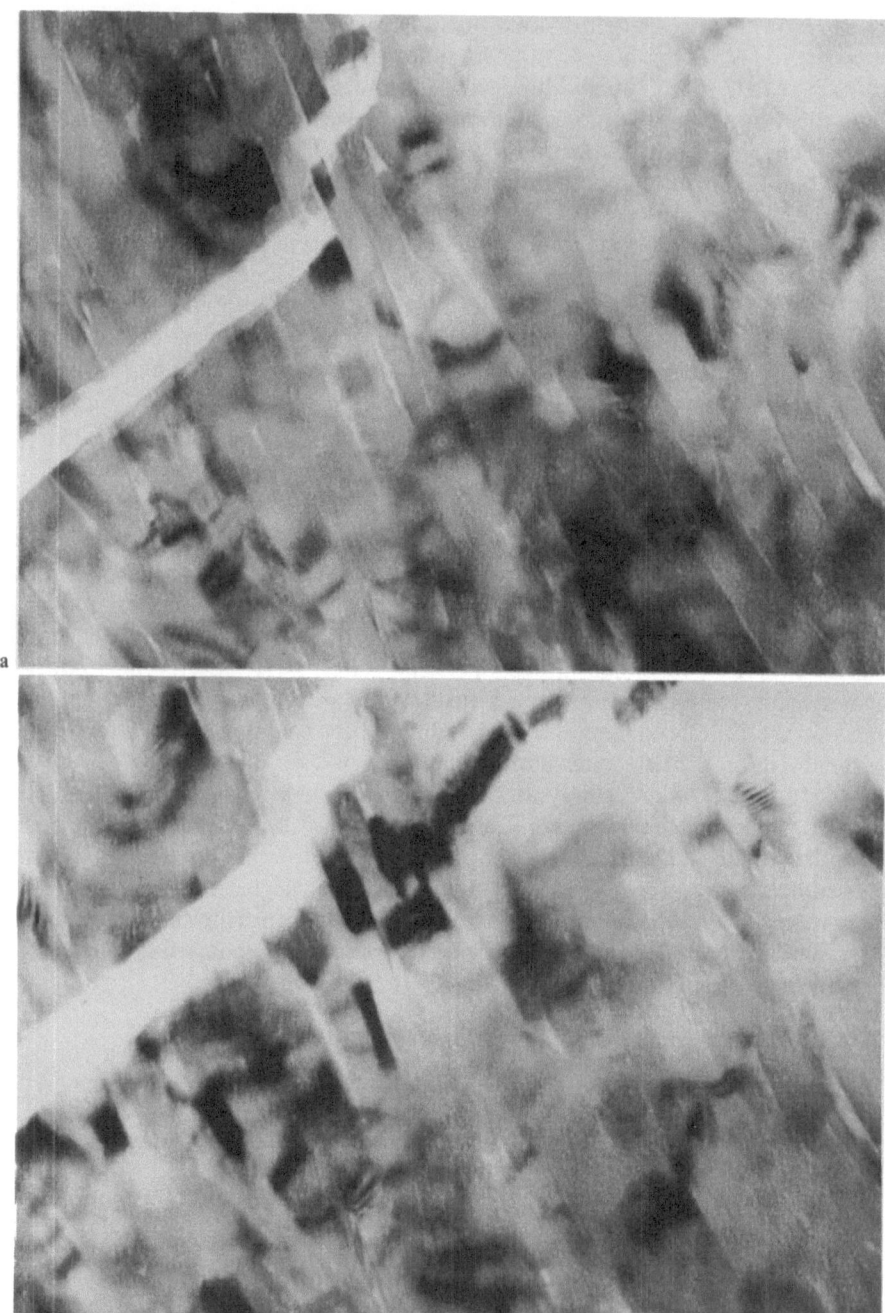

Fig. 30a, b. TEM (a) of ultrathin section of adult human permanent enamel prepared by ion-beam thinning. **b** The same part of the same field is imaged with a tilt-angle difference of 12°. The crystallite axis is from NNW to SSE. Note that Bragg extinction contours, most of which cross the crystallite axis, have changed dramatically between the two positions. The movement of these extinction contours can be followed continuously during tilting of the specimen at the microscope. Field width 1 μm

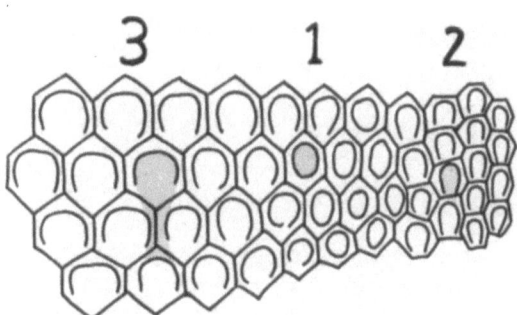

Fig. 31 a. The relationship between the secretory territories of the ameloblasts (*hexagonal outlines*) and the size and shape of prisms in mammalian dental enamel development. It is usually found that Pattern *3* prisms are made by the largest ameloblasts, and Pattern *2* by the smallest. The areas designated as prisms in each of the three patterned prisms are illustrated in one case with the *dotted overlay*

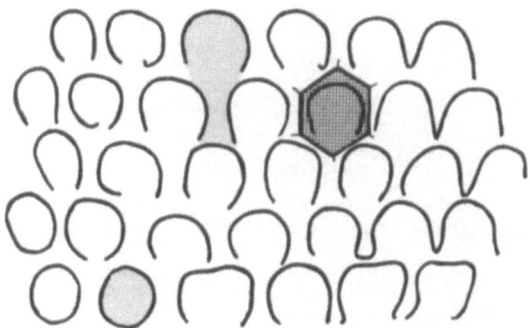

Fig. 31 b. A few of the varieties of form and packing pattern that may be found in human (mostly Pattern 3) enamel. Many other weird cross-sectional shapes may be encountered. Joined prism boundaries form arcades. *Dotted areas* are one Pattern 3 and one Pattern 1 prism. *Cross-hatched hexagon* indicates secretory territory of one ameloblast

tions based upon evidence obtained from TEM replica studies of fractured enamel. These authors considered that the cross-striations were due to a periodic variation in the width of the prism, with the crystals deviating more from the mean prism axis in the wider parts of the prisms. Their model gave an incorrect packing arrangement but contains an essential element which had not been previously identified.

Abundant circumstantial evidence suggests that these features represent daily growth lines in the enamel (ASPER 1916; GYSI 1931; SCHOUR and HOFFMAN 1939; BOYDE 1963, 1964; DEAN 1987). Some workers in this field have, however, doubted their reality (WEBER and GLICK 1975; WARSHAWSKY and BAI 1983; WARSHAWSKY et al. 1984a). In this respect, it must be stated that there is no doubt that the appearance of cross-striations may arise through the failure to recognise sections of rows of prisms, which, because they are so well aligned, can easily give rise to a misleading effect (Fig. 22; BOYDE 1964; GUSTAFSON

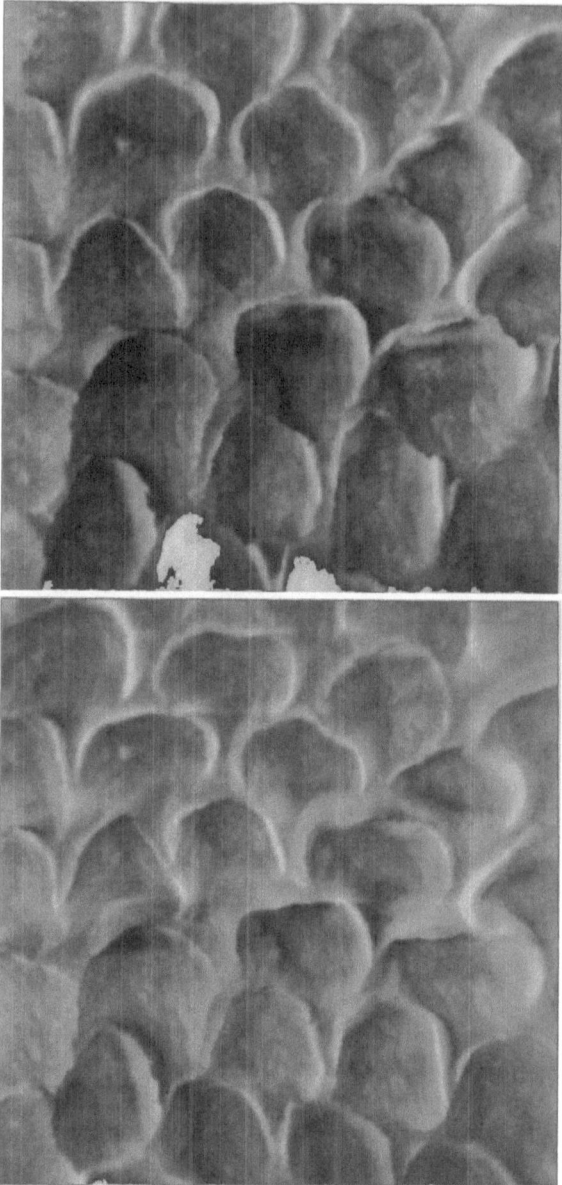

Fig. 32. Argon-ion-beam-etched human enamel viewed along the prism axis. Ion-etching causes selective removal of material close to the prism boundaries. Stereopair, tilt-angle difference 10°. Stereo-viewing will show čentral portion of prism near junction of head and tail to be very prominent. Field width 22 μm

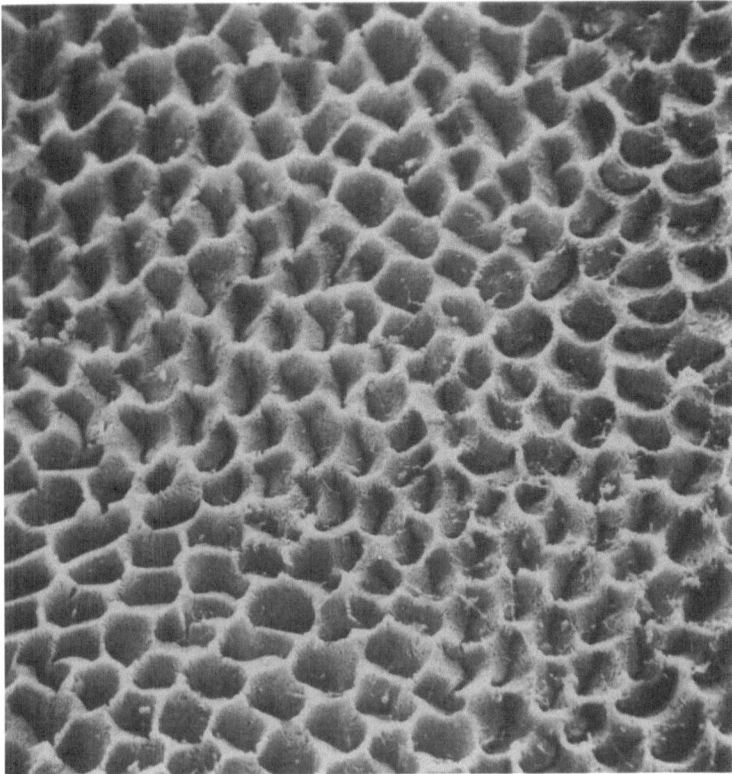

Fig. 33. SEM. Anorganic preparation of developing enamel surface of molar tooth of *Loxodonta africana* (African elephant). The exit directions of the prism boundaries which can be seen in the Tomes' process pits of this surface vary considerably within this one field of view, accounting for the irregular prism decussation found in this species. Arcade rows are also frequently seen. Field width 97 µm

1959; WEBER and GLICK 1975; GLIMCHER et al. 1965). On the other hand, there are several lines of microscopic data which are not so easily challenged. GLAS and NYLEN (1965) showed that periodic variations in microradiographic density along the length of the prisms did not change with the geometrical rotation of a rod-shaped sample, whereas the prominence of the sum images of the arcade-row boundaries did change. The appearance of varicosities in fractured surface preparations examined by SEM (Figs. 37, 38) leaves no room for such misidentification because the surface can be studied in 3-D, and the matched fracture-halves can be analysed. The appearance of cross-striations in micro-milled surfaces examined using BSE SEM similarly leaves no room for misidentification because the low-density prism boundaries appear clearly defined and at a different contrast than the variations in density imaging the cross-striations (Fig. 39). Although they are not always seen with every means of optical microscopy, the fact that they can be seen with optical microscopes which produce extremely thin optical sections – the confocal scanning microscopes – makes

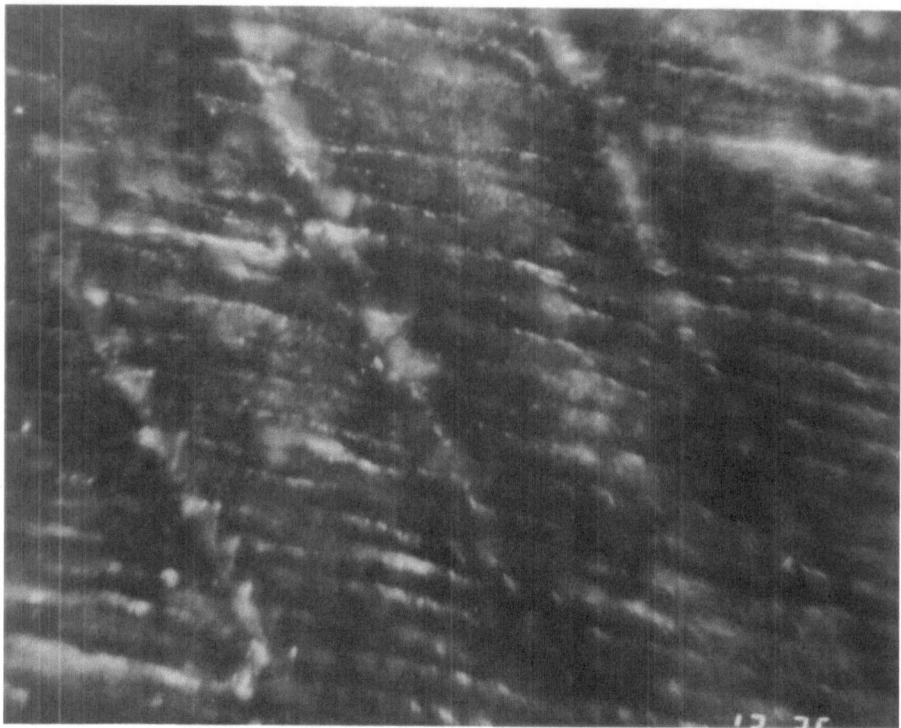

Fig. 34. Dark-field LM of longitudinal ground section of human premolar section, showing prisms, cross-striations and regular incremental line. Field height 190 μm

it the less likely that they are all artefactual (BOYDE 1987). Another line of reasoning which would contradict that view is that they are clearly parallel with and of the same nature as the more exaggerated growth lines occurring at regular and irregular longer-term (larger-distance) intervals (Figs. 30–41). These longer-period growth lines, generally called the brown striae of Retzius, have universally been accepted as growth lines showing the position of the forming surface at a sequence of intervals during tissue growth.

2. The Incremental Lines or Brown Striae of Retzius

In man, one variety of the incremental lines of Retzius occurs with particular regularity and is more marked in superficial layers of the enamel, and particularly in the lateral and cervical enamel (Figs. 42–44). The number of cross-striations between the succeeding lines of this nature is commonly in the range 7–9, suggesting an unknown rhythmicity in enamel formation at an interval of a week or slightly more (VOLLRATH et al. 1975). The close study of enamel histology utilising cross-striations and incremental lines can lead to a self-contained method for determining the age at death of young human skeletal remains (BOYDE

Fig. 35. Occlusal enamel region in longitudinal section of human premolar between crossed polars, showing a complicated fine cross-banding, which may be due to interference between cross-striations in adjacent layers and/or the course of the prism rows in the third dimension, perpendicular to the image plane. Field height 190 μm

1963, 1964), and this has had forensic, archaeological and palaeoanthropological applications (DEAN 1987; BEYNON and DEAN 1987; BEYNON and WOOD 1987).

Not all the brown striae of Retzius are regular, however (Fig. 45). Indeed, the pattern of these lines is the same in all those parts of the enamel that formed at the same time in a given dentition. Thus, the different teeth developing in one individual give the same pattern of incremental lines, which is distinct from that of another individual. The pattern is a kind of fingerprint of enamel development specific to the individual and may be used to determine whether two or more teeth derive from the same individual (AG GUSTAFSON 1955). These facts strongly suggest that the less regular class of incremental lines also have an origin in systêmic fluctuations of an unknown origin.

If we trace the incremental lines within the enamel out towards the tooth surface, they crop out at the surface in the troughs of the circular grooves of the perikymata pattern; that is, they correspond to regions in which the enamel thickness is slightly deficient (Figs. 46, 47). The tooth surface in the grooves shows characters of the bulk developmental stages. This is taken to indicate that a systemic phenomenon (which caused some disturbance in the

Fig. 36. Lateral enamel in longitudinal section of human premolar between crossed polars, showing undulating external profile of the prisms. Field height 190 μm

nature of the slowing-down of enamel production) caused ameloblasts which were soon to stop producing new enamel matrix to cease their matrix appositional phase slightly prematurely. Close examination to the prism cross section in the plane of the incremental lines within the enamel shows that the prism boundary shifts, towards the centre of the prism, shifting the ratio of pit-to-interpit enamel at that level. A hypothetical model of the circadian rhythm underlining the formation of the cross-striations of enamel prisms (described in the following section) indicates that this is to be expected as enamel formation slows down. This phenomenon is clearly illustrated many times in the literature and either has gone unrecognised or has been described as a change in the course of the prisms (GUSTAFSON 1959).

3. Neonatal Line

The neonatal line is a particularly marked incremental line corresponding to the disturbance in growth consequent upon parturition. It is possible for

Fig. 37. SEM. Human premolar lateral enamel, fractured in oblique transverse plane, showing the ▷ cross-sectional shape of the enamel prisms and the variations in external form of the prisms. (The fracture plane was pre-determined by grooving the specimen before breaking it: this plane of fracture does not often occur naturally). Field width 82 μm

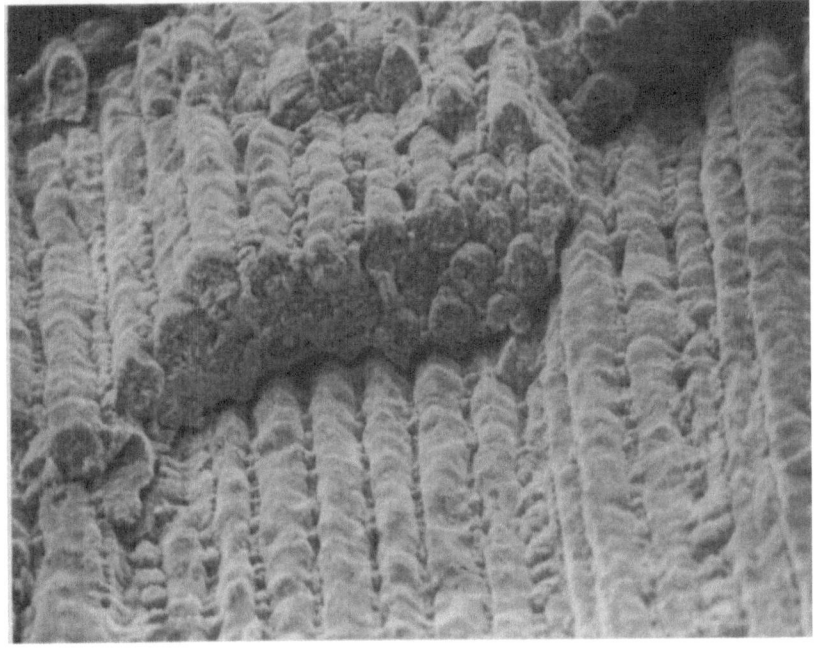

Fig. 38. Fractured human premolar lateral enamel, induced to break in the transverse oblique plane by notching the specimen, viewed from the cervical aspect, showing the impressions of the varicosities of the prisms in the overlying (cuspal) layer of prisms. Stereopair, tilt-angle difference 10°. Field width 90 μm

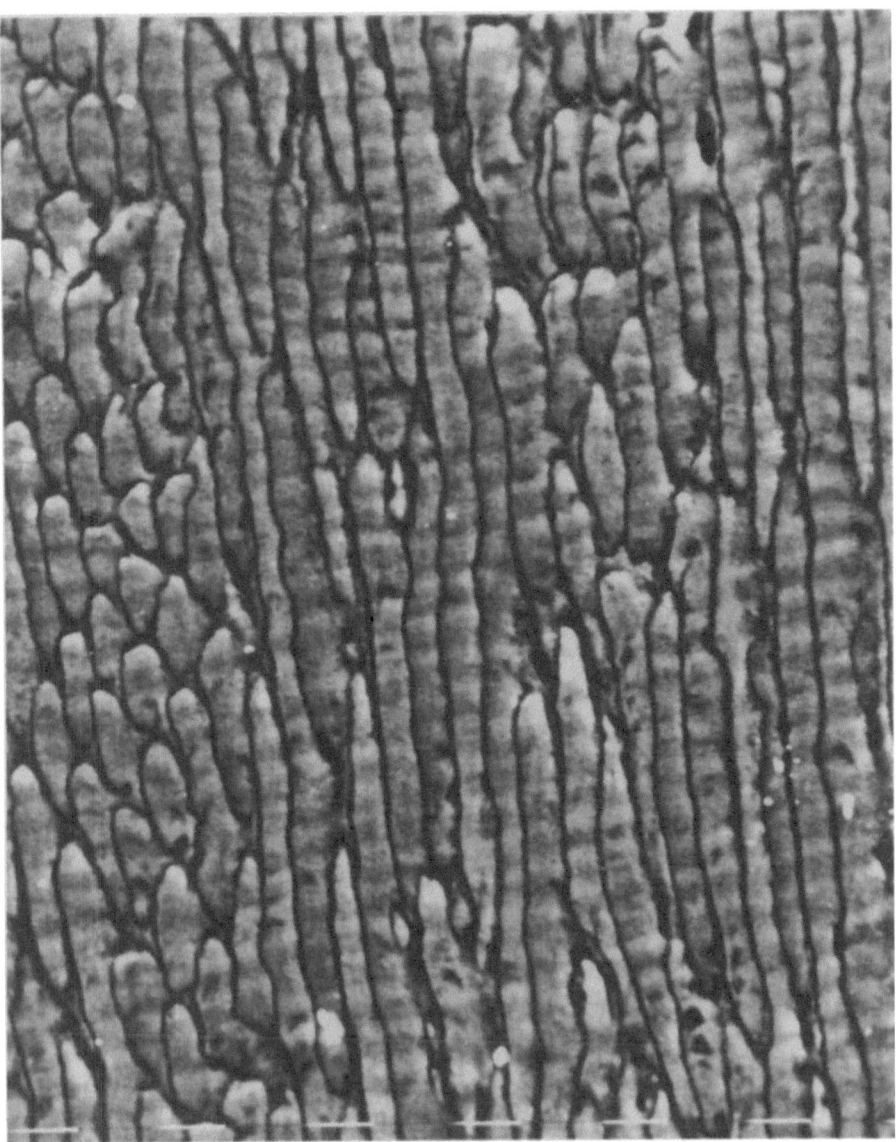

Fig. 39. BSE image of longitudinal section surface, diamond-micromilled, carbon-coated, showing density dependence of signal. Occlusal enamel region showing cross-striations of enamel prisms in more and less obliquely sectioned prisms. Scale ticks at lower border = 10 μm

this line to involve enamel that might be formed over a period of a week or more. It is always clearly marked, but its structure has not been clearly determined in many studies (WEBER and EISENMANN 1971). It normally occurs in human teeth in all the deciduous teeth and in the largest cusps of the first permanent molars.

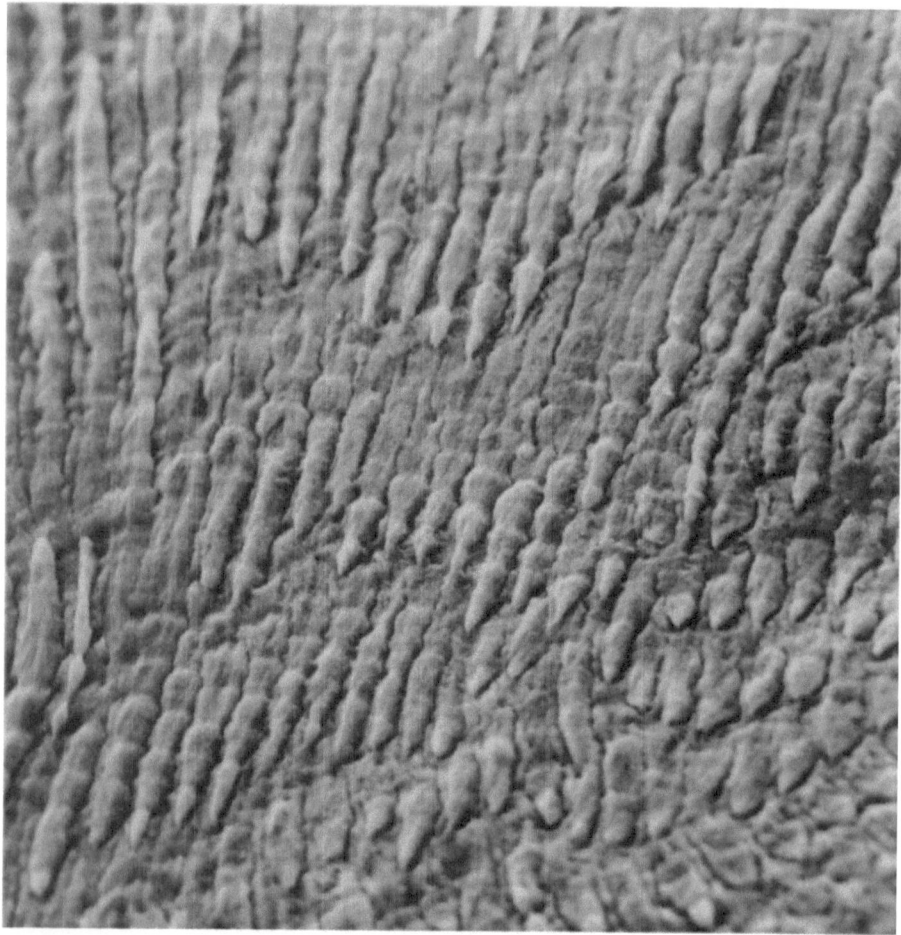

Fig. 40. Phosphoric acid-etched immature dog enamel, showing cross-striations of prisms having the same qualitative features in the same incremental layers. Field width 96 μm

4. Development of Cross-striations

A hypothesis due to BOYDE (1964, 1979) explains the morphological detail of the cross-striations and exaggerated cross-striations in the plane of the brown striae of Retzius (Figs. 48, 49). It starts with the notion, which can be deduced from the constancy of the overall shape of the Tomes' process pits, that there are favourable and unfavourable sites for the release of enamel matrix from the ameloblast. The favoured sites are the interpit and pit floor locations. The steep sides of the pits parallel with the prism direction (the general direction of progress of the ameloblast) represent near-zero growth sites. However, the degree to which the latter are unfavourable or unfavoured varies according to the rate of enamel secretion, which, we hypothesise, changes over a roughly 24-hour period. When the net rate of enamel secretion is reduced, the degree

Fig. 41. SEM. Human permanent molar enamel, cut in longitudinal section and acid-etched, showing selective dissolution of enamel in the plane of one incremental line (which may primarily have fractured, as they are prone to do). Field width 27 μm

of unfavourability of the wall sites is reduced, and more material is secreted in proportion. Conversely, during faster phases of secretion such sites are even less favoured, and even less material will be formed there than on the average. This will determine a change in the ratio of the amount of tissue secreted in the pit floor to that secreted in the interpit locations and in the walls of the pits before the final position of the prism boundary is dictated. The model thus essentially states that the ratio of interpit-phase enamel to pit-phase enamel at any one cross section or level is increased during the slow phases of amelogenesis and vice versa during fast phases.

Extra-slow phases would be marked by the more radical inward necking of the prism boundary seen in the striae of Retzius (HINRICHSEN and ENGEL 1966; WEBER et al. 1974). In longitudinal sections of the tooth, this would be seen as a kink in the prism boundary towards cervical. The shift would produce a greater statistical mix of crystals perpendicular to the incremental surface to those which formed in the interpit location, and this is clearly demonstrated by the shift in the extinction direction in the plane of the striae in polarised-light microscopy (GUSTAFSON and GUSTAFSON 1967).

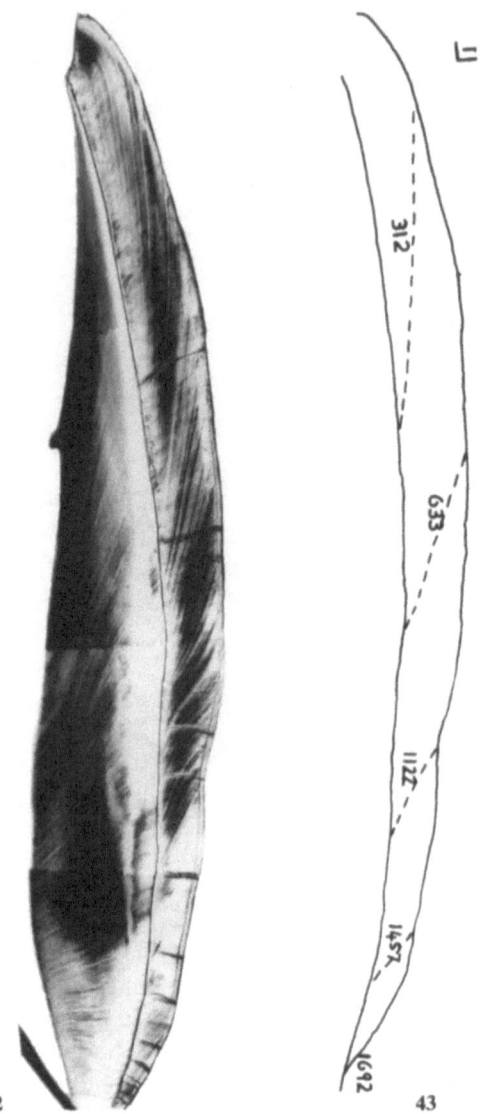

Fig. 42. LM montage of micrographs of labial fragment of human upper central incisor cut in longitudinal section, showing the incremental lines in the enamel

Fig. 43. Tracing of Fig. 42 with directions of a few incremental lines marked. The number written adjacent to each incremental line is the age (in days port partem) at which that incremental line formed. The data were derived by counting the number of cross-striations between incremental lines, transferring the count from the first permanent molar to this upper central incisor, and from one distinctive line to the next (BOYDE 1963)

Fig. 44. Higher magnification of one field of view from Fig. 42, showing regular incremental lines and cross-striations, in which seven or eight cross-striations can be counted between each pair of incremental lines. Field width 122 μm

Other facts concerning cross-striation can be explained as variations in the composition of the enamel secreted at different times during the 24-hour cycle. Cross-striations become more marked in enamel which has been subject to partial acid dissolution experimentally or in natural dental caries (Figs. 40, 41; DARLING 1958, 1961; GUSTAFSON 1945, 1957; GUSTAFSON and GUSTAFSON 1961; SIMMELINK and NYGAARD 1982). Earlier studies laid great emphasis on the

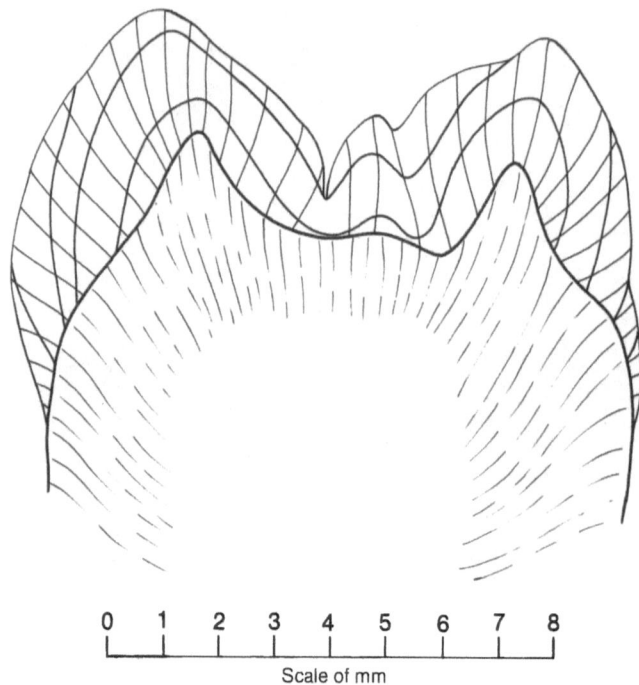

Fig. 45. Tracing of prism directions and incremental lines in a buccolingual longitudinal section through the mesial cusps of an upper permanent molar

heightening of the appearance of the phenomenon by gentle acid treatment of the ground section (Fig. 44; review in Boyde 1964), and explanations of this phenomenon simply assumed that the enamel retained more or less enamel matrix in correspondence with the cross-striations. A more likely explanation, however, is that the crystal chemistry of the enamel changes correspondingly, and that the more soluble locations represent a higher proportion of magnesium and carbonate in the enamel mineral phase. It might be expected that the carbon dioxide (carbonate/bicarbonate) available for incorporation in the enamel mineral component would be greater during those phases of the 24-hour cycle when metabolic activity was the greatest, i.e. when enamel was being secreted at its fastest (Boyde 1979). Considering that the final degree of mineralisation at any one level in the enamel is achieved years after its initial secretion (in the case of man), it must be an event which occurs at the time of secretion or initial mineralisation (crystal-center growlt) that is of importance.

That the centres of enamel crystallites are relatively magnesium- and carbonate-rich may be deduced from the fact that young enamel which is relatively rich in the centres of enamel crystallites is relatively rich in magnesium and carbonate (Robinson et al. 1979, 1984). In carious enamel, or in tissue subjected to partial dissolution in organic acid buffers, the centres of the crystallites are

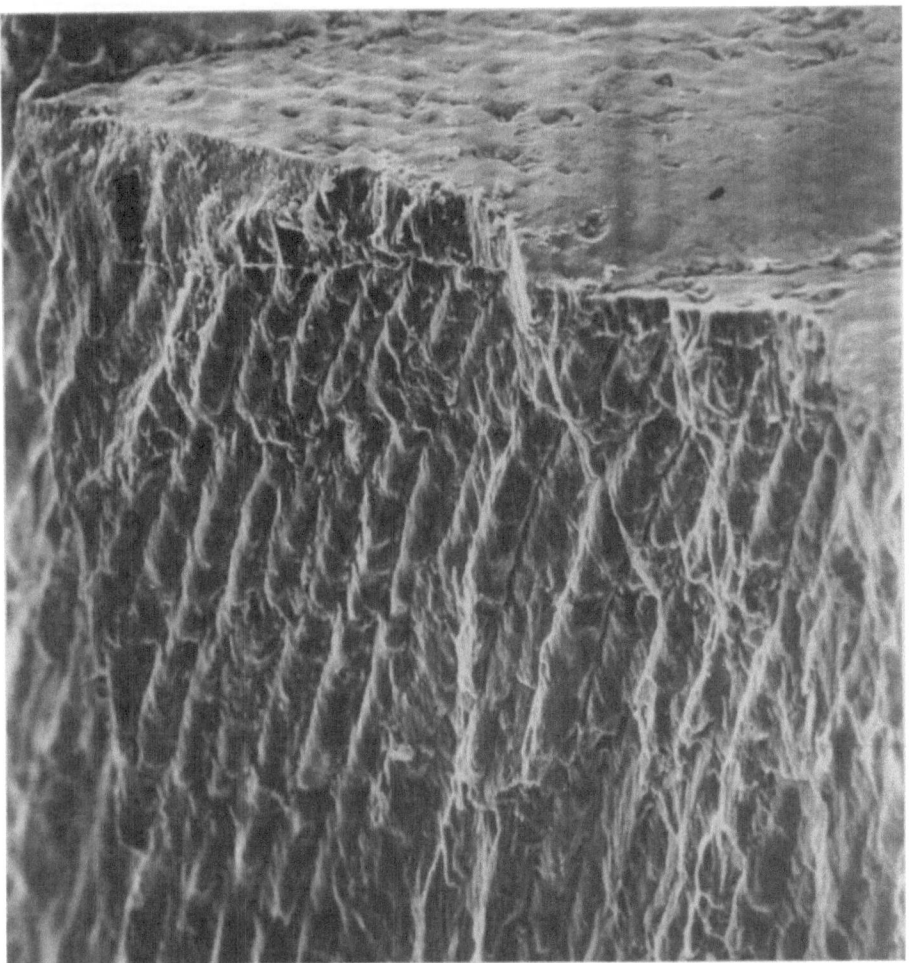

Fig. 46. SEM of human premolar, showing perikymata (*top*) and a longitudinally fractured surface. The incremental lines can be traced as discontinuities connecting with the overlapping layer line in the perikymata pattern. Field width 175 µm

selectively dissolved (ABRIGO 1972; JOHNSON 1967; JONGEBLOED et al. 1975; JOHANSEN 1964), and the concentration of magnesium and carbonate is markedly reduced against the average for the tissue (HALLSWORTH et al. 1972, 1973). A model explaining the cross-striations on this basis would probably have to assume that those parts of the crystals formed during fast phases of amelogenesis would contain proportionately more "core" of a carbonate-richer nature. Cross-striations are more prominent in hypocalcified enamel (HALS 1957).

The appearance of the cross-striations in microradiographs (GLAS and NYLEN 1965) and in images of micromilled surfaces of enamel visualised by

Fig. 47. Surface of unerupted human third permanent molar enamel damaged by pressure applied to the surface during extraction, showing fractures extending parallel with either the regular incremental lines or the prism direction. Field width 170 µm

back-scattered electrons in an SEM (Fig. 39) (BOYDE and JONES 1983) could both be explained as variations of the normal ratio of calcium and phosphate to magnesium and carbonate, a shift towards the latter being associated with a reduction in the mean atomic number or density. Such a small shift in crystal chemistry might also help to explain the enhanced visibility of cross-striations both in phase-contrast interference light microscopy (POOLE 1966), with the change in optical path difference being explained by the difference in density of the mineral phase, and in polarised-light microscopy, where the birefringence of the mineral phase would be dependent upon the crystal chemistry. However, the subtle balance in the proportion of crystals having different orientations because of their different growth sites (interpit and pit) would also explain the visibility in polarised light (GUSTAFSON 1945; GUSTAFSON and GUSTAFSON 1961, 1967).

For alternative, earlier explanations of the function and significance of cross-striations see AG GUSTAFSON (1959), GUSTAFSON and GUSTAFSON (1961, 1967, 1968), DARLING (1958, 1961) and HELMCKE et al. (1963).

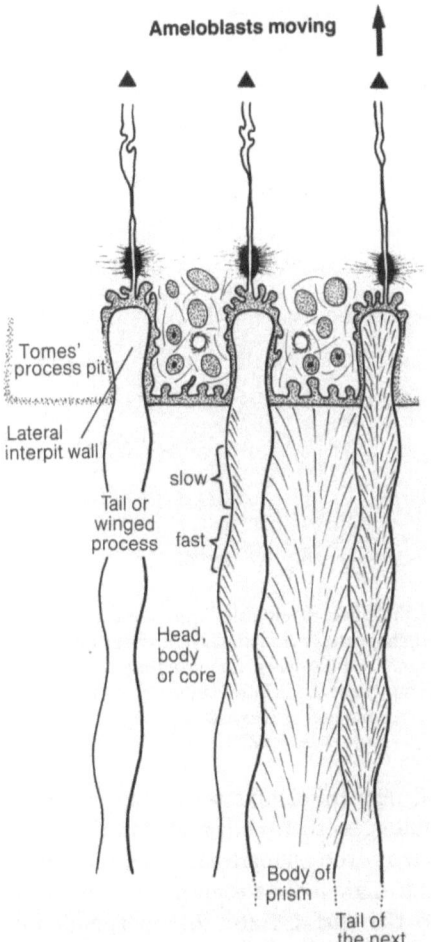

Fig. 48. A hypothesis concerning the development of the cross-striations of the enamel prisms. During the slow phases of circadian cycles of secretory activity of the ameloblasts, there is a reduction in the rate at which the surface of the ameloblast moves relative to the formative front. This movement is largely confined to those surfaces of the Tomes' process in relation to the circumdepression crystallites, i.e. in the interpit enamel. A change in the rate of the secretion of the ameloblasts would lead to a change in the balance of growth in pit floor as against interpit sites (BOYDE 1964)

V. General Directions of the Prisms as Seen in Low-Powered Light Microscopy

1. Prism Orientations

Prism directions (Fig. 45) reflect the course of ameloblasts during the entire period of amelogenesis and as such naturally run from the enamel-dentine juction to the surface of the enamel. In the cuspal regions of cheek teeth and in the incisive edges of anterior teeth, the mean prism orientation is perpendicu-

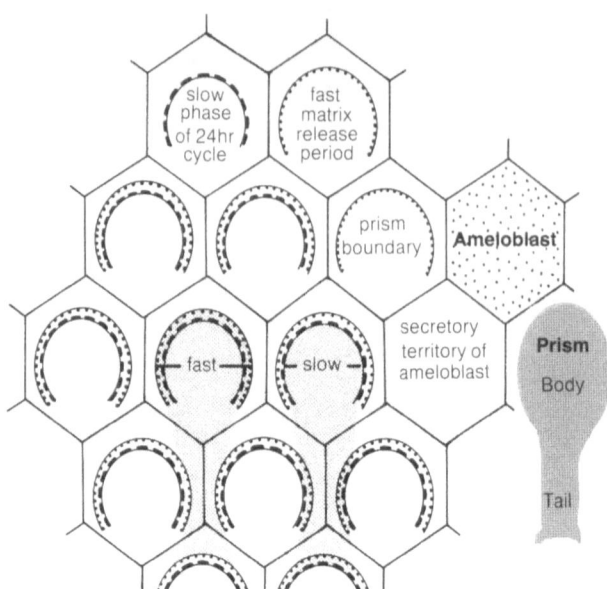

Fig. 49. The relative positions of the prism boundaries with respect to the secretory territories of individual ameloblasts (hexagonal shapes) as a function of time during the 24-hour secretory cycle. At fast phase of the cycle, the boundary is closer to the hexagon; at slow phase, it is more towards the centre of the hexagon. This is equivalent to a relative shift in the ratio of pit-floor- to interpit-wall-secreted material (from BOYDE 1976)

lar to the surface and also perpendicular to the incremental lines where these can be seen. On the lateral surfaces of teeth and on the lateral slopes of cusps, the prisms slope upwards from the enamel-dentine junction towards the biting surface of the tooth. In the high lateral regions of a human tooth, the angle is commonly about 60° to the surface, 30° to a perpendicular to the surface. In lower lateral regions of the human teeth, the prism orientation may be closer to a perpendicular to the mature tooth surface. Nevertheless, the prism orientation continues to run 45–60° to the plane of the incremental lines, i.e. to the surface of the enamel which prevailed during development. These details may be seen in any longitudinal section of any tooth: they are subject to great variability in detail, reflecting the individual developmental history of the tooth. The most general of these variations, however, depends upon the fact that neighbouring ameloblasts do not maintain the same relationships throughout the enamel development.

The course of the prisms alters and changes in both transverse and vertical senses on the route from the enamel-dentine junction to the surface. The alteration of the course in the transverse sense, considering lateral enamel, seems to be more important and will be considered in the next section.

In lower lateral regions of the tooth, a common finding is that prisms may deviate in a large curve towards cervical, recurving towards occlusal as the prism reaches the tooth surface (Figs. 50, 51). This curvature should not be confused that of the Hunter-Schreger bands.

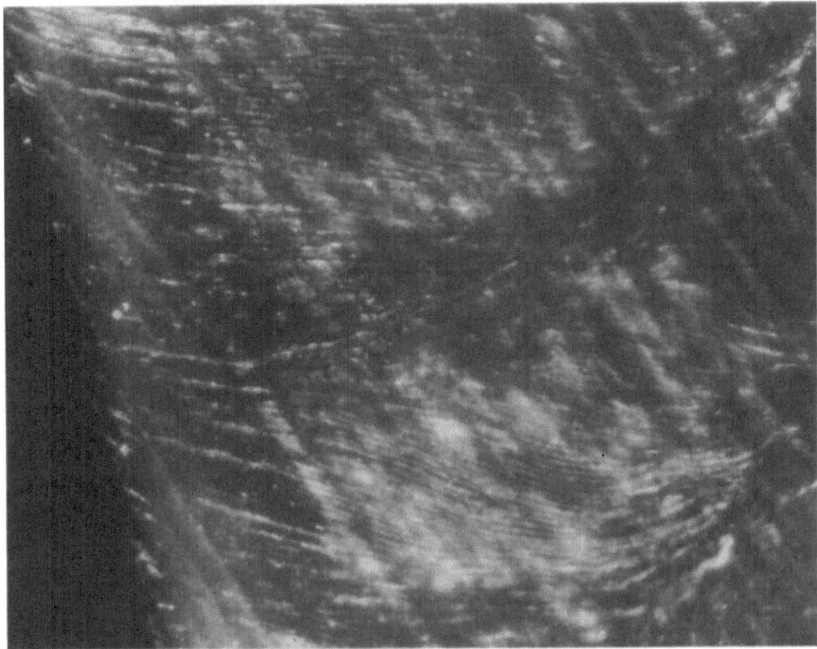

Fig. 50. Dark-field micrograph of longitudinal section of human premolar, showing cervical curvature of enamel prisms in a lower lateral region. The terminal segments of the prisms deviate towards cuspal in this field. Field height 540 μm

At the tooth surface proper, the prisms may undergo an alteration in their course in the last segment of their course in either sense. Some may deviate cervically from the last incremental line to the surface, for example, and in other regions in other teeth, one will find them deviating towards occlusal. Such sudden changes in the course of the entire prism sheet are uncommon in deep enamel. They always occur after a prominent incremental line: changes in the course of prisms are one of the underlying bases for the formation of incremental lines.

2. Prism Decussation

a) Movement of Ameloblasts

Certain features of enamel structure and development prove that ameloblasts are able to move their positions within the plane of the sheet of columnar cells to which they belong (WOLF 1942). The evidence for this viewpoint can be gleaned from any developing enamel surface preparation, bearing in mind that one prism boundary relates to one ameloblast. Tomes' process pits in the surface enter the surface in different directions in different parts of the field of view, indicating a different exit direction for the associated ameloblasts (WOLF 1942; BOYDE 1964). The slope of the prism (boundary) can be determined from

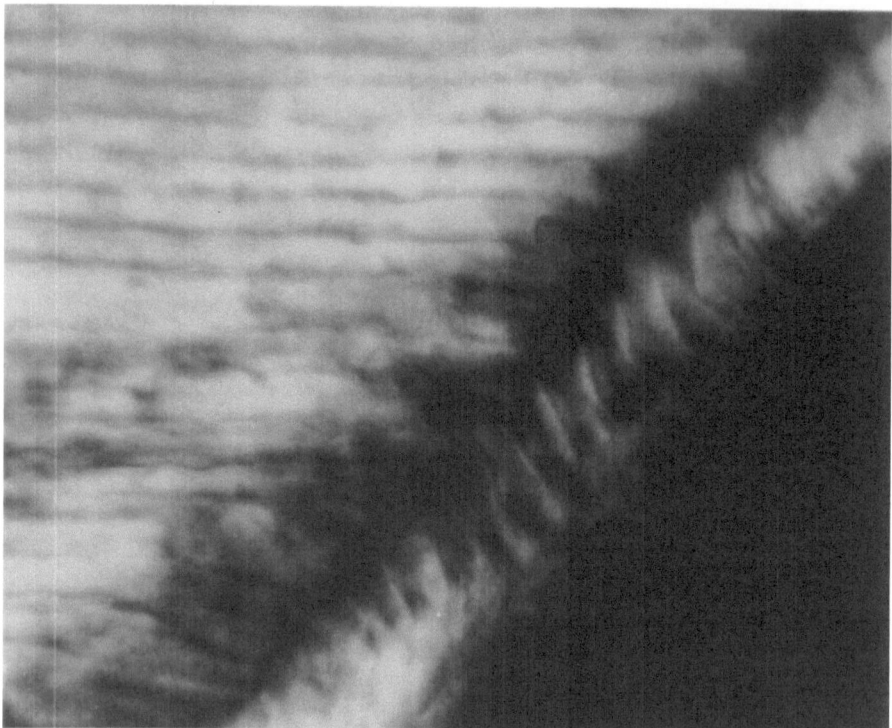

Fig. 51. Longitudinal section of human premolar in polarised light, cuspal to NE, showing cervical inclination of prisms in the last part of their courses to the surface. Field height 190 μm

the developing surface by tilting it to look along the steep lateral or cuspal, non-secretory, walls of the pits. [The author developed a real-time stereoviewing system for the SEM, which made this manoeuvre rather simple and reliable (BOYDE and MARTIN 1982)]. The three-dimensional course of the prisms is complex and sinusoidal: they have different neighbours at different parts of their lengths. However, at any one given depth within human enamel and within a small field of view, neighbouring prisms will have a nearly similar course. The bands which have matching courses have matching optical (LESTER 1965) and physical properties, including wear resistance (OSBORN 1965; SUNDSTROM 1966; BOYDE 1983). They are known as the Hunter-Schreger bands (APPLEBAUM 1960; OSBORN 1965, 1968a, b).

Fig. 52. Secretory ameloblasts of rat incisor during development of inner enamel showing alternate inclinations of proximal ends of cells in alternate transverse rows, which are sliding past each other at this stage of development. Papillary layer cells are seen in the top of field of view, enamel at the bottom. Field width 95 μm

Fig. 53. Developing enamel surface of rat incisor inner-enamel layer, showing alternate transverse rows of Tomes' process pits entering the surface with alternate orientations. Field width 56 μm

52

53

b) Extreme Decussation in the Rat Incisor

The development of the complex course of prisms may be understood from examination of developing enamel surface material and by contrasting the situation in different taxa, where prism decussation is developed to different extents (KAWAI 1955). The most extreme case of prism decussation is found in the inner enamel of the incisor teeth of myomorph and sciuromorph rodents (TOMES 1850; KORVENKONTIO 1934). Alternate transverse rows of ameloblasts slide past each other in opposing directions (Fig. 52), giving rise to Tomes' process pits at roughly 45° to the developing enamel surface (Fig. 53), with the prisms crossing each other at nearly 90° (WATSON and AVERY 1954; BOYDE 1964, 1969, 1975, 1978b; WARSHAWSKY 1968; KALLENBACH 1973).

The cellular mechanics involved in the movement of ameloblasts responsible for decussation have yet to be resolved, but some pertinent observations have been made. Ameloblasts contain actin filaments (REITH and ROSS 1973), an observation confirmed by the heavy-meromyosin studies of NISHIKAWA and KITAMURA (1983). The latter detailed study concluded that the role of the actin was most likely in the transport of secretory granules. In this respect, its involvement in translatory movement would be indirect, perhaps by controlling an eccentric secretion process. That secretion from ameloblasts is eccentric is clear from any analysis of the developing enamel surface (including that in human teeth), and it would be attractive to suggest this as a hypothetical driving force for translation of the cell across the forming enamel. However, one would then expect the pit to lead the cell and to make some kind of kink in the cell just outside the enamel. The morphology of the cells does not suggest that this is the case (WARSHAWSKY 1968; BOYDE 1969). KALLENBACH'S (1973) detailed study of the Tomes' process of the rat incisor inner-enamel ameloblast concluded that the cell led the prism. Images of the bodies of these cells (BOYDE 1975, 1978; SKOBE 1977) show a curvature along their length in the plane of the transverse row to which they belong, which again suggests that the cell body leads the way. The intercellular attachments between the cells along the sliding sides between transverse rows are clearly different from those between the cells in the same row (BOYDE 1964; KALLENBACH et al. 1965). KALLENBACH (1973) showed that the terminal web radiating inwards from the distal terminal bars extended further and was denser on the sliding faces.

c) Other Mammals

In all other mammals, the zones of prisms have common orientation property zones which are much wider. Zones 3, 4 and 5 prisms wide are found in the inner enamel of the incisors of the hystricomorph rodents (Fig. 54). Zones which

Fig. 54. Developing enamel surface of guinea pig lower incisor inner enamel, showing varying direction of exit of Tomes' process pits corresponding to direction of movement of ameloblasts in zones which are three to five pits (prisms) wide. Field width 112 μm

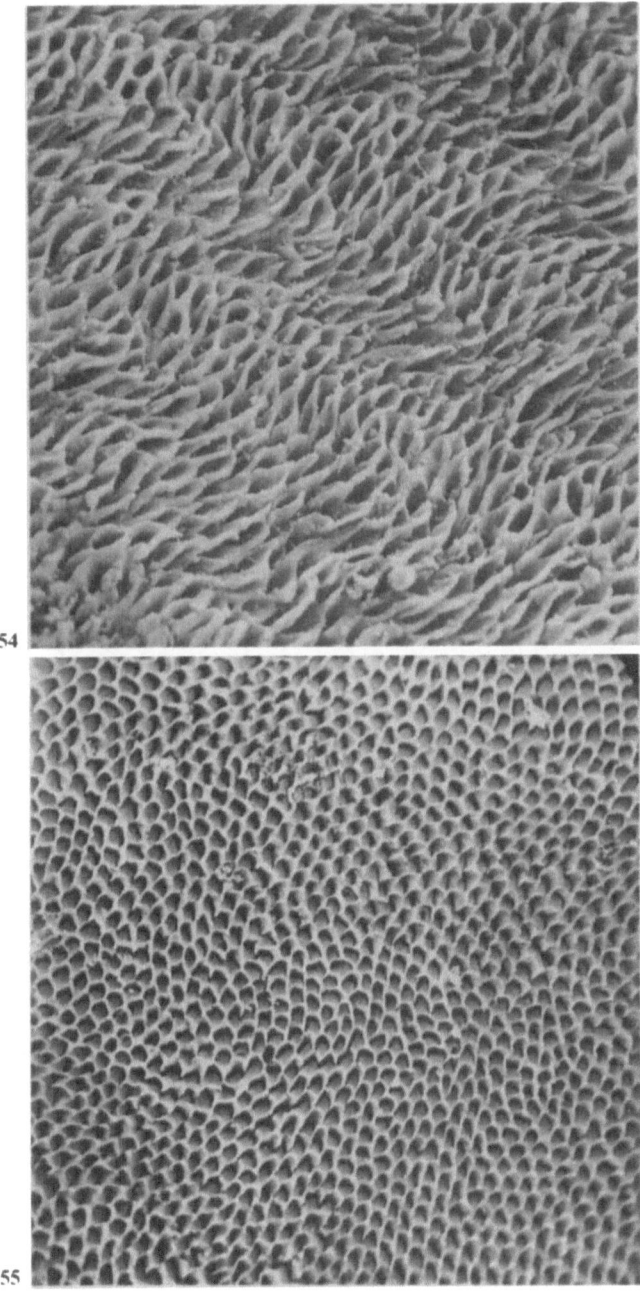

54

55

Fig. 55. Human upper lateral developing deciduous incisor: the enamel organ was removed by dissection followed by treatment with an NaOCl solution for 20 min to remove the residual debris before washing and freeze-drying. This field shows the mid-thickness developing enamel from the centre of the lingual side. Note the predominance of the Pattern 3 organisation of the pits (cuspal towards top) and the change in orientation corresponding to the development of the Hunter-Schreger bands.
Field width 196 μm

Fig. 56. Critical-point-dried developing enamel surface of orang-utan second deciduous molar. Lateral enamel surface showing strong variation in inclination of pits (and prisms), corresponding to the development of the Hunter-Schreger bands. Field width 166 μm

are wider, yet still clearly distinguished, may be found, for example, in the developing enamel of the Carnivora (KAWAI 1955; BOYDE 1969; OSBORN 1971).

Regarding primates, sharply demarcated zones (averaging 7–8 prisms wide) of decussating Pattern 2 prisms are found in the unique strepsirhine primate, *Daubentonia*, the aye-aye (SHELLIS and POOLE 1979). Decussating Pattern 1 zones have been reported in a lemur (BOYDE and MARTIN 1982; see also BOYDE and MARTIN 1984a, b; MARTIN 1983; MARTIN and BOYDE 1984). Wider, less sharply demarcated zones are found in the Pattern 3 enamel most commonly seen in Anthropoidea.

d) Human Enamel

In human enamel, zones of pits at the developing enamel surface are not so sharply distinguished because the changes in direction are not so important as in the examples we have just considered (Figs. 55, 56). The Hunter-Schreger bands are (very roughly said) about 100 μm wide, taking the term "band" or "zone" to mean a region in which the prisms are doing roughly the same

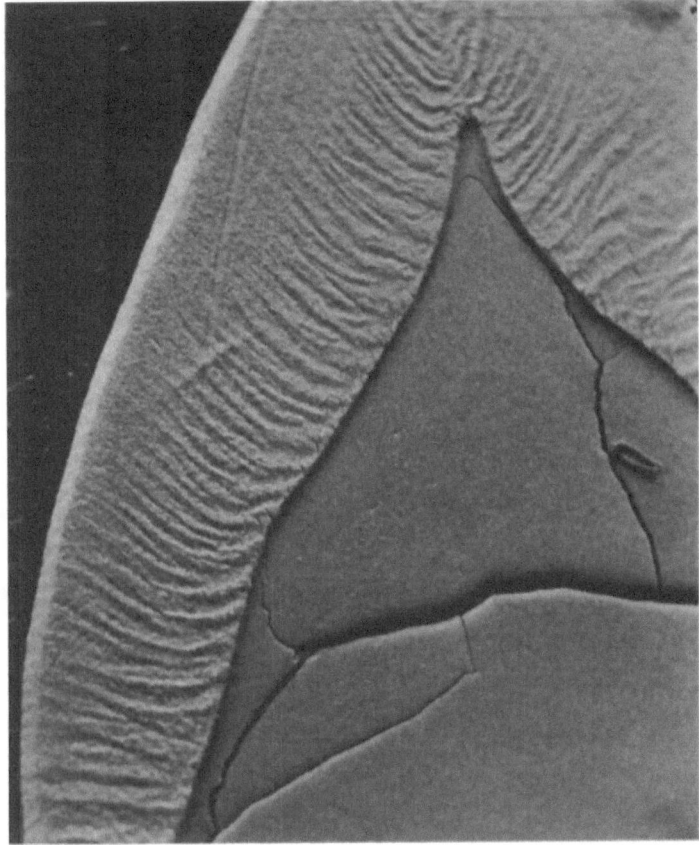

Fig. 57. Converted back-scattered electron image of longitudinal section of premolar (deep-etched with NaOCl to render it anorganic, followed by potassium citrate). The image shows contrasts due to the topographic relief of the Hunter-Schreger bands. (Specimen biased to +200 volts to prevent escape of secondary electrons. Electrons collected are secondary electrons due to BSE-releasing secondaries from the wall of the specimen chamber.) Field width 4500 μm

thing (LESTER 1965; OSBORN 1968). Moving in the longitudinal direction along the tooth from a location at which a prism has one characteristic, for example, deviating to the left, we will find that at a distance of some 10–13 prisms the opposite behaviour will be observed: in this example, the 10th–13th prism would be deviating to the right (Figs. 57–60). Good ways of observing this are to examine either the developing enamel surface or a surface ground parallel to the tooth surface ànd deep-etched in deeper layers of the enamel. Prisms may change their direction several times throughout the thickness of the enamel: their course may be sinusoidal or helicoidal. This can be seen directly in deep-etched enamel examined by SEM and in confocal LM (Figs. 61, 62). It can also be inferred from the details of prism orientations with respect to Hunter-Schreger bands seen in longitudinal sections observed by optical microscopy (OSBORN 1968a, b), as now explained (Fig. 59).

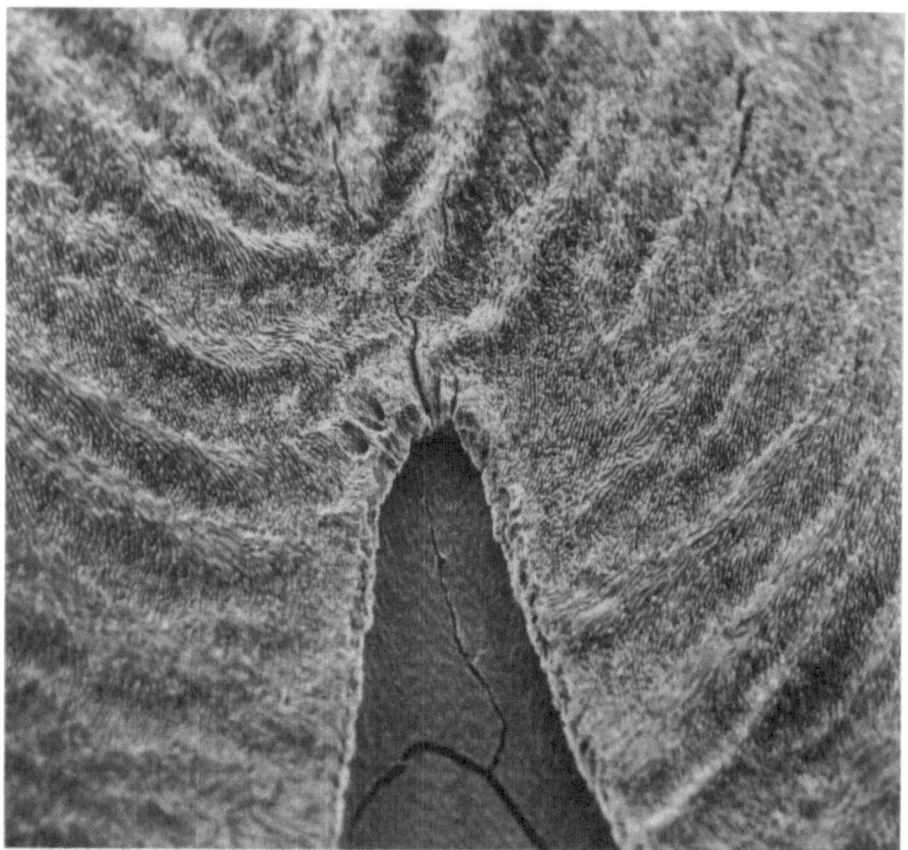

Fig. 58. Higher magnification of sample shown in Fig. 57, showing the changing orientation of the prisms in the Hunter-Schreger bands. Field width 827 μm

A diazone is a band of enamel seen in a longitudinal section in which the prism profiles indicate that these prisms have been sectioned more transversely than in the contrasting form of zone. In the parazones, the prisms appear to be more longitudinally sectioned. Both sets of zones cross the mean axis of the enamel prisms as projected in a longitudinal section plane. If the prism axis crosses the zone axis, then the prisms must be undulating up and down with respect to the plane of a longitudinal section, i.e. have a side-to-side component projected in the plane of an oblique transverse section which was nearly parallel with the prism direction in the locality.

The actual angles at which prisms cross in human lateral enamel are not as great as it looks from the appearance of diazones and parazones. (Actually, the prisms do not often really cross in human enamel since there is a grade in course from one to the next, as may be seen in the accompanying Figures.) OSBORN'S (1968) data showed that prisms deviate less than 20° from a mean radial course, giving a decussation angle of <40°. This is in agreement with

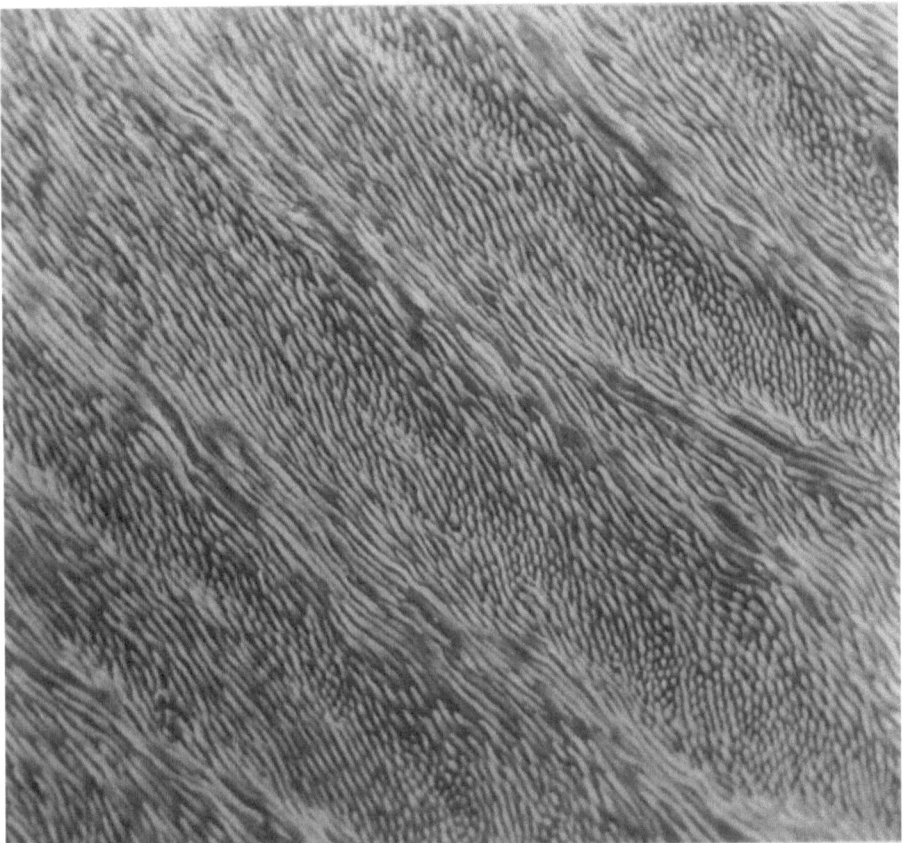

Fig. 59. Hunter-Schreger bands in deep-etched longitudinal section of human premolar, showing the enamel close to enamel-dentine junction, which is out of the right-hand side of the field of view. Note that prisms in the diazones apparently cross the zone axis more so than in the parazones. Field width 530 μm

values obtained by BOYDE (1976) using a different measuring strategy: the latter measured the deviation from the idealised mean path (which would be a perpendicular to the tooth surface in a transverse section plane) in a large number of locations in a large number of teeth, measuring the angle at $<>50$ μm deep to the tooth surface. This study emphasised the fact that prisms may deviate significantly from the "normal" path whilst close to the tooth surface, whereas Hunter-Schreger bands are only easily visible in the deeper enamel. Nevertheless, they often run up to the surface in human teeth. There is, however, an alternative explanation for group data obtained by pooling observations from many teeth (Fig. 63).

Human teeth are not circular in cross section, but have flattened surfaces. Ignoring their lateral, sinusoidal oscillations, prisms do not pass in mean straight lines from the enamel-dentine junction to the surface except where they are

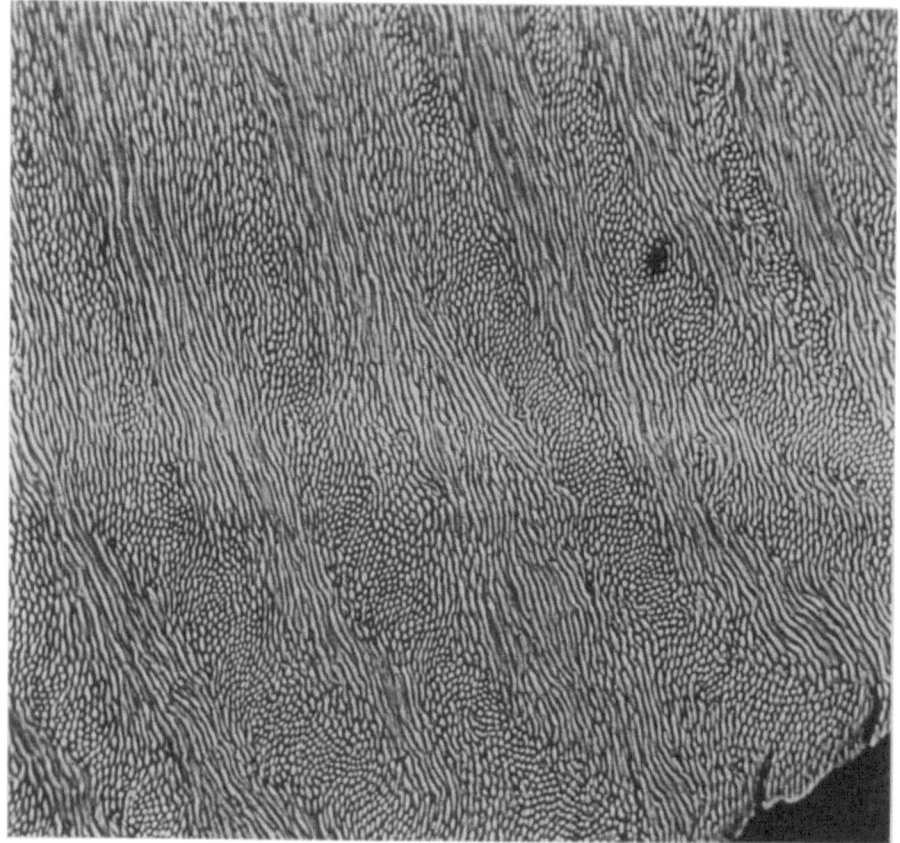

Fig. 60. BSE image of acid-etched (0.5% phosphoric acid) longitudinal section of human premolar, showing diazones and parazones. Field width 737 μm

in the centre of a flat tooth surface (Fig. 64). Others pass laterally at oblique angles to the tooth surface as projected into the transverse section view (BOYDE 1985).

Gnarled enamel is the name of the appearance of the enamel seen in incisive edges of anterior teeth and over cusps of posterior teeth (Fig. 65). It is probable that the gnarled enamel structure is not too different from that seen on lateral surfaces of teeth but merely that the usual histological section planes intercept the structure in a greater variety of directions, thus giving rise to the apparent complexity of structure (BOYDE 1964; OSBORN 1965). However, it is accepted in many textbooks that the gnarled enamel structure is more complicated. The appearances of the Hunter-Schreger bands in cuspal enamel depend upon the distance of the section from the ideal midline longitudinal section through the cusp centre (Fig. 66). If the section passes the centre, the bands appear to radiate from the enamel-dentine junction, much as they do along lateral surfaces. Sections well removed from the centre line show the bands as apparently curved

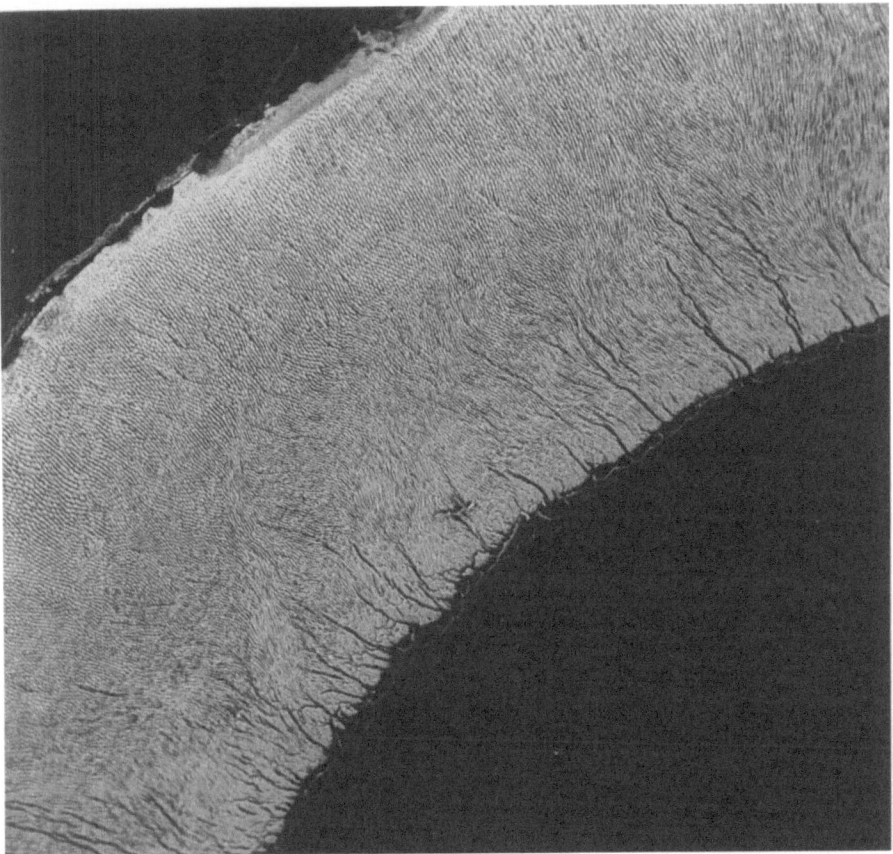

Fig. 61. Transverse section of human premolar polished and etched with 0.5% phosphoric acid, imaged with BSE, showing the course of enamel prisms as revealed in this plane of section. Deep clefts extending from enamel-dentine junction are tufts exaggerated by etching process. Field width 1870 μm

and tangential to the junction. The "gnarling" is due to the severe change in apparent orientation seen at zone boundaries, combined with the fact that the inclinations of the prisms to the incremental surface are shallower (i.e. the relative rate of translatory motion of the ameloblasts is greater) in the inner enamel of cuspal regions.

e) Functional Significance

There is a consensus in the literature that the complications of the course of the enamel prisms seen in the general phenomenon of decussation and the presence of Hunter-Schreger bands increase the strength of enamel, or at least its resistance to the propagation of fractures (OSBORN 1968). Enamel does tend to cleave via its prism boundaries (more details in a later section). The chances

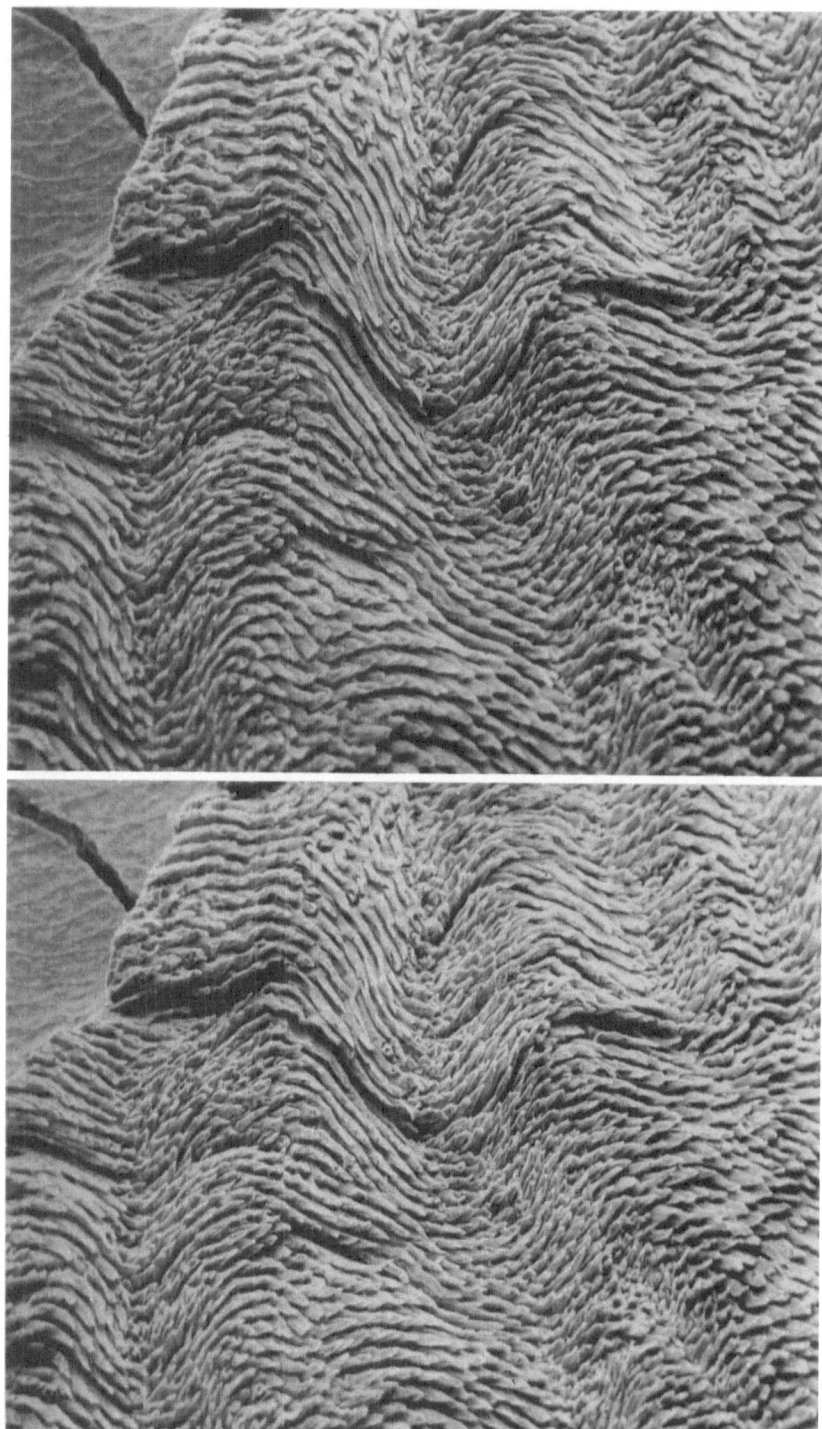

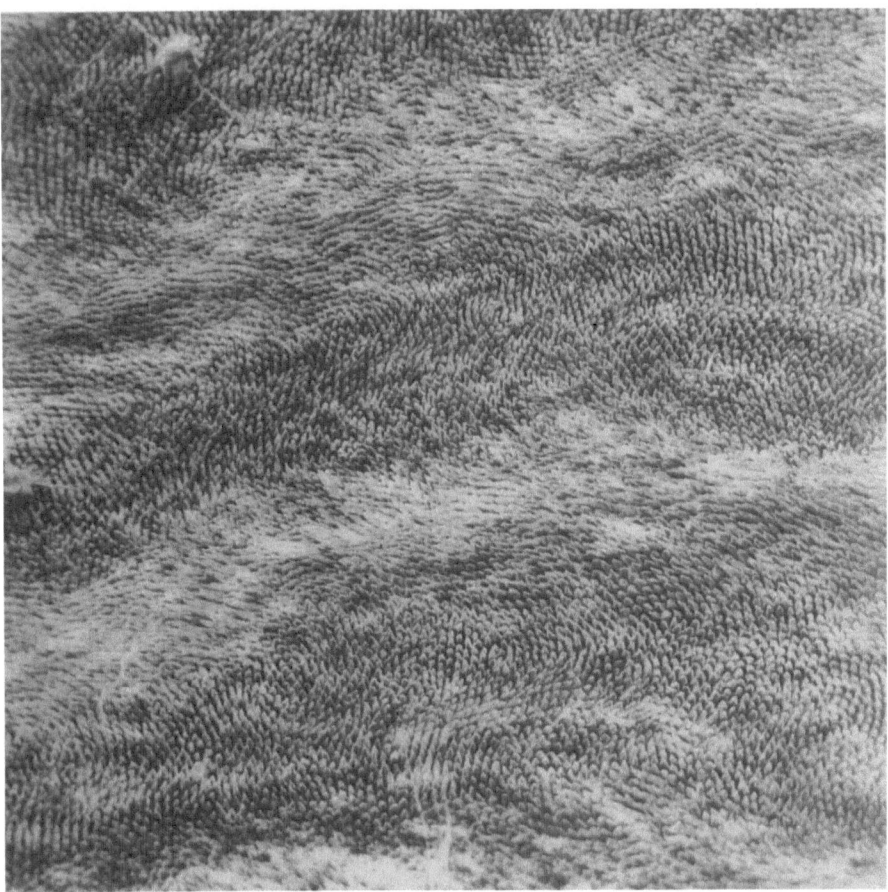

Fig. 63. Lateral enamel surface of human premolar deep-etched with EDTA to show the course of the enamel prisms at about 100 μm deep to the surface (from measurements made in close-by localities on the same sample). The zones of the Hunter-Schreger band pattern are here seen head-on. Note that one set of zones approaches the surface more nearly perpendicularly; the other has stronger sideways inclination. (A longitudinal section which would also show the same "dip" sense perpendicular to a surface like this would show parazones and diazones. One reason for the unequal angle of approach of decussating prisms to the surface is shown in Fig. 64.) Field width 684 μm

of a putative fracture propagating right through the tissue will obviously be reduced by the complicated sinusodial courses of the prisms. One would expect, therefore, that the Hunter-Schreger bands contribute towards the fracture and wear resistance of enamel.

◁ **Fig. 62.** Oblique transverse section of human premolar deep-etched with NaOCl and potassium citrate (stereopair, tilt-angle difference 10°), showing the gradual change in orientation of the prisms which is suggested by the developmental information. The type of surface shown here is strongly suggestive of a side-to-side bending of the prisms, which would imply that they progress from zone to zone during their development. Field width 37 μm

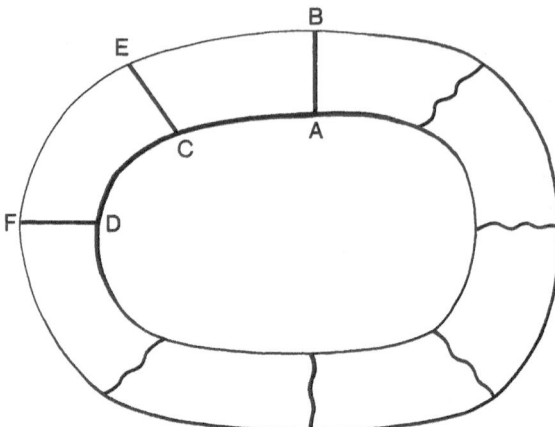

Fig. 64. Transverse section of a tooth indicating the embryological reasons why prisms do not approach the tooth surface at mean normal incidence at all portions of the perimeter. A prism running from *A* to *B* would be expected to approach the tooth surface perpendicularly. However, a prism running from a point like *C* (halfway betwen *A* and *D*) would reach the tooth surface at point *E* (halfway between *B* and *F*)

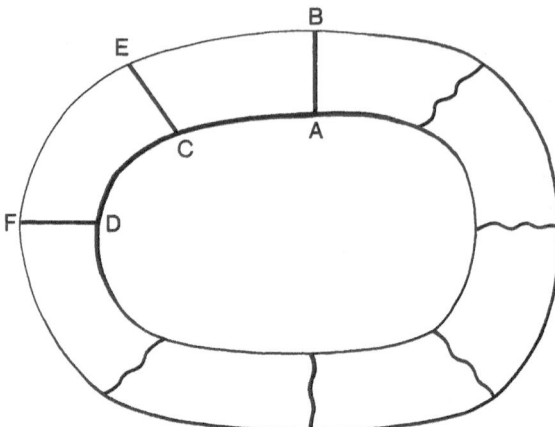

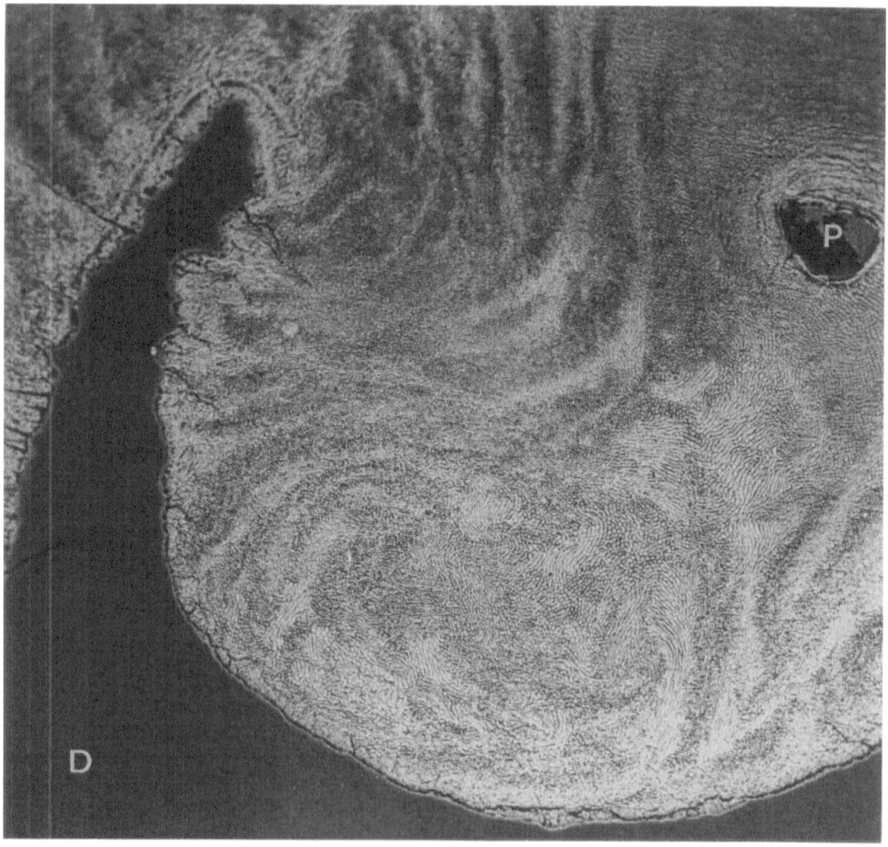

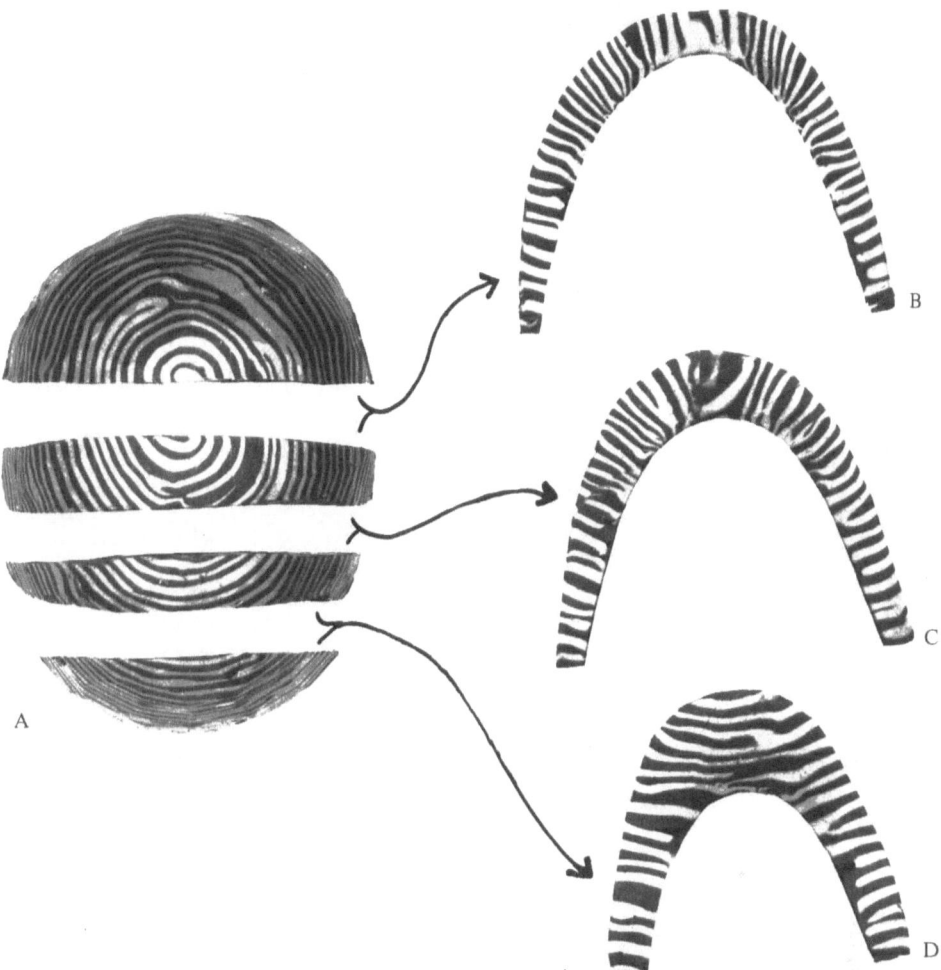

Fig. 66. A sectioned plasticine model of the arrangement of the Hunter-Schreger bands. The pattern begins as a spiral over the cusp tip. *A*, viewed from on top. *B, C, D*, the appearance of the bands in longitudinal section through the model cusp depends on how far the section lies from the centre of the cusp. At *B*, the bands apparently radiate nearly perpendicularly at the cusp centre; at *D*, very far from the cusp centre, the bands may appear to lie at a tangent to the enamel-dentine junction at the tip of the cusp

We shall examine two examples from comparative dental histology which may imply that the significance of the complicated courses of the enamel prisms is more than this. The enamel at the incisive edge of rodent incisors contains well-marked sheets of incisally inclined but transversely decussating prisms. Such

◁ **Fig. 65.** Transverse section of human premolar polished and etched to reveal prisms, and showing the complex changes of course of the enamel prisms in mature enamel in a cuspal region. *D* Dentine; *P* occlusal surface pit. Field width 1740 µm

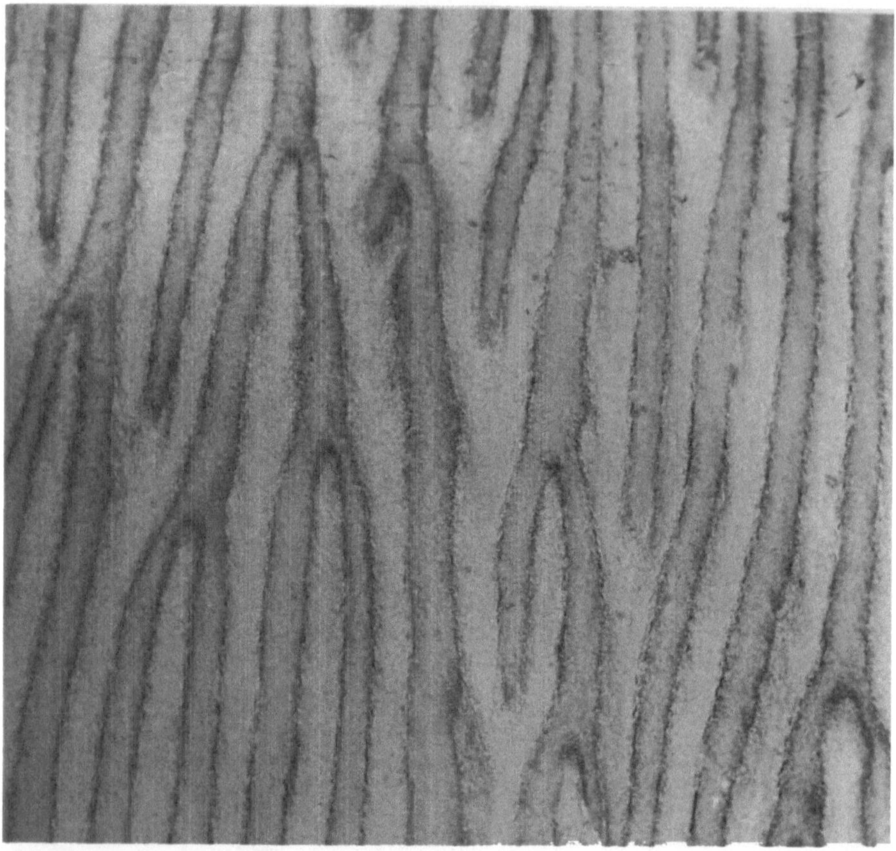

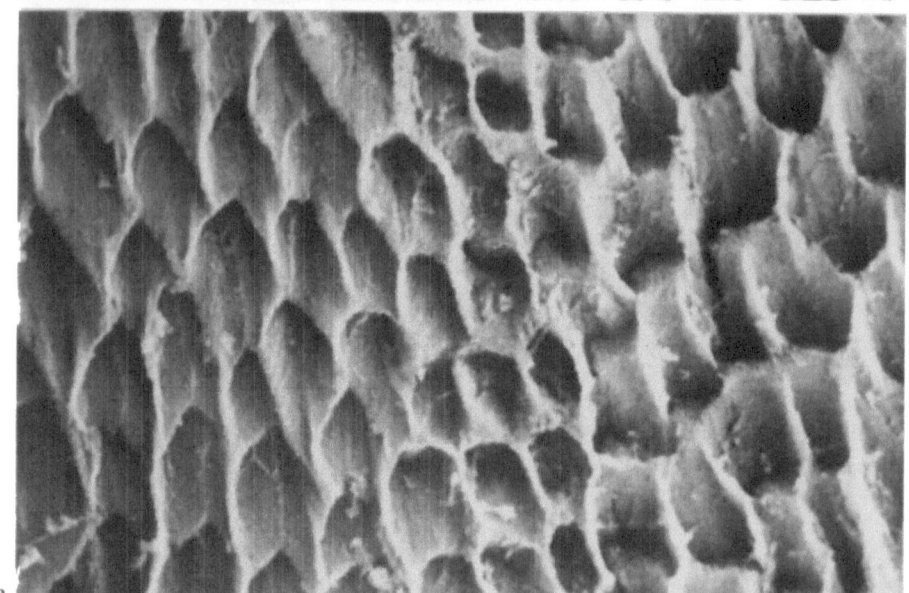

67

68

transverse sheets of prisms essentially ensure that prism boundaries are joined to a greater extent than in other arrangements. They may thus give rise to potential fracture planes, whereby entire sheets might be expected to cleave away from the functional surface. In this example, it may be the case that the decussation ensures rapid preferential functional wear in order to maintain the sharp biting edge of the tooth.

In the enamel of rhinoceroses (and in some extinct mammalian groups), the Hunter-Schreger bands are vertical rather than horizontal in the greater part of the enamel thickness (Figs. 66–68; RENSBERGER and VON KOENIGSWALD 1980; FORTELIUS 1984, 1985; BOYDE and FORTELIUS 1986). At the corresponding worn functional surfaces of these teeth, therefore, prisms either will approach the enamel surface perpendicularly or may lie in the plane of the functional surface. In the latter locations, the rate of wear of the tissue is preferentially increased. The enamel wears to produce a ridged surface corresponding to the distribution of the Hunter-Schreger bands. This microrelief may have a functional significance at the gross level in terms of the ability to cut through rough browsing. It is not known whether differential wear of parazones and diazones has any functional significance in man; it is, however, indirectly responsible for earlier views that Hunter-Schreger banding was associated with variations in the density of mineralisation since it leads to differential thickness of a ground section (OSBORN 1965; SUNDSTROM 1966; BOYDE 1983, 1985).

G. The Enamel-Dentine Junction

I. The Shape of the Junction

The enamel-dentine junction has a complicated shape at whatever level of magnification it is viewed. At the scales of magnification represented in light microscopy and SEM, it shows a scalloped profile in either longitudinal or transverse sections perpendicular to the junction (Fig. 2). The dentine surface is bayed with headlands projecting towards the enamel. At much higher magnifications, as seen in high-powered SEM and TEM, the junction is extremely complex and interdigitated (Fig. 69). The form of the junction would explain the difficulty in separating these two tissues for the purpose of making histologi-

Fig. 67. Low-magnification image of developing inner-enamel surface of rhinoceros permanent molar enamel, showing the distribution of the decussating zones. The thin darker lines show zone boundaries. "Uphill-going" zones representing pits with an occlusal inclination appear slightly lighter in this image than the cervically inclined zones. Cell columns moving in opposite directions collide at the Y junctions, and these portions of developing enamel surface are higher than surrounding areas as a consequence. Field width 2340 μm

Fig. 68. A zone boundary in developing inner-enamel surface of developing rhinoceros molar inner enamel, showing the opposing direction of the entry of pits in the vertically distributed zones. Longitudinal axis of tooth is vertical. Field width 51 μm

Fig. 69. Enamel-dentine junction exposed by osteoclastic resorption of lower surface of human deciduous molar prior to shedding, showing a thin zone in which prismatic detail is missing next to
the dentine (*top*), followed by a band in which prism boundaries are prominent from the resorbed
surface. Field width 27 μm

cal preparations and the strong resistance of enamel to cleave from the surface
of dentine in normal function. Fractures parallel with the enamel-dentine junction do not pass through the plane of the junction, but through the enamel
close to the junction (Fig. 70).

Preparations of the dentine side of the junction made by dissolving the
enamel away in acids or chelating agents (Figs. 71, 72) demineralise the surface
of the dentine matrix, which will then shrink and distort on drying for observation in SEM (WHITTAKER 1977). The embryological origin of the complex scalloped shape of the enamel-dentine junction in human teeth is not clear.

II. Features Originating at the Enamel-Dentine Junction

1. Spindles

Spindles are a class of low-density enamel feature, dark in BSE SEM images,
strongly reflecting light in optical microscopy, particularly concentrated at the

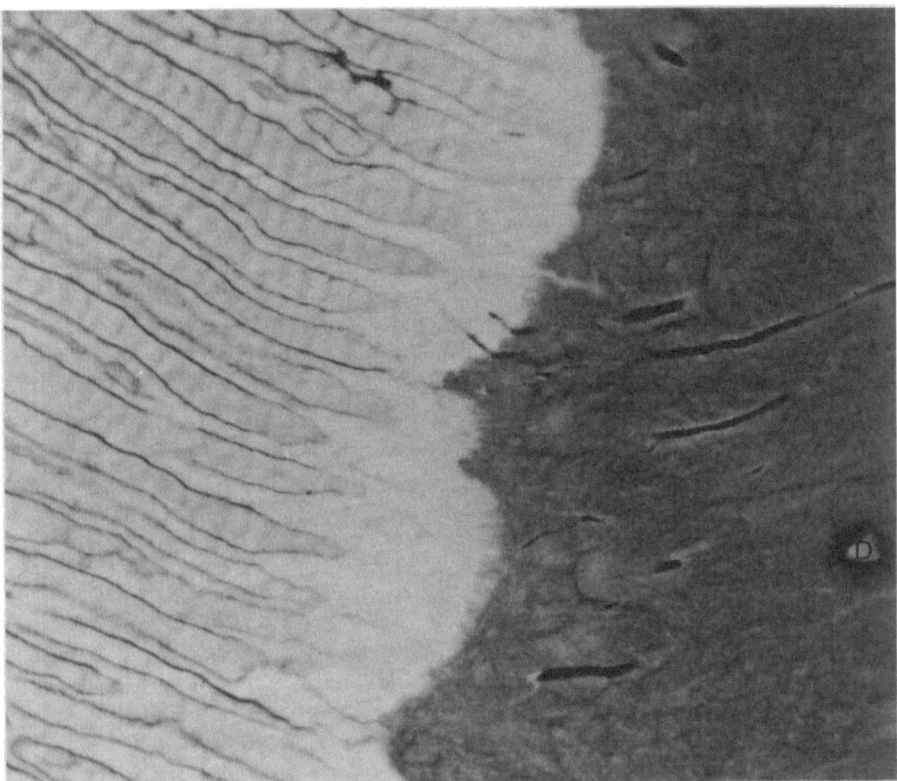

Fig. 70. Enamel-dentine junction in cuspal region of human molar prepared by sectioning longitudinally and diamond-micromilling, followed by carbon-coating and BSE imaging in an SEM. Contrasts
are mainly dependent upon density variations in the tissue. *D*, Dentine. Field width 95 μm

tips of cusps or biting edges. Concerning their embryological origin and nature,
the most commonly expressed view is that they are extensions of dentine tubules
in the enamel, no explanation being offered as to how this might occur. However,
the continuity of dentine tubules with enamel spindles can be demonstrated
in a good proportion of cases. Spindles stand perpendicular to the enamel-
dentine junction and are therefore not exactly parallel with the prisms in their
locality. They are roughly as long as a secretory ameloblast would have been
before the tissue was secreted. There is a possibility that they represent isolated,
dead ameloblasts incarcerated in enamel. Another special feature of the region
in which they are concentrated is that the large-diameter von Korff fibre bundles
of the collagen matrix of the dentine are more nearly perpendicular to the
junction: spindles may represent dentine matrix which projects across the general
line of the junction (Figs. 72, 73). However, BSE SEM imaging shows a rather
lower density for the spindle regions than the dentine matrix (Fig. 70). Nevertheless, some van Gieson-stained sections of tooth germs show projections of dentine matrix between ameloblasts prior to the onset of amelogenesis.

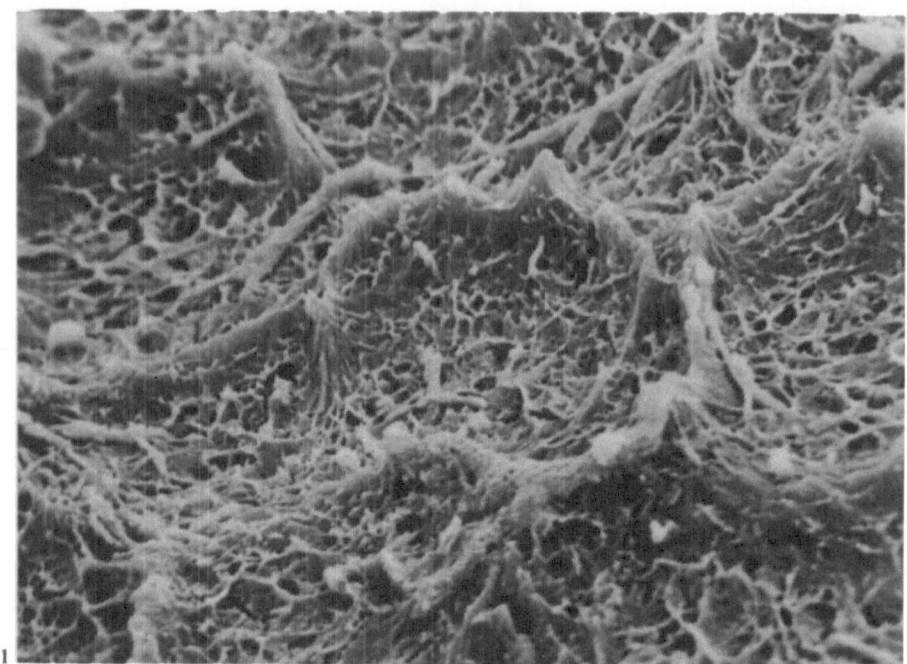

71

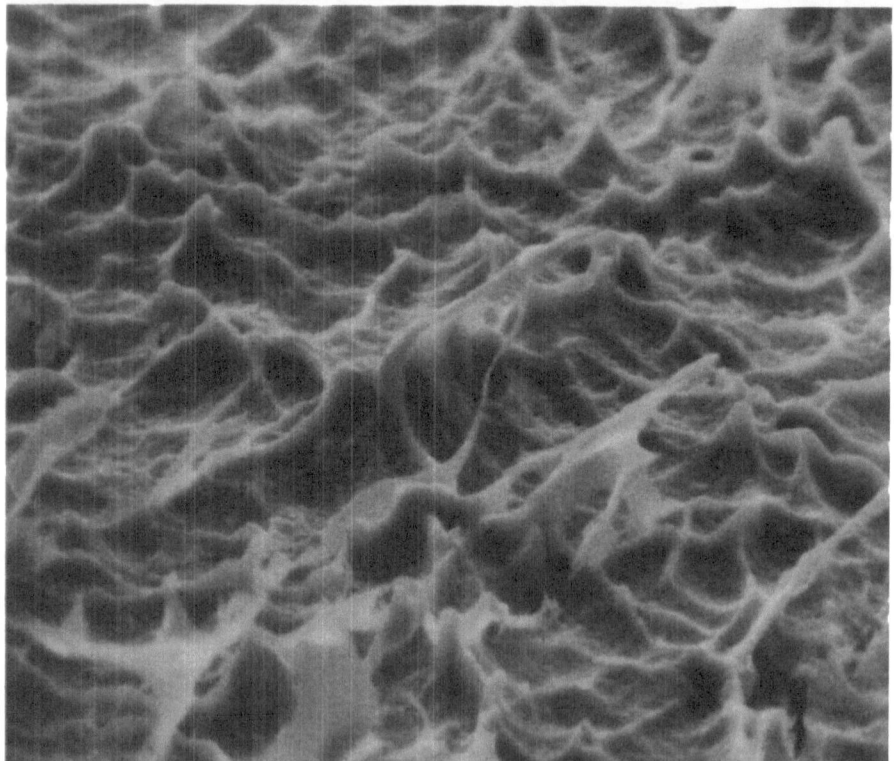

72

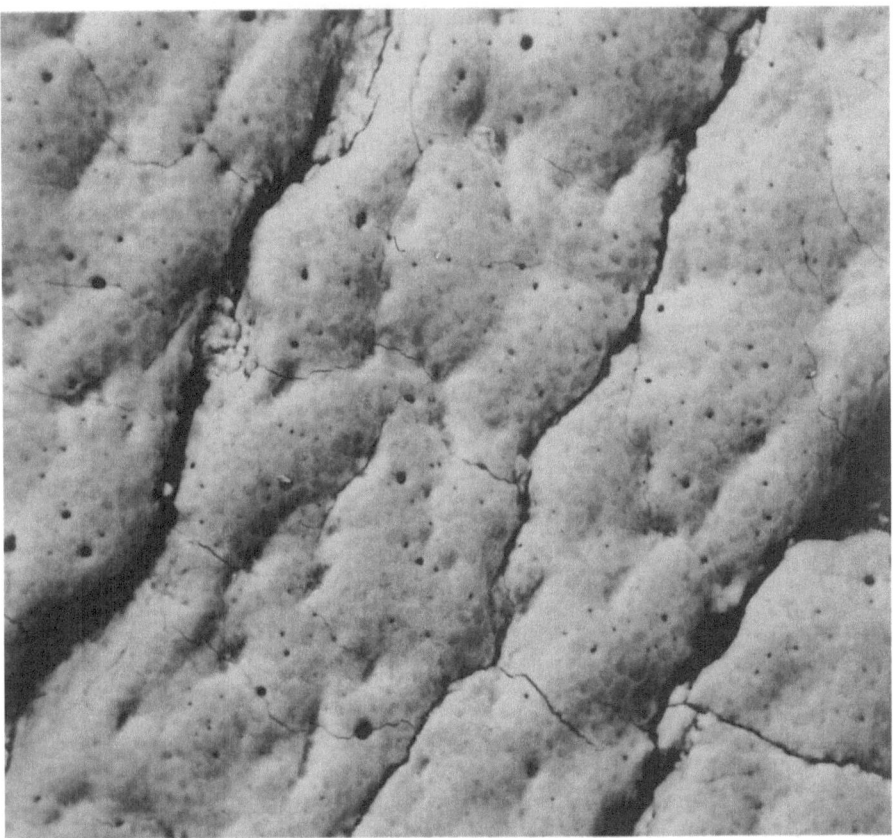

Fig. 73. Enamel side of enamel-dentine junction of lateral enamel in human molar exposed by first making the dentine anorganic by prolonged treatment with hot (50° C) 1% sodium peroxide solution, followed by air-polishing (using jet of sodium bicarbonate power in air surrounded by water). Prism cross-section profiles can be seen in the enamel. The depressions would correspond to dentine ridges on the enamel-dentine junction or the tuft planes. They are at the correct separation distance and in the correct orientation to be tuft planes. Cracks develop in the enamel due to the removal of its dentine support. Isolated black holes in enamel may be spindles or tubules. Field width 219 μm

2. Tufts

Tufts are a class of feature named for their appearance in low-powered optical microscopy of thick transverse sections of human teeth, where, owing to the great depth of field, several parts of these structures are projected into

Fig. 71. Enamel-dentine junction in human permanent tooth revealed by dissolving away the dentine with molar HCl, air-dried from absolute ethanol. Some of the detail in a preparation like this represents enamel matrix protein remnants dried down onto the dentine surface, the latter being deformed by the dehydration process as a consequence of also being demineralised. Field width 57 μm

Fig. 72. Enamel-dentine junction of human tooth exposed by removing overlying enamel with acid. Projections could be dentine collagen or enamel matrix components. Field width 45 μm

one image plane. The tufts are longitudinal fault planes (recurring at intervals of < >100 μm) in the inner part of the enamel, usually measuring one-fifth to one-third of the enamel thickness (this value depending upon the plane of the section with respect to the vertical level on the tooth), within which more enamel-matrix protein is retained in the mature tissue (Fig. 61). They project outwards from the enamel-dentine junction in the same direction as the prisms and therefore deviate from side to side in association with the Hunter-Schreger band pattern. The faulting in the tissue which gives rise to the tufts occurs predominantly in the longitudinal direction, joining one prism boundary to the next across the shortest distance in the interpit enamel. These less mineralised regions have a markedly lower refractive index than the surrounding enamel and therefore constitute strongly reflective planes in optical microscopy. This is the reason why they appear bright in reflected light and dark in transmitted light. A thin optical section reveals a fault following the prism direction. Focusing up and down, this fault plane can be seen to deviate from side to side in accordance with the direction of the prisms. In teeth from older individuals in which the enamel has acquired a high degree of translucency due to post-eruptive maturation, the tuft planes can be seen using low-powered stereobinocular light microscopy to look through the tooth surface.

3. Lamellae

Lamellae are faulting planes which run in the same direction as the tufts but extend throughout the entire thickness of the tissue from the enamel-dentine junction to the surface. It is difficult to distinguish lamellae from longitudinal cracks which develop in aged teeth in normal function. If there is a real distinction, then lamellae are cracks which develop prior to the eruption of a tooth, whereas cracks of the same nature are acquired during function. As with tufts, lamellae are normally studied in transverse sections of teeth, where they give rise to sufficient optical contrast when seen on edge. They may, however, be studied within the thickness of longitudinally sectioned enamel, using confocal microscopes such as the TSM (PETRAN et al. 1985).

4. Enamel Tubules

Enamel tubules are a class of feature seen only in the enamel closest to the enamel-dentine junction in human teeth. However, features of the same nature may be much more numerous and extensive in the teeth of other mammals. They are ubiquitous in marsupials, with the exception of the wombats (TOMES 1849; LESTER and BOYDE 1967a, b; LESTER 1971; PALAMARA et al. 1981; PHAKEY et al. 1984). They are well developed in many strepsirhine primates (CARTER 1922).

Enamel tubules are generally continuous with the dentine tubules at the enamel-dentine junction and have an analogous embryological origin. They are due to the retention within the developing enamel matrix for a limited period of a cell process extending from the Tomes' process region of the secretory ameloblast (LESTER 1971: Fig. 1; for a contrary opinion, see RISNES and FOSSE

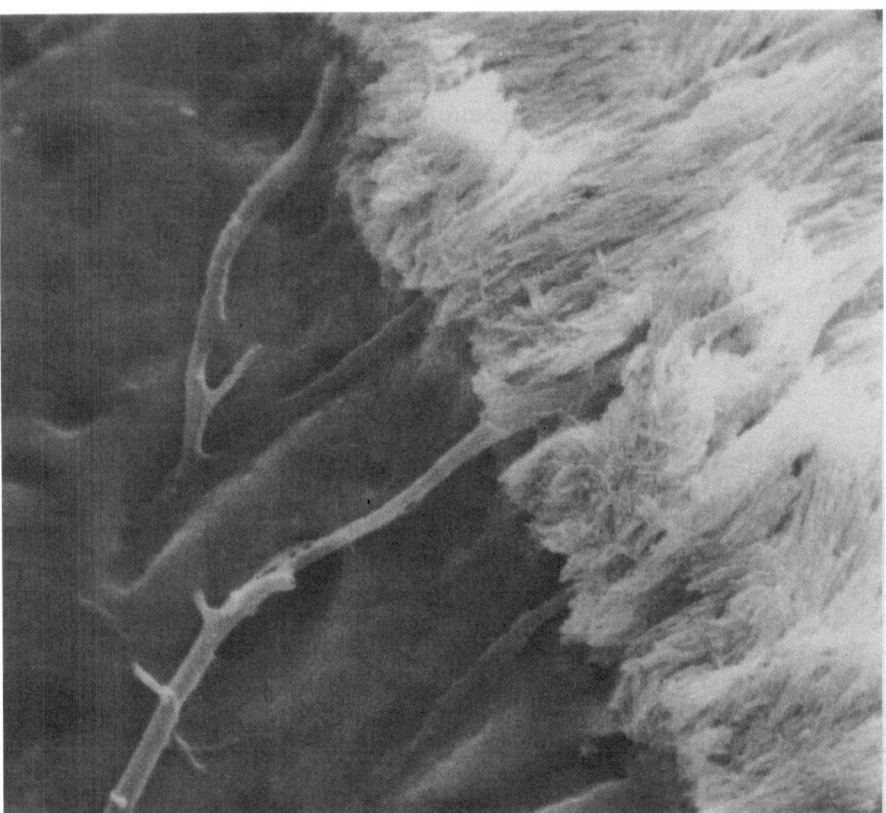

Fig. 74. Enamel-dentine junction in marsupial (*Trichosurus vulpecula*) prepared by filling open spaces in both tissues with methylmethacrylate monomer, polymerised in situ, and then by alternate treatment with NaOCl and acid-etching to dissolve away parts of the tissues leaving PMMA internal casts of enamel and dentine tubules. Field width 20 μm

1974). This cell process determines that there is a tubular deficiency within the developing enamel matrix, in which no enamel matrix is present and no enamel crystals grow. The distribution of tubules is thus well studied by making internal casts by polymerising methyl methacrylate monomer in situ (Fig. 74; BOYDE and LESTER 1984; LESTER et al. 1987).

In human material, enamel tubules are best recognised in longitudinal fractures replicated for TEM (LESTER and BOYDE 1967b). The detail which characterises these features is generally difficult to pick up in scanning electron microscopy of human teeth. Enamel tubules in man extend only a few microns from the dentine surface, but in doing so they penetrate the prism-free zone at the enamel-dentine junction: this zone is free of the boundary discontinuities which permit fluid flow through the bulk tissue (LINDEN 1968). Their patency in the mature tooth would also mean that dentine tubules were exposed when enamel was worn to the level where the enamel tubules were present. They might thus constitute a part of an advance warning system whereby pulp cells, and particularly odontoblasts, would be able to recognise impending exposure of dentine due to functional wear of the tooth.

Not even in the most extensive distribution of enamel tubules seen in marsupial mammals do they extend to the tooth surface (Tomes 1849). Their contents may be retained as a "fibril" in decalcified material, but the biochemical nature of the contents has not been identified. One would assume from their origin that they contain the remains of the disintegrated cytoplasmic processes of the ameloblast which caused them to form (Lester 1971).

H. Mature Enamel Surface Features

I. Fissures

Fissures between cusps or ridges are gross features of the occlusal surfaces of the posterior teeth, which originate at the borders or originally separate formative areas (the cusps). During crown development, dentine formation starts at several separate centres, with enamel development following and covering these dentine centres. As the plane of the enamel-epithelium-to-mesenchymal-papilla interface increases in extent, the distance between the original cusp centres increases. There comes a point when all the proliferative cells of the mesenchymal papilla have differentiated into odontoblasts and are producing dentine. The plane of the enamel-dentine junction of the occlusal surface of the tooth is then defined. At this time, however, the separate centres of enamel formation are essentially moving towards each other across the surface of the dentine sheet. If each enamel-formative centre has a rather steep side and the steep sides of adjacent centres are parallel, a deep fissure will form on the occlusal surface of the tooth. Enamel formation ceases when the width of the fissure approximates the length of the two sets of columnar ameloblasts and the associated papillary layer of the enamel organ.

Fissures may be as deep as the entire enamel thickness, so that dentine may possibly be exposed at the base of a deep fissure. Most commonly, however, a thin layer of enamel will have been formed by the contiguous ameloblasts at the base of the fissure.

There is a clear association between an excellent capillary blood supply to the maturation-stage enamel organ and normal enamel mineralisation (Tobin 1972; Iwaku and Ozawa 1979; Hodde et al. 1983). The dimensions of fissures would suggest that the fissure enamel organ may be relatively deprived of blood supply during the maturation phase, and that enamel abutting fissures reaches a lesser degree of maturity than at any other site on the crown surface.

II. Prism-Free Enamel and Smooth Areas

The greater part of the enamel surface of human teeth is smooth, particularly on those areas of the tooth surface which might occlude with teeth in the opposing arch. The prior distribution of ameloblasts can generally just be

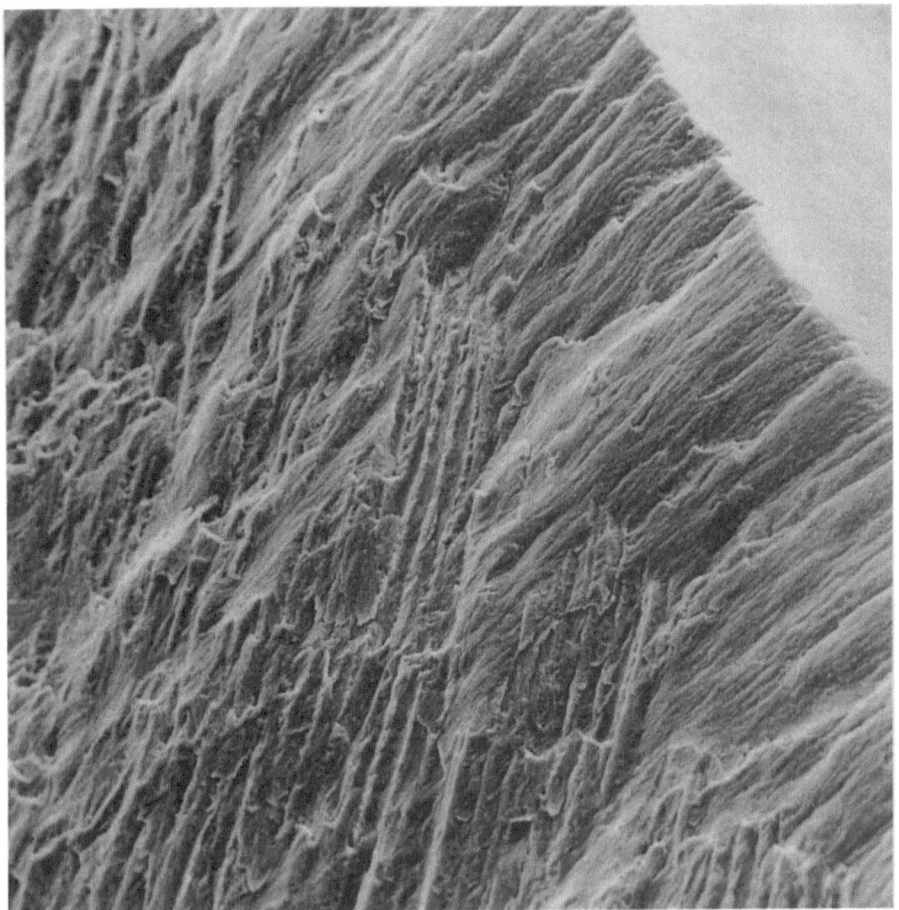

Fig. 75. Fractured human deciduous molar enamel, showing extensive prism-free surface zone enamel with a grain perpendicular to the finished enamel surface (seen in top right-hand corner), which contrasts with the prism direction seen in opposite corner. Field width 222 μm

recognised in the pattern of very shallow pits in the tooth surface which, at about half a micron deep (BOYDE 1971 a), are but a shadow of the former Tomes' process pits. Such shallow pitted enamel is prism free (RIPA et al. 1966, 1967). The limited changes in the orientation of such a surface would not be expected to produce any discontinuities in crystallite orientation (Figs. 75, 76).

III. Pits, Perikymata and Imbrication Lines

Further down on the sides of the tooth, the incremental lines crop out at the tooth surface. Here, smooth areas alternate with areas of pitted enamel, which in some instances may closely resemble a forming enamel surface; however, these pitted areas have a modal depth of 1.25 μm, compared with the

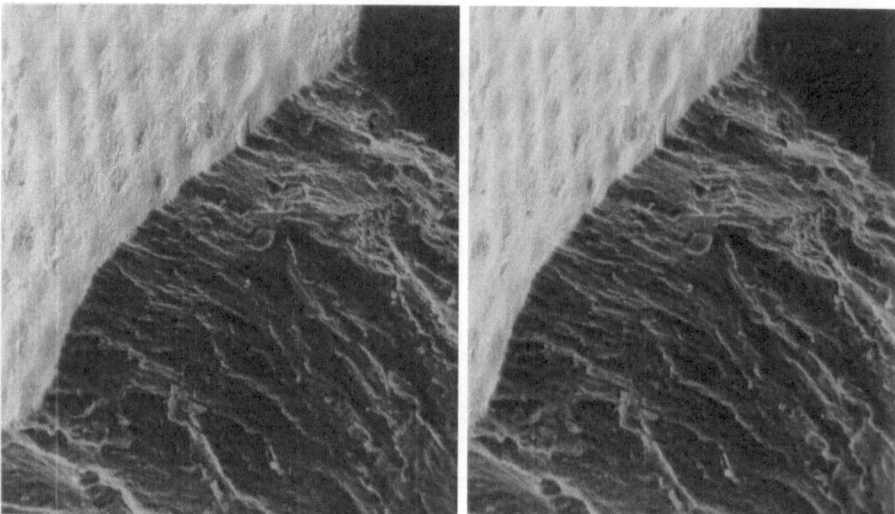

Fig. 76. Fractured human lower third permanent molar, fracture surface being nearly in the transverse direction of the tooth. The occlusal fragment is viewed from cervical. Stereopair, 10° tilt-angle difference. Stereoviewing will show the change in orientation from the direction of the prisms to the prism-free surface enamel. The shallow type of pit seen on the surface corresponds with a prism-boundary-free enamel layer underneath. Field width 89 μm

3- to 6-μm pits found during development. They also have less regular margins, sometimes very irregular where the walls of the pits have grown sideways across the floors (BOYDE 1971). Where smooth and pitted regions alternate in circular bands around the tooth, the smooth regions constitute the most peripheral regions of the tooth surface and overlap the pitted regions on their cervical sides. Projecting the pitted surface cervically and inwards into the enamel, we find that these planes correspond with the planes of the brown striae of Retzius. The overlapping edges of the enamel (formed by ameloblasts which were able to overcome the unknown influence which gave rise to the formation of the incremental lines) are known as imbrication lines. The appearance of (breaking) waves on the enamel surface gives rise to the name perikymata (Figs. 77–80).

Prism boundary discontinuities reach the enamel surface in the pits in the troughs of the perikymata. The smooth wave crests of the perikymata are composed of shallow pitted enamel in which the prism boundaries do not reach the surface: there is a thin layer of prism-free enamel (MANNERBERG 1960; NEWMAN and POOLE 1974; RIPA et al. 1966, 1967; BOYDE 1970, 1971 a, b, c).

Detail at the occlusal or incisal margin of imbrication lines indicates that the ability to recover to make prism-free surface-zone enamel is an all-or-none process for individual cells. One ameloblast stops secreting enamel, the next recovers to produce mineralisable matrix for a period of a day or a few days. The patches of enamel corresponding to individual cells can be identified in the detail of the edge of the imbrication line (BOYDE 1970; NEWMAN and POOLE 1974).

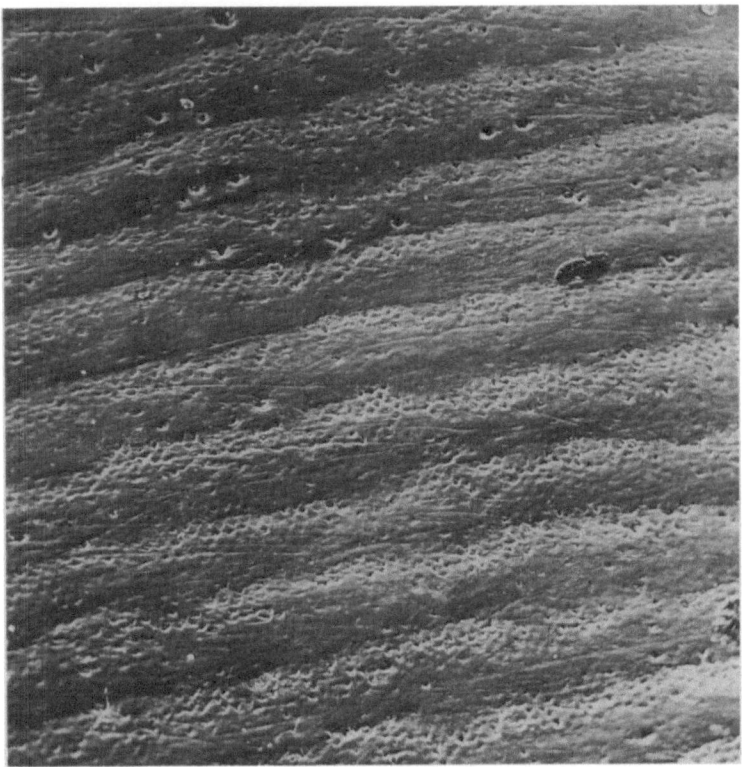

Fig. 77. Lateral enamel surface of human tooth which had been in function, showing the perikymata pattern of smooth bands of enamel alternating with pitted areas. Scratches have been caused by toothbrush abrasion on the tooth surface. Several isolated deep pits can be seen in the smooth areas. Field width 75 μm

IV. Cervical Margins

Perhaps the converse of the ability to recover to produce a little more enamel at the tooth surface is the ultimate cessation of the ability of any of the cells of the internal enamel epithelium sheet to produce enamel matrix. At the cervical margin of the tooth, there is a gradation of the time period over which ameloblasts are active and the rate at which they produce enamel in such a way as to lead to a tapered edge of the enamel which is steeply inclined towards the surface of the underlying dentine. It is difficult to recognise the prior location of individual ameloblasts. The ultimate end of the enamel is generally a smooth, well-defined line.

Cells of the internal enamel epithelium continue to function after the formation of enamel in inducing the differentiation of odontoblasts which form the peripheral layer of the root dentine (prior to its coverage by cementum). However, these internal enamel epithelium cells form part of an only two-layered sheet of epithelium, the stellate reticulum and the stratum intermedium being

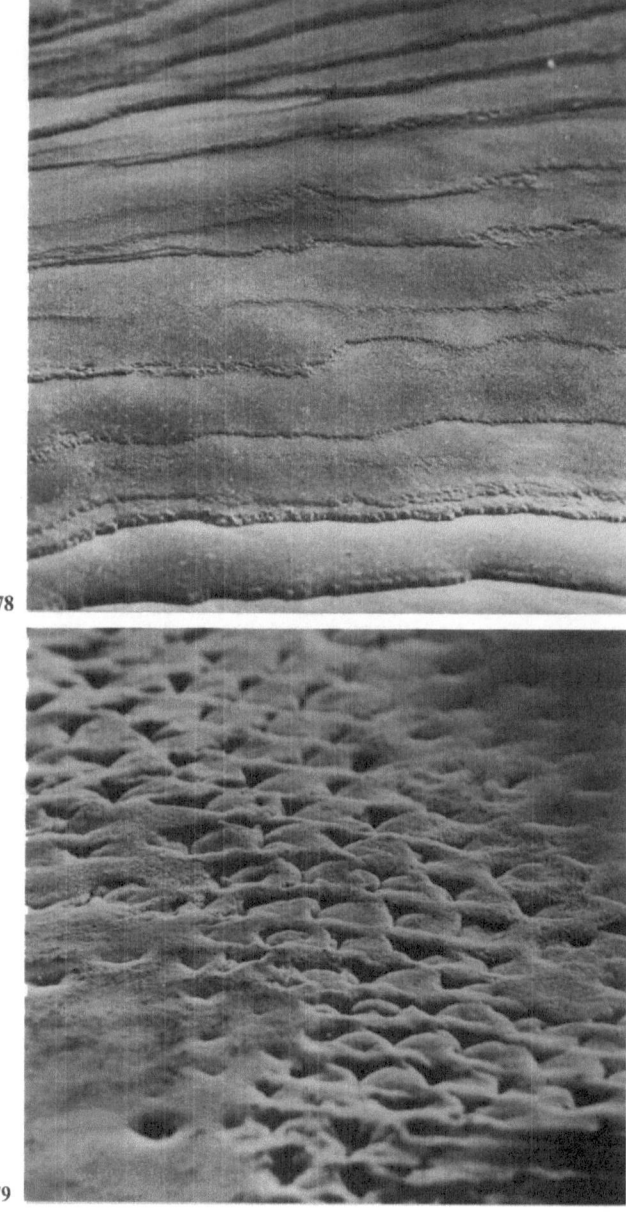

Fig. 78. Low-lateral region of human molar, showing imbrication lines (cervix top). Field width 2000 μm

Fig. 79. One groove of the perikymata pattern showing Tomes' process pits retained in a trough (cervix right). Field width 43 μm

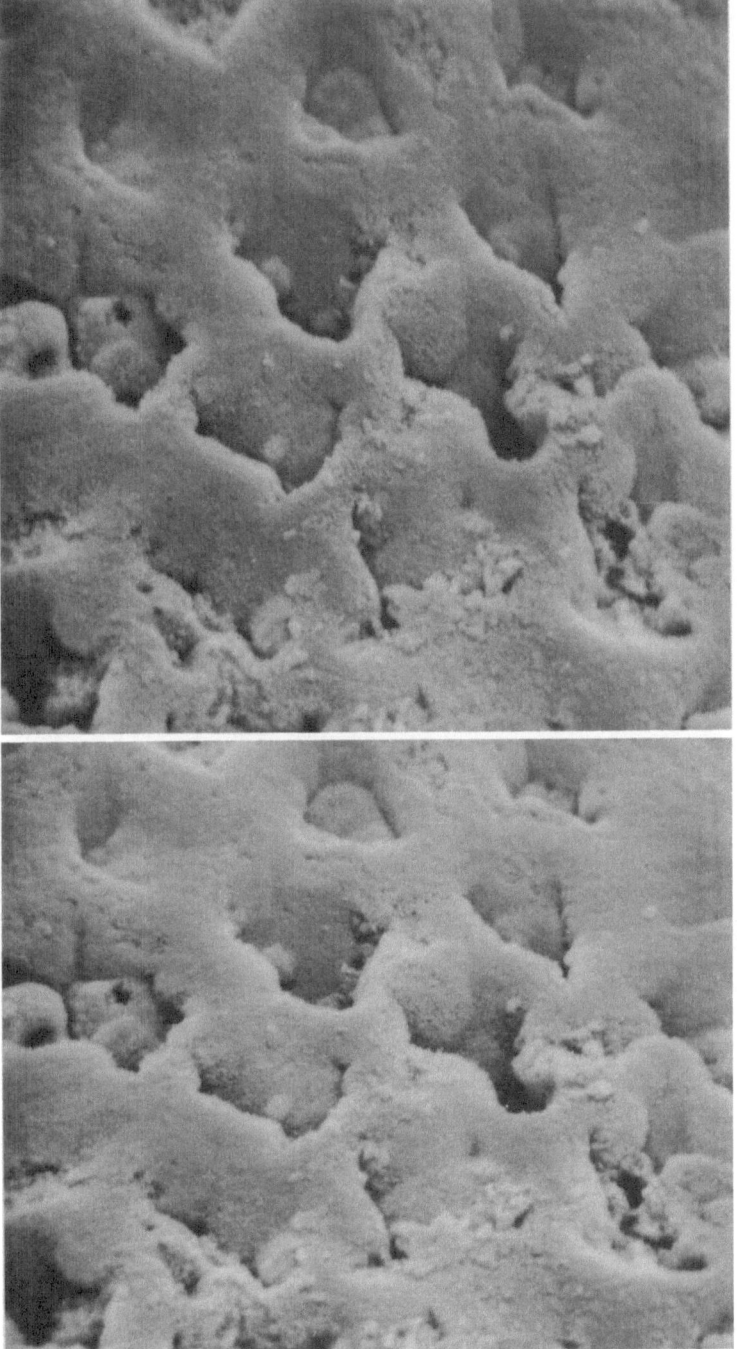

Fig. 80. Stereopair SEM image (tilt-angle difference 10°) of the groove of one perikyma in the mid-lateral region of a human premolar (cervix top). Field width 25 μm

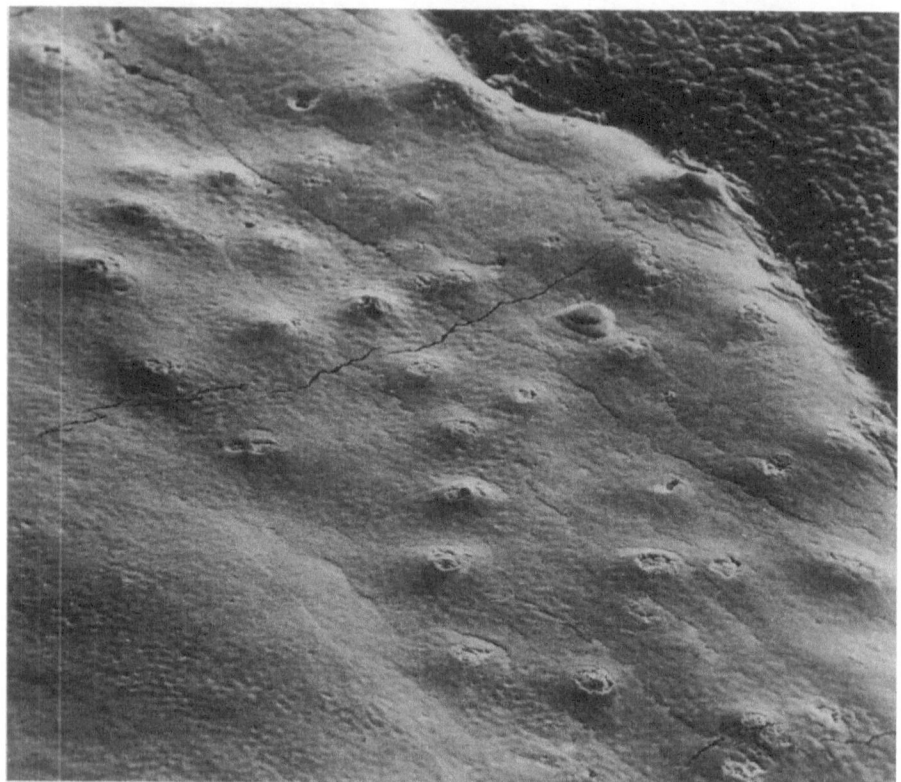

Fig. 81. A field of brochs in the cervical region of a human premolar. Cement surface in top right-hand corner. Normal low-lateral enamel surface in bottom left-hand corner. This is a relatively narrow field of rather sparsely positioned brochs. (Contrast with Fig. 82.) Field width 630 μm

missing from this part of the developing tooth organ. Occasionally, the down-growing epithelial enamel organ differentiates a stratum intermedium and a stellate reticulum, and the cells of the internal enamel epithelium sheet go on to become fully functional ameloblasts. This accounts for the origin of the so-called enamel pearls sometimes found separate from the enamel sheet on the crown proper. Enamel pearls are particularly common at the division of the roots of molars in man.

V. Brochs

Brochs (Figs. 81, 82) are a very common class of enamel surface feature, found in about 50% of premolars and in a lower proportion in other permanent tooth classes; they are sometimes found in deciduous teeth (JAYASINGHE 1987). They are located in bands about $^1/_2$ to 1 mm wide at the cervix (BOYDE 1970, 1971 b, c). The distribution is similar in contralateral teeth from one individual.

Fig. 82. Similar field in similar orientation to Fig. 81 in another human premolar, showing a wide broch field with densely positioned brochs and a few pits. Both Figs. 81 and 82 show that the superficial strata are missing at the tops of most of the brochs. Cementum can be seen just lapping onto the enamel edge in both Figs. 81 and 82. Field width 659 μm

When present, the enamel is generally less thick but reaches towards the same thickness as the preceding normal enamel surface at the most prominent parts of these features. Most commonly, superficial strata are missing at the most prominent portion of the brochs, but this is not always the case. There is a change in prism course in the underlying enamel associated with the overlying broch. Prisms tend to converge towards each other at the centre of the broch at the surface of the tooth in a type of wheat sheafing. It would therefore seem that ameloblasts move towards each other on the surface, perhaps concentrating their secretory efforts in a smaller locality and accounting for the projection of the brochs (JAYASINGHE 1987). The brochs are 30–50 μm in diameter (though it is difficult to define their peripheries) and 10–15 μm above the surrounding level of the enamel surface. They are most easily seen in SEM of the tooth surface but may also be identified with reflected-light microscopy (BOYDE 1970; JAYASINGHE 1987).

Although it is now known that brochs are projections of the tooth surface, these may be the same features as were described as micropits in studies of

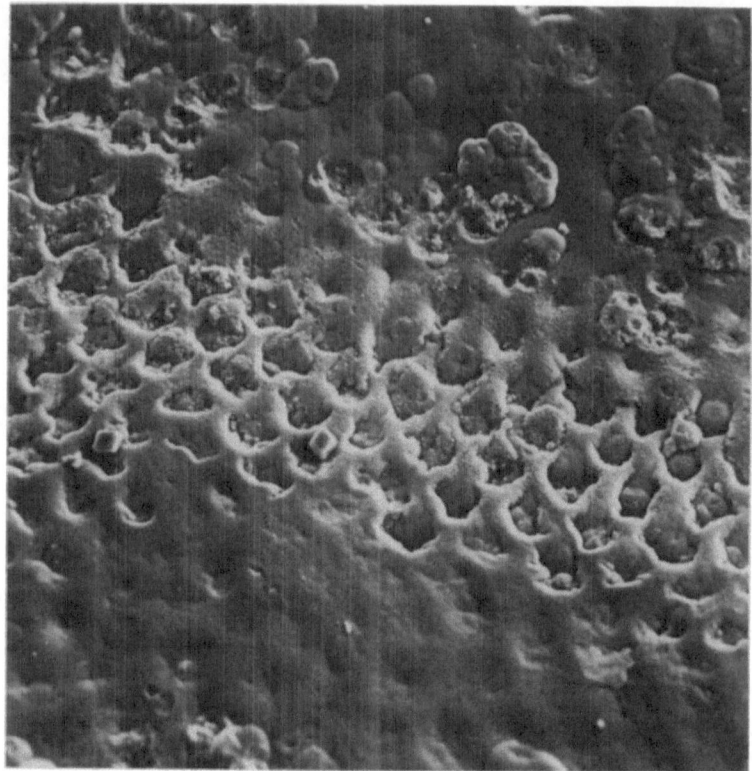

Fig. 83. Enamel surface from unerupted human lower permanent canine cleaned of overlying organic integuments by treatment with NaOCl (two salt crystals can be identified in two pits). Note Tomes' process pits in the groove of a perikyma; a smooth wave crest of a perikyma in lower portion of field of view; and surface-overlapping projections, some of which have broken away on the smooth perikyma band at the top of the field of view. Field width 99 μm

the enamel surface based upon the use of replicas prior to the availability of SEM (WOLF 1940; SCOTT 1952; MANNERBERG 1960, 1964, 1968). This misidentification of the surface relief is an obvious problem of the limitations of replica technology. There are, however, infrequent micropits in the same (cervical) region as the broch fields, and they may be mixed.

VI. Surface-Overlapping Projections and Isolated Deep Pits

Surface-overlapping projections (SOPs) and isolated deep pits (IDPs; or focal holes) are a class of feature also commonly found in human enamel, with a similar distribution pattern in contralateral teeth (Figs. 83–88). They are more common in lateral enamel. The SOPs may be as small as the amount of enamel produced by one ameloblast or may be produced by a cluster of many ameloblasts. SOPs usually form on the smooth wave crests of perikymata.

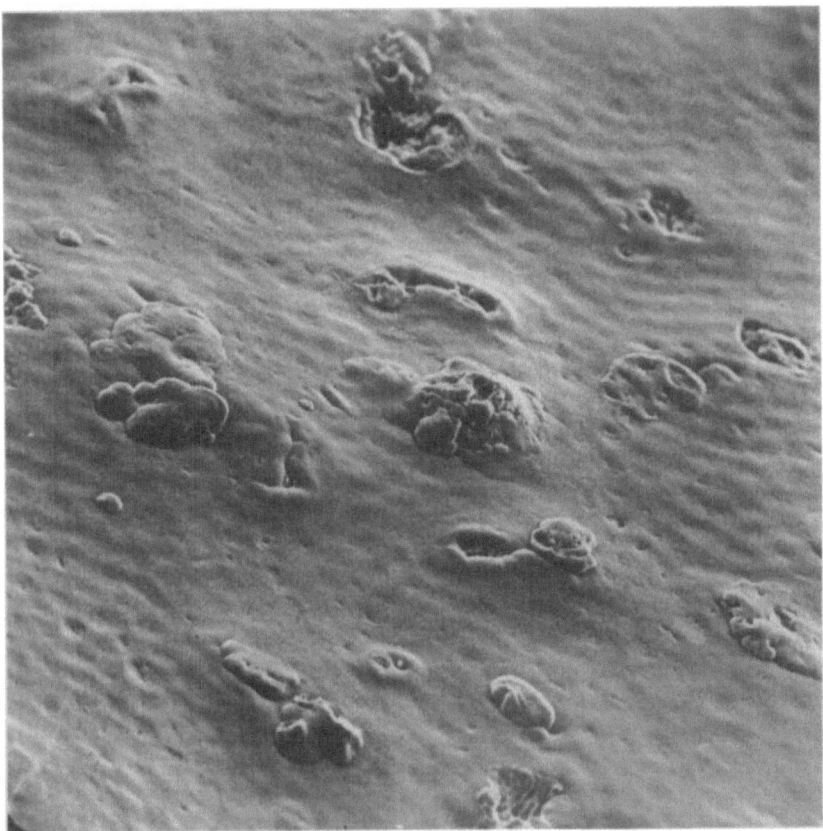

Fig. 84. Low-lateral region of labial face of human lower permanent canine, showing surface-overlapping projections and cleft-like, isolated deep pits (cervix to top right). Field width 204 μm

The association of IDPs with SOPs is well demonstrated in preparations which have been roughly handled. The SOPs can be made to break away to reveal the underlying IDPs by treatment with strong sodium hypochlorite solutions, as well as by rough mechanical handling. The embryological origin of this class of features has not been proven, but it seems reasonable to speculate that a "bubble" of gas or non-mineralisable secretion is included in the surface zone and then overlain by later formed, normal, mineralisable matrix, which therefore projects slightly above the surrounding plane of the surface (BOYDE 1971 b, c).

VII. Regional Differences in the Enamel Surface

Over the cuspal regions of human permanent teeth, shallow pits and a prism-free layer are the rule: perikymata are rare or absent. Deep, steep-sided, punched-out IDPs (Figs. 87–88) are commonly found in cuspal surface regions

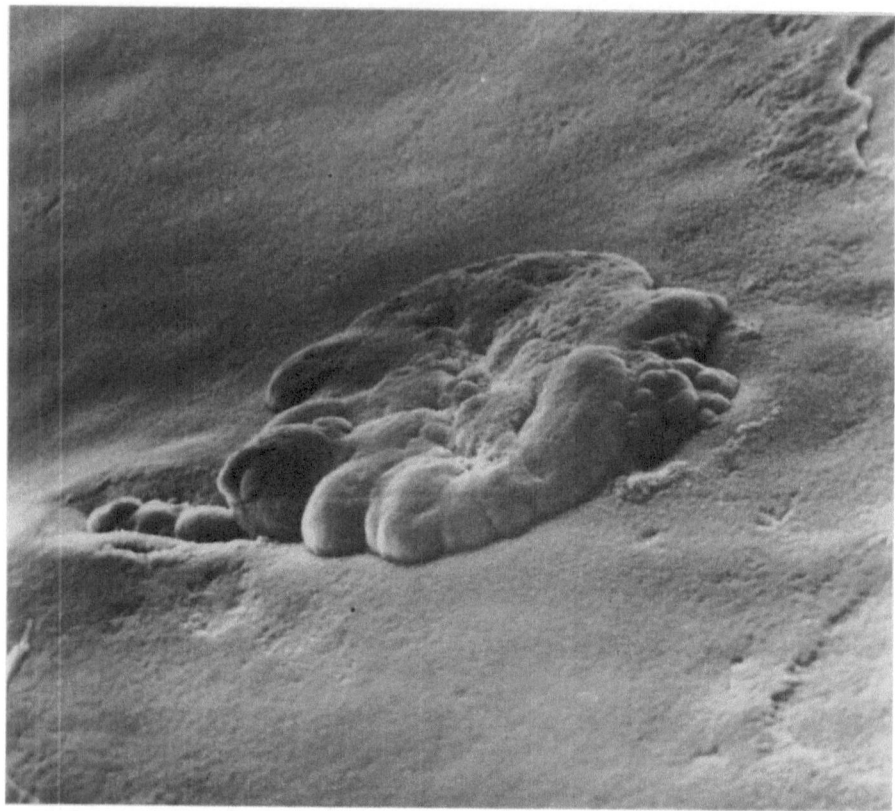

Fig. 85. High-magnification view of one surface-overlapping projection in same tooth as in Fig. 84. Field width 53 μm

without evidence of an association with SOPs (BOYDE 1970). Perikymata increase in frequency along the lateral surface and more frequently present as regular wave-like variations in enamel thickness. In low lateral and cervical regions, the perikymata become even more closely spaced and regularly present as over-lapping layer lines (imbrication lines). Usually, the entire surface is covered with shallow pits in a prism-free layer. Thus, the overlapping strata overlap regions in which there are none of the prominent "rod-end pits" with obvious prism boundary discontinuities typical of mid-lateral regions.

In human deciduous teeth, perikymata are insignificant in comparison with those seen in the permanent teeth. There is generally a thick layer of prism-free enamel over the entire tooth surface in deciduous teeth (RIPA et al. 1966, 1967) (Fig. 75).

VIII. Cement on the Enamel Surface

Cement on the enamel surface is rarely encountered and only at the cervical margin in human teeth. However, it is the normal situation over all the enamel surface in the cheek teeth of many mammals. In hypsodonty, the tooth may

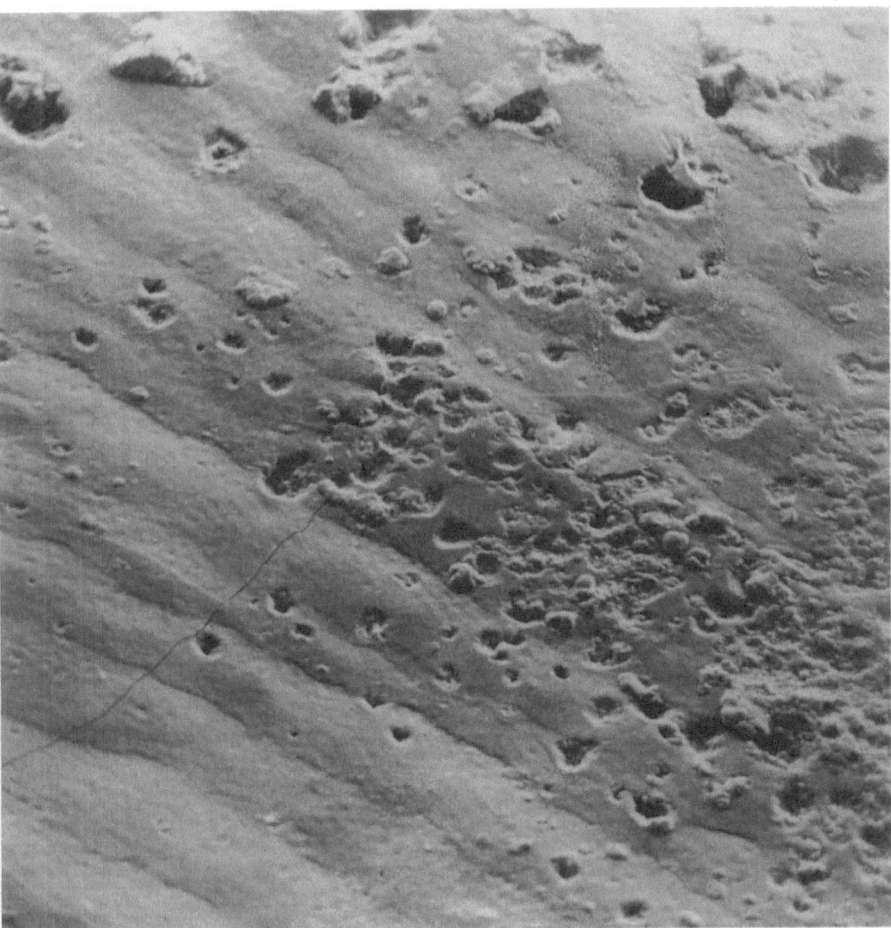

Fig. 86. Human premolar lateral enamel surface, showing isolated deep pits and surface-overlapping projections. The isolated deep pits arise where the surface-overlapping projections break away, which is accelerated by treatment of the unerupted tooth with sodium hypochlorite. Field width 510 μm

begin to function before the formation of an anatomical root. Cementum is formed on the outside of the enamel to attach the tooth to its socket via the periodontal ligament. Cementum is also formed to pack out the space which would otherwise be an open and deep fissure between the flattened cusps (lophs) of the occlusal surface. The cementum formed on the surface of the enamel is of two types: either attachment cement, which contains Sharpey fibres and has a histology resembling cellular cementum in human teeth, or packing and binding cement, which is a relatively coarse form of rapidly formed bone (JONES and BOYDE 1974). Some dental cementum attached to enamel is a form of calcified cartilage, e.g. in guinea-pig molars.

Two common strategies for the attachment of cement to enamel appear to have been adopted in mammals (JONES 1974). In the first, the end of appositional enamel formation must be rather more sudden than in man since shallower

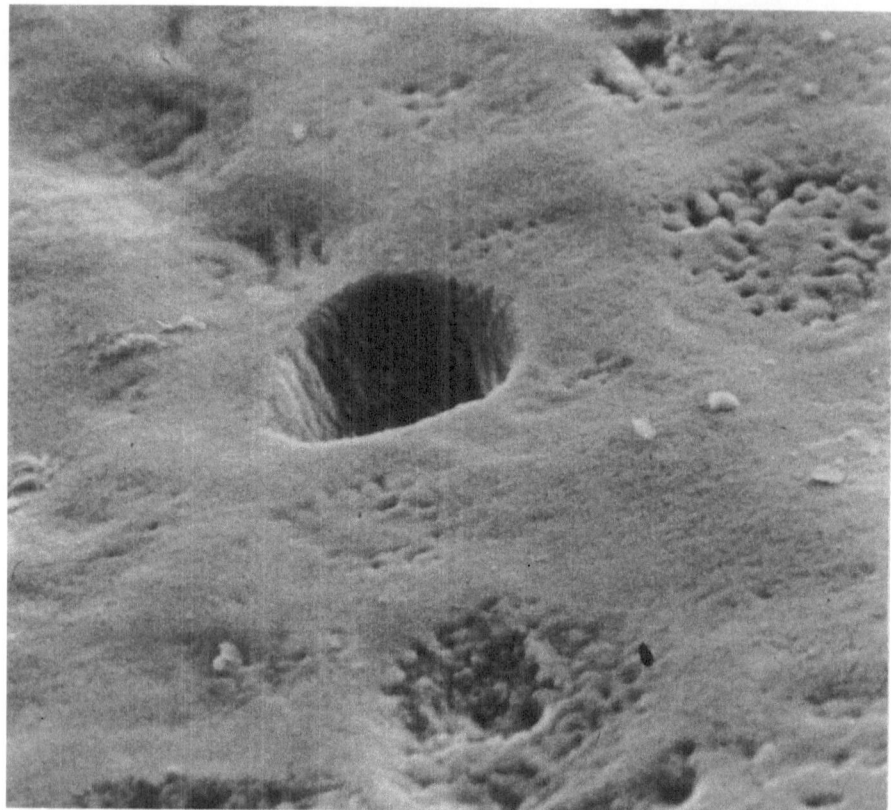

Fig. 87. Cuspal enamel of human premolar, showing a punched-out isolated deep pit. Field width 46 μm

but nevertheless significant Tomes' process pits are left all over the surface of the mature tissue. Following the maturation stage of enamel formation, the ameloblasts move away from the tooth surface to allow surrounding connective tissue cells to deposit cementum directly onto the enamel surface. Interdigitation between cement and enamel is achieved only through the complexities of the shape of the enamel surface at its completion (Fig. 89). In the Equidae (Figs. 90–91), the enamel surface is partially resorbed by osteoclasts before the deposition of cementum (JONES 1974; JONES and BOYDE 1974). This gives rise to a more complicated interdigitation between the two tissue types and probably a better functional bond. Certainly, it is much more difficult to separate these tissues in the dried teeth than in other mammals.

IX. Calculus on the Enamel Surface

Calculus – mineralised bacterial dental plaque – obviously cannot be classified as a tissue, but its occurrence on parts of human teeth is so common that it is a normal feature to be encountered in examining any fresh, extracted

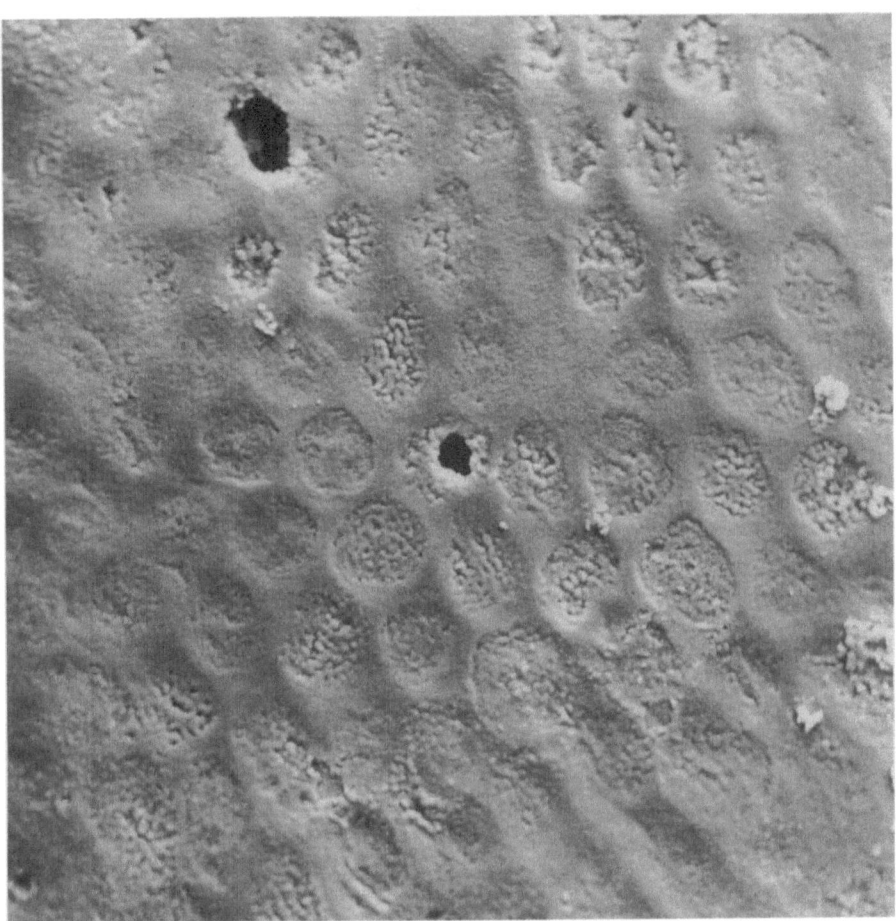

Fig. 88. Just below marginal ridge of human premolar, showing two punched-out isolated deep pits of different extents. Field width 78 μm

tooth (JONES 1972 a, b). [Unmineralised plaque is also ubiquitous but is usually removed in preparing the tooth for microscopical examination (JONES 1972 a, b)]. Calculus adapts exactly to replicate the finest details of the surface and is consequently firmly attached whilst wet. Differential shrinkage on drying may lead to its dislodgement. All the natural surface irregularities of the human tooth seem to encourage the initial growth of calculus, which may thus seem to grow out from individual pits (Fig. 92) in the troughs of the perikymata, as well as from the edges of imbrication lines and in broch fields (Fig. 93).

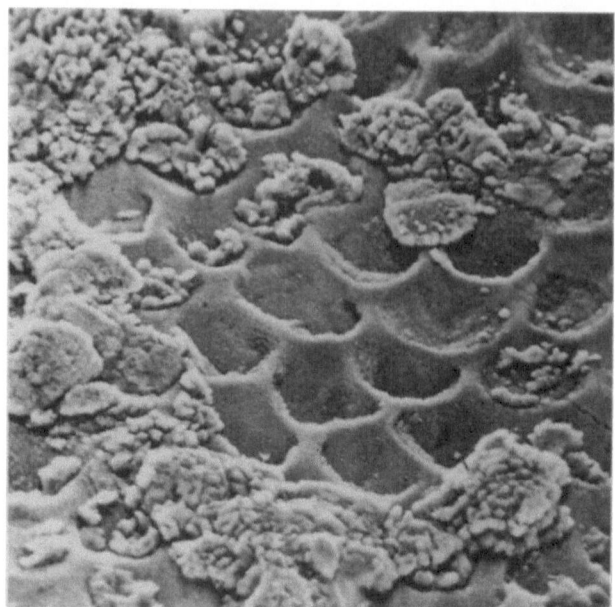

Fig. 89. Completed enamel surface of elephant molar, showing early stage of attachment of cementum to the pitted enamel surface. Field width 53 μm

J. Maturation of the Enamel and Eruption of the Crown

I. Maturation and Maturation Ameloblasts

Following the cessation of the secretory activity of the ameloblasts, they remain on the tooth surface until it acquires a high degree of mineralisation. [Although it is clear that there is a point at which ameloblasts produce no more mineralisable matrix, recent studies have shown that maturation ameloblasts still produce matrix proteins even in the maturation zone; these, however, are not assembled to make an extra layer of enamel (GRAVER et al. 1978; NANCI et al. 1984, 1988)]. In the case of human permanent teeth, the ameloblasts may be in this functional position for a period of years after secretion as such has stopped. During maturation, the concentration of amelogenins in the enamel is greatly reduced. It is not known how much enamelin is removed, if any. The amino acid composition changes, reflecting the net increase in the enamelin-to-amelogenin ratio. The mineral-to-organic-matrix ratio also increases, more due to the increase in mineral than to the removal of protein since the protein content is reduced in the earliest stages of maturation. The water content reduces dramatically. This is achieved through an increase in the size of the individual enamel crystallites and a reduction of the intercrystallite water and organic matrix-filled space (EASTOE 1982; ROBINSON et al. 1977, 1978, 1979, 1981 a, b; BELCOURT 1984; FINCHAM 1984; ROBINSON and KIRKHAM 1984; DEUTSCH et al. 1984).

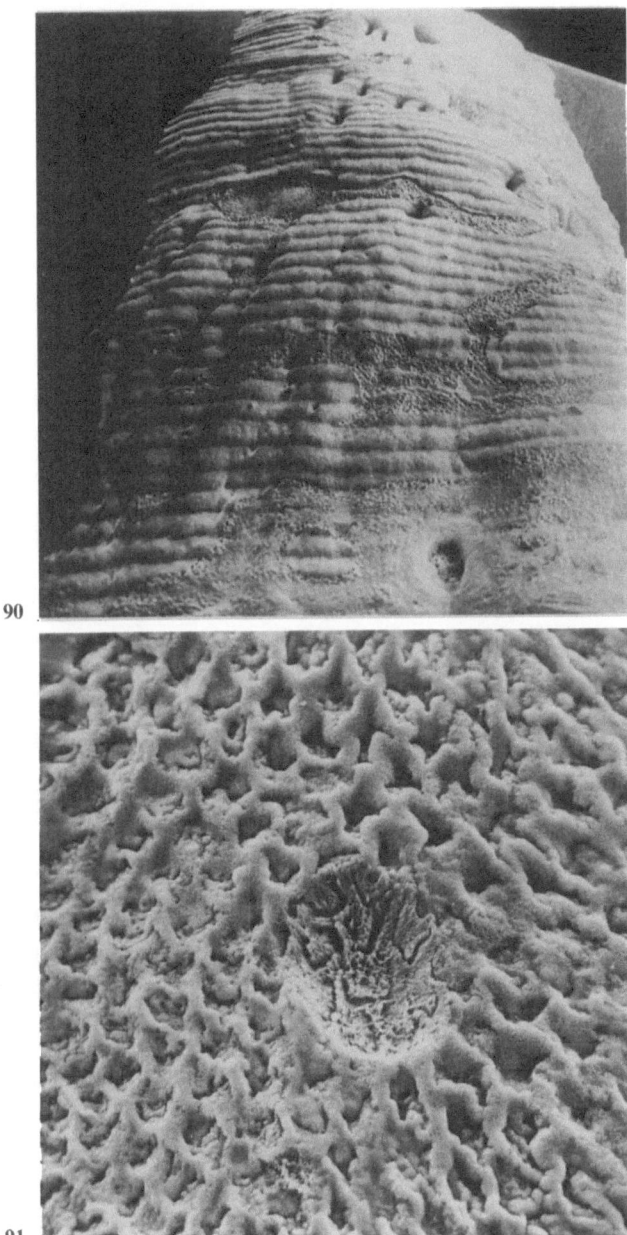

Fig. 90. Lateral view of horse deciduous molar, showing perikymata and bands of resorption caused by osteoclasts, mostly following the direction of the perikymata. Field width 5000 μm

Fig. 91. Completed enamel surface of horse molar, showing isolated osteoclastic resorption pit at centre of field of view. Field width 65 μm

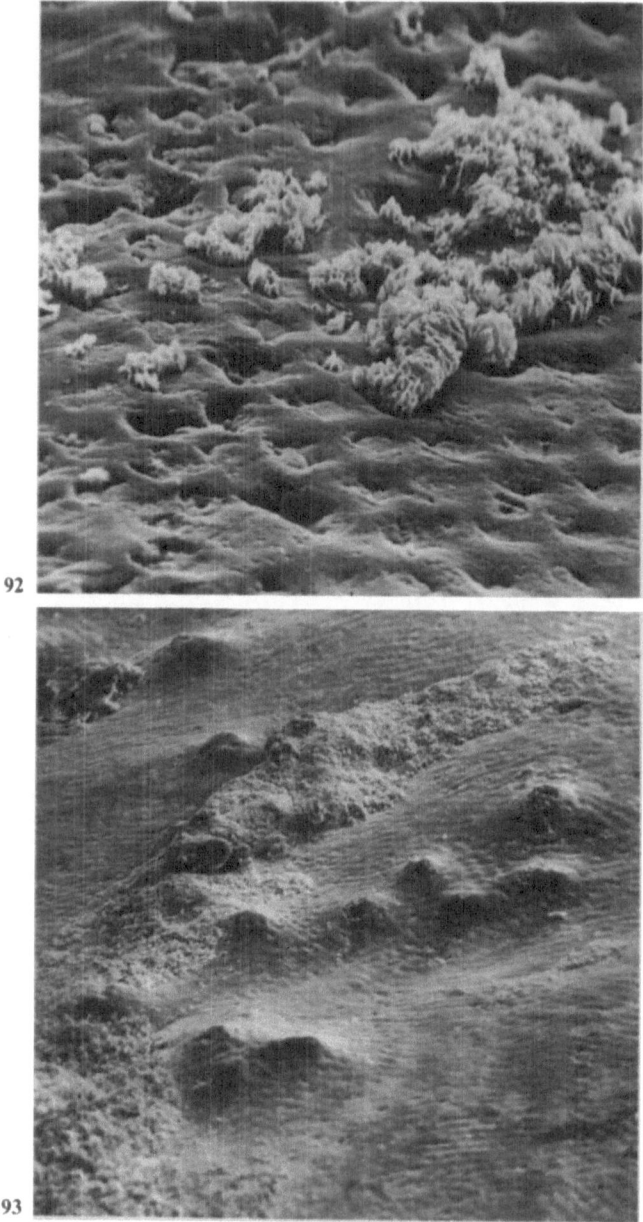

Fig. 92. Lateral enamel surface, showing development of mineralised bacterial dental plaque (calculus), spreading from Tomes' process pits in the trough of a perikyma. Field width 58 μm

Fig. 93. Field of brochs at the cervix of a human premolar, with calculus attached to the tooth surface. Field width 429 μm

II. Information from Rodent Incisor Studies

1. Main Features

As with so much else in the understanding of enamel cell biology, an overwhelming proportion of the available information derives from studies of the continuously growing rat incisor tooth: this is because any tooth in any animal of any age presents all the stages in the life history of enamel from the earliest stages of cell differentiation to the loss of tissue during normal function. Extensive observation and experimentation with the rodent (usually rat) incisor "system" have shown that maturation of the enamel is a complex process, involving rapid transport of calcium and phosphate (from a prodigious blood supply, via the enamel organ) and early removal of the organic matrix, especially the amelogenin component, with an accompanying change in the amino acid composition. There are active, repetitive cyclical changes in the function and in the microanatomical organisation of the maturation ameloblasts, in which two extreme morphologies can be recognised (REITH and COTTY 1962; REITH 1963; KALLENBACH 1968, 1970, 1974, 1980; KURAHASHI and MOE 1969; GARANT 1972; ROBINSON et al. 1977, 1981; SMITH and WARSHAWSKY 1977; JOSEPHSEN and FEJERSKOV 1977; BOYDE 1975; BOYDE and REITH 1976, 1977, 1978, 1981, 1982, 1983; IWAKU and OZAWA 1979; TAKANO and OZAWA 1980; TAKANO et al. 1982a, b; TAKANO et al. 1982; REITH and BOYDE 1981a, b, 1983, 1985; HODDE et al. 1982; JOSEPHSEN 1984; GARANT et al. 1983; GOLDBERG and SASAKI 1985; MOE 1971, 1981; REITH et al. 1982, 1984; SASAKI et al. 1983, 1984; SUGA et al. 1970; TAKANO and OZAWA 1984; TAKANO et al. 1982; NISKIKAWA and JOSEPHSEN 1987; SASAKI and GARANT 1987). The existence of two alternating morphologies of maturation ameloblasts can be traced back to SUGA's (1959) study, which showed that, during the maturation of enamel in rodent teeth, two ameloblast types could be identified by histological and histochemical criteria, and that these two types may appear several times alternately during the maturation process.

2. Transitional Zones

The change from secretory to maturation functions in rat incisor ameloblasts is associated with a cytological revolution, which involves the death and destruction of many cells, whose remains are phagocytosed by monocyte macrophages (KURAHASHI and MOE 1969; SMITH and WARSHAWSKY 1977). During this transitional process, the cells which had developed a temporarily complex interdigitation with the surface of the enamel matrix following the reduction and loss of their Tomes' processes (KALLENBACH 1968) become temporarily poorly attached to the enamel, as evidenced by their facility for detachment from the enamel in histological processing (FLEMING 1961). Following this, they produce a basal lamina and are well attached to the surface of the tissue. The surviving cells are much shorter than their secretory counterparts: because of the loss of numbers, they must also be marginally wider to cover the same surface. The distal portion (closer to the enamel) contains a concentration of mitochondria, and the plasma membrane facing the enamel is deeply infolded, considerably increasing the surface area of the cell opposite the enamel: both these fea-

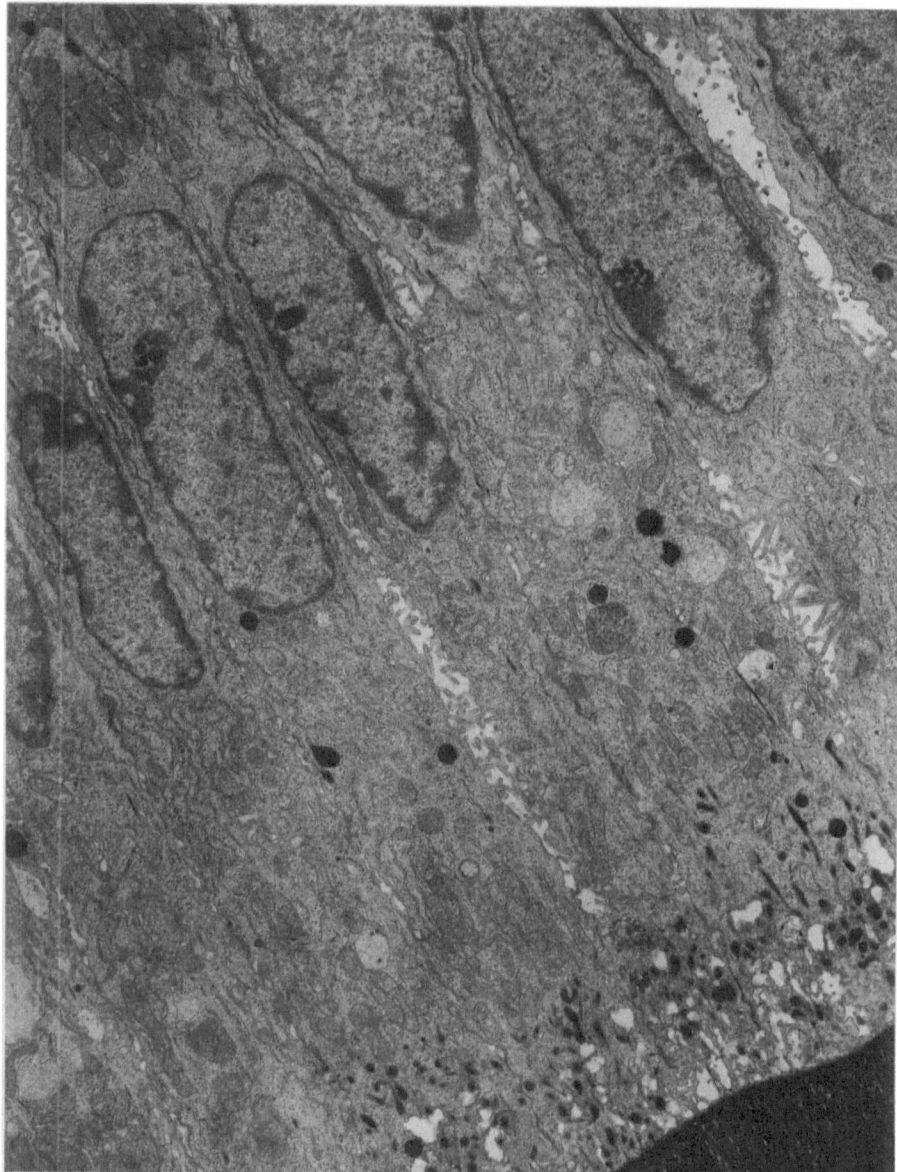

Fig. 94. Transmission electron micrograph of longitudinal section of rat incisor ameloblasts immediately after the transition from secretion to maturation. Note the shortening of the cells (the entire long axis of one cell being contained in the field of view: see Fig. 3) and the distal infolding. Dark features concentrated particularly in distal regions of cell are considered to be matrix material taken into the ameloblasts and undergoing "resorption". Field width 19 μm

tures are suggestive of water-pumping or salt-transporting cells (REITH 1963). The earliest maturation cells show the greatest distal convolution (Figs. 94, 95). There is a narrow intercellular space. Accumulations of material assumed to be matrix material en route into the ameloblast are seen in the distal compart-

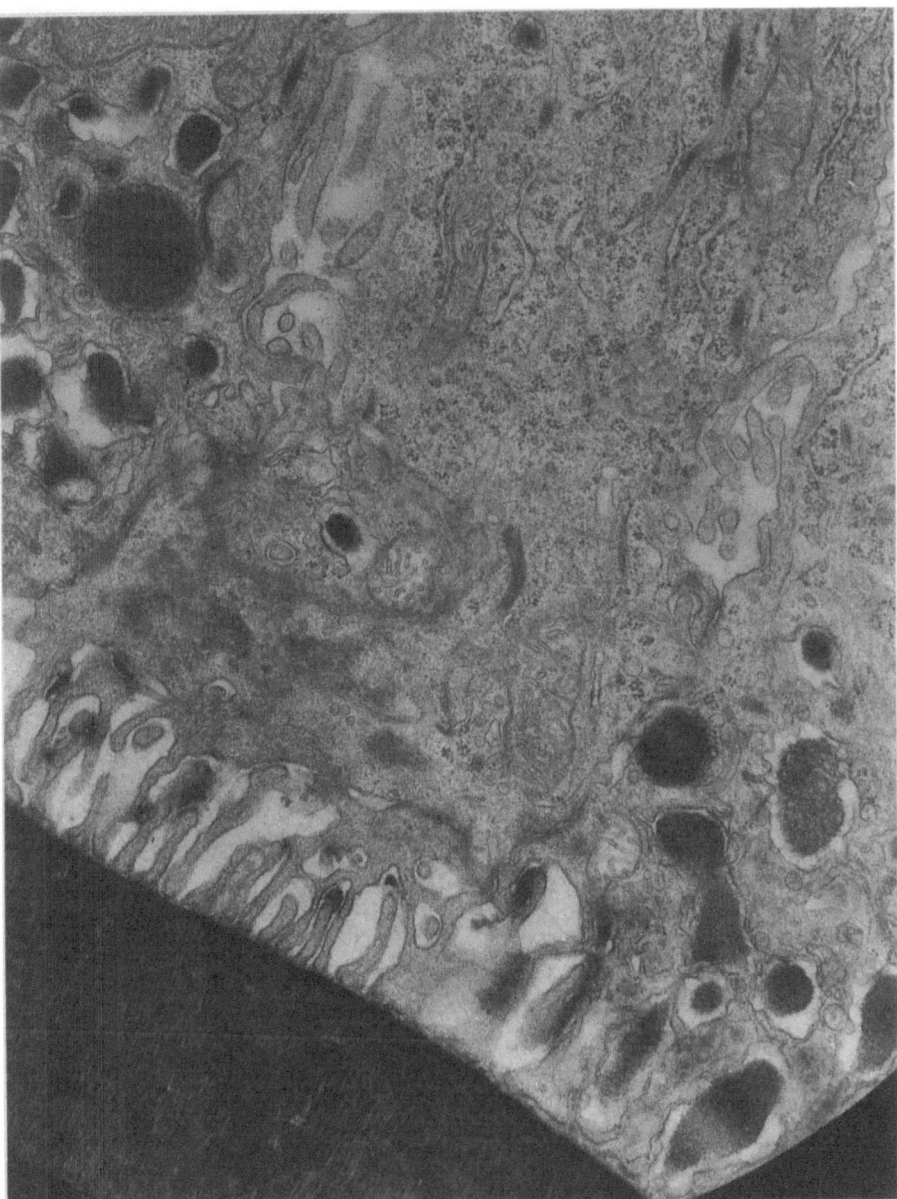

Fig. 95. Transmission electron micrograph of longitudinal section of rat incisor maturation ameloblasts from the early post-transitional region, showing the deeply infolded distal plasma membrane and dense patches within the ameloblasts, considered to be the enamel matrix material taken up into the cells and undergoing degradation. This is a ruffle-ended ameloblast of the earliest type. Field width 6 µm

ments and in apparently intracellular locations (REITH 1963; KALLENBACH 1968). Biochemical analysis of dissected fragments shows that a rapid uptake of phosphate occurs opposite the earliest, differentiated maturation ameloblast (ROBINSON et al. 1981 c).

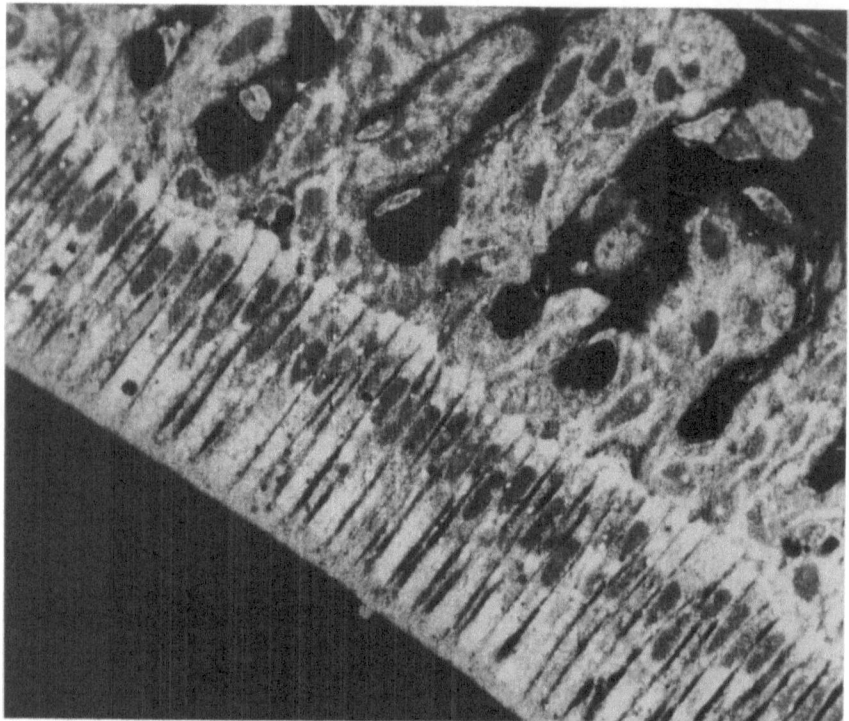

Fig. 96. Longitudinal section of rat incisor maturation-stage enamel organ perfusion-fixed with glutaraldehyde, post-fixed with osmium and embedded in PMMA. Longitudinal section surface ground and polished and imaged with back-scattered electrons in SEM. Field shows ruffle-ended maturation-stage ameloblasts. Field width 190 μm

3. Ruffle-Ended Ameloblasts

The majority of rat incisor maturation ameloblasts show a similar distal infolding and concentration of mitochondria, for which they have been designated as ruffle-ended ameloblasts (RA) by JOSEPHSEN and FEJERSKOV (1977). In the later RA, the cells are in close contact to one another in the ruffle-ended part of the cell, but over the greater part of the length of the cell there is a substantial intercellular gap (Figs. 96, 97, 98, 107). The ameloblasts may form tight junctions with each other at this distal region: the tightness of these junctions has been proven by the lanthanum tracer studies of JOSEPHSEN (1984) and the lanthanum and microperoxidase tracer studies of TAKANO and OZAWA (1984).

These cells are mostly "open" at their proximal ends, so that there is continuity of the extracellular space of the papillary layer with the lateral intercellular space of the ameloblasts.

4. Smooth-Ended Ameloblasts

The minority of maturation ameloblasts (Fig. 105) at any one stage of development show an extensive lateral intercellular space compartment which extends

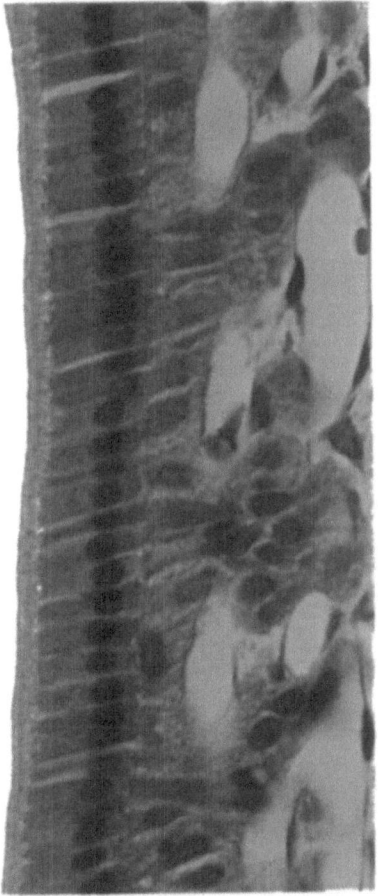
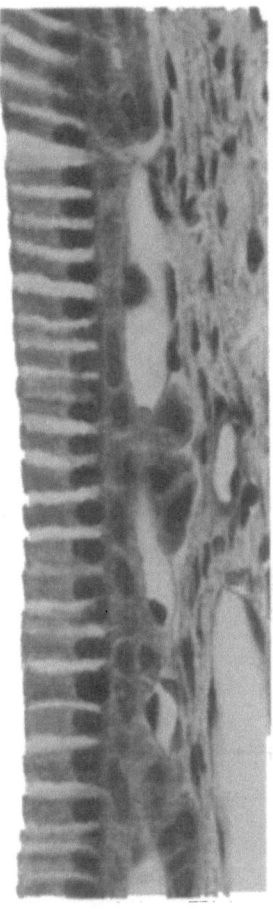

97   98

Figs. 97 and 98. Light micrographs of haematoxylin-and-eosin-stained, decalcified transverse section of rat incisor in maturation zone. Two fields within the same transverse section show ruffle-ended cells (Fig. 97) and smooth-ended ameloblasts (Fig. 98). (The maturation-cycle bands are not exactly transverse to the tooth axis; see Fig. 106a and b). Field width Fig. 97 = 88 μm, Fig. 98 = 66 μm

almost down to the enamel surface and in some cases seems to be in continuity with the enamel compartment (Figs. 99–104; TAKANO and OZAWA 1984). The distal infolding is strongly reduced or absent, so that these are known as the smooth-ended ameloblasts (SA), according to JOSEPHSEN and FEJERSKOV (1977). Cells of this type have tight junctions at their proximal ends, again proven by tracer studies (JOSEPHSEN 1984; TAKANO and OZAWA 1984). They may develop extensive gap-junctional complexes with their neighbours along their contacting, long-axis surfaces. The SA are distributed as narrow bands, usually roughly parallel with the general direction of differentiation of tooth-forming tissues. Single cells modulate backwards and forwards between these two morphologies, clearly spending much more time as ruffle-ended cells than as smooth-ended cells.

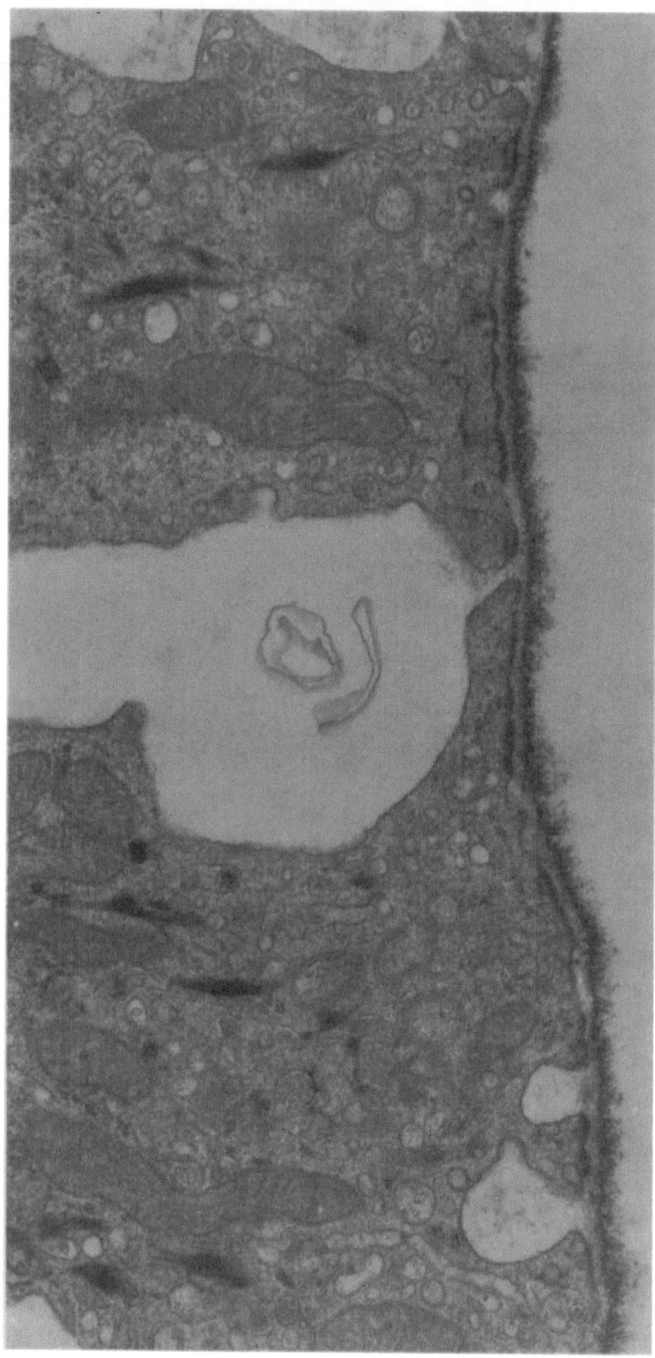

Fig. 99. Transmission electron micrograph of smooth-ended variety of rat incisor maturation amelo-blasts removed from the enamel surface by EDTA demineralisation. Centre of field of view shows the wide intercellular gap between two ameloblasts, extending as an open "space" through to the basal lamina still attached to the ameloblasts, which would have attached in turn to the maturing enamel surface. Field width 3 μm

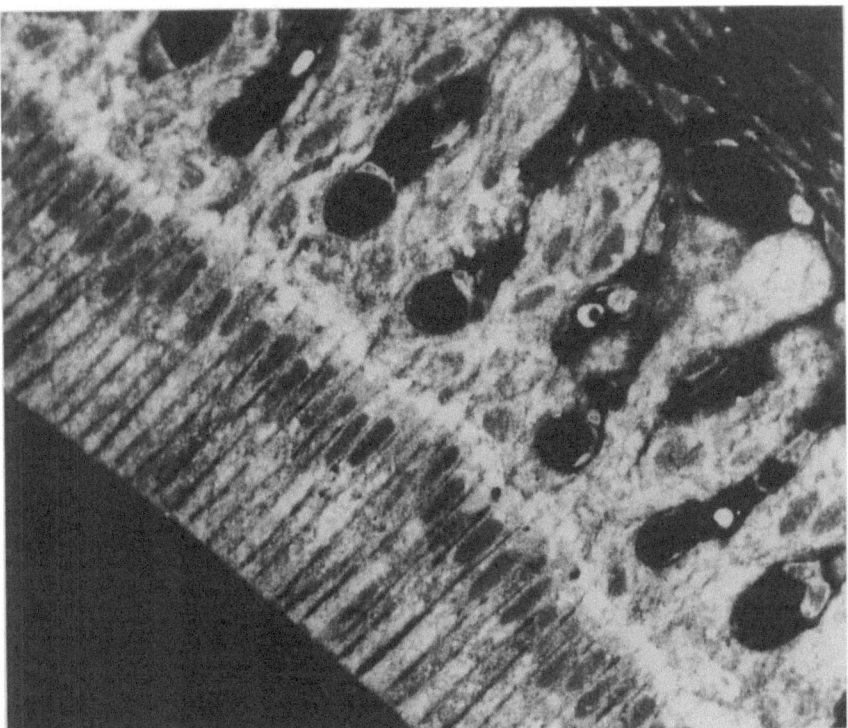

Fig. 100. Similar preparation to Fig. 96, showing field of smooth-ended ameloblasts. Field width
190 μm

5. Labelling Studies

The rate at which in vivo experimental markers can enter maturation enamel differs opposite these two types of cell. Radiocalcium, radiophosphate, tetracycline (Fig. 116), calcein and other fluorochrome markers all enter maturation enamel when injected into the bloodstream much more rapidly via the narrow, smooth-ended cell bands (BOYDE and REITH 1981; REITH and BOYDE 1981 b; DEUTSCH and PEER 1982; REITH et al. 1984; SMITH et al. 1988). More specifically, the greatest permeability from bloodstream to enamel matrix is at the border zones of change from ruffle- into smooth- and from smooth- into ruffle-ended ameloblasts, particularly the latter (which occurs at the incisal side of each narrow, SA band; BOYDE and REITH 1981), where labels may enter within seconds or one to two minutes. Here, it is important to note how rapidly changes can occur in the apparent distribution of injected marker substances in enamel (they cannot be used to study long-term growth processes in this tissue!). Radiocalcium and radiophosphate "shift" to lie opposite the RA cells if sampled at 30 minutes (TAKANO 1984). The distribution of tight junctions explains why this might be so (JOSEPHSEN 1984; TAKANO and OZAWA 1984). The most rapid route from the bloodstream to the enamel would be via the fenestrated capillaries of the papillary layer to the intercellular spaces of this layer (Fig. 105), which communicate with the lateral interameloblastic space compartment of the RA bands. The latter is in communication with the extensive lateral intercellular

438   A. BOYDE

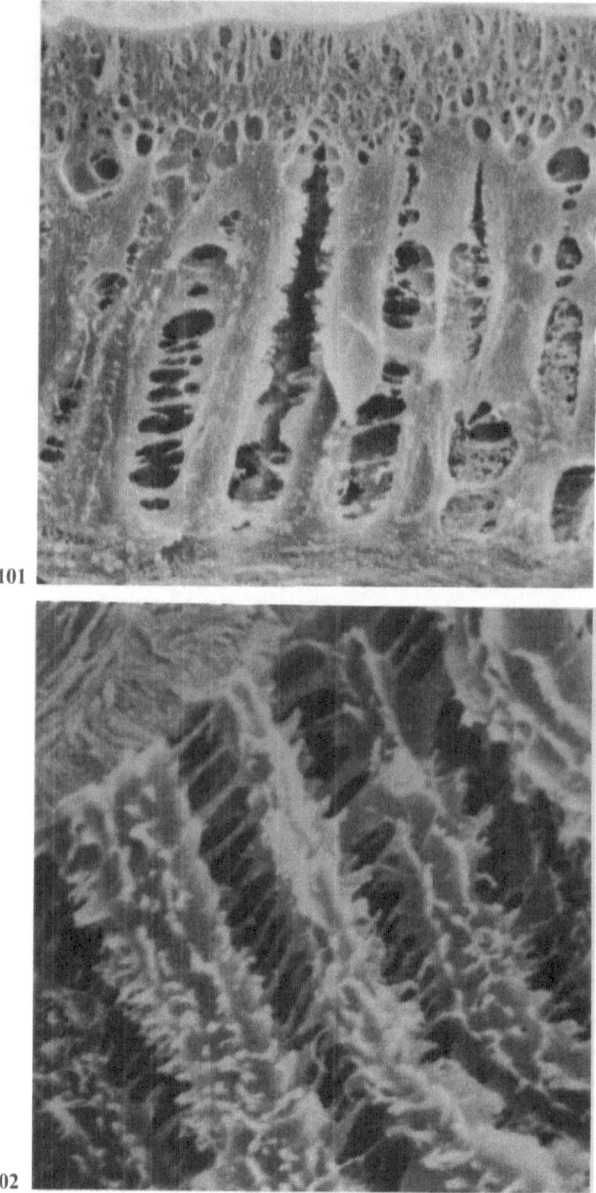

Fig. 101. SEM of ethanol-freeze-fractured and critical-point-dried rat incisor maturation ameloblasts, showing closed junctions at stratum intermedium ends of cell (bottom of field of view) and ruffle-ended specialisation at distal border adjacent to enamel (top of field). Field width 33 μm

Fig. 102. Lateral cell surface of smooth-ended rat incisor maturation ameloblasts, enamel at top left. Dry-fractured, critical-point-dried, maturation stage. Field width 20 μm

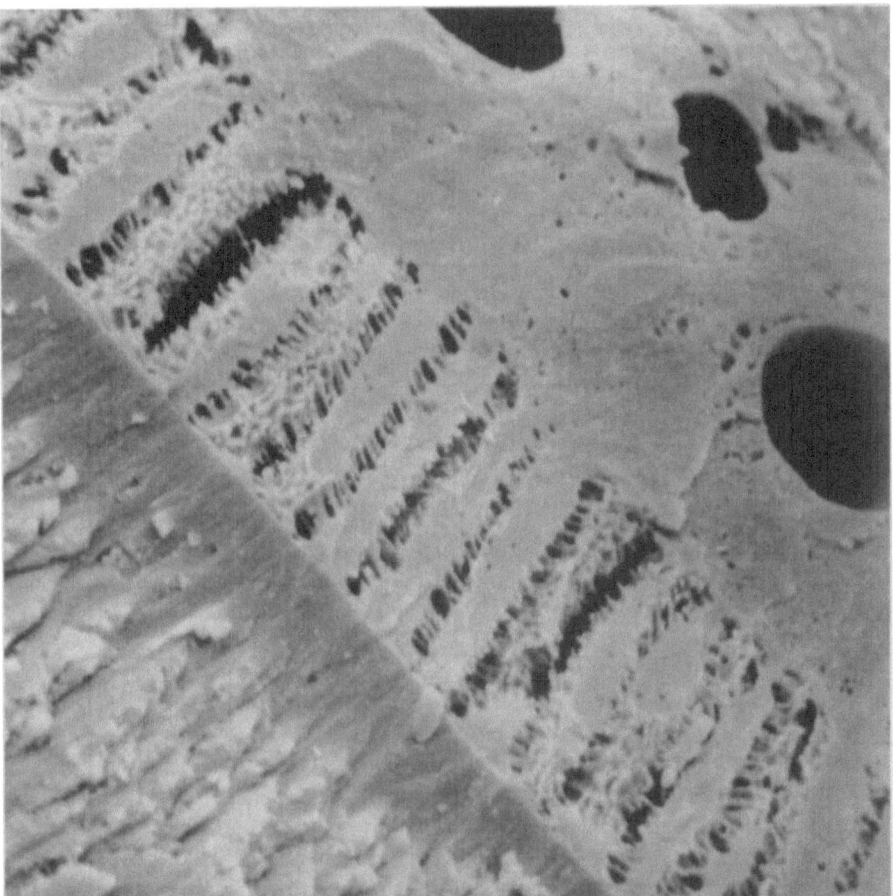

Fig. 103. Ethanol-freeze-fractured, critical-point-dried specimen, showing smooth-ended maturation-stage ameloblasts of rat incisor. Bottom left-hand corner shows enamel layer smooth-ended maturation ameloblasts, with papillary layer blood vessels seen in top right-hand corner. Field width 49 μm

compartment of the SA bands (otherwise shut off from the papillary layer extracellular space) only at the RA/SA borders. A similar "open route" is also preserved post mortem in glutaraldehyde perfusion-fixed material. Thus, in removing the enamel organs for processing for LM and TEM using EDTA demineralisation, it is found that the underlying enamel is etched first and most opposite the RA/SA and SA/RA borders, spreading to the tissue opposite the narrow SA bands and only later to that next to the RA (REITH et al. 1982).

6. Cyclical Phenomena

The fluorochrome marker studies show that waves of change sweep through the entire maturation ameloblast sheet, originating at the cervical end of the maturation portion of the enamel organ (BOYDE and REITH 1981). Using multiple

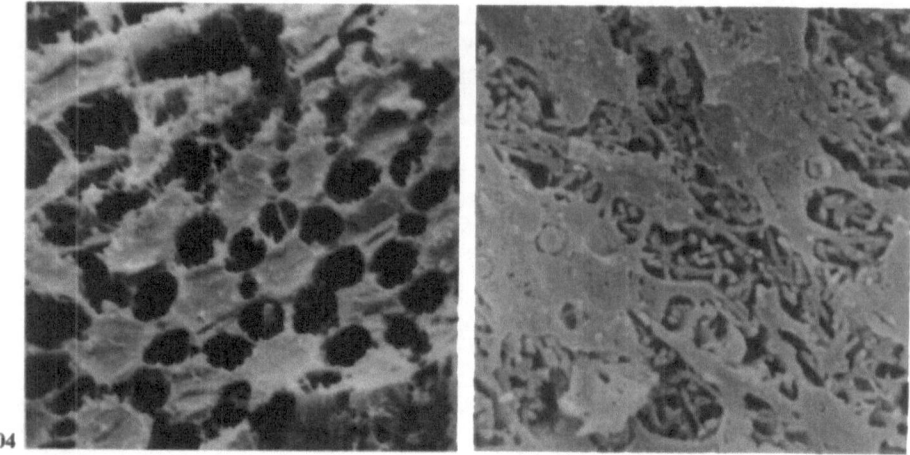

104   105

Fig. 104. Transverse freeze-fracture of rat incisor maturation-stage ameloblasts, showing the prominent intercellular gaps. Field width 19 μm

Fig. 105. Oblique freeze-fracture of papillary layer of maturation-stage enamel organ of rat incisor, showing the very complicated cell surfaces of the stratum intermedium cells. Field width 19 μm

markers at successive time intervals, SMITH et al. (1988) have been able to determine the rate of progression of these waves: several pass along the length of the rat incisor during one day, which proves that they are not standing waves moving down the tooth at the same rate that it erupts. Moreover, the "wave generator" lies towards the apical end of the (maturation) ameloblast sheet.

The enamel opposite smooth-ended ameloblasts has different properties than that adjacent to the ruffle-ended ameloblasts. The two types of band manifest different post-mortem staining reactions with a variety of stains (BOYDE and REITH 1982; TAKANO et al. 1982). They can even be identified in the drying properties of the enamel following the removal of the enamel organ (REID et al. 1984). Both phenomena may possibly represent a change in the hydrophobicity of the organic matrix. TAKANO et al. (1982) believed that the glyoxal bis(2-hydroxyanil) (GBHA) stain which they used to demonstrate the distribution of the narrow, SA, bands was specific for ionised calcium (which could not be the case since "ions" are not bound). If this reaction has any real connection with calcium, then it might at best demonstrate accessibility of the stain to the mineral component given the matrix chemistry in the narrow bands.

Post-mortem staining reactions of many types also clearly show an incremental banding pattern in both RA and SA zones superimposed upon reactions correlating with the distribution of the cell types. These incremental bands have the same orientation along the entire maturation region and are not parallel with the maturation bands, the orientation of which changes from band to band in a characteristic fashion (BOYDE and REITH 1982). The distance between such bands is in agreement with known values for daily growth increments

(SMITH and WARSHAWSKY 1977), leaving little room for doubt that they reflect a circadian rhythm imposed upon the enamel surface before maturation. GASSER et al. (1972) have shown a circadian rhythm in the rate of cell entry into the ameloblast layer in the rat incisor.

No changes corresponding to the maturation cyclical banding have yet been found in the chemistry of the organic matrix or in the degree of mineralisation of the corresponding enamel (ROBINSON et al. 1977, 1981; ROBINSON and KIRK-HAM 1984). It has been proposed that the cyclical changes represent a cycle in the activity leading to the removal of the surplus organic matrix from the enamel rather than a cycling in the input of the mineral component (NANCI et al. 1988). However, the greatest loss of matrix occurs next to the earliest maturation ameloblasts (ROBINSON et al. 1981), not next to the RA-SA cycling cells. It has also been suggested that the cyclical changes represent the rejuvenation or reviving or reactivation of the ameloblast in the SA bands (TAKANO and OZAWA 1980).

III. Studies on Rooted Teeth

Much less direct evidence has been accumulated for phenomena surrounding maturation in rooted teeth, i.e. those which are not of continuous growth and which might be better models for what happens in man. There are big differences on a temporal scale. A human permanent tooth may take 5 years to harden its enamel before eruption, and it may last for 100 years in function. A rat incisor forms and hardens its enamel in one month, and it is lost due to wear in the next (SMITH and WARSHAWSKY 1977). Human secretory ameloblasts engage in more maturation whilst secreting than do rat cells (ROSSER et al. 1967; DEUTSCH et al. 1984). Nevertheless, many phenomena are found in human and other rooted teeth which exactly match those seen in the rat incisor, so that it is tempting to believe that the mechanisms involved will be found to be the same. Thus, RA- and SA-type cells may be seen in LM and/or SEM preparations of human (Figs. 107, 108), monkey, cow, sheep, goat, cat and dog material (author's unpublished data); they have been described in TEM of the rooted rat molar (BOYDE and REITH 1979), and in cat (SASAKI 1984) and monkey (SKOBE, personal communication 1987). Tetracycline injected intravenously in kittens was found to label narrow bands of enamel opposite the maturation enamel organ of deciduous carnassials in a pattern similar to that described in rat incisors, i.e. first and most intensely at the SA-to-RA (incisal) transitional borders, followed by the RA-to-SA (apical) borders of the narrow bands, spreading thence to the entire narrow band (Fig. 106; BOYDE and REITH 1983, 1984). Post-mortem staining with GBHA has been shown to stain bovine maturation-stage enamel with bands resembling those produced in the rat incisor (TAKANO et al. 1982). The sequence of chemical changes in the organic matrix and the mineral component is similar in human (and other rooted teeth) and the rat incisor (ROBINSON et al. 1971, 1977, 1981a, b; DEUTSCH et al. 1984; COOPER 1967, 1968).

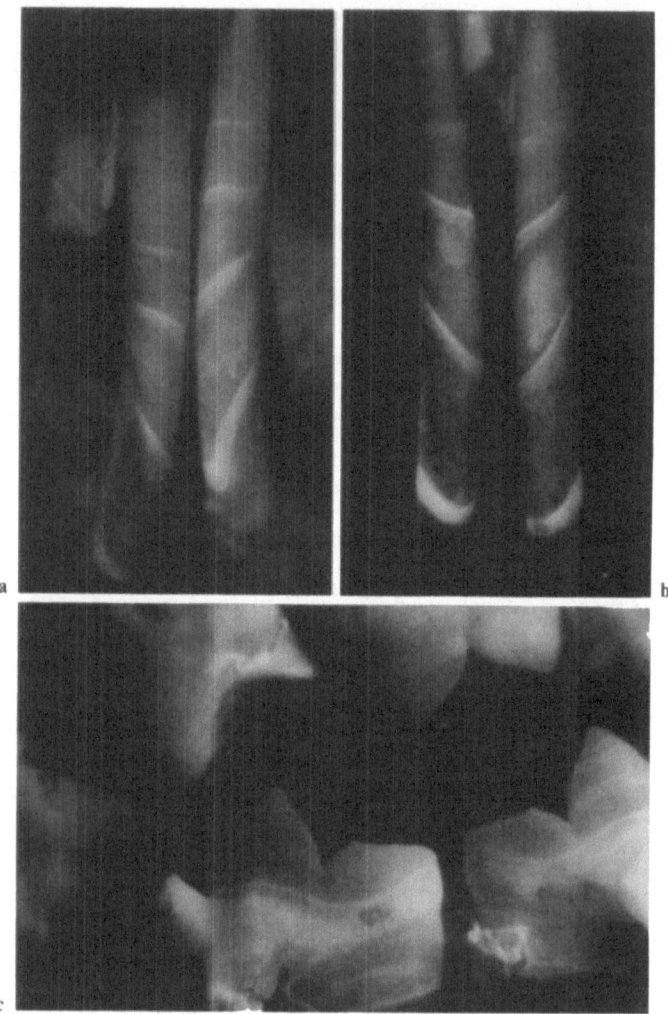

Fig. 106a–c. Yellow fluorescence under ultraviolet light of (**a, b**) rat incisors and (**c**) cat deciduous molars from animals given intravenous injections of tetracycline antibiotic shortly before being killed. Tetracycline has intensely stained narrow bands (opposite smooth-ended maturation ameloblasts). In the rat incisors, the most stained part of the band is on the incisal edge, and it is less intensely stained on the apical edge of each narrow band. The same was observed to be true for the cat carnassial teeth, but this is not so clearly shown in the recorded photograph. The pattern of distribution of bands is usually symmetrical, as is indicated in the rat incisors (**b**). The mandibular incisors (centre of **a**) of one animal show a markedly eccentric distribution. Field width **a, b** = 6 mm; **c** = 12 mm

IV. Mechanisms of Maturation and Matrix Removal

The literature contains divergent views on the mechanisms of maturation. Perhaps the simplest view is that the force generated by the growth of the crystals is sufficient to mobilise the unstable thixotropic protein matrix gel (EAS-TOE 1982). Teleologically speaking, this would explain the utility of employing the inherently unstable amelogenin as the principal component of the newly secreted matrix, together with evidence for the thixotropic nature of the gel formed at 37° C and the pH sensitivity, de-gelling at pH 6.8, which might be found at the surface of growing hydroxyapatite crystals (N. SIMMONS, personal communication). It would also explain how high degrees of maturation might be reached in the secretory stages of thick enamelled teeth, such as those of man. It does not explain why there is an apparently rather sudden loss of organic matrix which is not parallelled by a matching increase in mineral uptake in bovine and rat incisor enamel (ROBINSON and KIRKHAM 1984; ROBINSON et al. 1977, 1981a, b; DEUTSCH and PEER 1982; DEUTSCH et al. 1984). The hypothesis that crystal growth pressure leads to amelogenin extrusion would also not explain the active cyclical changes in the organisation of ameloblasts seen in the maturation zone, which strongly suggest active cellular participation. These cyclical changes may be correlated either with mineral ion input or organic matrix removal from the enamel, or with both.

Several mechanisms have been discussed concerning the transport of calcium and phosphate across the enamel organ: none have been proven, but the correct answer will have to explain the high transit rate from blood to enamel which is observed, much higher than transit rates into bone and dentine (BOYDE and REITH 1981).

Some authors believe that ameloblasts secrete proteases into the enamel matrix; no protease has yet been isolated (CRENSHAW and BAWDEN 1984). Clearly, however, the amount of residual matrix is reduced and the composition is changed during maturation. The amino-acid fingerprint of enamel protein changes from the amelogenin-rich newly secreted form to the enamelin-dominated mature tissue (ROBINSON et al. 1975, 1977, 1978, 1981; DEUTSCH et al. 1984; GLIMCHER et al. 1977; EASTOE 1982).

V. End of Maturation – Variations in Composition

Few data are available concerning the variations in the composition of enamel at the end of the maturation period and prior to eruption. Excellent information, on the other hand, is available for erupted human teeth, but here the factor of post-eruptive maturational changes must be taken into account. Enamel does not reach a uniform composition at the end of pre-eruptive maturation. The protein content is generally higher and the mineral content lower in the centre of the tissue and in the tuft regions close to the enamel-dentine junction (WEATHERELL et al. 1967, 1968; HALLSWORTH et al. 1972; COOPER 1967, 1968). The magnesium concentration is also higher in the more organic, less dense, central regions of the enamel (ROBINSON et al. 1981).

VI. Eruption or Emergence

At the completion of maturation and prior to eruption, the residual ameloblasts and the associated papillary layer cells of the enamel organ remain together as the reduced enamel epithelium (REE). The ameloblasts remain firmly

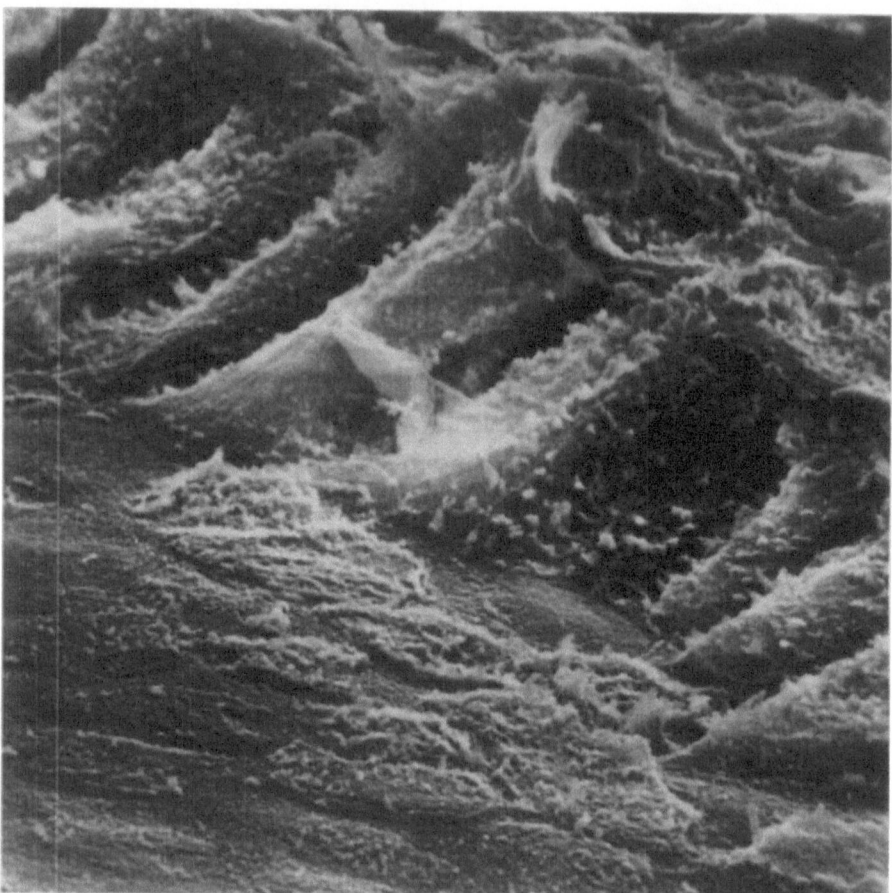

Fig. 107. Late maturation ameloblasts attached to the surface of a human third permanent molar, with part of the enamel organ stripped away. The amount of extracellular space between these cells and the continuity of the lateral intercellular space compartment down to the enamel surface would suggest that these are late smooth-ended ameloblasts. Field width 51 μm

attached to the tooth, and attempts to remove them from a tooth extracted prior to eruption lead to rupture through the cells leaving their distal portions attached, via a basal lamina, to the enamel (Figs. 107, 108). The tooth normally erupts with the accompanying REE cells attached, but some of these may be displaced by a migration of oral epithelium cells onto the enamel directly rather than over the reduced enamel organ. In either case, an epithelium is attached directly to enamel, and no connective tissue components are exposed to the oral cavity (LISTGARTEN 1966).

As the enamel emerges, its covering of REE or of oral epithelium is worn from the tooth surface. A recently erupted tooth may, however, be covered by a thin membrane of REE origin. This may be separated from the tooth by acid dissolution of the enamel, a procedure described by *Nasmyth* (1839).

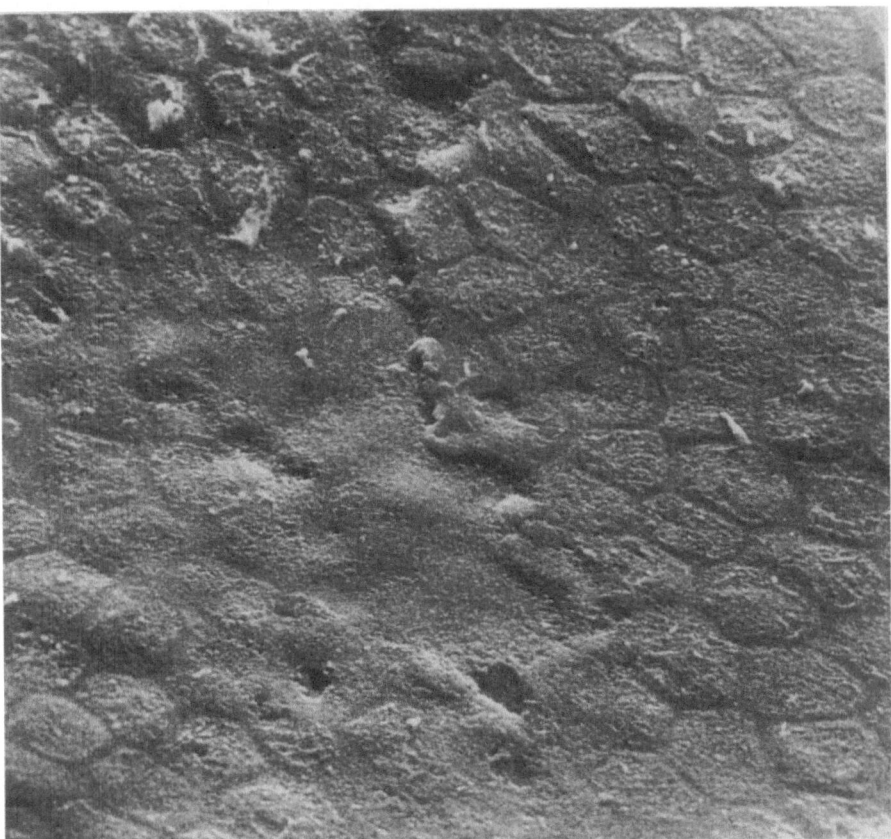

Fig. 108. Surface of human third permanent molar from which late maturation ameloblasts have been stripped. Note that a small plaque delineates the previous location of each cell: this is the distal portion of the ameloblast adherent to the basal lamina on the surface of the enamel. An imbrication line crosses the centre of the field of view cervical to right). Field width 90 μm

Nasmyth's membrane must be distinguished from pellicle, which is a material acquired by deposition onto the enamel surface in the mouth. Pellicle has as its origin salivary proteins which are bound to the hydroxyapatite surface of the enamel.

VII. Post-eruptive Maturation

After a tooth has erupted into the oral cavity, it is possible for the density of the tissue to be increased by the continued deposition of mineral within the tissue (GOLDBERG et al. 1979, 1987). (There is also the possibility that the quality of the tissue may be degraded by demineralisation within the tissue deep to the surface, leading to the development of carious change). Post-eruptive maturation of enamel can lead to an apparent colour change in the tooth,

greater translucency of the enamel bringing the tooth colour closer to that
of the underlying yellowish dentine. However, any improvement in the quality
of the enamel is offset by the generation of cracks in the tissue developing
lamella-like features. Cracking or crazing of the enamel is supposed to be a
frequent occurrence in arctic weather conditions, where one would presume
that repetitive thermal stressing is the underlying cause.

K. Functional Changes in Enamel

I. Wear

Enamel is remarkably resistant to wear, but it does wear, and the wear
is mostly unrelated to enamel structure unless chipping-out of prisms and incremental lines occurs (RENSBERGER 1978; WALKER et al. 1978). Excessive rates
of wear are found either where there is an abrasive content in the diet or
where the teeth contact each other in non-masticatory function, for example,
in individuals who habitually grind their teeth together for substantial periods
(bruxism) or who grit their front teeth or bite their nails. The modern Western
human diet contains little abrasive material – no sand mixed in with the food!
Studies of archaeological human populations have all shown an extent of occlusal wear far greater than is encountered at the present time. This can be
accounted for by assuming that small, sharp rock particles crushed between
the opposing teeth will gouge grooves in the occlusal surface enamel, leading
to a much greater rate of wear than would be possible from tooth-to-food
or tooth-to-tooth contact.

Tooth-to-tooth wear is normally distinguished by the term "attrition". Its
purest form is found in contact surface wear between teeth in the same dentition
at approximal facets. The occurrence of wear in this location depends upon
the degree of flexibility in the attachment provided by the periodontal ligament
apparatus. Heavy wear implies repeated lateral relative movement between adjacent teeth. Even in modern Western humans, it is exceptionally rare not to
find a small contact surface facet which has polished away all the mature enamel
surface detail. Such contact facets may reach several square millimetres in individuals who also show extensive occlusal wear. Wear accelerated by contact
of other foreign materials with the teeth (for example, holding tools, hairgrips
or a smoking pipe in the front teeth) is properly called abrasion in English
usage.

In examining the relationship between the microscopical structure of enamel
and wear phenomena, two principal structural factors should be noted. So long
as the wear facet is perpendicular to the prism and crystallite direction (as
it is to a fair approximation in the as-developed condition over the entire crown
surface, with a particularly close approximation to this simplification at occlusal
and incisal edges), there are no structural discontinuities which can be exploited
in breaking away largish fragments from the wear surface. Enamel surfaces
which present prisms perpendicular to the surface are wear resistant. On the

other hand, if a surface is produced in which prisms lie more nearly parallel with the worn surface, there is a strong likelihood that groups of prisms may cleave away from this surface by fractures exploiting the prism boundary discontinuities. The regular and prominent incremental lines (brown striae of Retzius) of the more superficial lateral enamel also constitute planes within the tissue along which fractures may tend to propagate, the more so the less mature the tooth. The consequences for wear in normal usage of the human incisor teeth are the following. Once the enamel covering the dentine at incisive edges has worn away, the prism direction for the labial enamel of the upper incisors is more parallel with the worn surface than the lingual. The incisive edge now also intercepts brown striae. Pieces of enamel may break away parallel with both the prisms and the brown striae, leaving a rough, chipped and sharp edge at the labial surface of the tooth. The opposite set of circumstances in normal occlusal relationships apply for the lower incisor teeth.

II. Erosion

Erosion of the enamel results from chemical dissolution of the mineral phase of the tissue. It is a common clinical circumstance and has two major origins. The first is the consumption of excessive amounts of citrus fruits, in which the citrus acid functions as both an acid and a chelating agent in the removal of enamel mineral. Secondly, if stomach contents are regurgitated, the acid stomach fluid is able to erode those surfaces of the teeth with which it comes into contact. Acid-etching of the enamel is currently in widespread use in clinical dental practice.

L. Pathology

I. Acquired and Inherited Disorders

The major acquired pathologies of dental enamel are its destruction internally by caries; its amplification externally by the deposition of mineralised bacterial dental plaque (calculus), which has hardly any direct consequences for the enamel itself; and excessive wear and erosion.

Several inherited disorders affect the enamel directly or indirectly (SAUK et al. 1972; HALS 1957). Of these, the commonest are the groups distinguished by the terms "dentinogenesis and/or amelogenesis imperfecta". In dentinogenesis imperfecta, the inadequately mineralised dentine support to the enamel leads to the cracking and flaking of the enamel from the surface of the dentine, underscoring the normal function of dentine in enamel function. Amelogenesis imperfecta refers to a group of conditions in which the main disorder of enamel formation seems to be that of maturation, so that the tissue, although formed, remains with a high organic matrix and low mineral content. It is, therefore, very soft and easily abraded, flaking off from the surface of what may be a relatively normal dentine.

II. Linear Hypoplasias

Linear hypoplasias result where generations of ameloblasts developing from the differentiation of the internal epithelium cells at one particular longitudinal address on the developing tooth circumference are unable to form enamel or the correct thickness of enamel. The most likely explanation is development of a mutation in the internal enamel epithelium at the cervical loop of the tooth germ in the corresponding location. Other hypoplasias are not of genetic origin.

III. Hypoplastic Grooves

Hypoplastic grooves may commonly be related to a severe systemic disturbance, such as one of the exanthematous fevers. All teeth forming enamel at the same time form the same pattern of rings of reduced enamel thickness. Such hypoplasias are essentially exaggerated perikymata. In very severe cases, they may involve almost the entire thickness of the enamel. In such cases, a wide band of enamel which was forming prior to the systemic disturbance is exposed at the completed, mature tooth surface. This enamel is generally hard and sound yet shows a typical developing-enamel surface morphology (Fig. 109) (BOYDE 1970). These circumstances indicate that secretory ameloblasts are particularly sensitive to such insults, but that they may recover and differentiate to become successful and competent maturation ameloblasts.

IV. Hypoplastic Pits

Hypoplastic pits are localised areas of defective formation of enamel. They may be related to local infectious disease affecting the related portion of the enamel organ. This may result from severe carious involvement of a deciduous predecessor, with the ensuing bacterial infection of the deciduous pulp eventually spreading to affect the enamel organ of the permanent tooth germ.

V. Dental Fluorosis

Dental fluorosis is a consequence of oversupply of fluoride in drinking water and foodstuffs. It is endemic in human populations in certain circumscribed geographical localities. In mild forms, large patches of the enamel will be seen to be excessively white, i.e. to scatter light abnormally. In severe cases, the discoloured patches are orange or brown. Such teeth cause an aesthetic disfigurement, but there is little indication that they are functionally incompetent. The explanation of the discoloration is possibly that the fluoride leads to the formation of fewer larger crystals, leading to an alteration in the light-scattering properties of the enamel. It may also have direct and indirect effects on the

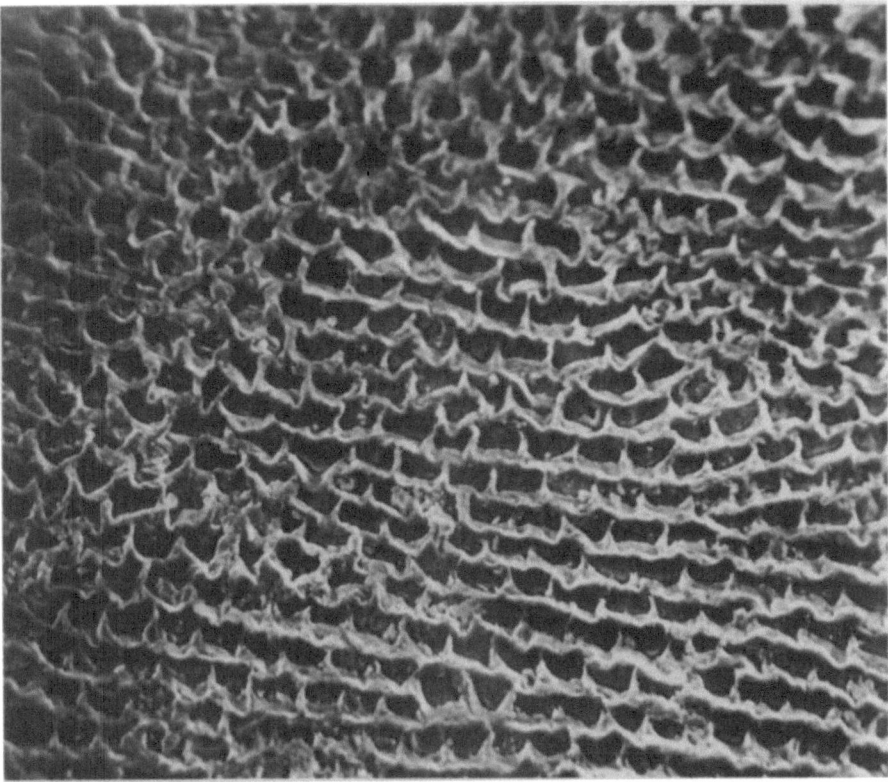

Fig. 109. Surface of enamel in a deep hypoplastic groove in a human premolar (cervix to top). The details of this surface are typical of active bulk-enamel secretion: in this case, the arrangement of the pits suggests arcades of prism boundaries in some locations. Field width 177 μm

synthesis and processing of the enamel matrix proteins and/or on the removal of matrix proteins during maturation (TRILLER 1984; MANNERBERG 1968).

VI. Enamel Caries

Enamel caries is a disease of modern human populations. It is caused by the fermentation of disaccharide sugars by certain species of bacteria in the plaque inhabiting the tooth surface to produce organic acids. The net result is a subsurface demineralisation of the enamel. Most, if not all, features of natural enamel caries can be imitated in artificial systems in which organic acid buffers are supported in gels to restrict diffusion. This subsurface demineralisation phenomenon can be imitated in sintered hydroxyapatite blocks and is therefore not explained by enamel structural features (LANGDON et al. 1980). However, during the progress of a natural enamel caries lesion, certain features

Fig. 110. Back-scattered electron image of longitudinally sectioned, PMMA-embedded molar with a proximal surface carious lesion. (Tooth surface seen in bottom right-hand corner.) Contrasts in this image are mainly due to density variations, though there is a significant contribution from topography, owing to the uneven way in which carious enamel polishes. Note the prominence of the cross-striations and the regular incremental lines in the cariously affected enamel. Field width 491 µm

of the enamel structure are exploited or accentuated. Thus, the enamel prism boundaries may be widened and traced with, for example, fluorescent dye tracers, further than the unaltered prism boundaries. Both the cross-striations of the enamel prisms and the brown striae of Retzius are also exaggerated in the carious tissue (Fig. 110). This probably relates to the fluctuating carbonate content of these structures, with the more carbonate-rich layers being selectively dissolved by the dilute organic acids. Demineralisation of the subsurface enamel in caries results in its softening, which can be demonstrated by measurement or experiment. Probing the relatively harder surface which covers the underlying lesion may cause cavitation. A cavity results where the surface enamel has caved in due to the lack of underlying support. Literature reviews and more

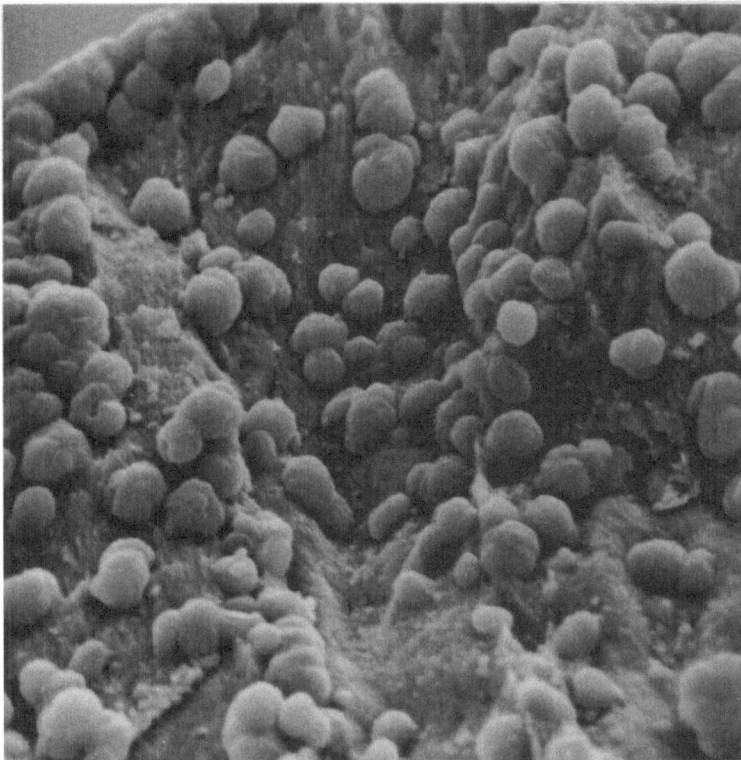

Fig. 111. Surface inside carious cavity in enamel in human tooth. Unmineralised bacterial dental plaque has been removed by prolonged extraction with 1,2-ethane diamine, leaving shapes in the cavity wall which may be either mineralised bacteria or whitlockite, caries crystals. Field width 11 μm

information concerning the histopathology of caries may be found in SHELLIS and HALLSWORTH (1987), JONES and BOYDE (1987), LENZ (1956), DARLING (1958, 1961), GUSTAFSON (1957), MANNERBERG (1964), JOHNSON (1967), VAHL et al. (1966), and VAHL and PLACKOVA (1967).

Caries develops in locations where thick plaque layers may remain relatively undisturbed on the tooth surface, allowing longer periods for the build-up of the necessary acidic conditions. These locations are at the depths of the fissures of the occlusal surfaces of posterior teeth; just below contact surface facets for teeth in the same arch; and around the necks of teeth, particularly on buccal and labial faces. Carious lesions may "heal" to a limited extent by remineralisation within the porous enamel of the subsurface lesion. The remineralisation occurs with a different crystal species than was originally present in the enamel and a different crystal form. There is a substantial whitlockite component in the remineralisation crystals. Some of the bacteria which may penetrate the surface of a cavitated lesion may themselves become mineralised during the reversal phase of the carious process (Fig. 111).

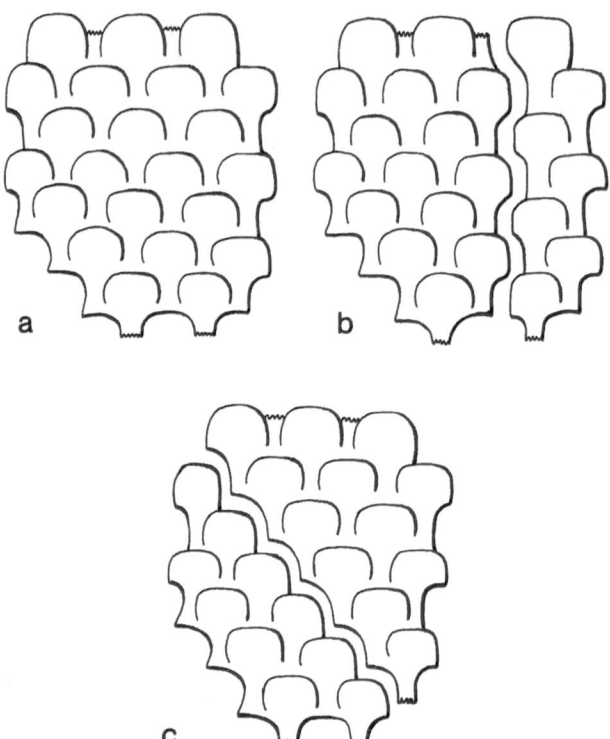

Fig. 112a–c. The most common directions of easily generated fractures in human enamel as visualised from a head-on view of the prisms. Prism boundaries tend to join to each other across the shortest interval so long as it is not necessary to break crystallites. This criterion is satisfied for the fracture lines shown here

VII. Iatrogenic Cutting and Fracturing

Restorative dentistry aims to arrest the progress of an extensive caries lesion and/or to replace the lost dental tissue and restore the tooth to its original functional shape. In the preparation of cavities for restoration by prosthetic materials, enamel is subject to influences which are not part of the normal range of experience for the teeth of mammals other than man.

The need to understand how enamel behaves when it is cut by the dental surgeon and the respects in which its continued function is compromised by absence of the surrounding enamel and/or the underlying dentine support pro-

Fig. 113. Acid-etched human premolar lateral enamel prepared by grinding away material from ▷ the dentine side of a slab of enamel and etching it from "inside out". (A stirred acid system was used to prevent the deposition of new acid calcium phosphate species within the enamel structure; thus, the prism boundary spaces have been preferentially attacked.) A discontinuity runs across the field of view joining prism boundaries in the sense indicated in Fig. 112. Lamellae and tufts also frequently run in this direction. Field width 29 μm

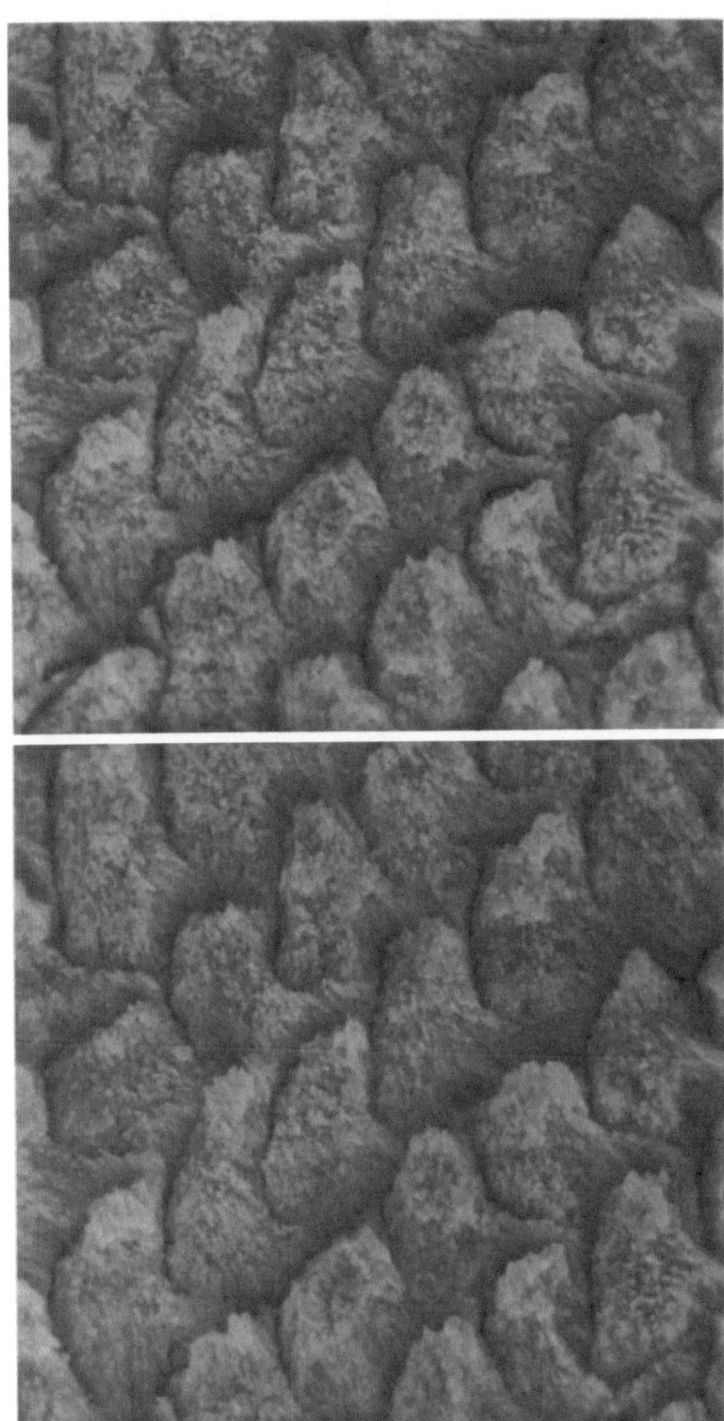

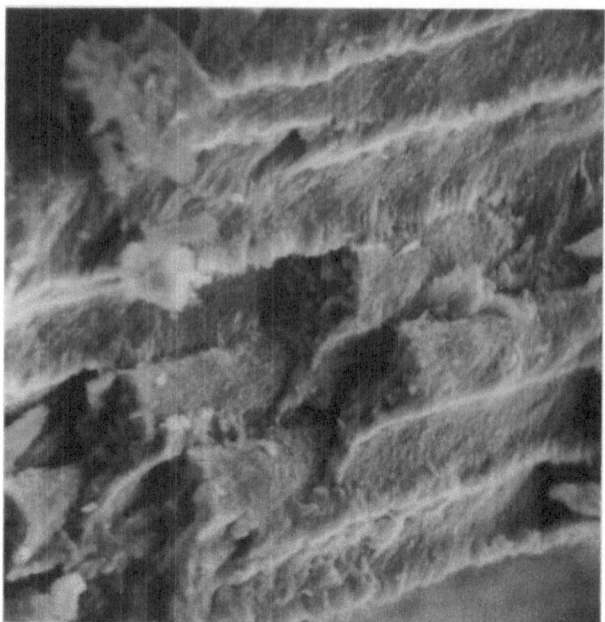

Fig. 114. Fractured human enamel in which the fracture faces either are running longitudinally (as in the directions indicated in Fig. 112) or have fractured the prisms transversely (except at one location where the tail of a prism has fractured across at centre right of field of view). Field width 55 μm

vided the motivation for studies which have led to a generally improved understanding of the functional relationships in enamel (BOYDE 1976a, 1978a).

Mature enamel cannot be cut except as an exceptionally thin slice, using a diamond ultramicrotome knife. Cutting involves a process of plastic deformation. If very thin sections of enamel are cut, this deformation may be confined mostly to the organic matrix between adjacent crystallites, so that small broken fragments of crystals may be held together by the original organic matrix but are no longer in their original relationships. Cutting enamel under the clinical circumstances of cavity preparation involves removing very substantial amounts of enamel per cutting event compared with what we understand from ultramicrotomy. Under these circumstances, the enamel fractures exploiting its own internal discontinuity system. However, an examination of the cut surface will reveal no detail of the enamel structure due to the smearing which is the last event in any cutting process. Smearing involves the plastic deformation of a layer of enamel material over the surface of cut enamel due to locally high pressures existing underneath the points at which the cutting instrument contacts the

Fig. 115. Enamel which has been induced to fracture in the transverse-oblique direction (the most ▷ difficult direction, breaking across the tail portions of the prisms), viewed from the tail (cervical) side. The sample was prepared and notched to make this particular fracture plane possible and inevitable. Field width 29 μm

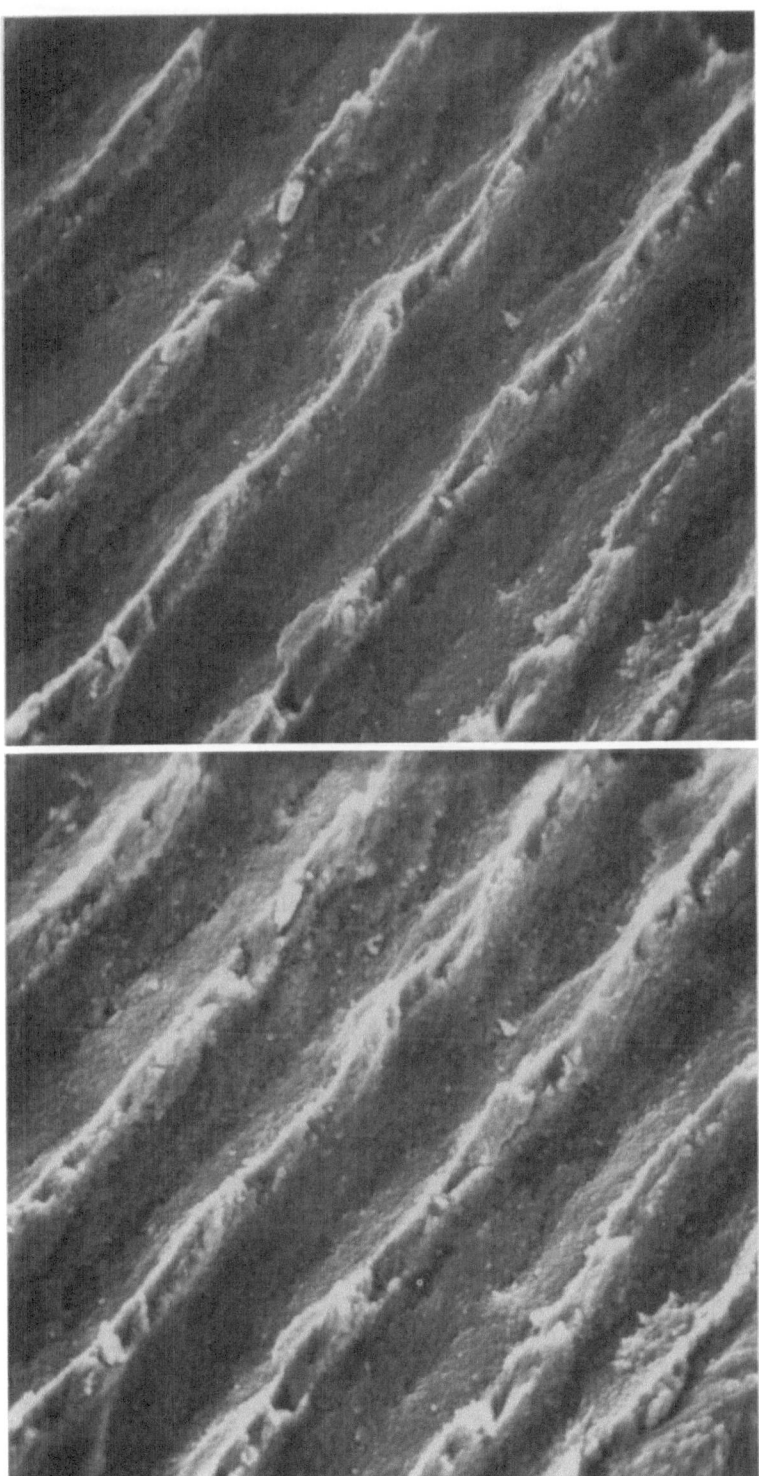

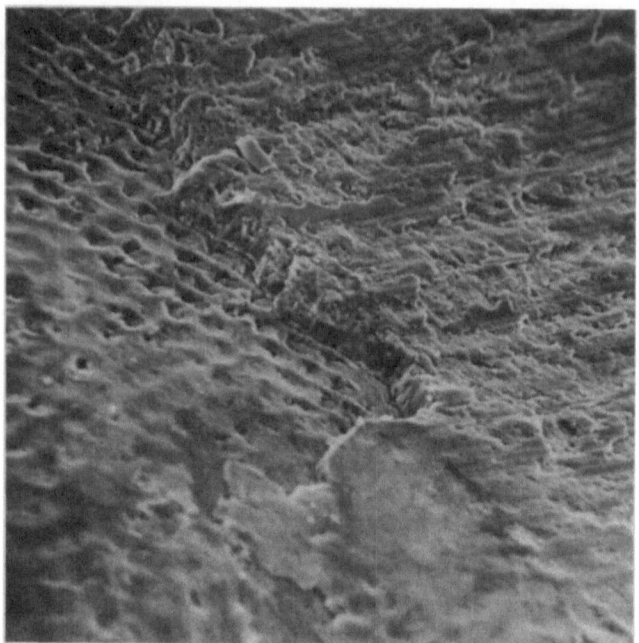

Fig. 116. SEM of low-lateral enamel surface of human tooth scaled with tungsten carbide scaler before extraction. Scaling has caused fracture of the enamel parallel with incremental lines exposing a pitted surface (resembling the troughs of normal perikymata), which was the profile of the plane of the incremental surface inside the tooth. Field width 112 μm

enamel. The smeared layer is generated both by rotary cutting instruments moving at high speed and by hand instruments moving at a very low speed (BOYDE 1976a, 1978).

Trying to view the processes of fracturing in enamel must involve thinking in three dimensions. Enamel crystallites are long, thin and strong. Enamel cleaves by breaking straight across the crystallites or by separating in the organic matrix "glue" joining the crystals side to side. The organic matrix is thickest and weakest at the prism boundary discontinuities, and where possible, a fracture will separate enamel at this discontinuity. However, the discontinuities themselves are not continuous, so that a process of fracturing enamel has to involve cleavage through continuous enamel structure. The least distance in which such cleavage has to run will be from the end of one end of a horseshoe- or arch-shaped prism boundary to the nearest point of the next neighbouring prism boundary on the cervical side (Figs. 112–114). This fracture plane only involves the necessity to separate crystals side from side and is therefore energetically favoured. A potential fracture plane which would join prism boundaries in the same transverse row has to overcome the obstruction caused by the fact that the crystals in this vicinity have a strong component of orientation across the desired fracture line. Breakages in this direction are therefore unfavoured (Fig. 115). When and where they do occur, they give rise to a relatively

Fig. 117. Short-term, acid-etched (stirred system) enamel surface in which the prism boundary discontinuities have been expanded by the etching process, making the prisms more prominent. Field width 112 µm

rough texture on the fractured surface where the interpit-phase enamel crystallites have had to be ruptured transversely. No extensive fracture through human enamel can avoid the necessity to break across some prisms. Such breaks are commonly transverse to the prism axis, so that a fractured surface consists of a series of longitudinally exposed prisms with small areas of transversely broken prisms. Intermediate fracture orientations with respect to the prism long axis are uncommon.

The prominent incremental lines in the subsurface enamel of lateral regions of the tooth also constitute planes of relative weakness in the enamel, which are exploited in the fracturing process (BOYDE 1976a) and in rough manipulations of the enamel surface [as in clinical scaling and polishing with abrasive instruments (Fig. 116)]. The fact that they are weak may be explained by the necking of the prism boundary which occurs in this plane, leading to the shift in the balance of crystallites growing in the long axis of the prisms to the interpit crystallites running at, say, 45° to the prism axis. This necking of the prism boundary must also mean that there is an increased proportion of prismatic crystallites ending at a prism boundary at that location, and that prism boundaries may either be joined or at least come into a closer range than normal. There is also a greater proportion of "prism boundary" in the plane of the brown striae. Thus, the transverse break across the prism will have to break across fewer crystallites if it occurs in the plane of the incremental line. It is also possible that the crystallites themselves may be less perfect in the plane of the incremental line. The coloration of the brown striae itself implies

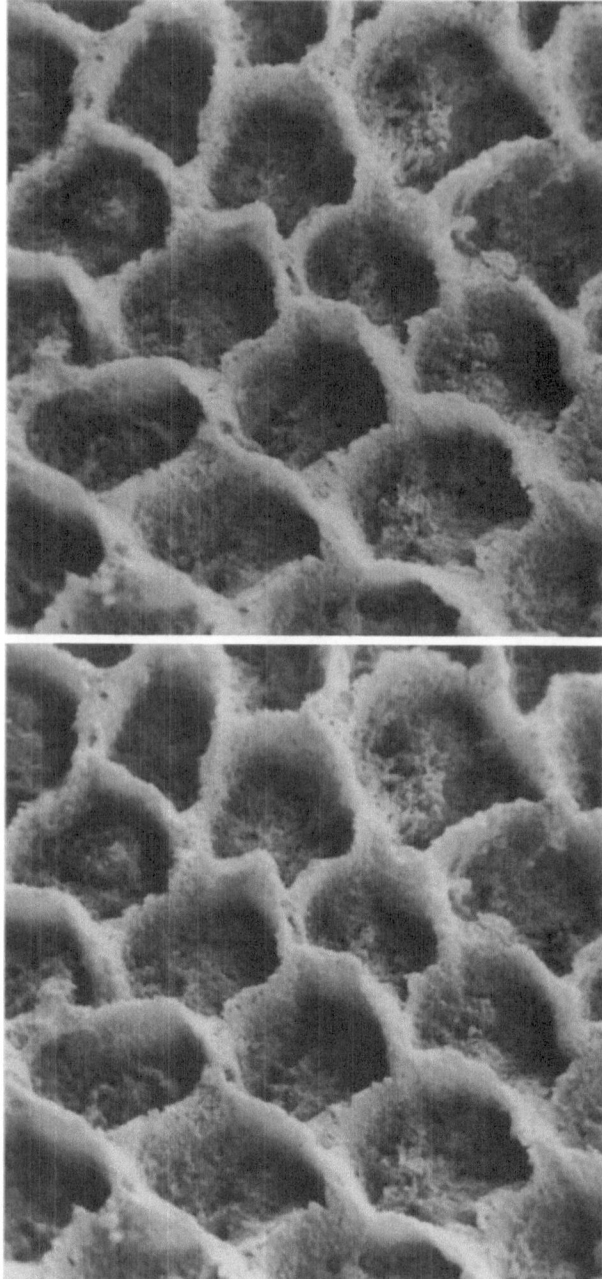

Fig. 118. Acid-etched human premolar enamel from a level 250 μm deep to facial surface, showing continuous honeycomb of material in juxta-prism-boundary region: non-stirred system. Stereopair, 10° tilt. Field width 22 μm

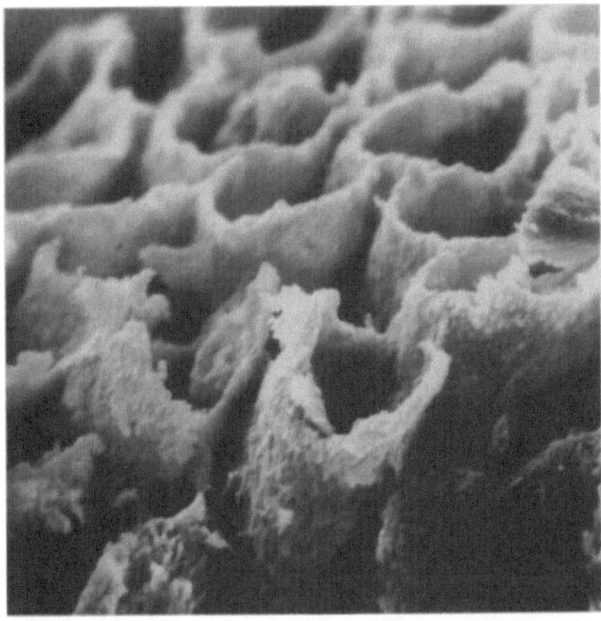

Fig. 119. Oblique view of acid-etched enamel surface, fractured perpendicular to the etched surface in the lower part of the field of view to show the considerable extent of the artefactual continuous honeycomb formed in the juxta-prism-boundary regions by a remineralisation phenomenon prior to the exposure of this surface: non-stirred system. Field width 23 μm

a change in the Rayleigh scattering properties of the enamel, an explanation of which would involve a statistical shift in the balance of crystallite size and intercrystalline space, which may be relevant.

VIII. Acid-Etching

The subject of acid-etching of enamel as employed in clinical dentistry needs some mention because of the large volume of uncritical literature generated in the clinical dental field. Acid-etching is used both to remove the smeared layer which results from cutting a cavity in the tooth (because the smeared layer may not be well attached to the underlying cut enamel) and to increase the surface area of contact, wettability and the degree of interdigitation of the enamel with the adhesive resin restorative material, orthodontic bracket bonding adhesive or fissure-sealant material which will be applied. Images of enamel surfaces which have been etched and dried ready for examination in an SEM may give rise to misleading information about enamel structure if the etching procedure was of the type used in clinical practice, in which concentrated phosphoric acid is used and no stirring is possible. The interpretation of an etched surface involves a knowledge of the relative rates of removal of what has been removed, yet we are left with material that has not yet been

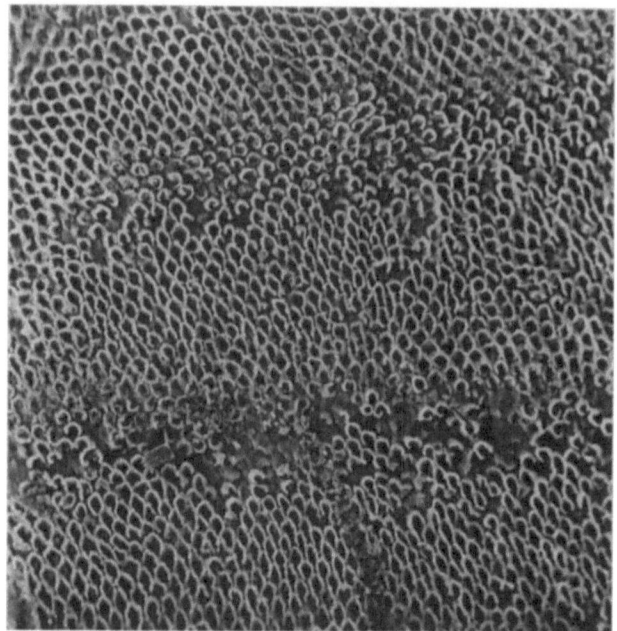

Fig. 120. Phosphoric acid-etched human enamel parallel with facial surface, showing development of continuous honeycomb artefact due to changes in the juxta-prism-boundary region prior to exposure of this surface. Note also the change in etching pattern in the planes of two incremental lines roughly one-third and two-thirds of the way down. Field width 221 μm

removed, and which has not in itself influenced the progress of the etching process. The early stages of acid-etching of enamel (and particularly with concentrated phosphoric acid) involve the partial dissolution of juxta-prism-boundary enamel with a solution which is exceptionally likely to deposit relatively insoluble acidic calcium phosphate species. (In the normal clinical acid-etching situation, the solution is not stirred, and there is an exceptionally high solid-to-solution ratio.) Such crystals may be deposited in the juxta-prism-boundary space to be exposed by the advancing total demineralisation front, thus eventually remaining proud of the surface (Figs. 117–120). These features may be, and generally are, interpreted as "prism sheaths", implying a more organic phase located at enamel prism boundaries. However, they are not prism boundary material. They are not even in the location of the original prism boundaries, but are distributed on both sides of the original prism boundary location. That they are more extensive than the original prism boundaries is indicated by the fact that the resulting honeycomb is usually or often continuous.

Acid-etching is used to improve the strength of the bond of adhesive dental restorative materials to cavities in enamel. The structure of enamel dictates that this bond must be much weaker in enamel regions etched parallel with the prism long axis (sideways-on etching; BOYDE 1985) than where etching proceeds along the prism boundary direction (head-on etching; BOYDE 1985). Differential etching is also related to incremental phenomena in the enamel. Deeper

Fig. 121. Phosphoric acid-etched natural enamel surface of human premolar, showing bands in which prism sheaths stand proud of the surface. These bands correspond to regions in which the prism boundary discontinuities were exposed in the troughs of the perikymata. Deposition of acidic calcium phosphate species could have happened in these exposed prism boundaries ahead of the initial etching process. An alternative explanation is that they may have been exposed to fluoride in the mouth, making the juxta-prism-boundary regions less soluble during the etching process. The regions where the etch pattern is less prominent lay beneath the smooth, prism-free regions of the wave crests of the perikymata before the etching process began. Field width 153 μm

within the enamel, a different etch structure is produced in the plane of the brown striae of Retzius (Fig. 120). Etching from the enamel surface inwards in unstirred systems, it is found that the prism boundaries deep to the troughs of the perikymata (in which the prism boundaries reach the enamel surface) remain proud of the etched surface to a much greater extent than those which previously had underlain the prism-free surface zone. Several explanations can be offered for this. The reason why the prisms deep to the prism-free surface zone seem to be less prominent may be that these prism boundaries are less marked, or simply that they began to etch later since the prism-free layer has first to be removed. The "proud sheath" phenomenon may be due to the generation of acidic calcium phosphate species in the juxta-prism-boundary regions at an early stage of the etching process. It may also be explained by the exposure of these prism boundaries to fluoride in the mouth, creating a conversion of the hydroxyapatite to fluorapatite, rendering the juxta-prism-boundary material less soluble.

Acknowledgements. My studies of enamel have been supported by grants from the Medical Research Council and The Science Research Council. I am particularly grateful for the long-term support and assistance of Dr. S.J. Jones and for many years of technical assistance from Elaine Maconnachie. Of recent years, I have received particular assistance from T.F. Watson and R. Radcliffe. Writing this chapter proved difficult after the loss of my dear friend and scientific collaborator of many years, Edward J. Reith. In finally doing so, I have been reminded constantly of his unique insights and contributions to our understanding of amelogenesis, and I most humbly dedicate this chapter to his memory. Several of the figures used in this chapter were made by Professor Reith and his excellent technical assistant Margie Schmid, to whom I am also much indebted.

My long-term interest in enamel was helped and fuelled by the availability of the new method of scanning electron microscopy. It is a pleasure to acknowledge the help of Roy Switsur, who in turn introduced me to A.D.G. Stewart, and thus to high-resolution scanning electron microscopy early in my own studies of enamel. Finally, I would like to acknowledge the help and friendship of Prof. Ronald W. Fearnhead, who made everything possible for me to start studying this interesting tissue.

References

Abbott F (1889) Growth of enamel. Dent Cosmos 31:749–762

Abrigo SC (1972) Enamel crystal development in the rat incisor. Thesis, Case Western Reserve University

Allan JA (1967) Maturation of enamel. In: Miles AEW (ed) Structural and chemical organisation of teeth. Academic, New York, pp 467–492

Allan JH (1959a) Investigation into the mineralisation pattern of human dental enamel. I. Polarised light studies. J Dent Res 38:1096–1107

Allan JH (1959b) Investigation into the mineralisation pattern of human dental enamel. II. X-ray absorption studies. J Dent Res 38:1109–1118

Allan JH (1959c) Investigation into the mineralisation pattern of human dental enamel. III. Decalcified section studies. J Dent Res 38:119–1128

Angmar B, Carlstrom D, Glas JE (1963) Studies on the ultrastructure of dental enamel. IV. The mineralization of normal human enamel. J Ultrastruct Res 8:12–23

Angmar-Mansson B (1971) A quantitative microradiographic study on the organic matrix of developing human enamel in relation to the mineral content. Arch Oral Biol 16:133–145

Aoba T, Tanabe T, Moreno EC (1988) Function of amelogenins in porcine enamel mineralisation during the secretory stage. Adv Dent Res (1988)

Applebaum E (1960) The arrangement of the enamel rods. NY State Dent J 26:185–188

Arends J, Jongebloed WL (1979) Ultrastructural studies of synthetic apatite crystals. J Dent Res [Suppl B] 58:837–843

Asper H (1916) Über die braune Retzius'sche Parallelstreifung im Schmelz der menschlichen Zähne. Schw Vrtljschr Zahnheilk 26:277–314

Avery JK, Visser RL, Knapp DE (1961) The pattern of mineralisation of enamel. J Dent Res 40:1004–1019

Belanger LF (1957) The mineralization of rat enamel in the light of Ca45 autoradiography and microincineration. J Dent Res 36:595–601

Belcourt AB (1984) High molecular weight proteins in developing enamel their probable interactions. In: Fearnhead RW, Suga S (eds) Tooth enamel IV. Elsevier, Amsterdam, pp 540–542

Beynon AD, Dean MC (1987) Crown-formation time of a fossil hominid premolar tooth. Arch Oral Biol 32:773–780

Beynon AD, Wood BA (1987) Patterns and rates of enamel growth in the molar teeth of early hominids. Nature 326:493–496

Boyde A (1963) Estimation of age at death of young human skeletal remains from incremental lines in dental enamel. Third International Meeting, in Forensic Immunology, Medicine, Pathology, and Toxicology (London April 16–24 1983) Plenary Session IIA. Copies may be obtained from the author

Boyde A (1964) The structure and development of mammalian enamel. Thesis, available from Senate House Library, University of London

Boyde A (1965) The structure of developing mammalian dental enamel. In: Stack MV, Fearnhead RW (eds) Tooth enamel [1964]. Wright, Bristol, pp 163–171, 192

Boyde A (1967) The development of enamel structure. Proc Roy Soc Med 60:923–928

Boyde A (1968) Correlation of ameloblast size with enamel prism pattern: Use of scanning electron microscope to make surface area measurements. Z Zellforsch 93:583–593

Boyde A (1969) Electron microscopic observations relating to the nature and development of prism decussation in mammalian dental enamel. Bull Group Int Rech Sci Stomatol Odontol 12:151–207

Boyde A (1970) The surface of the enamel in human hypoplastic teeth. Arch Oral Biol 15:pp 897–898

Boyde A (1971a) Scanning electron microscopy of the completed enamel surface. In: Fearnhead RW, Stack MV (eds) Tooth enamel II [1969]. Wright, Bristol, pp 39–41

Boyde A (1971b) The tooth surface. In: Eastoe JE, Picton DC, Alexander A (eds) The prevention of periodontal disease [1970]. Kimpton, London, pp 46–63

Boyde A (1971c) New surface features of human dental enamel. J Anat 109:343–344

Boyde A (1973) Quantitative photogrammetric analysis and qualitative stereoscopic analysis of SEM images. J Microsc 98:452–471

Boyde A (1975) A method for the preparation of cell surfaces hidden within bulk tissue for examination in the SEM. SEM/1975/I:295–303

Boyde A (1976a) Enamel structure and cavity margins. Operative Dentistry 1:13–28

Boyde A (1976b) Amelogenesis and the structure of enamel. In: Cohen B, Kramer IRH (eds) Scientific Foundations of Dentistry. Heinemann, London, pp 335–352

Boyde A (1978a) Cutting teeth in the SEM. Scanning 1:157–165

Boyde A (1978b) Development of the structure of the enamel in the incisor teeth in the three classical subordinal groups in the Rodentia. In: Butler PM, Joysey KA (eds) Development, function and evolution of teeth [1976]. Academic, London, pp 43–58

Boyde A (1979) Carbonate concentration, crystal centers, core dissolution, caries, cross striations, circadian rhythms, and compositional contrast in the SEM. J Dent Res [Spec Suppl B] 58:981–983

Boyde A (1980) Histological studies of dental tissues of odontocetes. Sci Rep Int Whaling Comm (Special Issue) 3:65–87

Boyde A (1983) Airpolishing effects on enamel, dentine and cement. Br Dent J 156:287–291

Boyde A (1985) Anatomical considerations relating to cavity preparation. In: Vanherle G, Smith DC (eds) Posterior composite restorations. 3M, Minneapolis, pp 377–403

Boyde A (1987) Applications of the tandem scanning reflected light microscope and three dimensional imaging. Ann NY Acad Sci 483:426–440

Boyde A, Fortelius M (1986) Development, structure and function of rhinoceros enamel. Zool J Linn Soc 87:181–214

Boyde A, Jones SJ (1983) Backscattered electron imaging of dental tissues. Anat Embryol (Berl) 168:211–226

Boyde A, Lester KS (1984) Further SEM studies of marsupial enamel. In: Fearnhead RW, Suga S (eds) Tooth Enamel IV, Elsevier, Amsterdam, pp 442–446

Boyde A, Martin L (1982) Enamel microstructure determination in hominoid and cercopithecoid Primates. Anat Embryol (Berl) 165:193–212

Boyde A, Martin L (1984a) The microstructure of primate dental enamel. In: Chivers DJ, Wood BA, Bilsborough A (eds) Food acquisition and processing in primates. Plenum, New York, pp 341–367

Boyde A, Martin L (1984b) A non-destructive survey of prism packing patterns in primate enamels. In: Fearnhead RW, Suga S (eds) Elsevier, Amsterdam, pp 417–421

Boyde A, Martin L (1987) Tandem scanning reflected light microscopy of primate enamel. Scan Electron Microsc 1:1935–1948

Boyde A, Pawley JB (1975) Transmission electron microscopy of ion erosion thinned hard tissues. In: Pors Nielsen S, Hjorting-Hansen E (eds) Calcified tissues. FADL's, Copenhagen, pp 117–123

Boyde A, Reid SA (1983) New methods for cathodoluminescence in the scanning electron microscope. SEM/1983/IV:1803–1814

Boyde A, Reith EJ (1976) Scanning electron microscopy of the lateral surfaces of rat incisor ameloblast. J Anat 122:603–610

Boyde A, Reith EJ (1977) Scanning electron microscopy of rat maturation ameloblasts. Cell Tissue Res 178:221–228

Boyde A, Reith EJ (1978) Electron probe analysis of maturation ameloblasts of the rat incisor and calf molar. Histochemistry 55:41–48

Boyde A, Reith EJ (1979) A correlated scanning and transmission electron microscope study of maturation ameloblasts in developing molar teeth of rats. J Anat 197:421–431

Boyde A, Reith EJ (1981) Display of maturation cycles in rat incisor enamel with tetracycline labelling. Histochemistry 72:551–561

Boyde A, Reith EJ (1982) In vitro histological and tetracycline staining properties of surface layer rat incisor enamel also reflect the cyclical nature of the maturation process. Histochemistry 75:341–351

Boyde A, Reith EJ (1983) Cyclical uptake pattern of tetracycline in post-secretory maturation phase enamel demonstrated in rooted teeth. Calcif Tissue Res 35:762–766

Boyde A, Stewart ADG (1962) A study of the etching of dental tissues with argon ion beams. J Ultrastruct Res 7:159–172

Boyde A, Switsur VR, Fearnhead RW (1961) Application of the scanning electron-probe X-ray microanalyser to dental tissues. J Ultrastruct Res 5:201–207

Boyde A, Jones SJ, Reynolds PS (1978) Quantitative and qualitative studies of enamel etching with acid and EDTA. SEM/1978/II:991–1002

Boyde A, Petran M, Hadravsky M (1982) Tandem scanning reflected light microscopy of internal features in whole bone and tooth samples. J Microsc 132:1–7

Braden M (1976) Biophysics of the tooth. In: Kawamura M (ed) Frontiers of oral physiology, vol 2. Karger, Basel, pp 1–37

Brown WE, Chow LC, Siew C, Gruninger S (1984) Acidic calcium phosphate precursors in formation of enamel mineral. In: Fearnhead RW, Suga S (eds) Tooth enamel IV. Elsevier, Amsterdam, pp 8–13

Carlson SJ, Krause DW (1985) Enamel ultrastructure of multituberculate mammals: An investigation of variability. Contr Mus Paleontol Univ Michigan 27:1–50

Carlstrom D (1963) Polarization microscopy of dental enamel with reference to incipient carious lesions. Adv Oral Biol 1:255–295

Carlstrom D, Glas JE (1963) Studies on the ultrastructure of dental enamel. J Ultrastruct Res 8:1–11

Carter JT (1922) On the structure of the enamel in the primates and some other mammals. Proc Zool Soc (Lond) 2:599–608

Cooper JS, Poole DFG (1973) The dentition and dental tissues of the agamid lizard, *Uromastyx*. J Zool 169:85–100

Cooper WEG (1967) A microchemical investigation of the mineralisation of dental enamel in the pig. Caries Res 1:174–184

Cooper WEG (1968) A microchemical, microradiographic and histological investigation of amelogenesis in the pig. Arch Oral Biol 13:27–44

Crabb HSM (1959) The pattern of mineralisation of human dental enamel. Proc Roy Soc Med 52:118–122

Crabb HSM (1968) Structural patterns in human dental enamel revealed by the use of microradiography in conjunction with two dimensional microdensitometry. Caries Res 2:235–252

Crabb HSM, Darling AI (1960) The gradient of mineralisation in developing enamel. Arch Oral Biol 2:308–318

Crabb HSM, Darling AI (1962) The pattern of progressive mineralisation in human dental enamel. Pergamon, Oxford

Crenshaw MA, Bawden JW (1984) Proteolytic activity in embryonic bovine secretory enamel. In: Fearnhead RW, Suga S (eds) Tooth enamel IV. Elsevier, Amsterdam, pp 109–113

Crenshaw MA, Takano Y (1982) Mechanisms by which the enamel organ controls calcium entry into developing enamel. J Dent Res 61:1574–1579

Daculsi G, Kerebel B, (1978) High resolution electron microscope study of human enamel crystallites: size, shape and growth. J Ultrastructure Res 65:163–172

Daculsi G, Menanteau J, Kerebel LM, Mitre D (1984) Enamel crystals: shape, length, and growing process; high resolution TEM and biochemical study. In: Fearnhead RW, Suga S (eds) Tooth enamel IV. Elsevier, Amsterdam, pp 14–18

Darling AI (1958) Studies of the early lesion of enamel caries. Br Dent J 105:119–135

Darling AI (1961) The selective attack of caries on the dental enamel. Ann R Coll Surg Engl 29:354–369

Dauphin Y (1987a) Implications of preparation processes on the interpretation of reptilian enamel structure. Paläont Z 61:331–337

Dauphin Y (1987b) Premier bilan de l'étude de la structure de l'émail dentaire chez les reptiles fossiles et actuels. CR Acad Sci (Paris) 305/II: 1217–1219

Dean MC (1987) Growth layers and incremental markings in hard tissues; a review of the literature and some preliminary observations about enamel structure in *Paranthropus boisei*. J Human Evolution 16: 157–172

De la Hire (1699) Sur les dents. Hist Acad Roy des Sci (Paris) 41–43

Deutsch D, Peer E (1982) Development of enamel in human fetal teeth. J Dent Res 61: 1543–1551

Deutsch D, Shapira L, Alayoff A, Leviel D, Yoeli Z, Arad A (1984) Protein and mineral changes during prenatal and postnatal development and mineralization of human deciduous enamel. In: Fearnhead RW, Suga S (eds) Tooth enamel IV. Elsevier, Amsterdam, pp 234–239

Dostal A (1987) Rasterelektronenmikroskopischer Vergleich der Zahnschmelzprismen höherer Primaten. Dissertation, Universität Wien

Eastoe JE (1982) Enamel protein chemistry. Past, present and future. J Dent Res 58 B: 753–763

Ebner V von (1903) Über die Kittsubstanz der Schmelzprismen. Dtsch Monatschr Zahnheilk 21: 505–529

Ebner V von (1906) Über die histologischen Veränderungen des Zahnschmelzes während der Erhärtung, insbesondere beim Menschen. Arch Mikr Anat 67: 18–81

Elliott JC (1965) The interpretation of the infra-red absorption spectra of some carbonate containing apatites. In: Fearnhead RW, Stack MV (eds) Tooth enamel [1964]. Wright, Bristol, pp 20–22 and 50–58

Elliott JC (1969) Recent progress in the chemistry, crystal chemistry and structure of the apatites. Calcif Tissue Res 3: 293–307

Elwood WK, Bernstein MH (1968) The ultrastructure of the enamel organ related to enamel formation. Am J Anat 122: 73–94

Fearnhead RW (1960) Mineralisation of rat enamel. Nature 188: 509–510

Fearnhead RW (1961) Electron microscopy of forming enamel. Arch Oral Biol 4: pp 24–28

Fearnhead RW, Elliott JC (1962) Observations on the relationship between the inorganic and organic phases in dental enamel. In: Fifth International Congress for Electron Microscopy. Academic, New York, paper QQ7

Fincham A (1984) Amelogenins: progress and problems. In: Fearnhead RW, Suga S (eds) Tooth enamel IV. Elsevier, Amsterdam, pp 114–119

Fleming HS (1961) Transitional ameloblastic activity zone in mice teeth. J Dent Res 40: 268–281

Fortelius M (1984) Vertical decussation of enamel prisms in lophodont ungulates. In: Fearnhead RW, Suga S (eds) Tooth enamel IV. Elsevier, Amsterdam, pp 427–431

Fortelius M (1985) Ungulate cheek teeth: developmental, functional, and evolutionary interrelations. Acta Zool Fenn 180: 1–76

Fosse G (1968a) A quantitative analysis of the numerical density and the distributional pattern of prisms and ameloblasts in dental enamel and tooth germs. Acta Odontol Scand 26: 285–336

Fosse G (1968b) A quantitative analysis of the numerical density and the distributional pattern of prisms and ameloblasts in dental enamel and tooth germs. Acta Odontol Scand 26: 409–433

Fosse G (1968c) A quantitative analysis of the numerical density and the distributional pattern of prisms and ameloblasts in dental enamel and tooth germs. Acta Odontol Scand 26: 501–603

Frank RM (1968) Etude ultrastructurale de la dentinogenèse et de la amélogenèse. Thèse présentée pour le doctorat en chirurgie dentaire, Faculté de Médecine de Strasbourg, France

Frazier PD (1968) Adult Human Enamel: An electron microscopic study of crystallite size and morphology. J Ultrastruct Res 22: 1–11

Gantt DG (1979) A method of interpreting enamel prism patterns. SEM/1979/II: 975–981

Gantt DG, Pilbeam DR, Steward GP (1977) Hominoid enamel prism patterns. Science 198: 1155–1157

Garant PR (1972) The demonstration of complex gap junctions between the cells of the enamel organ with lanthanum nitrate. J Ultrastruct Res 40: 333–348

Garant PR, Nagy A, Cho MI (1983) A freeze fracture study of ruffle ended post secretory ameloblasts. J Dent Res 63: 622–628

Garant PR, Sasaki T, Colflesh DE (1988) Na – K – ATPase in enamel organ: Its location and roles in enamel maturation. Adv Dent Res (in press 1988)

Gasser RF, Scheving LE, Pauly JE (1972) Circadian rhythms in the cell division of the inner enamel epithelium and in the uptake of 3H-thymidine by the root tip of rat incisors. J Dent Res 51: 740–746

Glas JE (1962) The orientation of the apatite crystallites as deduced from X-ray diffraction. Arch Oral Biol 7:91–104

Glas JE, Nylen MU (1965) A correlated electron microscopic and microradiographic study of human enamel. Arch Oral Biol 10:893–908

Glimcher MJ, Daniel EJ, Travis DF, Kamhi S (1965) Electron optical and X-ray diffraction studies of the organisation of the inorganic crystals in embryonic bovine enamel. J Ultrastruct Res 7:1–77

Glimcher MJ, Brickley-Parsons D, Levine PT (1977) Studies of enamel proteins during maturation. Calcif Tissue Res 24:259–270

Goldberg M, Sasaki T (1985) Intramembrane particle distribution on the plasma membrane of ruffle-ended and smooth ended maturation ameloblasts of the rat incisors. J Biol Buccale 13:251–260

Goldberg M, Genotelle-Septier, Molon-Noblot M, Weill R (1979) Maturation tardive de l'émail dentaire humain. J Biol Buccale 7:353–363

Goldberg M, Carreau JP, Arends J (1987) Biochemical and scanning electron microscope study of lipids chloroform methanol extracted from unerupted and erupted human tooth enamel. Arch Oral Biol 32:765–772

Graver HT, Herold RC, Chung TY, Christner PJ, Pappas C, Rosenbloom (1978) Immunofluorescent localisation of amelogenins in developing bovine teeth. Dev Biol 63:390–401

Grine FE, Krause DW, Fosse G, Jungers WL (1987) Analysis of individual, intraspecific and interspecific variability in quantitative parameters of caprine tooth enamel structure. Acta Odontol Scand 45:1–23

Grove CA, Judd G, Ansell GS (1972) Determination of hydroxyapatite crystallite size in human dental enamel by dark-field electron microscopy. J Dent Res 51:22–27

Gustafson AG (1955) The similarity between contralateral pairs of teeth. Odont Tidskr 63:245–248

Gustafson AG (1959) A morphologic investigation of certain variations in the structure and mineralisation of human dental enamel. Odont Tidskr 67:361–472

Gustafson G (1945) The structure of human dental enamel. A histological study by means of incident light, polarized light, phase contrast microscopy, fluorescence microscopy and micro-hardness tests. Odont Tidskr [Suppl] 53:1–150

Gustafson G (1957) The histopathology of caries of human dental enamel. Acad Odont Scand 15:13–55

Gustafson G, Gustafson AG (1961) Human dental enamel in polarized light and contact microradiography. Acta Odont Scand 19:259–287

Gustafson G, Gustafson AG (1967) Microanatomy and histochemistry of enamel. In: Miles AEW (ed) Structural and chemical organisation of teeth, vol 2. Academic, New York, pp 75–134

Gustafson G, Gustafson AG (1968) A new concept of dental enamel structure and formation. Odont Revy 19:265–270

Gustavsen F, Silness J (1969) Crystal shape in the prism sheath region of sound human enamel. Acta Odont Scand 27:617–629

Gysi A (1931) Metabolism in adult enamel. The Dental Digest 37:661–668

Hallsworth AS, Robinson C, Weatherell JA (1972) Mineral and magnesium distribution within the approximal carious lesion of dental enamel. Caries Res 6:156–168

Hallsworth AS, Weatherell JA, Robinson C (1973) Loss of carbonate during the first stages of enamel caries. Caries Res 7:345–348

Hals E (1957) Hypocalcification of the enamel. Investigation of three cases. Acta Odont Scand 15:177–198

Hammarlund-Essler E (1958) A microradiographic-microphotometric and X-ray diffraction study of human developing enamel. Trans Roy Schools Dentistry Stockholm & Umea 4:15–25

Helmcke JG (1953) Atlas des menschlichen Zahnes im elektronenmikroskopischen Bild. 1. Teil, Histologie des normalen Zahnes. Berlin, Transmare Photo GMBH

Helmcke JG (1955) Elektronenmikroskopische Strukturuntersuchungen an gesunden und kranken Zähnen. Dtsch Zahnärztl Z 10:1461–1478

Helmcke JG, Schulz L, Scott DB (1963) Querstreifung der menschlichen Schmelzprismen. Dtsch Zahnärztl Z 18:569–637

Hinrichsen CFL, Engel MB (1966) Fine structure of partially demineralised enamel. Arch Oral Biol 2:65–93

Hodde KC, Boyde A, Reid SA, Reith EJ, Schmid MJ (1983) The vascular architecture of the rat incisor enamel organ. Beitr Elektronemikroskop Direktabb Oberfl 16:431–443

Hunter J (1770) The natural history of the human teeth: explaining their structure, use, formation, growth and diseases. Johnson, London

Huszar G (1971) Observations sur l'épaisseur de l'émail. Bull Group Int Rech Sci Stomatol Odontol 14:155–167

Ishiyama M (1984) Comparative histology of tooth enamel in several toothed whales. In: Fearnhead RW, Suga S (eds) Tooth enamel IV. Elsevier, Amsterdam, pp 432–436

Ishiyama M (1987) Enamel structure in odontocete whales. Scanning Microsc 1:1071–1079

Iwaku F, Ozawa H (1979) Blood supply of the rat periodontal space during amelogenesis as studied by the injection replica SEM method. Arch Histol Jpn 42:81–88

Jayasinghe JAP (1987) A study of the incidence and the histology of brochs. M.Sc. Thesis, University College London

Johansen E (1964) Microstructure of enamel and dentin. J Dent Res 6:1007–1020

Johnson NW (1967) Some aspects of the ultrastructure of early human enamel caries seen with the electron microscope. Arch Oral Biol 12:1505–1521

Jones SJ (1972a) The tooth surface in periodontal disease. The Dental Practitioner and Dental Record 22:462–473

Jones SJ (1972b) Calculus on human teeth. Apex (Journal of the Dental School of University College Hospital, London) 6:1–5

Jones SJ (1974) Ph.D. Thesis, available from Senate House Library, University of London

Jones SJ, Boyde A (1974) Coronal cementogenesis in the horse. Arch Oral Biol 19:605–614

Jones SJ, Boyde A (1987) Scanning microscopic observations on dental caries. Scanning Electron Microsc 1:1991–2002

Jongebloed WL, Molenaar I, Arends J (1975) Morphology and size distribution of sound and acid-treated enamel crystals. Calcif Tissue Res 19:109–123

Josephsen K (1984) Lanthanum tracer study of permeability of ameloblast junctional complexes in maturation zone of rat incisor enamel organ. In: Fearnhead RW, Suga S (eds) Tooth enamel IV. Elsevier, Amsterdam, pp 251–255

Josephsen K, Fejerskov O (1977) Ameloblast modulation in the maturation zone of the rat incisor enamel organ. A light and electron microscope study. J Anat 124:45–70

Kakei M, Nakahara H (1983a) A light microscopic study of the localization of carbonic anhydrase activity in the developing dentin and enamel of the rat lower incisor. Jap J Oral Biol 25:374–377

Kakei M, Nakahara H (1983b) Immunological relationship between carbonic anhydrase isoenzyme C and immature enamel matrix proteins of the rat incisor. Jap J Oral Biol 25:1125–1128

Kallenbach E (1968) Fine structure of rat incisor ameloblasts during enamel maturation. J Ultrastruct Res 22:90–119

Kallenbach E (1970) Fine structure of rat incisor enamel organ during late pigmentation and regression stages. J Ultrastruct Res 30:38–63

Kallenbach E (1973) The fine structure of Tomes' process of rat incisor ameloblasts and its relationship to the elaboration of enamel. Tissue Cell 5:501–524

Kallenbach E (1974) Fine structure of rat incisor ameloblasts in transition between enamel secretion and maturation stages. Tissue Cell 6:173–190

Kallenbach E (1977) Fine structure of secretory ameloblasts in the kitten. Am J Anat 148:479–512

Kallenbach E (1980) Access of horseradish peroxidase (HRP) to the extracellular spaces of the maturation zone of the rat incisor enamel organ. Tissue Cell 12:165–174

Kallenbach E, Clermont Y, LeBlond CP (1965) The cell web in the ameloblasts of the rat incisor. Anat Rec 153:55–70

Katchburian E, Holt SJ (1972) Studies on the development of ameloblasts. 1. Fine structure. J Cell Sci 11:415–447

Kawai N (1955) Comparative anatomy of the bands of Schreger. Okajimas Folia Anat Jpn 27:115–131

Kollar EJ, Fisher C (1980) Tooth induction in chick epithelium: Expression of quiescent genes for enamel synthesis. Science 207:993–995

Korvenkontio A (1934–35) Mikroskopische Untersuchungen an Nagerincisiven unter Hinweis auf die Schmelzstruktur der Backenzähne. Histologisch-phyletische Studie. Annal Zool Soc Zool-Bot Fenn Vanamo (Helsinki) 2:1–274

Kurahashi Y, Moe H (1969) Electron microscopy of the ameloblasts in the later stage of the matrix formation stage and in the maturation stage of the enamel in rat. In: Araya S (ed) Hard Tissue Research. Ishiyaku, Tokyo, pp 256–285

Langdon DJ, Elliott JC, Fearnhead RW (1980) Microradiographic observation of acidic subsurface decalcification in synthetic apatite aggregates. Caries Res 14:359–366

Lester KS (1965) The bands of Schreger, the role of reflexion. Arch Oral Biol 10:361–371

Lester KS (1971) On the nature of "fibrils" and tubules in developing enamel of the opossum, Didelphis marsupialis. J Ultrastruct Res 30:64–77

Lester KS (1988 – submitted) Procerberus enamel: a missing link

Lester KS, Boyde A (1967a) Some observations on the enamel dentine junction. J Dent Res 46:1286 Abst

Lester KS, Boyde A (1967b) The structure and development of marsupial enamel tubules. Z Zellforsch 82:558–576

Lester KS, Boyde A (1987) Relating developing surface to adult ultrastructure in chiropteran enamel by SEM. Adv Dent Res 1:181–190

Lester KS, Boyde A, Gilkeson C, Archer M (1987) Marsupial and monotreme enamel structure. Scan Electron Microsc 1:401–420

Lenz H (1956) Elektronenmikroskopische Untersuchungen bei beginnender Schmelzcaries. Zahnärztl Rundschau 65:285–289

Linden LA (1968) Microscopic observations of fluid flow through enamel in vitro. Odont Revy 19:1–17

Listgarten MA (1966) Phase-contrast and electron microscopic study of the junction between reduced enamel epithelium and enamel in unerupted human teeth. Arch Oral Biol 11:999–1016

Lumsden AGS (1979) Pattern formation in the molar dentition of the mouse. J Biol Buccale 7:77–103

Lyon DG, Darling AI (1957) Orientation of the crystallites in human dental enamel. Br Dent J 102:483–488

Mannerberg F (1960) Appearance of tooth surface as observed in shadowed replicas. Odont Revy [Suppl] 11:6

Mannerberg F (1964) The incipient carious lesion as observed in shadowed replicas (en face pictures) on the same teeth. Acta Odont Scand 22:343–363

Mannerberg F (1968) Appearance of the tooth surface of teeth showing dental fluorosis as observed by shadowed replicas. Odont Revy 19:271–291

Martin LB (1983) The relationships of the later miocene Hominoidea. Ph.D. Thesis, available from Senate House Library, University of London

Martin L, Boyde A (1984) Rates of enamel formation in relation to enamel thickness in hominoid primates. In: Fearnhead RW, Suga S (eds) Tooth enamel IV. Elsevier, Amsterdam, pp 447–451

Martin L, Boyde A, Grine F (1988) SEM of primate enamel. Scanning Microsc (in press)

Meckel AH, Griebstein WJ, Neal RJ (1965) Structure of mature human dental enamel as observed by electron microscopy. Archs Oral Biol 10:775–783

Moe H (1971) Morphological changes in the infranuclear portion of the enamel producing cells during their life cycle. J Anat 108:43–66

Moe H (1981) Adaptation of arterioles to moving capillaries. Acta Anat (Basel) 109:369–377

Mummery JH (1916) On the structure and arrangement of the enamel prisms, especially as shown in the enamel of the elephant. Proc Roy Soc Med 9:121–138

Nalbandian J, Frank RM (1962) Microscopie électronique des gaines, des structures prismatiques et interprismatiques de l'émail foetal humain. Bull Group Int Rech Sci Stomatol 5:523–542

Nanci A, Warshawsky H (1984a) Characterization of putative secretory sites on ameloblast of the rat incisor. Am J Anat 171:163–189

Nanci A, Warshawsky H (1984b) Relationship between the quality of fixation and the presence of stippled material in newly formed enamel of the rat incisor. Anat Rec 208:15–31

Nanci A, Bendayan M, Slavkin HC (1984) Distribution of enamel protein antigens during mouse incisor amelogenesis as revealed by high resolution immunocytochemistry. In: Fearnhead RW, Suga S (eds) Tooth enamel IV. Elsevier, Amsterdam, pp 141–145

Nanci A, Slavkin HC, Smith CE (1988) Application of high resolution immunocytochemistry to the study of the secretory, resorptive and degradative functions of ameloblasts. Adv Dent Res (in press 1988)

Nasmyth A (1839) Researches on the development, structure and diseases of the teeth. Churchill, London, 165 pp

Newman HN, Poole DFG (1974) Observations with scanning and transmission electron microscopy on the structure of human surface enamel. Arch Oral Biol 19:1135–1143

Nishikawa S, Josephsen K (1987) Cyclic localisation of actin and its relationship to junctional complexes in maturation ameloblasts of the rat incisor. Anat Rec 219:21–31

Nishikawa S, Kitamura H (1983) Actin filaments in the ameloblast of the rat incisor. Anat Rec 207:245–252

Nylen MU, Eanes ED, Omnell KA (1963) Crystal growth in rat enamel. J Cell Biol 18:109–123

Orams HJ (1976) Ultrastructural study of human dental enamel using selected area argon ion beam thinning. Arch Oral Biol 21:663–675

Osborn JW (1965) The nature of the Hunter-Schreger bands of enamel. Arch Oral Biol 10:929–933

Osborn JW (1968a) Evaluation of previous assessments of prism directions in human enamel. J Dent Res 47:217–222

Osborn JW (1968b) Directions and interrelationship of prisms in cuspal and cervical enamel of human teeth. J Dent Res 47:395–402

Osborn JW (1969) The 3-dimensional morphology of the tufts in human enamel. Acta Anat (Basel) 73:481–495

Osborn JW (1971) The relationship between the optical density of prism borders in dog tooth enamel and the angle from which they are viewed. Arch Oral Biol 16:1055–1059

Osborn JW, Roberts AM (1971) Optical fringe effects at prism borders in human tooth enamel sections. J Microsc 93:123–128

Palamara J, Phakey PP, Rachinger WA, Orams HJ (1981) Electron microscope study of the dentine enamel junction of kangaroo teeth using selected area argon ion beam thinning. Cell Tissue Res 221:405–419

Petran M, Hadravsky M, Boyde A (1985) The tandem scanning reflected light microscope. Scanning 7:97–108

Phakey PP, Orams HJ, Palamara J, Rachinger WA (1984) Macropodinae enamel ultrastructure. In: Fearnhead RW, Suga S (eds) Tooth enamel IV. Elsevier, Amsterdam, pp 452–456

Poole DFG (1956) The structure of the teeth of some mammal-like reptiles. Quart J Microsc Sci 97:303–312

Poole DFG (1966) The use of the microscope in dental research. Br Dent J 121:71–79

Poole DFG, Brooks AW (1961) The arrangement of crystallites in enamel prisms. Arch Oral Biol 5:14–26

Poole DFG, Johnson NW (1967) The effects of different demineralizing agents on human enamel surfaces studied by scanning electron microscopy. Arch Oral Biol 12:1621–1634

Reid SA, Boyde A, Reith EJ (1984) Cyclical phenomena occurring during the maturation of the enamel of rat incisor teeth. Histochemistry 81:521–524

Reith EJ (1961) The ultrastructure of ameloblasts during matrix formation and the maturation of enamel. J Biophys Biochem Cytol 9:825–840

Reith EJ (1963) The ultrastructure of ameloblasts during early stages of maturation of enamel. J Cell Biol 18:691–696

Reith EJ (1970) The stages of amelogenesis as observed in molar teeth of young rats. J Ultrastruct Res 30:111–151

Reith EJ, Boyde A (1981a) The arrangement of ameloblasts on the surface of maturing enamel of the rat incisor tooth. J Anat 133:381–388

Reith EJ, Boyde A (1981b) Autoradiographic evidence of cyclical entry of calcium into maturing enamel of the rat incisor tooth. Arch Oral Biol 26:983–987

Reith EJ, Boyde A (1985) The pyroantimonate reaction and transcellular transport of calcium in the rat molar enamel organ. Histochemistry 83:539–543

Reith EJ, Cotty VF (1962) Autoradiographic studies on calcification of enamel. Arch Oral Biol 7:365–372

Reith EJ, Ross MH (1973) Morphological evidence for the presence of contractile elements in the secretory ameloblast of the rat. Arch Oral Biol 18:445–448

Reith EJ, Boyde A, Schmid MJ (1982) Correlation of rat incisor ameloblasts with maturation cycles as displayed on enamel surface with EDTA. J Dent Res 61:1563–1573

Reith EJ, Schmid MJ, Boyde A (1984) Rapid uptake of calcium in maturing enamel of rat incisor. Histochemistry 80:409–410

Rensberger MJ (1978) Scanning electron microscopy of wear and occlusal events in some small herbivores. In: Butler PM, Joysey KA (eds) Development, function and evolution of teeth. Academic, London, pp 416–438

Rensberger JM, Koenigswald WV (1980) Functional and phylogenetic interpretation of enamel microstructure in rhinoceroses. Paleobiology 6:477–495

Retzius A (1836) Mikroskopiska undersokningar ofver tandernas sardeles tandbenets struktur. Kongl Vetenskaps Acad Handlinger (Stockholm) [Ar] 1836:52–140

Ripa LW, Gwinnett AJ, Buonocore MG (1966) The prismless outer layer of deciduous and permanent enamel. Arch Oral Biol 11:41–48

Ripa LW, Gwinnett AJ, Buonocore MG (1967) The prismless enamel surface. Microscopy with polarized light. Dent Radiogr Photogr 40:38–39

Risnes S, Fosse G (1974) The origin of marsupial enamel tubules. Acta Anat (Basel) 87:275–282

Robinson C, Kirkham J (1984) Enamel matrix components: alterations during development and possible interactions with the mineral phase. In: Fearnhead RW, Suga S (eds) Tooth enamel IV. Elsevier, Amsterdam, pp 261–265

Robinson C, Weatherell JA, Hallsworth AS (1971) Variations in the composition of dental enamel within thin ground tooth sections. Caries Res 5:44–57

Robinson C, Lowe NR, Weatherell JA (1975) Amino acid composition, distribution and origin of tuft protein in human and bovine dental enamel. Arch Oral Biol 20:29–42

Robinson C, Lowe NR, Weatherell JA (1977) Changes in amino-acid composition of developing rat incisor enamel. Calcif Tissue Res 23:19–31

Robinson C, Fuchs P, Deutsch D, Weatherell JA (1978) Four chemically distinct stages in developing enamel from bovine incisor teeth. Caries Res 12:1–11

Robinson C, Briggs HD, Atkinson PJ, Weatherell JA (1979) Matrix and mineral changes in developing enamel. J Dent Res [Spec Issue B] 58:871–880

Robinson C, Briggs HD, Atkinson PJ (1981a) Histology of enamel organ and chemical composition of adjacent enamel in rat incisors. Calcif Tissue Int 33:513–520

Robinson C, Briggs HD, Atkinson PJ, Weatherell JA (1981b) Chemical changes during formation and maturation of human deciduous enamel. Arch Oral Biol 26:1027–1033

Robinson C, Weatherell JA, Hallsworth AS (1981c) Distribution of magnesium in mature human enamel. Caries Res 15:70–77

Robinson C, Hallsworth AS, Kirkham J (1984) Distribution and uptake of magnesium by developing bovine incisor enamel. Archs Oral Biol 29:479–482

Romaniuk K, Shroff FR (1965) The relationship of directional variation in the terminal portions of enamel rods to their cross sectional appearance. NZ Dent J 61:94–99

Ronnholm E (1962) An electron microscopic study of the amelogenesis in human teeth. J Ultrastruct Res 6:229–303

Rosser H, Boyde A, Stewart ADG (1967) Preliminary observations of the calcium concentration in developing enamel assessed by scanning electron probe X-ray emission micro analysis. Arch Oral Biol 12:431–440

Ruch JV, Karcher-Djuricic V (1971) Mise en evidence d'un role specifique de l'epithelium adamantin dans la differenciation et le maintien des odonblastes. Ann Embryol Morph 4:359–366

Sahni A (1984) The evolution of mammalian enamels: evidence from multituberculata (Allotheria, extinct); primitive whales (Archaeocete Cetacea) and early rodents. In: Fearnhead RW, Suga S (eds) Tooth Enamel IV. Elsevier, Amsterdam, pp 457–461

Sasaki T (1983) Ultrastructure and cytochemistry of the Golgi apparatus and related organelles of the secretory ameloblasts of the rat incisor. Arch Oral Biol 28:895–905

Sasaki T (1984a) Morphology and function of maturation ameloblasts in kitten tooth germs. J Anat (Basel) 138:333–342

Sasaki T (1984b) Endocytotic pathways at the ruffled borders of rat maturation ameloblasts. Histochemistry 80:263–268

Sasaki T (1984c) Tracer, cytochemical, and freeze-fracture study of the mechanisms whereby secretory ameloblasts absorb exogeneous proteins. Acta Anat (Basel) 118:23–33

Sasaki T, Garant T (1987) Mitochondrial migration and Ca–ATPase modulation in secretory ameloblasts of fasted and calcium loaded rats. Am J Anat 179:116–130

Sasaki T, Higashi S (1983) Scanning and transmission electron microscopy of developing enamel surfaces in the kitten tooth germs. J Electron Microsc (Tokyo) 32:163–171

Sasaki T, Higashi S, Tachikawa T, Yoshiki S (1981) Morphogenesis of gap junctions in rat amelogenesis. J Electron Microsc (Tokyo) 30:191–197

Sasaki T, Higashi S, Tachikawa T, Yoshiki S (1983) Thin section, tracer and freeze fracture study of the smooth ended maturation ameloblasts in rat incisors. Acta Anat (Basel) 117:303–313

Sasaki T, Higashi S, Tachikawa T, Yoshiki S (1984) Absorptive and digestive functions of maturation ameloblasts in rat incisors. In: Fearnhead RW, Suga S (eds) Tooth enamel IV. Elsevier, Amsterdam, pp 266–270

Sasaki T, Colflesh DE, Garant PR (1988a) Regulation of Ca transport by Ca−ATPase-calmodulin complex in the enamel organ. Adv Dent Res (in press 1988)

Sasaki T, Yeh JH, Takagi T (1988b) Isolation of two amelogenin peptides and their amino acid sequences. Adv Dent Res (in press 1988)

Sauk JJ, Vickers RA, Copeland JS, Lyon HW (1972) The surface of genetically determined hypoplastic enamel in human teeth. Oral Surg Oral Med Oral Pathol 34:60–68

Schour I, Hoffman MM (1939) Studies in tooth development. II. The rate of apposition of enamel and dentin in man and other animals. J Dent Res 18:161–175

Schreger D (1800) Beiträge zur Geschichte der Zähne. Beitr vergleich Zergliederungskunst 1:1–20

Scott DB (1952) Microscopic studies of dental tissues. II. Optical microscopy of tooth surface. Oral Path 8:638–645

Scott DB, Wyckoff RWG (1947) Electron microscopy of tooth structure by the shadowed collodion replica method. US Pub Hlth Reps (Washington) 62:1513–1516

Selvig KA (1973) Electron microscopy of dental enamel: analysis of crystal lattice images. Z Zellforsch 137:271–280

Selvig KA, Halse A (1972) Crystal growth in rat incisor enamel. Anat Rec 153:453–468

Shellis RP, Poole DFG (1979) The arrangement of prisms in the enamel of the anterior teeth of the aye-aye. Scan Electron Microsc II:497–506

Shellis RP (1984a) Inter-relationships between growth and structure of enamel. In: Fearnhead RW, Suga S (eds) Tooth enamel IV. Elsevier, Amsterdam, pp 467–471

Shellis RP (1984b) Variations in growth of the enamel crown in human teeth and a possible relationship between growth and enamel structure. Arch Oral Biol 29:697–705

Shellis RP, Hallsworth AS (1987) The use of scanning electron microscopy in studying enamel caries. Scan Electron Microsc 1:1109–1123

Shobusawa M (1952) Vergleichende Untersuchungen über die Form der Schmelzprismen der Säugetiere. Okajimas Folia Anat Jpn 24:371–392

Shroff FR, Romaniuk K (1964) A preliminary investigation of the surface structure of the enamel of erupted deciduous teeth. NZ Dent J 60:298–305

Silness J (1969) Some variations in the microradiographic appearance of human deciduous enamel. Odont Revy 20:93–110

Simmelink JW, Nygaard VK (1982) Ultrastructure of striations in carious human enamel. Caries Res 16:179–188

Skobe Z (1976) The secretory stage of amelogenesis in rat mandibular incisor teeth observed by scanning electron microscopy. Calcif Tissue Res 21:83–103

Skobe Z (1977) Enamel rod formation in the monkey observed by scanning electron microscopy. Anat Rec 187:329–334

Skobe Z, Prostak K, Stern D (1981) Ultrastructure of secretory ameloblasts in a monkey Macaca mulatta. Arch Oral Biol 26:1075–1090

Slavkin CH (1970) Epithelial mesenchymal interactions related to periodontal disease. J Periodont Res 41:5–13

Slavkin HC, Brownell AG, Bringas P, MacDougall M, Bessem C (1983) Basal lamina persistence during epithelial mesenchymal interactions in murine tooth development in vitro. J Craniofac Genet Dev Biol 3:387–407

Smith CE, Warshawsky H (1977) Quantitative analysis of cell turnover in the enamel organ of the rat incisor. Anat Rec 187:63–98

Smith CE, McKee MD, Nanci A (1988) Rapid ameloblast modulation cycles. Adv Dent Res (in press)

Smreker E (1905) Über die Form der Schmelzprismen menschlicher Zähne und die Kittsubstanz des Schmelzes. Arch Mikr Anat 66:312–331

Suga S (1959) Amelogenesis. Histological and histochemical observations. Int Dent J 9:394–420

Suga S, Murayama Y, Musashi T (1970) A study of the mineralization process in the developing enamel of guinea pigs. Arch Oral Biol 15:597–612

Sundstrom B (1966) Schreger bands and their appearance in microradiographs of human dental enamel. Acta Odont Scand 24:179–194

Sundstrom B, Zelander T (1968) On the morphological organisation of the organic matrix of adult human enamel after decalcification by means of a basic chromium (III) sulphate solution. Odont Revy 19:1–15

Swancar JR, Scott DB, Njemirovskij Z (1970) Studies on the structure of human enamel by the replica method. J Dent Res 49:1025–1033

Takano Y (1984) Remarks in discussion and fig 2:3.9. In: Fearnhead RW, Suga S (eds) Tooth enamel IV. Elsevier, Amsterdam, p 312

Takano Y, Crenshaw MA (1980) The penetration of intravascularly perfused lanthanum into the ameloblast layer of developing rat molar teeth. Arch Oral Biol 25:505–511

Takano Y, Ozawa H (1980) Ultrastructural and cytochemical observations on the alternating morphologic changes of the ameloblasts at the stage of enamel maturation. Arch Histol Jpn 43:385–399

Takano Y, Ozawa H (1984) Autoradiographic and tracer experiments on the exit route for the resorbed organic matrix of the enamel at the stage of maturation. In: Fearnhead RW, Suga S (eds) Tooth enamel IV. Elsevier, Amsterdam, pp 271–275

Takano Y, Crenshaw MA, Reith EJ (1982a) Correlation of Ca incorporation with maturation ameloblast morphology in the rat incisor. Calcif Tissue Int 34:211–213

Takano Y, Crenshaw MA, Bawden JW, Hammarstrom L, Lindskog S (1982b) The visualisation of patterns of ameloblast modulation by the glyoxal bis (2-hydroxyanil) staining method. J Dent Res 61:1580–1586

Thesleff I, Stenman S, Vaheri A, Timpl R (1979) Changes in the matrix proteins, fibronectin and collagen, during differentiation of mouse tooth germ. Dev Biol 70:116–126

Tobin CE (1972) Correlation of vascularity with mineralisation in human fetal teeth. Anat Rec 174:371–380

Tomes J (1849) Structure of the dental tissues of marsupial animals. Philos Trans R Soc Lond 139:403–412

Tomes J (1850) On the structure of the dental tissues of the order Rodentia. Philos Trans R Soc Lond 140:529–567

Triller M (1984) Fluorosis: A model to study enamel lesions. In: Fearnhead RW, Suga S (eds) Tooth enamel IV. Elsevier, Amsterdam, pp 368–372

Vahl J, Plackova A (1967) Elektronenoptische Untersuchungen von braunen Schmelzflecken (arretierte Karies). Dtsch Zahnärztl Z 22:629–630

Vahl J, Höhling HJ, Plackova A, Bures H (1966) Elektronenmikroskopische Ultradünnschnittuntersuchungen an Zähnen mit Schmelzflecken, herrührend von initialer Karies, artifizieller Karies und Mineralisationsstörungen. Dtsch Zahnärztl Z 21:983–989

Vollrath L, Kantarjian A, Howe C (1975) Mammalian pineal gland: 7 day rhythmic activity? Experientia 31:458–460

Vrba ES, Grine FE (1978) Australopithecine enamel prism patterns. Science 202:890–892

Wakita M, Hinrichsen K (1980) Ultrastructure of the ameloblast-stratum intermedium border during ameloblast differentiation. Acta Anat (Basel) 108:10–29

Wakita M, Kobayashi S (1983) The three dimensional structure of Tomes' processes and the development of the microstructural organization of tooth enamel. In: Suga S (ed) Mechanisms of tooth enamel formation. Quintessence, Tokyo, pp 65–89

Wakita M, Tsuchiya H, Gunji T, Kobayashi S (1981) Three-dimensional structure of Tomes' processes and enamel prism formation in the kitten. Arch Histol Jpn 44:285–297

Walker A, Hoeck HN, Perez L (1978) Microwear of mammalian teeth as an indicator of diet. Science 201:908–910

Warshawsky H (1968) The fine structure of secretory ameloblasts in rat incisors. Anat Rec 161:211–230

Warshawsky H (1979) Radioautographic studies on amelogenesis. J Biol Buccale 7:105–126

Warshawsky H, Bai P (1983) Knife chatter during thin sectioning of rat incisor enamel can cause periodicities resembling cross-striations. Anat Rec 207:533–538

Warshawsky H, Josephsen K, Thylstrup A, Fejerskov A (1981) The development of enamel structure in rat incisors as compared to the teeth of monkey and man. Anat Rec 200:371–399

Warshawsky H, Bai P, Nanci A (1984a) Lack of evidence for rhythmicity in enamel development. INSERM 125:241–256

Warshawsky H, Bai P, Nanci A, Josephsen K (1984b) Morphological visualisation of two categories of enamel proteins in relation to the crystals of rat incisor enamel. In: Fearnhead RW, Suga S (eds) Tooth Enamel IV. Elsevier, Amsterdam, pp 177–182

Watson ML (1960) The extracellular nature of enamel in the rat. J Biophys Biochem Cytol 7:489–492

Watson ML, Avery JK (1954) The development of the hamster lower incisor as observed by electron microscopy. Am J Anat 95:109–161

Weatherell JA, Weidmann SM, Hamm SM (1967) Density patterns in enamel. Caries Res 1:42–51

Weatherell JA, Robinson C, Hiller CR (1968) Distribution of carbonate in thin sections of dental enamel. Caries Res 2:1–9

Weber DF (1973) Sheath configurations in human cuspal enamel. J Morphol 14:479–480

Weber DF, Eisenmann DR (1971) Microscopy of the neonatal line in developing human enamel. Am J Anat 132:375–392

Weber DF, Glick PL (1975) Correlative microscopy of enamel prism orientation. Am J Anat 144:407–420

Weber DF, Eisenmann DR, Glick PL (1974) Light and electron microscopic studies of Retzius lines in human cervical enamel. Am J Anat 141:91–104

Weill R, Tassin MT (1965) Etude histochimique de la matrice de l'émail histogenèse chez le rat. Acat Histochem 22:259–282

Weinstock A, Leblond CP (1971) Elaboration of the matrix glycoprotein of enamel by the secretory ameloblasts of the rat incisor as revealed by radioautography after galactose-H^3 injection. J Cell Biol 51:26–51

Whittaker DK (1977) The enamel-dentine junction of human and *Macaca irus* teeth: a light and electron microscopic study. J Anat 125:323–335

Wolf J (1940) Plastische Histologie der Zahngewebe. Deutsche Zahn Mund und Kieferheilkunde 7:507–538

Wolf J (1942) Der Einfluss der Ameloblastenverschiebungen auf die Gestalt und den Verlauf der Schmelzprismen. Deutsche Zahn Mund und Kieferheilkunde 9:488–515

Young MF, Shimokawa HS, Sobel ME, Termine JD (1988) A characterisation of amelogenin messenger RNA in the bovine tooth germ. Adv Dent Res (in press 1988)

Special Aspects of Biomineralization of Dental Tissues

H.J. Höhling

A. Does a Correlation Exist Between the Fundamental Processes of Enamel, Dentine, and Cementum Mineralization Which Might Lead to a General Calcification Theory?

As seen in earlier contributions to this volume concerning enamel, dentine, and cementum formation (and also considering bone and cartilage mineralization in the epiphyseal growth plate), it is clear that distinct differences exist in hard tissue formation in these various systems. These arise, e.g., from changes in the expression of the enamel cells (the ameloblasts) and the dentine cells (the odontoblasts) – though both are prismatically shaped, and densely arranged in parallel. Morphology at the electron-microscopical level shows important variations in crystal nucleation, growth, and arrangement of the crystallites into larger groups. In enamel so-called prisms exist with a diameter in the range of 4–6 µm and long, prismatically shaped crystallites perfectly arranged in parallel. In dentine, however, there is not such a precise parallel arrangement of crystals over such large areas; the much smaller needle- and ribbonlike elongated crystallites are arranged in parallel only on and in the 50–100-nm-thick collagen fibers of type I. Over wide reaches, while the collagen fibers in dentine form an irregular network, the collagen fibers in cementum are partly arranged in parallel to form bundles, thus leading to a parallel arrangement of the crystallites. (Since it is more important to clarify the relationship and correspondences between enamel and dentine, the collagen mineralization of cementum will not be discussed here, although it shows great similarities to that of dentine.)

Those who deny important correlations between the fundamental processes of enamel and dentine mineralization would point at the cellular differences and at those in the crystal arrangement, and at the so-called matrix vesicles (MV; collagen-free, membrane-surrounded entities, $\varnothing \sim 0.1$ µm) as primary sites of crystal formation in the collagen-rich, hard tissues. The MV appear in the predentine of the mantle dentine in the first stage of mineral formation, as was discussed in preceding papers, and are assumed to induce the Ca-phosphate mineralization. In enamel, mineralization starts directly in the vicinity of the distal ends of the ameloblasts and their processes, the so-called Tomes' processes, and MV do not arise.

This primary mineralization function of the MV has been regarded as so fundamental that BERNARD and MARVOSO (1981) classify hard-tissue systems into those which contain MV (*primary* hard-tissue formation, including mantle dentine, woven bone, growth plate cartilage) and those which do not (*secondary,* subsequential mineralization, e.g., circumpulpal dentine, cementum, enamel, lamellar bone).

At this point it should be stressed that the MV are important structures for original mineral nucleation in those hard-tissue systems possessing a relatively broad, unmineralized matrix in the vicinity of the hard-tissue cells (e.g., predentine, osteoid, the longitudinal septa in the epiphyseal growth plate). The MV may be regarded as important "local factors", markers at which mineralization starts and from which it spreads in all directions so that large compartments gradually become systematically mineralized.

In spite of this there are many specialists who are of the opinion that the fundamental extracellular processes leading to crystal nucleation and crystal growth are so alike that this similarity is more important than any differences in the larger-scale arrangement. They believe that one of the most significant tasks for the future is to understand the process of crystal nucleation and growth in these different hard tissues. Afterwards one would be able to develop a general calcification theory encompassing the fundamental events identical in the development.

It will thus be the aim of this chapter to describe and compare the crystal nucleation, growth, and arrangement of enamel and dentine (also in relation to cementum, bone, and growth plate cartilage as far as is necessary). To try to formulate a basis for a theory, it will be necessary to discuss transport of the Ca and phosphate groups, the main constituents of the prevailing mineral apatite, and the nature of the important matrix macromolecules in enamel and dentine. By matrix are meant those macromolecules which have the capacity to bind Ca and phosphate groups into unstable Ca-phosphate clusters from which nuclei will develop and subsequently form into stable entities. They might continue to grow and coalesce with neighboring entities to form the final crystallites visible under the electron microscope. Nuclei are the smallest entities of a developing crystal which are still in an unstable equilibrium with their surroundings, attaining stability as further ions are added.

Further, the structural arrangement of the different matrix macromolecules will have to be discussed, i.e., what is their crystalline or paracrystalline structure? Paracrystals are crystals in which the periodically repeating atoms – or groups of atoms – are displaced from the average atomic distance in such a way that long-range order is destroyed.

The amino-acid sequence analysis of the proteinaceous part of the matrix macromolecules must be searched for special regions existing in a periodic manner along the protein chain. These could be regarded as active sites for Ca-phosphate nucleation, that is, microregions along the chain which preferentially bind Ca and/or phosphate groups to induce nucleation. From the crystalline structure one could obtain rough or even relatively precise values for the spatial distances between such active sites. If the crystalline structure has not yet been evaluated the nature of the amino-acid sequence might give an idea of the arrangement between assumed active sites.

B. Fundamental Aspects of Enamel Mineralization

I. Mineralization in an Ordered, Ionotropic Gel

Figure 1 is a cross section through prisms of the "key hole" type of young enamel (see also Fig. 2). It is difficult or impossible to cut ultrathin sections of embedded, mature, sound enamel without "shattering," i.e., without changing the original arrangement and orientation of the crystallites. It is also difficult to obtain unshattered, ultrathin sections of young enamel. Nevertheless, the thin crystallites nicely outline the characteristic shape of prisms. One can see that the prisms possess characteristic borders over most areas while remaining "open" to one side.

The real arrangement of the crystallites in relation to the distal ends of the cells, the Tomes' processes with their asymmetric, "noselike" appearance, is shown in Fig. 2. The left-hand drawing (A) presents a cross section through the "hexagonally" drawn ameloblasts (Am; dotted lines) and through the corresponding enamel prisms (Pr) (quadratic hatching). One has to consider that

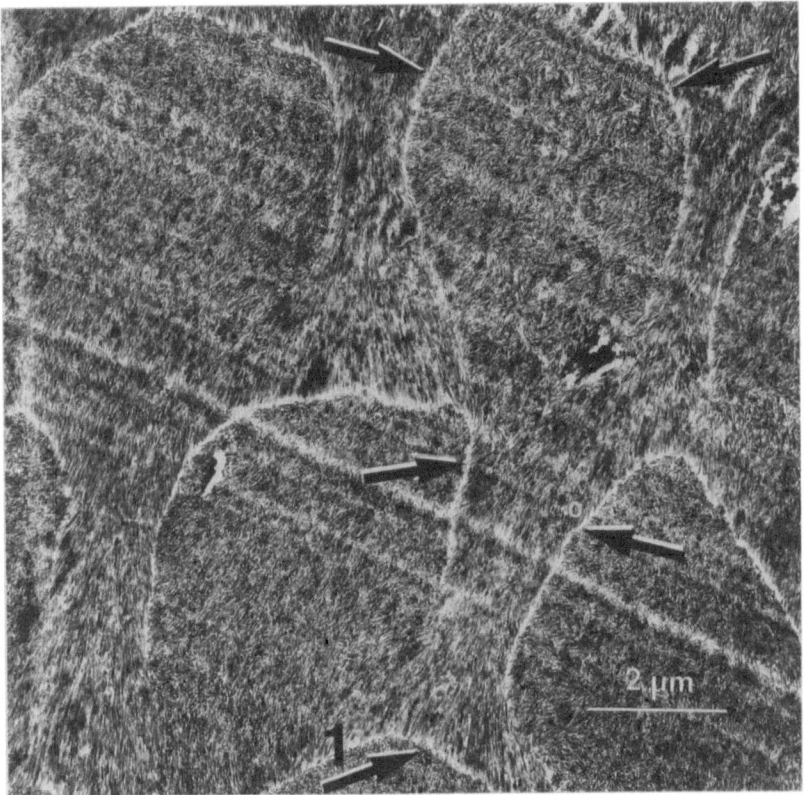

Fig. 1. Electron micrograph of a cross section through enamel prisms of the key-hole type (Fig. 12 of HÖHLING 1966). The *arrows* point to the border of a key-hole prism; *arrow 1* points to the "open" end of the key-hole prism

the ameloblasts lie above and the enamel prisms below the drawing plane (as can be seen in the side view, right-hand drawing B).

In the cross section (A) of drawing (a) the ameloblasts (AM) are arranged in a so-called hexagonal dense packing. They "produce" or contribute to the formation of a large enamel prism (Pr) in the shape of a key-hole, known as the "key-hole" type, with an open border to the bottom. This type is visible in Fig. 1. In drawing (b) the AM are arranged in parallel rows from top to bottom, producing the shape of a horseshoe, known as the "horseshoe" type, with an open border to the bottom. In Fig. 2c a roundish, "closed" type is shown, which, roughly speaking, may develop when the Tomes' processes have not yet reached their asymmetric form at the very beginning of enamel formation, or after they have lost it at the end of enamel formation.

As mentioned before, the right-hand drawing (Fig. 2B) gives the side view or longitudinal section through the "noselike" Tomes' processes (AF) and the enamel prisms (Pr). This side view is not given in Fig. 2c since the asymmetric Tomes' process has not yet developed. One can see that the enamel prisms are lying beneath the Tomes' processes, which protrude into the developing enamel prisms (Pr). The apatite crystallites which are arranged in parallel are shown as lines (visible only under the electron microscope). A characteristic morphological correlation exists between the outlines of the Tomes' processes and those of the enamel prisms (as found by BOYDE 1964), and the elongated hexagonal apatite crystallites with their crystallographic c-axis in the direction of the lines by which they are represented in Fig. 2B. On the elongated concave side (arrow 2) of the Tomes' processes in Fig. 2a, b, the long crystallites are arranged more or less perpendicular to the process membrane (AF); on the shorter "convex" side (arrow 1) the long crystallites are arranged more obliquely. These variations in orientation produce the enamel prism borders, which follow the elongation of the "tip" of the noselike Tomes' processes (dotted line, arrow 3). As mentioned before, one can see this border line in the cross section in Fig. 1.

While BOYDE (1964) clarified the morphological correspondence between the outlines of the Tomes' processes and the developing enamel prisms, HÖHLING (1966) tried to explain why the characteristic orientation of the elongated crystallites in the long direction of the enamel prisms occurs in this special orientation. Figure 3 shows a group of ameloblasts (A) with their asymmetric noselike Tomes' processes (AP) which retract in direction peripheral to the later tooth surface (arrow 1). They excrete in the opposite direction (arrow 2), partly via the "ameloblastic bodies" of FRANK and NALBANDIAN (1967), the fiberlike matrix macromolecules (O). In the concave region (K) these macromolecules are oriented more or less perpendicularly to the surface of the Tomes' processes, and in the convex region (F) they are oriented in an oblique manner.

HÖHLING (1966) and HÖHLING et al. (1982) assumed that the Ca^{2+} ions (possibly also the Na^+ ions) being pumped out of the cells would stream into the recently extruded gel of the macromolecules. By this influx (and the retraction of the cells in the opposite direction, arrow 1) the fibrous macromolecules would be oriented in the streaming direction, leading to an ionotropic gel (THIELE 1964) by reacting with corresponding negatively charged groups of

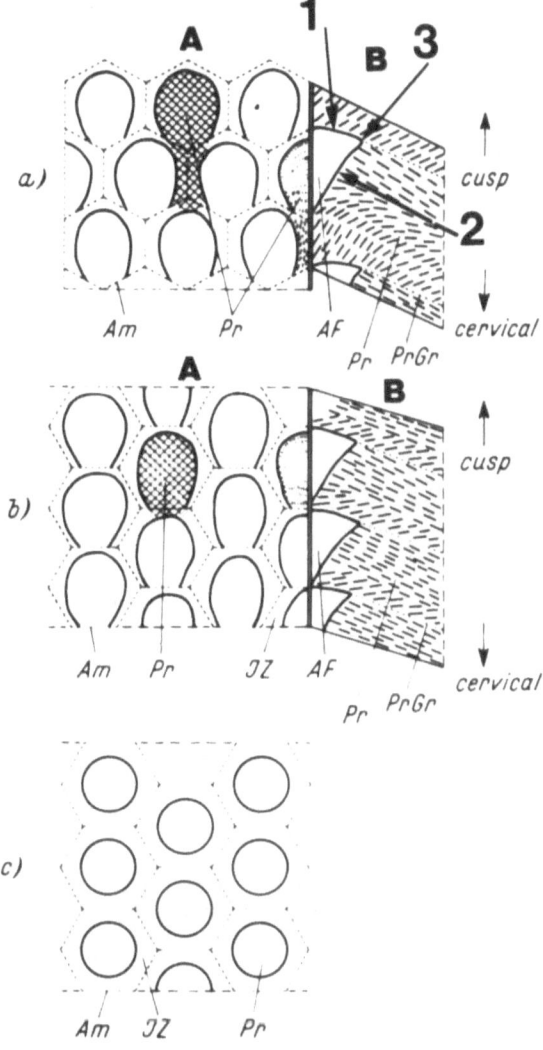

Fig. 2a–c. The local relation of the ameloblastic processes to the enamel prisms and the arrangement of the crystallites (Fig. 20 of Höhling 1966; mainly according to Boyde 1964). **a** The key-hole type of enamel prisms in transverse and longitudinal section. *A* cross section: the cross sections of the cells are shown as hexagons (*broken lines*); *Am*, ameloblast; *Pr*, key-hole enamel prisms. *B* Longitudinal section: the long prismatic crystallites, shown as lines, are mainly arranged perpendicular to the cell border. *AF*, noselike ameloblast process, Tomes' process; *PrGr*, prism border caused by the different orientation of the crystallites due to the noselike, asymmetric ameloblast process. *Arrow 1*, convex side of Tomes' process; *arrow 2*, concave side of Tomes' process; *arrow 3*, tip of Tomes' process. **b** The horseshoe type of enamel prisms in transverse and longitudinal section. *A* Cross section: the ameloblasts (*Am*) are arranged in rows; *Pr*, horseshoe enamel prism; *JZ*, interprismatic region between horseshoe prisms. *B* Longitudinal section: an arrangement of the crystallites corresponding to that in **a**; they are less inclined towards the prism long axis at the open side of the prism. **c** Representation of a roundish, "closed" enamel prism (*Pr*) in cross section. The ameloblasts (*Am*) have not yet developed the noselike Tomes' processes (*AF*) shown in **a** and **b**

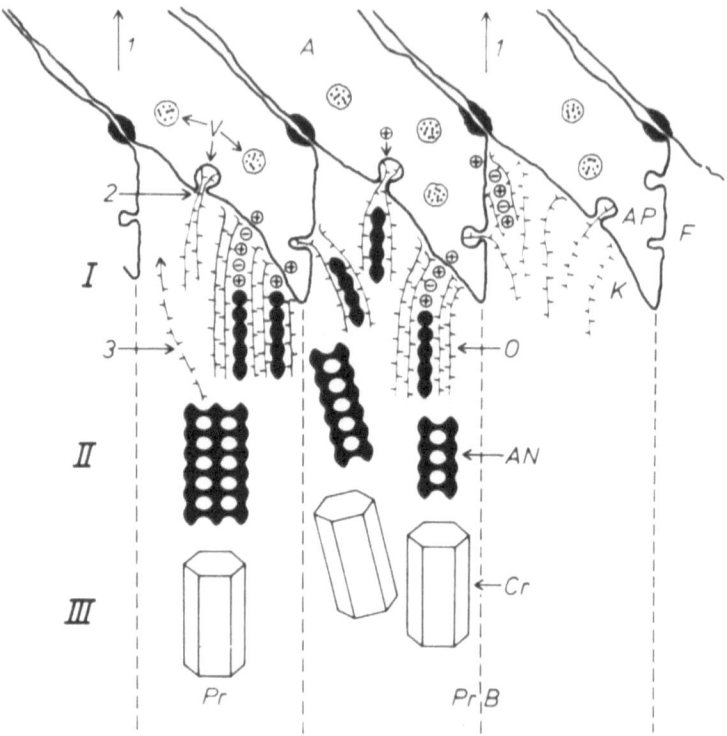

Fig. 3. Approximate scheme of the formation of an ordered ionotropic gel in the enamel matrix and the subsequent ordered mineralization (Fig. 6 of HÖHLING et al. 1982). Ameloblasts (*A*) with their asymmetrical, noselike (*K*, concave; *F*, front) process (*AP*) retract to the later tooth surface (*arrow 1*) and extrude proteins (*O*) in the opposite direction by exocytosis via ameloblastic bodies (*V*), according to Frank and NALBANDIAN (1967) (*arrow 2*). Ca$^+$ ions ($\oplus$) are pumped out by the Ca-ATPase. The ions react with the charged groups of the proteins, leading to parallel arrangement of the molecules; this is followed by ion binding and subsequent crystal formation (region *I*). After degradation and removal (*arrow 3*) of most of the organic substance, the mineral strands (*AN*) may coalesce with neighbors, (region *II*) and finally form larger crystallites (*Cr*, region *III*). The crystallites shown (*Cr*) are about 500–1000 times too thick compared with the diameter of the enamel prisms (*Pr*). *PrB*, prism border

the macromolecules. At the same time, by a process not yet clarified, the inorganic phosphate groups also enter this streaming process and bind to the newly bound Ca groups, to which further Ca ions could bind. This process begins first with unstable Ca-phosphate clusters; these develop into nuclei, which grow into stable Ca-phosphate entities, e.g., the dotlike islands. The islands can form chains (Fig. 3, region I).

The long Ca-phosphate needles and/or ribbons can not develop in one continuous process from one nucleus; instead repetitive active sites must exist along the matrix macromolecule that produce several nuclei from which multiple dotlike islands develop. These grow and coalesce with neighbors along the direction of the crystallographic c-axis to form the long chains or needles (region I). Further, after extensive decomposition and transport of the broken-down macromolecules to the ameloblasts (arrow 3), the apatite chains can also grow perpendicularly to the c-axis and finally coalesce with neighboring apatitic chains to form larger ribbons (dark structures in region II).

However, the question of crystal nucleation and growth has to be discussed later in a separate chapter, since these ideas are not in full agreement with the prevailing ones.

II. Calcium (and Phosphate) Transport

HÖHLING (1966) was the first to propose that the formation of an ordered, ionotropic gel in the extracellular matrix is brought about by the influx of Ca^{2+} ions into the extracellular space (and the retraction of the ameloblasts). At that time, it had not yet been clarified that Ca^{2+} ions are transported through the ameloblasts and pumped out at the distal end in the region of enamel formation, so this concept remained rather hypothetical. During the past few years, however, different studies have supported the assumption that the ameloblasts control enamel mineralization and transport Ca and phosphate to the extracellular matrix to induce Ca-phosphate mineralization. For example, ^{45}Ca autoradiography (e.g., NAGAI and FRANK 1975) has shown that Ca travels through the cells, passing several organelles, such as mitochondria, endoplasmic reticulum, and the Golgi complex. MUNHOZ and LEBLOND (1974), using this method, calculated a transport time of only 30 s from the blood vessels to the enamel matrix, which is in agreement with ionic transport speed. The results of ^{45}Ca autoradiography were supported by cytochemical precipitation of Ca, e.g., with Na-pyroantimonate (DEPORTER 1977; EISENMAN et al. 1979). However, the pyroantimonate technique is not without problems (VAN IREN et al. 1979). Further confirmation comes from electron-probe X-ray microanalysis, which has been applied to ameloblasts by EISENMAN et al. (1984), BOYDE and REITH (1977), and REITH and BOYDE (1978).

Besides ^{45}Ca autoradiography, the finding of the Ca pump, a Ca-ATPase at the distal region of the ameloblasts, would be more direct proof that Ca is extruded into the matrix. SASAKI and GARANT (1986) have localized the Ca-ATPase at the electron-microscopical level by applying the one-step lead method at alkaline pH, after transcardiac perfusion with a formaldehyde-glutaraldehyde mixture. Levamisol was added to block the activity of the unspecific alkaline phosphatase (APase). While studying the secretory ameloblasts, it was found that Ca-ATPase is present above all in the proximal and distal regions of the Tomes' processes, from which the matrix macromolecules are also extruded. A scheme of this Ca-ATPase distribution is indicated by elliptic dots in Fig. 4. The localization of Ca-ATPase is thought to support the idea (Fig. 3) that Ca^{2+} ions are pumped out of the ameloblasts in the region of the Tomes' processes and stream into the gel of the matrix, producing a parallel arrangement of the macromolecules by reaction with the negatively charged groups.

It should be mentioned that the details of the Ca transport through the ameloblasts have not been clarified. SASAKI and GARANT (1987) have found, with the application of the calmodulin blocker trifluoperazine (TFP) to rats,

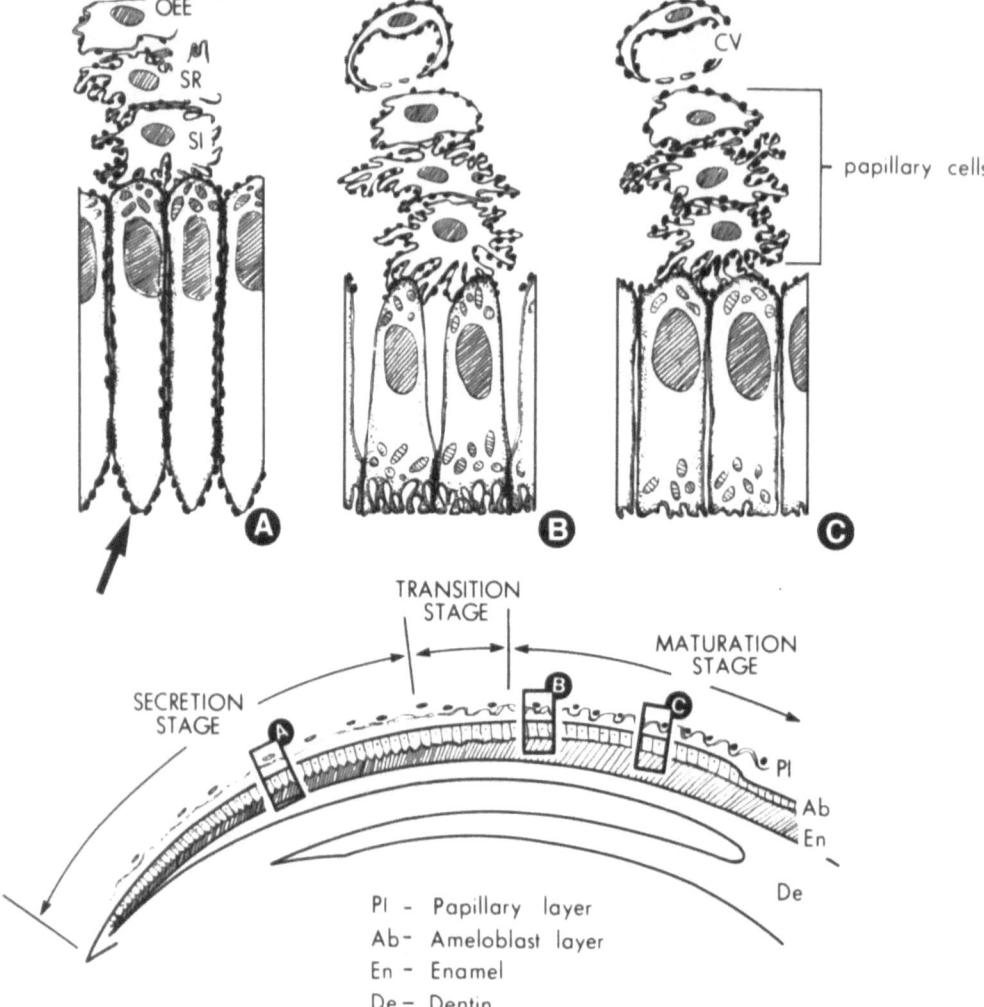

Fig. 4. Ca-ATPase localization in the cells of the enamel organ (Fig. 18 of SASAKI, GARANT 1986). The Ca-ATPase is represented by *dotted lines;* the *arrow* points to the Ca-ATPase in the region of the Tomes' process, in secretory *stage A*

that the enzymatic reaction of Ca-ATPase was drastically decreased at the plasma membrane, including the region of the Tomes' processes. In addition, the maturation ameloblasts of the ruffle-ended and smooth-ended type did not show any important Ca-ATPase activity (SASAKI and GARANT 1986), suggesting that at this stage an active Ca transport through the ameloblasts no longer exists (Fig. 4B, C). As mentioned by GARANT (1986), at this stage Ca may also reach the mineralization front by paracellular diffusion. The details of the Ca transport process through the secretory ameloblasts are still unknown. REITH (1983) assumes that Ca may enter the ameloblasts at the basal side (coming from the blood vessels), involving an APase at the outer cell membrane as Ca porter (possibly also as phosphate porter). This was assumed by WARNER et al. (1983) to explain the entrance of Ca and phosphate into the matrix vesicles (as will be discussed later). The Ca^{2+} ions would then be bound to transport proteins located at the inner side of the ameloblast membrane. According to REITH (1983), these would transport the Ca^{2+} ions very quickly, along the fluid region of the membrane to the distal ends of the ameloblasts (the Tomes' processes). The Ca^{2+} ions would then be pumped into the extracellular space by the Ca-ATPase. It should be mentioned that this hypothesis has not yet received sufficient experimental support.

III. Current Ideas About the Morphological Aspects of Crystal Nucleation and Crystal Growth in Enamel

1. The Prevailing Theory on the Morphology of Crystal Nucleation and Crystal Growth in Enamel

It is widely accepted that the first crystallites which can be observed under the electron microscope in the vicinity of the Tomes' processes are long plates, that is crystallites with their length in the direction of the c-axis; they have a width of more than 10 nm but a thickness of only about 2 nm (Fig. 5c) (DACULSI and KEREBEL 1978; WEISS et al. 1981). DACULSI and KEREBEL (1978) and WEISS et al. (1981) used high-resolution electron microscopy to distinguish the different lattice planes, such as the (100) (010) (110) planes, and measured the width and thickness of the crystallites with high precision by counting the lattice planes, as the exact distance between them is known from X-ray diffraction work. DACULSI and KEREBEL (1978) have measured the dimensions of the developing crystallites at various maturation stages by relating the size to the distance from the cell surface. Increasing distance corresponds to more advanced maturation. For instance, within a distance of about 10 µm from the cell surface they observed only two lattice planes parallel to the c-axis, making it simple to determine their thickness. Within about 25 µm there were three lattice planes, etc. By determining their width and thickness up to some 100 µm, they concluded that crystal development or, better, crystal growth takes place in a two-stage process. In the first stage (up to 200 µm from the surface) crystal formation starts with very thin, long plates or ribbons, about 2 nm thick, which grow mainly widthwise. The increase in width is much greater than that in thickness. This process has already been described by RÖNNHOLM (1962) (see Fig. 5c, left). DACULSI and KEREBEL (1978, Chart V) define the second stage as starting when the crystal width has reached a value of about 60 nm at the end of the fetal stage. The crystallites grow mainly in thickness (Fig. 5c, right). NYLEN et al. (1963) had also assumed such a secondary process in which the crystallites grow mainly in thickness. At the end of the embryonic stage, DACULSI and KEREBEL (1978) measured crystallites and found a width of about 61.3 nm and a thickness of about 16 nm, with a width-to-thickness ratio of about 3.87. The corresponding values for crystallites from mature enamel were 68.3 and 26.3 nm,

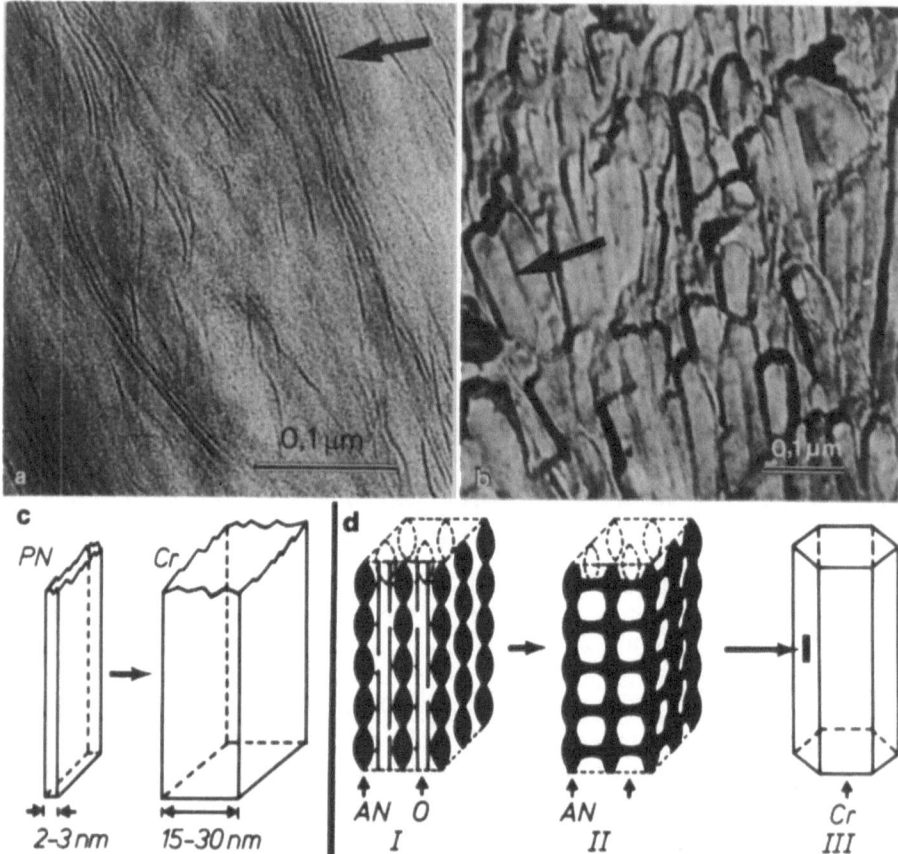

Fig. 5a–d. Electron micrographs of developing crystallites in fetal enamel and of mature crystallites, and scheme of crystal formation in enamel. **a** Early crystal formations in fetal enamel; the *arrow* points to apatitic strands lying in a group of four, showing high intrinsic contrast. **b** Carbon replica of the parallel apatite crystallites of mature enamel. We believe that many hexagons with only a low degree of flattening are visible (*arrow*). **c** Crystal formation according to prevailing ideas (e.g., DACULSI and KEREBEL 1978). It is assumed that first a thin crystallite (*PN*) is formed (*left*) with a thickness of only about 2 nm and a width of more than 10 nm. After growth in width, growth in thickness occurs in a second stage (*right*; *Cr*, crystallite). A possible final stage of hexagons is not shown. **d** The concept of HÖHLING et al. (1982). For stage I we assume that parallel Ca phosphate strands, needles, possibly with a dotlike substructure, develop (*AN*) on the parallel protein chains (*O*). They may develop by very rapid coalescence of dotlike islands (see **a**). In stage II these strands have grown and coalesced with neighbors to form larger entities, e.g., ribbons. They then grow, heal crystal defects, and finally form (stage III) the mature crystals, e.g., hexagons (*Cr*)

respectively, with a width-to-thickness ratio of 2.59. This implies that the mature crystallites are flattened hexagons (not shown in Fig. 5c; isodiametric hexagons are shown in Fig. 5d, III). DACULSI and KEREBEL (1978) point out that besides growth in thickness, the growing crystallites also fuse to form bigger entities. This is shown in Fig. 5d and will be discussed in more detail in the following chapter. I am surprised that most authors (e.g., DACULSI and KEREBEL 1978;

WEISS et al. 1981) assume that the crystallites of mature enamel are such *flat* hexagons. I have observed a more or less *pencil-like* shape, i.e., more isodiametrical crystallites, often hexagons (Fig. 5b). At present this discrepancy can only be mentioned without solution. As it is almost impossible to obtain cross sections of the crystals, perhaps the authors studied obliquely cut sections. WARSHAWSKY et al. (1987) also conclude that sections of crystallites which are generally described as cross sections of flattened hexagons are not, or not always, really cross sections. They describe them as elongated parallelepipeds yielding rhomboidal cross sections.

This is not the place to discuss the nature of the primary mineral formed in enamel, whether it starts as an apatite or as an octocalciumphosphate (OCP). Nevertheless it should be mentioned that WEISS et al. (1981) have correlated the thickness values of early crystal formations with the dimensions of the crystal-elementary cell of apatite as well as of octocalciumphosphate. They have come to the conclusion that characteristic values of crystal thickness correspond nicely to the dimensions of the lattice cell in the direction of the a-axis for the apatite and octocalciumphosphate or for both. So they propose that a layer of octocalciumphosphate may develop first and may be transformed rapidly into that of apatite; subsequently, apatite may develop on the primary OCP layer during the further process of crystal growth.

2. Aspects of Our Own Ideas Concerning the Morphology of Crystal Nucleation and Crystal Growth in Enamel

The prevailing theory of crystal nucleation morphology and crystal growth in enamel has just been discussed. It is assumed that in the first stage a "two-dimensional" crystallite develops with a dimension along the crystallographic c-axis reaching the μm range, with a primary width of more than 10 nm and a thickness of about 2 nm. It should be stressed that this primary crystallite already has such large dimensions that it cannot be the primary formation, the primary seed. There must be smaller primary formations which develop secondarily to this "two-dimensional" crystallite.

As seen in Fig. 3 and Fig. 5d, Ca-phosphate chains – needles with dotlike substructures (probably apatitic chains, Fig. 5a, arrow) – seem to exist in a stage which precedes the stage of the primary "two-dimensional" crystallite. General proof cannot be given of their existence since such dots, islands in the Ca-phosphate needles, lie at the border of the resolution for ultrathin sections in the electron microscope. Nevertheless, in correlation to primary crystal formations in collagen (as will be discussed in the chapter on dentine) and in other organic matrices the conclusion which is presented schematically in Fig. 5d was reached. The dotlike substructure of the apatite chains leads one to assume (Fig. 5d, I) that in a primary stage different nuclei develop along the matrix macromolecules which rapidly grow and form stable dotlike islands. They rapidly make contact with neighboring dots, fuse with them, and form needlelike strands in which the dotlike substructure is normally no longer visible (Fig. 5a, arrow). Also, the Ca-phosphate strands make contact laterally with neighboring strands arranged in parallel, coalesce with them, and form ribbon-

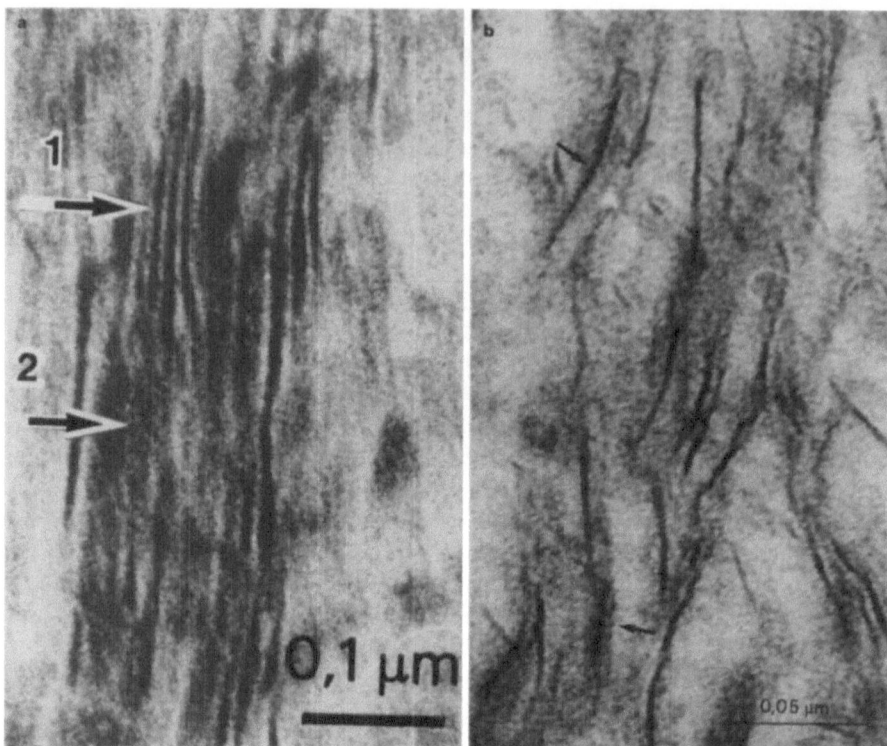

Fig. 6a, b. Electron micrographs of Ca-phosphate strands of developing enamel with high intrinsic contrast. **a** *Arrow 1* points to three Ca-phosphate strands with high contrast; *arrow 2* points to a larger entity with less contrast. We assume that the three strands have coalesced to form this larger crystallite (see HÖHLING et al. 1982, Fig. 5). **b** The *arrows* point to Ca-phosphate strands in which dotlike substructures, islands, are visible

like structures (Fig. 5d, II). These may correspond to the thin long platelets which DACULSI and KEREBEL (1978) described as primary crystallites. Deliberations here suggest they are secondary formations. Following Fig. 5d, II these ribbons grow in thickness, reach neighboring ribbons, and also coalesce with them. During the coalescence process prevailing lattice defects are healed progressively so that the final crystal contains relatively few lattice defects. This has been demonstrated by high-resolution electron microscopy. As mentioned before, many pencil-like crystallites have been observed with more isodiametric cross sections (Fig. 5b). The hexagonal crystallite shown in Fig. 5d (III) represents this. Using scanning electron microscopy on mature enamel crystallites ARENDS and JONGEBLOED (1978) also observed that pencil-like hexagons prevail with a much lesser degree of flattening than that mentioned by DACULSI and KEREBEL (1978). The thrust of the present argument (Fig. 5d, I–III) is that the final, mature crystal, with only few lattice defects, develops from several small entities in different steps.

Using the electron microscope it is difficult to prove whether real dotlike structures exist in the apatite chains or ribbons or whether they must be regarded as artifacts. Strands in which dotlike substructures are visible can be seen in Fig. 6b (arrows). Further, whether needlelike strands exist or whether they are ribbons lying on their short edge as most authors describe is also uncertain (Fig. 5c). Because these strands have a substructure consisting of dots (Fig. 6) and often show a curved course (Fig. 6), I have concluded that they cannot really be platelike, ribbonlike crystallites, which would have straight outlines.

These early apatitic strands normally have a high intrinsic electron contrast (Figs. 5a, 6). Most authors concluded that this contrast is due to the fact that the ribbons are lying on their short edge. Thus, more mass would be irradiated by the electron beam. In comparison I have concluded that the high contrast is due to the so-called amorphous contrast. On electron-transparent foils of different composition REIMER (1962) found that the regions in which the substance is "amorphous", or to be more exact, in which the crystallites are very small and have many lattice defects, show a much higher electron contrast than well-crystallized regions with the same mass thickness. On this basis it was concluded that the early apatite strands still have many lattice defects producing the high contrast (Fig. 6a, arrow 1, Fig. 6b). The larger apatite crystallites have less contrast (except the zones of Bragg reflexion) since most of the lattice defects healed during growth and coalescence. In the region indicated in Fig. 6a, arrow 2, the contrast is already less than that indicated by arrow 1 since greater entities with less defects already exist (see also HÖHLING et al. 1982, Fig. 5).

3. Does the Enamel Crystal Develop from One Nucleus or from Several Nuclei?

From the discussion above, the assumption is made that several apatite nuclei develop at several active sites along the matrix macromolecules and that they grow to stable islands which fuse to needles and ribbons (Fig. 3; Fig. 5d; 7b). This concept, however, is not the prevailing one. Another, widely accepted theory is shown in Fig. 7a. The Ca-phosphate nucleus develops some distance from the cell and grows to a stable crystallite. It continues to grow by adding and incorporating further ions at the top of the crystal. This represents the growth in length along the c-axis. This idea of a continuous growth in length from only one original entity indicates that active sites at the matrix macromolecule are necessary only at the beginning of crystal formation. During further growth in length, possible active sites (Fig. 7a) would no longer be needed since the growing crystallite would directly incorporate the arriving Ca^{2+} and phosphate ions. (In the chapter on collagen mineralization in dentine we shall also show that for this matrix both the ideas illustrated in Fig. 7 may apply: on the one hand that several active sites per 67-nm period exist and on the other that only one active site per 67-nm period is used for apatite nucleation. Since the complete amino-acid sequence for type I collagen is known, and since a paracrystalline model for collagen exists, more deliberations are possible for this system.)

As in the case of collagen, it is likely that several active sites along the matrix macromolecules exist and induce a nucleation for enamel mineralization

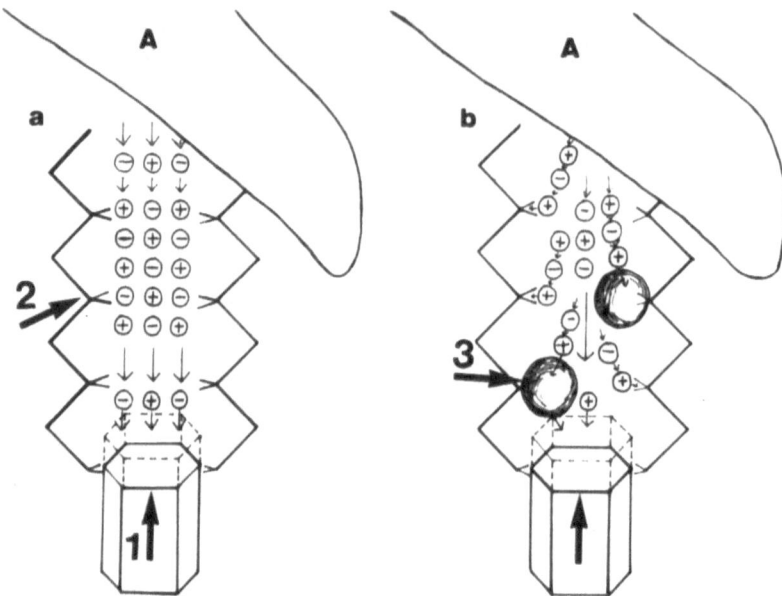

Fig. 7a, b. The concept of longitudinal crystal growth in enamel. **a** (Prevailing concept: a crystallite (*arrow 1*) grows continuously in length by incorporating Ca² ions (+) and phosphate ions (−); the influence of further possible nucleating sites on the matrix macromolecule strand is not considered (*arrow 2*). *A*, Ameloblast. **b** Our concept: the crystallite (*arrow*) grows in length by incorporating Ca^{2+} and phosphate ions and by coalescing with preformed Ca-phosphate islands (*arrow 3*). The unmineralized region is not drawn to scale

(Fig. 7b). It is unlikely that all the Ca^{2+} and phosphate ions which enter the microchannels between the matrix macromolecules (see also Fig. 3) can diffuse to the developing crystal and become incorporated without being bound by the charged groups of macromolecules in the channel to form clusters and afterwards nuclei. At this point it is necessary to describe the nature of the enamel macromolecules and to discuss how far the amino-acid sequence analysis has reached to see whether such possible active sites for Ca-phosphate nucleation do exist.

IV. Description of the Matrix Macromolecules in Enamel

1. General Chemical Characterization

Since this contribution appears in a morphologically oriented series, chemical aspects of the matrix will be treated only briefly. While the content of the organic substance per mass is about 20% in developing enamel, it is reduced to much less than 1% in mature enamel. This means that a large amount of organic material is lost in the process of enamel formation and maturation. Further, amino-acid analyses have shown that a characteristic change of the prevailing amino acids occurs in the transition to mature enamel. Chemical

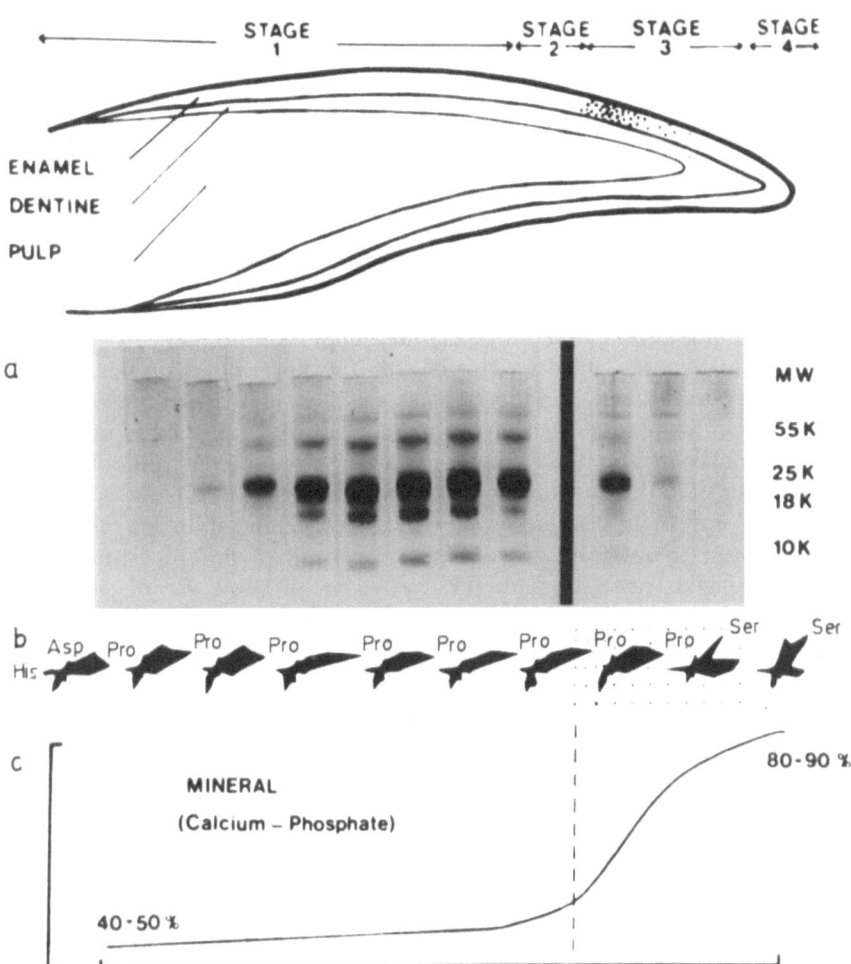

Fig. 8a–c. The protein pattern and the amino-acid composition of the proteins from developing enamel to mature enamel (Fig. 1 of ROBINSON et al. 1983). *Top:* The growing rat incisor, from stage 1, developing enamel, to stage 4, mature enamel. **a** Diagrams of SDS-acrylamide gel electrophoresis of the proteins from stage 1 to 4, with amelogenins and enamelins. **b** Rose diagram of the amino acid composition of the proteins at the different stages of development. **c** Representation of the increasing mineral content (Ca-phosphate) with increasing maturation

separation methods, such as gel chromatography and gel electrophoresis, have revealed several components with different mol weights for developing as well as for mature enamel. Consequently, it is of great interest to clarify whether it is possible to classify these different components. EASTOE (1965) divided enamel proteins and glycoproteins into two groups, the amelogenins and the enamelins, a classification which has been generally accepted. It is possible to analyze the development from the early to the mature enamel (Fig. 8, top) and to show the existence and the fate of these two protein groups by proceeding with the analyses from stage 1 to stage 4 along the surface of the continuously growing rat incisor. The pattern of the distribution of the proteins separated by gel electrophoresis and the corresponding amino-acid composition presented in the rose type are also illustrated in Fig. 8.

Amelogenins: The amelogenins constitute up to 90% of the developing enamel matrix protein; they comprise the bulk of material produced and secreted by the ameloblasts. It is widely assumed that they fill space or form the right structure for the backbone of the matrix, and/or determine the size of the developing crystallites, but do not directly serve as the nucleation-inducing matrix. I believe that spacefilling is not a sufficient function and do not understand what "determining the size of the crystallites" actually means. Only in the stage of nucleation are the crystallites under the influence of the matrix macromolecules; when they grow to crystallites they will fill up the free microspaces. Since the growth in length is not really hindered and since the length of the crystallites reaches the μm range or even mm range, it is unlikely that macromolecules influence the growth in length. How the macromolecules influence the growth in thickness is more difficult to explain.

Further, it must be considered how the amelogenins and enamelins are spatially arranged in relation to each other to fulfill a different function. The molecular weights according to SDS gel electrophoresis are in the range of 25000–30000 Daltons (Fig. 8a). FINCHAM (1984) indicates nearly the same range (25000–27000 Daltons) according to the latest, appropriate separation methods. It is not yet clear whether these amelogenins are primary gene products. There is evidence that they may result from the degradation of larger molecules (ROBINSON et al. 1979). Fig. 8a shows that amelogenins with an even lower molecular weight exist (e.g., in the range of 5000 Daltons). This suggests that a slow degradation of the amelogenins takes place during enamel maturation. Most of the degraded molecules would be transported to the ameloblasts and degraded further (Fig. 3, arrow 3). Originally, the amelogenins were characterized by their special amino-acid composition. Figure 9 represents the amino-acid composition in the rose diagram. The proteins which contain a large amount of amelogenin are characterized by a relatively high content of proline (Pro), glutamic acid (Glu), leucine (Leu), and histidine (His).

SASAKI et al. (1984) isolated an amelogenin (mol weight 19350 Daltons) from developing bovine incisor enamel and determined the complete amino-acid sequence of all the CMBr-cleaved peptides using automated Edman degradation. The molecule consists of 170 amino acids. Noteworthy is that the sequence is characterized by a periodically appearing Pro, the prevailing amino acid, which should have a determining influence on the tertiary–quaternary structure of the amelogenins. Further, a periodical, clustered appearance of charged amino acids which may act as nucleating sites cannot be detected; this may strengthen the concept that the amelogenins do not represent the nucleating matrix.

Enamelins: This glycoprotein group constitutes the remaining 10% of the developing enamel matrix. They differ markedly in amino-acid composition (ROBINSON et al. 1975; TERMINE et al. 1979). Their molecular weight appears to be higher than that of the amelogenins, lying in the range 50000–70000 Daltons (Fig. 8a). They contain much more serin (Ser), phosphate (as Ser-phosphate), and aspartic acid (Asp) than the amelogenins. The Pro, Glu, and

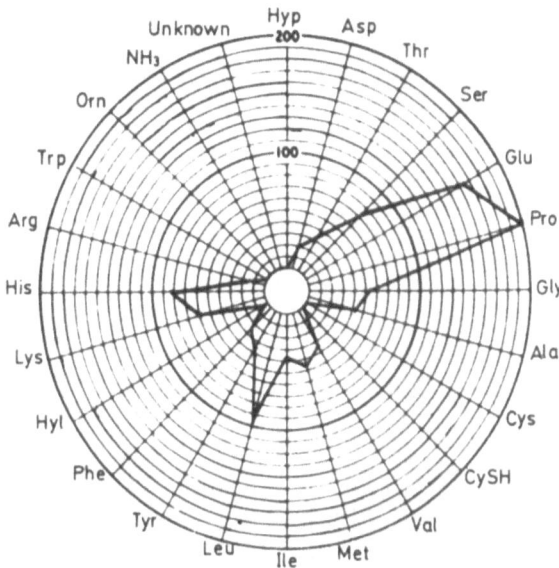

Fig. 9. Rose diagram of the amino-acid composition of pooled developing enamel protein; *Pro*, *His*, and *Leu* are predominant (Fig. 2 of ROBINSON et al. 1983)

Leu content is greatly reduced when compared with the amelogenins. The fate of the enamelins during the process of mineralization and maturation is even less clear than that of the amelogenins. However, since they are connected with the mineral phase and can be extracted only after mineral dissolution with EDTA, it is assumed that they act as a matrix to induce nucleation. This latter function seems to be supported by the immunohistochemical demonstration of enamel reported by HAYASHI et al. (1986). As Fig. 10 shows, the gold particles of the electron-microscopical immunogold method are spatially related to the filamentous and ribbonlike structures, the original sites of these crystallites, on decalcified ultrathin sections.

Up to now the question of how the amelogenins and the enamelins (if they really exist as different entities) must be spatially arranged has not been discussed. The arrangement would have to be such that the amelogenins – which would not act as matrix, but contribute to the backbone structure – can be decomposed and transported to the cell while the nucleation-inducing enamelins stay mainly at the original site, bound to the crystallites. One solution to this problem might be that the amelogenins form a central core, while the enamelins are arranged around it. This may indicate that the two proteins possess a different paracrystalline three-dimensional structure.

2. Possible Crystalline, Paracrystalline Structure of the Enamel Proteins

Considering that the protein composition of the developing and mature enamel is heterogeneous, it would be difficult to clarify experimentally in which way the protein or proteins are crystallographically structured. It might be

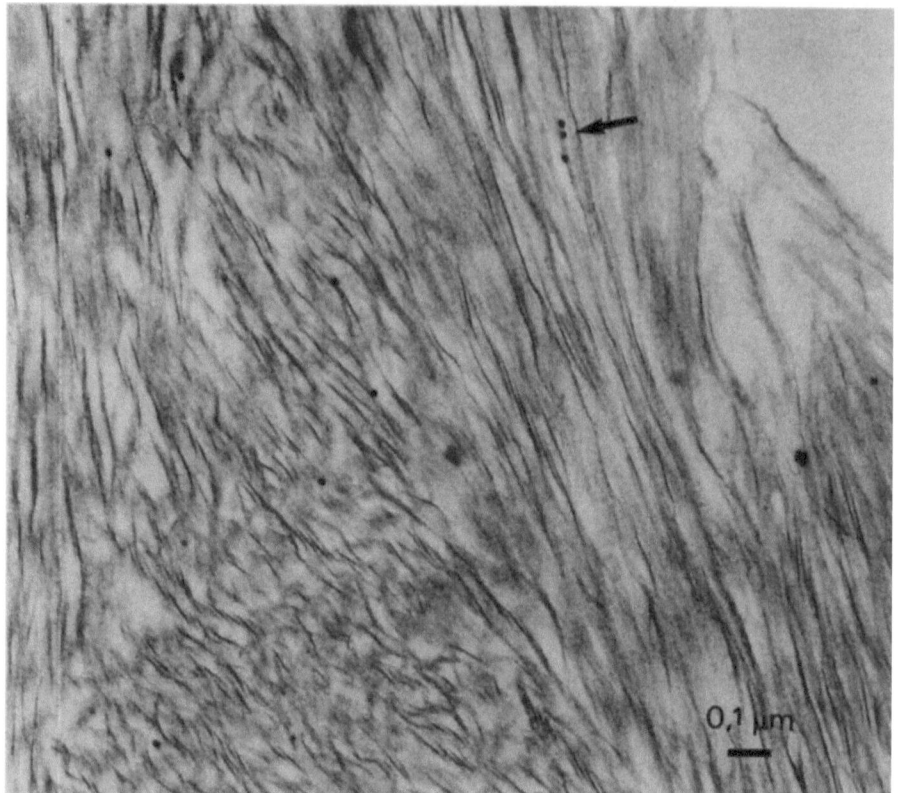

Fig. 10. Electron micrograph showing the enamelins stained by the protein A-gold immunohistochemical method (*arrow*). (With permission, from Prof. E. Bonucci 1985)

useful to analyze the primary structure, the amino-acid sequence, and to compare it with proteins which have a corresponding amino-acid sequence (if such proteins exist) and whose three-dimensional arrangement is known. Not enough sequence analyses, which would make such a comparison possible, have been obtained yet (Sasaki et al. 1984). Another way to clarify the structure is to apply more indirect methods, such as optical rotary dispersion, to the macromolecules in solution to obtain information on the main aspects of the tertiary structure. According to Bonar et al. (1965), optical rotary dispersion analysis has shown that a small portion of the enamelins exists in the α-helix form.

The most important direct method for the determination of a crystalline and paracrystalline structure is X-ray diffraction analysis. It can be applied to the enamel proteins only after demineralization since the X-ray reflections of the crystallites dominate and conceal the more or less diffuse reflections of the proteins.

In the early 1960s, several groups applied X-ray diffraction to demineralized fetal and mature enamel. At that time amelogenins and enamelins had not yet been classified. HÖHLING et al. (1963), GLIMCHER et al. (1961), and PERDOK and GUSTAFSON (1961) obtained two diffuse reflections, with d spacings in the range of 1.08 and 0.46 nm (HÖHLING et al. 1963), 0.98 and 0.465 nm (GLIMCHER et al. 1961), and 2 nm, 0.95 nm, 0.7 nm, and 0.55 nm (PERDOK and GUSTAFSON 1961). A weak diffuse reflection in the range 0.378–0.393 nm was also found by HÖHLING (1966).

PERDOK and GUSTAFSON (1961) concluded that a special keratin exists, which they called δ-keratin. GLIMCHER et al. (1961) suggested that a protein with the rare cross β-structure exists – a structure in which the protein chains are believed to be folded after a supercontraction (see HÖHLING 1966, Fig. 28). This structure was proposed by ASTBURY et al. in 1959, but it is no longer certain whether this structure really exists. Since the X-ray reflections of HÖHLING et al. (1963) and HÖHLING (1966) were in agreement with those of a disoriented β-keratin, these authors discussed the possibility of the existence of such a β-keratin for the main portion of the enamel protein. However, since denatured proteins also show these reflections, and since two or three diffuse reflections seemed to be too few as a basis from which to derive the existence of a real paracrystalline structure, HÖHLING (1966) considered simply that an ordered, ionotropic gel exists with the fiber ordered in the direction shown by Fig. 3 and suggested that a prestage of β-keratin may exist.

Recently, JODAIKIN et al. (1986) have dissected enamel portions from the different zones of enamel maturation along the continuously growing rat incisor (see Fig. 8, top). The specimens were fixed and demineralized according to many different procedures. The demineralized specimens were kept humid during X-ray diffraction. The authors obtained two diffuse rings with d values 0.8–1.2 nm, 0.35–0.5 nm, and a relatively sharp one at 0.467 nm. This relatively sharp reflection always appeared, irrespective of the preparation. It was more intense in the zone of advanced mineralization than in the zone of incipient mineralization and is, in addition to other reflections, an important feature of X-ray patterns received from proteins with a β-pleated sheet structure. The authors therefore assume that this structure prevails in the enamel proteins, and since it is more intense in the stage of advanced mineralization of mature enamel, where the enamelins prevail, they conclude that it is mainly the enamelins which have this configuration. This conclusion seems to be supported by X-ray and infrared studies on chemically isolated amelogenin and enamelin fractions. They show that only the enamelins have the features of the β-pleated sheet structure.

Since the enamelins are believed to be the main matrix components which induce the Ca-phosphate nucleation, this β-pleated sheet structure would have to be discussed in connection with the Ca-phosphate nucleation. HÖHLING et al. (1963), and HÖHLING (1966) have already discussed, on the basis of their few X-ray diffraction rings, such a β configuration but hesitated finally to assert the existence of a fully developed β configuration, as mentioned before. According to LEE et al. (1977) and WEINER and TRAUB (1984), this β-pleated sheet structure also exists as a structural basis in other mineralizing systems, e.g., under certain conditions in the dentine as phosphoproteins, and in the matrix macromolecules of mollusc shells. So the β-pleated sheet structure should be regarded with interest in the future as the structure of the matrix.

Since the organic substance in enamel is heterogeneous as assessed on the basis of chemical separation techniques, more structural analyses are necessary to clarify whether other structural protein configurations also exist. As mentioned earlier, amino-acid sequence analyses may also contribute to clarifying the structure, and they may show whether characteristic regions exist along the protein chain which might function as active sites for nucleation.

3. Amino-Acid Sequence Analysis for Active Sites for Ca-Phosphate Nucleation

Amino-acid sequence data are important to clarify whether groupings of charged amino acids exist which may act as active sites for Ca-phosphate nucleation. It is widely assumed that the amelogenins do not act as a nucleating matrix. We have already mentioned that SASAKI et al. (1984) were able to analyze the full sequence of an amelogenin with a molecular weight of 19 000 Daltons. In this amino-acid sequence, we could not detect microsites with a neighborhood of more than one charged amino acid which might function as active sites to bind, for example, Ca ions for the process of nucleation.

SEYER and GLIMCHER (1969) have analyzed amino-acid sequences of two peptides, consisting of 46 and 43 residues in the enamelins. According to preliminary information, these amino-acid sequences may represent an NH_2 terminal section of a larger matrix macromolecule with a molecular weight of 25 000–30 000 Daltons. Since differences exist between these and some amino-acid sequences obtained earlier, and since the precise location and degree of phosphorylation is not entirely certain, these results should be regarded as preliminary. SEYER and GLIMCHER (1969) found that the two isolated peptide components with 46 and 43 amino acids contain the sequences Glu-Ser(P)-Leu and Glu-Ser(P)-Tyr respectively three times. In accordance with these authors we assume that the grouping Glu-Ser(P) [Ser(P): serine phosphate] may act as a Ca-phosphate nucleating site. Since the one Glu-Ser(P) group is located at the NH_2 terminal, and assuming that the other two are located in the middle and at the other end of this peptide, the distance between these Glu-Ser(P) sites would be about 6 nm, if the distance between two neighboring amino acids were about 0.3 nm.

The question arises whether results already exist which would support such a distance between active sites. Concerning collagen as matrix, the full amino-acid sequence of the collagen type I molecule, which exists in dentine, has been elucidated; the distances of polar regions, rich in charged amino acids, lie in this range. We have concluded from center-to-center distances between dots, which represent islands in the apatitic chains in collagen, that nucleating sites in this distance range do indeed exist. We believe we have observed dotlike substructures in apatitic chains in developing enamel (Fig. 6).

Preliminary, approximate measurements of the center-to-center distances between such dots in the apatitic chain place them mainly in the range of 3.5–5.5 nm (HÖHLING et al. 1982, Fig. 2d). It would be important to carry out more systematic work on these early apatitic strands to find out whether they frequently exhibit a dotlike substructure. If they do, more measurements of center-to-center distances should be carried out to clarify whether this distance range, mainly 3.5–5.5 nm, really exists. With the greatest caution we may say that the preliminary amino-acid sequence analysis of a small peptide allows one to calculate a rough distance of about 6 nm between Glu-Ser(P) dimers which might act as Ca binding sites. Our first center-to-center distance measurements between islands in apatitic chains result in such a distance range, possibly also indicating the distance between active sites. That these comparisons are the subject of so much speculation indicates that we are only just beginning to clarify the nature and local distribution of active, nucleating sites at the matrix macromolecules.

V. Crystal Structure of the Developing and Mature Enamel Crystallites

We have often used the term "apatitic chains" for the developing crystallites in enamel. This is not totally correct. While it is generally accepted that mature enamel crystallites possess the apatite structure, $Ca_{10}(OH)_2(PO_4)_3$, there is no consensus about the crystal structure of the early developing crystallites. Analysis of the crystal lattice planes by high-resolution electron microscopy has revealed that the mature crystallites have a high degree of lattice perfection. Only a few lattice distortions are found. The so-called dark lines should be mentioned as possible sites of lattice distortion (NAKAHARA and KAKEI 1984). The apatite of the mature crystallites can be described as a carbonate apatite in which 2%–3% of the phosphate groups in the lattice are replaced by carbonate groups. Two OH groups can also be replaced by a carbonate group. We shall not describe the nature of this apatite structure but only say that the elongation of the crystallites in the direction of the c-axis correlates with rows of Ca atoms and OH atoms in this direction. Since there is no agreement on the crystal structure of the developing Ca-phosphate mineral in enamel (as in dentine, bone, etc.) we shall briefly discuss the different theories.

1. Existence of a Primary Ca-Carbonate Mineral

CASCIANI et al. (1979) have applied laser Raman microspectrometry to the earliest mineral formations in enamel. They have come to the conclusion that a Ca-Mg carbonate (huntite) develops first. Since these results have not been supported by other groups, they should be regarded with caution. Nevertheless it is assumed by several authors that the primary Ca-phosphate formations, e.g., the Ca-phosphate chains, which we have described, are rich in carbonate. This carbonate content would cause a lattice distortion and possibly cause the so-called central dark lines which are normally observed in enamel mineral formations (NAKAHARA and KAKEI 1984).

2. Existence of a Primary Apatite with Strong Lattice Distortions

We have come to the conclusion, as have several other groups, that the primary Ca-phosphate chains, needles and ribbons have already developed the apatite lattice, containing, however, many lattice distortions. This can be derived from several observations. Electron diffraction of these formations in the vicinity of the cells shows a lower number of diffraction rings, which are more diffuse than those of mature enamel. However, they are in agreement with the apatite diffraction rings. The diffuse character would indicate the small size and the lattice distortions of the crystallites. Further, like other groups, we have found that these early formations show a high intrinsic electron contrast (in the unstained state). Other groups conclude that this electron contrast is due to the existence of platelets, ribbons lying on their short edge, which cause a higher mass thickness and thus a higher contrast. As already mentioned, we assume that this high contrast is due to the low degree of crystallinity. REIMER (1962) has observed that in electron-transparent foils of different substances, such as Sb_2S_3, the so-called amorphous regions show a much higher contrast than the crystalline regions (except the zones of Bragg reflection), both possessing the same mass thickness. The "amorphous" regions do not contain a fully amorphous substance, but a substance with small crystallites containing many lattice distortions. We assume that the mineral strands and ribbons have an analogously low degree of crystallinity. As they grow in thickness and coalesce with neighbors, the defects in the lattice would be more and more healed (as is indeed shown by high-resolution electron microscopy for the more mature crystallites); they also lose the high intrinsic contrast (see Fig. 6).

3. Primary Existence of an Octocalciumphosphate

Octocalciumphosphate (OCP), $Ca_8H_2(PO_4)_6 \cdot 5H_2O$, is a mineral which is structurally nearly identical to apatite, $Ca_{10}(OH)_2(PO_4)_6$. The only structural difference is an extra H_2O sheet, which exists as an interlayer between the (100) planes of the apatite elementary cell. Expulsion of these H_2O sheets from OCP would lead to hydrolysis to apatite. While OCP was found in several pathological mineralizations such as calculi and aortic mineralizations (e.g., HÖHLING et al. 1968), BROWN et al. (survey 1984) have reported that it also exists in an early transitory stage of enamel formation. According to BROWN et al. (1984) the question is not whether an acidic precursor exists but what the nature of the transitory phase is. They conclude that the precursor is OCP and not brushite, which theoretically might also exist. Regarding the appearance of OCP during enamel formation, they propose two possible mechanisms: (1) Growth takes place by formation of OCP crystals in toto, which then hydrolyze to apatite. (2) A unit elementary cell layer of OCP would be formed at one time on (100) apatite faces and transformed to apatite by hydrolysis. The authors speculate that the first process would take place during early mineral formation and the second during enamel maturation. If our belief is correct that Ca-phosphate strands appear first and ribbons develop from them, clarification is needed of how this might be consistent with the theory of BROWN et al. (1984).

C. Fundamental Aspects of Dentine Mineralization and, in Part, of Cementum, Bone and Cartilage

I. General Aspects of Enamel and Dentine Mineralization

In the preceding section on enamel mineralization, it was shown that mineralization starts in an ordered, ionotropic gel directly at the distal cell surface, at the Tomes' process. The crystallites are in a strictly parallel arrangement in enamel prisms, and crystal formation starts with very small needlelike or ribbonlike crystallites (probably with dotlike substructures) and proceeds to large crystallites (the largest ones to appear in the organism) with a high degree of lattice perfection. We shall see in this section that in dentine formation a zone of pure organic substance, the predentine, exists with a width of about 20 μm, indicating that the mineralization lags behind the formation and extrusion of organic substance. Enamel formation of organic substance and its mineralization take place almost simultaneously. Further, we shall see that for dentine mineralization, so-called matrix vesicles (MV) play a fundamental role for the induction of the primary crystal formation. It seems to be established that MV mineralization precedes the mineralization of the collagen fibrils and that of the intervening ground substance. First we will have to discuss the MV mineralization.

II. Mineralization of the Matrix Vesicles

1. Development and Morphology of Matrix Vesicles

BONUCCI (1967) and ANDERSON (1967) deduced from electron microscopical investigations that the primary crystal formation in mineralizing cartilage and bone starts in MV (Fig. 11). Later it was found that the primary mineralization in the predentine, in the so-called mantle dentine (without a preexisting mineralized zone), also starts in MV (EISENMANN and GLICK 1972; SISCA and PROVENZA 1972; KATCHBURIAN 1973) (Fig. 12). The MV are round or oval-shaped formations surrounded by a membrane and contain no collagen. Four major opinions exist concerning their formation (RABINOVITCH and ANDERSON 1976). The prevailing opinion is that, above all in the chondrocyte, MV form by a budding from the processes of the cells. According to a second hypothesis (RABINOVITCH and ANDERSON 1976), the MV appear in the vicinity of the cell membrane as preformed structures and are extruded, possibly via the cell processes. It was not clarified in which cell organelle they are formed and in what way they are transported to the membrane.

For the prevailing hypothesis of a budding from cell processes one may argue that the matrix of the MV is identical with that of the cell process of this region. We would like to point out that the MV contain neither exclusively nor mainly cytoplasm of the cell process, but do, in particular, contain specially synthesized macromolecules, which play a specific role in the induction of crystal nucleation.

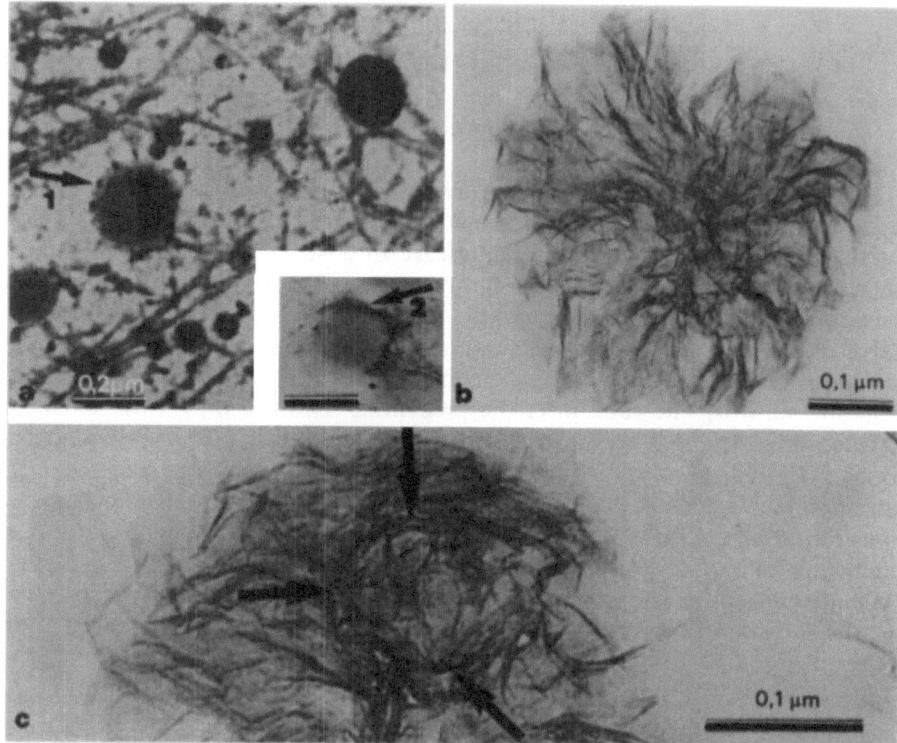

Fig. 11 a–c. Electron micrographs of matrix vesicles and mineralized noduli of the epiphyseal growth plate (Fig. 1 of Höhling et al. 1980b). **a** Stained matrix vesicles (*MV*) with dotlike staining islands (*arrow 1*). *Insert: arrow 2* points to a bent Ca-phosphate needle which developed at the MV membrane. **b** Ca-phosphate nodule with radiating Ca-phosphate needles and ribbons. **c** Nodule at higher magnification. The *arrows* point to bent Ca-phosphate needles which probably reinforce the original matrix vesicle membrane

The release of MV by the budding from cell processes can only be assumed for the chondrocytes in the epiphyseal growth plate. The dentinal tubules, the odontoblastic processes, from which MV are extruded into the predentine (Fig. 12), have no such special processes. In this case the second general assumption for MV extrusion is relevant, namely that preformed structures are blebbed off the cell wall. This process was proposed, for example, by KATCHBURIAN (1973). We shall come back to this point when we discuss the development of the MV on the basis of staining with bismuth ions. The third opinion proposed by RABINOVITCH and ANDERSON (1976) is that MV develop from degradation products of the chondrocytes. We have indications that MV can develop in this way. However, since such cell degeneration does not happen in dentine, we need not discuss this process of formation in connection with the dentine mineralization. The fourth proposal by RABINOVITCH and ANDERSON (1976) is the assumption that MV develop secondarily by an assembly of macromolecules in the extracellular space. Since it is difficult to prove this mode of development we shall not discuss it.

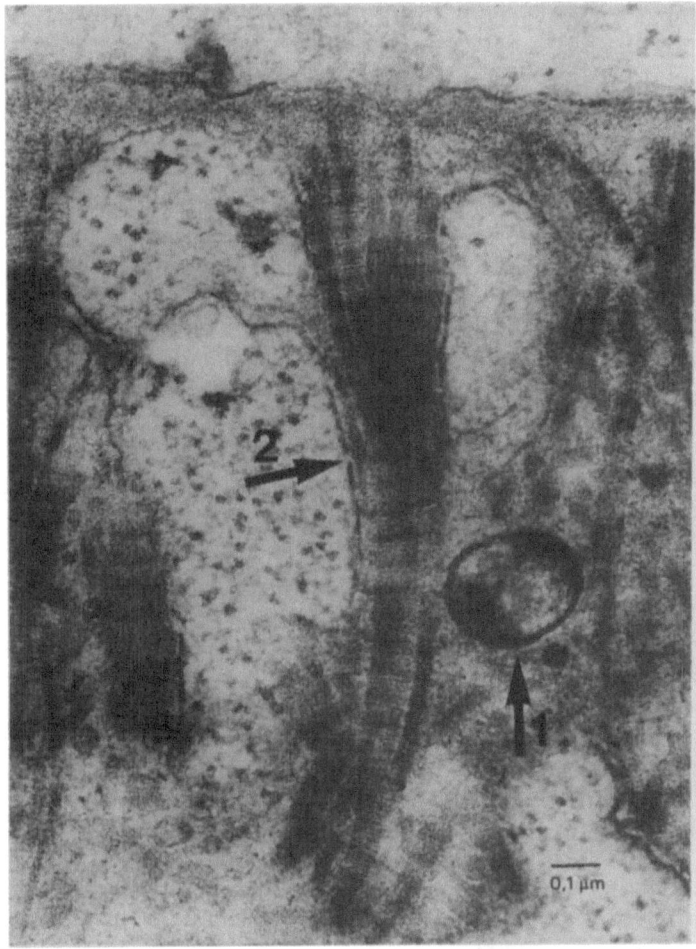

Fig. 12. Electron micrograph of predentine with a matrix vesicle (*arrow 1*) and collagen fibers (*arrow 2*). (With permission of Dr. P.L. GLICK, from EISENMANN and GLICK 1972)

Most of the studies on the development of the MV and their extrusion have been carried out in the cartilage of the growth plate. For several years it was assumed that, after extrusion, the MV are transported into the middle of the longitudinal septum and mineralized in that region (e.g., BONUCCI 1967). According to this assumption the mineralization of the MV would take place a long way away from the cells. However, cryopreparation and observation using transmission electron microscopy and electron probe X-ray microanalyses on dry thin sections have shown that vesiclelike structures already appear at a distance of $^1/_2$–1 μm from the cell border (BARCKHAUS et al. 1981). They enrich Ca and phosphate groups already in that region. We do not know whether the full mineralization of the MV is completed near the cells or whether it occurs gradually, e.g., when they may be transported into the middle of the longitudinal septa.

2. Histochemical and Chemical Characterization of the Components
of the Matrix Vesicles

To understand the process of crystal nucleation and growth inside and at the inner surface of the MV it is important to know the nature of the macromolecules and the way in which they induce crystal nucleation. We shall see that relatively little is known of the components of the MV. We know that, as expected, the MV membrane is rich in lipids, mainly phospholipids such as phosphatidylserin (WUTHIER et al. 1977). As early as in the first years of the description of the MV it could be shown that they are rich in alkaline phosphatase (APase) (e.g., ALI et al. 1970).

Against the general assumption that APase activity exists at the MV membrane and is connected with the Ca and phosphate transport or nucleation process, THYBERG and FRIBERG (1978) believe that two types of MV exist. They describe a round (type I) MV – corresponding morphologically to the MV described by most authors as the normal MV – which represents lysosomal vesicles being extruded into the extracellular space. Released lysosomal enzymes would make the surrounding matrix "calcifiable" for a primary mineralization (while the vesicles would become mineralized in a second step). The authors do not discuss the function of type II vesicles, which are described as vesicles with an irregular shape. Although VÄÄNÄNEN and KORHONEN (1981) have found, by gradient centrifugation, some lysosomal activity for a certain fraction, they have also isolated a MV-rich fraction which contains only APase and pyrophosphatase activity; so APase activity predominates which would induce the primary mineralization. Thus there is not a sufficient basis for discussing two types of MV. We shall concentrate on the MV which are believed to induce Ca-phosphate mineralization.

Attempts have been made to enrich the MV chemically by density ultracentrifugation at $300000\,g$ and to characterize the enriched fractions biochemically (STEIN et al. 1981; MULRAD et al. 1981). Before ultracentrifugation the tissue was treated with proteolytic enzymes such as trypsin and collagenase in order to separate the MV from the ground substance. On the enriched MV fraction various biochemical separation methods such as SDS gel electrophoresis were applied to clarify whether components of different molecular weight exist and to characterize them. STEIN et al. (1981) and MULRAD et al. (1981) have separated relatively pure fractions of the MV and carried out SDS polyacrylamide gel electrophoresis. They have obtained several protein bands in the range 30000–240000 Daltons.

So far, only a few macromolecules of the MV have been characterized. Alkaline phosphatase was found to be present, which had already been demonstrated histochemically. F-actin was also found, comprising about 11% of the total amount of the protein (STEIN et al. 1981; MULRAD et al. 1981). Considering that the organic substance inside the MV is active as a nucleating matrix and that it seems unlikely that actin is a strong nucleating substance, one may question whether F-actin is really present inside the MV. Electron microscopy of the enriched MV fractions shows a relatively high percentage of extravesicular substance, so F-actin might also be an extravesicular component. Immunohistochemistry at the electron microscopical level may clarify whether F-actin is a component of the MV themselves. Neutral peptidases have also been detected in the MV fraction (HIRSCHMAN et al. 1983). Probably it will take several more years before all components of the MV have been determined and before the details of the mechanism of crystal formation in relation to matrix macromolecules can be reasonably discussed. We believe that matrix macromolecules with a dominance of negatively charged groups are present in the MV, such as glycoproteins and/or proteoglycan monomers. This assumption stems from our staining experiments with Bi^{3+} ions applied to the MV and corresponding structures.

3. Analysis of the Fine Structure and the Pathway of the Matrix Vesicles by Means of Staining with Bi^{3+} Ions

Since little is known about the components of the MV and their structural arrangement, we asked whether ultrastructural studies may contribute indirectly to a clarification of this question. We looked for a dye which would stain characteristic negative groups and whose staining nuclei would be so small that their spatial arrangement could give information on the macromolecular arrangement, i.e., on center-to-center distances between active sites of the staining nuclei. These center-to-center distances might correlate to active sites for the Ca-phosphate nucleation of the developing hard tissue. BARCKHAUS et al. (1986) have found that the staining of fixed ultrathin sections with bismuth nitrate would fulfill these requirements. This staining method had been used by SERAFINI-FRACASSINI and SMITH (1966). We have not yet applied this method to dentine, but we have used it on tissue of the epiphyseal growth plates of young black-headed sheep and young guinea pigs. The fixed ultrathin sections were stained with a 0.5% solution of bismuth nitrate in 0.1 M nitric acid in water at pH 1.2. It could be expected that mainly negative groups of the macromolecules (such as carboxylate and sulfate groups) would be stained. BARCKHAUS et al. (1986) found elliptic staining products of different sizes at the cell processes and at the flat regions of the cell. The long diameter lies in the diameter range of the MV. Corresponding staining products in the extracellular region often show a round configuration. BARCKHAUS et al. (1986) concluded that the large elliptic formations are the MV in the prestage of extrusion. This would mean that they can be blebbed off from the cell processes as well as from flat regions of the cell, as was discussed before. Perhaps the small staining products represent mainly aggregated macromolecules with prevailing negative groups (such as the proteoglycan aggregates or monomers) or are partly prestages of the MV, which would form by an extracellular aggregation. We assume that the larger round ·staining products correspond to the MV since their size is identical to that of the MV, with diameters around 0.1 μm.

BARCKHAUS et al. (1986) studied the Bi^{3+} staining pattern of the MV at higher magnification (Fig. 13) and measured the center-to-center distances between the staining nuclei, which may eventually be of importance for assessing distances between active sites for nucleation, when more is known about the nature of the macromolecules of the MV. We found parallel chains of such

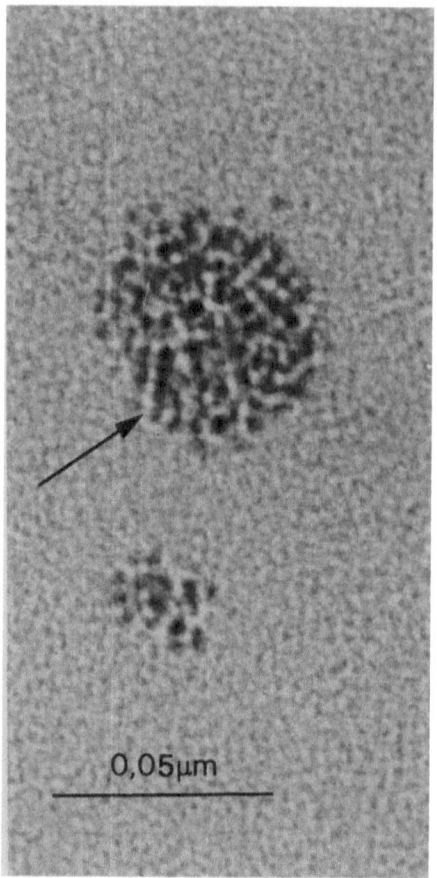

Fig. 13. Electron micrograph of a Bi^{3+}-stained round structure, probably a matrix vesicle in the epiphyseal growth plate; the *arrow* points to strands with a dotlike substructure which probably make indirectly visible macromolecular strands

staining dots which might represent the existence and arrangement of fibrous macromolecules (Fig. 13, arrow). The center-to-center distances between such chains lie mainly in the range of 5–6 nm. They may roughly represent the distances between fibrous macromolecules. The center-to-center distances between dotlike staining products inside such chains, i.e., fibrous macromolecules, might roughly represent the center-to-center distances between active sites for the nucleation of the staining substance. Assuming that such fibrous macromolecules act as a nucleating matrix for the Ca-phosphate crystallization, we also measured the distances between apatite chains and their dotlike islands inside the mineralized MV and in the so-called noduli (Fig. 11b), the mineralized compartments which comprise not only the MV but also some of their immediate neighborhood (Fig. 11c). We found that the center-to-center distances between these Ca-phosphate chains and between the dotlike islands inside these mineral chains are mainly around 5 nm. The question arises whether these distances correspond to those of the Bi^{3+}-staining chains and dots (Fig. 13).

4. Mineralization of the Matrix Vesicles and Their Immediate Neighborhood

a) Mineralization in the Matrix Vesicles

Unmineralized and unstained MV show some high intrinsic contrast (HÖHLING et al. 1976b, Fig. 4), perhaps indicating a Ca and phosphate enrichment in the prestage of nucleation. Figure 4 from HÖHLING et al. (1976b) also shows early mineralized microcompartments, the noduli, which already contain mineralized extravesicular regions (Fig. 11b, c). This means that the MV mineralization does not stop in a first stage at the MV membrane but proceeds directly beyond the MV border.

It is not known where in the MV the mineral nucleation starts; only some indications exist that it might begin at the inner MV membrane (BOSKEY 1981, survey). Further, no conclusive results exist concerning the way in which the Ca and phosphate ions are transported into the MV. If the MV blebbed off from the cell membrane, their membranes would mainly correspond to the cell membrane and contain a Ca-ATPase. If such a Ca-ATPase were active it would pump Ca out of the MV instead of into it, the opposite of what would be necessary for Ca and phosphate enrichment of the MV.

The early assumption that the APase at the MV membrane would split phosphate esters and transport inorganic phosphate (Pi) into the MV was given up by several authors (survey, e.g., WUTHIER 1982), as it would not explain the transport of Ca into the MV. Since we do not believe simply that Ca and phosphate diffusion can explain the high influx of Ca and phosphate and the intense mineralization inside the MV, we are still open to the assumption that an APase splits phosphate esters and transports the Pi produced into the MV. We can imagine that, in an as yet unknown way, the Ca transport is coupled to this Pi transport system. Ca ions may enter the active site of the enzyme or be bound to the phosphate esters and translocated into the MV together with the Pi. WARNER et al. (1983) carried out transport studies with ^{45}Ca and ^{32}Pi for the isolated MV fraction. They came to the conclusion that the APase at the MV membrane does not act as a phosphohydrolase but as a Pi-binding and transporting protein. The Ca transport into the MV would be coupled directly or indirectly to this enzyme, since they found that ^{45}Ca and ^{32}Pi are transported together, although the Ca/P ratios vary over time. In view of these results we cannot see why they reject the idea of the above-mentioned function of a phosphohydrolase being connected with such a transfer, since Pi must be produced before it can be transported.

Concerning the early stage of MV mineralization, WUTHIER (1982) believes that the first mineral stage is a so-called amorphous Ca-phosphate (ACP). Though the latest X-ray diffraction studies have shown that a real ACP does not exist, or only at a level of less than 1% in developing hard tissue (GRYNPAS et al. 1984), we are open to the opinion that in a very early, brief stage such "amorphous" prestages of crystallites do exist. However, such an ACP would rapidly be transformed to apatitic formations with a low state of crystallinity.

Some authors believe that the primary crystallites develop at the inner membrane of the MV (Fig. 11a insert, arrow 2). According to BOSKEY (1981, survey), the phosphate groups of the phospholipids in the membrane would first bind

Ca. Thus calcium-phospholipid-phosphate (Ca-Pl-PO$_4$) complexes would be formed in a first stage with a molar Ca/P ratio of about 1. Such complexes have been found in various hard tissues as well as in in vitro mineralization studies of the MV-rich fractions. The Ca-Pl-PO$_4$ complex would attract further Ca^{2+} and phosphate ions, so that unstable Ca-phosphate clusters may develop, followed by nuclei and subsequently stable Ca-phosphate islands. Parallel to this phospholipid mineralization, or directly afterwards, the mineralization starts and proceeds in the interior of the MV. Figure 11 b, c shows mineralized noduli at a fairly high magnification, with the probable outlines of the original MV marked in the center (Fig. 11 c, arrow). Apatitic strands and needles are visible. The magnifications are too low clearly to show strands with dotlike substructure. At the periphery of the noduli the apatitic needles are radially arranged (Fig. 11 b), and needles have partly coalesced with neighbors to form ribbons. The nodules (including the MV) are densely mineralized (Fig. 11 b, center). Very small crystal formations are densely packed. This dense mineralization, i.e., the high amount of mineral substance per unit volume, can be compared with that in peritubular dentine, which is like the nodules in having no collagen as matrix (HÖHLING et al. 1976b). Afterwards, when the whole extracellular volume in the longitudinal septum of the growth plate (like the whole predentine) has become mineralized, these mineralized noduli are still recognizable. It was determined by electron probe X-ray microanalysis that their amount of mineral substance per unit volume is much higher than that of the intervening regions, which are mineralized secondarily and which also contain collagen as matrix (BARCKHAUS and HÖHLING 1978).

b) Extravesicular Mineralization; Radial Arrangement of the Elongated Crystallites

The peripheral region of the noduli, which mainly comprises the region surrounding the original MV, is characterized by a radial arrangement of the apatitic strands (Fig. 11 b). From the beginning of our analysis on this nodule mineralization we said that this radial arrangement is caused by radially arranged matrix strands, and HÖHLING et al. (1976b) concluded that these matrix strands probably are the proteoglycan monomers (Fig. 14). Later, HUN-ZIKER et al. (1981), BONUCCI and REURINK (1978), and SCHERFT and GROOT (1981), using electron microscopy, observed that such fibrous strands do indeed radiate from the surface of the MV membrane and it is assumed that they are proteoglycan monomers. Since we could not focus the electron beam for electron probe X-ray microanalysis exactly onto the MV, and since such a small volume would not emit sufficient X-rays for an elemental analysis, we do not know whether the recorded S$_K$ radiation stems from the MV alone or also from the surrounding region (BARCKHAUS et al. 1981). It was concluded that the sulfur X-rays registered are indicative of the sulfate groups of sulfated proteoglycans. Gradually the nodule mineralization would proceed, in part via the radiating proteoglycan strands, and finally reach the collagen fibrils and the surrounding ground substance. It has not been fully clarified whether the collagen mineralization starts after the full mineralization of the nodules

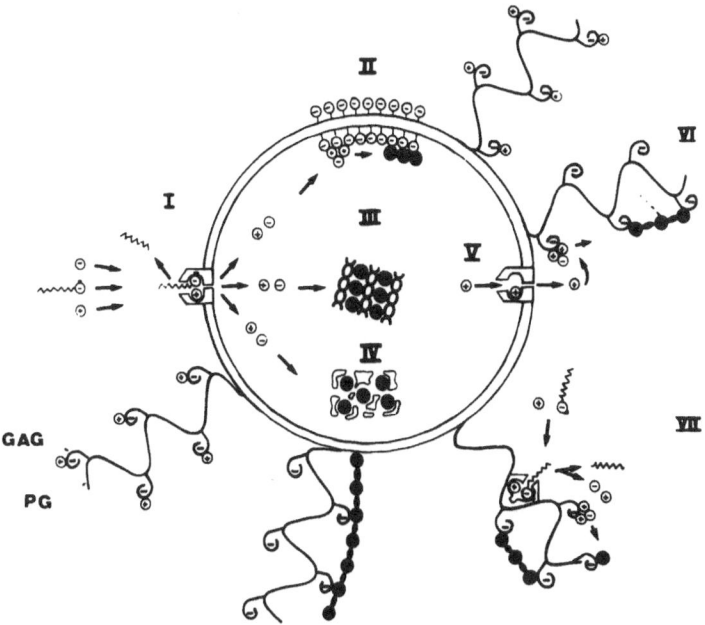

Fig. 14. Approximate scheme of Ca and phosphate entry into the matrix vesicles (*MV*) and MV mineralization. *I*, Possible splitting of phosphate esters and incorporation of Ca and inorganic phosphate into the MV. *II*, Ca-phosphate mineralization at the phospholipid membrane. *III*, *IV*, Ca-phosphate mineralization on macromolecular strands: *III*, longitudinal view; *IV*, cross-sectional view. *V*, Secondary extrusion of Ca^{2+} ions into the extravesicular space; this is unlikely in our opinion. *VI*, *VII*, Dotlike mineralization of the radiating *PG* macromolecules. *GAG*, glycosaminoglycan chain

(Fig. 11 b, c) or whether it starts during nodule mineralization. Nevertheless, it is widely accepted that chronologically, the MV are mineralized before the collagen fibrils. We shall see that the process of collagen mineralization is more complex than may be assumed at first glance. Above all, one has to differentiate between mineralization at the collagen fiber surface, which is connected with attached noncollagenous proteins, and mineralization inside the fibrils, which is mainly connected with the proteins of the collagen fibers.

III. Mineralization of the Collagen Fibers

1. General Aspects

Collagen is the dominant fibrous protein not only in connective tissue but also in hard tissues, bone, dentine, cementum, and even the mineralizing cartilage of the epiphyseal growth plate. It comprises about 80%–90% (by weight) of the organic substance in demineralized dentine and bone. Its relative content in the predentine and osteoid is somewhat less since the process of mineralization is connected with a loss of noncollagenous proteins, including the sulfated proteoglycans (PUGLIARELLO et al. 1970; BAYLINCK et al. 1972; ENGFELDT and

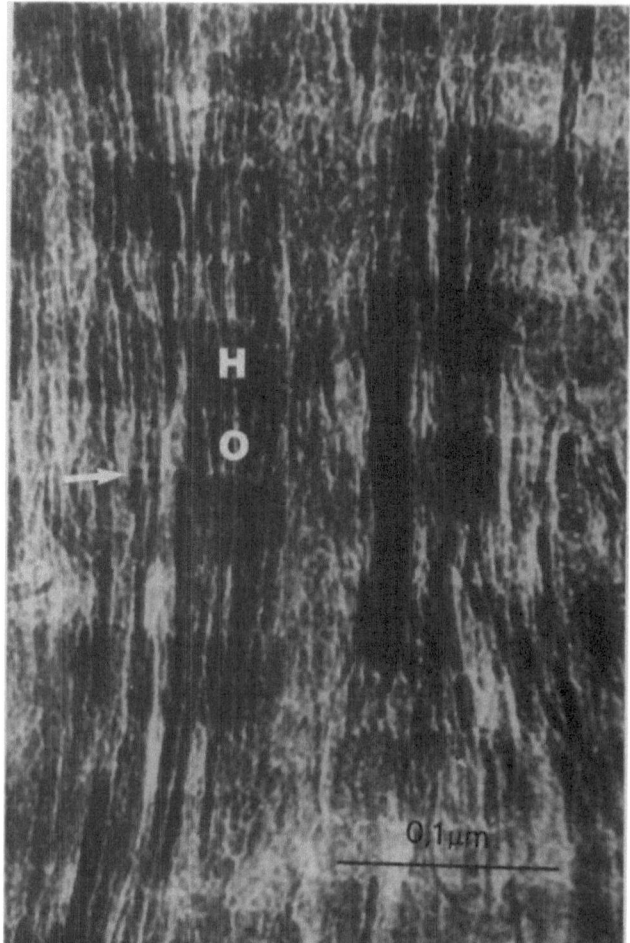

Fig. 15. Electron micrograph of mineralized collagen fibers arranged in parallel in mineralizing turkey tibia tendon. Some groups of crystallites are bound to the hole zone (*H*) and overlapping zone (*O*); others have grown beyond these zones. Often a dotlike substructure of the needles can be observed (*arrow*)

HJERPE 1972; HÖHLING et al. 1981a). While it was sometimes suggested that only the surface or the outer region of the collagen fiber is mineralized, it has to be noted that normally the whole collagen fiber in these hard tissues is filled with mineral substance. When the collagen fibers are arranged in parallel to form thicker bundles, as in lamellar bone and cementum, interior regions may be less mineralized.

It is generally assumed that the arrangement of the crystallites reflects and reinforces the macromolecular structure of collagen, as seen in Fig. 15. However, this reflection of the collagen structure by the crystallites is apparent only in relatively thick collagen fibers (diameter more than 50 nm), with the fiber bundles preferably arranged in parallel, but not in the thin collagen fibers of the

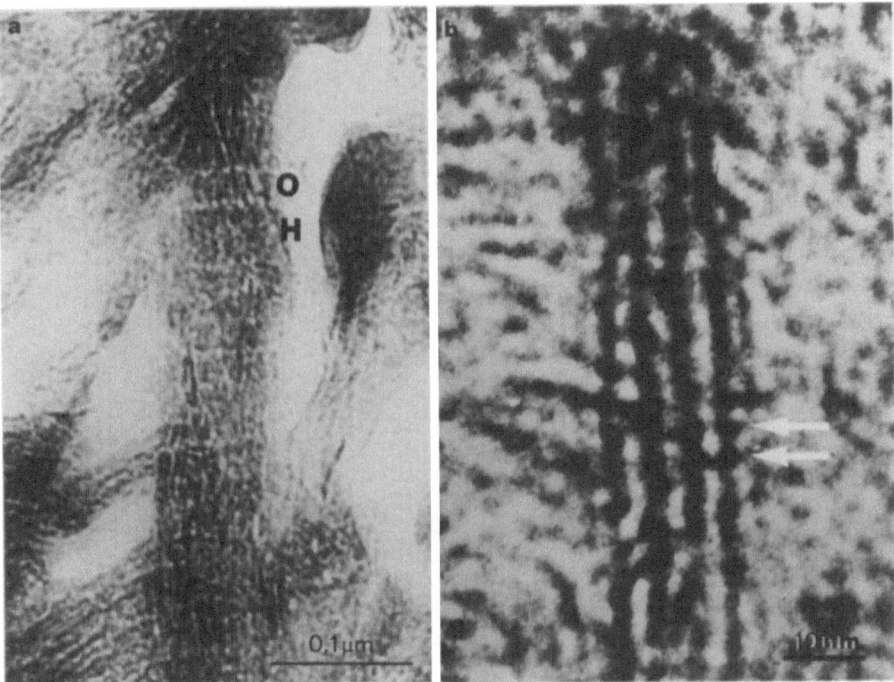

Fig. 16a, b. Electron micrographs of mineralized collagen fibers in developing dentine. **a** The collagen hole zone (*H*) and overlapping zone (*O*) are just visible as tiny light holes; Ca phosphate strands with dotlike substructure are partly visible (Fig. 3 of HÖHLING et al. 1971). **b** Higher magnification of Ca phosphate strands with a dotlike substructure (*arrows*), probably inside a collagen fiber

mineralizing cartilage (diameter 10–20 nm). In the dentine the collagen fibers form a network and are not arranged in parallel, in contrast to lamellar bone and cementum. The parallel arrangement of the crystallites is normally also visible in the singly lying fibrils of mature dentine, which have a thickness in the range 50–100 nm. However, strands of apatitic formation also appear in developing dentine, for which one can only assume that they are mineralized collagen fibers, because of the diameter of these strands and of a parallel arrangement of the crystallites over long distances. In this case, the arrangement of the crystallites in characteristic parallel lines is missing. In a few characteristic cases we observed tiny lines, running perpendicularly over the mineralized strand at distances reproducing the characteristic 67-nm period of collagen (Fig. 16a). This periodicity will be discussed in the chapter on collagen structure. The crossbanding in Fig. 16a is produced by tiny microholes, in which no mineral substance is present.

A frequent characteristic of mineralized collagen fibers is that the outer or surface region seems to be more densely mineralized than the inner region (Fig. 17). We believe that this is not only due to the fact that more Ca^{2+} and phosphate ions are present at the fiber surface for mineral formation, but that the ion diffusion into the interior of the fibril would be restricted. It is known that noncollagenous macromolecules are bound in a characteristic way

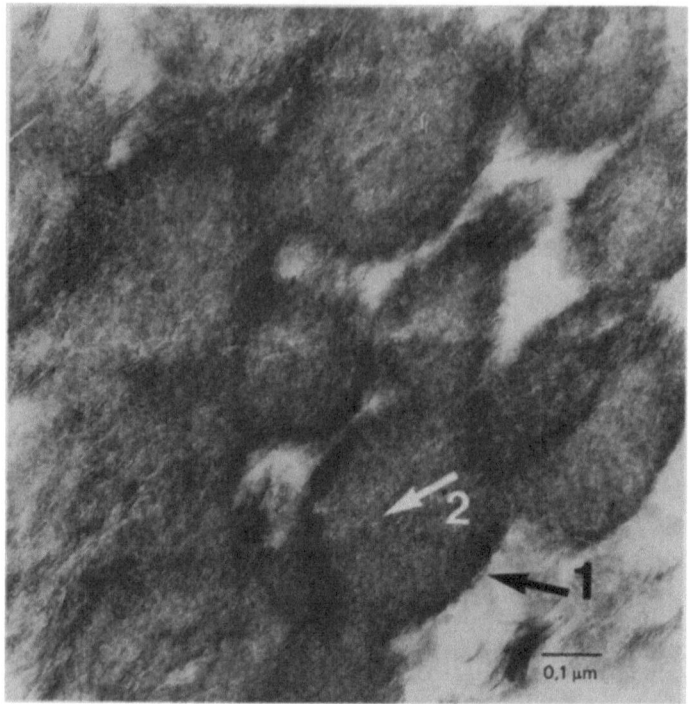

Fig. 17. Electron micrograph of mineralized collagen fibers in cross section. *Arrow 1* points to the periphery, where there seems to be more mineral substance per unit volume than in the interior (*arrow 2*)

to the fiber surface; and we conclude that they are more intensely nucleating matrix macromolecules than collagen itself. With this comment we do not want to discuss the "old" and frequently discussed question whether collagen is a well or a poorly nucleating macromolecule. Such a discussion is not fruitful since, in the course of evolution, collagen has become the dominant matrix macromolecule; in the mineralized state it fulfills important biomechanical functions. Hence it is more important to discuss the structural relationship between the collagen structure and the Ca-phosphate mineral.

2. Relation to the Paracrystalline Collagen Structure

a) Morphological Observations and Measurements of Mineralized Collagen

When we closely study electron micrographs of mineralized collagen fiber bundles at higher magnification, we can see apatitic strands and needles, often with a dotlike substructure, i.e., composed of islands (Fig. 16b). Often ribbon-plate-like crystallites prevail (Fig. 15), which are arranged in parallel lines perpendicular to the mineralized fiber, the distance from line to line being about 67 nm. This is the so-called collagen macroperiod, D. Often two lines of crystallites exist, the total distance $D = 67$ nm being subdivided into two parts of 0.6

and 0.4 D. Further, groups of crystallites exist (mainly needles) which are much longer than the macroperiod D. Their length seems to be independent of the collagen macroperiod (Fig. 15).

HÖHLING et al. (1974) and KREFTING et al. (1980) measured the center-to-center distances between parallel needlelike apatite crystallites in collagen of bone, dentine, and mineralizing turkey tibia tendon. They found that they lie mainly in the range of 4–5 nm. HÖHLING et al. (1971) also attempted to determine the number of dots (islands) per collagen macroperiod D. They did not find apatitic strands exhibiting a dotlike substructure along the full length of the macroperiod; however, from the mean center-to-center distances between these islands and from the counted number of dots for a certain length along the macroperiod D they concluded that the number of such islands per macroperiod D must be 9–13. While the distance between the apatitic lines (Fig. 15) corresponds to the collagen macroperiod, the question arises whether the results of the measurements of center-to-center distances between the apatitic formations can also be related to the collagen structure. To answer this question a short description of collagen structure is necessary. A review of the relationship between mineral formation and collagen structure has been given by HÖHLING et al. (1980a).

b) Collagen Structure

While two decades ago collagen was regarded as a single entity, research over the last 15 years has revealed that at least 10 different collagen types exist as different gene products. Furthermore, complex cellular processing of the final extracellular collagen fibers was detected. It is beyond the scope of this chapter to describe the different types of collagen. It suffices to mention that the most frequent form of collagen, type I, which prevails in the skin and the tendon, is also the dominant collagen in bone, dentine, and cementum. Collagen type II prevails in the mineralizing epiphyseal growth plate, where, in addition, collagen type X has been found in the hypertrophic zone (e.g. AYAD et al. 1987); it may also be connected with the mineralization process. We shall discuss the aspects of the paracrystalline structure of collagen type I necessary to understand collagen mineralization.

Collagen type I consists of two identical α1 (I) protein chains and one α2 chain whose amino-acid sequence has been almost completely determined (e.g., for survey see FIETZEK and KÜHN 1975). At this point we should mention that glycine (Gly) amounts to 33% of all amino acids, appearing at every third position in the triplet Gly-x-y, while the amino acids proline (Pro) and hydroxylproline (Hyp) make up nearly another third (22%). Knowledge of the high percentage of Gly was important in clarifying the paracrystalline structure of the collagen molecule by X-ray diffraction analysis. RAMACHANDRAN and KARTHA (1954) and RICH and CRICK (1955) concluded from the X-ray diffraction data that the molecule exists as a triple helix (Fig. 18). The three chains in the triple helix are mainly held together by hydrogen bonds. TRAUB et al. (1969) continued the X-ray diffraction analysis on collagen and proposed a model which is similar to that of RICH and CRICK (1955) and contains fewer

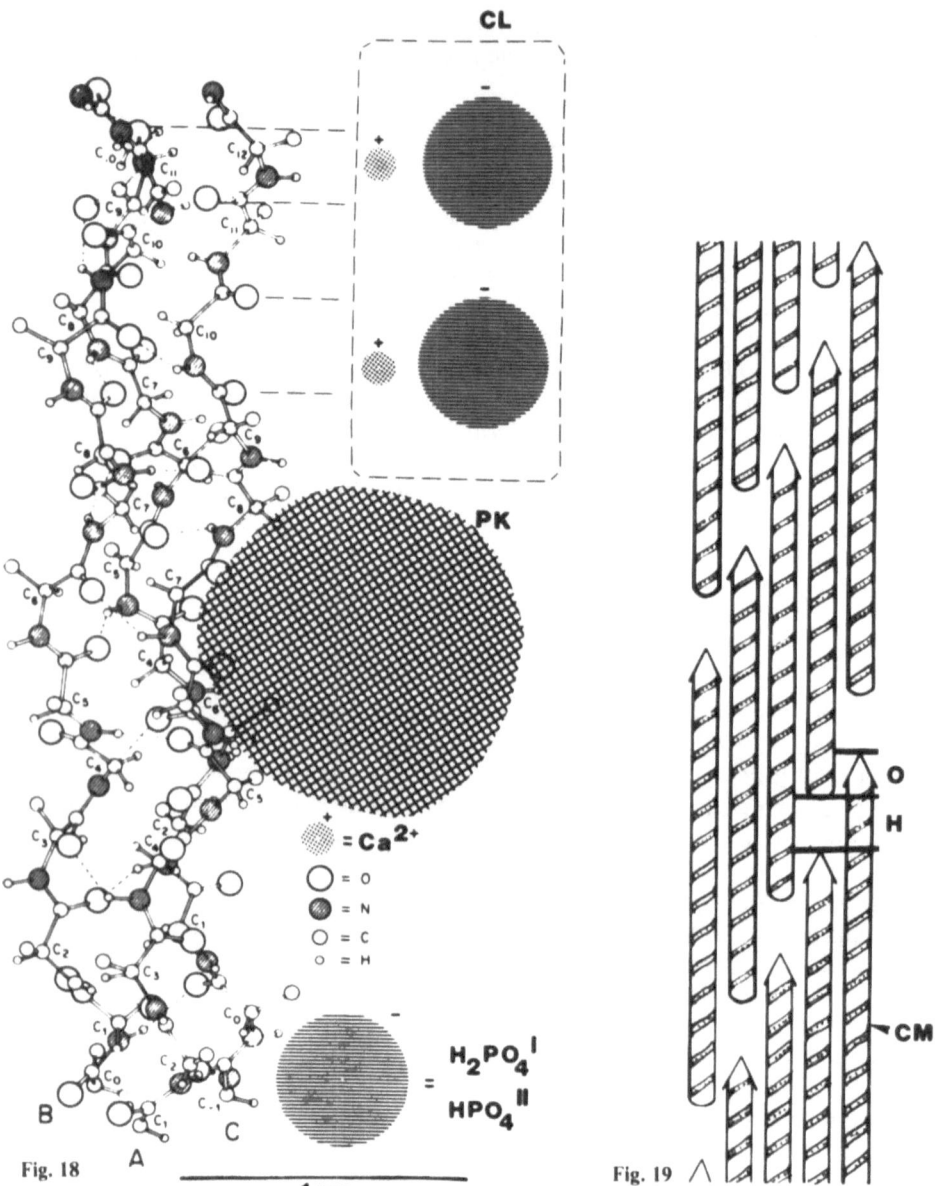

Fig. 18. Triple helix of Ramachandran (1963); a schematic representation of Ca^{2+} ions ($+$) and phosphate ions ($-$) bound to form clusters *Cl* (*top*) and nuclei *PK* (*below*)

Fig. 19. Scheme of the lateral aggregation of collagen triple helices (*CM*) according to Hodge and Petruska (1963). A terminal overlapping zone (*O*) and a zone with microholes (*H*) exist

hydrogen bonds than the RAMACHANDRAN model (Fig. 18). The triple helix has a diameter of about 1.4 nm and a length of about 290 nm. As Fig. 18 shows, the triple helix has a regular periodicity of extrusions and invaginations, Gly always being situated in the invaginations. Here the question arises of whether the binding of Ca^{2+} and phosphate ions (Fig. 18, top) by charged amino acids takes place mainly in the extrusions or in the invaginations, or whether neighboring extrusions and invaginations function as an active site (Fig. 18, bottom). Before we can discuss this point we have to explain the "hierarchy" for the development of helices, leading to the formation of visible fibers.

HODGE and PETRUSKA (1963) developed a two-dimensional model for the lateral aggregation of the helices (Fig. 19). As Fig. 19 shows, the neighboring helices are arranged together with an axial shift of $D = 67$ nm. This arrangement produces a zone with end-overlapping of the molecules, the "overlapping" zone (0) with a length of 0.4 D, and a zone without overlapping regions, which produces elongated spaces between neighboring molecules, the so-called hole zone (H) (length 0.6 D). At the end of the 1960s and the beginning of the 1970s, models were developed describing a three-dimensional arrangement of this two-dimensional structure, to form microfibrils. The existence of such microfibrils with a diameter of about 4.0 nm had been postulated using X-ray small angle diffraction; later they were directly observed by an electron microscope using negative staining (HOSEMANN and NEMETSCHEK 1973). According to SMITH (1968) and MILLER and PARRY (1973), five molecules (triple helices) form a five-stranded rope, the microfibril. MILLER and PARRY (1973) proposed that they aggregate in a tetragonal array (Fig. 20d, e). This model is now less favored than the original hexagonal model of KATZ and LI (1974) as somewhat modified by TRUS and PIETZ (1980) and by the pseudohexagonal model of MILLER and TOCHETTI (1981). This hexagonal structure would not greatly change the principal aspects of crystal nucleation and growth (as far as the existence of the 4-nm microfibrils is assumed) which we have demonstrated originally in the tetragonal model (Fig. 20d, e) and which we will discuss later.

For several years it was assumed that these microfibrils aggregate continuously (e.g., shown in Fig. 20d, e) to form the thick collagen fibers with the typical cross striation. However, HÖHLING et al. (1981 b) presented evidence that the large collagen fibers in mineralizing turkey tibia tendon with a diameter in the range 100–400 nm are not only composed of microfibrils with a diameter of 4 nm. There are also somewhat thicker fibrils with a diameter of about 20 nm in the thick tendon fiber. ITOH et al. (1982) obtained corresponding results for collagen of soft tissue. The hierarchy for the development from the small to the large entity would involve four steps: (1) the molecule (triple helix 1.4 nm); (2) the microfibril (with a diameter of 4 nm); (3) a somewhat thicker microfibril with a diameter depending upon the type of the tissue and thus the diameter of the thick fiber; and (4) the thick collagen fiber with cross striation, exhibiting a width of 50–100 nm in bone and dentine and up to 400 nm in mineralizing turkey tibia tendon.

Concerning the active sites for apatite nucleation in the interior of the collagen fiber, it is relevant to note that the arrangement of the microfibrils into

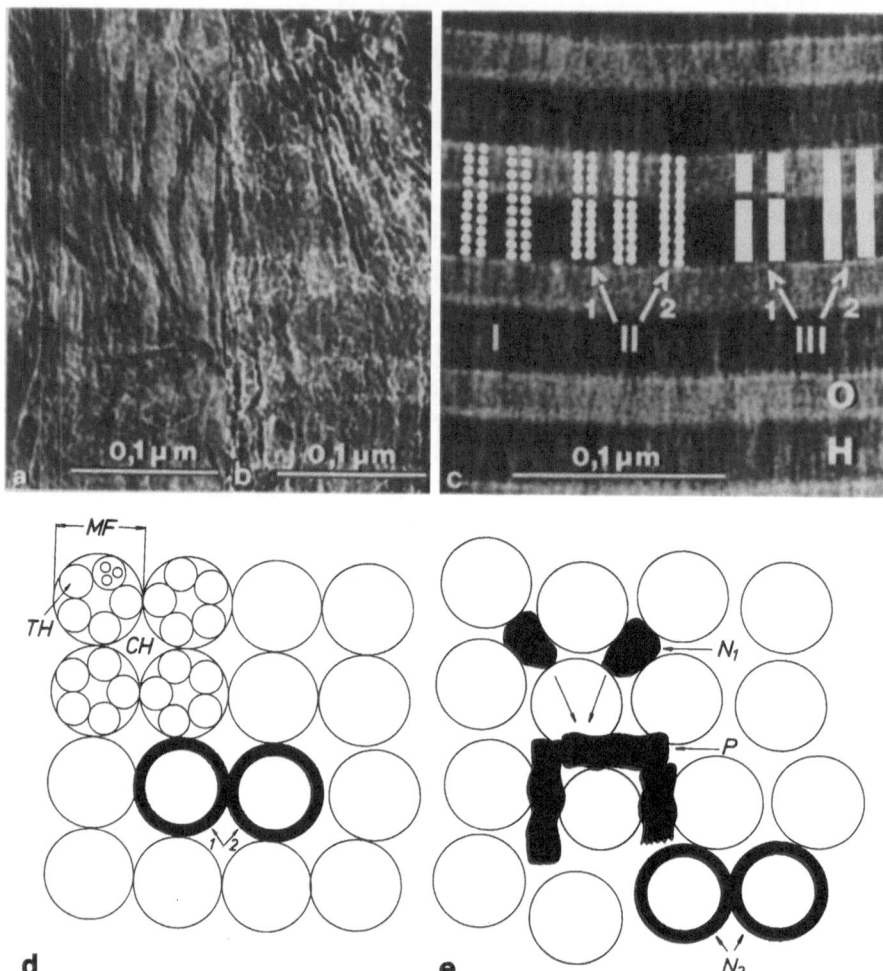

Fig. 20 a–e. Electron micrographs of mineralized turkey tibia tendon and of negatively stained collagen fibers, and a cross section through the MILLER and PARRY (1973) collagen model. **a, b** Longitudinal section through mineralized turkey tibia tendon showing the hole zone (*H*) and overlapping zone (*O*). **c** Negatively stained collagen fiber with stages of crystal formation (*I, II, III*). *I → III*, transition from apatitic islands to ribbonlike crystallites (*III*). **d, e** A cross section through the microfibrils (*MF*) in the tetragonal model of MILLER and PARRY (1973). *CH*, microholes between the microfibrils; *N₁* dotlike islands developed in the microholes; *N₂*, direct embedding of a microfibril with mineral substance according to BOYDE (1974); *P*, Ca-phosphate strands which have coalesced to form ribbons; *TH* triple helices

4-nm-thick structures produces microchannels with a diameter of about 1.8 nm. Within the channels apatitic nucleation and crystal growth can take place (Fig. 20 d, e). Two of the protein chains of the triple helix protrude into the channels and contribute to the nucleation (Figs. 18, 20c, d). Amino-acid sequence analysis, especially in relation to the cross striation, representing the so-called polar regions, has shown that polar, charged amino acids prevail in these regions (VON DER MARK et al. 1970). As they induce the nucleation of the positively or negatively charged staining substance (uranyl acetate, phosphotungstic acid) they may represent sites for a preferential apatitic nucleation. The number of cross striations per macroperiod D and their distance correspond to those of the islands in the apatitic chains in collagen discussed before. Since negatively charged amino acids (such as Glu, Asp) are often neighbors of positively charged amino acids, up to four polar amino acids may appear in the two protein chains of such an extrusion of the triple helix (Fig. 18). The question of whether one or more Ca^{2+} ions, in connection with accompanying phosphate ions, would be bound to such an extrusion to form a mineral cluster probably cannot be answered. In the near future it will also not be possible to clarify whether a Ca-phosphate cluster of one such extrusion or those of a few extrusions are necessary to form a crystal nucleus and subsequently a stable entity (Fig. 18). Our conception is depicted in Fig. 20. Figure 20a, b represents longitudinal sections through mineralized collagen fibers, with the hole (H) and overlapping zone (O) and Fig. 20c is a negative staining micrograph of a collagen fiber of the same magnification. On the left side of Fig. 20c (I) apatitic islands are shown (as dots) in the hole zone (H) and overlapping zone (O). The islands (N_1) would develop at active sites in the microchannels between the microfibrils as shown in Fig. 20d, e, which is a cross section through a collagen fiber with its microfibrils (MF). The islands in Fig. 20c would rapidly coalesce to form apatitic strands (region II); arrow 1 shows the crystallites which are confined to the hole and overlapping zone, and arrow 2 those which grow beyond the border of both zones. Both sorts of crystallites can be seen in Fig. 20a, b. Region III of Fig. 20c shows that the apatitic strands and needles coalesce with neighbors to form ribbonlike crystallites. The same is shown in the cross section of the collagen fiber (Fig. 20e, P). Arrow 1 in region III of Fig. 20c shows that the ribbonlike crystallites are confined to the hole (H) and overlapping zone (O), and arrow 2 that they have grown beyond the borders of both zones.

Concerning the appearance of lines of crystallites in the hole and overlapping zone, we assume that the growth in length is hindered by cross bridges between the molecules and microfibrils in the border region of the hole and overlapping zone. Concerning the longer crystallites in Fig. 20a, b, which are not confined to the hole and overlapping zone, two explanations are possible: on the one hand they might have developed on the surface of the thicker type of microfibrils (thicker than 4 nm) so that their growth in length was not hindered by the collagen structure; on the other hand, the cross bridges which would hinder the growth in length may not be evenly distributed or in certain cases the growing crystallite might break them, thus reaching a greater length.

Since the data on the collagen structure, on the aggregation to fibers, and on the amino-acid sequence analysis are in agreement with our observations on crystal development and with our measurements of their arrangement, we believe that our ideas (presented in Fig. 20) contain elements of the true matrix-crystal relationship in collagen. However, details of this relationship will probably have to be corrected. For instance, while we have concluded that a crystal of the length of about 67 nm has developed from several islands which have formed at different active sites, BERTHET-COLOMINAS et al. (1979) have developed a different hypothesis. They assume that only one nucleus develops in the region of the collagen d band and grows continuously in length to form a crystallite of the length of about 67 nm. It would grow independent of any other active sites it might pass during growth. This is only one of the different theories that are possible. Further intense analyses are necessary to clarify the details of the process of this heterogeneous nucleation in collagen, enamel, and the noncollagenous matrices.

c) Mineralization on the Surface of the Collagen Fibers

Figure 17 shows that a denser mineralization, i.e., a higher amount of mineral substance per unit volume, appears to exist at the surface regions of the collagen fibers. Electron probe X-ray microanalysis has demonstrated that a higher degree of mineralization indeed exists in the surface region, extending into the interfibrillar space. It can reach values of up to 100% more mineral substance per volume (HÖHLING et al. 1976a). The crystallites in this region are smaller and more densely packed; and their parallel arrangement does not reflect the macromolecular collagen structure. It is impossible to clarify whether this region of dense mineralization is confined to the surface of the fiber (from there proceeding to the interfibrillar space) or whether it also penetrates into the interior of the thick fiber. We believe that this dense mineralization is mainly due to the nucleating activity of noncollagenous proteins, e.g., phosphoproteins and proteoglycans. They are probably bound in a characteristic way to collagen fiber. For instance, SCOTT and HAIGH (1985) have applied the electron-dense dye cupromeronic blue to nonaqueous demineralized bone and shown that proteoglycan filaments are oriented parallel to the fiber axis, while in soft tissue they are circularly arranged around the collagen fiber in the region of the d band. They may have a strong influence on crystal nucleation. This example suggests that the nucleation process at the collagen surface is complex since noncollagenous proteins are probably involved. We assume that the collagen mineralization starts somewhere at the surface of the collagen fiber, from where it might proceed into the interior of the collagen fiber (and possibly also into the extracollagenous space).

IV. Extracollagenous Mineralization

1. General Aspects

We have just discussed that the degree of mineralization is higher in the surface regions of the collagen fiber, which probably comprises part of the intercollagenous space (the region of the ground substance). It is our impression that the extracollagenous region is more densely mineralized. The extracollagenous regions were more highly mineralized, for example, in the mineralizing turkey tibia tendon (HÖHLING et al. 1976a), the nodules comprising the MV which contain no collagen (BARCKHAUS and HÖHLING 1978), the peritubular dentine, which also contains no collagen, and the dense regions of osteopetrotic bone which also contain less collagen (HÖHLING et al. 1972). The question arises as to which sort of macromolecules exist outside the collagen fibers and induce crystal formation.

2. Types of Macromolecules of the Noncollagenous Regions

Special biochemical separation methods have been used to gain access to all the noncollagenous proteins in the different hard tissues. TERMINE et al. (1979) proposed as the first step preliminary dissociative extraction of the hard tissue in nondemineralizing agents to dissolve all the proteins not strongly bound to the mineral; in a second step the extraction-denaturing agent is combined with EDTA to bring the mineral into solution and thus to gain access to the mineral-bound organic substance. It was found that the prevailing groups of noncollagenous proteins in bone and dentine are phosphoproteins (phosphoryns), proteoglycans, glycoproteins, and proteins containing γ-carboxyglutamic acid, the so-called Gla-proteins. I shall discuss the nature and possible function of these proteins mainly in relation to dentine mineralization. BUTLER (1984) has surveyed these noncollagenous proteins in dentine and bone, and this survey will be used here. The most important class of noncollagenous proteins in dentine is that of the phosphoproteins (phosphoryns).

a) Phosphoproteins

DIMUZIO and VEIS (1978) call the dentine phosphoproteins phosphoryns. They are proteins with a high organic phosphate content, this being due to serine phosphate alone in dentine (LEE et al. 1977), while in bone a phosphothreonine has also been found (COHEN-SOLAL et al. 1978). In the dentine of bovine teeth probably only a single component exists, with a molecular weight of 155000 Daltons. It contains 46% phosphoserine and 45% aspartic acid (Asp). It is thought that in rat incisors three families probably exist, these having molecular weights in the region of 90000–95000 Daltons and differing phosphate content (BUTLER et al. 1983). Autoradiographic studies (WEINSTOCK and LEBLOND 1973) have shown that the phosphoprotein is secreted after transport through the odontoblastic processes at the dentine mineralizing front. It is rapidly incorporated into dentine. Since the phosphoryns bind Ca strongly (STETLER-STEVENSON and VEIS 1987), it is assumed that they induce nucleation by binding calcium.

b) Proteoglycans

The proteoglycans (PGs), i.e., the proteoglycan monomers, consist of a central protein chain to which disaccharide chains are attached. In cartilage these PG monomers are often attached to long hyaluronate chains and form PG aggregates. There is reason to assume that they bind Ca but prevent Ca phosphate nucleation and/or crystal growth. Probably a special structural change and deaggregation of the hyaluronate aggregates in the hypertrophic zone is necessary to induce mineral nucleation (e.g., BUCKWALTER et al. 1987). Furthermore, the PG subunits are smaller in this state (e.g., CAMPO and RAMANA 1986). In bone and dentine only the PG monomers and not the hyaluronate aggregates exist. Hence the PGs are much smaller. A small PG with a core protein of molecular weight of 38000 Daltons and one or two glycosaminoglycan (GAG) chains (containing mainly chondroitin sulfate 4, less 6) with an average molecular weight of 40000 Daltons was found by FISHER et al. (1983) in developing bone. The structure of the PGs in predentine and dentine has not yet been so well characterized (RAHEMTULLA et al. 1984). GOLDBERG et al. (1983, 1987) applied the electron-dense dye cupromeronic blue in the critical electrolyte concentration mode with $MgCl_2$ to predentine and demineralized dentine. They found elongated, ribbonlike PG staining products 60 nm long and 14 nm thick in spaces between the collagen fibers. As with bone collagen (SCOTT and HAIGH 1985), the collagen-bound PGs were positioned with their long axes parallel to the axis of the collagen fiber. The amount of staining substance was less in the dentine. These results show that part of the PG is bound to collagen and part contributes to the intercollagenous ground substance, as is the case for the phosphoryns. That there is less PG in dentine than in predentine supports the results of chemical and microprobe tests showing that a loss of PGs takes place with the process of mineralization. This loss has been taken by many authors to indicate that the PGs hinder mineralization and must be removed before biomineralization can start and continue. However, I believe that this problem is more complex and that the PG monomers have a partly crystal nucleation function and might partly act as Ca-transporting macromolecules. After the transport the Ca-transporting molecules would decompose and be transported in a partly decomposed state to the cells to undergo further decomposition. Also, there may be a surplus of some of the nucleation-inducing PG molecules and these may be decomposed and transported to the cells for further decomposition. As discussed in Sect. II 4b, we assume that the fibrous strands radiating from the matrix vesicles are PG monomers and Ca-phosphate nucleators with a center-to-center distance of the Ca-phosphate nucleating sites in the range of 5–6 nm.

c) Glycoproteins

In the early 1960s HERRING (survey 1968) had already characterized a bone-specific glycoprotein rich in sialic acid with a molecular weight of about 23000 Daltons. Recently, extraction and separation methods with added protease inhibitors to prevent breakdown of the molecules have been used. TERMINE et al. (1981) found three glycoproteins from fetal calf bone associated with the mineral phase. The 32000-Dalton component was called osteonectin by TERMINE. Bone contains a relatively large amount of it. Since it has a strong affinity to collagen and apatite, it was assumed that it initiates crystal formation of the collagen fiber. For dentine BUTLER et al. (1981) and DIMUZIO et al. (1983) isolated a sialic acid-rich glycoprotein with a molecular weight of 95000 Daltons. Many authors assume that these glycoproteins also contribute to the induction of biomineralization.

d) γ-Carboxyglutamic Acid-Containing Proteins and Gla-Proteins,
e.g., Osteocalcin

HAUSCHKA et al. (1975) and PRICE et al. (1976) have isolated a special protein from bone which they called osteocalcin. It contains 49–51 amino acids, with two or three Gla residues. It was found that the protein chain folds back on itself with NH_2 and COOH regions juxtaposed; two α-helical regions are separated by a β-turn, stabilized by a disulfide bridge. The three Gla residues project outwards of the molecule and might bind Ca. HAUSCHKA et al. concluded that osteocalcin developed from a higher molecular weight precursor. The Gla-protein would split off and is assumed to be causally related to biomineralization. PRICE et al. (1976) believe that the Gla-protein has developed independently and that it is not a causative factor for biomineralization. Meanwhile PRICE et al. (1985) have isolated a second Gla-protein which they assume is connected with the beginning of mineral formation. LINDE et al. (1980) also found Gla-proteins in dentine. Immunohistochemical localization studies by BRONCKERS et al. (1985) have shown that the dentine Gla-protein is demonstrable only in dentine and not in predentine. This may mean that the protein is extruded at the predentine/dentine border and rapidly incorporated into the mineralizing dentine, as is the case for the phosphoryns.

D. Conclusion: Differences and Similarities in Mineralization in Enamel, Dentine, and Other Hard Tissues, and a General Calcification Theory

In Sect. A the question was already raised of whether similarities exist between mineralization in enamel, dentine, and the other hard tissues such as cementum, bone, and mineralizing cartilage. Before this question could be discussed it was necessary to describe the main processes of enamel mineralization (Sect. B) and dentine mineralization (Sect. C). It became evident that, at the electron microscopical level in the μm range, important differences exist concerning the arrangement of the crystallites. Furthermore, the long prismatic crystallites in enamel are thicker by a factor of about 10 than the needle- and ribbonlike crystallites in dentine, cementum, bone, and cartilage (diameter 30–50 nm) (Fig. 5a, b) and are also much longer. While the crystallites in enamel are arranged precisely in parallel in relatively large compartments, the enamel prisms (Fig. 1) having a diameter of about 5 μm, the compartments of the parallel arrangement in dentine are much smaller. They normally have the same diameter as the collagen fibers, lying roughly in the range of 50–100 nm. In cementum and bone, the diameter of these compartments is somewhat larger since collagen fibers are arranged in parallel in bundles; the c-axes of the ribbon- and needlelike crystallites follow the direction of these bundles. Furthermore, while the long prismatic crystallites in the enamel prisms are arranged precisely in parallel, the crystallites of the noduli, the mineralization of which started in the matrix vesicles (Fig. 11 b), are normally radially arranged.

In spite of such differences in the arrangement and size of the crystallites between the different hard-tissue systems, the question should finally be posed of whether similarities also exist. When the early stages of mineral formation for enamel (Fig. 5a, arrow) and dentine (e.g., Fig. 16) are compared, one can see that they have a similar appearance. They consist of mineral strands in the form of needles with diameters in the range 1.5–5 nm. While most authors who have analyzed enamel mineralization (e.g., DACULSI and KEREBEL 1978) describe the earliest enamel crystallites as long thin ribbons, we believe that mineral strands exist, which become ribbonlike at a secondary stage. We have concluded that the same sequence occurs for the first stages of mineral formation in collagen-rich systems such as dentine. So, according to these deliberations, initial mineral formation in the various hard tissue systems should show similarities in morphological expression and in diameter range. Further, we assume that the crystallites do not grow continuously, e.g., from one dotlike structure to a long needle, but develop from several nuclei which grow into islands which rapidly coalesce.

The question arises of whether other authors have also reached similar conclusions with respect to fundamental similarities in the process of crystal formation for the different hard-tissue systems.

BONUCCI's idea of the existence of "crystal ghosts" seems to be in agreement with our ideas (review, BONUCCI 1985). BONUCCI used "postembedding demineralization" (PEDS), e.g., with acidic phosphotungstic acid or with acidic bismuth nitrate, and ultrathin sections to study various hard tissues (e.g., cartilage, bone, enamel). He found that contrast-rich strands appear at the same site at which the mineral strands had existed before. He concluded that the strands represent the original matrix strands. They appear in noduli of epiphyseal cartilage as well as in those of bone. He also found crystal ghosts with similar morphology in developing dentine and enamel. He assumed that these crystal ghosts are the morphological expression of the same fundamental process of mineral formation in the various different hard tissues which were analyzed using this method. The fibrous strands would induce crystal nucleation in the various hard tissues. They may partly contain similar basic components.

He described the crystal ghosts as "glyco-lipo-proteic" components. However, every hard-tissue system may contain specific macromolecules in its crystal ghosts such as osteonectin for bone, phosphoryns for dentine, and enamelin for enamel.

The "crystal ghosts" require further consideration. The question arises of whether the staining in ultrathin sections connected with the demineralization represents the matrix substance at its original site. If microchannels with a diameter in the range of that of the early crystallites (~ 2 nm) exist between the matrix macromolecules, the latter need not be dislocated during early crystal formation (Figs. 3, 7). In this case the stained matrix might be located at its original site. However, if the diameter of the channels were smaller, the early growing crystallites would dislocate the matrix macromolecules by distances of about 1–2 nm since they cannot incorporate much organic substance.

The crystal ghosts only demonstrate the existence of matrix substance. Electron microscopy of ultrathin sections has insufficient resolution to show morphological details of the macromolecular structure. If islands do exist in the early mineral strands, it would be interesting to clarify whether the crystal ghosts also contain such a substructure as a possible indication of active sites. It has been mentioned that knowledge of the amino-acid sequence and of the paracrystalline structure of the matrix macromolecules would be necessary to obtain information on the spatial distances between active sites for crystal nucleation. Had this information been obtained, it would not mean that the distances between active sites which are really active during in vivo mineralization were known. It might be that an active site attracts all the Ca^{2+} and phosphate ions of the neighborhood at the expense of an adjacent active site which, for example, might have started to bind ions at a later stage. The explanation of the process of nucleation at the macromolecular level becomes even more complex when one considers that development of the crystal nuclei depends on the distribution and activity of the APase molecules along the matrix macromolecules. They deliver the Pi, and possibly the Ca^{2+} ions for crystal nucleation. An attempt should be made to clarify whether the distances between the dots (the islands) in the early crystallites (e.g., Fig. 16b) really indicate the distances between active sites that occur during in vivo mineralization. One way that such clarification might be obtained derives from purified matrix macromolecules such as collagen molecules or microfibrils (not the thick fiber) being systematically mineralized in vitro. The mineralized macromolecules should be sprayed as monolayers onto films to get a higher resolution in the electron microscope. Such in vitro experiments should be used to study whether crystal islands indeed develop first and then coalesce to form needles. If they do exist the center-to-center distances between such islands should be determined to obtain information on the distances between active sites. Should the contrast of Ca-phosphate islands in the electron microscope be too low, Sr^{2+} ions should be used. However, we do not know whether such in vitro experiments will really solve this problem.

Only the question of the nucleus-matrix relationship has been discussed here; we have omitted to discuss the way in which the Ca and phosphate could be transported to the active site without being nucleated as Ca-phosphate on their way, i.e., at the wrong site. We have also omitted to discuss the way in which the enzyme machinery, e.g., the APase, must function to induce the mineral nucleation. If one tries to understand fully the details of the nucleation process, which obviously needs several controlling steps, it becomes obvious that we are still a long way from having completely clarified this multistep process.

Acknowledgements. We thank Mrs. Anita Möllers and Mrs. Marita Müller for the careful typing of the article.

References

Anderson HC (1967) Electron microscopic studies of induced cartilage development and calcification. J Cell Biol 35:81–101

Ali SY, Sajdera SW, Anderson HC (1970) Isolation of calcifying matrix vesicles from epiphyseal cartilage. Proc Natl Acad Sci USA 67:1513–1520

Arends J, Jongebloed WL (1978) Crystallite dimensions of enamel. J Biol Buccale 6:161–171

Astbury WT, Beighton E, Parker KD (1959) The cross-β configuration in supercontracted proteins. Biochem Biophys Acta 35:17–25

Ayad S, Kwan APL, Grant ME (1987) Partial characterization of type X collagen from bovine growth plate cartilage. FEBS Lett 220:181–186

Barckhaus RH, Höhling HJ (1978) Electron microscopical microprobe analysis of freeze dried and unstained mineralized epiphyseal cartilage. Cell Tissue Res 186:541–549

Barckhaus RH, Krefting ER, Althoff J, Quint P, Höhling HJ (1981) Electron microscopic microprobe analysis on the initial stages of mineral formation in the epiphyseal growth plate. Cell Tissue Res 217:661–666

Barckhaus RH, Greinke F, Goebeler M, Höhling HJ (1986) Analysis of the fine structure and pathway of extrusion of the matrix vesicles (MV) and related compounds in the epiphyseal growth plate. In: Ali SY (ed) Cell mediated calcification and matrix vesicles. Elsevier Science Publishers, Amsterdam New York Oxford, pp 39–44

Baylink J, Wergedal J, Thompson E (1972) Loss of proteinpolysaccharides at sites where bone mineralization is initiated. J Histochem Cytochem 20:279–292

Bernard G, Marvoso V (1981) Matrix vesicles as an assay for primary tissue calcification in vivo and in vitro. In: Ascenzi A, Bonucci E, de Bernard B (eds) Matrix vesicles. Wichtig Editore, Milano, pp 5–11

Berthet-Colominas C, Miller A, White SW (1979) Structural study of the calcifying collagen in turkey leg tendons. J Mol Biol 134:431–445

Bonar LC, Mechanic GL a Glimcher MJ (1965) Optical rotary dispersion studies of neutral soluble proteins of embryonic bovine enamel. J Ultrastruct Res 13:296–307

Bonucci E (1967) Fine structure of early cartilage calcification. J Ultrastruct Res 20:33–50

Bonucci E (1985) Crystal matrix relationship in calcifying organic matrices. In: Belcourt AB, Ruch JV (eds) Tooth morphogenesis and differentiation. Les Editions INSERM, Paris, pp 459–471

Bonucci E, Reurink J (1978) The fine structure of decalcified cartilage and bone. A comparison between decalcification procedures performed before and after embedding. Calcif Tissue Res 25:179–190

Boskey AL (1981) Current concepts of the physiology and biochemistry of calcification. Clin Orthop 157:225–257

Boyde A (1964) "The structure and development of mammalian enamel". PhD Thesis, Dept Anatomy, The London Hospital Medical College

Boyde A (1974) Transmission electron microscopy of ion beam thinned dentine. Cell Tissue Res 152:543–550

Boyde A, Reith EJ (1977) Qualitative electronprobe analysis of secretory ameloblasts and odontoblasts in the rat incisor. Histochemistry 50:347–354

Bronkers ALJJ, Gay S, DiMuzio MT, Butler WT (1985) Immunolocalization of γ-carboxyglutamic acid containing proteins in developing molar tooth germs of the rat. Coll Relat Res 5:17–22

Brown WE, Chow LC, Siew C, Gruninger S (1984) Acidic calcium phosphate precursors in formation of enamel mineral. In: Fearnhead RW, Suga S (eds) Tooth enamel IV. Elsevier Science Publishers, Amsterdam New York Oxford, pp 8–13

Buckwalter JA, Rosenberg LC, Ungar R (1987) Changes in proteoglycan aggregates during cartilage mineralization. Calcif Tissue Int 41:228–236

Butler WT (1984) Matrix macromolecules of bone and dentin. Coll Relat Res 4:297–307

Butler WT, Bhown M, DiMuzio MT, Linde A (1981) Noncollagenous proteins of dentin. Isolation and partial characterization of rat dentin proteins and proteoglycans using a three-step preparative method. Coll Relat Res 1:187–199

Butler WT, Bhown M, DiMuzio MT, Cothran WC, Linde A (1983) Multiple forms of rat dentin phosphoproteins. Arch Biochem Biophys 225:178–186

Campo RD, Romano JE (1986) Changes in cartilage proteoglycans associated with calcification. Calcif Tissue Int 39:175–184

Casciani FS, Doty S, Etz E (1979) The Raman spectra of mineralizing rat incisor enamel. J Dent Res 58A:231 (Abstr No 551)

Cohen-Solal L, Lian JB, Kossiva D, Glimcher MJ (1978) The identification of O-phosphothreonine in the soluble non collageneous phosphoproteins of bone matrix. FEBS Lett 89:107–110

Daculsi G, Kerebel B (1978) High resolution electron microscope study of human enamel crystallites: Size, shape, growth. J Ultrastruct Res 65:163–172

Deporter DA (1977) The early mineralization of enamel. Calcif Tissue Res 24:271–274

DiMuzio MT, Veis A (1978) Phosphoryns – major noncollageneous proteins of rat incisor dentin. Calcif Tissue Res 25:169–178

DiMuzio MT, Bhown M, Butler WT (1983) The biosynthesis of γ-carboxyglutamic acid-containing proteins by rat incisor odontoblasts in organ culture. Biochem J 216:249–257

Eastoe JE (1965) The chemical composition of bone and tooth. Adv Fluor Res Dent Caries Prevention 3:5–17

Eisenmann DR, Glick PL (1972) Ultrastructure of initial crystal formation in dentin. J Ultrastruct Res 41:18–28

Eisenmann DR, Ashrafi SH, Neumann A (1979) Calcium transport and the secretory ameloblast. Anat Rec 193:403–422

Eisenmann DR, Ashrafi SH, Zaki AE (1984) Calcium distribution in freeze dried enamel organ tissue during normal and altered enamel mineralization. Calcif Tissue Int 36:596–603

Engfeldt EB, Hjerpe A (1972) Glycosaminoglycans of dentine and predentine. Calcif Tissue Res 10:152–159

Fietzek PP, Kühn K (1975) Information contained in the amino-acid sequence of the α1-chain of collagen and its consequences upon the formation of the triplehelix of fibrils and crosslinks. Mol Cell Biochem 8:141–157

Fincham AG (1984) Amelogenins, progress and problems. In: Fearnhead RW, Suga S (ed) Tooth enamel IV. Elsevier Science Publishers, Amsterdam New York Oxford, pp 114–119

Fisher LW, Termine JD, Dejter SW Jr, Whitson SW, Yanagishita M, Kimura JH, Hascall VC, Kleinman HK, Hassell JR, Nilsson B (1983) Proteoglycans of developing bone. J Biol Chem 258:6588–6594

Frank RM, Nalbandian J (1967) Ultrastructure of amelogenesis. In: Miles AEW (ed) Structural and chemical organization of teeth, I. Academic, New York, pp 399–462

Garant PR (1986) personal letter July 30

Glimcher MJ, Bonar LC, Daniel EJ (1961) The molecular structure of the protein matrix of bovine dental enamel. J Mol Biol 3:541–546

Goldberg M, Septier D (1983) Electron microscopic visualization of proteoglycans in rat incisor predentine and dentine with cuprolinic blue. Arch Oral Biol 28:79–83

Goldberg M, Septier D, Escaig-Haye F (1987) Glycoconjugates in dentino-genesis and dentine. In: Graumann W, Lojda Z, Pearse AGE, Schiebler TH (eds) Progress histochem and cytochem 17, no 2. Fischer, Stuttgart New York

Grynpas MD, Bonar LC, Glimcher MJ (1984) Failure to detect an amorphous calcium phosphate solid phase in bone mineral: A radial distribution function study. Calcif Tissue Int 36:291–301

Hauschka PV, Lian JB, Gallop PM (1975) Direct identification of the calcium-binding amino-acid γ-carboxyglutamate in mineralized tissue. Proc Natl Acad Sci USA 72:3925–3929

Hayashi Y, Bianco P, Shimokawa H, Termine JD, Bonucci E (1986) Organic-inorganic relationships, and immunhistochemical localization of amelogenins and enamelins in developing enamel. Basic Appl Histochem 30:291–299

Herring GM (1968) The chemical structure of tendon, cartilage, dentine and bone matrix. Clin Orthop 30:261–299

Hirschman A, Deutsch D, Hirschman M, Bab IA, Sela J, Mulrad A (1983) Neutral peptidase activities in matrix vesicles from bovine fetal alveolar bone and dog osteosarcoma. Calcif Tissue Int 35:791–797

Hodge AJ, Petruska JA (1963) Recent studies with the electron microscope on ordered aggregates of the tropocollagen macromolecule. In: Ramachandran GN (ed) Aspects of protein structure. Academic, London New York, pp 289–300

Höhling HJ (1966) Die Bauelemente von Zahnschmelz und Dentin aus morphologischer, chemischer und struktureller Sicht. Hanser, München, S 9–127

Höhling HJ, Frank RM, Harndt R (1963) Röntgenographische Untersuchungen an der organischen Matrix von foetalem und jugendlichem menschlichen Schmelz. Das Deutsche Zahnärzteblatt 17:77–82

Höhling HJ, Fearnhead RW, Lotter G (1968) Kombination von Röntgen- und Elektronenbeugung sowie Elektronenmikroskopie zur Bestimmung der Mineralkomponenten bei der Aorten-Verkalkung. German Medical Monthly 13:135–138 (Thieme, Stuttgart)

Höhling HJ, Kreilos R, Neubauer G, Boyde A (1971) Electron microscopy and electron microscopical measurements of collagen mineralization in hard tissues. Z Zellforsch 122:36–52

Höhling HJ, Steffens H, Heuck F (1972) Untersuchungen zur Mineralisierungsdichte im Hartgewebe mit Proteinpolysaccharid bzw. mit Kollagen als Hauptbestandteil der Matrix. Z Zellforsch 134:283–296

Höhling HJ, Ashton BA, Köster HD (1974) Quantitative electron microscopic investigations of mineral nucleation in collagen. Cell Tissue Res 148:11–26

Höhling HJ, Barckhaus RH, Krefting ER, Schreiber J (1976a) Electron microscopic microprobe analysis of mineralized collagen fibrils and extracollageneous regions in turkey leg tendon. Cell Tissue Res 175:345–350

Höhling HJ, Steffens H, Stamm G, Mays U (1976b) Transmission microscopy of freeze dried unstained epiphyseal cartilage of the guinea pig. Cell Tissue Res 167:243–263

Höhling HJ, Ashton BA, Fietzek PP (1980a) Kollagenmineralisierung, in: Kuhlencordt F, Bartelheimer H (eds) Handbuch der inneren Medizin VI/1A Knochen-Gelenke-Muskeln. Springer, Berlin Heidelberg New York, S 59–80

Höhling HJ, Barckhaus RH, Krefting ER (1980b) Hard tissue formation in collagen rich systems, calcium phosphate nucleation and organic matrix. Trends Biochem Sci 5:8–11

Höhling HJ, Althoff J, Barckhaus RH, Krefting ER, Lissner G, Quint P (1981a) Early stages of crystal nucleation in hard tissue formation. In: Schweiger HG (ed) International Cell Biology 1980–1981. Springer, Berlin Heidelberg New York, pp 974–982

Höhling HJ, Barckhaus RH, Krefting ER, Althoff J, Quint P, Niestadtkötter R (1981b) Relationship between the Ca-phosphate crystallites and the collagen structure in turkey tibia tendon. In: Veis A (ed) The chemistry and biology of mineralized connective tissues. Elsevier, North Holland New York Amsterdam Oxford, pp 113–117

Höhling HJ, Krefting ER, Barckhaus RH (1982) Does correlation exist between mineralization in collagen rich hard tissues and that in enamel? J Dent Res 61:1496–1503 (sp Issue)

Hosemann R, Nemetschek TH (1973) Reaktionsabläufe zwischen Phosphorwolframsäure und Kollagen. Kolloid Z u Z Polymere 251:53–60

Hunziker EB, Herrmann W, Schenk RK, Marti T, Müller M, Mohr H (1981) Structural integration of matrix vesicles in calcifying cartilage after cryofixation and freeze substitution. In: Ascenzi A, Bonucci E, DeBernard B (eds) Matrix vesicles. Wichtig Editore, Milano, pp 25–31

Itoh T, Klein L, Geil PH (1982) Age dependence of collagen fibrils and subfibril diameters revealed by transverse freeze-fracture and etching technique. J Microsc 125:343–357

Jodaikin A, Traub W, Weiner S (1986) Protein conformation in rat tooth enamel. Arch Oral Biol 31:685–689

Katchburian E (1973) Membrane bound bodies as initiators of mineralization of dentine. J Anat 116:285–302

Katz EP, Li ST (1974) Structure and function of bone collagen fibrils. J Mol Biol 80:1–15

Krefting ER, Barckhaus RH, Höhling HJ, Bond P, Hosemann R (1980) Analysis of the crystal arrangement in collagen fibrils of mineralizing turkey tibia tendon. Cell Tissue Res 205:485–491

Lee SL, Veis A, Glonek T (1977) Dentin phosphoprotein: An extracellular calcium-binding protein. Biochemistry 16:2971–2979

Linde A, Bhown M, Butler WT (1980) Noncollageneous proteins of dentin. A reexamination of proteins from rat incisor dentin utilizing techniques to avoid artifacts. J Biol Chem 255:5931–5942

Miller A, Parry DAD (1973) The structure and packing of microfibrils in collagen. J Mol Biol 75:441–447

Miller A, Tochetti D (1981) Calculated X-ray diffraction pattern from a quasi hexagonal model for the molecular arrangement in collagen. Int J Biol Macromol 3:9–18

Mulrad A, Sela J, Deutsch D, Bab I (1981) Actin in extracellular matrix vesicles obtained from bone and cartilage. In: Ascenzi A, Bonucci E, DeBernard B (eds) Matrix vesicles. Wichtig Editore, Milano, pp 215–220

Munhoz COG, Leblond CP (1974) Deposition of calcium phosphate into dentine and enamel as shown by radioautography of sections of incisor teeth following injection of ^{45}Ca into rats. Calcif Tissue Res 15:221–235

Nagai M, Frank RM (1975) Transfer du ^{45}Ca par autoradiographie en microscope électronique au cours de l'amélogenèse. Calcif Tissue Res 19:211–221

Nakahara H, Kakei M (1984) Central darkline and carbonic anhydrase, problems relating to crystal nucleation in enamel. In: Fearnhead RW, Suga S (eds) Tooth enamel IV, Elsevier, Amsterdam New York Oxford, pp 42–46

Nylen MU, Eanes ED, Omnell KA (1963) Crystal growth in rat enamel. J Cell Biol 18:109–123

Perdok WG, Gustafson G (1961) X-Ray diffraction studies of the insoluble protein in mature enamel. Arch Oral Biol 4:70–75

Price PA, Otsuka AS, Poser JW, Kristaponis J, Raman N (1976) Characterization of a γ-carboxyglutamic acid-containing protein from bone. Proc Natl Acad Sci USA 73:1447–1451

Price PA, Williamson MK, Otawara Y (1985) Characterization of matrix Gla protein. A new vitamin K-dependent protein associated with the organic matrix of bone. In: Butler WT (ed) The chemistry and biology of mineralized tissues. Ebsco Media, Birmingham Alabama USA, pp 163–195

Pugliarello MC, Vittur F, DeBernard B, Bonucci E, Ascenzi A (1970) Chemical modifications in osteons during calcification. Calcif Tissue Res 5:108–114

Rabinovitch AL, Anderson C (1976) Biogenesis of matrix vesicles in matrix vesicles in cartilage growth plates. Fed Proc 35:112–116

Rahemtulla F, Prince CW, Butler WT (1984) Isolation and partial characterization of proteoglycans from rat incisor. Biochem J 218:877–885

Ramachandran GN (1963) The triple helical structure of collagen. In: Ramachandran GN (ed) Aspects of protein structure. Academic, London New York, pp 39–55

Ramachandran GN, Kartha G (1954) Structure of collagen. Nature 174:269–279

Reimer L (1962) Änderung des elektronenmikroskopischen Bildkontrastes beim Übergang amorphkristallin und flüssig-kristallin. Naturwissenschaften 49:297

Reith EJ (1983) A model for transcellular transport of calcium based on membrane fluidity and movement of calcium carriers within the more fluid microdomains of the plasma membrane. Calcif Tissue Int 35:129–134

Reith EJ, Boyde A (1978) Histochemical and electronprobe analysis of secretory ameloblasts of developing rat molar teeth. Histochemistry 55:17–26

Rich A, Crick FHC (1955) The structure of collagen. Nature 176:915

Robinson C, Lowe NR, Weatherell JA (1975) Amino acid composition, distribution and origin of 'Tuft' protein in human and bovine dental enamel. Arch Oral Biol 20:29–42

Robinson C, Briggs HD, Atkinson PJ, Weatherell JA (1979) Matrix and mineral changes in developing enamel. J Dent Res 58 B:871–880

Robinson C, Weatherell JA, Höhling HJ (1983) Formation and mineralization of dental enamel. Trends Biochem Sciences 8:284–287

Rönnholm E (1962) An electron microscopic study of the amelogenesis in human teeth. I. The fine structure of the ameloblasts. J Ultrastruct Res 6:229–248

Sasaki T, Garant PR (1986) Ultracytochemical demonstration of ATP dependent Ca-pump in ameloblasts of rat incisor enamel organ. Calcif Tissue Int 39:86–96

Sasaki T, Garant PR (1987) Calmodulin blocker inhibits Ca^{++}-ATPase activity in secretory ameloblasts of rat incisor. Cell Tissue Res 248:103–110

Sasaki S, Takagi T, Suzuki M, Baba T, Minegishi K (1984) Aminoacid sequence of developing bovine enamel protein. In: Fearnhead RW, Suga S (eds) Tooth enamel IV, Elsevier Science Publishers, Amsterdam New York Oxford, pp 151–155

Scherft JP, Groot CG (1981) The development of matrix vesicles into bone nodules, studied with colloidal thorium dioxide. In: Ascenzi A, Bonucci E, DeBernard B (eds) Matrix vesicles. Wichtig Editore, Milano, pp 173–177

Scott JE, Haigh H (1985) Proteoglycan-type I collagen fibril interactions in bone and noncalcifying connective tissues. Biosci Rep 5:71–81

Serafini-Fracassini A, Smith JW (1966) Observations on the morphology of the protein-polysaccharide complex of bovine nasal cartilage and its relationship to collagen. Proc R Soc Lond [Biol] 165:440–449

Seyer J, Glimcher MJ (1969) The amino-acid sequence of two O-phosphoserine containing tripeptides isolated from the organic matrix of embryonic bovine enamel. Biochem Biophys Acta 181:410–418

Sisca RF, Provenza DV (1972) Initial dentine formation in human deciduous teeth. Calcif Tissue Res 9:1–5

Smith BJ (1968) Molecular pattern in native collagen. Nature 219:157–158

Stein RM, Hsu HHT, Anderson HC (1981) Protein profiles of isolated fetal calf and rachitic rat matrix vesicles by polyacrylamide gel eletrophoresis. In: Ascenzi A, Bonucci E, DeBernard B (eds) Matrix vesicles. Wichtig Editore, Milano, pp 117–122

Stetler-Stevenson WG, Veis A (1987) Bovine dentin phosphoryn: Calcium ion binding properties of a high molecular weight preparation. Calcif Tissue Int 40:97–102

Termine JD, Torchia OA, Conn KM (1979) Enamel matrix: Structural proteins. J Dent Res 58B:773–778

Termine JD, Belcourt AB, Conn KM, Kleinman HK (1981) Mineral and collagen binding proteins of fetal calf bone. J Biol Chem 256:10403–10408

Thiele H (1964) Ordnung von Fadenmolekülen durch Ionendiffusion, ein Prinzip der Strukturbildung. Protoplasma 58:318–341

Thyberg J, Friberg U (1978) The lysosomal system in endochondral growth. In: Graumann W, Lojda Z, Pearse AGE, Schiebler TH (eds) Progress in histochemistry and cytochemistry 10, no 4. Fischer, Stuttgart

Traub W, Yonath A, Segal DM (1969) On the molecular structure of collagen. Nature 221:914–917

Trus BL, Piez KA (1980) Compressed microfibril models of the native collagen fibril. Nature 286:300–301

Väänänen HK, Korhonen LK (1981) Phosphatases of matrix vesicles. In: Ascenzi A, Bonucci E, DeBernard B (eds) Matrix vesicles. Wichtig Editore, Milano, pp 111–116

Van Iren F, van Essen-Joolen L, van der Duyn Schouten P, Boers-van der Sluijs P, De Bruijn WC (1979) Sodium and calcium localization in cells and tissues by precipitation with antimonate: a quantitative study. Histochemistry 63:273–294

Von der Mark K, Wendt P, Rexrodt F, Kühn K (1970) Direct evidence for a correlation between amino-acid sequence and cross striation pattern of collagen. FEBS Lett 11:105–108

Warner GP, Hubbard HL, Lioyd GC, Wuthier RE (1983) ^{32}Pi and ^{45}Ca metabolism by matrix vesicle enriched microsomes prepared from chicken epiphyseal cartilage by isosmotic percoll density gradient fractionation. Calcif Tissue Int 35:327–338

Warshawsky H, Bai P, Nanci A (1987) Analysis of crystallite shape in rat incisor enamel. Anat Rec 218:380–390

Weiner S, Traub W (1984) Macromolecules in mollusc shells and their function in biomineralization. Philos Trans R Soc Lond [Biol] 304:421–434

Weinstock M, Leblond LP (1973) Radioautographic visualization of the deposition of a phosphoprotein at the mineralization front in the dentine of the rat incisor. J Cell Biol 56:838–845

Weiss MP, Voegel JC, Frank RM (1981) Enamel crystallite growth: width and thickness study related to the possible presence of octocalcium phosphate during amelogenesis. J Ultrastruct Res 76:286–292

Wuthier RE (1982) A review of the primary mechanism of endochondral calcification with special emphasis on the role of cells, mitochondria and matrix vesicles. Clin Orthop Rel Res 169:219–242

Wuthier RE, Majeska RJ, Collins GM (1977) Biosynthesis of matrix vesicles in epiphyseal cartilage. I. In vivo incorporation of ^{32}P orthophosphate into phospholipids of chondrocyte, membrane and matrix vesicle fractions. Calcif Tissue Res 23:135–139

Author Index

Page numbers in *italics* refer to bibliography

Subject Index

Numbers in italics refer to pages upon which the subject in question is discussed in length

A. Oksche, L. Vollrath (Eds.)

Handbook of Microscopic Anatomy

Volume I/3
H. G. Schwarzacher
Chromosomes
in Mitosis and Interphase
1976. 116 figures, 3 tables. VIII, 182 pages.
ISBN 3-540-07456-2

Volume II/6
H. Schmalbruch
Skeletal Muscle
1985. 129 figures. XI, 440 pages. ISBN 3-540-15608-9

Volume II/Part 7
E. D. Canale, G. R. Campbell, J. J. Smolich,
J. H. Campbell
Cardiac Muscle
1986. 101 figures. XIII, 318 pages. ISBN 3-540-16379-4

Volume VI/7
L. Vollrath
The Pineal Organ
1981. 190 figures. XVII, 665 pages. ISBN 3-540-10313-9

Volume VI/8
P. Böck
The Paraganglia
1982. 61 figures. XV, 315 pages. ISBN 3-540-10978-1

Volume VII/6
G. Aumüller
Prostate Gland and Seminal Vesicles

Springer-Verlag
Berlin Heidelberg New York
London Paris Tokyo Hong Kong

1979. 142 figures (some in color) in 181 separate illustrations. X, 380 pages. ISBN 3-540-09191-2

A. Oksche, L. Vollrath (Eds.)

Handbook of Microscopic Anatomy

Volume 5/Part 5

H. E. Schroeder, University of Zürich

The Periodontium

With Technical Assistance of M. Amstad-Jossi,
R. Kröni, W. Scherle

1986. 127 figures. VI, 418 pages.
ISBN 3-540-4-16604-1

Contents: Introduction. - History and Nomenclature: Glossary of Current Terminology for the Healthy Periodontium. - Periodontium, a Developmental and Functional Unit. - Development, Structure, and Function of Periodontal Tissues: Cementum. Alveolar Process and Alveolar Bone. Periodontal Ligament. Gingiva. - Current Trends. - References. - Author Index. - Subject Index.

The Periodontium is the first and only extensive and fully integrated summary of present knowledge on the development, structure, and function of the four tissues (cementum, alveolar bone, periodontal ligament, gingiva) constituting the periodontium. It covers the most essential literature including some 1800 references, and is furnished with numerous outstanding illustrations.

The text focuses on the human periodontium, though data on animals is also included, and describes it as it relates to clinical dentistry. This description utilizes both information derived from the application of various morphological research tools, such as light microscopy, histochemistry, scanning and transmission electron microscopy, stereology and most of the recent experimental evidence from in vivo and in vitro studies. In particular, the various types of root cementum and their development are reviewed for the first time in such a manner, and thus have been treated with emphasis on their role in supporting teeth.

This new, wide-ranging treatment will be valuable for all those either theoretically or clinically interested in the structure and function of the periodontium.

Springer-Verlag
Berlin Heidelberg New York
London Paris Tokyo Hong Kong

Springer